EXPLORING NATURAL DISASTERS

Natural Processes and Human Impacts

Preliminary Edition

By Gabi Laske
University of California – San Diego

COPYRIGHTED MATERIAL — DO NOT DUPLICATE, DISTRIBUTE, OR POST

Bassim Hamadeh, CEO and Publisher
Michael Simpson, Vice President of Acquisitions
Jamie Giganti, Managing Editor
Jess Busch, Graphic Design Supervisor
Seidy Cruz, Acquisitions Editor
Sarah Wheeler, Senior Project Editor
Brian Fahey, Licensing Associate

First published in the United States of America in 2014 by Cognella, Inc.

Cover Image Credits: Copyright © 2009 by iStockphoto.com/Nadya Lukic, Copyright © 1980 by Getty Images/Science Source/USGS, Copyright © 2007 by iStockphoto.com/Brasil2, Copyright © 2008 by iStockphoto.com/milehightraveler, Copyright © 2010 by iStockphoto.com/RapidEye, Copyright © 2008 by Getty Images/Stocktrek Images, Copyright © 2010 by iStockphoto.com/HoltWebb, Copyright © 2008 by iStockphoto.com/DPhoto, Copyright © 2009 by iStockphoto.com/Byronsdad, Copyright © 2007 by iStockphoto.com/luoman, Copyright © 2011 by Depositphotos Inc./Akihito Yokoyama, Copyright ©2009 by Depositphotos Inc./Aleksandr Kuznetsov

Printed in the United States of America

ISBN: 978-1-60927-135-0 (pbk) 978-1-62661-651-6 (br)

Contents

COPYRIGHTED MATERIAL — DO NOT DUPLICATE, DISTRIBUTE, OR POST

COPYRIGHTED MATERIAL — DO NOT DUPLICATE, DISTRIBUTE, OR POST

COPYRIGHTED MATERIAL — DO NOT DUPLICATE, DISTRIBUTE, OR POST

Preface

This book is a work in progress. It was borne out of the need to give my UCSD course SIO15 "Natural Disasters" a suitable textbook. Some chapters are finished texts but others are still in draft version, which are marked as such. The careful reader will recognize that the chapters in crudest draft version closely resemble the lecture notes on the class website, on which this book is based: http://quakeinfo.ucsd.edu/~gabi/sio15

To keep the purchase price of the book down, I provide all case studies and the study guide (including key terms, key concepts and study questions) on my website at http://quakeinfo.ucsd.edu/~gabi/book. I highly recommend that SIO15 students access and use this additional material. Apart from pricing issues, an advantage of removing case studies from a printed book is that databases of case studies are easier to update.

Many natural disasters textbooks are written by experts in geology or engineering, and are therefore weighted heavily toward classical disasters such as earthquakes, landslides and floods, with a few shorter chapters on volcanoes, hurricanes and tornadoes. In a warming climate, the impact from extreme weather events is predicted to increase. This book therefore contains more than the usual number of chapters on "the atmosphere" and tries to strike a balance to previous textbooks. This book also contrasts with the typical engineering textbook in that much emphasis is put on conveying a scientific framework for understanding the processes involved in natural disasters. The book therefore includes chapters on basic physical principles, the solar system, plate tectonics and on basic processes in Earth's atmosphere. Finally, several chapters are dedicated to how human action not only mitigates the extent of damage from natural disasters but also contributes to them, and how human action is changing the planet in profound ways.

The vast majority of figures chosen for this book comes from the public domain, e.g. from Wikipedia and from various government agencies and public institutions, such as the US Geological Survey (USGS), the National Atmospheric and Oceanographic Administration (NOAA), the National Aeronautics and Space Administration (NASA) and the Smithsonian Institute. By choosing images from the public domain, I am hoping to stimulate the remembrance of the material presented in this book. In addition, I aim to raise awareness that much material that commercial entities convey, such as weather channels, depend on data collected and provided by government agencies, and not lastly public tax dollars. Though the book uses a good amount of scientific jargon the aim here is to convey key words with an easy-to-follow style of writing. Hopefully, the casual science reader and information seeker will find a wealth of useful information as much as my students do, while still providing an entertaining and stimulating book for my expert colleagues.

References are currently given at the end of each chapter. Repeat references are often given as two- to four-letter codes. The complete reference is as follows:

AB: Abbott, P.L., 2009. Natural Disasters. 7th edition. McGraw-Hill, New York, pp. 504.

BH: Best, D.M. and Hacker, D.B., 2010. Earth's Natural Hazards: Understanding Natural Disasters and Catastrophes. Kendall/Hunt Publishing, pp. 310.

EDS: Hamblin, W.K. and Christiansen, E.H., 2001. Earth's Dynamic Systems. 9th edition. Prentice Hall, pp. 735.

EW: Burt, C.C., 2004. Extreme Weather, 1st edition. Norton&Company, New York, pp. 298.

EIA: Energy Information Agency of the DOE, 2009. Annual Energy Review 2009. www.eia.gov/aer

IEA: International Energy Agency (IEA), 2010. Key World Energy Statistics. Downloadable on the IEA website www.iea.org

IPCC: Intergovernmental Panel on Climate Change, 2007. Climate Change 2007: Synthesis Report (Fourth Assessment Report), print and online ppt material.

KB: Keller,E.A. and Blodgett, R.H., 2008. Natural Hazards, 2nd edition. Pearson Prentice Hall, Upper Saddle River, NJ, pp.478.

LIFE: Purves, W.K., Sadava, D., Orians, G.H. and Heller, H.C., 2001. Life, The Science of Biology. 6th edition. W.H. Freeman and Co., Gordonsville, VA.

ME: Marshak, S.,2005. Earth: Portrait of a Planet. 2nd edition. Norton & Company, New York, pp. 748.

MT: Ahrens, C.D., 2003. Meteorology Today, 7th edition. Brooks/Cole-Thomson Learning, Pacific Grove, CA, pp. 544.

RO: Robinson, A., 1993. Earthshock. 1st edition. Thames and Hudson, London, pp. 298.

RU: Ruddiman, W.F., 2000. Earth's Climate: Past and Future, W.H. Freeman and Company, 450 pp.

TG: Garrison, T., 1999. Oceanography. 3rd edition. Brooks/Cole Thomson Publishing, Pacific Grove, CA, pp. 541.

TT: Trujillo, A.P. and Thruman, H.V., 2005. Essentials of Oceanography. 8th edition. Pearson Prentice Hall, Upper Saddle River, NJ, 518 pp.

UE: Press, F. and Siever, R., 2003. Understanding Earth, 4th edition. W.H. Freeman and Company, 573 pp.

WW99: Platt McGinn, A., 1999. Safeguarding the Health of Oceans. Worldwatch Institute.

WW06: Halweil, B., 2006. Catch of the Day. World Watch Institute, publication #172, Worldwatch Institute. http://worldwatch.org

Chapter 1: Introduction to Natural Disasters and Human Impact (draft)

1.1 Natural Disasters and the Human Component

In Southern California, people typically think of earthquakes first when asked which type of natural disaster comes to mind. Indeed, there have been some very destructive earthquakes in the past that also caused many casualties. Larger earthquakes are likely to claim more lives than smaller earthquakes. But the size of an earthquake is not the only primary factor. While a magnitude-6.7 earthquake in 2003 destroyed much of the city of Bam in Iran (Case Study 1) and claimed tens of thousands of lives, a very similar earthquake in 1994 in the Los Angeles area caused extensive damage to the freeway system but claimed only 65 lives. Submarine earthquakes can cause devastating tsunami, but while upward of 200,000 people perished in the 26 December 2004 Sumatra-Andaman earthquake, less than 20,000 people perished in the larger 11 March 2011 Tohoku, Japan earthquake. Clearly, local geology plays an important role, but also how people prepare for earthquakes and how buildings are constructed. Much of the damage from an earthquake can result not from the shaking itself but from other processes triggered by the earthquake. Two of these are tsunami and landslides. But there are several other "secondary disasters". For example, the 18 April 1906 San Francisco earthquake was so destructive not primarily by the shaking but rather by the subsequent ranging fires and the inability to fight these fires effectively (Case Study 2).

Earthquakes, volcanic eruptions and processes making landscapes vulnerable to landslides, are fueled by energy provided through Earth's internal processes. This book attempts to provide a glance into the physical processes involved. It explains why earthquakes and volcanic eruptions occur at some places on the globe but not at others, and why both are devastating in some places but not so in others. The book also discusses the mechanical principles behind masses moving on a slope, the basic principles of streams and rivers along their journey from spring to the oceans, and why some rivers are more prone to flooding than others.

Large earthquakes and humongous volcanic eruptions can claim lives counting into the tens if not hundreds of thousands, but other types of natural disasters can be just as devastating. Large hurricanes can produce a huge storm surge that wipes out everything in its path. As occurred during 2005 Hurricane Katrina, low-lying areas are particularly vulnerable to devastating flooding. Fatality numbers become large in densely populated areas or when large portions lie below sea level, such as is the case for Bangladesh (Table 1.2). Hurricanes, other severe storms and tornadoes are fueled by

COPYRIGHTED MATERIAL — DO NOT DUPLICATE, DISTRIBUTE, OR POST

energy supplied in Earth's atmosphere, as opposed to Earth's internal energy. Hurricanes, which fuel on warm tropical marine air, occur in certain regions on the globe but not in others. And while tornadoes can accompany nearly every severe thunderstorm, the confluence of various air masses and the jet stream in the U.S. Midwest sets the stage for particularly frequent and devastating tornadoes.

As devastating as single hurricanes and tornadoes can be, it is perhaps surprising that of all natural disasters in the U.S., the #1 killer is none of the above. In fact, it is none of the severe weather that comes to mind first. Instead, on average in the last two decades, heat waves have been emerging as the silent but most serious killer in the U.S., more so than exposure to cold and winter storms. Heat waves are "silent killers" because they are not of the sudden, spectacular kind that creates news hype. The 1995 "Chicago Heat wave", an event that easily makes the list of 10 most fatal natural disasters in U.S. history, claimed nearly 1000 lives (Case study 1 in Chapter 13). But rarely anybody remembers this event, not because it occurred 10 years before Hurricane Katrina. Heat waves are often underestimated in their potential to do harm to the human body, which makes it somewhat difficult to keep accurate records. The suffering is slow, and a "natural cause" rather than the heat wave may be declared as the cause of death. Nevertheless, during heat waves, fatality rates increase, particularly among the elderly and poor who cannot afford air conditioning in their homes or move around to escape the heat, as dramatically came to light in the 2003 European heat wave that claimed some 30,000 lives (Case Study 2 in Chapter 13). In a warming climate, with more extreme weather events, we need to get prepared for more such events.

Though all these natural disasters are caused by natural processes, human action dramatically influences the outcome of a disaster. A functioning tsunami warning system and resulting evacuations brings down fatality rates by orders of magnitudes. No tsunami system was in place to warn people living around the Indian Ocean that the devastating 2004 Sumatra-Andaman tsunami would be coming even though they would have had many hours to reach safer elevations. Shaking from earthquakes can be exacerbated by inadequate building construction. While humans have no control over the onset of earthquakes and volcanic eruptions, we do have an increasing influence on the start and extent of landslides, floods and even wildfires. In a natural environment, the #1 cause for wildfires is lightning. However, 85% of all wildfires are nowadays started by human action, from arson over land clearance to that carelessly discarded cigarette butt in "bone-dry" grassland.

The natural Earth provides awesome amounts of energy to fuel natural disasters but the increasing human population and daily action that changes Earth's surface is reaching an equally awesome and frightening level. We do not only exacerbate the extent of natural disasters, but we contribute to them on a massive scale, and even create new ones. Oil and gas exploration gone awry may seem anecdotal, but the 2006 Lusi mud volcano in Indonesia (see Chapter 6) will spew mud for decades to come. Mining, housing and infrastructure development, and farming profoundly change Earth's surface. In numerous cases, these new landscapes are vulnerable to flooding and accelerated erosion. Human action has removed as much as half of Earth's original forests. Many natural forests have been replaced by managed forests, which often support less biodiversity. Removal of the tropical rain forest is particularly devastating as the nurturing but thin humus layer is easily eroded, leaving behind badlands that take thousands of years to re-populate with a mature forest. Earth has experienced mass extinctions in the past, triggered by humongous volcanic eruptions and asteroid impacts. But experts fear that if the current rate of the human-induced disappearance of species continues for another 200 years, the planet will face the greatest mass extinction of all times.

Many local problems may be exacerbated remotely. So is evidence mounting that the devastating drought in Africa's Sahel in the late 20th century was enhanced, if not caused, by fossil fuel burning in North America. Farming and industry not only profoundly affect the chemical balance on land, but now also tracks into the vast oceans. Poisons such as mercury, emitted from coal burning power plants, are not only taken up by small fish but climb up the food chain and ultimately harm our own lives.

Humans exploit Earth's resources, from oil, to minerals to fresh water as if there was no tomorrow, often without being aware that many of these resources are non-renewable, meaning that regeneration of these resources typically takes longer than a human lifespan. Oil and Earth's mineral resources are formed on geological timescales, often on the order of millions of years. But even renewable resources are often not harvested on a sustainable level, and this process has now reached the oceans. Within only a few decades, the vast majority of the ocean's fisheries that are relevant to the fishing industry, and our dinner table, are now either overexploited to the point of near-extinction or to an unsustainable level, thereby jeopardizing our own future food supply. While natural disasters have profoundly changed Earth's surface and life on it, humans have the means to dwarf some of Earth's most awesome powers. This book not only describes Earth's natural powers and processes but aims to raise awareness of the equally awesome powers humans now have, and of the responsibility to leave a planet behind that is suitable to raise our own future generations and that of life on Earth in general.

1.2 What is a Natural Disaster?

- event of a natural cause with high death rate or extensive destruction
- event of a natural cause that has a negative impact on the environment
- event of a natural cause with low death rate but occurring often
- a natural trend that has negative long-term consequences

An example of a recent devastating natural disaster was Hurricane Ike, and Atlantic hurricane in 2008 that reached well into the continent. In more recent times, in 2011, relatively weak Hurricane Irene traveled along the U.S. east coast and affected millions of people.

Not all large hurricanes make the news. An example of a truly remarkable hurricane that did not make the news (because it did not kill nor destruct human construction) was Ioke in 2006. Ioke crossed the northern Pacific Ocean in August 2006, passing Johnston Atoll and Wake Island but did not affect permanently inhabited areas. Some residents of Wake Island evacuated. It was the first category 5 and thereby the largest hurricane that ever formed in the Central Pacific Ocean. See the back end of a news clip to get info and link to satellite image.

Some Obvious Natural Disasters

These include mostly sudden events or events

- earthquakes
- volcanic eruptions
- landslides and other mass movements
- floods (though some floods may build up over a longer time)
- fires
- hurricanes, tornadoes and severe storms
- impacts of space objects

Some Not so Obvious Natural Disasters

These events may happen slowly

- droughts and extreme weather
- global warming and cooling
- changing sea level
- coastal erosion
- shift in resident species

1.3 Why Study Natural Disasters? – Some Numbers

Effects of Natural Disasters

- natural disasters kill, destroy, cost (prevention, relief efforts and rebuilding)
- costs have increased exponentially over last 50 years
- the gap between losses and insured losses increases

Natural Disasters and Human Interaction

- humans can mitigate disaster risks (e.g. build EQ proof houses)
- humans can increase disaster risks (e.g. build along EQ faults)
- humans can enhance effects of natural disasters (e.g. build wrong flood channels)
- humans can trigger natural disasters (e.g. inappropriate drainage)

The Need to Understand Natural Disasters

- need to understand process involved
- improve identification of risks
- improve prevention of losses (lives and structures)
- improve forecast and evacuations
- increase effectiveness of disaster relief (fight rising costs)

Table 1.1 Worldwide Cumulative Fatalities due to Natural Disasters between 1947-1980[1]

	Earthquakes	Tsunami	Volcanic Eruptions	Floods	Hurricanes	Tornadoes	Other Severe Weather	Mass Movements	Total
# Fatal Events	180	7	18	333	210	119	147	45	1059
Fatalities by Area									
North America	77	60	96	1633	1997	4568	5003	323	13,757
Caribbean and Central America	30,613	----	151	2575	16,541	26	510	260	50,676
South America	38,837	----	440	4396	----	----	340	5262	49,275
Europe	7750	----	2000	11,199	250	39	6816	640	28,694
Asia	354,521	4459	2805	170,664	478,574	4308	34,403	4356	**1,054,090**
Africa	18,232	----	----	3891	864	548	5	----	23,540
Oceania	18	----	4000	77	290	----	117	----	4502
Totals	**450,048**	4519	9492	194,435	**498,516**	9489	47,194	10,841	1,224,534

most frequent event: floods (333 events)

highest number of fatalities: hurricanes (498,516 deaths) and earthquakes (450,048 deaths)

COPYRIGHTED MATERIAL — DO NOT DUPLICATE, DISTRIBUTE, OR POST

lowest number of fatalities: tsunami (4,519)

but then the Dec 26, 2004 Sumatra earthquake tsunami killed 245,000 people, making tsunami more fatal than floods in general (194,435 deaths from 1947-1980)

area with highest death rate: Asia (1,054,090) (*high population density*)

area with lowest death rate: Oceania (4,502); North America (13,757)

somewhat surprisingly, Africa has the third lowest number of fatalities from natural disasters though starvation due to droughts is not included in this comparison

Table 1.2 The Deadliest Events between 1970-2011

Area	Time	Event	# Fatalities
Bangladesh	11/14, 1970	Tropical Cyclone[1]	300,000
China	07/28, 1976	Tangshan Earthquake	255,000
Haiti	01/12, 2010	Earthquake	222,570
Indian Ocean	12/26, 2004	Sumatra-Andaman Earthquake	220,000 (most from tsunami)
Myanmar (Burma)	05/02, 2008	Tropical Cyclone Nargis	138,300
Bangladesh	04/30, 1991	Tropical Cyclone Gorky	138,000
China	05/12, 2008	Sichuan Earthquake	87,449
Pakistan	10/08, 2005	Earthquake	73,300
Peru	05/31, 1970	Nevados Huascaran Earthquake and volcanic mudslide	66,000
Russia	06/15, 2010	Heat wave and peat fires	55,630
Iran	06/21, 1990	Gilan Earthquake	40,000
Europe	06/01, 2003	Heat wave	35,000
Iran	12/26, 2003	Bam Earthquake	26,271
Armenia	12/07, 1988	Earthquake	25,000
Iran	09/16, 1978	Tabas Earthquake	25,000
Colombia	11/13, 1985	Nevado del Ruiz volcanic mudslide	23,000
Guatemala	02/04 1976	Earthquake	22,084
India	01/26 2001	Gujarat Earthquake	19,737
Japan	03/11 2011	Tohoku Earthquake	19,184
Turkey	08/17 1999	Izmit Earthquake	19,118

[1]A cyclone in the Indian Ocean is the same phenomenon as a hurricane in the Atlantic or East Pacific.

This table was compiled from data provided by Swiss Reinsurance Company.

Things that we learn from this table:

- single-event hurricanes/cyclones and earthquakes are major killers
- high fatality rates in areas with high population density
- this is not necessarily a developing country (e.g. Africa is missing from this table)
- To put this into perspective of what humans can do: WW II claimed 62 Mio lives, according to Wikipedia (http://en.wikipedia.org/wiki/World_war_2). By January 2008, the Iraq War claimed more than 1,000,000 lives, according to the Opinion Research Business survey (http://en.wikipedia.org/wiki/Casualties_of_the_Iraq_War).

Table 1.3 The Most Costly Events between 1970-2011 (insured losses)

Area	**Time**	**Event**	**Cost in Billion [billion U.S. $]**	**Fatalities**
USA	08/25 2005	Hurricane Katrina	74.7	1,836
Japan[1]	03/11 2011	Tohoku Earthquake	35	19,184
Pakistan[2a]	08/29 2010	Floods	43	??
USA	08/23 1992	Hurricane Andrew	43	25.6
USA	09/11 2001	terrorist attack[3]	21.2	2,982
USA	01/17 1994	Northridge Earthquake	21.2	61
USA	09/13 2008	Hurricane Ike	21.1	136
USA	09/02 2004	Hurricane Ivan	15.4	124
USA	10/19 2005	Hurricane Wilma	14.5	35
USA[2b]	08 1993	Mississippi Flood	12	48
Thailand	07/27 2011	Monsoon Floods	12	813
New Zealand	02/22 2011	Earthquake	12	181
USA	09/20 2005	Hurricane Rita	11.6	34
USA	08/11 2004	Hurricane Charley	9.6	24
Japan	09/27 1991	Typhoon Mireille	9.3	51
USA	09/15 1989	Hurricane Hugo	8.3	71
Chile	02/27 2010	Earthquake	8.2	562
Europe	01/25 1990	Winter Storm Daria	8.0	95

Europe	12/25 1999	Winter Storm Lothar	7.8	110
USA	04/22 2011	severe storms/tornadoes	7.3	354
USA	05/20 2011	severe storms/tornadoes	7.1	155
Europe	01/18 2007	Winter Storm Kyrill	6.6	54

[1] Total losses exceed $200 billion, making this the costliest natural disaster.

[2a] not listed at Swiss Re. Total economic impact.

[2b] not listed at Swiss Re. Majority not insured.

[3] not a natural disaster but puts things into perspective, regarding costs and fatalities.

This table is compiled from data provided by Swiss Reinsurance Company, and more recent data.

Things that we learn from this table:

- until 2010, the 10 costliest events occurred in the U.S., one of them man-made
- 5 of these 10 events occurred within only 15 months (Aug 04 - Oct 05); the year 2006 was relatively "quiet"
- most of the costliest disasters were hurricanes though major flood events were also very costly
- Again, to put this into perspective, as of 2010, the war in Iraq has cost over $ 3 Trillion (3000 Billion or 3 Million Million).

NB: The most costly natural disaster in S.D. county was probably the 26 October 2003 Cedar Fire Paradise. It was started by a lost hunter who ignited a signal fire. The Cedar Fire, together with the Paradise and Otay fires that burned on the same weekend was the second largest wildfire in the history of CA (after the Great Fire of 1889). These three fires burned 350,000 acres, claimed 15 lives and destroyed 3500 buildings worth $1.2 billion; these fires were closely followed in significance by the Witch Creek, Rice and Poomacha fires in 2007.

1.4 Severe Weather and Heat Waves

Table 1.4 Average Annual Death Rates Due to Severe Weather in the U.S.[1]

Type	(1940-2011)	(1982-2011)	(1992-2011)	(2002-2011)
	72 yr ave	30 yr ave	20 yr ave	10 yr ave
Lightning	128	54	45	37
Flood	104	93	83	78
Tornado	102	73	82	108

Hurricane	46	47	64	114
Heat	-	137*	169	119
Cold	-	29*	27	27
Winter Storm	-	40*	37	23
Wind	-	-	51*	45
Rip currents	-	-	-	46

[1]This table shows long- and short-term averages.

*Data not available over complete time span.

The numbers are compiled using Hazstats data from the National Weather Service (NWS) (http://www.weather.gov/os/hazstats.shtml). Note that the #1 killer depends on the time span over which data are available, i.e. lightning has been the long-term top killer (72 yr average), but the recent short-term killer is heat (20 yr average). This is because of a change in the long-term behavior in lighting deaths, see below.

Statistics of certain types of disasters

The data displayed here are the same annual numbers that are compiled into Table 1.4.

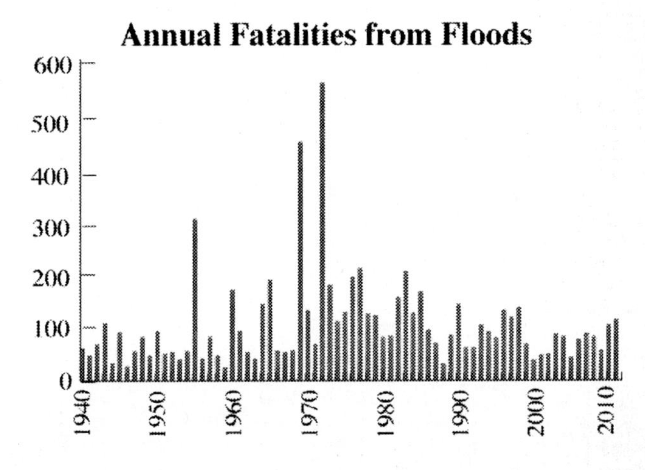

Floods: The impact of single events in the mid-1950s and around 1970 has been declining and so the average fatality rate from floods has been declining. The background yearly fatality rate is not declining, however, and may in fact be increasing.

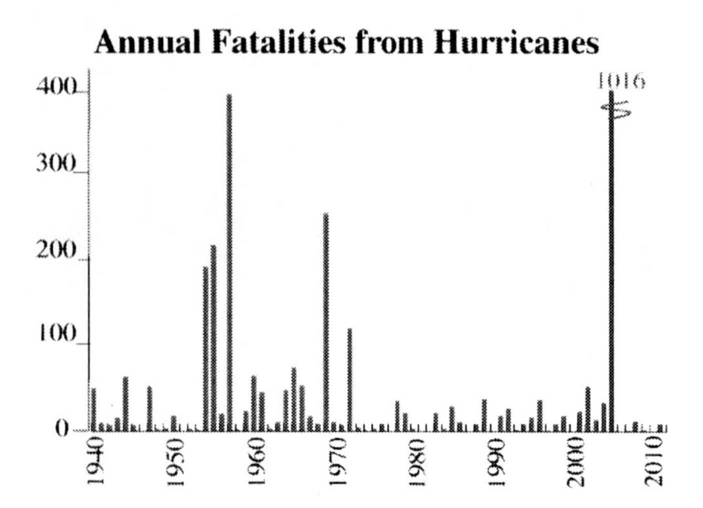

Hurricanes: Recent averages of deaths from hurricanes are much higher than the 72-year long-term average, because they are dominated by the fatalities of Hurricane Katrina in 2005. On the other hand, the 6000 fatalities of the 1900 Galveston, TX Hurricane increase a 1900-2011 long-term annual death rate by 54! It is therefore important to consider over what time period the data have been collected and if there have been single major events. A long-term average is usually a best estimate for a "typical" average because a single catastrophic event does not dominate the statistics.

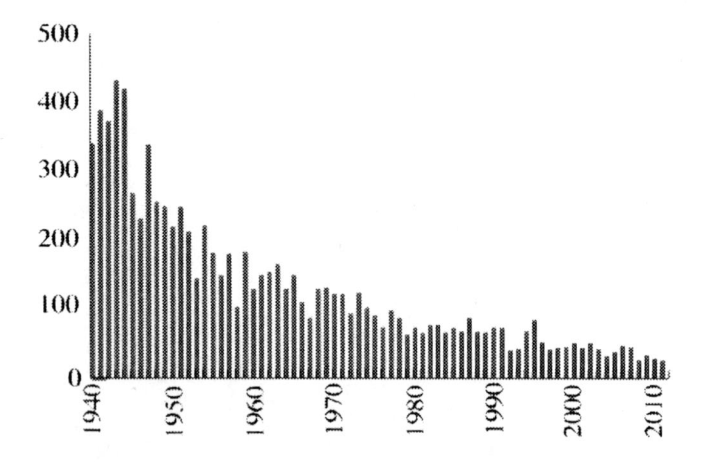

Lightning: The death rate due to lightning strikes is intriguing. The typical annual death rate has actually gone down by a factor of 10 (from 400 to 40) since the 1940s. The long-term average is therefore controlled by high death rates in the earlier 1900s. This is presumably because there were many more people outdoors during thunderstorms than nowadays. Taking single events not into account, it is still true that more people die from lightning strikes than from hurricanes but, perhaps surprisingly, heat waves are now the #1 killer in the U.S..

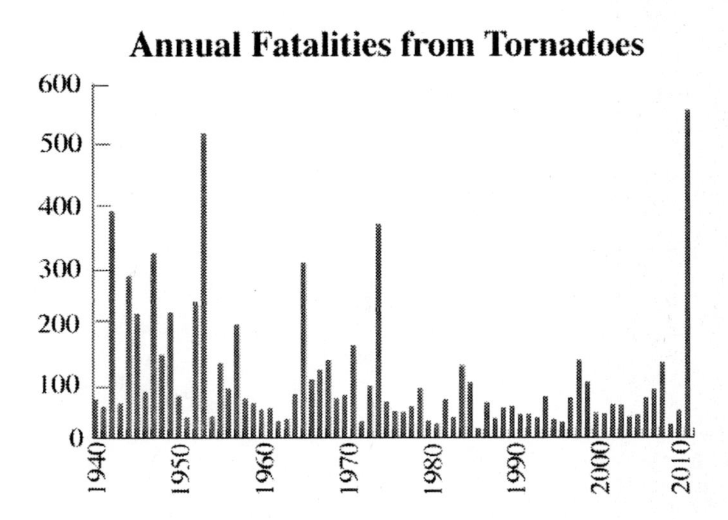

Tornadoes: A rather disturbing fact is that the number of tornado deaths have not declined over the past 30 years, after a marked decline in the late 1970s. In fact, the average fatality rate has increased over the last 3 decades (see table) though this is due to the fact that the impact of the particularly deadly year 2011 increases the average! A subtle increase in annual fatalities (see figure) could be caused by an increase in more powerful tornadoes though there is currently no convincing evidence to support this idea. But another reason could be that people have become complacent and do, in fact, no longer build tornado shelters. One reason being that such shelters add to the construction price of construction price of a new home.

The Tri-State Tornado in 1925 has been the deadliest single tornado in U.S. history, with 695 people losing their lives. An additional 52 people lost their lives in other tornadoes during the 18 March 1925 tornado outbreak.

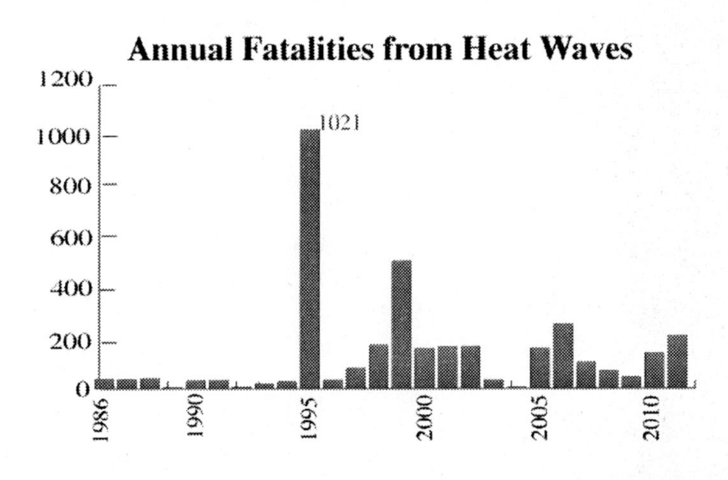

Heat waves: Records on fatalities from heat waves go back only to 1986. A single event in 1995, the Chicago Heat Wave, dominates recent statistics. However, more people die from heat in recent years than 20 years ago. The big worry is that the trend may be increasing. Over the last 30 years, heat waves have been the #1 killer in the U.S., with floods being a distant second. In the last 20 years, more people died from heat waves than from tornadoes and floods combined.

Table 1.5 Billion-Dollar Losses in the U.S. (1980-2003)

Heat waves/droughts:	144
Hurricanes:	102 (before Katrina)
Floods:	55
Cold-weather storms:	28
Wildfires:	13
Tornadoes:	7

This table was compiled from data provided by Munich Reinsurance Company.

Lightning, Heat Waves, Tornadoes and Statistics - Things that we learn from all this

- in the long term, lightning has been the #1 killer in the U.S. but the fatality rate has declined exponentially
- single catastrophic events, such as Hurricane Katrina, may dominate short-term averages, causing an apparent increase in average fatality
- more people in the U.S. die from heat than from severe cold (at least 5 times more!)
- **over the last 30, 20 and 10 years, heat waves have been the #1 killer**
- even not counting the particularly fatal year 2011 deaths from tornadoes have increased over the last 20 years!
- **not counting Hurricane Katrina as single event, on average, heat waves and droughts are the costliest natural disasters in the U.S.**

Heat Waves and the Elderly[2]

In 2004, a year without heat waves the 60-90 age group accounted for 21% of the people dying from severe weather. In 2006, a year with heat waves, they accounted for 34%. Elderly people cope with heat less well than younger people. Older people, especially those living on a limited budget, also may not have air conditioning in their houses.

A particularly troubling heat wave occurred in 1995, almost exactly 10 years before Hurricane Katrina caused more than 1000 fatalities. The **Chicago Heat Wave July 12-16, 1995** also cost more than 1000 lives and there are remarkable parallels to the Katrina disaster:

- increased fatalities among elderlies
- increased fatalities among poor and blacks who lived in sub-standard housing
- lack of preparedness/overwhelmed emergency facilities

Aggravating factors included:

- urban heat island effect: paved ground retains heat during night (vegetation would release cooling water at night during respiration)
- temperature inversion trapped heat, humidity and polluted air near ground
- power failures
- poor people had no A/C or no money to use A/C
- people afraid of crime did not open doors and windows (during the heat wave of the 1930s, people slept outdoors) Wikipedia entry at http://en.wikipedia.org/wiki/Chicago_Heat_Wave

Heat waves may always have been a major killer but no long-term data are available at the NWS to test this. However, typical annual numbers have increased by a factor of four since the late 1980s (from 40 to 160). One could argue that the increase in heat-related deaths is a consequence of global warming but this hypothesis is difficult to test due to the lack of earlier data. Climate also has not change uniformly over the last 100 years. In fact, the 1970s were unusually cold. We will revisit this topic in lectures 15 and 23 on global climate change.

Severe Weather - An all too common natural disaster

On average, severe weather causes the most fatalities of all natural disasters. Tragically, some fatalities may be avoidable.

For example, the weekend of September 23, 2006 brought severe weather to the Midwest (Missouri) and South (Kentucky and Arkansas), with 8 fatalities by Sept 24 (12 by Sep 25). Of these 8, 1 was killed by a tornado, 1 was traffic related in heavy rain, 1 person was killed by lightning and 5 people died by flooding. All of these 5 people died in cars while the driver tried to cross a creek or other flooded regions. *Trying to cross flowing water is dangerous!* This includes creeks and flash flood areas in the desert.

- The water may be deeper than expected.
- The current of even shallow water can be strong and can carry cars away.
- Once the water reaches the exhaust pipe of a car, the engine stops running and people are trapped.

NOAA State of the global climate, issued since 1997 (http://www.ncdc.noaa.gov/sotc/global/)

References and Websites

1. Abbott, P.L., 2009. Natural Disasters. 7th edition. McGraw-Hill, New York, 504 pp.
2. Klinenberg, E. 2003. Heat Wave, a social autopsy of disaster in Chicago, University of Chicago Press, 328 pp. *This book describes the weather conditions during the July 1995 "Chicago Heat Wave" and what factors contribute to the disasters like this one. It discusses why so many people died and provides demographical insight into this particular event. It also tries to convey what it was like during this crisis.*
3. Wikipedia Webpage on the 1906 San Francisco Earthquake

Ellsworth, W.L., 1990. Earthquake history, 1769-1989. In: Wallace, R.E., The San Andreas Fault System, California: U.S.G.S. Professional Paper 1515, 152-187, online excerpts at http://earthquake.usgs.gov/regional/nca/1906/18april/index.php

Other Recommended Reading

Natural Disasters in General:

Robinson, A., 2002. Earthshock: Hurricanes, Volcanoes, Earthquakes, Tornadoes, and Other Forces of Nature, revised edition, Thames and Hudson Publ., 304 pp.

Chapter 2: Energy Sources of Earth Processes and Disasters

Natural disasters can have awesome powers of devastation. An asteroid impacting at high speed causes widespread destruction. Earthquakes are characterized by strong shaking, sometimes sending an all-engulfing tsunami across oceans. A volcano ejects debris in powerful eruptions . Landslides race down the hillsides. Flash floods advance with force and tear off everything in their path. Severe storms are accompanied by fierce winds and heavy precipitation. Tornadoes have the most violent and destructive winds on Earth, and the winds in fierce tropical storms are not far behind. Wildfires produce immense amounts of heat. All these processes fuel on large amounts of different types of energy. There are four basic types of energy relevant to natural disasters:

1. kinetic energy (energy of motion) can be used directly to drive destruction
2. potential energy (gravitational and elastic) can potentially be used to destruct
3. rotational energy plays a role in some types of mass movements and air circulation
4. thermal energy (**heat**) fuels Earth's internal and external processes, including volcanism and severe weather

Each of these types of energy can be transformed into other types of energy. Potential energy has to be transformed into kinetic energy to cause destruction, e.g. in mass movements and earthquakes. Potential energy is also transformed into rotational energy when a round body rolls down a slope. Kinetic energy can be transformed into potential energy when a lava bomb is ejected from a volcano and gains height. And kinetic energy can be transformed into heat when a falling space object hits Earth's surface at high speed. This chapter examines the different types of energy in more detail.

Earth tides deform Earth's surface everywhere on the planet. But tides manifest themselves best as the daily observable changes in sea level along the coast. Daily sea-level changes can reach several meters, and seawalls needs to be erected accordingly to protect buildings and other structures from flooding. The situation becomes particularly perilous when large waves from strong winter storms or from tropical storms add to the height of daily high tides. On top of this, we also have to be aware that twice a year the high tides become particularly high. The importance to understand tides and distinguish them from other effects on sea level changes has increased recently after it is becoming abundantly clear that global warming is having a measurable effect on sea level rise. But even if we do not have large-scale natural disasters in mind, a person walking along the beach needs to know the tide calendar. Beaches seemingly wide at low tide can shrink dramatically and even disappear by the time of the following high tide. This includes some spots along the beach of the Scripps Institution of Oceanography (SIO) in La Jolla, CA. Hikers walking from SIO north toward Del Mar may be cut off upon their return and may have to seek an alternative route back to SIO. Hikes on extremely wide beaches that extend for miles can become death traps because there will be no way to go to higher elevation when the tide comes in. A prime example of this is the Wadden Sea that stretches for 500 km (310 mi) along the Dutch, German and Danish shores of the North Sea[1]. This very shallow intertidal zone is a UNESCO World Heritage site and attracts 8 – 10 million people each year for mudflat hiking and bird watching. During low tide, the ocean can often not be seen from the shore, luring hikers miles out into the flats. But when the tide comes in only few hours later unwary tourists may be overcome by the rush of water and have to be rescued by lifeboat. Tourists are therefore strongly encouraged to venture out only with experienced, licensed guides.

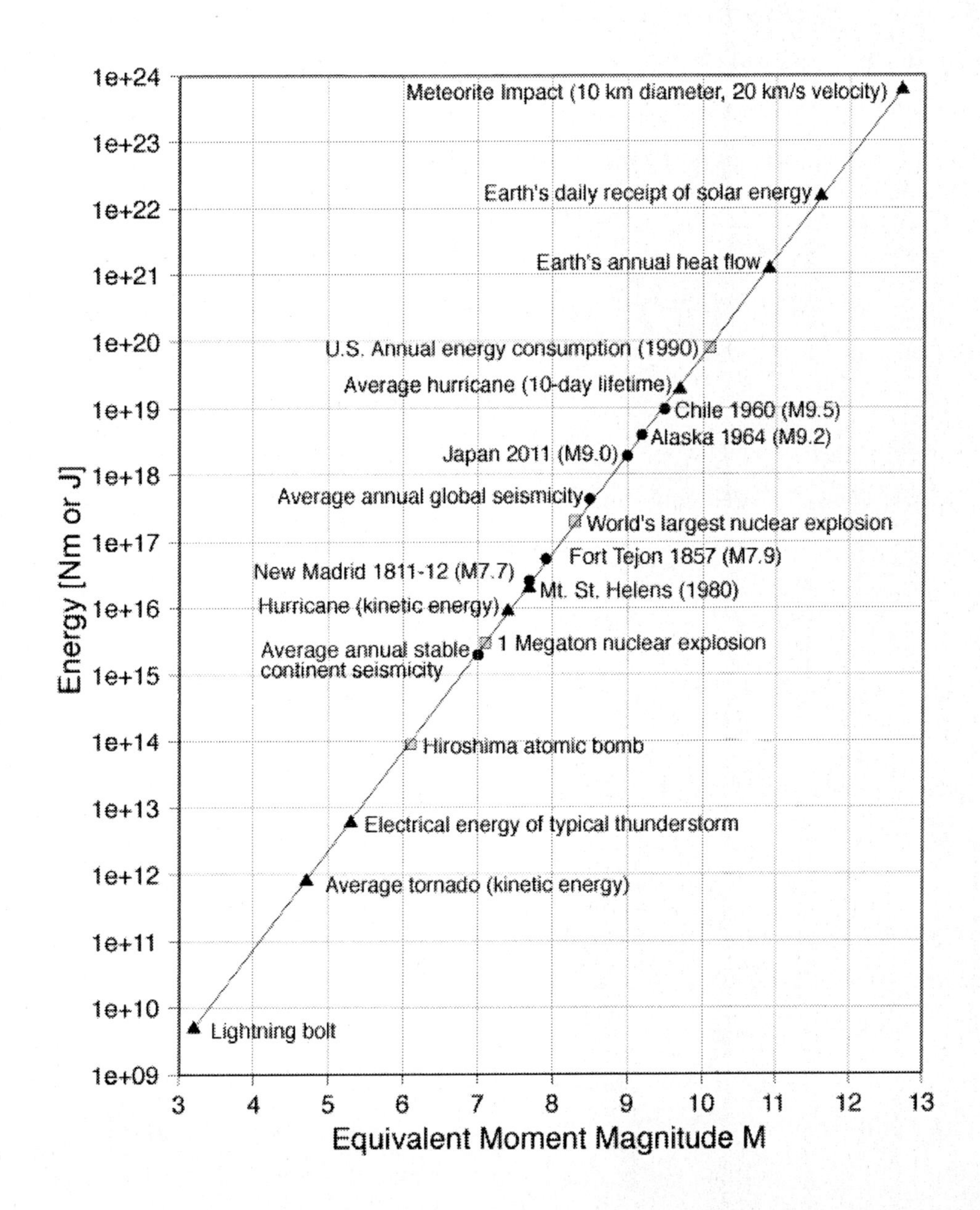

Figure 2.1 An approximate comparison of energy released by earthquakes of a certain magnitude with that of other disasters, natural and human processes. (data modified from Johnston[2])

2.1 Forces, Energy and Power

Earthquakes can causes extensive damage and destruction. So do volcanic eruptions. Large hurricanes are the most powerful storms in Earth's atmosphere. But no single event on Earth has been as powerful as a large asteroid impact. How do we measure the size of an event and how do we compare these events? We can evaluate a process by the amount of energy it releases. It turns out that this is not an easy task, for two reasons: different processes may release different kinds of energy and even one event many may release different kinds of energy. For example, an earthquake releases kinetic energy that results in shaking. Some energy is spent on generating earthquake waves to propagate through Earth. But the earthquake also spends energy to heat and crush the rock in the earthquake zone when two blocks scrape past east other and experience friction. Estimating this energy is quite difficult. In a volcanic eruption much energy is used to expel material but large amounts of heat are also released. Hurricanes fuel on heat that is released during condensation of water vapor. Earthquake specialists compare the size of an earthquake, and the energy released, through the moment magnitude. The moment magnitude scales logarithmically with energy so that 30 times more energy is released for every step on the moment magnitude scale. Seismologist Arch C. Johnston made an attempt at comparing the energy released by different processes (Fig. 2.1).

With a 9.5 magnitude, the 1960 Valdivia, Chile earthquake that sent a devastating tsunami across the Pacific Ocean has so far been the largest earthquake ever recorded. The devastating 11 March 2011 Tohoku, Japan earthquake had a 9.0 magnitude and released a little less than 1/3 of the energy of the 1960 quake. Even an earthquake as small as a 5.3 or so can cause significant damage even though it releases 1 million times less energy. How does this compare to the energy of other natural disasters? By the time it impacts on Earth's surface, a good-size asteroid has a huge amount of kinetic energy, maybe 100,000 times more energy than the 1960 Chile earthquake. The 1980 eruption of Mount St. Helens, the largest in the contiguous U.S. in at least 100 years, released energy comparable to that of each of the two magnitude-7.7 earthquakes near New Madrid, MO. The strong winds in a hurricane are indicative of large amounts of kinetic energy that compare to a 7.4-magnitude earthquake. But a hurricane lasts a lot longer than an earthquake, sometimes up to 2 weeks. During this time, the hurricane releases more energy than the largest recorded earthquake! In fact, if we could harness the energy of such a hurricane, we could cover more than a tenth of our annual energy consumption, in 1990 that is. Of course, this is not feasible. But this comparison gives us an idea of how much energy just one hurricane really packs, or on the other hand, how energy hungry our lifestyle has become. And we release some of this energy in bombs. The world's largest nuclear bomb test (1961 Russian "Tsar Bomba") released an equivalent of 50 megatons TNT and was comparable to a magnitude 8.4 earthquake, surpassing the largest earthquake in the contiguous U.S. (1857 Fort Tejon, CA)! Even an average tornado packs as much energy as a 4.8-magnitude earthquake. Though this seems a relatively small earthquake that would not cause much damage, a tornado uses all its energy in a very small area, often causing total destruction to everything in its path.

The Kinetic Energy of a Moving Body

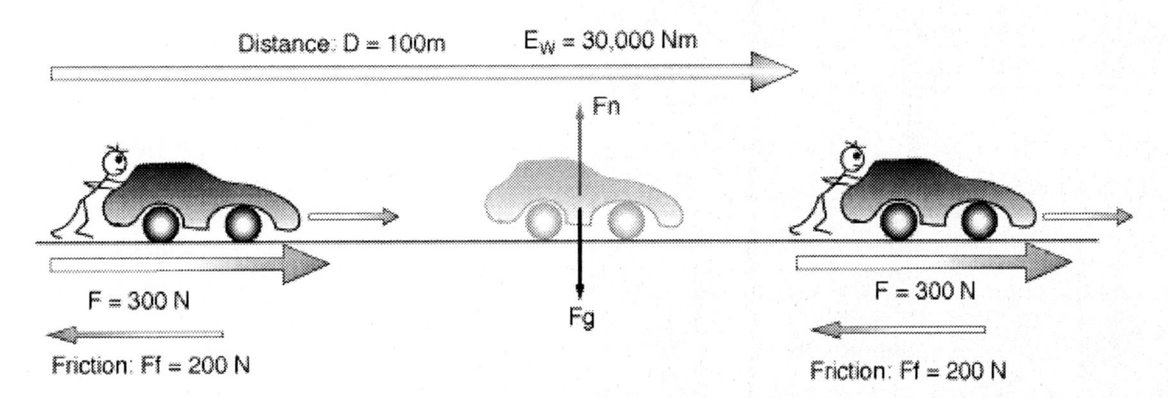

Figure 2.2a Concept diagram of how work is done on a body, i.e. energy invested, by exerting a force on it. A person pushes a car over a distance of 100 m, using a force of 300 N. The person does work on the car in the amount of 30,000 Nm.

Why do different processes have different types of energy, and how is this energy generated, stored and spent? Starting with mechanical processes, kinetic energy is the energy of motion. Any moving object, such as a landslide, a lava flow racing down a volcano, an asteroid approaching Earth, a building shaken by an earthquake but also winds in a storm have kinetic energy. Kinetic energy is related to **forces** that act on a body. Energy then is understood as **work** done on a body or the capacity to do work. An expression often used in the context is **power** which is the energy spent in a certain time. More power is invested on a process if a certain amount of energy is spent in less time than during a longer time period. For example, two runners of the same weight participate in a 10-mile walk. One finishes in 3 hours but the other in only 2 hours. Both runners spend the same amount of energy, but the second runner used more power.

In the context of discussing forces acting on a body, it is useful to recall **Newton's three laws of motion**:

1. An object continues in a state of rest or in a state of motion at a constant speed along a straight line, unless compelled to change that state by net force.
2. When a net external force, F, acts on an object of mass, m, the acceleration, a, that results is proportional to the force and has a magnitude that is inversely proportional to the mass. The direction of the acceleration is the same as the direction of the force.
3. Whenever one body exerts a force on a second body, the second body exerts an oppositely directed force of equal magnitude on the first body.

To understand the significance of Newton's laws of motion, we consider a person pushing a car (Fig. 2.2a). The first law explains why the car starts moving after it is pushed. The person pushing it exerts a force, F, and applies work, E_W, on the car. The work done is proportional on the distance, D, over which the car is pushed: $E_W = F \cdot D$. If the person exerts a force of 300 N on the car, over a distance of 100 m, then the work done is 30,000 Nm. The SI unit (International System of Units) for mechanical energy is 1 Nm = 1 J, where the N stands for "Newton" and the J stands for "Joule". In scientific texts, F and a would be bold-faced to signify that forces and accelerations have a direction. To keep things simple, we omit this style of writing but use italics for all symbols.

The second law explains how the state of the car changes. A body that experiences a force changes its current state. If the person in Fig. 2.2a keeps pushing the car with the same force, it accelerates until the person stops pushing. The force is related to acceleration and mass of the car through $F = m \cdot a$. The SI unit for force is $1\ N = 1\ kg{\cdot}m/s^2$. If we ignore friction for now, and assume that the car has a mass of 800 kg, and the person keeps pushing with the same force, then the acceleration would be $a = F / m = 0.375\ m/s^2$. But friction actually slows down the progress. The person pushing the car has to spend 200 N just to overcome the resistance of the car to being pushed across the ground, so that only a force of 100 N is actually spent on accelerating the car. This gives $a = F_{100} / m = 0.125\ m/s^2$.

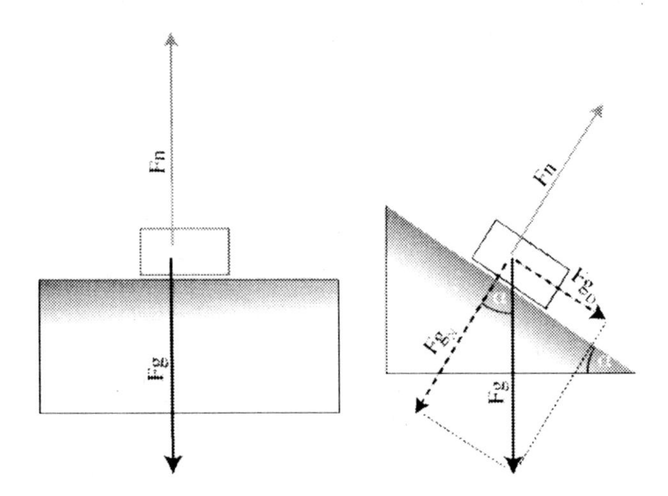

Figure 2.2b Concept diagram of the normal force acting on a body sitting on a flat surface (left) and one sitting on a slope (right). According to Newton's third law, the ground pushes back with the normal force, Fn, that has the same strength as the part of the body's gravitational force that pushes perpendicular to the surface. For horizontal surfaces, $Fn = - Fg$. But for a sloped surface, Fn is smaller. In this case, Fg is decomposed into a component perpendicular to the surface, Fg_N (N stands

for normal), and one parallel to the surface, Fg_D (D stands for downslope). The normal force is then smaller: $Fn = -Fg_N = -\cos a \bullet Fg$, where a is the slope angle.

The third law signifies that when the person pushes the car, the car actually pushes back with the same force, i.e. 300 N, and the person pushing the car feels this force. Newton's third law also explains the somewhat odd-appearing concept of the normal force, Fn, a surface exerts on a body (Fig. 2.2a). If the surface is horizontal, then Fn points straight up in response to the gravitational force of the body, $Fg = m \bullet g$, where g is Earth's gravitational acceleration $g = 9.81$ m/s^2. In our case, the normal force is $Fn = -7{,}848$ N. The minus sign signifies that Fn points in the opposite direction of Fg. If the surface is tilted, e.g. when a body sits on a slope, then the normal force counteracts the component of Fg that is perpendicular to the slope (Fig.2.2b). The frictional force, Ff, is proportional to Fn and the friction coefficient, μ, which essentially expresses how much resistance a rough surface provides against moving a body across it. Note that, since Fn is proportional to the mass of the body, the frictional force is as well.

We can also answer the question how fast the car moves after the person has pushed the car over distance D. For this, we use the fact that the car has gained kinetic energy during its acceleration. The kinetic energy, E_{kin}, is proportional to a body's mass and the square of its velocity, v, through $E_{kin} = \frac{1}{2} \bullet m \bullet v^2$. Recall that the work done on the car was 30,000 Nm, but we ignored friction to get this number. The actual energy that was available to the car to accelerate was $E_{W100} = 100$ N • 100 m = 10,000 Nm. At the end of the 100 m, this energy is transferred into kinetic energy of the car. We can now determine the final velocity as

$$v = \sqrt{\frac{2 \bullet E_{W100}}{m}} = 5 \text{ m/s (18 km/h; 11.2 mph).}$$

Of course, this is only a thought experiment because it is unlikely that a person can push a car continuously with such a force over such a long distance.

COPYRIGHTED MATERIAL — DO NOT DUPLICATE, DISTRIBUTE, OR POST

Potential Energy and Energy Transfer

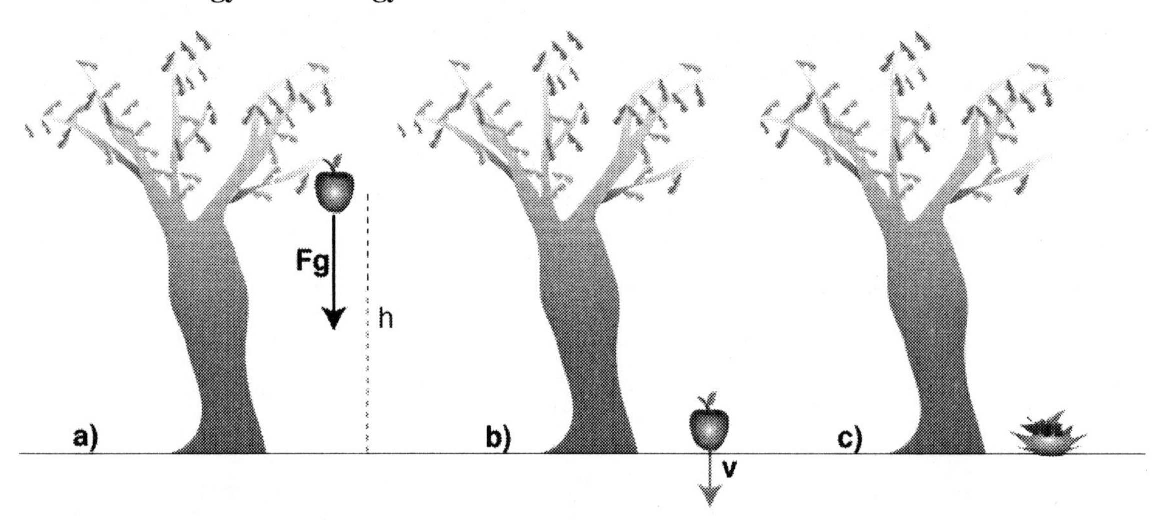

Figure 2.3 The free-falling apple. An example of how potential energy is transformed into kinetic energy. This concept applies to falling and impacting objects such as an asteroid.

The second type of energy relevant to natural disasters is **potential energy**. This energy is essentially stored energy that could potentially be used in a process. Here, again, we concentrate on mechanical processes. Fig. 2.3a shows an apple that hangs on a tree. It is at rest, at height h above the ground, say $h = 5$ m. However, it is only at rest because the tree holds it there while Earth's gravity exerts a gravitational force on the apple where $Fg = m \cdot g$. If the mass, m, of the apple is 100 g, then the force acting on the apple is 0.98 N. Newton's third law applies here because the tree exerts a force with the same amount but in opposite direction to hold the apple in place. The gravitational force could potentially be used to apply work on the apple over the height, h: $E_W = Fg \cdot h$. This work is the potential energy of the apple, with $E_{pot} = m \cdot g \cdot h = 4.9$ Nm. The apple will fall toward the ground once detached. In fact, it will accelerate because Fg continues to act on it until it hits the ground. While it is falling, it loses potential energy (because h becomes smaller) and it gains kinetic energy, because its velocity increases. Potential energy is transferred into kinetic energy, and the apple loses potential energy at the same rate as it gains kinetic energy. In the split-second before the apple hits the ground, all potential energy is transferred into kinetic energy, and we can easily compute the final velocity as

$$v = \sqrt{\frac{2 \cdot Epot}{m}} = \mathbf{9.9\ m/s\ (35.6\ km/h;\ 22.1\ mph).}$$

At such a high speed, it would be no wonder that the apple does not survive the fall intact. Especially on a hard surface that does not give, all the kinetic energy is now used to crush the apple. Increasing the mass of the impacting body and the distance from Earth's surface, an asteroid or a piece of it impacting on Earth has so much kinetic energy that Earth's surface no longer withstands. Kinetic energy is then used to also crush Earth's surface to form an impact crater. Some energy is also transferred to heat upon impact to melt rock in a characteristic way. So-called tektites are tell-tale signs of impacts on Earth and help distinguish an impact crater from a volcano crater.

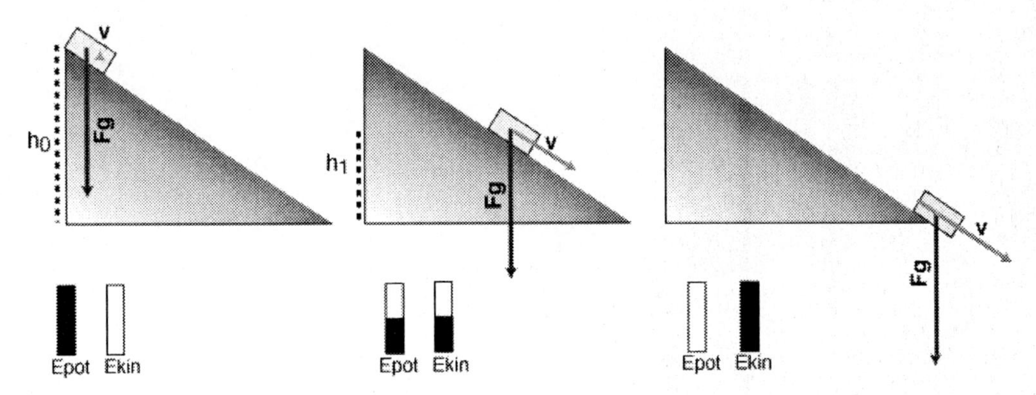

Figure 2.4 A block sliding down a slope. An example of how potential energy is transformed into kinetic energy. This concept applies to landslides.

A similar exchange between potential and kinetic energy occurs when a body moves down on a slope (Fig. 2.4), such as occurs during a landslide. Without going into too much detail until Chapter 8, a mass initially at rest (left) has the potential energy $E_{pot} = Fg \cdot h_0$ and the kinetic energy $E_{kin} = 0$. After it starts moving down the slope, it loses potential energy and gains kinetic energy. Halfway down the slope, half the potential energy is transferred into kinetic energy and $E_{pot} = Fg \cdot h_1 = E_{kin}$. At the bottom of the slope, the mass spent all its potential energy and now reached the highest velocity before it deforms and comes to rest. If there was no friction, potential energy is transferred only into kinetic energy and the height at which both energies would equal is $h_1 = ½ \cdot h_0$. But friction actually held the mass on the slope to start with, and so some potential energy is lost on the way down to overcome friction. In this case, the point at which $E_{pot} = E_{kin}$ is farther downslope and the final E_{kin} near the bottom is smaller than the original E_{pot} when the mass was at rest near the top. Once the mass reaches the bottom, the kinetic energy is spent to disintegrate the mass, similar to the case with the falling apple.

The Potential Energy of a Spring and along an Earthquake Fault

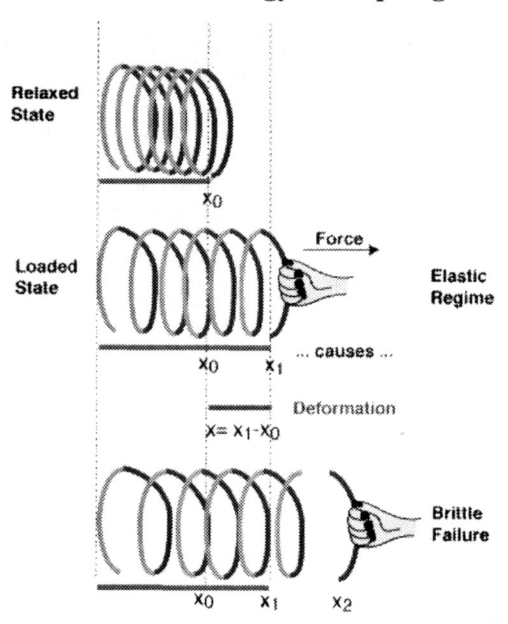

Figure 2.5 Loading a spring that was initially in a relaxed state (above) is an example of how elastic potential energy can be stored. The range within which a loaded spring returns to its initial state after releasing the end is the elastic regime. If we pull hard enough, the spring will eventually break during brittle failure. We use this concept to study earthquakes.

The potential energy just described in also called **gravitational potential energy** because it involves Earth's gravitational potential, or simply put, the ability of Earth to attract other bodies, and we will get back to this later. Potential energy can also be stored through other means. When we pull on a coil spring such as shown in Fig. 2.5, energy is stored in the spring in form of **elastic potential energy**. Again, this is called potential energy because it can be used and transformed into other kinds of energy. For example, when we let go of the spring, it will move and the potential energy transforms to kinetic energy. The energy is called elastic energy because the spring goes back to the shape it had before we started pulling (see Box 1). In this thought experiment, we assume that the left end is attached to a solid frame. In the relaxed state of the spring, its right end rests at location x_0. After we pull with force Fx, the end moves to location x_1, so the end moved a distance x which is also called the **deformation**. Note that **force causes deformation**. The force with which we have to pull on the end is proportional to the stiffness of the spring. This stiffness is expressed by the **spring constant**, k. It is greater the stiffer the spring is. The force also increases as we pull farther. The force then is defined as $Fx = k \cdot x$, and the elastic potential energy is $E_{pot} = \frac{1}{2} \cdot k \cdot x^2$. We can pull on the spring to a certain extend before it stops to return to its original state after release. This is called the **elastic regime**. If we pull beyond this point, the spring will either experience permanent, **plastic**

deformation (see Box 1). If we pull farther yet, the spring will eventually break. This is also called **brittle failure.** Plastic deformation and brittle failure are irreversible processes.

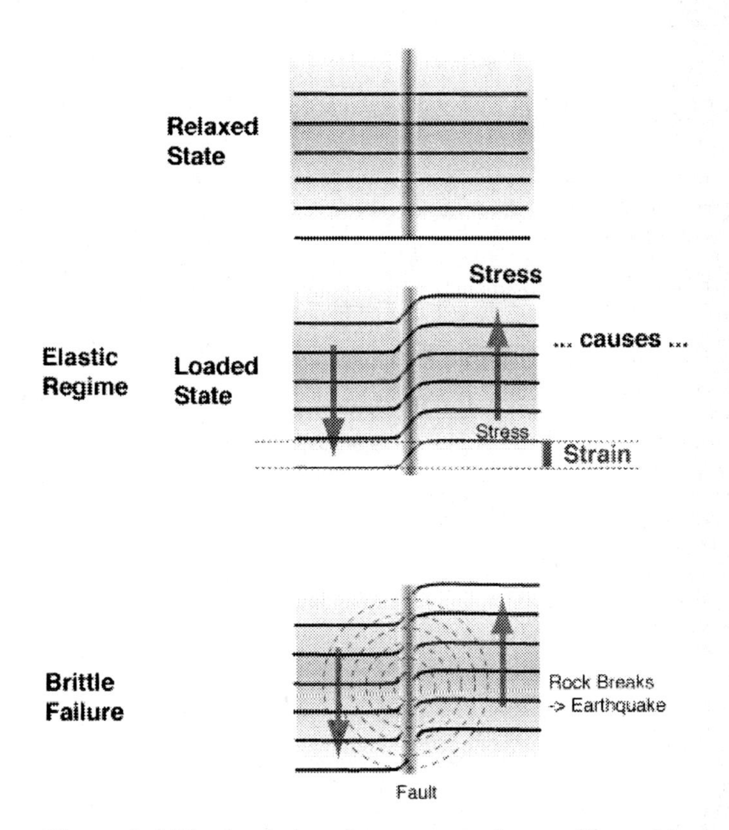

Figure 2.6 The loaded-spring concept also applies to blocks sliding past each other. This concept describes the process of loading an earthquake fault.

The loaded-spring concept can be applied to a slightly more complex situation of two blocks being attached along a deformable contact zone (Fig. 2.6) that we call the earthquake fault. If the blocks hold tight to each other along the fault, the **fault is locked**. This is accomplished by protrusions (**asperities**) along the rough edges of the two blocks that hook into each other. Just as a coil spring can be loaded by exerting a force on one end, the fault can be loaded by pulling the blocks in opposite directions along the fault. Here, the cumulative force exerted on the fault is called the **stress**. In correspondence to the deformation of a spring, the relative movement of the blocks with respect to each other is the **strain**. Note that the **stress causes strain**. In the elastic regime, the blocks go back to their original position if the pull on the blocks stops. Similar to the spring example, there exists a regime where so much stress is applied that the fault experiences permanent plastic deformation. The fault then **creeps** without producing and earthquake. Brittle failure, or an earthquake, occurs when the stress is so great that the friction in the fault is overcome, the protrusions

COPYRIGHTED MATERIAL — DO NOT DUPLICATE, DISTRIBUTE, OR POST

break and the blocks start to glide past each other instantaneously. In the real world along an earthquake fault, not all protrusions are the same and the equivalent of a spring constant may be different. Hence, with the same amount of stress, a segment along the same fault may be creeping, while another segment may be locked, thereby increasing the likelihood of a larger earthquake. The pulling stress described above is also called **shear stress**. When the blocks are pulled apart, the acting forces cause **tensional stress**. If the blocks are pushed together, the fault experiences **compressional stress**. In a tensional and compressional stress regime, the motion along a fault may occur in a direction perpendicular to the acting forces, i.e. along the fault but not in the same direction as the stresses act.

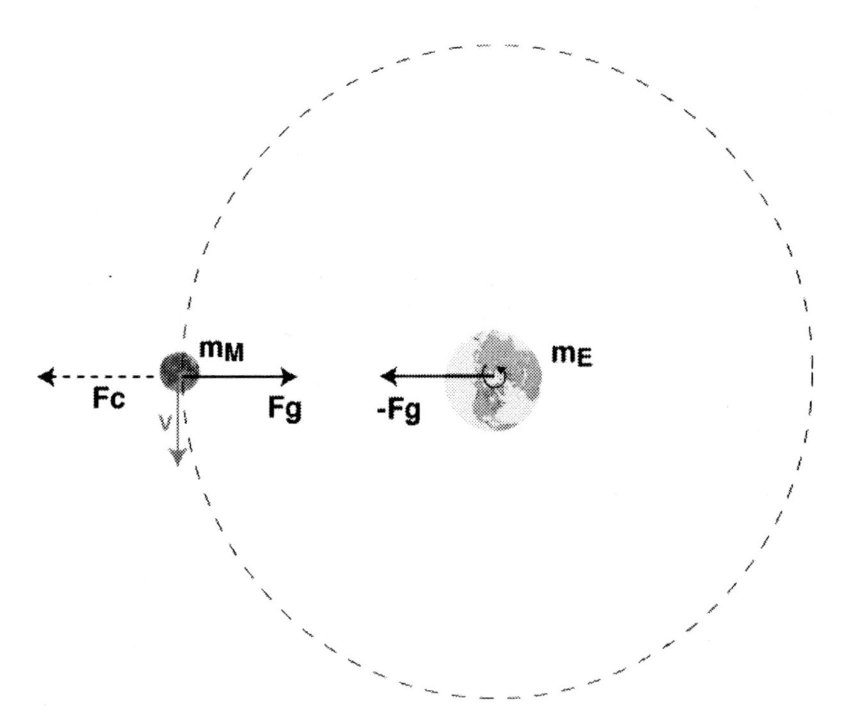

Figure 2.7 The gravitational force on the Moon exerted by Earth, *Fg*, is compensated by the centrifugal force, *Fc*, holding the Moon in orbit around Earth. This is an example for a body having both potential and kinetic energy. The spinning Earth (once a day) is an example of a body having rotational energy.

Gravitation and Rotational Energy

Above, we learned that gravitational potential energy is a result of the fact that Earth attracts bodies such as the apple falling from a tree. Because of Newton's third law of motion, it is not only true that Earth attracts the apple, but the apple also attracts Earth with the same force. The formulae for the gravitational force and the potential energy in that case were simple because Earth is such a huge body compared to the small apple. The more general formula to express the mutual attracting force between two bodies, such as Earth and the Moon (Fig.2.7), follows Newton's Law of Universal Gravitation: every particle in the universe exerts an attracting force on every other particle. The formula to describe this is:

$$Fg = \frac{G \bullet m_E \bullet m_M}{r_M^{\;2}} \quad \textbf{(2.1)}$$

where the symbols are defined as follows:

G: gravitational constant; 6.67384×10^{-11} m^{3}kg^{-1}s^{-2}

m_E: mass of Earth; 5.9722×10^{24} kg

m_M: mass of Moon; 7.3422×10^{22} kg

r_M: distance between the centers of mass of the two bodies;

(average Earth-Moon distance: 384,399 km)

First, if we plug in all the numbers, we obtain the attracting force between Earth and Moon as $Fg_{(E\text{-}M)} - 1.98 \times 10^{20}$ N. We can also calculate the attracting force between Earth and Sun. With $m_S =$ 1.9891×10^{30} kg and the mean Earth-Sun distance $r = 1.496 \times 10^8$ km, $Fg_{(E\text{-}S)} = 3.54 \times 10^{22}$ N. These are very large numbers, compared to the 0.98 N the falling apple experienced. The question is whether the formula above resembles the formula $Fg = g \bullet m$ that we used for the apple. Indeed, it does, because

$$g = \frac{G \bullet m_E}{r^2} = \mathbf{9.81\ ms^{-2}}$$

when we plug in (we substitute) Earth's radius as $r = 6{,}371$ km. The 5 m the apple is above Earth's surface can be ignored here because it is so much smaller, but to be more mathematically correct, $r =$ 6,371,005 m. This shows, however, that the same physics applies to the Earth-apple system as to the Earth-Moon system. For objects that are much closer to Earth's surface compared to Earth's radius, we can use the simple formula. For everything else the formula above is applied.

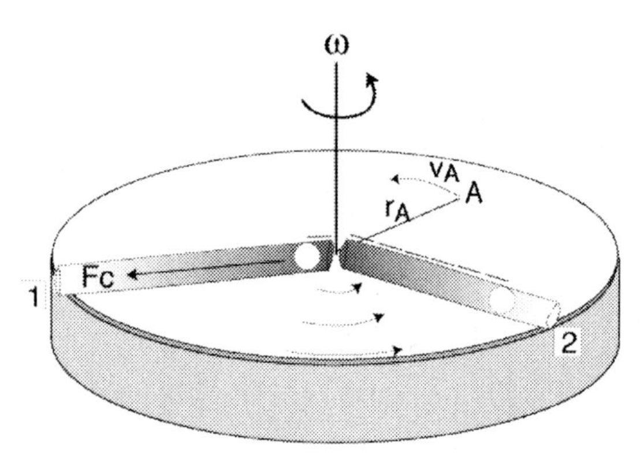

Figure 2.8a Visualization of the centrifugal force, *Fc*. We place a freely moving ping-pong ball into an open tube that is fixed to a rotation disk. When the disk starts to spin, the ping-pong ball will move outward, pulled by the centrifugal force, *Fc*. By the time the disk moves from point 1 to point 2, the ping-pong ball moved in the tube by 5 units (dashed line). The centrifugal force is an outward-pointing inertial force that initially resting bodies experience in a rotating system. The strength of the force increases with the square of the angular speed, □ = $2\pi f$, where f is the rotation frequency. At point A, the angular speed is related to the velocity of the point through v_A = □ • r_A.

Figure 2.8b American hammer throw athlete John Flanagan competing at the Summer Olympics 1908 in London, UK. The athlete accelerates a 16-lb (7.3 kg) ball that is attached to a wire by swinging the ball in circles above his head and by turning a few times himself. When the ball gained enough speed, the athlete releases the ball. The name "hammer throw" comes from older versions of the sport when a sledge hammer was thrown. Hammer throw is an Olympic sport still today. (source: Wikipedia).

The question now is, if the Earth-Moon attracting force is so large, why does the Moon not fall onto Earth? The answer is because the Moon orbits around Earth at a distance at which the centrifugal force, *Fc*, exactly counteracts the gravitational force, *Fg* (Fig. 2.7). The centrifugal force is not a real force but a fictitious force. Without emerging into deep discussion on reference frames, the centrifugal force is an inertial force that feels like a force to the orbiting body as a result of the orbital movement. We can "visualize" the centrifugal force, by watching objects move on a rotating disk (such as a record player from the old days) (Fig. 2.8a). The tube is needed in the experiment to keep the ping-pong ball on the disk long enough for the experiment. Without the tube, the ping-pong ball would quickly move sideways off the disk, in the direction of the spinning disk. On a faster spinning disk, *Fc*, is strong enough to push the ball out of the tube. The outward acceleration increases with the square of the angular speed of the disk and with distance from the center: $a_c = \omega^2 \bullet r$; where $Fc = a_c \bullet m$; and m the mass of the ping-pong ball. With $v = \square \bullet r$ the acceleration can also be written as $a_c = v^2/r$. To keep the ball on the disk, a force counteracting *Fc* needs to come to play. For example, we could attach a string to the ball that is fixed near the center of the disk. If we spin the disk fast enough that the string experiences tension, the ball will experience *Fc* pointing outward, but also an inward-pointing force exerted by the string that counteracts and is in balance with *Fc*. This is the centripetal force, the same force with which an athlete performing a hammer throw holds the revolving ball back with a wire before releasing it (Fig.2.8b). Getting back to the Earth-Moon system, the gravitational force between Earth and the Moon counteracts, and is in balance with *Fc* and $Fc = -Fg$ where the minus sign expresses that they point in opposite directions (Fig. 2.7). With its center rising $h = 378{,}028$ km above Earth's surface (384,399 – 6371 km), the Moon has potential energy that we can calculate using the formulae above. The Moon also has kinetic energy $E_{kin} = ½ \bullet m_M \bullet v_M^2$, where $v_M = 1.022$ km/s is the velocity along Moon's orbit around Earth. We can also verify that $Fc = a_c \bullet m_M = v_M^2 / r_M \bullet m_M = Fg$ (ignoring the minus sign here).

Any spinning body, such as the revolving Earth has rotational energy (Fig 2.7). Without getting into too much detail, Earth's rotational energy is $E_{rot} = ½ \bullet I_E \bullet \square_E^2$, where I_E denotes Earth's inertia and $\square_E$ denotes Earth's rotation rate. Rotational energy is also spent in slumping mass movements where a mass tears off the top and slides along a curved surface. The toe of the slump (bottom end) is then lifted up relative to its position before the slumping movement. Rotational energy also plays an important role in destructive tornadoes and large storm systems.

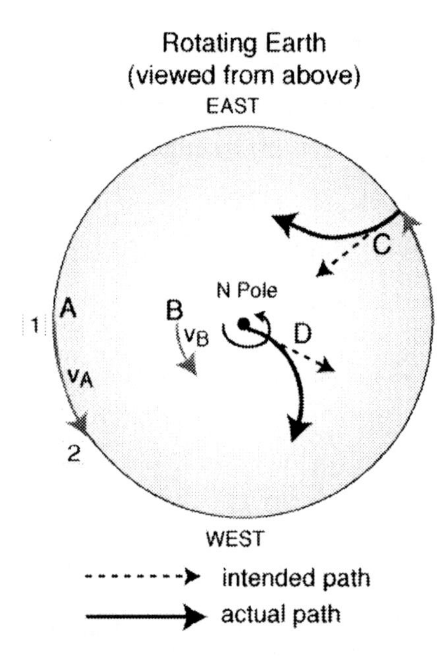

Figure 2.9 Movements on Earth as seen from the North Pole. From this perspective, movements can be compared to those on a rotating disk such as an old music record.

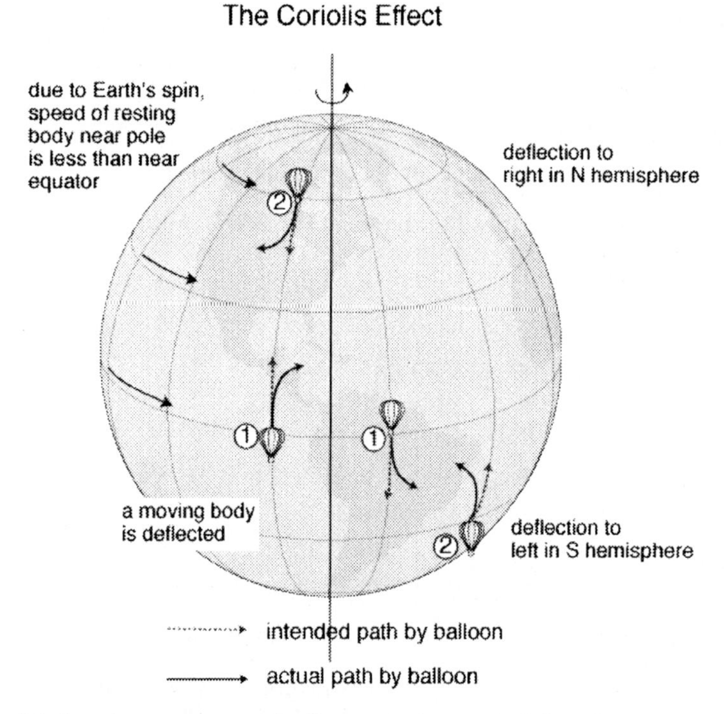

Figure 2.10 Movements on Earth, now considering its spherical shape. The Coriolis Effect deflects moving objects in the Northern Hemisphere to the right, and in the Southern Hemisphere to the left.

2.2 The Coriolis Effect

The spinning-disk experiment in Fig. 2.8 reveals another very important concept. As mentioned, the tube on the disk prevented the ping-pong ball to move sideways. We now repeat the experiment without the tube. This experiment is equivalent to observing objects moving on the spinning Earth, when we look down onto the North Pole (Fig. 2.9). An important point to observe here is that, as the disk moves from point 1 to point 2, locations closer to the center of the disk (B) travel a shorter path than points closer to the edge of the disk (A). Consequently the speed at which point B travels is less than that of point A, i.e. $v_B < v_A$. If a body initially attached to the disk now moves radially inward (C) or outward (D), it will retain its initial speed at which it moved around the center of the disk. If we now release the ping-pong ball from point A and send it straight to the center, it will advance relative to the intended path because it has an excess speed v_A by the time it reaches point B (displayed as scenario C). Similarly, a ball released radially outward from close to the center will stay behind because now it has a lower speed by the time it reaches an outer point (displayed a scenario D). In both cases, the ball will deflect to the right. If we wanted to simulate movements with view on Earth's South Pole, then we would spin the disk in the other direction and objects on the disk would be deflected to the left.

This same effect is observed on a rotating sphere, such as Earth (Fig. 2.10). A balloon starting northward near the equator is equivalent to sending the ball on the spinning disk from the edge toward the center. On Earth, the balloon will eventually advance and be deflected to the right. If the balloon starts at higher latitudes and moves south, then it will stay behind but also be deflected to the right. Objects moving away from equator are deflected eastward, while objects moving toward equator are deflected westward. Objects in the **Northern Hemisphere** are therefore always deflected to the **right**, while objects in the **Southern Hemisphere** are deflected to the **left**.

The effect just described is the **Coriolis Effect**. It explains how objects are deflected on a spinning body. The Coriolis Effect comes into play when studying air movements such as winds. We will also learn that air movement is cyclonic in large storm systems, as a result of the Coriolis Effect. Ocean currents are also affected by this, and even coastal upwellings and downwellings have something to do with the Coriolis Effect. However, one exception should be noted here already, despite what is commonly said: in contrast to storm systems, the way in which tornadoes rotate is **not** a result of the Coriolis Effect but mostly where in a thunderstorm a tornado develops (more on this in Chapter 14).

2.3 Earth's Principal Types of Energy

We will now bring the knowledge we gained on different types of energy into the context of energy sources on Earth and of natural disasters in more detail. So far, we have discussed three principal forms of energy but we have left out the forth one, heat. Depending on where the heat comes from, it fuels different processes on Earth. Hence we will distinguish between Earth's internal and external heat sources. We then distinguish between five types of energy.

1: kinetic energy: the energy of motion. During Earth's early history, large amounts of kinetic energy from impacting space objects (asteroids, comets, space debris) were converted to heat upon impact. In fact, there were so many impacts that this was one of the principal ways to heat up the planet internally. The amount of energy the space object brings with it increases with its speed and mass. Volcanic debris that is ejected from a volcano at high speed has large amounts of kinetic energy. The debris can race down the slopes of the volcano, thereby gaining even more kinetic energy by transforming some potential energy, only to cause great destruction at the foot of the volcano and surrounding area. With increasing amounts of initial kinetic energy, the debris can flow for longer distances, thereby affecting areas farther away from the volcano. Kinetic energy is also relevant to processes in the atmosphere. Wind storms with stronger winds have more kinetic energy that is spent on moving and destroying anything from small objects to large structures.

2: potential energy: through Earth's gravitational field, elevated objects are pulled toward Earth's surface. The attracting force, and the amount of potential energy, increases with the mass of the attracted body and with its distance from Earth's surface. A space object has more potential energy the farther away it is from Earth. If it is on collision course with Earth, it can transform more potential energy into kinetic energy to gain more speed upon impact, causing more destruction. Even processes less dramatic than impacts are fueled by potential energy, such as landslides, snow avalanches and other downhill movements. A slide initiated at low elevation, with less distance to travel down a slope will likely cause less destruction than one starting higher up. The volcanic debris mentioned above also is given potential energy when it exits the crater of a volcano, with more energy as the elevation of the crater increases. Processes in Earth's atmosphere are also fueled by potential energy. Precipitation such as rain and hail are formed high above Earth's surface. Large hailstones therefore have large amounts of potential energy to transform into kinetic energy. Since energy scales with the mass, a grapefruit-size hailstone can cause a lot of damage to a brand new car. But even inside Earth, gravitation and therefore potential energy plays a major role to layer and turn over Earth's mantle (called mantle convection). In fact, gravity is the primary cause for plate tectonics. We also learned that earthquake processes are related to the release of potential energy, elastic potential energy in this case.

3: rotational energy: we touched on this type of energy only briefly. In the big picture of Earth processes, it may not play such an important role. But understanding the details in some natural disasters requires the consideration of rotational energy. In some types of slumps, the moving mass actually stays pretty much were it is. Nevertheless, it can cause great destruction. As mentioned above, in a rotational slump, the mass breaks off near the top and moves along a curved surface beneath. The head of the slump drops, causing damage to structures that once stood on the top. Structures that were located on the toe of the slump will likely be destroyed as they cannot respond to the deformation experienced by the toe. The destruction is caused by the release of rotational energy in the slump. Strong cyclonic storms have large amounts of rotational energy. Tornadoes, the atmospheric phenomena with the highest surface wind speeds on Earth, are ferocious because of rotating winds. Rotational energy also plays a great role in disaster prevention. For example, to prepare tall buildings or elevated freeways for earthquake shaking, they are often put on protective bearings that pretty much resemble bearings in the wheels of an inline skate or skateboard. When the ground shakes, the kinetic energy from the shaking is taken up by the bearings that transform the kinetic into rotational energy that is spent internally in the protective bearings. So, locally, rotational energy plays an important role.

4: Earth's internal heat: this is a very important energy source to fuel processes in the solid Earth. This heat is mostly a relic of Earth's early formation. Together with gravity, Earth's internal heat drives mantle convection and ultimately plate tectonics. Earth's internal heat is released during volcanic eruptions, but also in hot springs. Through plate tectonics, Earth's internal heat is also responsible for earthquakes. NB: the heat released during earthquakes is insignificant. So, where did the internal heat during Earth early formation come from? The heat resulted from four basic processes:

a) **impact energy during formation**: large impacting space objects spent huge amounts of kinetic energy upon impact to deform and/or destroy both itself and Earth's surface layers. Much of this energy, however, was transformed to heat. During Earth's early history, much more space objects impacted on Earth than nowadays, so impacts were a significant source of heat.

b) **energy from initial gravitational compression**: initially, when the body later to become Earth formed, the particles were held together rather loosely, like in some of today's smaller asteroids. But when the planet grew, so did its gravitational forces. Eventually, the pressure on Earth's interior closer to the center exerted by the material above grew so large that the inside started heating up.

c) **early differentiation**: during this process, gravity was at play. When Earth grew through the capture of more space debris, the heavier elements in the captured material, such as iron, started to sink toward Earth's center. During sinking, the elements lost potential energy that was eventually turned into heat.

d) **decay of radioactive elements**: this process is still ongoing but was much more prominent in Earth's early history. During radioactive decay, the atoms of an element are split to form atoms of new, lighter elements. For example, uranium and potassium are naturally occurring radioactive elements that experience decay. During the splitting of atoms, some energy is also released. The process of splitting atoms is called **fission**.

5: Earth's external heat: as noted above, impacting space objects provide energy to heat Earth's interior, but they also heat its surface. This was relevantly particularly throughout Earth's very early history, but impacts are rare today. And hopefully, this will not occur during our lifetime. Through radiation, solar energy is now Earth's primary external heat source and drives processes in the atmosphere, including convection. When Earth's surface air is warmed by solar heating, it tends to rise, thereby replacing colder air aloft that then may sink. Ultimately, this convection drives weather, sometimes severe weather and great storms. Solar heat also drives evaporation. We will see below that evaporation costs energy, and solar heat provides this energy. This energy is not lost. Rather, it is stored as latent heat that can later be regained during condensation. This very important process fuels some of the greatest storms on Earth, tropical cyclones. The amount of heat Earth receives from the Sun is vastly greater than the amount of heat Earth releases. Earth's surface received 5300 more heat from the Sun than from within. As radioactivity is one of the sources for Earth's internal heat, radioactivity also produces Sun's energy. However, the process in the Sun is **fusion**, not fission. In fusion, atoms are fused together to form new, heavier elements. As with fission, energy is released during this process. Currently, hydrogen atoms fuse to form helium atoms in the Sun. The Sun's core is heating up as fusion continues. Eventually, a long time from now (in about 5 billion years), helium fusion will start to produce carbon.

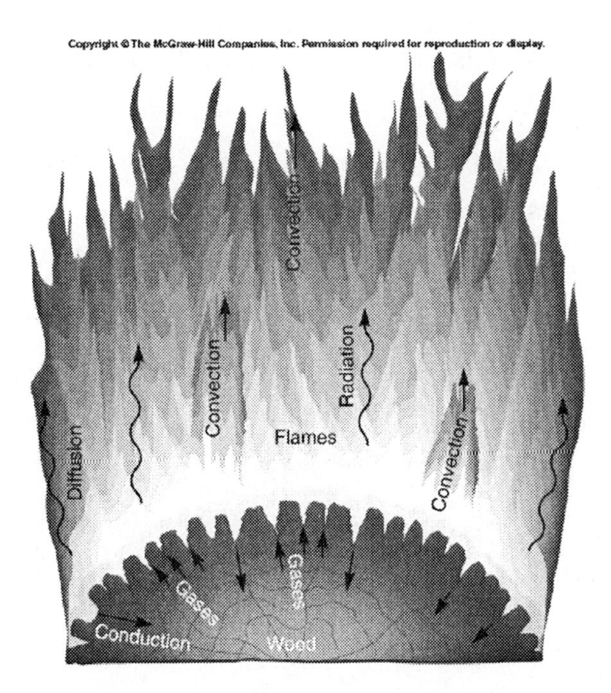

Figure 2.11 The four ways to transport heat, illustrated on a burning wood log: conduction, radiation, diffusion and convection. (source: AB)

The Transfer of Heat

Heat can be carried through a body such as a burning wood log or Earth, and even through Earth's atmosphere in four basic ways (Fig. 2.11): by conduction, radiation, diffusion and convection. During conduction, radiation and diffusion, heat is always transported along a temperature gradient, i.e. between locations with different temperatures, when heat is carried from warmer areas to cooler areas. During **conduction**, heat flows through a body. Energy is passed on as kinetic energy between electrons or atoms within the body. The atoms within a body are fixed in place but they can still vibrate, and transfer energy, depending on the temperature of the body. In the air, the conduction of heat takes place through the collision of molecules. In good thermal conductors such as metals, the heat transfer happens rapidly while in poorly conducting solids such as wood, heat transfers very slowly. Wood is therefore a relatively good insulator against heat, but also against cold.

Thermal **radiation** occurs in air. Here, only electromagnetic energy is emitted without any particle movement involved in the transfer. Strictly speaking, particles (atoms) moving in the hot log emit this radiation but they are not involved in the actual heat transport. A person sitting close to a fire log feels the heat (which is infrared electromagnetic waves) radiating from the log even though the surrounding air may be cold.

In thermal **diffusion**, single particles carrying the heat actually move along with the heat instead of colliding with other particles as is the case with conduction, although the boundary between diffusion and conduction is somewhat fuzzy. This process is mostly relevant in gases, such as air, and liquids.

In **convection**, entire portions of material move with the heat, i.e. mass transfer occurs along with heat transfer. This mode of heat transport is the most effective one. Since material is moved, this mode of heat transfer has a mechanical aspect as well. In Earth's interior this process is extremely important to transport heat from the center to the surface and is mostly responsible for Earth's loss of heat since the planet's early formation.

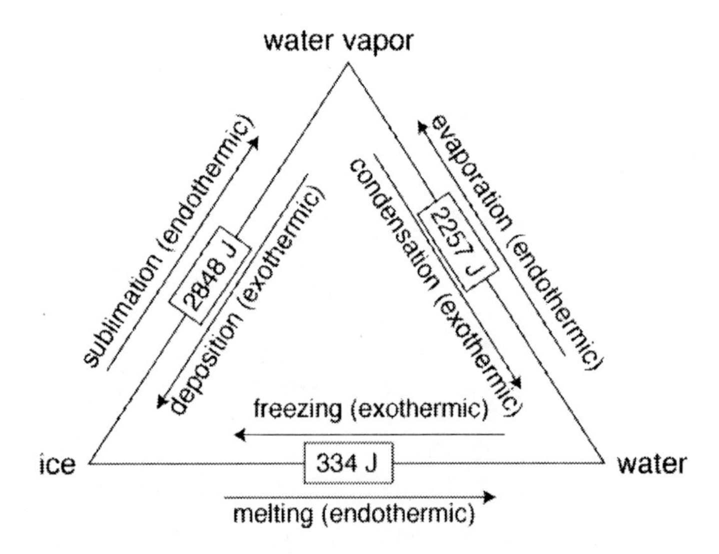

Figure 2.12 The three phases of H_2O and the naming of phase changes. Processes from solid to liquid or gas are endothermic (energy is required), as is the change from liquid to gas. The reversed phase changes are exothermic (energy is released). Numbers along the paths indicate how much energy is required or released during the phase change of 1 g of H_2O.

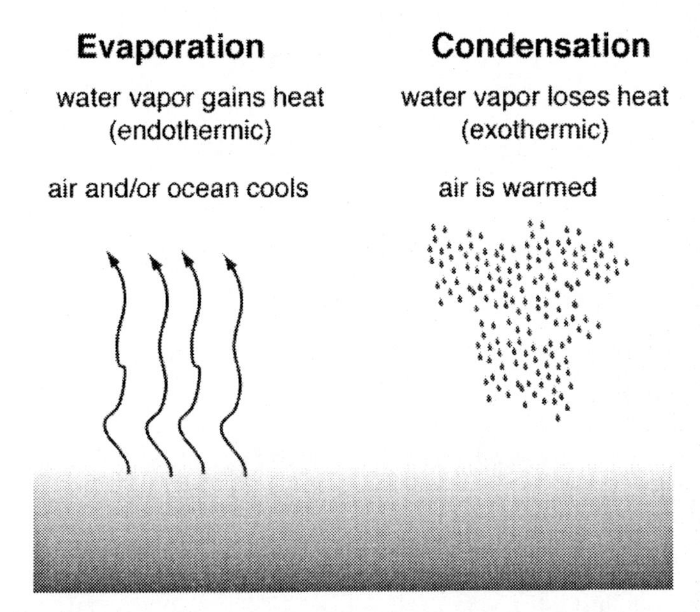

Figure 2.13 The uptake and release of latent heat in the air. During **evaporation,** heat is taken up in the process. The air therefore gains heat while the water from which evaporation takes place loses heat. During **condensation**, heat is released from the air. The released heat can be used to fuel other processes.

2.4 The Concept of Latent Heat

The concept of latent heat is an extremely important concept to understand how big storms such as tropical cyclones can amass immense amounts of energy to fuel violent winds and weather. These storms are fueled by **latent heat** stored in the water vapor in the atmosphere. The term comes from the Latin word *latere* for "lying hidden". Latent heat cannot be used immediately to fuel any process. It is therefore said to be stored away, hidden. But it can be released in a so-called **phase change** and then be used up.

To fully understand this concept, we need to familiarize ourselves with two principal properties of H_2O. First of all, H_2O exists in 3 basic different forms called **phases**. The 3 phases of H_2O are:

1) ice, H_2O is a solid
2) water, H_2O is a liquid
3) water vapor, H_2O is a gas

Every time H_2O converts from one phase to another, energy is either taken up or released during the process (Fig. 2.12). At its melting point at 0°C (32°F) H_2O turns from a solid (ice) to a liquid (water).

Energy has to be invested to make this phase change happen. This is an **endothermic phase change**. The energy is called the **latent heat of fusion,** and 334 J, or 80 cal, are required to melt 1 g of ice. At the boiling point at 100°C (212°F), H_2O turns from a liquid to a gas (water vapor). This phase change is also endothermic, and the energy of 2,257 J, or 539 cal (**latent heat of vaporization**) is required for the phase change of 1 g of water. During these phase changes, energy is taken up without changing the temperature of the medium. If either phase change occurs in the other direction (water to ice; water vapor to water), then latent heat is released, and the phase change is **exothermic**.

Evaporation and **condensation**, the change from the liquid to the gaseous phase and vice versa, can occur at temperatures other that 100°C. These changes are relevant to meteorological processes (Fig. 2.13). The energy required for evaporation or released during condensation depends on the temperature of the medium. At 0°C, 2500 J (600 cal) of heat is required to evaporate 1 g of water (**latent heat of condensation**). This same amount of energy is released during condensation. Condensation causes the formation of clouds, fog and rain. At temperatures more than 1°C lower or higher, less energy is required for these processes. The water from which evaporation takes place loses this energy and cools in the process. This could be any body of water, including oceans, lakes and rivers. The heat taken up by the water vapor is **latent heat**. During **condensation**, the air containing the water vapor gains the latent heat which is now available to warm the air and drive convection and winds, and ultimately to fuel severe storms.

Just how significant are these numbers? The **calorie** (cal) was first defined by Nicolas Clément in 1824 as a unit of heat. It approximates the amount of energy needed to raise the temperature of 1 g of water by 1°C (1.8°F), or 4.19 J. The word "approximate" expresses the fact that the exact amount of energy required for this process depends slightly on the temperature of the water and on pressure. But this is irrelevant for the discussion here because the changes are within 0.01 J. Recalling now that the condensation of just 1 g of H_2O releases 600 cal of heat, this energy can be used to raise the temperature of 600 g (1.32 lb) of water by 1°C, which is quite a lot, considering that 600 g of water at room temperature is about 20.25 fluid oz.

Table 2.1 Heat Capacity of Some Materials

Material	**Volumetric Heat Capacity [J/cm^3/°C]**	**(mass) Specific Heat Capacity [J/g/°C]**
elements		
Al at 25°C	2.42	0.90

Cu at 25°C	3.45	0.39
Au at 25°C	2.49	0.13
air and H_2O		
air[(2)]	0.0013	1.0035
steam at 100°C		2.08
Water at 25°C	4.18	4.18
Water at 100°C	4.22	4.18
ice at -10°C	1.94	2.11
Human Body	3.35	
building materials		
asphalt		0.92
brick		0.84
Glass (silica)	2.09	0.84
Glass (pyrex)		0.75
glass (flint)		0.50
Quartz Sand	1.30	0.84
Granite	2.17	0.79
marble		0.88
Concrete		0.88
Wood		1.2–2.3

(2) effects of variations in temperature and humidity are on the order of $4x10^{-5}$ J/cm^3/°C

The other principal and very important property of water is its **heat capacity**. By definition, the heat capacity of a substance is the amount of heat required to change the temperature of the substance. More specifically and in units, this would be the number of Joules per °C (or physicists use the Kelvin instead of the degrees Celsius where the temperature change by 1 K is the same as the change by 1°C). Often, this measurable quantity is given as heat capacity per mass. This is the (mass) specific heat capacity, or specific heat (in units of J/°C/g). The specific heat then gives the amount of energy needed to heat the mass of 1 g of a substance by 1°C (5/9°F). The heat capacity can also be given as the volume-specific heat capacity or volumetric heat capacity which carries the unit J/°C/cm^3.

Table 2.1 lists a few materials and their specific heat as well as the volumetric heat capacity. The reason why both are listed, and used, is because the density (mass per volume) of different materials is different and some prefer one quantity to the other, e.g. the volumetric heat capacity is used primarily in engineering but the specific heat by physicists. Regardless of the preference, water has the highest heat capacity of the materials listed here. This means that water can take up large amounts of heat before the temperature rises significantly. Or conversely, water can release large amounts of heat before its temperature sinks. Water also has a much higher heat capacity than rock or building materials. This is particularly relevant in **climate science**. The high heat capacity of the oceans explains why coastal climates are much milder than climates in the interior of continents at the same latitude. Air has a heat capacity that is several orders of magnitude smaller than that of water. Solar heating causes the air temperature to rise quickly, but the ocean temperatures rise very slowly. Temperatures in rock also climb more quickly than that in the oceans. Continental interiors, therefore heat up more quickly, and cool more quickly, than land in the vicinity of an ocean. Continental interiors also experience more extreme temperatures than coastal areas. Oceans therefore function as **climate moderator**.

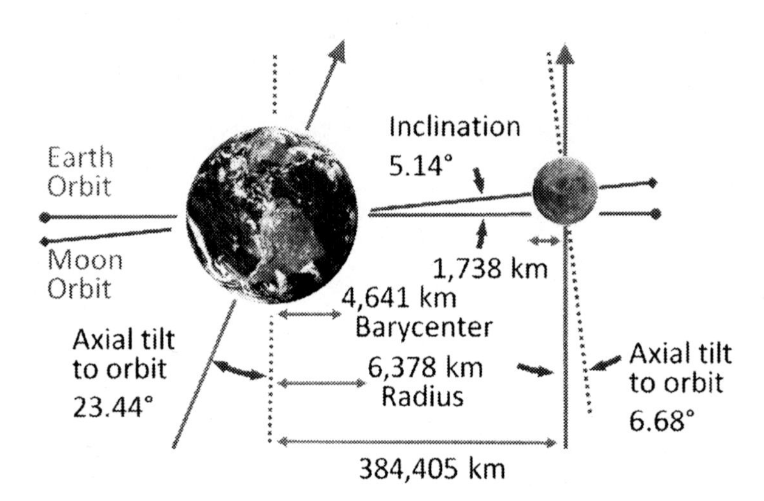

Figure 2.14 Details of the Earth-Moon system. In addition to Earth's radius at the equator, 6,378 km, and the distance between the centers of Earth and the Moon (384,405 km; not to scale), the **barycenter** (the center of mass of the Earth-Moon system) is also shown. The Moon's orbital plane is tilted with respect to Earth's orbital plane around the Sun. This is why we do not observe a solar eclipse every month when the Moon is between Earth and the Sun. (source: Wikipedia).

2.5 Earth Tides

We now return to the Earth-Moon system (Fig. 2.14) to discuss in more detail tidal forces and their impact on Earth's surface. Section 2.1 introduced the concept of gravitation in the most general sense. The attracting gravitational force of the Moon on Earth's surface causes it to deform (Fig. 2.15a).

Before examining details of this deformation, it is important to study Moon and Earth as a revolving two-body system. In a simplified scenario Earth is regarded as the (fixed) center, while the much smaller Moon orbits around Earth. But even though the Moon is much smaller (the Moon's radius is only 0.273 of Earth's radius) and has much less mass (the Moon's mass is only 0.0123 of Earth's mass), the center of mass of the Earth-Moon system, the **barycenter**, is 4,641 km away from Earth's center (note that this is still within Earth though) (Fig. 2.14).

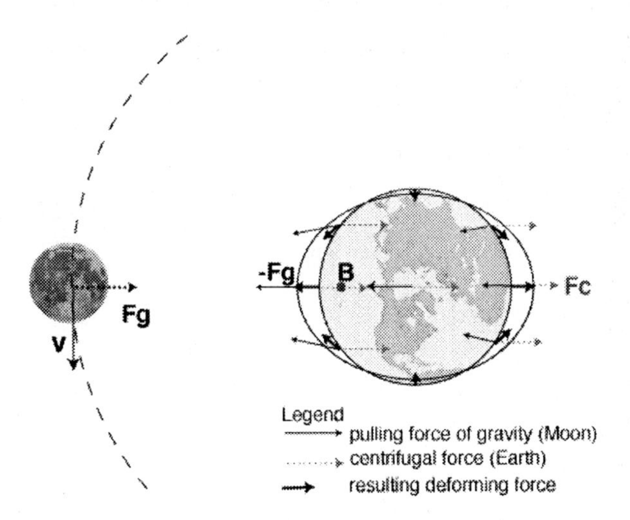

Figure 2.15a The Earth-Moon system viewed from above and onto the Moon's orbital plane around Earth. Also shown are the acting forces on Earth's surface and on Earth's center as well as the resulting deformation of Earth's surface. Two bulges form, one facing the Moon and the other facing away from the Moon.

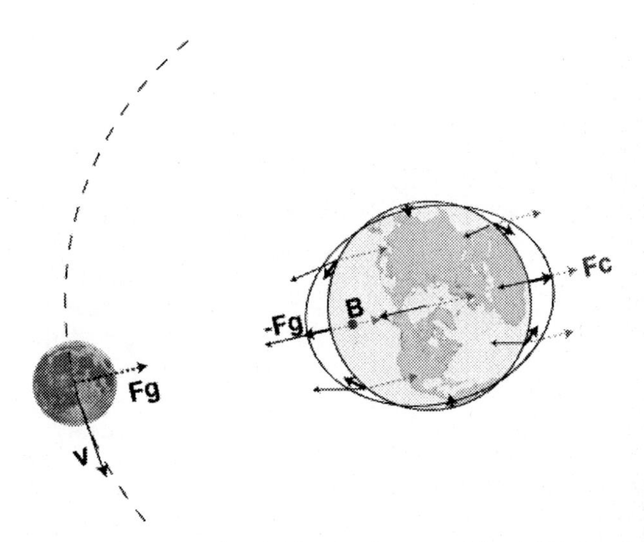

Figure 2.15b After 24 h, Earth rotated by 360°. At the same time the Moon advanced. So Earth has to rotate a bit further for the same spot to experience high tide as the day before (about 50 min).

The Earth-Moon system then revolves around the barycenter, once every 27.3 days (corresponding to the Moon's orbital period around Earth, relative to the position of the stars; **sidereal month**). Two forces now come into play (Fig. 2.15). The Moon tugs on every point on and inside Earth. At each of these points, the gravitational force exerted by the Moon points back to the center (of mass) of the Moon. At each point, the magnitude of this force depends hereby on the distance to the Moon's center. Points not facing the Moon experience a smaller force than points facing the Moon. So, since all these forces point toward the Moon, why does Earth not fall into the Moon? Because Earth revolves around the barycenter. During the course of 27.3 days, every point on Earth then follows the path of a circle. This concept is easy to verify by tracing this motion on a piece of paper. Recall that circular motion induces a centrifugal force. Every point on Earth experiences the same centrifugal force, pointing in the same direction and having the same magnitude. We obtain the final resulting force tugging on Earth's surface by adding the two vectors for the gravitational force and the centrifugal, at each point on Earth. This concept explains why Earth has **two bulges**, one facing the Moon but also one that points away from the Moon.

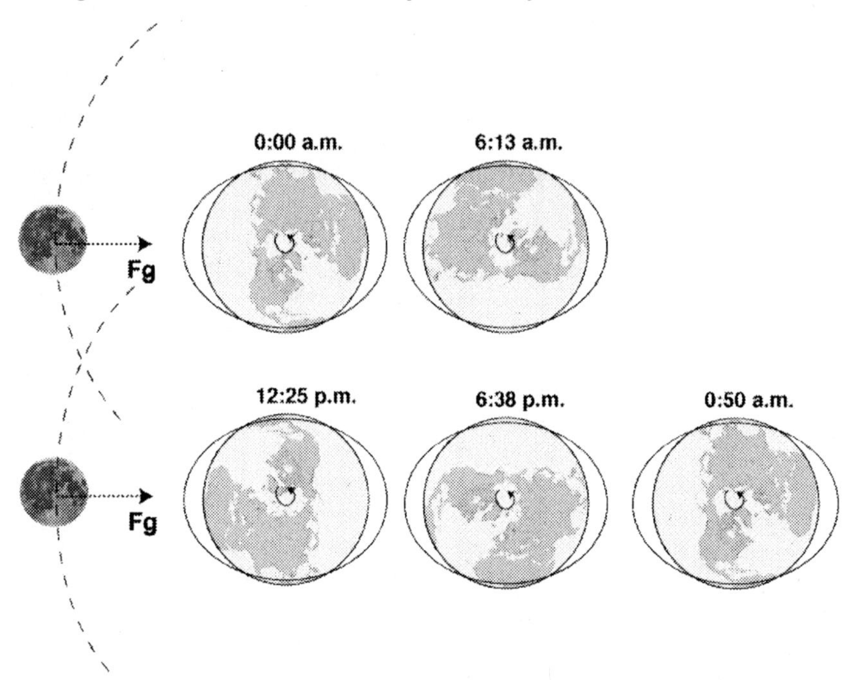

Figure 2.15c Position of Earth relative to the Moon for a full cycle between two high tides (24 hours and 50 min). Here, we assume that the first bulge occurs at 0:00 o'clock.

Now consider that within this temporal framework of the Earth-Moon system, Earth spins around its axis once a day. The reference frame now becomes important (Fig. 2.15b). While Earth spends 24 hours for one rotation around its axis, the Moon has already advanced along its orbit. So in order to align the same point on Earth with the Moon, Earth has to rotate a bit more, for about 50 min more. During the course of 24 h 50 min, every point on Earth now goes through two bulges (Fig. 2.15c). These are the two times each day when the points experience a **high tide**. **Twice a day**, each points also goes through a "squeeze" which indicates a **low tide**. The term tide comes from the low-German word *tiet* for "time". The figures here are only schematic to illustrate the concept. Fig. 2.15 ignores the fact Earth's rotation axis is tilted with respect to the Moon's orbital plane around Earth (Fig. 2.14). Taking this into account explains why the two high tides each day at a point on Earth are not equally high. It also affects the timing of high and low tides.

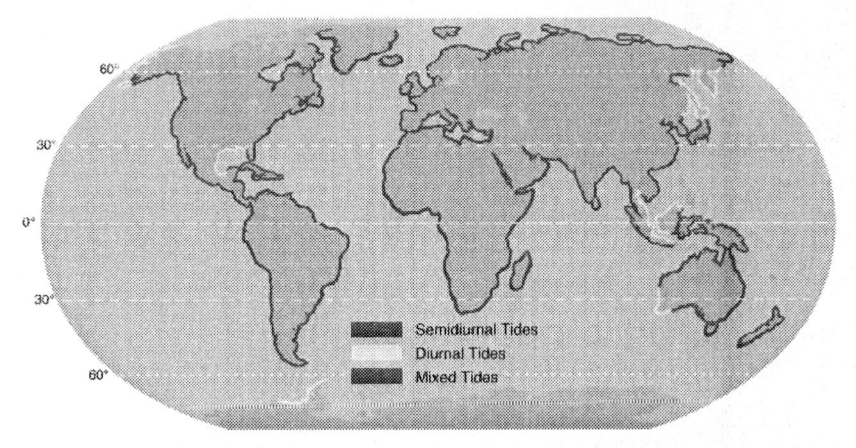

Figure 2.16 Coastal areas with semidiurnal, diurnal and mixed semi-diurnal tides. The majority of coasts experience semidiurnal tides. Tidal forcing depends on many factors, including water body dimensions, coastline as well as continental shelf geometry. (source: Wikipedia)

Most points on Earth experience two high tides and two low tides every day (Fig. 2.16) (**semidiurnal tides**). Though few, at some locations only one high tide and one low tide occurs (**diurnal tides**). These locations include some stretches of Alaska's west coast, Kamchatka, northern Japan, the northern coast of Indonesia and the west coast of Borneo, the southwestern tip of Australia and the west coast of the Antarctic Peninsula. Many locations experience mixed semi-diurnal tides that have some signature of semidiurnal tides as well as diurnal tides. La Jolla, CA is such a location. It can experience two high tides but the level of the two high tides is different, and so is the level of the two low tides.

The National Oceanographic and Atmospheric Administration (NOAA) Tides & Currents website provides an excellent tutorial on Earth tides and observations[3]. The changes in sea level are measured with **tide gauges**. In the old days, these were 12-in wide pipes (the "stilling" wells) that reached into the water from a pier. A float within the pipe was attached through a wire with a recording device and rose and fell with the tides. In a modern-day tide gauge, the "stilling" well only

COPYRIGHTED MATERIAL — DO NOT DUPLICATE, DISTRIBUTE, OR POST

functions as protective pipe for a sophisticated acoustic and electronics system. The system sends an audio signal down the tube and measures the it takes for a reflection off the water surface to return, much like an echo sounder[4]. Tide gauges do not only measure the tides but any changes in sea level. This includes storm surges, tsunami, and even the very slow rise in sea level caused by global warming.

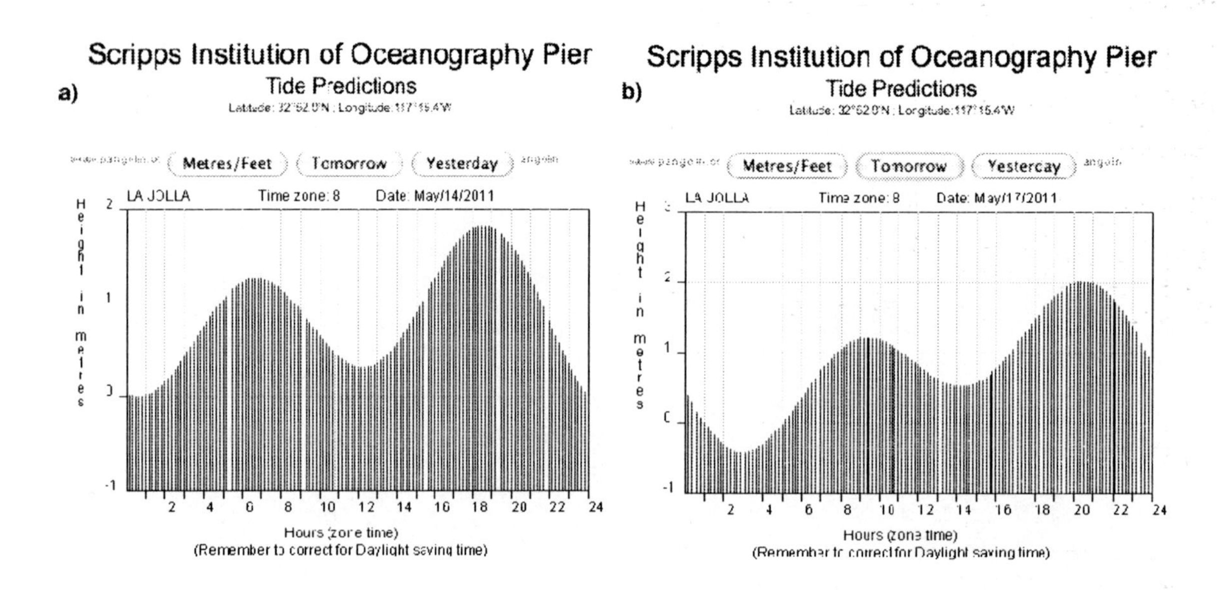

Figure 2.17 The tide calendar for La Jolla, CA: a) for 14 May 2011; b) 17 May 2011. The highest high tide is at 19:31 PDT on 14 May and at 21:30 PDT on 17 May. The tidal range is 1.9 m on 14 May and 2.3 m on 17 May, the day of Full Moon. Note that the height scale is different in both plots. Also note that these are predicted tides, not measured tides. Tides are measured using tide gauges. The tide predictions at the SIO pier are broadcast at http://ocean.peterbrueggeman.com/piertide.html. Peter Brueggeman is the head librarian of the SIO library.

Since the forces playing a role in the Sun – Earth – Moon system are well know, the tides on Earth can be predicted fairly accurately and published in so-called **tide calendars**. A typical course of tides throughout the day is shown in Fig. 2.17. We observed two high tides and two low tides in La Jolla, CA in mid-May 2011. The later high tide is the higher high tide, characteristic of mixed semi-diurnal tides. Over the course of the three days between 14 May and 17 May 2011, the timing of the low and high tides was delayed by 150 min (the two low tides and the lower high tide) and 120 min (the higher high tide). The difference between the lowest low tide and the highest high tide each day is the **tidal range**. In mid-May 2011, the tidal range in La Jolla was about 2 m (6.6 ft).

Figure 2.18 Low and high tide in Alma, New Brunswick in the Bay of Fundy, Canada. At 16 m (52 ft), the tidal range in this bay is the largest well-documented in the world. (source: Wikipedia).

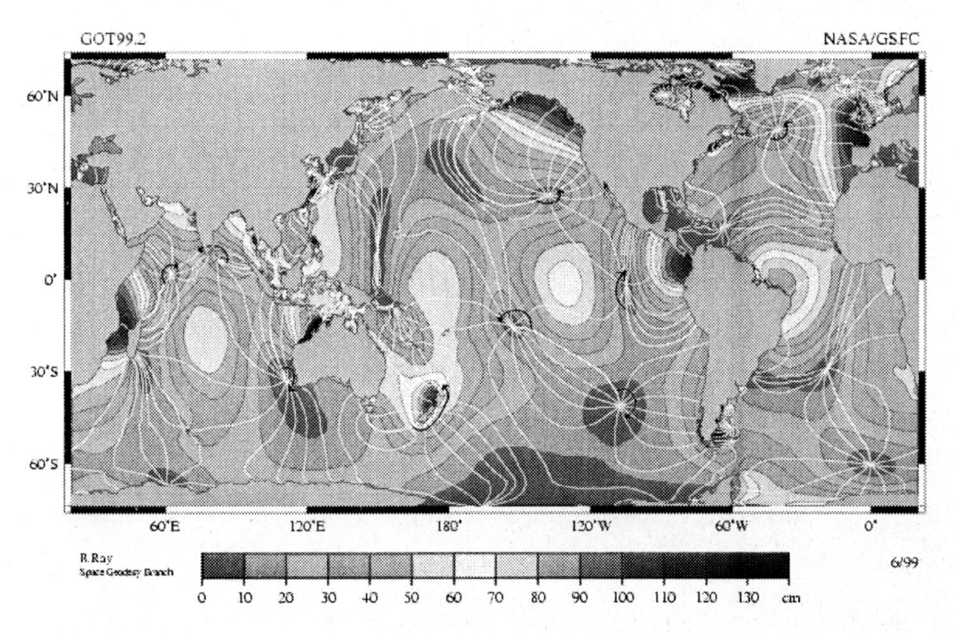

Figure 2.19 The tidal range as a result of the semidiurnal lunar tide (shaded). The white lines are lines of the same tide, in 1-hour increments. Curved arcs around the amphidromic points (points with no tides) show the direction of the tides. (source: Wikipedia/NASA).

The tidal range changes throughout the month, and also throughout the year but in La Jolla, it rarely exceeds 2.5 m (8.2 ft). Some areas, however, experience much larger tidal ranges. The location with the largest tidal ranges in the world is found in the Bay of Fundy (Fig. 2.18). A boat docked at the pier during high tide may lie dry during low tide. Other locations include coasts along the Channel between France and the UK, with a tidal range of up to 9 m (30 ft). A top 50 list can be found on the NOAA Tides & Currents website.[5] Figure 2.19 illustrates the tidal range as a consequence of the semidiurnal tide that is dominant over the diurnal tide at most locations on Earth. The Bay of Fundy and the Channel stand out but large tidal ranges are also found in Davis Strait joining the Labrador Sea, along the west coast of Panama and northern Colombia, northwestern Australia, East Africa to Madagascar, western New Zealand and the Gulf of Alaska.

New Moon

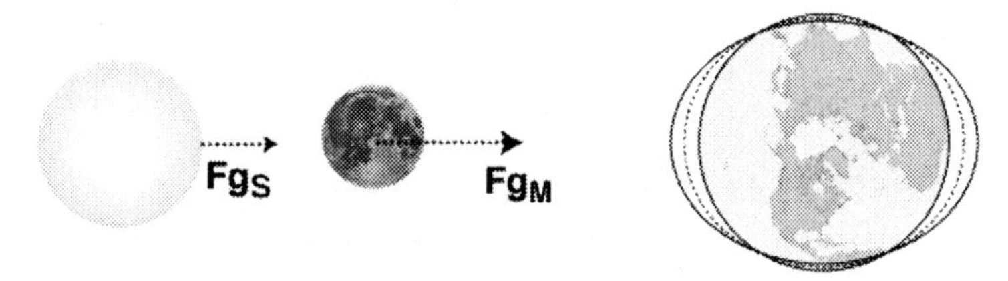

Figure 2.20a Position of Earth, Moon and Sun during New Moon. Sun's and Moon's tidal forces add to a spring tide.

First Quarter Moon

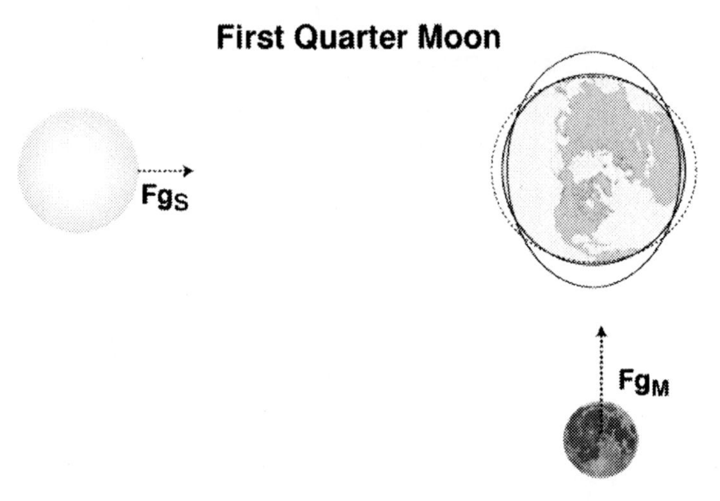

Figure 2.20b Position of Earth, Moon and Sun during a First Quarter Moon. Sun's and Moon's tidal forces even out to form a neap tide.

Full Moon

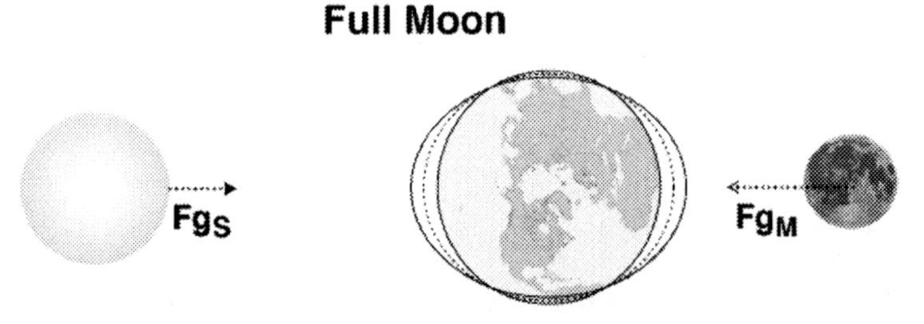

Figure 2.20c Position of Earth, Moon and Sun during Full Moon. Sun's and Moon's tidal forces add to a spring tide.

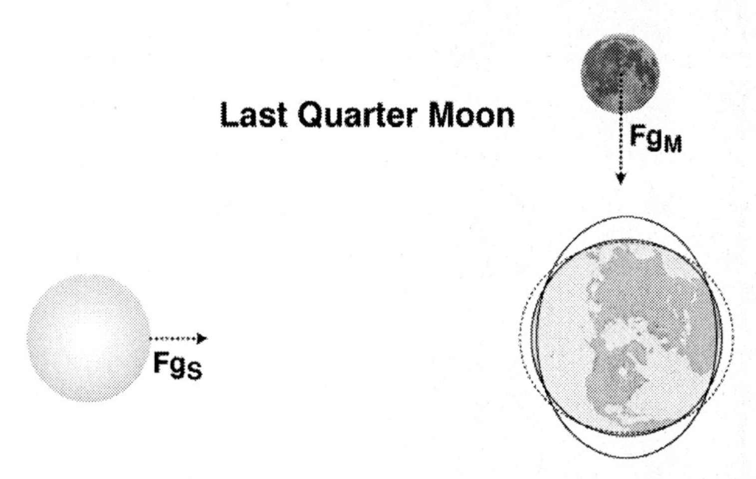

Figure 2.20d Position of Earth, Moon and Sun during a Last Quarter Moon. Sun's and Moon's tidal forces even out to form a neap tide.

In Fig. 2.17 we can see that the structure of high and low tides changes rather rapidly over the course of only a few days. This is due to the fact that the Moon is not the only celestial body exerting tidal forces on Earth. Even though the Sun is much farther away from Earth than the Moon, the Sun contributes to the tides on Earth with about 30% (i.e. the lunar tidal force on Earth surface is about twice as large as the solar tidal force). Whenever the Sun and the Moon are aligned with Earth (this alignment is also called **syzygy**) (Fig. 2.20a and 2.20c) the lunar and solar tidal forces add up and the high tides are particularly high (Fig. 2.21). This occurs when the Sun and the Moon are on the same side of Earth and pull from the same side (Fig. 12.20a) but also when Earth is between the Sun and the Moon and they pull in opposite directions (Fig. 12.20c). At the same time, the low tides are particularly low. The tidal range is at a maximum during the course of a month. These are the times of **spring tides** that occur **twice a month**. The alignment of Earth, the Sun and the Moon are accompanied by characteristic Moon phases. A **New Moon** is observed when the Moon is between the Sun and Earth (Fig. 2.20a) and a **Full Moon** is observed when Earth is between the Sun and the Moon. So why do we not observe a lunar eclipse every time we observe a Full Moon? Recall from Fig. 2.14 that the Moon's orbit around Earth is tilted with respect to Earth's orbit around the Sun. During New Moon, when the Moon is between Earth and the Sun, we do not observe a lunar eclipse for the same reason. Even though the Moon is not illuminated during New Moon, this is **not** an eclipse. How can you tell the difference? A New Moon lasts a day and the area of illumination on the face of the Moon changes only slowly from day to day. An eclipse lasts only a few hours, and the area of illumination changes throughout the same night.

As the Moon moves out of the New Moon phase, the area on the Moon illuminated by the Sun is increasing and we observe a waxing Moon. If the Moon and the Sun are at right angle (Fig.

12.20b), on the **First Quarter Moon**, the lunar and solar tidal forces pull in different directions and partially cancel each other. The tidal range is at a minimum, with the smallest difference between high tide and low tide. This also occurs when the Moon moves out of the Full Moon into the **Last Quarter Moon** (Fig. 2.20d). These are the times of **neap tides** which also occur **twice a month**. The changing phase of the Moon after the Full Moon but before the New Moon is also called the waning Moon.

The time between two Full Moons is about 29.5 days (**synodic month**; period of the Moon's revolution with respect of the Sun – Earth line). The Moon's orbit is actually slightly elliptical, with an eccentricity of 0.055. If the Moon is at its **perigee** (point on orbit closest to Earth) then the tidal range is higher. The highest spring tides occur when the passage of the Moon through the perigee coincides with a Full Moon or New Moon (every 7.5 synodic months/221 days). Twice a year, in June and in December, Earth experiences particularly large spring tides when Earth, the Sun and the Moon are aligned at perigee (point along orbit closed to Earth) and perihelion (point along orbit closest to the Sun). The term **King tide** is a popular term to describe a particularly high spring tide that occurs only a few times a year. However, this is not a scientific term.

On March 6, 1962, a catastrophic **perigean spring tide** struck the entire Atlantic coastline of the U.S. from the Carolinas to Cape Cod in the darkness of predawn, and for the following 65 hours. This disastrous flooding event resulted in a loss of 40 lives and over $0.5 billion in property damage.

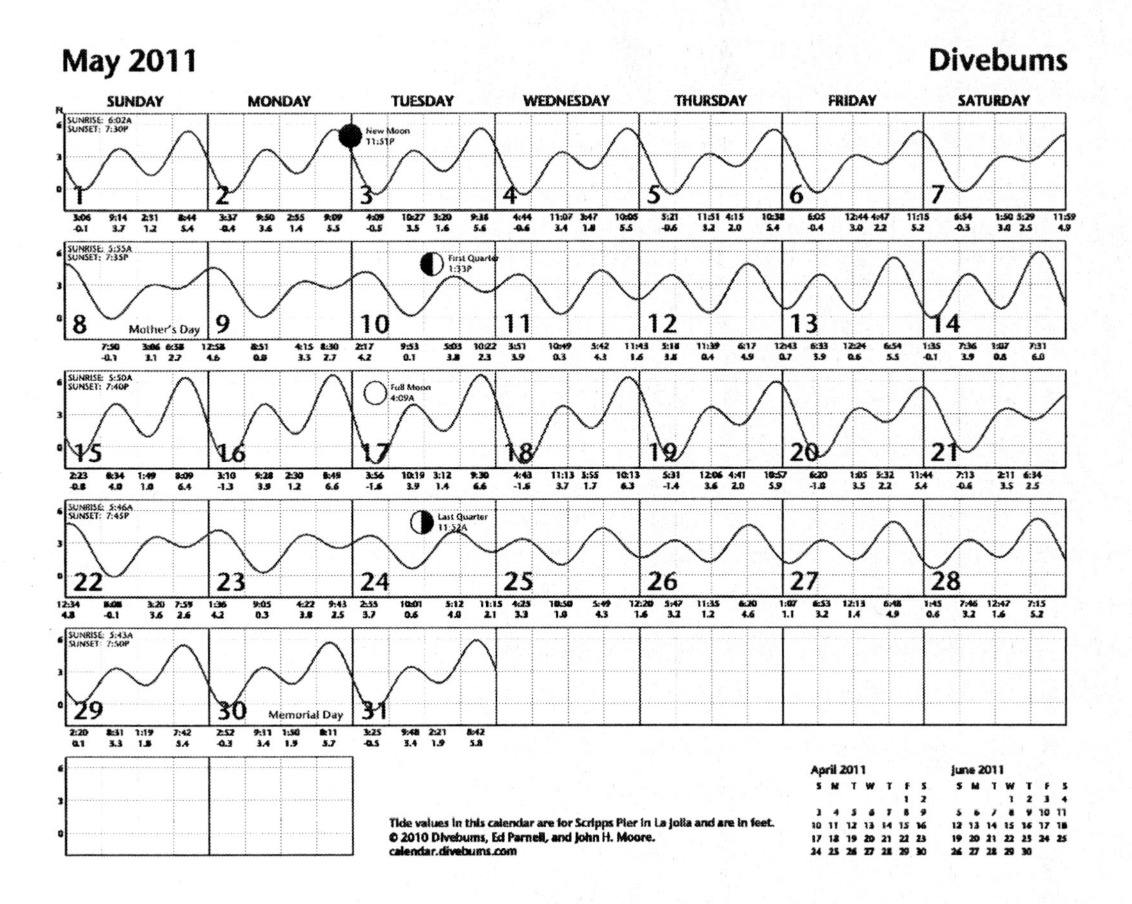

Figure 2.21 The tide calendar for La Jolla, CA for the month of May 2011. New Moon is near midnight on 2 May 2011, Full Moon is on 17 May, while the First and Last Quarter Moons are on 10 May and 24 May. Spring tides occur on New Moon and Full Moon while neap tides occur on the Quarter Moons. Note that the scale is in feet. The calendar can be downloaded at http://divebums.com/Tides2011

BOX 1: Elastic, Ductile and Brittle Material

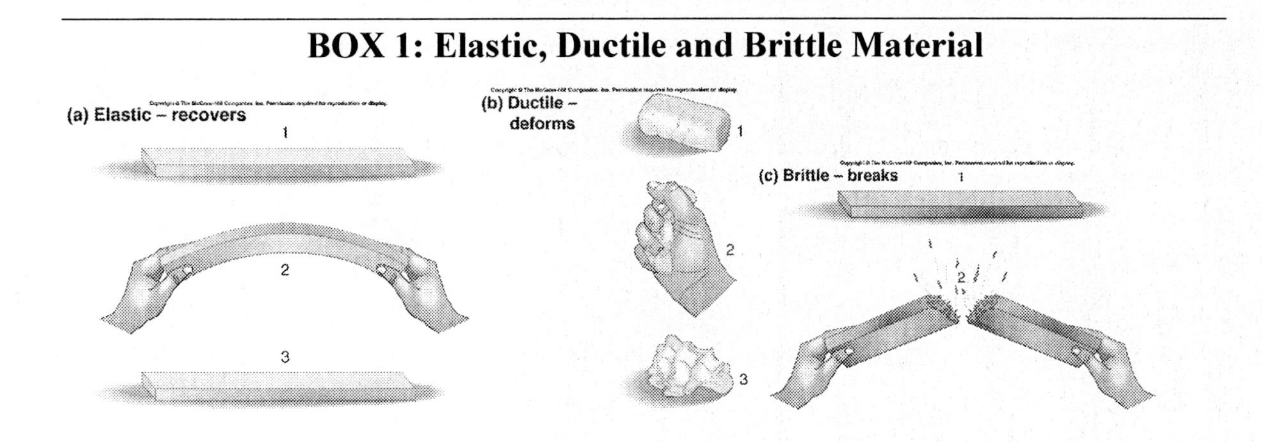

COPYRIGHTED MATERIAL — DO NOT DUPLICATE, DISTRIBUTE, OR POST

Figure 2.22 Different material behavior in response exerting forces. (a) Elastic material recovered after the force is removed. (b) Ductile or plastic material deforms permanently. (c) Brittle material breaks. (source: AB)

Earth scientists use the terms elastic, ductile and brittle to describe some of the mechanical properties of Earth's layers and rocks (Fig.2.22). In addition to these three, one layer (Earth's outer core) behaves like a liquid. In particular, elastic, ductile and brittle describe a material's reaction when a force (or stress) is applied to it.

An **elastic material** can be bent without breaking. After the deforming force is removed, the material returns to its original state. Elastic materials store potential energy when forces are exerted on them. All this energy is used to bring the material back to its original form after the forces are removed. An example of an elastic material that we discussed is the coil spring in Fig. 2.5. A tall palm tree bending in the wind is elastic because it goes back to its upright position after the winds die down. Wood boards are elastic, as are popsicle sticks and plexiglas. Some materials are easier to bend than others, e.g. a foam board is easier to bend than a wood board of the same thickness, and both are considered elastic materials.

A **ductile material** can also bend without breaking when forces are applied. It is pliable. But the deformation is plastic, not elastic. This means that the material does **not** return to its original state after the bending forces are removed. Instead, it remains in the form that it was last when forces were applied. Examples of ductile material include plumber's putty, play dough, cookie dough, modeling clay, soft wax and other soft, pliable materials. There is no storage of potential energy, so no energy that could be spent on bringing the material back to its original form.

A **brittle material** breaks when forces are applied, particularly when forces seem relatively weak. Brittle materials include glass, peanut brittle, graphite pencils, nutshells. Some materials may behave like an elastic material if forces are weak enough or if forces are exerted slowly. Conversely, elastic materials behave like brittle materials if the applied forces exceed a threshold given by the elasticity of the material (see. Figs. 2.5 and 2.6). Brittle materials may store some potential energy when forces are acting slowly to allow elastic behavior. The potential energy is released instantaneously when the material breaks. It is used to damage the material.

References and Websites

1. Wind, wide spaces and mud. September 2009 issue of the Atlantic Times. Retrieved in March 2011 at http://www.atlantic-times.com/archive_detail.php?recordID=1919

2. Johnston, A.C., 1990. An earthquake strength scale for the media and the public. Earthquakes & Volcanoes, 22, 214-216.
3. NOAA websites on Earth tides: tide and current data available at http://oceanservice.noaa.gov/facts/find-tides-currents.html and NOAA educational page on tides at http://oceanservice.noaa.gov/education/tutorial_tides/
4. NOAA educational webpage on old and new tide gauges at http://oceanservice.noaa.gov/education/kits/tides/tides10_oldmeasure.html and http://oceanservice.noaa.gov/education/kits/tides/tides11_newmeasure.html
5. the top 50 list of locations with the largest tidal range on NOAA's Tides & Currents website at http://www.co-ops.nos.noaa.gov/faq2.html#26

The NOAA Tides & Currents website gives and excellent introduction to everything you ever want to know about Earth tides: http://tidesandcurrents.noaa.gov/restles1.html

Other Recommended Reading

Evans, B. and Jameson, M., 2008. Category 7, Tor Books.

This is a science-based FICTION story about a scientist-gone-crazy who tries to manipulate the size and path of a hurricane. The story is about how a hurricane can be fueled, i.e. its latent heat increased by a secret laser weapon. The author is a meteorologist and so is familiar with the science of weather. The book also discusses past real-life efforts in the U.S. to manipulate and control weather, and the consequences of an attempt in 1947 to divert a hurricane before it made landfall in Georgia (Project Cirrus). Project Cirrus was superseded in 1992 by Project Stormfury which was considered to be a failure. More recent cloud seeding attempts in China to increase rainfall are thought to have contributed to unusually severe snowstorms.

Chapter 3: The Solar System and the Layered Earth

Figure 3.1 Montage of planetary images taken by spacecraft managed by the Jet Propulsion Laboratory in Pasadena, CA. From top to bottom: Mercury, Venus, Earth (and Moon), Mars, Jupiter, Saturn, Uranus and Neptune. (source: NASA/JPL/wikipedia)

Earth is nearly unimaginably old, when it comes to a human lifetime, and it took this long time for Earth to develop into the planet we know today. It has a unique place in our Solar System. Our planet has just the right distance from the Sun for livable temperature conditions. Water in liquid form is abundant, one of the most important factors to facilitate "life as we know it". Earth has a powerful sustained magnetic field that protects life on the surface from harmful cosmic particles. But, unlike that of some other planets such as Jupiter, Earth's field is not too strong to be harmful by itself. Our atmosphere is also unique. Geological processes removed much of the initially present CO_2. This allowed the development of a mild but not too oppressive atmospheric greenhouse, unlike that of

Venus. Yet, it is the geological processes that are sometimes responsible for catastrophes on Earth. Large volcanic eruptions can devastate and destroy landscapes for miles on end, and humongous eruptions affect an entire continent, if not the entire globe. Large earthquakes cause widespread, destructive shaking, and large earthquakes in the oceans may send off monster tsunami across the widest oceans. Processes in the atmosphere, too, can release huge amounts of energy to fuel damaging winds and powerful storm systems.

But why should we review the universe and the Solar System in a natural disasters book? It turns out that the more we explore the other bodies in our Solar System, the more we learn about the processes on our own planet. We also learn that some of the processes we thought were unique to Earth actually happen elsewhere, be it in modified form. Mars has the tallest volcanoes in the Solar System, not Earth. And Io, a moon of Jupiter, has a unique type of active volcanism that sends plumes of material much farther into space than volcanoes on Earth currently do. Cryovolcanism, a type of volcanism that erupts volatiles such as ammonia and methane instead of lava, was discovered on the Neptune moon Triton. Venus may have plate tectonics just like Earth, but the style is much more catastrophic than on Earth. Europa, another moon of Jupiter, may have a type of ice plate tectonics, facilitated by a water "ocean" beneath that may also harbor primitive life. By studying the atmosphere of other bodies in the Solar System, we understand, and appreciate, how special our own atmosphere is. On the other hand, there are weather phenomena on other planets that are similar to those on Earth. Jupiter has a persistent gigantic storm visible as the Giant Red Spot. Unimaginably strong winds blow on Neptune. Despite a thin atmosphere, Mars has strong dust storms that we can observe from Earth. They allow us to ultimately comprehend how a sand storm starting in the Sahara desert may be responsible for diseases in the Caribbean coral reefs.

Earth's position as one of many bodies in the Solar System makes it vulnerable to impacts from time to time. It has become very clear in the last 3 decades that Earth experienced at least several very large impacts that wiped out entire branches in the family tree of life. Such large impacts are rare, but even much smaller impacts can potentially wipe out a large city. The recent launch of the Spaceguard program illustrates the urgency of better understanding how real the risk for such an impact is and how we could respond to it.

The combination of sustaining a magnetic field and plate tectonics requires unique properties in a layered body. At the end of this chapter, we will review Earth's internal structure, and the fundamental differences between oceans and continents. We will also understand how Earth's strong outer shell, that is the region of earthquakes, can be held up by a weaker soft layer that reacts to the growth of ice sheets and flows away.

Figure 3.2 Distant galaxies in deep space in a Hubble Telescope Ultra Deep Field "photograph". The image consists of data collected from September 2003 through January 2004 in a small region in the constellation Fornax. The patch of sky in which the galaxies reside was chosen because it had a low density of bright stars in the near field. (source: NASA/wikipedia)

Figure 3.3 One of the most famous images taken by the Hubble Space Telescope, Pillars of Creation shows stars forming in the Eagle Nebula. The photograph is a composite of 32 different images from four separate cameras and was taken 1 April 1995. The Pillars formed from the gas (mostly hydrogen) and dust of eroding stars and formed the breeding ground for new stars (within the dark dots). The left Pillar was several times larger in length than the Solar System is across. The Eagle Nebula is 7,000 light years from Earth. In 2007 it was announced that the Pillars were destroyed by a supernova's shockwave about 6,000 years ago. Observers on Earth currently see the shockwave approaching the

Pillars, but will not see the actual destruction for about another 1,000 years. (source: NASA/wikipedia)

3.1 The Universe and Our Solar System

The universe is commonly defined as the totality of everything that exists, including all matter (e.g. planets, stars, interstellar matter) and energy. Its age is currently estimated at 13.75 ± 0.11 billion years which can be determined in several ways. The first way is based on the **Big Bang theory** which is the prevailing cosmological model that explains the formation of the universe from a single point, and its early history. Measurement of the microwave background radiation helps estimate the cooling of the universe (since the Big Bang). And together with current observations of the expansion of the universe, extrapolations can be made backward in time to the Big Bang. Alternatively, the age of the universe can be estimated from other measurements, such as the age of chemical elements, the oldest star clusters and the oldest white dwarf stars[(1)]. The estimated age from these varies between 14.5 and 11.5 billion years. The age of protons in the universe, which are considered stable, is currently also 13.7 billion years. The region that is visible from Earth (the observable universe) is a sphere with a radius of 46 billion light years (the distance that light travels in 46 billion years; 10×10^{12} km). In the 1920s observations by astronomer Edwin Hubble showed that the **Milky Way**, the galaxy we live in, was just one of around 200 billion **galaxies** in the observable universe. Launched into space in 1990, the Hubble Space Telescope (HST) was named after him. One of the most captivating images from the HST is its Ultra-Deep Field image that for the first time gave us comprehensive visible-light images of galaxies in the deep-space universe (Fig. 3.2).

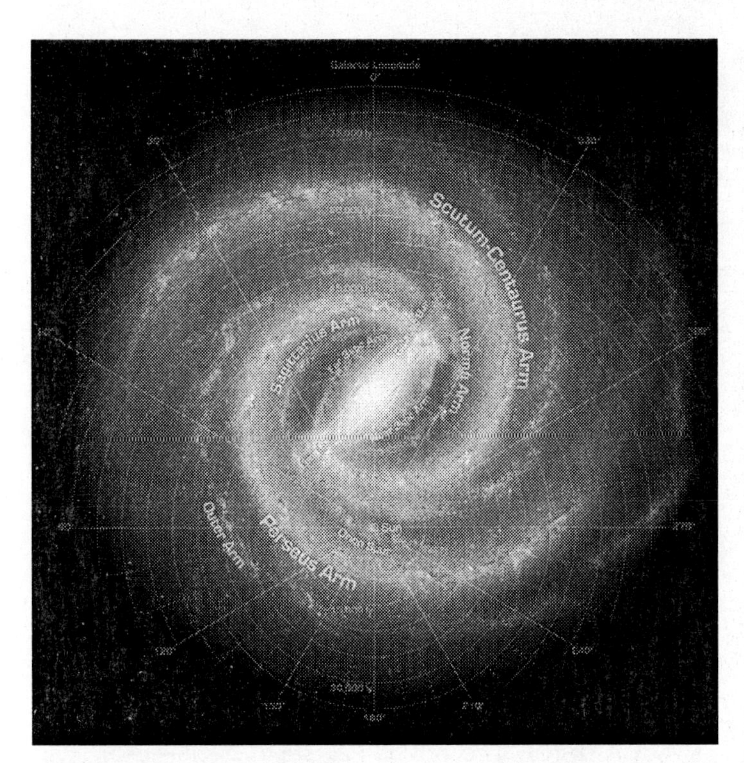

Figure 3.4 Artist's conception of the spiral structure of the Milky Way, a barred spiral galaxy. Four spiral arms emerge from the ends of a central bar of stars. Scientists recently discovered that two of these arms, the Perseus Arm and the Scutum-Centraurus Arm, are much more dominant than the others. The Solar System is located in the Orion-Cygnus Arm, which is one of the minor arms. (source: NASA/wikipedia)

A galaxy contains anywhere between a few ten million to a hundred trillion (10^{14}) stars. Earth, along with our Solar System, is located in the Milky Way. This galaxy is 100,000-120,000 light years in diameter and 1,000 light years thick, and so resembles a rather flat disk, with several spiral arms (Fig. 3.4). The Milky Way contains 200 – 400 billion stars, where the oldest known star is 13.2 billion years old. The entire galaxy rotates once every 15 – 50 million years. It is a barred spiral galaxy rather than an ordinary spiral galaxy, which means that it has a bar-shaped core region at its center that is surrounded by a disk of gas, dust and stars in four distinct arm structures. Our **Solar System** is located in the Orion-Cygnus spiral arm and orbits about 28,000 light years from the center of the galaxy. It is currently about 20 light years above the galaxy's equatorial plane.

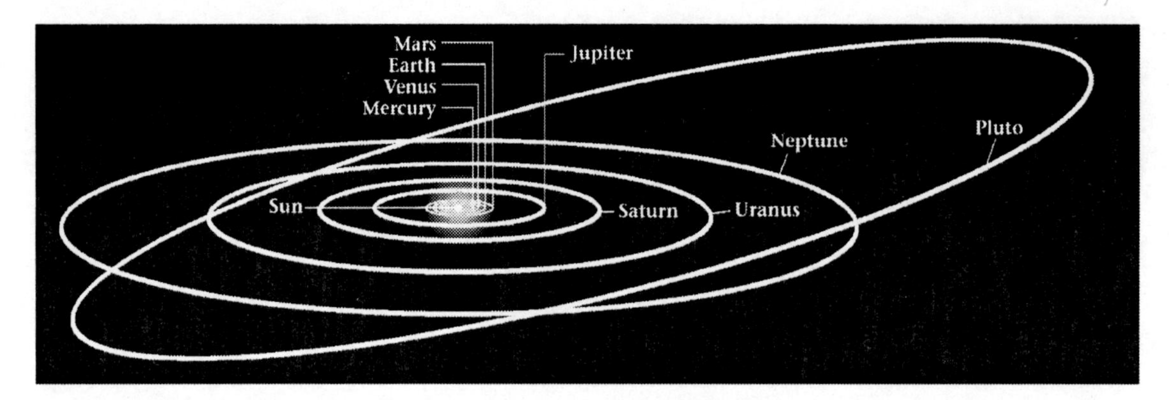

Figure 3.5 The orbits of the 8 planets of our Solar System around the Sun, together with that of Pluto. The orbits of all planets lie in the same plane, but Pluto's is highly inclined. (source: ME)

The bulk of the matter in the Solar System (Fig. 3.5) consists of hydrogen and helium created in the Big Bang. Other chemical elements, which are less than one-thousandth as abundant as hydrogen, were formed by nuclear reactions in the interiors of earlier generations of stars that existed between the Big Bang and the origin of the Solar System 4.55 billion years ago. Stellar explosions and winds distributed these heavier elements throughout the universe. The age of the Solar System is estimated from the study of meteorites. Earth is about 4.5 billion years old and the Moon about 100,000 years younger.

The Formation of the Solar System

As far as how the Solar System formed, no single hypothesis can currently explain all observations adequately. Among many observations, two of the most fundamental ones are: 1) the Sun carries 99.9% of the mass of the solar system; 2) the planets orbiting the Sun carry 99% of the angular momentum. The angular momentum essentially describes the capability and vigor with which the Solar System and its members rotate. Initially postulated by German philosopher Immanuel Kant in 1755, and later updated, the **modified nebular hypothesis** is currently the most widely used model. It supports the idea that the Sun and the planets formed at about the same time from a huge rotating spherical cloud (nebula) of ice, gas and debris, perhaps somewhat similar to what is being observed in the Pillars of Creation nowadays (Fig. 3.3).

In the modified nebular hypothesis, the spinning could contracts. To conserve angular momentum, the speed at which the cloud spins increases, much like a figure skater performing a pirouette spins faster when tugging in the arms. The rotating cloud eventually flattens to a disk, the **protoplanetary disk**. The exact reason for the flattening is poorly understood but one hypothesis is that particles from above and below the disk collide near the disk and propel each other farther out - this is needed to explain why the planets carry most of the angular momentum in the Solar System.

The immense density at the center of the spinning disk leads to gravitational collapse and the formation of the Sun, where the immense pressure and temperature during gravitational collapse start thermo-nuclear processes in the Sun (i.e. the fusion of hydrogen). At the same time, bodies start to collide in the rotating disk and stick together (**accretion**). Exactly how this happens is still debated among scientists. Two possible models are called gravitational instability theory and core-accretion theory. A publication in the journal American Scientist likens the core-accretion process to what happens to dust particles under one's bed after a lengthy time of not vacuuming[(2)]: dust starts to grow together and clump (accrete) into dust bunnies. Accretion first leads to **planetesimals** (a few km radius), then to **protoplanets** (planetary embryos) and eventually to **planets** (> 1200 km radius). Protoplanets are large enough to have nearly rounded into spheres, and have undergone some degree of melting and subsequent differentiation, where heavier elements (such as iron) sink toward the center and lighter elements (such as oxygen) rise toward the surface.

Figure 3.6 The 8 planets in our Solar System, along with the currently 5 dwarf planets, including former 9th planet Pluto (as of March 2012). Sizes are to scale, but relative distances from the Sun are not. (source: NASA/wikipedia)

The Solar System – A Summary

Our Solar System contains a multitude of celestial bodies, but the most prominent ones are without doubt the Sun and the planets. The word planet comes from the Greek word *planes* for "wanderer". The old Greek had noticed that the planets were not fixed in the night sky, like the other stars, but moved around over the course of months and years. Many cultures viewed them as divine, which explains why they have names of Greek or Roman gods. Of the eight current planets, six were recognized as planets since ancient times. Uranus and Neptune were discovered relatively late, in the 18th and 19th centuries, after telescopes became available. Until 2006, we had nine planets. They were, with growing distance from the Sun: Mercury, Venus, Earth, Mars, Jupiter, Saturn, Uranus, Neptune and Pluto. And many of us remember rhymes learned in school to remember the order of the planets (also called mnemonic phrases), for example "My Very Educated Mother Just Served Us Nine Pickles" (another rhyme substitutes "Pizza" for "Pickles") or "My Very Easy Method Just Speeds Up Naming Planets". At a conference in Prague, Czech Republic, the International Astronomical Union (IAU) defined on 24 August 2006 that a planet has to meet four conditions:

- has its own orbit around the Sun
- has sufficient mass to assume a nearly round shape
- is not a satellite of a planet or other nonstellar body
- has cleared the neighborhood around its orbit

The forth criterion means that a planet has to have "scooped up" all the debris along its orbit as a result of its large mass and gravitational force. Pluto fails to meet this criterion and became a dwarf planet. Our rhymes lost their last words, but they are still in use, otherwise unchanged. The planets of the Solar System now fall into two groups. The four planets closest to the Sun are the **inner planets**, also called the **rocky planets**. They are made of solid rock and are more Earth-like than the other planets, which is why they are also called **terrestrial planets**. The four other planets, the **outer planets**, are much larger than the terrestrial planets, and they consist mostly of gas (with perhaps a rocky core). These are the **gas giants**. The **Asteroid Belt** separates the terrestrial planets from the gas giants. All planets orbit the Sun with the same sense of rotation, counterclockwise when viewed from above the Sun's north pole. All planets have nearly circular orbits that nearly lie within the same plane, the **ecliptic**. The revolution of a planet around the Sun can take from 80 days (Mercury) to 165 years (Neptune). All planets also rotate about a spin axis, where a spin can take from 9h (Jupiter) to 243 days (Venus). Venus and Uranus (and Pluto) have a **retrograde rotation**, which means that the spin direction is opposite the revolution direction around the Sun. The tilt of the rotation axis of

Uranus and Pluto is so large that they essentially lie on their side, living Venus as the only planet with a true retrograde rotation. Except for Mercury and Venus, all planets have **moons**, also called **natural satellites**.

Table 3.1 Significant Celestial Bodies in the Solar System (sorted by Distance from the Sun)

Body	Type	Radius (ER)	Mass (EM)	Density [g/cm^3]	Distance from Sun (AU)	# Satellites
Sun	star	109.25	332,946.8	1.408	0	-
Mercury	planet	0.383	0.055	5.427	0.4	0
Venus	planet	0.950	0.815	5.24	0.7	0
Earth	planet	1.0	1.0	5.515	1.0	1
Moon	Earth moon	0.27	1/81.3	3.346	(a)	0
Mars	planet	0.53	0.107	3.94	1.5	2
Ceres[e]	dwarf planet	0.074	1.6×10^{-4}	2.08	2.77	0
Jupiter	planet	10.7	317.8[b]	1.33	5.2	63[c]
Io	Jupiter moon	0.29	0.015	3.52		
Europa	Jupiter moon	0.25	0.008	3.02		
Ganymede	Jupiter moon	0.413	0.025	1.94		
Callisto	Jupiter moon	0.378	0.018	1.84		
Saturn	planet	9.0	95.15	0.70	9.5	60+3?[d]
Titan	Saturn moon	0.404	0.023	1.88		
Rhea	Saturn moon	0.12	3.9×10^{-4}	1.24		
2060 Chiron	asteroid/comet	0.03	$\approx 1.7\times 10^{-6}$	?	13.7	
Uranus	planet	4.0	14.54	1.30	19.6	27
Neptune	planet	3.882	17.15	1.76	30	13
Triton	Neptune moon	0.21	2.14×10^{-4}	2.06		
Pluto	dwarf planet	0.19	0.002	2.1	39	4
Charon[f]	Pluto moon	0.095	2.5×10^{-4}	1.7	39	0
Haumea	dwarf planet	0.12	6.77×10^{-4}	3.0	43	2
Makemake	dwarf planet	0.11	4.92×10^{-4}	2.0?	46	2
Eris[g]	dwarf planet	0.196	0.0025	2.5	68	1

Table annotations:

ER, Earth radius = 6371 km

EM, Earth mass = 5.974×10^{24} kg

AU, astronomical unit = 1.4960×10^{8} km

[a] distance to Earth: 3.844×10^{5} km

[b] this is 2.5 times the mass of all the other planets taken together

[c] Ganymede, Callisto, Io and Europa are similar to terrestrial planets; Ganymede is larger than Mercury

[d] Titan (largely made of ice) is the only satellite with an atmosphere and is larger than Mercury

[e] Ceres is located in the Asteroid Belt. It was discovered on 1 January, 1801, 45 years before Netpune. It was considered a planet for half a century before reclassification as an asteroid.

[f] Charon is currently considered Pluto's satellite but Pluto/Charon are really a binary dwarf planet system

[g] Eris is larger and more massive than Pluto!

Aristarchos of Samos (310 – ca. 230 B.C.), a Greek scientist, was the first to suggest a model of the Solar System where the Sun is at the center (**heliocentric model**). But his model was rejected for nearly 1800 years. The **geocentric model** (in which the Sun and the planets revolve around Earth), introduced by Aristotle in 350 B.C. and further developed by Ptolemy just seemed more plausible. It was hard to understand at the time, how Earth could move and rotate without leaving the birds and clouds behind. Of course it was also hard to understand in general why we would not live in the center of the universe. The 16th-century Polish astronomer Nicolaus Copernicus is credited with our modern heliocentric view. Copernicus' revolutionary book "De revolutionibus orbium coelestium" (On the revolutions of heavenly spheres) was published just before he died in 1543. The book explains how the observed motion of celestial objects can be explained without putting Earth at rest in the center of the universe. Following Copernicus' publication, Johannes Kepler developed the laws of motion of the planets in 1595, and Isaac Newton developed the laws on gravitational attraction and dynamics in 1687.

The Sun

The Sun is a star at the center of the Solar System. Its radius and diameter are about 109 times larger than that of Earth (Table 3.1; Fig. 3.6). It is made of **plasma,** a highly ionized gas. Even though it consists mainly of hydrogen (about 75%) and helium (nearly 25%), and to only about 1.3% of other elements (incl. oxygen, carbon, neon and iron), the Sun contains 99.86% of the mass of the Solar System. The Sun's mass can be determined approximately on the basis of Newton's law of universal gravitation (equation 2.1), considering that Earth's mass is negligible compared to the Sun's mass, and knowing the Earth – Sun distance. Sunlight takes about 8 minutes to reach Earth (1.496×10^{11} m / 3×10^{8} m/s = 499 s = 8 min 18.6 s). Knowing the Sun's radius, its density can be determined to be about 1.4 that of water.

As described in Chapter 2, the Sun undergoes fusion of hydrogen, with a core temperature reaching nearly 15.7 million °C (28.3 million °F). About 40% of the original hydrogen has been used up. The Sun's energy reaches Earth in the form of electromagnetic waves. Much of it is in the visible range of the spectrum (light), but contributions also come from shorter wavelengths (incl. ultraviolet, X-rays and gamma rays) and longer wavelengths (infrared, microwaves and radio waves). The temperature of the intensely bright surface of the Sun (**photosphere**) is 6,000 K (5,715°C; 10,340°F). The temperature of the **corona** above the photosphere is 1 – 2 million °C (1.8 – 3.6 million °F) though the corona can reach 20 million °C (36 million °F) in its hottest regions. The Sun's corona extends into outer space to form the **solar wind**. Exactly why the corona is so much hotter than the photosphere is an area of intense research. A current idea is that so-called Alfvén waves, low-frequency traveling oscillations of ions and the magnetic field, provide the energy to heat the corona. There are 3 other zones between the photosphere and the corona: the temperature minimum, the chromosphere, and the transition zone. Beyond the corona is the heliosphere that is blown by the solar wind. It reaches across the entire Solar System and past Pluto.

The Sun also maintains a magnetic field that is some factor 5 stronger than Earth's magnetic field. The Sun rotates, where its plasma rotates faster near the equator (rotation period of 25 days) than near the poles (about 34.5 days). This differential rotation causes the magnetic field lines to become twisted over time, causing magnetic field loops to erupt from the Sun's surface, and the formation of **sunspots** and **solar prominences**. This magnetic activity has an 11-year (solar) cycle as the Sun's magnetic field reverses about every 11 years (i.e. the polarity switches).

Figure 3.7 The four terrestrial planets: Mercury, Venus, Earth and Mars (from left). (source: NASA/wikipedia)

3.2 The Terrestrial Planets

Figure 3.8 The first photograph ever taken of an "Earthrise". The photo was taken on 24 December 1968 by Apollo 8 astronauts. (source: NASA/wikipedia)

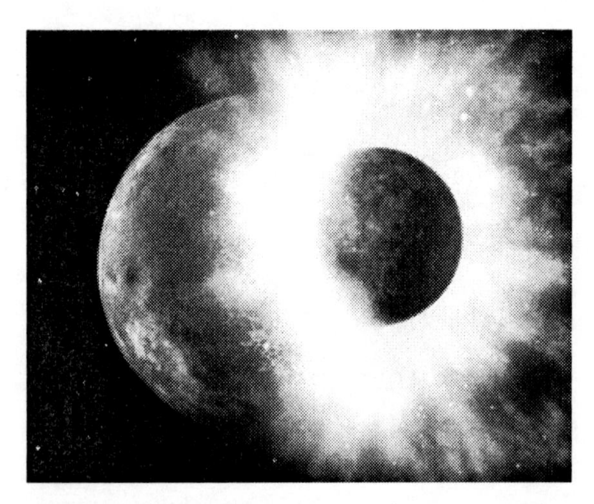

Figure 3.9 Artist's rendering of a Mars-sized body colliding with the early Earth. This collision was responsible for the formation the Moon. (source: NASA/wikipedia)

The Earth – Moon System

Earth is the third planet from the Sun. We also refer to it as the "Blue Planet" because the liquid water on Earth gives it a bluish color when viewed from space. With a radius of 6371 km (3957 mi), Earth is the largest of the inner planets. It orbits the Sun in 365 days and 6 hours (365 d 6 h) and rotates around its own axis in 23 h 56 min, as viewed from space, or relative to the stars, which is called the **sidereal rotation period**. Earth's rotation axis is tilted by 23.5° with respect to its orbital plane around the Sun (Fig. 2.13). This tilt is responsible for seasonal climate variations on Earth. The planet has a unique atmosphere in the Solar System that allows life as we know it to thrive. An ozone layer protects life on land from harmful ultraviolet (UV) radiation, and the oxygen allows animals to breathe. Earth's early atmosphere was very similar in composition to that of Venus and Mars, with CO_2 as a major constituent. But photosynthesis allowed the accumulation of oxygen and geological processes removed much of the CO_2. Earth's current atmosphere has about 78% nitrogen, 21% oxygen and 1% in other constituents. The solid Earth has a crust, a mantle and a metallic core, and details are discussed near the end of this chapter. Earth also has a strong dipolar magnetic field (with 2 poles) whose shape resembles that of a bar magnet. It protects Earth's surface from the harmful solar wind. The field is maintained by a **dynamo** caused by fluid flow of iron in Earth's outer core. We will learn much about Earth in later chapters, so the summary here is only brief. Like other bodies in the Solar System, Earth has been bombarded with impacting objects, primarily during its early formation, but also later during its history after life first developed. Some of these impacts were large enough to cause major mass extinctions. Impacts are therefore discussed later in this chapter. Here, we continue with Earth's satellite, the Moon.

Even though the **Moon** is larger than Pluto and other dwarf planets, it is a satellite, not a planet (the Moon violates the first criterion for being a planet). However, the Moon distinguishes itself by being the largest planetary satellite in the Solar System regarding its size relative to the planet it orbits. The Moon orbits Earth in 27 d 7 h 44 min. It is in **synchronous rotation** with Earth meaning that the same hemisphere faces Earth all the time. The Moon's orbital plane is inclined relative to Earth's orbital plane around the Sun by only 5.14° (Fig. 2.13), and its rotation axis is tilted with respect to its orbital plane by 6.68°. Though seemingly small, the tilt of the Moon's orbital plane in sufficient to explain why we do not observe a solar eclipse every 29.5 days when New Moon occurs. However, depending on where along its orbit the Moon is during a New Moon, a solar eclipse can be observed occasionally. At certain locations on Earth, they may even be total solar eclipses when the Moon obstructs the Sun completely. The next total solar eclipse observable along a narrow swath across the contiguous U.S. will occur on 21 August 2017.

The Earth-facing hemisphere is marked with dark volcanic **maria** among bright ancient crustal highlands and **impact craters**. With a radius of 1,737 km (1,080 mi), its width is about ¼ that of Earth, and its mass is about 1/81 that of Earth. The Moon's gravitational acceleration on the surface is 16.7% of that on Earth, so objects weight only 16.7% on the Moon. This explains why astronauts of the Apollo program could leap in big jumps. The Apollo missions returned 382 kg (842 lb) of lunar rock that helped determine the Moon's surface and internal makeup, and develop models for its origin. The lunar rocks are very old, about 3.16 to 4.5 billion years old, and they resemble basalts, volcanic rocks on Earth. The Moon's density of 3.35 g/cm^3 is significantly less than that of Earth (5.52 g/cm^3). This indicates that much of the Moon's internal composition primarily resembles that of Earth's mantle. Nevertheless, the Moon is a differentiated body, with a distinct crust, mantle and core. The core has a radius of about 500 km, which is less than 1/3 of that of the entire satellite (the radius of Earth's core is nearly ½ of Earth's total radius). Several lunar orbiters in recent years have confirmed the existence of lunar water ice in shadowed craters near the poles. Finding water on the Moon is a significant find as it may allow humans to set up a Moon base in the future.

The Moon has a weak magnetic field, where the field strength is less than 1/100 of that of Earth's magnetic field. Its non-dipolar shape argues against a current generation of the field in the Moon's core. It is thought that the crustal magnetization is either the remnant of an ancient field when the Moon still generated a field, or the signature of temporary fields that were generated during large impact events. The Moon also has **no atmosphere** to speak of, which means that there is no erosion by wind and precipitation. This explains why ancient impact craters have been preserved on the Moon but not on Earth. Smaller falling bodies also do not burn up in the atmosphere and so can impact on the Moon. This gives scientists an opportunity to study impact craters and hypothesize how they form, on the Moon as well as on Earth. Scientists inferred that the last event creating basaltic lava flows occurred about 1.2 billion years ago. The Apollo missions in the 1960s revealed that the Moon experiences **moonquakes** that can be very deep, up to 1000 km from the Moon's surface. But their cause is very different from that of earthquakes. As discussed in Chapter 2, the Moon and the Sun exert tidal forces on Earth, which causes Earth's surface to deform. Similarly, Earth exerts strong tidal forces on the Moon. The resulting stress buildup within the Moon causes the quakes.

At the time of the Apollo missions, much debate existed about how the Moon formed. The four models were:

1. Fission: the Moon formed from the material that formed Earth, separating somehow from Earth in the process
2. Capture: the Moon originally formed far away from Earth and was eventually captured through Earth's gravitational attraction

3. condensation: the Moon formed near to Earth, simultaneously but independently from the early accretion disk
4. giant impact: a planetesimal named Theia, perhaps the size of Mars, hit the young proto-Earth 4.5 billion years ago, some 30 – 50 million years into the formation of the Solar System

The first three models have been rejected by research in the last few decades. Model #1 would require Earth to have an unrealistically fast initial spin. Model #2 would require Earth to have an unrealistically large atmosphere to slow down the Moon during capture. Model #3 does not explain why the Moon's interior is depleted of metallic iron. These models also cannot explain the high angular momentum of the Earth-Moon system. Model #4 is currently the most compelling model. Scientists still debate, however, how much material came from the **impactor** (the impacting body) and how much from the proto-Earth. Most recent chemical analyses of Moon rocks indicate a closer resemblance to Earth than previously thought, so much of the Moon should have come from Earth [(3)]. In September 2009, news media reported that scientists measured the temperature in shadowed craters near the poles of the Moon as being -238ºC (-397ºF)[(4)], and thereby 10ºC lower than the -230ºC (-382ºF) measured in 2006 on the dayside of Pluto. The scientists speculated that the Moon craters could be one of the if not the coldest place in the Solar System.

Mercury

Mercury is the planet nearest to the Sun. It is also the smallest planet (Fig. 3.7; Table 3.1). Of the eight planets, its orbit around the Sun is the most elliptical. Due to its close proximity to the Sun, Mercury can be observed only shortly before sunrise or just after sunset. Ancient cultures viewed the elusive and quickly moving planet as a swift messenger in their myths and religions. The Romans named the planet after their messenger god, Mercury (the equivalent to the Greek God Hermes). Slowed down by the Sun's immense tidal forces, Mercury's rotational period is 2/3 of its orbital period (58.7 days and 88.0 days). Due to the large rotational period compared to the orbital period, the night side does not receive any sunlight for months at a time, and the surface temperature can drop to -183ºC/-297ºF, ranking among the coldest spots in the Solar System. On the other hand, the side facing the Sun can reach temperatures of 467ºC/872ºF (some of the highest in the Solar System) when the planet is near the perihelion.

Mercury has no satellites. Its surface resembles that of the Moon, being heavily cratered and with regions of smooth plains. The planet has no substantial atmosphere. Its high density suggests that an iron core takes up 42% of its volume (Earth's core takes up only 17% of Earth's volume). The large iron core generates a magnetic field that has about 1% the strength of Earth's magnetic field.

Mercury may be the remnant of a giant early collision with a planetesimal 1/6 of its size that nearly stripped it to its core. This would explain why Mercury has a high concentration of iron compared to Earth. Another theory is that Mercury formed from the solar nebula before the Sun's energy output stabilized. Temperatures on the planet would have been so hot that the rock layers vaporized and were carried away by the solar wind. A third hypothesis is that as Mercury was accreting, the solar nebula caused drag on the accreting particles, thereby removing lighter material and leaving behind only heavier material from which Mercury formed. Mercury is currently visited by NASA's MESSENGER spacecraft to map the magnetosphere, surface topography, gravity field and chemical composition of volatiles in the shadowed craters near the poles.

Venus

After Mercury, Venus is the only other planet closer to the Sun than Earth is. As a consequence, Venus remains within 48° from the Sun, as seen from Earth. It is brightest shortly before sunrise or shortly after sunset. For this reason, this planet is also known as the Morning Star or the Evening Star (though Venus is not a star!). Venus is named after the Roman goddess of love and beauty. It is sometimes called Earth's sister planet, because the two planets have several things in common. They are about the same size and the surface gravity is about the same. Venus is slightly less dense than Earth but the bulk composition is estimated to be very similar to that of Earth. Venus is the only other planet in the Solar System that is thought to have some process resembling plate tectonics and/or volcanism. This is inferred from the fact that Venus has much less craters than the other terrestrial planets and the Moon, and so its surface must be relatively young. Venus' plate tectonics is thought to be catastrophic and/or episodic where the whole planet is resurfaced at once. The last such event is thought to have occurred 500 million years ago.

Venus has a retrograde rotation. The orbital period is 224.7 days while the rotation period (sidereal period) is 243 days. The orbital and the retrograde spin rotation combine so that a Venus day (the time when the Sun reappears at the same spot in the sky) is 117 (Earth) days. It is currently unknown why Venus is the only large body in the Solar System that has a true retrograde rotation. One idea to explain Venus' "wrong spin direction" is that a large clump of material struck the young accreting planetesimal at an angle during the early stage of the Solar System, which caused the emerging planet to spin backward. Venus has no satellites. Its weak magnetic field is induced by the interaction between the ionosphere and the solar wind. The lack of an intrinsic magnetic field provides little protection from cosmic radiation and caused the loss of hydrogen that has been swept away by the solar wind. Like Earth, Venus has an atmosphere though it is very dense, with a surface pressure 92 times that on Earth. Venus' atmosphere is composed primarily of CO_2 (96.5%; with most of the remaining 3.5% being nitrogen) causing a very strong, runaway greenhouse effect. As a result, the surface temperature of 460°C/860°F is so high that initially existing oceans would have evaporated and surface rocks have a permanent red glow. The slow rotation of Venus' solid surface contrasts with the rapid rotation of its clouds. The tops of the clouds rotate with the same sense as the surface but 60 times faster, causing strong winds aloft (300 km/h, circling Venus every 4 – 5 Earth days).

Venus' dense atmosphere permanently shrouds its surface. It takes powerful radio telescopes, such as the giant 300-m (1000-ft) dish at Arecibo, Puerto Rico, to map the surface of Venus, and even then, we reach only small areas. In the 1970s the Soviet spacecrafts Venera 7 and 8, confirmed high temperatures, atmospheric pressure and CO_2 content. Venera 8 actually survived on the surface for 50 min. NASA's Pioneer, Venera 15/16 and, in the early 1990s, Magellan were in orbit around Venus and mapped its surface topography using radar imaging. Almost 60% of its surface is covered by a rolling plain with topography less than ±1 km. Only 16% of its surface lies below this plain, which is much less compared to Earth's 2/3 of ocean floor. A northern "continent", Terra Ishtar, is about the size of the continental U.S.. Its highest mountain chain, Maxwell Montes, reaches 11 km (6.8 mi) in height, about 2 km (1.2 mi) more than Earth's Mt. Everest. An equatorial "continent", Aphrodite Terra, is about twice as large as the northern one and is much rougher.

Mars

Mars is the forth planet from the Sun and the first planet farther away from the Sun than Earth is. The planet is named after the Roman god of war. It is often called the "Red Planet" because the abundance of iron oxide on its surface gives it a reddish appearance. Even though Mars is quite a bit smaller than Earth (the diameter is a little more than half of Earth's), the two planets are quite similar. Several recent missions to Mars confirmed that the planet must have had an abundance of water earlier in its history. This hypothesis is based on surface features (morphology) that resemble erosional processes on Earth that involve water. In addition, certain minerals that form in the presence of water were found on Mars. Some of this water has been confirmed to exist as ice sheets near the poles. In 2007, the Mars rover Spirit sampled rock that contained water molecules and the 2008 Phoenix lander directly sampled water ice in shallow Martian soil. It is thought that the presence of water facilitates primitive life and the search for bacteria is on, after scientists first claimed in 1996 that they discovered evidence of bacterial live on the Martian meteorite ALH 84001 found in Antarctica in 1984 (and name after the Allan Hills where it was found). But others contest that the wormlike "biogenic features" seen only under the scanning electron microscope may have formed in a geochemical process or may have been the result of contamination on Earth.

Compared to other planets, surface temperatures on Mars are quite Earth-like. During the polar winter, temperatures reach as low as -140°C (-220°F), while the lowest temperature ever measured on Earth is -90°C (-130°F). During the summer, temperatures can reach balmy 20°C (68°F), though the average temperature on Mars is only -63°C (-81°F) (that on Earth is 16°C/61°F). Mars' obliquity (the tilt of the spin axis relative to the orbital plane) is currently 25.2° and so very similar to that of Earth. Throughout a Martian year, the planet goes through 4 seasons just like Earth does. However, Mars can have much larger obliquities, which is thought to be responsible for extreme climate changes. Mars rotates around its axis in 1 d 37 min, but with 687 days, the orbital period of Mars is nearly twice as long as that of Earth. This allows winters to become relatively colder than on Earth and summers relatively warmer.

Mars has two tiny, irregularly shaped satellites, Deimos and Phobos, that are about 20 km (12.5 mi) in diameter and were discovered in 1877. Their origin is uncertain but they may be captured asteroids. In Greek mythology, Phobos (panic/fear) and Deimos (terror/dread) accompanied their father, Ares (the Greek god of war) into battle. Mars has a metallic core made of iron and nickel, and its size relative to the rest of the planet is comparable to Earth's core. But since the overall density of Mars is much lower than that of the other terrestrial planets, the core of Mars likely has a significant amount of lighter elements. Mars does not appear to have a global magnetic field that is maintained by a dynamo like that on Earth. But a strong local magnetic field indicates that the crust has been magnetized locally. On the other hand, the magnetized rocks exhibit "stripes" in the lateral magnetization pattern similar to those found on the ocean floor near mid-ocean ridges on Earth. On Earth, these symmetric "stripes" are evidence for past magnetic field reversals and seafloor spreading. An alternative explanation for Mars' magnetized rocks is that Mars initially did have a magnetic dynamo, some 4 billion years ago, with magnetic field reversals occurring similar to those on Earth. But the planet's dynamo ceased when the core at least partially froze, and much of the magnetized crust was reworked by ancient plate tectonics. Mars also has volcanoes, with 25-km (15.5 mi) tall Olympus Mons being the tallest volcano in the Solar System. Olympus Mons is a shield volcano and thought to be extinct, having produced the last lava flow some 2 million years ago and the last caldera (large collapsed volcano crater) some 150 million years ago. These estimates come from counting and comparing impact craters.

Mars also has an atmosphere though it is very thin and not very effective as protection against the solar wind. The surface pressure on Mars is less than 1% of that on Earth. It is thought that Mars had a thicker atmosphere during the first 1 billion years of its existence. The atmosphere consists to 95% of CO_2, 3% nitrogen, 1.6% argon and traces of oxygen and water. Methane occurs in localized traces suggesting volcanoes or comets as source. However, a microbiological or geological process has also been suggested. Despite its thin atmosphere, Mars is known to have strong dust storms the observation of which helps us understand the global impact of relatively local dust storms on our own planet.

Figure 3.10 The four gas giants: Jupiter, Saturn, Uranus and Neptune (from left). (source: NASA/wikipedia)

3.3 The Gas Giants

Jupiter

Jupiter is the 5th planet from the Sun and the first planet on the other side of the Asteroid Belt, i.e. the first of the **outer planets** (Fig. 3.10). With a mean radius of 69,911 km (43,423 mi), it is roughly 10.7 times as wide as Earth (Table 3.1) and the largest planet in the Solar System. Jupiter is therefore named after the principal god of Roman mythology, in analogy with the Greek deity Zeus. Its brightness in the night sky sometimes rivals that of Venus, as it may leave reflections on a lake or calm ocean in a dark night. Having 2.5 times the mass of all the other planets in the Solar System taken together, Jupiter is indeed a giant. Jupiter takes up 60% of the mass and 70% of the angular momentum of the planets of the Solar System. However, the planet's density is only 1.33 g/cm^3 and therefore nearly 4 times lower than that of Earth, and barely larger than that of water. The structural make-up of Jupiter, and the other outer planets, therefore has to be very different from that of the terrestrial planets. In fact, Jupiter's makeup is closer to that of the Sun and so Jupiter's origin can be traced back directly to the solar nebula. Jupiter has no crust but is thought to have a rocky core that is larger and perhaps 10 times more massive than planet Earth. But much of Jupiter is made of gas that becomes denser with depth, turning liquid at about 1000 km (620 mi) depth, and turning metallic some 14,000 km (8,700 mi) and 20% of the way down. Most of Jupiter's interior is therefore in liquid form, consisting to 71% (by mass) of hydrogen, 24% helium and 5% other elements, including iron and silicates. The atmosphere is made of 90% hydrogen (by volume) and 10% helium.

Jupiter's orbital period is 11 y 317.6 d. Its surface rotates in a dizzying 9 h 52 min though different latitudes move at different speeds, forming structural bands. The bright-colored bands are called **zones** and are regions of rising gas, while the darker bands are called **belts** and are regions of falling gas. The tops of the dark belts are about 20 km (12.4 mi) lower and 10°C (18°F) warmer than the tops of the brighter zones. The most prominent feature of Jupiter's surface is its **Great Red Spot**, being 2 – 3 times larger in diameter than Earth. The Red Spot is the vortex of a giant, violent and long-lasting storm, similar to storms on Earth. It drifts slowly relative to the clouds as the planet rotates and has been around for at least the last 150 years. From the sense of rotation, we know that the storm is a high-pressure system rather than a low as we are used to on Earth. Why has this storm lasted so long while systems on Earth break up much faster? Heat from below fuels the Red Spot. It has more mass, which makes it more stable. Also, Jupiter has no continents that tend to break up storms such as hurricanes on Earth.

Jupiter radiates 1.6 times as much heat as it receives from the Sun. It therefore has to have a heat source inside. This heat is produced through gravitational contraction where the planet shrinks by about 2 cm (roughly 1 inch) each year. Contrary to some popular accounts though, Jupiter is far away from being "almost" a star generating energy through nuclear processes. Jupiter has an intense magnetic field that is generated by currents in the metallic hydrogen layer and is 14 times stronger than Earth's magnetic field. At the height of Jupiter's clouds, the field strength is 10 times that of Earth's magnetic field, which itself is a strong field. As a result, Jupiter has giant aurorae. Jupiter also has wispy rings of material around it that are thought to be made of ejected material from Jupiter's four innermost moons, Adrastea, Matis, Thebe and Amalthea. They were first observed in 1979 by the Voyager 1 space probe and more thoroughly investigated in the 1990s by the Galileo orbiter.

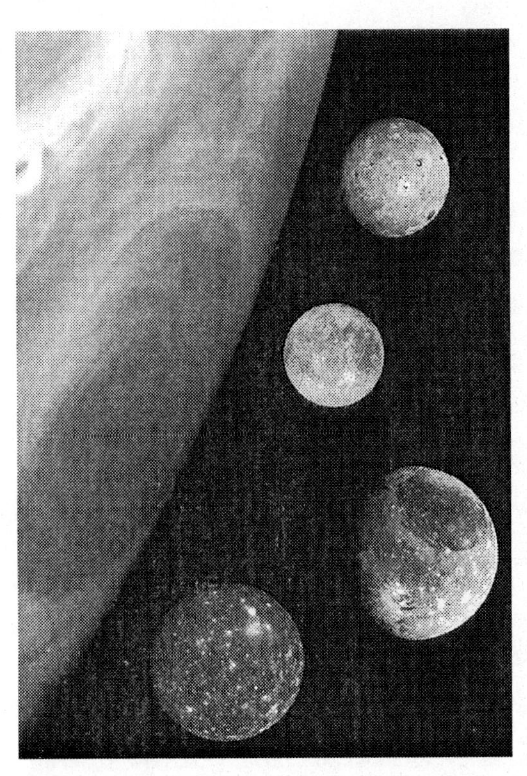

Figure 3.11 Composite image of Jupiter and the four Galilean satellites: Io, Europa, Ganymede and Callisto (from top, and with distance from Jupiter). (source: NASA/wikipedia)

Jupiter has a large number of satellites, 63 or so, and counting. The four largest moons, Io, Europa, Ganymede and Callisto are called the **Galilean satellites**, with Io being the closest and Callisto the most distant from Jupiter (Fig. 3.11). The Galilean satellites are large enough to be viewed easily through a small telescope. **Ganymede** is the largest and, with a radius of 2631 km (1634 mi), is larger than planet Mercury but less dense. Ganymede is the largest moon in the Solar System. It completes an orbit around Jupiter in about 7 days. It is composed of approximately equal amounts of silicate rock and water ice, where the ice surrounds a silicate mantle and a metallic core. Dark regions with many ancient craters (4 billion years old) cover about 2/3 of the moon's surface. The other 1/3 is covered by lighter-colored regions, with slightly younger, intriguing grooved terrains. The origin of these is not well understood but they are thought to have formed during tectonic activity that was brought on by tidal heating exerted by Jupiter. Ganymede is the only satellite in the Solar System with a magnetosphere thought to be generated through convection within a liquid core. It has a thin oxygen atmosphere.

Callisto is the second largest Galilean satellite and almost as large as Mercury, and the third-largest moon in the Solar System. It orbits Jupiter in 16.7 days. Unlike the other three Galilean satellites, it does not participate in an orbital resonance. But it rotates synchronously with its orbital period, meaning that the same hemisphere of Callisto always faces Jupiter (similar to the Moon's synchronous rotation around Earth). Being farthest away, Callisto does not experience tidal heating. It has a density similar to that of Ganymede but it did not differentiate early in its history. The heavily cratered surface of Callisto must be very old, and there is no evidence of tectonic movement. Like Europa, it is covered with ice, with a water ocean proposed at 100 km depth. This leaves the question open whether this satellite may harbor life. Callisto is surrounded by an extremely thin atmosphere made of CO_2. Molecular oxygen has been proposed as minor constituent but not yet confirmed.

Europa is the sixth-closest moon of Jupiter and the smallest of the Galilean satellites. Slightly smaller than Earth's Moon, Europa is primarily made of silicate rock and probably has an iron core, which would explain its higher density compared to Ganymede and Callisto. Its surface is composed of ice and is one of the smoothest in the Solar System. Yet, its intriguing grooves clearly show evidence of tectonic activity. The lack of craters also suggests that the surface must be very young, at about 20 – 180 million years of age. It is proposed that heat energy from tidal flexing allows the existence of a 100-km (62 mi) thick water "ocean" that surrounds the rocky interior beneath an icy crust that undergoes plate tectonics. It is thought that Europa's water ocean could harbor life. Europa orbits Jupiter in just 3.5 days. Due to its proximity to Jupiter, the radiation level at the surface of Europa is equivalent to a dose of about 540 rem per day, an amount considered to cause illness or death in humans. Europa has a nearly non-existent atmosphere composed mainly of oxygen. The oxygen is not biogenic but results from the interaction of solar UV radiation with the icy surface.

Io is Jupiter's innermost Galilean satellite, the fifth moon out from Jupiter and the forth-largest moon in the Solar System. With over 400 active volcanoes, Io is the most geologically active object in the Solar System. This extreme activity is the result of immense tidal heating exerted by Jupiter. Several volcanoes produce plumes of sulfur and SO_2 that climb as high as 500 km (300 mi) above the surface. Io's surface is also dotted with more than 100 mountains, some taller than Mt. Everest, that have been uplifted by strong compression at the base of the moon's silicate crust. Having a relatively high density that is larger than that of Earth's Moon, Io's crust surrounds a silicate mantle and a molten iron or iron sulfide core that takes up 20% of Io's mass. Depending on its composition, the radius of the core could make up half of Io's radius, or a little as 1/5. Some of Io's volcanism can be observed on Earth's-based telescopes, such as the Keck telescope in Hawaii, and the Hubble Space Telescope. Io is located within one of Jupiter's most intense radiation belts and receives about 3,600 rem per day.

Saturn

Named after the Roman god of agriculture, Saturn is the second-largest planet in the Solar System. With its rings, Saturn is arguably the most beautiful planet to observe with a telescope. Saturn completes its orbit around the Sun every 29 y 174.2 d. It rotates with a period of 10 hr 34 min. The density of Saturn is only 0.7 g/cm^3 and so lower than that of water. So, as the joke goes, Saturn would float in a bathtub if we found one big enough to hold it. The bulk of Saturn is made of hydrogen and helium, reflecting Saturn's formation directly from the solar nebula. It is thought to have a rocky core made of silicates and metals (iron and nickel) but it cannot make up more than 20% in terms of radius. The core is surrounded by metallic hydrogen, an intermediate layer of liquid hydrogen and liquid helium, and an outer gaseous layer. Electrical currents within the metallic hydrogen layer are thought to generate the planet's magnetic field that is about 20 times weaker than that of Jupiter.

Saturn's atmosphere exhibits color bands similar to that of Jupiter though they are much fainter. Occasionally, long-lived oval and other features known from Jupiter's atmosphere can also be seen on Saturn. Once every Saturn year, near the northern hemisphere's summer solstice, short-lived Great White Spots occur which are associated with storms. They have been observed since at least 1876. Peaking at 1800 km/h (1120 mph), the winds on Saturn are among the fastest winds on the Solar System's planets (second only to those on Neptune).

Probably most fascinating are Saturn's extensive rings which extend from 6,630 to 120,700 km above Saturn's surface (4120 to 74950 mi). The rings are about 20 m (65 ft) thick and are composed of 93% water ice and 7% carbon. The particles that make up the rings range from dust particle size to 10 m (33 ft). Currently, two theories exist to explain the ring's formation: 1) the rings are remnants of a destroyed moon of Saturn; 2) the rings are left over from the original nebular material from which Saturn formed.

Saturn has at least 62 moons, with 53 having formal names. **Titan** is the largest moon, having more than 90% of the mass in orbit around Saturn. Having a mean radius of 2576 km (1600 mi), Titan is about 50% larger than Earth's Moon and is the second-largest moon in the Solar System. It is the only moon known to have a dense atmosphere, and the only object other than Earth with clear evidence of stable bodies of surface liquid. Discovered by Dutch astronomer Christriaan Huygens in 1655, Titan was the first known moon of Saturn. The satellite is composed mainly of water ice and rocky material, where the dense atmosphere conceals the surface and nothing was known until the arrival of the Cassini-Huygens spacecraft in 2004. This mission discovered the liquid hydrocarbon lakes (composed of methane and ethane). The surface is geologically young, with few impact craters. The atmosphere is largely composed of nitrogen, with minor components leading to methane and ethane clouds and nitrogen-rich organic smog. Wind and rain on Titan create surface features similar to those found on Earth, including sand dunes, rivers and lakes. With its liquids and robust nitrogen atmosphere, Titan's methane cycle is viewed as an analog to Earth's water cycle, although at much lower temperatures. The satellite is also thought as a possible host of microbial extraterrestrial life. **Rhea**, Saturn's second-largest moon, may have a ring system on its own and an atmosphere. Its radius of 764 km (475 mi) is about half that of Earth's Moon. Many of the other moons of Saturn are very small, most of them having a diameter of 50 km (31 mi) or less.

Saturn radiates about twice as much energy as it absorbs from the Sun, so Saturn must still "generate" heat, like Jupiter but more so. It is thought that only 2/3 of Saturn's internal heat remains from its early formation and that the other 1/3 is produced by continuing differentiation and contraction.

Uranus

Uranus is named after the ancient Greek god of the sky, the father of Cronus (Saturn) and grandfather of Zeus (Jupiter). Uranus can be seen with the naked eye but it is so dim and its orbital motion is so slow that it was not recognized as a planet until William Herschel announced its discovery on 13 March 1781. It was the first planet discovered with a telescope. Uranus is similar in composition to Neptune, and both are different from the other two gas giants that are more than twice in diameter. Astronomers sometimes put them in their own category as **ice giants**. Uranus' atmosphere is similar to Jupiter's and Saturn's in its primary composition of hydrogen and helium. But Uranus' atmosphere contains more ices of water, ammonia, methane and traces of other hydrocarbons. Uranus is thought to have a rocky core surrounded by an icy mantle. With a minimum of -224°C (-371°F), Uranus is the coldest planet in the Solar System. Uranus also has a ring system, a magnetosphere, and numerous moons.

Uranus orbits the Sun in 84 y 139 d, and it rotates in a retrograde manner around its spin axis in 17 h 14 min. Unique to Uranus, however, is that its rotation axis is tilted 97.8° with respect to its orbital plane, so the retrograde movement is very different from that of Venus. Because the rotation axis is nearly aligned with the orbital plane, Uranus experiences seasons very different from the other planets. Unlike the other gas giants, Uranus does not produce any excess heat. A possible explanation for both the unusual rotation as well as the lack of heat is explained by the hypothesis that Uranus collided very early with a protoplanet that threw it into its odd rotation. The collision would also have expelled the primordial heat (original heat), leaving behind a cooler core.

Neptune

Neptune is the most distant planet in the Solar System and received its name from the Roman god of the sea. It was first observed on 23 September 1846 by Johann Galle, after scientists observed unexpected changes in the orbit of Uranus and predicted that this is caused by the tug of another planet. Neptune is similar in composition to Uranus. Traces of methane in the outermost regions account for Uranus' and Neptune's blue-greenish appearance. In contrast to Uranus' atmosphere, however, Neptune's exhibits active and visible weather patterns. For example, at the time of the 1989 Voyager 2 flyby, Neptune had a Great Dark Spot resembling Jupiter's Great Red Spot. These weather patterns are sustained by the strongest winds on any planet in the Solar System, with recorded wind speeds as high as 2,100 km/h (1300 mph). In comparison to Uranus, Neptune's more varied weather is believed to be driven by internal heating. Neptune radiates about 2.6 times as much energy as it receives from the Sun. Explanations for this excess heat includes radiogenic heating from the core (fission of heavy elements), conversion of methane under high pressure into hydrogen, diamond and longer hydrocarbons and differentiation. Neptune orbits the Sun in 164 y 288 d and rotates around its axis in 16 h 6 min. With temperatures reaching -218°C (-360°F) at the cloud tops, Neptune's outer atmosphere is one of the coldest places in the Solar System. Neptune also has a faint and fragmented ring system that was detected in 1960 and confirmed during the Voyager 2 flyby.

With a radius of 1353 km (840 mi), **Triton** is Neptune's largest moon. It was discovered 17 days after Neptune's discovery. It is the only large moon with a retrograde orbit. Since its composition is similar to that of Pluto, it is thought that Triton was captured from the Kuiper Belt (see below). Triton has a surface of mostly frozen nitrogen, a crust composed of mostly water ice, an icy mantle and a substantial core composed of rock and metals. It is geologically active, revealing **cryovolcanism** (ice volcanism) and tectonic terrains. Geyers are believed to erupt nitrogen.

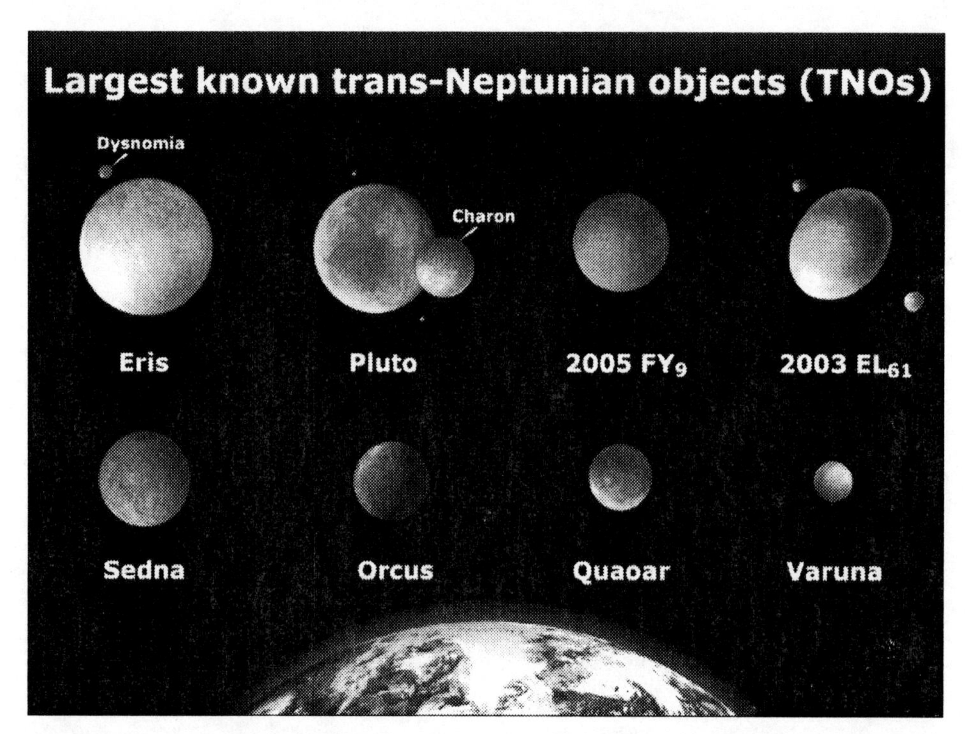

Figure 3.12 Comparison of Eris and other trans-Neptunian objects with Pluto and Charon. For scale, Earth is also shown. Pluto and Eris are dwarf planets, while Sedna, Orcus and Quaoar may be declared as such in the future. 2005 FY_9 and 2003 EL_{61} are dwarf planets Makemake and Haumea. Relative sizes are somewhat uncertain. Sedna and Makemake are thought to have a density similar to that of Pluto, while Haumea and Quaoar are thought to be denser, with 3 g/cm^3. (source: NASA/wikipedia)

3.4 Other Members in the Solar System

Pluto and the Dwarf Planets

Pluto is one of the trans-Neptunian objects (TNOs), meaning they lie beyond the orbit of Neptune. Pluto was discovered relatively late, on 13 February 1930 by Clyde W. Tombaugh. For nearly 8 decades, it has been viewed as the 9th planet in our Solar System even though Pluto did not quite fit in. With an inclination of 17°, Pluto's orbit around the Sun is strongly inclined relative to the orbits of all other planets (Fig. 3.5). With an eccentricity of nearly 0.25, Pluto's orbit is also much more elliptical than that of the other planets. Pluto orbits the Sun with a period of 248 y 33 d and rotates about its own axis in 6 d 9 h 17 min. The spin axis has a tilt of 120° meaning that Pluto lies on its side and its rotation is retrograde, similar to that of Uranus. Pluto is very small and, being located beyond the gas giants, would need its own category of Solar System planets, if the term "inner planets" were synonymous with "terrestrial planets". Having a diameter less than half of that of Mercury, Pluto is only 2/3 as wide as the Moon and is smaller than other celestial objects in the Solar System. Pluto also has a very large moon, Charon, discovered in 1978 and half as wide as Pluto. Two more satellites were found in 2005, Nix and Hydra, and another one in 2011.

In 1977, Charles T. Kowal discovered **Chiron** at the Palomar Observatory in San Diego county, CA. This 230 km long and 140 km wide body belongs to a new class of objects between the orbits of Saturn and Uranus (named **centaurs**). Initially assumed to be an asteroid, Chiron also exhibits the properties of a comet. Though much smaller than Pluto, its discovery rang in a lively debate on whether Pluto really should be regarded as a planet or not. Following the discovery of Chiron, the Palomar Observatory was instrumental at discovering other larger celestial bodies. On 4 June 2002, Chad Trujillo and Mike Brown found **Quaoar**, then thought to be about half as wide as Pluto (Fig. 3.12). Quaoar is about 43 AU from the Sun. Its mean density is estimated at around 3 g/cm^3 and so is much larger than Pluto's. Then came **Sedna,** with a density similar to that of Pluto. Discovered by Mike Brown and collaborators on 14 November 2003, this TNO has a very elongated orbit, with a perihelion at 76 AU and an aphelion at 937 AU. Being 1200 to 1600 km wide (7.5 to 10 mi), Sedna is larger than Quaoar. Discovered in 2004 by Mike Brown's group, **Haumea** (named after the Hawaiian goddess of childbirth) is at the same distance as Quaoar but about 1.5 times wider. It has about the same density as Quaoar. On 31 March 2005, Mike Brown's team found **Makemake** (named after the Rapanui god of fertility and creator of humanity), a body nearly as large as Haumea and only a little farther out from the Sun, and having a lower density (perhaps $2g/cm^3$). A similar number of asteroid objects (Ceres, Pallas, Juno and Vesta) have at one time been considered for planet status but then rejected.

Pluto's ultimate demise came with **Eris**, discovered in January 2005 also by Mike Brown's group. Initially named Xena, Eris is slightly larger than Pluto, denser, and nearly twice as far from the Sun as Pluto is. The name comes from the Greek goddess of strife and discord and adequately stands for the cause of the debate on Pluto's status. Eris has one known moon, Dysnomia. Since Eris is larger than Pluto, scientists had to decide whether it should become the 10th planet in the Solar System or whether Pluto should finally be declared something other than a planet. The IAU therefore decided in 2006 that a planet has to meet four criteria. It turns out that Pluto has only 7% of the mass of other objects in its orbit, and so Pluto does not meet the 4th criterion. It was therefore put into a new class of celestial bodies, the **dwarf planets** that have to meet the following four criteria:

- has its own orbit around the Sun
- has sufficient mass to assume a nearly round shape
- is not a satellite of a planet or other nonstellar body
- has NOT cleared the neighborhood around its orbit

These criteria are the same as those for a regular planet, except for the last one that allows other objects, other than a moon, to reside in a dwarf planet's orbit. The second criterion prevents the declaration of just any object to become a dwarf planet. As of fall 2011, the IAU has recognized five dwarf planets in the Solar System: Ceres, Pluto, Haumea, Makemake and Eris. Only Ceres and Pluto have been observed in enough detail to demonstrate that they meet the definition of a dwarf planet. Eris has been accepted because it is larger and more massive than Pluto. Haumea (rather egg-shaped but having 2 moons) and Makemake qualified because their brightness of +1 or larger suggests that their diameter is 838 km or larger, a recommendation by the IAU. Sedna, 2007 OR_{10}, Orcus and Quaoar are still under consideration to become a dwarf planet. Being nearly as large as Haumea, Sedna may have the best chances. Larger than Quaoar, Orcus was discovered on 17 February 2004 by Mike Brown and collaborators. It has about the same distance from the Sun as Pluto, but is always on the other side. Also having a large moon, Vanth, it is sometimes viewed as the anti-Pluto. 2007 OR10 is thought to be as large as Quaoar, or maybe even as Sedna, and is currently the largest body in the Solar System without a name. **Ceres** is currently the only asteroid that is a dwarf planet, while the much smaller Pallas, Juno and Vesta fall under the category "small Solar System bodies".

Pluto's struggle for status is not over as Charon does not really orbit Pluto, raising doubt that Pluto has its own orbit around the Sun, a criterion to qualify as a dwarf planet. The fact that the barycenter (center of mass) of the Pluto-Charon system does not lie within Pluto, but also not in Charon makes this system unique, and there is hope for those who hang on to Pluto as something special. Some scientists proposed that Pluto and Charon should be name a binary dwarf planet system. The jury on this is still out.

Figure 3.13 Artist's rendering of the Asteroid Belt between the orbits of Mars and Jupiter. (source: NASA/wikipedia)

Figure 3.14 Asteroid Ida and its moon Dactyl. This photo was taken by NASA's spacecraft Galileo on 28 August 1993, from a range of 10,870 km (6755 mi). Ida is irregularly shaped and about 56 km (35 mi) long. Dactyl (the little dot to the right) is only about 1.5 km across. (source: NASA/wikipedia)

The Asteroid Belt

The **Asteroid Belt** is located between Mars and Jupiter (Fig. 3.13). In fact, it is located where scientists predicted another planet. In 1766, before Uranus was confirmed as a planet, German scientist Johann D. Titius and later Johann E. Bode discovered that the distance of the planets to the Sun follow a simple arithmetic rule quite closely: the distance can be determined in AU according to the formula $a = (n + 4) / 10$; where n is a number taken from the series $n = 0, 3, 6, 12, 24, 48, 96, 192$ …, in the order of the planets' position from the Sun. For example, $n = 6$ for Earth as the third planet from the Sun and the distance is computed as 1 AU. The formula is also known as Titius-Bode Law, or also as Bode's Law. Though it is not entirely understood why this formula works, it predicts the planets' distances quite well, up to Uranus. A possible explanation why it works involves that the planets orbit the Sun in resonance. The Titius-Bode Law does not work for Neptune and Pluto. In fact, Pluto's position is predicted as the next position after Uranus, and Neptune's position remains unexplained (there should not be a planet there). Another curiosity is that there is no planet for $n = 24$.

COPYRIGHTED MATERIAL — DO NOT DUPLICATE, DISTRIBUTE, OR POST

Ceres was original declared as this 5th planet from the Sun (between 1801 and the 1860s) but was then declared not a planet, after other bodies were found along its orbit that we now know as asteroids.

There are at least two theories why there is no real planet in the Asteroid Belt: 1) a large impactor destroyed a planet that initially formed there; 2) due to the strong tug of Jupiter, a planet could never form. There are millions of asteroids in the Asteroid Belt, and 1 – 2 million are larger than 1 km (0.6 mi) in diameter and qualify as planetesimals. Most asteroids are oddly shaped, some like potatoes, and too small to become spherical, except for Ceres that is now a dwarf planet. There are two more objects larger than 500 km in diameter: Pallas and Vesta. Due to their size, Ceres, Pallas and Vesta fall under the category protoplanet. Some asteroids even have a moon, such as asteroid Ida (Fig. 3.14). Asteroids are low-density, rocky objects, and usually have one of three compositions: carbon-rich (C-type), stony (S-type) or metallic (M-type).

The danger for Earth (and Mars) is that the asteroids do not follow the same, nearly circular orbit. In fact, many asteroids have elliptical orbits that reach into the orbits of Mars and Earth and some reach well beyond Jupiter. Asteroids with so-called Mars-crossing orbits are called **Amors** and those with Earth-crossing orbits are called **Apollos**. The latter will eventually impact on Earth, if we wait long enough so that both the asteroid and Earth are at the same place at the same time. Some previously existing Apollos already impacted on Earth, and the very large ones caused major mass extinctions. Some 7,000 Amors are also considered near-Earth asteroids because their orbits take them close to Earth. Between 500 and 1000 of these have a diameter of 1 km or larger.

Figure 3.15 Comet Hyakutake photographed in Arizona, with a Saguaro cactus in the foreground. The photo was taken on 27 March 1996 when Hyakutake was closest to the Sun, inside Mercury's orbit. Polaris, the north star, is the bright star seen just to the upper right of the comet's head. Hyakutake will not return to the Sun for another 70,000 years. (source: NASA)

Figure 3.16a Hale-Bopp was probably the most widely observed comet of the 20th century, and one of the brightest seen for many decades. It was discovered on 23 July 1995 and could be seen with the naked eye for a record 18 months, mostly in 1997. A photo taken on 29 March 1997 near Pazin in Istria, Croatia, shows the yellowish dust tail and a second, fainter bluish gas tail. (source: Philipp Salzgeber, Wikipedia)

Figure 3.16b Hale-Bopp with its two tails. It passed its perihelion on 1 April 1997. The photo was taken on 4 April 1997 at the Johannes-Kepler-Observatorium, Linz, Austria. (source: Wikipedia)

Comets

Compared to the stars and planets, comets are fascinating but perhaps also frightening bright objects in the sky. They may stay in the sky for months, or even a whole year. In many cultures, comets have been seen as harbinger of fortune but also disaster. While the former cannot be proven scientifically, the latter is certainly true when they crash on Earth. We have not witnessed a comet impact on Earth but at least two impacts have been observed on Jupiter since 1994.

Comets come from the outer reaches of the Solar System, some on a regular basis. **Halley's Comet**, for example, has a period of about 75 years. Some scientists and theologians assume that Halley is also the "Star of Bethlehem". The comet's last return was observed in 1986 and its next visit is predicted for 2061. Halley's visits to the inner Solar System have been observed and recorded since at least 240 B.C. though subsequent observers did not recognize that they observed the same object. The comet's period was first determined in 1705 by English astronomer Edmond Halley. In 1986, the comet was the first to be observed in detail by spacecraft, giving scientists the opportunity to learn more about the comet's composition and structure.

Before reaching into the inner Solar System, comets are simply rocky objects that contain dust particles embedded in an icy crust where the ice is made of H_2O, CO_2 and NH_3 (ammonia). They

COPYRIGHTED MATERIAL — DO NOT DUPLICATE, DISTRIBUTE, OR POST

are therefore also referred to as "**dirty snowballs**". A comet is typically a few km to tens of km across. Elongated Halley's comet is relatively small, only 15 km long and 8 km wide. Upon approach to the Sun, solar energy causes the comet to eject icy volatiles that vaporize and begin to glow as they react with the sunlight. Dust particles are torn away with the ejected volatiles. At this stage, the comet consists of three parts, the original rocky part, now called the core or **nucleus**, the bright **dust tail**, and the bluish **coma**. The tail and the coma make for spectacular displays in the night sky. The dust tail is curved away from the Sun as the dust particles are slightly blown away by the pressure of the sunlight hitting them. In the gas tail, the volatiles are ejected with high speed, always away from the Sun, in a different direction from the dust tail. Event though the nuclei of comets are usually smaller than 50 km across, their tails can be several AU long.

After a comet left the inner Solar System, its ejected dust remains behind along its orbit. If the comet's orbit was an Earth-crossing orbit, Earth will fly through this dust at the same time every year to make for bright meteor showers. The most visible meteor showers are the Perseid shower in August and the Leonid shower in November. The dust for the Perseid shower was left behind by the 27-km wide **109P/Swift-Tuttle comet** that was discovered in 1862 and reappeared in 1992 (Table 3.2). After its 1992 rediscovery, scientists were first concerned that it may impact on Earth or the Moon upon its reappearance in August 2126. Of particular concern was its large size relative to the 10-km wide impactor that is thought to have wiped out the dinosaurs some 65 million years ago. But subsequent more detailed studies indicate that the comet, which was first observed by the Chinese at least as far back as 69 B.C., is in a stable orbit and will not impact on Earth for at least another 2000 years. The dust for the Leonid shower comes from the **55p/Tempel-Tuttle comet** that was discovered in 1865. It was also observed in 1366 and 1699 but not recognized as a recurring (periodic) comet. It last appeared on 28 Feb 1998, and it will reappear on 20 May 2031. With the nucleus being only 3.6 km across, it is relatively small. The comet's orbit intersects Earth's orbit nearly exactly so that the dust does not have to spread much over time to cause meteor showers on Earth. Meteor showers observed in 2009 were caused by dust from passages in 1466 and 1533.

Most comets are too faint to be observed without a telescope but a few comets each decade are visible with the naked eye. A comet's first appearance (after its absence) in the sky is also called **apparition**. Two of the most spectacular recent comets that were observable from Earth were Hyakutake in 1996 (Fig. 3.15) and Hale-Bopp a year later. **Hyakutake**, also called the Great Comet of 1996, was unknown until Japanese amateur astronomer Yuji Hyakutake sighted in on 31 January 1996. Only 2 km (1.2 mi) across, relatively small Hyakuake became visible to the naked eye in early March. By 24 March, the comet was one of the brightest objects in the sky, with a long tail that stretched 35° across the sky. With a minimum distance of 0.1 AU (14,960,000 km), Hyakutake's passage carried it very close to Earth on 25 March in one of the closest approaches of any comet in 200 years. Comets usually move too slowly to watch their actual advance with the naked eye. But upon close approach, comets can change their position markedly within minutes. On 25 March, Hyakutake covered 1/2° (diameter of the Full Moon) in 30 min. The coma reached up to 80° across the sky and was up to 2° wide. With a length of 3.8 AU, Hyakutake's tail surpassed the record of 2.2 AU which was held by the tail of the Great Comet of 1843. Unfortunately, the comet's quick passage and unfavorable weather in Europe did not allow observers there to fully appreciate its might. Hyakutake last reached into the inner Solar System 17,000 years ago. Due to the gravitational perturbation of the giant planets, its period has increased, and it is not expected to return for another 70,000 years. On 1 May, the comet reached its perihelion at 0.23 AU, well inside Mercury's orbit. By that time, the comet also developed a dust tail. But it was not seen easily from Earth because of its proximity to the Sun. After its perihelion passage, Hyakutake faded quickly and was lost to the naked eye before the end of May.

Hale-Bopp was probably the most widely observed comet in the 20th century (Fig. 13.16). It was visible for a record 18 months, twice as long as the previous record holder, the Great Comet of 1811. Hale Bopp was discovered on 23 July 1995 by two independent observers, Alan Hale and Thomas Bopp in the U.S.. The comet was discovered when it was still at a great distance from the Sun, raising hopes that it would be very bright by the time it reached the inner Solar System. Being nearly 40 km (25 mi) across, it is a monster of a comet. Holding true to this promise, it was so bright that it could be watched even in a light-polluted big city when it reappeared in January 1997. Before that, the comet became first visible to the naked eye in May 1996, just when Hyakutake moved away from the Sun, but the brightening was initially slow. Its alignment with the Sun later in 1996 meant that it could not be observed continuously before it reappeared in January 1997. By February, it had "grown" two tails, a bluish gas tail pointing straight away from the Sun and a yellowish dust tail curving away along its orbit. In fact, it had grown a third, very faint sodium tail that is thought to have emerged from the inner coma, but it could be observed only with powerful instruments. Hale-Bopp had its closest approach to Earth on 22 March at a distance of 1.32 AU. The comet passed the perihelion on 1 April 1997 at which point it developed into a spectacular sight. It shone brighter than any star in the sky except for Sirius and was visible well before the sky got fully dark each night. Its dust tail stretched 40-45° across the sky. At this point, Hale-Bopp was visible all night to observers in the Northern Hemisphere. After it passed the perihelion, it moved into the southern celestial hemisphere, escaping observers in the Northern Hemisphere. The last naked-eye observation in the Southern Hemisphere was reported in December 1997. Ten years later, in October 2007, astronomers were still able to track it at a distance of 25.6 AU from the Sun, and they expect to track it with a large telescope until perhaps 2020.

Hale-Bopp likely visited us last about 4,200 years ago and there is evidence that it was observed by the ancient Egyptians. Its orbital plane is nearly perpendicular to Earth's orbital plane which means that a close approach to planets, or even and impact, should be extremely unlikely. However, in April 1996, Hale-Bopp passed Jupiter at 0.77 AU, close enough to be affected by the planet's gravity. This shortened the comet's period by more than a factor 2 so that Hale-Bopp is expected to reappear in about 2400 years, in 4385 to be precise. Scientists also speculate that Hale-Bopp nearly crashed on Jupiter on its last visit in the year 2215 B.C., thereby dramatically changing its orbit. In fact, that visit may have been its first to the inner Solar System.

Being the brightest comet in 40 years, **McNaught** was easily visible to the naked eye in January and February 2007, after it was discovered by British-Australian Astronomer Robert H. McNaught on 7 Aug 2006. On 12 January, the comet reached its perihelion at a distance of 0.17 AU, and it was visible worldwide in broad daylight. McNaught's tail reached 35° across the sky. Never heard of it? You are not alone. The comet was visible only in the Southern Hemisphere.

It is thought that comets entering the inner Solar System for the first time have brighter displays than returning comets because more material can interact with sunlight. But the brightness of the display of comets is hard to predict. Some returning comets can have bursts, producing brighter tails than anticipated, and others are nearly complete busts. **Kohoutek** that returned after 150,000 year in 1973 was such a comet. Kohoutek comes from the Oort cloud (see below) on the outer reaches of the Solar System, and it was believed that its appearance in 1973 was its first approach to the Sun. Kohoutek was hailed in the media as "the comet of the century". It came within 0.14 AU of the Sun at which point it could be observed with the naked eye but it was nowhere near what scientists predicted. The comet subsequently became synonymous with spectacular duds but nevertheless inspired numerous artists and musicians. In the aftermath, scientists agreed that Kohoutek is actually a much closer Kuiper Belt comet (see below), which would explain the rocky makeup of the comet and the lack of outgassing otherwise responsible for a bright coma and tail.

Table 3.2 Orbital Parameters for Selected Comets

Name	Perihelion [AU]	Eccentricity	Inclination	Orbital Period [y]	Last perihelion	Next perihelion
Hale-Bopp	0.91	0.9951	89.4°	4,200* 2520 – 2533	1 Apr 1997	~ 4385

Hyakutake	0.23	0.9999	124.9°	17,000 ~70,000	1 May 1996	~ 72,000
Halley	0.59	0.967	162.3°	75.3	9 Feb 1986	28 Jul 2061
Kohoutek	0.14	> 1.0	14.3°	~150,000 ~75,000	7 Mar 1973	~ 77,000
McNaught	0.17	1.000	77.8°	~92,600	12 Jan 2007	?
Swift-Tuttle	0.96	0.96	113.5°	133.3	11 Dec 1992	12 Jul 2126
Temple-Tuttle	0.98	0.906	162.5°	33.2	28 Feb 1998	20 May 2031
Encke	0.33	0.847	11.76°	3.3	6 Aug 2010	21 Nov 2013

* two periods indicate a change in the period during the last perihelion. The top number is the previous period, while the bottom number is the new one.

The Origin of Comets

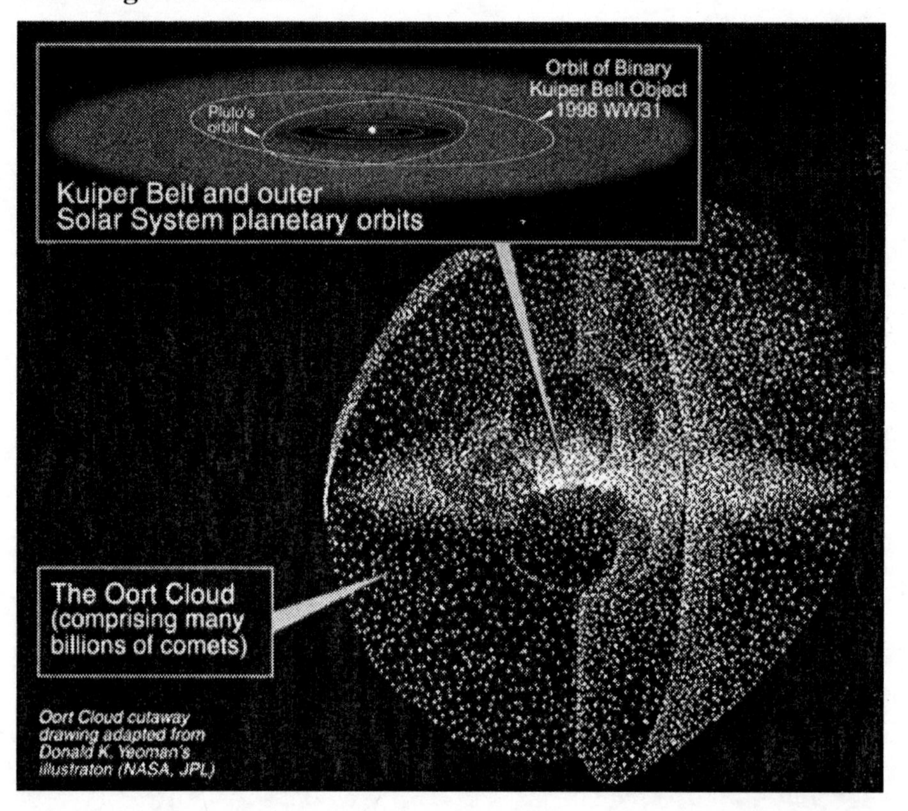

Figure 3.17 Artists rendering of the Kuiper Belt and Oort Cloud. (source: Wikipedia/NASA)

Scientists now agree that millions of (tail-less) comets reside in a vast, diffuse spherical cloud surrounding and reaching into the Solar System (Fig. 3.17). This cloud, the **Oort Cloud**, extends out to 50,000 AU, about 1 light-year. According to an out-dated model, scientists assumed that long-period comets with a recurrence time (time between two visits) of more than 200 years exclusively originate from the Oort Cloud. In this model, the long-period comets are categorized in "new comets", if the semimajor axis (the longer of the two axes in an ellipse) is more than 10,000 AU. "Returing comets" have a shorter semimajor axis. Short-period comets, with a recurrence time of less than 200 years were regarded as originating from a region closer to the Sun. This region is the **Kuiper Belt**. It aligns with the ecliptic of the Solar System and extends from the orbit of Neptune outward to 55 AU. The problem with this comet model is that comets like Swift-Tuttle do not fit in because the period is short but its inclination carries it out of the plane of the Kuiper Belt. We also now know that the tug of the outer planets, especially Jupiter, can change the periodicity of comets dramatically (e.g. that of Hyakutake and Kohoutek but also Hale-Bopp), and the period can shorten but also lengthen. In a more recent model, the comets are therefore classified into those that come from nearly any direction in the Solar System (nearly isotropic; essentially from the Oort Cloud) and those that come from the ecliptic, essentially from the Kuiper Belt. In the new model, the distinction between Oort Cloud and Kuiper Belt comets essentially remains but the constraints on periodicity are somewhat relaxed. The new model has an additional layer of complexity/categories that is irrelevant for our purposes. Under the new model, Kohoutek is a Kuiper Belt comet, despite its long period, but classified as such because of its physical properties. Short-period comets are still likely to come from the Kuiper Belt, except for those with a large inclination. Hale-Bopp was originally a Kuiper Belt comet but then migrated outward into the Oort Cloud. Halley started out as a Oort Cloud comet but then transformed into a Kuiper Belt comet. Hyakutake and McNaught are Oort Cloud comets. Temple-Tuttle and Encke, an Earth-crossing comet, are classical Kuiper Belt comets. With a period of 3.3 years, **Comet Encke** is currently known as having the shortest period. Its diameter is about 5 km (3 mi) and its orbit is frequently perturbed by the inner planets. Encke is thought to be the originator of several meteor showers in November and June/July known as the Taurids.

3.5 Impacts

Figure 3.18 A meteor or shooting star (upper right) during the Perseid meteor shower observed by Joe Westerberg in Joshua Tree National Park, CA on 11 August 2007. Meteors cut across the circular paths around Polaris on which stars and planets appear to travel. (source: Joe Westerberg, NASA)

Figure 3.19 The Willamette Meteorite shown here in 1920 was "found" in Oregon in 1902, but it was apparently known to Native Americans before that. The iron-nickel meteorite weighs about 15.5 tons. (source: Wikipedia)

Table 3.3 Some Notable Meteorites[5]

Name	Location	Date Found	Weight Dimension	Comment
Allende	Chihuahua, Mexico	8 Feb 1969	2 tons	largest known carbonaceous chondrite
Willamette	Oregon, US	1902	15.5 tons	largest meteorites found in U.S.
Cape York	Greenland	1894	58 tons	fell 10,000 yrs ago; fragments; world's second largest meteorite; 34-ton fragment on display
Hoba	Namibia	1902	66 tons	largest known (iron) meteorite; fell 80,000 years ago
Nogata	Japan	19 May 861	0.5 kg	oldest recorded fall of (stony) meteorite
Carancas	Peru	15 Sep 2007	> 3 m	chondrite; crater 4.5 m deep, 13 m wide; fireball 1000 m above ground;
Allan Hills 81005	Antarctica	1982	?	first meteorite determined as lunar
Allan Hills 84001	Antarctica	1984	2 kg	Mars meteorite; with Martian life?

Meteoroids, Meteors and Meteorites

The popular use of the term "meteor" cuts across a wide range of objects and phenomena. A meteor often stands for an impacting body, such as in the name of famous Meteor Crater in Arizona. But this is really a misnomer because a meteor is not an object, and scientists call the crater Barringer Crater instead. Scientists make the following distinctions between meteoroid, meteorite and meteor:

- **meteoroid**: a junk of matter such as a fragment of an asteroid or a comet that orbits the Sun. A meteoroid may intersect Earth's orbit and enter the atmosphere and, if it is large enough to survive, impact on Earth's surface. A stony meteoroid with a diameter of 10 km (6.2 mi) can produce an explosion of around 20 kilotons (comparable to the Hiroshima bomb).
- **meteor**: (Fig. 3.18) also termed "shooting star", a meteor is the visual phenomenon when a meteoroid burns up in Earth's atmosphere. Some meteors are also accompanied by acoustic phenomena such as a crackling, hissing or swishing sound.
- **meteorite**: (Fig 3.19) remaining piece of a meteoroid after it impacts on Earth's surface. A meteorite is the fragment that survives the travel through the atmosphere and the impact on

the ground. Strictly speaking, a meteorite is the remnant on the ground but the term is often also used for the still falling object.

Meteoroids are often called **bolides** though this term is also used for exceptionally bright, fireball-like meteors. A **meteor shower** is observed when meteors are caused by streams of space debris, such as the remnants of a comet's dust tail, that enter Earth's atmosphere at high speed and on parallel trajectories. The meteors then appear to radiate from one point in the sky.

Space objects that enter Earth's atmosphere primarily come from intact or fragmented asteroids and comets. Meteoroids enter Earth's atmosphere every day, at speeds of more than 10 km/s, and most particles are smaller than a grain of sand. Most of these burn up in the atmosphere and never reach the ground. However, some meteoroids are larger. The fact that Earth's surface does not look as cratered as the surface of some other planets shows just how instrumental the atmosphere is at protecting us, setting aside for now that erosion on Earth's surface also plays a major role at wiping out impact craters. The magnitude of the atmosphere's protecting power became tragically clear when the Space Shuttle Columbia burned up upon re-entry on 1 Feb 2003, after the heat shield on one wing was damaged upon take-off.

Some larger meteoroids survive the burn-up in the atmosphere and impact on Earth's surface. The fact that 2/3 of Earth's is covered by oceans means statistically that only 1/3 of the meteoroids actually impact on land. Some scientists therefore speculated that impacts in the oceans go nearly unnoticed. This is probably true for smaller impacts. But large impacts cause humongous tsunami (catastrophic ocean waves that are discussed in Chapter 6) that can devastate coastlines around the entire ocean. The surviving meteorites on land are categorized into two groups, **iron meteorites** and **stony (or rocky) meteorites**. The former are thought to resemble Earth's core, while the latter are thought to resemble Earth's mantle. Stony meteoroids are less likely to survive the journey through the atmosphere. Nevertheless, the vast majority of falling meteorites (94%) are stony meteorites, and only 6% are iron meteorites. But the fraction of actually recovered stony meteorites is much less. Stony meteorites may easily be mistaken for ordinary rocks and not recognized as meteorites. Often, a stony meteorite is only recognized as such when its fall is actually observed. Once on the ground, stony meteorites also weather more quickly than iron meteorites. Stony meteorites are classified as **chondrites** and **achondrites**. Chondrites contain small round particles (chondrules) that are composed of silicate minerals that appear to have been melted. Achondrites do not contain these nodules. 86% of the falling meteorites are chondrites, while 8% are achondrites.

Most meteorites come from asteroids and most iron meteorites are thought to come from the cores of asteroids that were initially large enough to melt and differentiate. The larger meteorites are believed to have their origin in the Asteroid Belt. Some of the smaller meteorites have been identified as Moon rock, while still others have been identified as pieces from Mars. The **Willamette Meteorite** (Fig. 3.20) is a 15-ton iron meteorite and the largest meteorite found in North America, and the 6th-largest in the world. Since there was no impact crater at the site where it was found, it is believed that it impacted somewhere in Canada or Montana and was then transported as a glacial erratic (a boulder that was initially transported by a glacier) during the Missoula Floods at the end of the last Ice Age (about 13,000 years ago).

The **Hoba Iron Meteorite** lies on the farm "Hoba West" near Grootfontein, Namibia. It is the largest known single meteorite. It was found and uncovered in 1920 but has never been moved due to its large weight of an estimated 66 tons. It measures 2.7 × 2.7 m (8 ft 9 in) × 0.9 m (3 ft). Part of the meteorite has rusted away, therefore its original weight may have been as much as 100 tons! It is also the most massive naturally-occurring piece of iron known on Earth's surface. The meteorite is thought to have fallen less than 80,000 years ago, and it left no preserved crater.

In January 1982, an expedition in Antarctica found the **Allan Hills 81005** meteorite, named after the hills where it was found. The meteorite was recognized as unlike any other known meteorite. Subsequent comparison with lunar rocks from the Apollo program at the Smithsonian Institute in Washington, D.C. confirmed that the meteorite must have come from the Moon. As of October 2010, about 134 lunar meteorites have been discovered in the deserts of Antarctica, northern Africa and Oman. How do these meteorites come to Earth? It is thought that impacts on the Moon making craters of a few km in diameter or less launches meteoroids that reach Earth's surface. Similarly, meteorite **Allan Hills 84001**, that was found two years later, is thought to come from Mars and impacted on Earth about 13,000 years ago.

An estimated 500 meteorites survive a meteoroid's journey to Earth's surface each year, only 5 – 6 of which are recovered and made known to scientists. The chance of being hit by a meteorite is extremely small, and surviving it is even less likely. Nevertheless, one such incident was reported in 1954. A woman in Sylacauga, AL was slightly injured when a 4-kg (8.8 lb) stony meteorite crashed through the roof of her home and bounced off her radio before hitting her.

The most notable recent event was the **Carancas impact event** near Lake Titicaca, Peru on 15 September 2007. Around noon local time, a chondrite crashed near the village of Carancas. The impact left a crater 4.5 m (15 ft) deep and 13 m (43 ft) wide. Witnesses reported that boiling water and noxious gases spilled from the crater. A bright fireball with a smoky tail was seen 1000 m (3300 ft) above the ground. The strong explosion shattered windows 1 km (0.6 mi) away. It is estimated that the impactor must have been at least 3 m across before breaking up. Initial speculations that this was a volcanic event were ruled out by scientists 3 days after the event because minerals found in some of the rocks in the crater occur naturally only in meteorites. People reported to become sick soon after the impact, experiencing nausea, headaches, diarrhea and vomiting. Radiation was soon rejects as possible cause for the illnesses that were instead caused by sulfur and arsenic poisoning. The ground water in the area is known to contain arsenic compounds. It is now believed that the arsenic vapors that people inhaled did not come from the meteorite but from the ground water that was boiled during the impact.

Comet Impacts

Figure 3.20a Photo of the Comet Shoemaker-Levy 9 taken on 17 May 1994 by the Hubble Space Telescope. (source: Wikipedia/NASA)

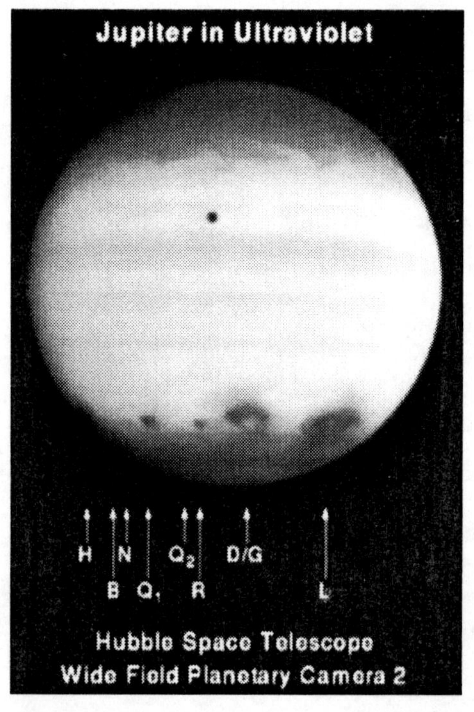

Figure 3.20b Ultraviolet image of Jupiter taken on 21 July 1994 by the Hubble Space Telescope. Dark spots near the bottom mark the impact sites of Shoemaker-Levy 9 fragments. The spots are very

dark in UV because large amounts of dust were deposited high in Jupiter's stratosphere, and the dust absorbed sunlight. The dark dot near the top comes from a Galilean moon. (source: Wikipedia/NASA)

That the ultimate fate of comets can end in impacts became drastically clear in July 1994 when observers on Earth trained their telescopes to watch the Comet **Shoemaker-Levy 9** crash on Jupiter. It was the first such event observed from Earth. Several space observatories, such as the Galileo, Ulysses and Voyager 2 spacecrafts, also provided a wealth of data to study details of the impact but also the recovery of Jupiter's gaseous surface. The impact also demonstrated how little time sometimes lapses between discovery and impact. It also raised awareness of how important Jupiter may be at scooping up space objects, thereby protecting the inner Solar System from impacts. The comet was discovered by astronomers Carolyn and Eugene M. Shoemaker and David Levy at the Palomar Observatory, CA on 24 March 1993. Shoemaker-Levy 9 was the 9th short-period comet discovered by the team (hence the name). Oddly, the comet orbited Jupiter rather than the Sun, unlike all other comets known at the time. Tracing back the comet's orbital motion, the scientists found that the comet was probably captured by Jupiter 20 – 30 years earlier. Before the capture, the comet likely orbited the Sun between just inside Jupiter's orbit and the interior of the asteroid belt. Prior to 1993, a previous very close approach in July 1992 took the comet into the orbit of Jupiter's closest moon. This likely tore the comet apart into several fragments as a result of Jupiter's strong tidal forces (Fig. 3.20). From the fragments, scientists estimated the original size of the comet to be 5 km (3.1 mi) across, about the size of Comet Hyakutake.

The fragments, up to 2 km (1.2 mi) in diameter, impacted on Jupiter's southern hemisphere between 16 and 22 July 1994, at a speed of approximately 60 km/s (134,200 mph). The resulting "fireballs" reached a temperature spike of 24,000ºC (43,000ºF). Much of this dissipated within a minute or so but temperature anomalies persisted for one or two weeks. The plumes from the fireballs reached a height of 3,000 km (1,850 mi). Over 6 days, 21 distinct impacts were observed, with the largest coming on 18 July when fragment G struck the planet. The impact created a giant dark spot, over 12,000 km (7,500 mi) across (nearly the size of Earth!). It was estimated to have released energy equivalent to that of 6 million megatons of TNT (600 time the world's nuclear arsenal). Two impacts on 19 July created similar-sized impact marks. The prominent scars from the impacts (Fig. 3.20b) were more visible that the Giant Red Spot and remained for many months.

After the impact, scientists reevaluated and raised the probability of fragmented impacts on Jupiter. During the Voyager missions, scientists had identified 13 "mysterious" chains of craters on Callisto and 3 chains on Ganymede. After witnessing Shoemaker-Levy 9 impacting on Jupiter, it became clear that these chains were the result of fragmented impacts as well. On 19 July 2009, a new black spot about the size of the Pacific Ocean appeared in Jupiter's southern hemisphere. The site was warm, with the same anomalies of hot ammonia and silica-rich dust as observed in the 1994 impact. Scientists therefore concluded that another impact occurred. This time, the impactor was smaller (200 – 500 m or 650 – 1650 ft across) but stronger, which led scientists to conclude that it was a previously unknown asteroid instead of a comet. On 3 June 2010, another, smaller impact event was observed.

These impacts highlight Jupiter's role as "cosmic vacuum cleaner" for the inner Solar System. Due to the strong gravitational pull, it is thought that 2,000 – 8,000 times more objects crash into Jupiter than into Earth. Astronomers argue that without the presence of Jupiter, impacts on Earth that cause mass extinctions would have been more frequent to the point that they would have prevented complex life from evolving to the level that we enjoy today.

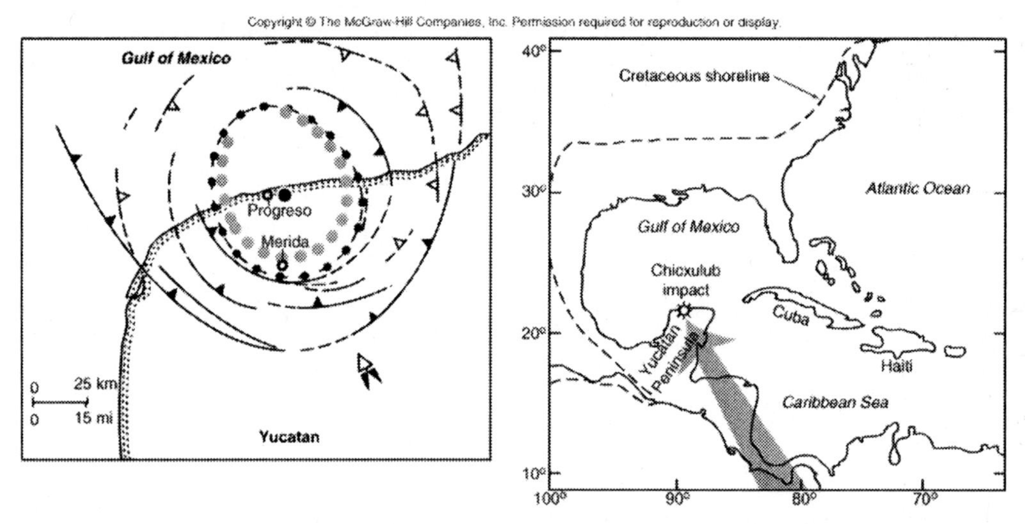

Figure 3.21 Left: Contour lines of a gravity survey of the buried Chicxulub impact crater near the edge of the Yucatan Peninsula, Mexico. The contour lines appear to open up toward the northwest suggesting that the asteroid came from the southeast. Right: The approximate path of the asteroid. Its impact would have caused a humongous tsunami and sent a superhot cloud of gases and debris into the atmosphere and over North America. (source: AB)

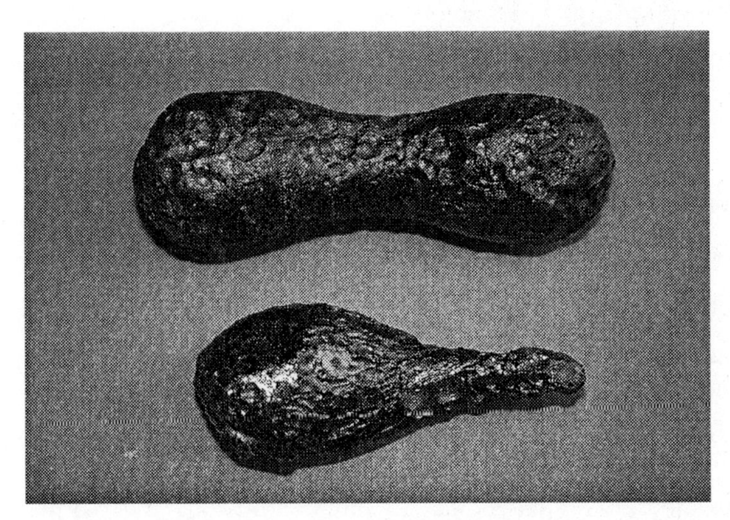

Figure 3.22 Tektites. These glassy, often tear-shaped rocks form under immense pressure and subsequent instantaneous cooling, such as during a meteoroid impact. (source: Wikipedia)

Table 3.4 The Ten Largest Impact Craters on Earth

Name	**Location**	**Diameter [km]**	**Diameter [mi]**	**Age [10^6 yrs]**
Vredefort	Free State, South Africa	300	185	2020
Sudbury	Ontario, Canada	250	155	1850
Chicxulub	Yucatán, Mexico	170	105	65
Manicouagan	Quebec, Canada	100	62	214
Popigai	Siberia, Russia	100	62	35.7
Chesapeake Bay	Virginia, U.S.	90	56	35.5
Acraman	South Australia, Australia	90	56	590
Puchezh-Katuni	Nizhny Novgorod Oblast, Russia	80	50	167
Morokweng	Kalahari Desert, South Africa	70	43	145
Kara	Nenetsia, Russia	65	40	70

Large Impacts

Large meteorites that survive the passage through the atmosphere crash onto Earth and leave craters like those found on the Moon. From the geometry of the crater, scientists estimate the size of the impactor as well as the speed upon impact and the angle from which it struck. The world's largest known impact crater is the 300-km wide (190 mi) **Vredefort crater** in South Africa (Table 3.4). The impacting asteroid must have been approximately 5 – 10 km (3.1 – 6.2 mi) across. The crater is estimated to be 2 billion years old, and is the second oldest, after the Suavjarvi crater in Russia that is thought to be 300 million years older. The Vredefort crater was long assumed to have formed during a volcanic explosion but evidence from so-called shatter cones found in the 1990s identified the crater as an impact crater. Shatter cones are cone-like features in the sediments beneath the crater that are formed by the shock waves released in meteorite impacts or underground nuclear explosions.

In North America, the largest such crater is evidenced in the **Sudbury Basin** in Ontario, Canada. The Basin is 62 km (39 mi) long, 30 km (19 mi) wide and 15 km (9.3 mi) deep though the modern surface is much shallower. It is thought that the basin is only part of the initially much wider impact crater with a diameter of 250 km (155 mi). The Sudbury crater is the second-largest known impact crater on Earth and also one of the oldest. It is thought that a 10 to 15-km wide (6.2 – 9.3 mi) body impacted on Earth's surface 1.8 billion years ago, scattering debris as far as 800 km (500 mi) away, over an area of 1,6 million km^2 (620,000 sq mi). Ejected rock fragments have been found as far as Minnesota. About 37.2 million years ago, another impact created the 8.4-km wide (5.2 mi) Lake Wanapitei crater to the northeast.

Not far behind in size is the **Chicxulub crater** in Yucatán, Mexico (Fig. 3.21). At a diameter of 170 km (110 mi), it is one of the largest known craters on Earth. The impacting bolide must have been at least 10 km (6 mi) in diameter. The crater is actually buried by sediments, and half of it is submerged in the Gulf of Mexico. It was discovered by petroleum geophysicist Glen Penfield in the 1970s. Subsequently scientists gathered more evidence that the crater was indeed the results of an impact and not a volcanic crater. This included anomalies in the gravity field but also samples of **shocked quartz** (minerals formed under instantaneous overpressure) and rocks called **tektites** (Fig. 3.22). Tektites are glassy objects, typically a few cm across. They form under immense pressure and temperature, when a large meteorite hits Earth's surface and melts the surrounding rock. Tektites are not found around volcanoes. But Penfield's discoveries were not made public until 1981 at a conference of the Society of Exploration Geophysicists.

Parallel to this work, physicist **Luis Walter Alvarez** at UC Berkeley and his son Walter, a geologists, developed their hypothesis that a large extraterrestrial body had struck Earth about 65 million years ago and was responsible for the demise of the dinosaurs. They based their theory on the discovery of an intriguing dark layer of **iridum** they found in 65-million-year old sediments in Italy, belonging to the K-T boundary between the Cretaceous and Tertiary periods (K stands for "Kreide", the German word for Cretaceous). Iridium is rare on Earth's surface and the only two sources are thought to be either extraterrestrial or very extensive volcanism that brings magma to the surface from very deep within Earth. Adding to the puzzle, the iridium layer was also found elsewhere on the planet. The 65-million-year old soil in Italy also contained a 1cm-thick layer of clay. The clay contained soot, glassy spherules, shocked quartz crystals, microscopic diamonds and other minerals that only form under extreme pressure and temperature. The scientists considered if Earth could have passed a giant nebular cloud to explain the iridium, but this would not explain the strange high-pressure/high-temperature minerals in the clay layer. They finally settled on an asteroid impact but their initial publication in 1980 was met by great skepticism. Only Penfield's discovery of the Chicxulub crater allowed scientists to connect the dots. Still, there was a large group of scientist that argued that very extensive volcanism related to flood basalts in India, the Deccan Traps, was an alternative explanation for Alvarez' geological finds. But a panel of 41 scientists agreed on 4 March 2010 that the Chicxulub impact indeed triggered the K-T boundary mass extinction. The soot in the clay layer is believed to be the result of humongous wildfires that where ignited after the impact.

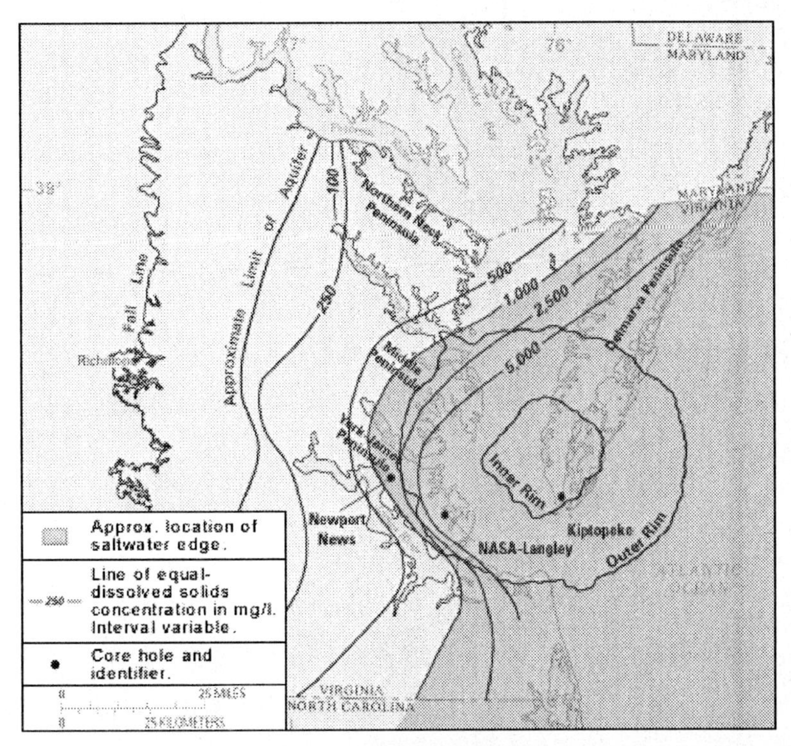

Figure 3.23a Map of the Chesapeake Bay Impact Crater. (source: USGS, wikipedia)

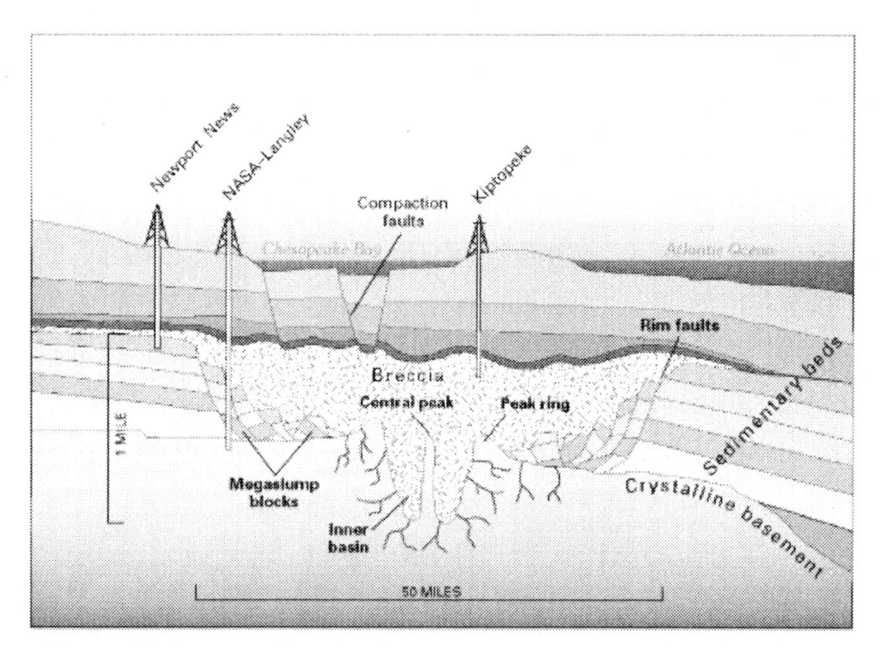

Figure 3.23b Cross-section of the Chesapeake Bay Impact Crater. The crater is buried by sediments, but the fragmented crust beneath shows the characteristics of an impact crater. As the Chixculub crater, this crater was discovered by seismic imaging techniques. (source: USGS, wikipedia)

With a diameter of about 90 km (56 mi), the largest impact crater in the U.S. is the Chesapeake Bay crater (Fig. 3.23). A bolide struck Earth here about 35 million years ago with a speed exceeding 11 km/s (24,600 mph). The bolide was completely vaporized by the impact. At the time, the climate was much warmer than today. A tropical ocean reached farther inland. The impact devastated marine as well as terrestrial habitats, sending a humongous tsunami ashore, possibly even overtopping Blue Ridge Mountains. The impact was so powerful that the basement rock beneath the crater fractured to a depth of 8 km (5 mi). The crater rim fragmented, forming several terrace-like rings that slump along so-called ring faults (Fig. 3.23b). Evidence for an impact crater arose in 1983 when drilling off-shore Atlantic City, NJ revealed a 20-cm (8 in) thick layer of material ejected from the crater. The layer also contained shocked quartz and tektites. But it was not until seismic imaging for oil exploration revealed the actual crater that has been buried over time by several 100 m of sediments. The continuing slumping of the crater and the compaction of the sediments above have controlled the course of rivers and shaped Chesapeake Bay. It has also affected aquifers that are an important source of drinking water to the region. The present-day freshwater aquifers are located above deep salty brine, making the entire lower Chesapeake Bay area vulnerable to groundwater contamination (Fig. 3.23a).

Figure 3.24 The Barringer Meteorite Crater near Winslow, AZ. (source: Wikipedia/USGS)

Meteorite impacts do not have to be as large and fatal as the Chicxulub impact in order to leave craters on Earth's surface. A spectacular crater and popular tourist attraction is the **Barringer Crater** (or Meteor Crater) near Winslow, AZ (Fig. 3.24). It also carried the name Canyon Diablo crater, named after the closest community in the late 19th century. In 1903, engineer and businessman Daniel M. Barringer suggested that the crater was the result of the impact of an iron meteorite. The crater is about 1,200 m (4,000 ft) in diameter, some 170 m (570 ft) deep, and is surrounded by a rim that rises 45 m (150 ft) above the surrounding area. The center of the crater is filled with about 230 m (750 ft) of debris. The crater is believed to have formed about 49,000 years ago by the impact of a 300,000-ton bolide, about 50 m across (165 ft). The final speed of the impactor, the Canyon Diablo iron meteorite is currently estimated at about 13 km/s (29,000 mph), after earlier estimates placed the speed at 20 km/s (45,000 mph). It is believed that about half of the Canyon Diablo meteorite vaporized above the ground before it impacted. Fragments of the meteorite have been collected around the rim since Native Americans inhabited the area. The origin of the crater was long disputed and some argued for a volcanic origin. After all, the San Francisco volcanic field is located only 65 km (40 mi) away, and 'intriguing' cracks in the ground, perhaps resembling volcanic dikes, radiate from the crater. The ultimate proof for the impact hypothesis did not come until the 1960s when Eugene M. Shoemaker found shocked quartz near the crater that cannot be the result of volcanism. Shoemaker's discovery is considered the first definite proof of an extraterrestrial impact.

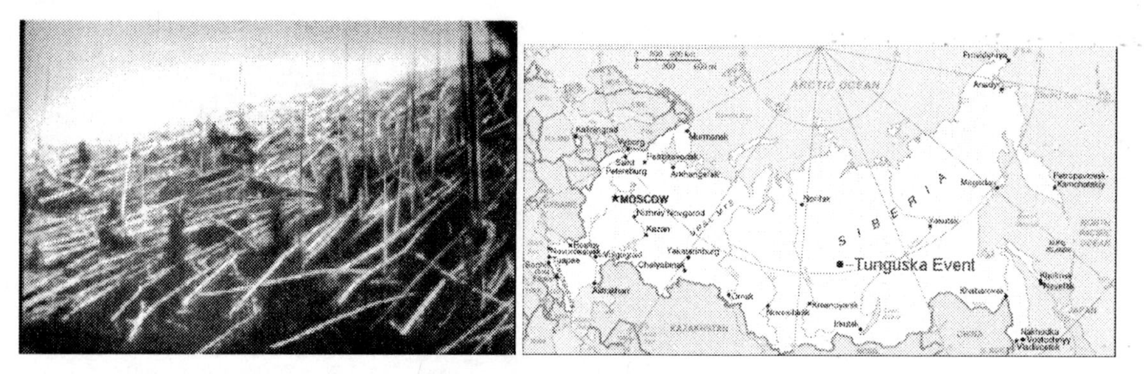

Figure 3.25 The 1908 Tunguska event. Left: Trees knocked over by the Tunguska blast. Photograph from the Soviet Academy of Science 1927 expedition led by Leonid Kulik. Right: Approximate location of the Tunguska event. (source: Wikipedia)

Bolides falling toward Earth's surface do not always leave a crater behind but can still cause an enormously powerful explosion. Such events are estimates to occur once every 300 years (an estimate by Eugene Shoemaker). The last one happened on 30 June 1908 near the **Podkamennaya Tungaska** River in Russia (Fig. 3.25) and is considered the largest recorded impact event over land in Earth's recent history. An explosion of about 1/3 the power of the Tsar Bomba explosion, the largest nuclear explosion in the world (Fig. 2.1) knocked over an estimated 80 million trees in an area covering 2,150 km^2 (830 sq mi). People near Lake Baikal some 700 km (435 mi) away observed a column of bluish light, nearly as bright as the Sun, moving across the sky. About 10 min later, there was a flash and a sound similar to that of artillery fire. The sounds were accompanied by a shock wave that knocked people off their feet and broke windows hundreds of kilometers away. The powerful explosion sent seismic waves to instruments across Eurasia. At the Mt. Wilson Observatory, CA, a decrease in atmospheric transparency was observed, caused by suspended dust, that lasted for months. At the time, there was little scientific interest in this event, which is why it took nine years for the first expedition to visit the area in 1927. No crater was found. The only significant piece of evidence were the blasted trees. Later, in the 1950s and 1960s, silicate and magnetite spheres with extraterrestrial signature were found that are consistent with debris from a meteorite airburst. The bogs in the area also exhibited an unusually high concentration of iridium. The leading scientific explanation for this is now the airburst of a meteoroid 6 – 10 km (4 – 6 mi) above Earth's surface. With advancing research, scientists now know that such events, though much weaker, occur about once a year in the upper atmosphere. In such events, a stony meteoroid of about 10 m (30 ft) in diameter can produce explosions similar to that of the Fat Man bomb dropped on Nagasaki in 1945.

Near-Earth Objects and the Torino Scale

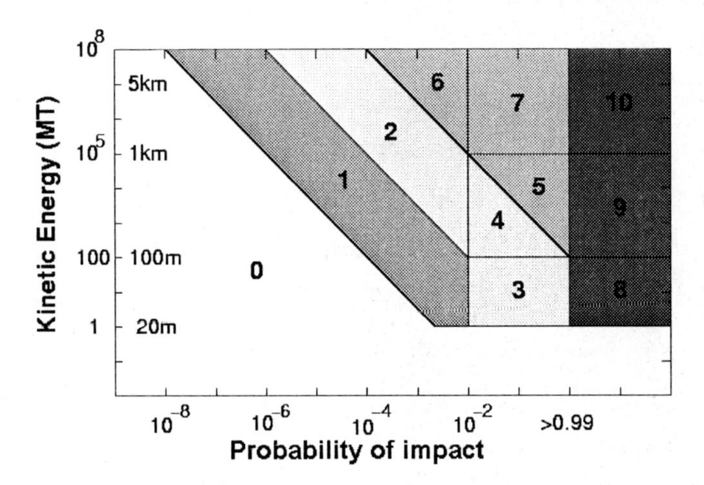

Figure 3.26 The Torino Scale uses a scale from 0 to 10 to categorize the impact hazard associated with near-Earth objects (NEOs). A larger number indicates a more serious hazard. The hazard depends on the probability of impact (x-axis) as well as the size and the kinetic energy of the NEO (y-axis). The probability of an impact depends on the orbital parameters of the NEO. (source: wikipedia)

In previous sections, we have learned that objects with orbits that come near Earth's orbit or, worse, cross Earth's orbit have an increased chance of impacting on Earth at some point in their life. Objects with a perihelion of 1.3 AU or less but an aphelion greater than this fall under this category. These objects are called **near-Earths object** (NEOs) and include a few thousand near-Earth asteroids (NEAs), near-Earth comets, a number of Earth- or solar-orbiting spacecraft and meteoroids large enough to be tracked in space before striking Earth. Most NEAs have a perihelion between 0.983 and 1.3 AU. Scientists now agree that previous large impacts have fundamentally controlled the path of life on Earth. Consequently, awareness of the potential dangers of NEOs has increased since the 1980s. Experts agree that the U.S. and China carry the greatest risk of being struck by an impactor. The U.S., the EU and other nations have therefore launched an effort called **Spaceguard**(6) to catalog and monitor NEOs. In the U.S., NASA has a congressional mandate to catalog all NEOs that are at least 1 km (0.6 mi) wide. As of October 2008, about 980 such objects have been detected. But this contrasts with an estimate in 2006 that the detection success is probably only 20%, i.e. 4900 objects remain to be found, many of them being in the southern celestial hemisphere. To help close the gap, a telescope in Australia will help cover about 30% of the sky that has no yet been mapped.

Clearly, a NEO's potential danger depends on its size but also on the velocity upon impact (the **terminal velocity**). The latter can be estimated by observing an object's current velocity in the sky provided its orbit is known. Scientists express the impact hazard on the **Torino Scale** (Fig. 3.26). There is also a more complex scheme, the Palermo Technical Impact Hazard Scale. For the Torino Scale, the hazard of an object is determined by estimating the impact energy and the probability of an impact. Recalling the formula for kinetic energy from Chapter 2, it is easy to see that larger bodies and faster bodies have a greater impact energy than smaller or slower bodies. The probability of an impact depends on the orbital parameters of the object. If the perihelion is outside of Earth's orbit (> 1.0 AU) then the probability is less than if the perihelion is inside of Earth's orbit (< 1.0 AU). The orbital inclination also plays a role. Kuiper Belt objects have a greater probability than Oort Cloud objects with a large inclination. Finally, the period has to be taken into account. For a given highly eccentric orbit, comets with a shorter orbiting period are much more likely to reach the timing of an impact on Earth than comets with a very long period. The Torino Scale then gives an estimate for a given probability (essentially orbital parameters) on one hand, and size and velocity on the other. So pose small and slow objects with a low probability little risk (0 on the Torino Scale). But the hazard increases when that same object has a higher probability (e.g. the orbit takes the object close to Earth). Conversely, for objects of the same probability, the hazard increases with the size and speed of the object.

On 25 December 2004, minor planet MN_4 (a NEA later named 99942 Apophis) was assigned a 4 on the Torino Scale, the highest rating so far. But the difficulty of precisely forecasting impacts was demonstrated by what happened during the days after this initial assignment. Experts estimated that there was a 2.7% chance of an impact in 2029. A slightly smaller chance would have placed its Torino rating at 2. But then scientists examined the asteroid's orbit closer and found that a resonance changed its closest approach from 2029 to 2036. Subsequently, the Torino hazard went to 1. Apophis is about 270 m (885 ft) across and, as of October 2009, the probability of an impact in 2036 is estimated to be 1 in 250,000 (4×10^{-6}). This brings the asteroid's Torino hazard back to a 0. As of September 2011, the NEO with the highest Palermo rating is 1950 DA, a NEA with a perihelion of 0.84 AU and an orbital period of 2 y 79 d. It is about 1.3 km (0.8 mi) across and has a 0.33% chance of impacting on Earth in 2880. By this time, scientists will either have refined predictions to more precisely determine the probability of an impact and/or found ways to deflect the asteroid.

3.6 Earth's History and Structure

Summary of Earth's History

The formation of Earth is tied very closely to the formation of the rest of the Solar System, which started from the solar nebula about 4.55 billion years ago. As we have learned at the beginning of this chapter, the planets along with Earth, first formed through random collisions of particles to grow into planetesimals, then accreted into larger protoplanets, and ultimately into the planets we know today. The process of accretion occurred rather violently through collisions that provided the energy to heat the growing planet. Additional heat from gravitational contraction during growth eventually provided enough energy to at least partially melt Earth's interior. This allowed the process of differentiation where heavier elements (e.g. iron) sank toward the center to form Earth's metallic core. In the process, the lighter elements ended up near the surface to form a low-density but still soft outer crust.

It is thought that differentiation was already happening when a Mars-sized body struck Earth 4.5 billion years ago, and only 50 million years into the formation of the Solar System. The impact could not have been later because the Apollo program brought Moon rocks of that age back to Earth. After the impact, Earth cooled to form a thicker solid shell, the lithosphere. And by 3.9 billion years ago, Earth already had small continents, oceans and a dense atmosphere composed primarily of CO_2. Exactly where the water came from is still a matter of debate but current thinking is that much of it was contributed by impacting icy comets. It took Earth a full 1 billion years before life first appeared on the planet about 3.5 billion years ago. Life was extremely primitive, in the form of bacteria. Another 1 billion years later, 2.5 billion years ago, Earth had developed large continents. We do not know exactly when plate tectonic started but scientists hypothesize that it has been occurring for at least the last 1.5 billion years. The surface of Earth has been changing since then, with supercontinents such as Pangaea forming only to break up a few tens to hundreds of million years later. This is discussed in more detail in Chapter 4.

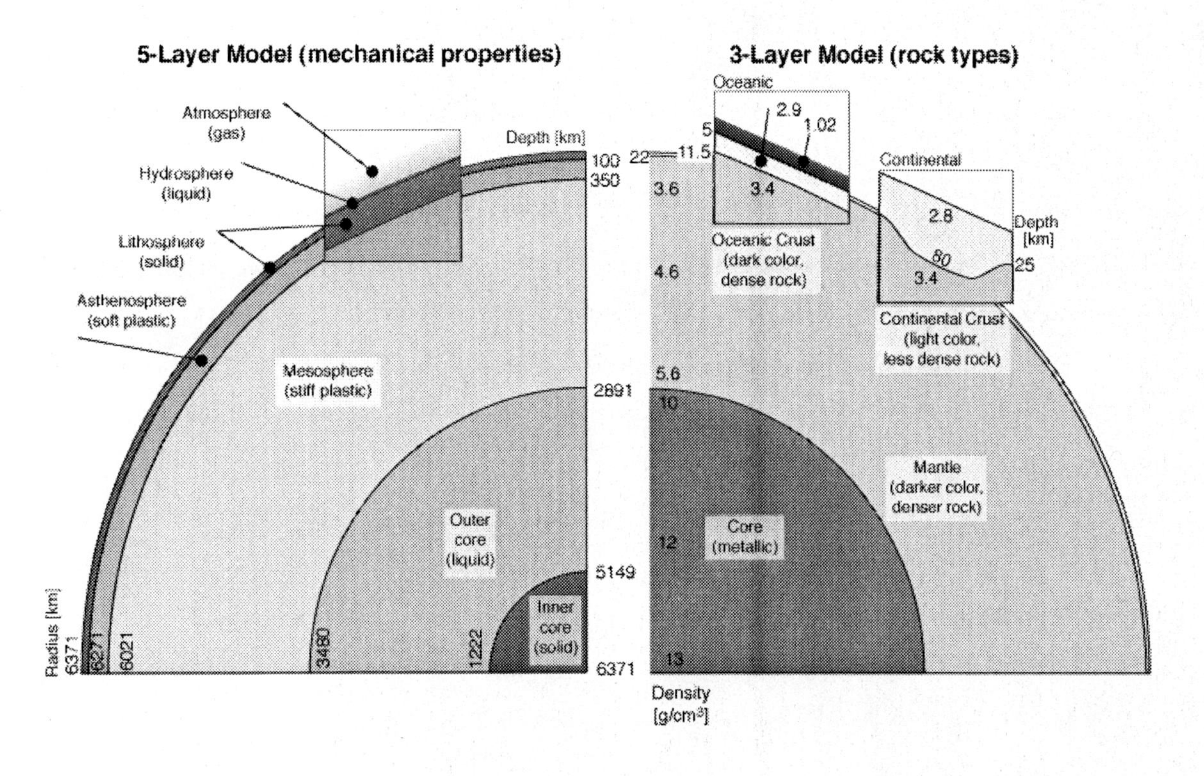

Figure 3.27 Two ways to describe the layering of Earth. On the left, differences in physical properties divide Earth's solid part into 5 layers. On the right, different material types divide it into 3 layers.

The Layered Earth

From astronomical studies, Earth's orbital parameters, dimensions and moment of inertia have been known quite accurately for a long time. Earth's mean radius is 6,371 km (3,957 mi), not taking into account the atmosphere. This gives a diameter of 12,742 km (7,914 mi), and a circumference of roughly 40,000 km (24,850 mi). Earth is slightly oblate, a bit wider near the equator (the equatorial radius is 6,378 km) and a bit flattened near the poles (the polar radius is 6,357 km). Even though the difference between the equatorial and the polar radius of 21 km (13 mi) seems large, from space, Earth still looks like a perfect sphere.

Oddly, Earth's inner structure has remained inaccessible for a long time. Most of the planets were known to the ancient Greeks. Space was explored with telescopes, and Neptune, the last planet to be discovered, was known by 1846. On the other hand, Earth's interior was assumed into the 19th century to be a nearly amorphous mass, with some pockets of molten rock to feed volcanoes. And Jules Verne's 1864 "Journey to the Center of the Earth" remained a fantasy. At the beginning of the 20th century, sketchy images of Earth's interior had developed a denser core because it was necessary to bring into agreement Earth's low-density surface rocks with the overall moment of inertia and its mass (i.e. Earth cannot be made of the same surface rock through and through because it would be too light).

In the late 19th century, in 1883 to be more precise, English seismologist John Milne working in Japan speculated that waves radiated from an earthquake could travel through Earth and be recorded at the other end, if instruments were sensitive enough to measure them. He likened this idea with exploring the interior of the human body with X-rays. Six years later, German geophysicist Ernst von Rebeur-Paschwitz recorded the motions from the great 18 April 1889 earthquake in Tokyo, Japan on two pendula in Potsdam and Wilhelmshaven, Germany, some 9000 km (5,600 mi) away. Scientists now knew that earthquake waves exist and can be used essentially like X-ray to study Earth's interior. In 1897, Richard D. Oldham, a British scientist working at the Geological Survey in India, identified different types of seismic waves in recordings of the great magnitude-8.1 Assam earthquake. Models of Earth's upper layering were determined and refined. It was quickly determined that seismic velocities increase with depth and seismic rays (the path along which seismic waves travel) curve upward rather than going straight through Earth. In 1909, Croatian meteorologist and seismologist Andrija Mohorovičić discovered a boundary at about 50 km (31 mi) depth that separates the lighter crust from the much denser mantle. This discontinuity is now known as the **Moho discontinuity**, one

COPYRIGHTED MATERIAL — DO NOT DUPLICATE, DISTRIBUTE, OR POST

of the most profound boundaries inside Earth (Fig. 3.27). Subsequently, scientists found another boundary that separates the mantle from the **core**. At 3,820 km (2,375 mi), its depth estimate was initially somewhat off. Working in Germany at the time, German-American seismologist Beno Gutenberg refined this number in 1914, and his estimate of 2,900 km (1,800 mi) remains within a few km of our current best estimate of 2,891 km (1,795 mi). However, seismic signals behaved rather oddly in that they formed a shadow-zone for waves that were supposed to penetrate into the core. This indicated that seismic velocities in the core are markedly lower in the "weaker" core than in the mantle above. Waves with shearing motion (shear waves, see Chapter 5) were missing, indicating that the core behaves like a **liquid**. In 1936 followed a publication by Danish seismologist Inge Lehmann who provided the first evidence for an **inner core**(7). Its **solid** state was hypothesized early on but the ultimate proof was not delivered until relatively late in the 20th century. In 1971, Freeman Gilbert at the Scripps Institution of Oceanography (SIO) and Adam Dziewonski at UT Dallas determined its rigidity from the observations of Earth's resonance frequencies, also called normal modes or free oscillations(8).

Modern models of Earth's interior are comprised of 3 or 5 principal layers depending on what parameters are described. The *3-layer model* distinguishes between different principal types of rock: the **core** is metallic and is surrounded by a **mantle** of dense rock. The metallic core is thought to resemble metallic meteorites. Near the surface, above the Moho discontinuity is the **crust** that consists of less dense rock that is typically lighter in color, particularly on continents. Crustal rocks in the oceans are slightly denser than those on continents. And the oceanic crust is markedly thinner than continental crust that can be up to 80 km (50 mi) thick beneath the large and high-reaching Himalayan and Andean mountain belts. The oceanic crust is only 6.5 km (4 mi) thick, beneath about 4 to 5 km (2.5 – 3.1 mi) of ocean water.

The **5-layer model** describes the different mechanical properties of the layers. Here, the **solid inner core** is surrounded by a **liquid outer core**. Fluid flow in the outer metallic core generates and maintains Earth's magnetic field. The core-mantle boundary separates the liquid outer core from the **solid mantle** above. The mantle consists of peridotite rock, and its composition is thought to be similar to that of chondritic, stony meteorites. In a zone between about 125 km (78 mi) and about 350 km depth (217 mi), the mantle is relatively weak and responds plastically to pressure. This is the **asthenosphere**. The **lithosphere** above comprises a strong, brittle layer.

Scientists make a further distinction between the upper and lower mantle, with a transition zone between 410 to 660 km depth (255 – 410 mi). We will learn in Chapter 4 that earthquakes occur from the surface into the transition zone, but not beyond, for physical reasons. The transition zone is a zone where mantle rock undergoes phase transformations where the atoms in the minerals are rearranged to take up less space, in response to the increasing pressure with depth. On geologic time scales, the mantle behaves like a ductile material and convects, and there is evidence for two styles of convection. In one style, the mantle convects as a whole. In the other style, the transition zone acts as a barrier so that a convection cell forms in the upper mantle and another one in the lower mantle.

The Asthenosphere, the Lithosphere and the Principle of Isostasy

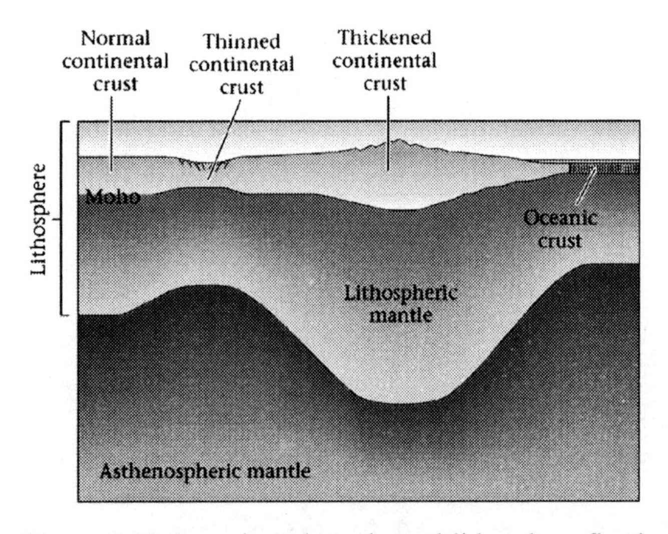

Figure 3.28 Oceanic and continental lithosphere floating in the weaker asthenosphere. Oceanic crust is thinner but denser than continental crust. (source: modified from ME)

The mechanical properties of the asthenosphere and the lithosphere profoundly determine the evolution of Earth's surface. Its interplay is ultimately responsible for the drift of continents and plate tectonics. The term **lithosphere** is based on the Greek word *lithos* for rock. It comprises Earth's crust and portions of the uppermost mantle beneath (Fig. 3.28). It is cool and mechanically strong. But it is also brittle and breaks under the stresses occurring on Earth. In the upper 70 km (43 mi), the lithosphere consists of rocks that are rich in silicates, which makes it particularly brittle. The lithosphere is broken up into 12 major plates and a few minor plates that all move about along Earth's surface. Oceanic lithosphere is typically 100 km (60 mi) thick and about 150 km (95 mi) thinner than the typically 250-km (155 mi) thick continental lithosphere. But oceanic lithosphere is denser than its continental counterpart. Earthquakes occur within the lithosphere.

The term **asthenosphere** is based on the Greek word **asthenes** for "weak". This layer immediately beneath the lithosphere consists of mantle rock, but it is weak, soft and ductile. In fact, some parts of the rocks may be molten. Under stress, the asthenosphere deforms rather than breaks.

The asthenosphere is denser than the lithosphere, which is why it is able to hold up the lithosphere in a stable fashion.

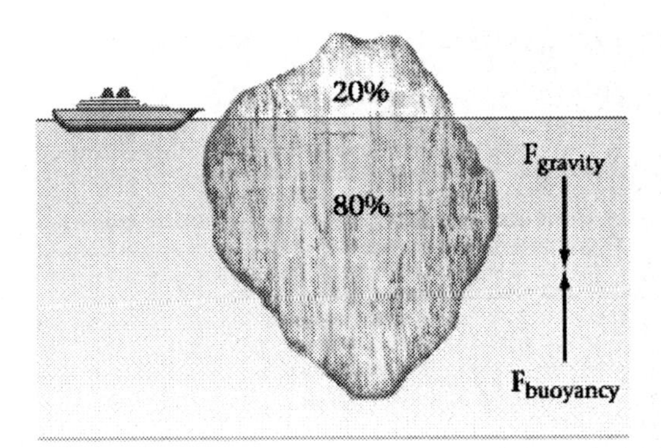

Figure 3.29 An example of isostasy. A low-density iceberg floats in the higher-density water. (source: from ME)

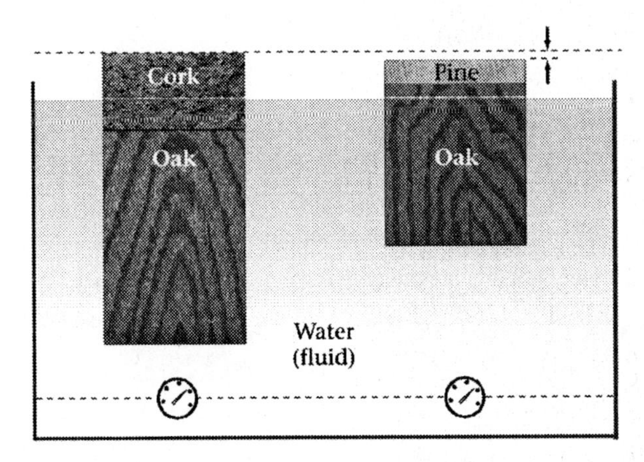

Figure 3.30 An example of isostasy of different materials. Here, oak wood represents the mantle part of Earth's lithosphere, cork represents continental crust and the denser pine wood the oceanic crust. The hydrostatic pressure (pressure from the column of mass above) is in balance at the bottom of the water basin. The thicker oak on the left compensates for the lighter cork compared to the pine wood on the right. Nevertheless, the column on the left sticks out of the water (the asthenosphere) a little farther than the column on the right. (source: from ME)

Obeying the principles of **isostasy**, the lithosphere "floats" on the asthenosphere like a "strong" iceberg in the ocean water (Fig. 3.29). In **isostatic equilibrium**, the gravitational force acting on the floating body (pointing toward the center of Earth) is in perfect balance with the buoyancy force exerted by the density contrast between the body and the medium it floats in (pointing upward). The body then displaces the medium it floats in by the equivalent of its mass. If the gravitational force is larger than the buoyancy force, the body sinks. And if the gravitational force is smaller, the body rises.

As a result of slightly different rock properties in the crust, the oceanic lithosphere is much thinner than the continental lithosphere even though the mantle beneath the crust is thought to be the same beneath continents and oceans. This can be understood better when considering the experiment in Fig. 3.30. The light cork resembles the continental crust, while a denser layer of pine wood represents the oceanic crust. In order for the tops of the two block to stick out of the water basin by similar amounts, much less of the oak wood beneath (resembling the uppermost mantle in the lithosphere) is required to pull down the pine wood into the water than is required to pull down the cork. This principle explains why continental interiors typically have deep-reaching roots, or keels.

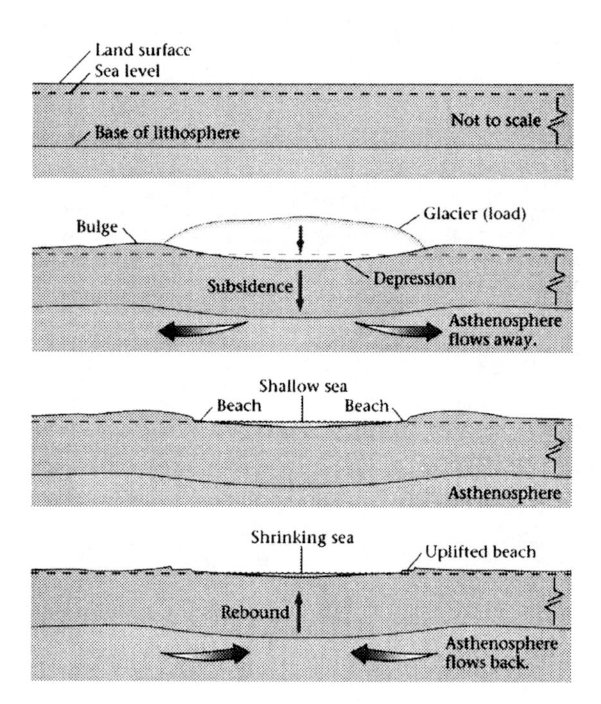

Figure 3.31a Cross sections illustrating the concept of post-glacial rebound. Before glaciation or the beginning of an ice age, the surface of the lithosphere is flat. The weight of a glacier or ice sheet pushes down on the lithosphere. The ductile asthenosphere flows sideways allowing the lithosphere to subside and pushing the perimeter up to form a bulge on the surface. After the end of an ice age, the ice melts to form a lake. Eventually the asthenosphere flows back, the bulge recedes, the water flows elsewhere and the lithosphere rebounds. (source: ME)

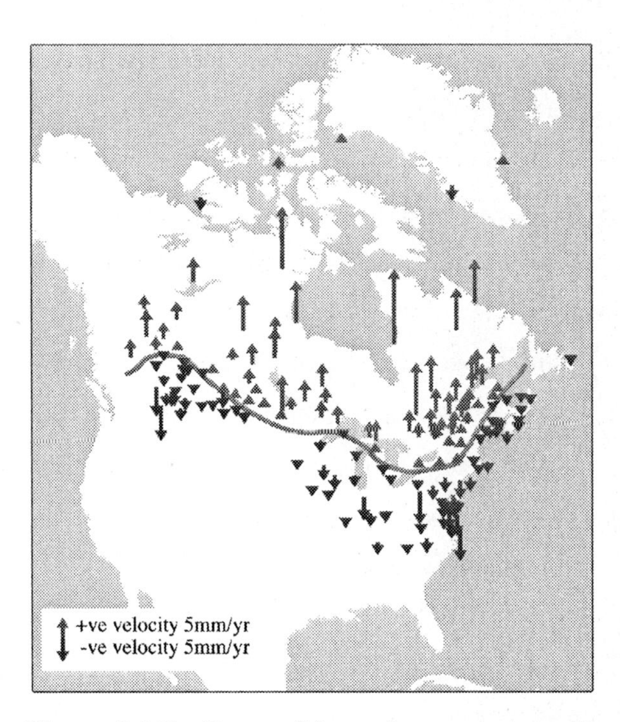

Figure 3.31b Rates of isostatic movement still taking place in North America in response to the melting of the ice sheet from the last ice age, as indicated by satellite data. (source: ME)

Like water, the asthenosphere reacts to imbalances in the gravitational and buoyancy forces. If weight is therefore added to the lithosphere, the asthenosphere beneath will react by flowing away sideways (Fig. 3.31). The lithosphere subsequently sinks. If weight is taken away, the asthenosphere will flow back and the lithosphere rises. This cycle occurs when Earth goes into an ice age when glaciers and ice sheets grow (**glaciation**) and when ice ages end. During glaciation, the precipitation does not drain through rivers into the ocean but instead stays to grow into ice sheets. This adds weight to the top of the lithosphere. The asthenosphere then reacts and the lithosphere sinks. Depending on the thickness of the ice sheet, the top of the rocky crust may therefore be pushed below sea level. Since the asthenosphere cannot flow away endlessly, it will pile up near the outer edges of the ice sheet and push up the lithosphere there, and the lithosphere flexes in response. At the end of the ice age, the ice sheet melts, and the water can find its pathways back into the oceans. The lithosphere loses weight on top and the asthenosphere beneath can flow back. The lithosphere rises beneath the previous location of the ice sheet, but sinks where the piled up asthenosphere had pushed up the lithosphere in the interim.

That this is a measurable effect still today, some 10,000 years after the last ice age is demonstrated in Fig. 3.30b. The northern part of North America is still uplifting, by up to a few cm per year, near Hudson Bay where the core of the ice sheet was located. Farther south, beyond the edges of the ancient ice sheet, the continent sinks by a few mm per year. This rebounding movement is also called **post-glacial rebound**. It is important to consider this concept when interpreting recent data of sea level changes. A current local measurement of a fall in sea level at a particular costal station along Baffin Bay does not at all indicate that the global sea level drops! Instead, scientists have to be very careful at correcting their data for post-glacial rebound. It then turns out that even these stations indicate that the global sea level is indeed rising, an indicator for a warming climate.

The principles of isostasy as just described do not only apply to post-glacial rebound. The formation of any larger lake and reservoir causes the asthenosphere to react to the changes in gravity. So has the surface near the center of Lake Mead in Nevada and Arizona dropped by 170 mm (6.7 in!) between 1935 and 1950 in response to the rising water level in this reservoir behind Hoover Dam. Accordingly, the surface immediately northeast of Lake Mead experienced uplift by 20 mm. The lithosphere beneath new reservoirs also reacts to the new stresses imposed by the added weight of the water. Cases of so-called **induced seismicity** have recently added to the negative impacts of building large reservoirs that also include the problems of displacing people and destroying habitats.

References and Websites

1. Wright, E.L., 2009. Age of the Universe, http://www.astro.ucla.edu/~wright/age.html. retrieved Feb 2012.
2. Laughlin, G.P., 2006. Observations of distant worlds are beginning to reveal how planetary systems form and evolve. Am. Scientist, 94, 420 – 429.
3. Discovery News, 26 Mar 2012. Moon's creation questioned by chemistry. At http://news.discovery.com/space/moon-creation-magma-meteor-ancient-space-120326.html
4. New Scientist, 29 Sep 2009. Moon is coldest known place in the solar system. At http://www.newscientist.com/article/dn17810-moon-is-coldest-known-place-in-the-solar-system.html
5. Wikipedia webpage on meteorites. Accessed on 30 March 2012. At http://en.wikipedia.org/wiki/Meteorites also Pasachoff and Filippenko, 2006 (see below)
6. the central website of the Spaceguard Foundation is located at https://www.cfa.harvard.edu/~marsden/SGF/
7. Lehmann, Inge, 1936. P'. Publications du Bureau Central Séismologique International A 14(3), 87 – 115.
8. Gilbert, F. and Dziewonski, A.M., 1971. Solidity of the Inner Core of the Earth inferred from Normal Mode Observations, Nature, 234, 465 – 466.

NASA home page of Lunar and Planetary Science. The site has comprehensive fact sheets and other information on every planet, Pluto, asteroids and comets.
http://nssdc.gsfc.nasa.gov/planetary/

The planets homepage at JPL (Jet Propulsion Laboratory):
http://pds.jpl.nasa.gov/planets/

Other Recommended Reading

A general astronomy textbook:

Pasachoff, J.M. and Filippenko, A., 2006. The Cosmos: Astronomy in the New Millenium, Thomson-Brooks Cole, Pacific Grove, CA, 500. This is a somewhat pricy but very entertaining college textbook. The authors include themselves in the narrative. Much information in Chapter 3 comes from this book.

Two textbooks on the Solar System:

Beatty, J.K, Petersen, C.C. and Chaikin, A., 1999. The New Solar System, Cambridge Univ. Press, 421 pp. Somewhat dated, pre-2006 IAU redefinition, but very informative, with very nice images.

Wilkinson, J., 2009. Probing the New Solar System. CSIRO Publishing, 312 pp. This book is from the post-2006 International Astronomical Union redefinition of what constitutes a planet.

A coffee-table book:

Giunti Editorial Group, 2005. Atlas of the Skies, Taj Books, 240. This is an appealing, very affordable introductory paperback book on astronomy, instruments and the heavenly bodies. A warning: beware of (lots of) typos! The written English translated from the original Italian is oftentimes unconventional. Some figure labels are not translated. Consultation of other references is recommended.

B. Morrison, 2010. Rogue Wave, Pocket Books, 416 pp. This science-fiction story describes the effects of a bolide impact in the Pacific Ocean, causing a humongous tsunami. The story tracks a scientist at the Pacific Tsunami Warning Center who races against time and desperately tries to save the population of Hawaii.

Chapter 4: Continental Drift and Plate Tectonics

Earth is the only planet in the Solar System that currently has plate tectonics. Since no old craters can be found on Venus, scientists speculate that Earth's sister planet experiences episodic catastrophic resurfacing, the last occurring about 500 million years ago. Evidence for volcanism as recent as 2 million years ago is dramatically underlined on Mars by Olympus Mons. Being over 25 km (82,000 ft) high, Olympus Mons is nearly three times taller than Mt. Everest (8848 m; 29,029 ft), and Mauna Loa which rises more than 9 km above the seafloor. However, there are no known current eruptions on Mars. Plate tectonics and associated mantle convection transport heat effectively from Earth's interior to the surface, thereby providing the energy to generate devastating earthquakes and volcanic eruptions. To understand why earthquakes, volcanic eruptions, tectonic uplift and mass movements occur at some places but not at others, it is necessary to recall the basic principles of plate tectonics. These are outlined in this chapter. After the review of this chapter, it will become clear why some areas are prone to much larger earthquakes than other areas. It will also become clear why it is impossible that the catastrophic, huge Tohoku, Japan earthquake on 11 March 2011 cannot occur in Germany even though news media posed the question "what if?". We will learn that the most devastating earthquakes occur along a particular type of plate boundary, the subduction zone. The U.S. has no modern memory of such an earthquake ever to occur along their shores. But recent research has revealed that a stretch on the West Coast along the Cascadia Trench has the potential for a Tohoku-type earthquake. In fact, such an earthquake occurred a few hundred years ago.

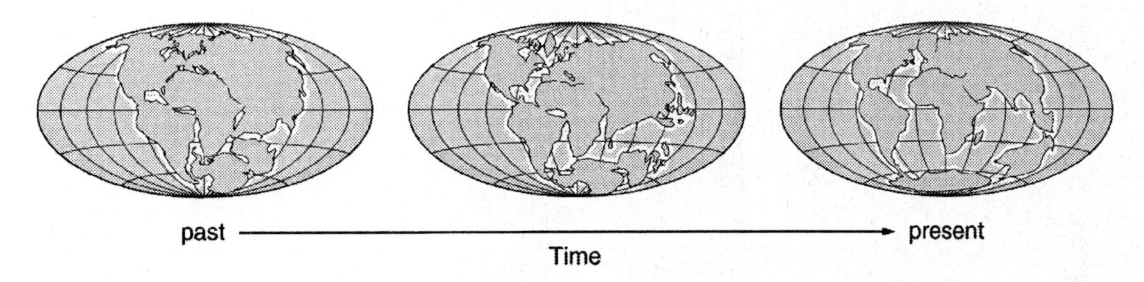

Figure 4.1 A concept figure illustrating how current continents fit together to form a supercontinent, not unlike the pieces of a jigsaw puzzle. If continents drift, then such a supercontinent may have existed in the past. (source: ME)

COPYRIGHTED MATERIAL — DO NOT DUPLICATE, DISTRIBUTE, OR POST

Figure 4.2 Alfred Wegener, the German meteorologist who proposed a comprehensive model of continental drift and presented the geological evidence to support his idea. (source: NISEE; UC Berkeley)

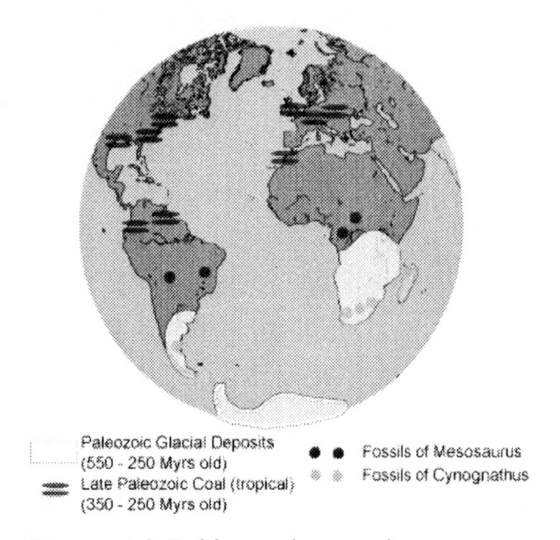

Figure 4.3 Evidence that continents were once joined: certain fossil groups and geological features are found on both sides of the Atlantic Ocean.

Figure 4.4 Supercontinent Pangaea existed in the Permian Period from about 300 to 250 million years ago before it started to break up about 200 million years ago. (source: wikipedia)

4.1 Developing and Explaining the Idea of Drifting Continents

Initial Observations

Some continents on both sides of oceans have outlines that suggest that they can be put together to form larger pieces in a jigsaw puzzle, if the continents are moved back together. Depending on whether we use the outline of the continents or also include the **continental shelves** (regions with thinned, altered continental crust in shallow oceans), they fit better or worse but in principle, we can move all pieces together to form a large supercontinent (Fig. 4.1). In 1912, German meteorologist Alfred Wegener (Fig. 4.2) took up this idea and formulated the theory of **continental drift** in which continents move across the globe over geological timescales.

Apart from the continents fitting together, there are also other pieces of evidence that support continental drift. Fossils of a certain kind often match at continental boundaries across oceans (Fig. 4.3). Without continental drift, one would have to explain how a certain species could independently occur at exactly these locations, across an ocean, but not at other locations nearby on the same continent. There are also geological units that seem to continue across continental borders, such as mountain belts and coal and glacial deposits. Earth's magnetic field also helped provide evidence for continental drift through **apparent polar wander curves** that plot the geographic time evolution of the poles of Earth's magnetic field. The concept of these curve is quite complex but crucial to support the continental drift hypothesis. A more detailed treatment therefore follows in the next section.

All these pieces of evidence suggest that there have been points in time in Earth's history when all continents merged to form large supercontinents, only to break up some time later. The most well-known supercontinent is probably **Pangaea** that existed in the Permian Period between 300 and 250 million years ago, and started to break up shortly after that (Fig. 4.4). Pangaea was not arranged in the way that Figure 1 suggests. This is because continents move not only sideways along lines of latitude but also toward and away from the poles, and even rotate on Earth's surface. Pangaea was aligned mainly in the north-south direction. It influenced global ocean circulation in a much different way than today's continents do. For example, it inhibited latitudinal (along a line of latitude) west-east ocean circulation that exists today in the Southern Oceans. The arrangement of continents therefore influences the transport of heat throughout the oceans, and ultimately Earth's climate. Pangaea likely caused one of the coldest climatic periods in Earth's history.

Subsequent Supporting Observations and Plate Tectonics

The concept of continental drift seems obvious to us today but Alfred Wegener's theory was hotly debated and mostly rejected because Wegener could not provide a mechanism for why continents drift. What forces make continents move and where would the energy come from to drive the drifting? The breakthrough came not before scientists were able to map the seafloor. In 1960, Harry Hess first reported on volcanically active **mid-ocean ridges** that he first proposed in the 1950s. The idea was born that new lithosphere could form along these ridges, and the new material could push plates apart. The ultimate proof of this concept came with the discovery of **magnetic stripes** on the seafloor later in the 1960s. Contrary to what the name suggests, these stripes are not caused by different coloration in the rocks but rather describe variations in magnetic properties. The magnetic stripes pattern was symmetric to the mid-ocean ridge axis, thereby supporting the idea of symmetric **seafloor spreading** in which new oceanic lithosphere is formed along the ridges and then moves away from the ridge (see section "Seafloor Spreading" below). But if new seafloor forms along the mid-ocean ridges, does this imply that Earth "grows"? We know from other geophysical observations that Earth is not expanding. For example, scientists have been measuring the length of a day very accurately. On an expanding Earth, the days would become longer but there is no evidence that this is happening. But where does the lithosphere go if Earth does not expand? It is consumed somewhere else, in so-called **subduction zones**! The idea of subduction zones was born when the deep sea drilling project (DSDP) and dredging (picking up rocks in a basket lowered to the seafloor) did not find any oceanic lithosphere that is older than 200 million years. The theory of **plate tectonics** includes all these concepts. It explains Wegener's observations and how lithosphere can be produced and consumed so that Earth does not change its size (expands or shrinks).

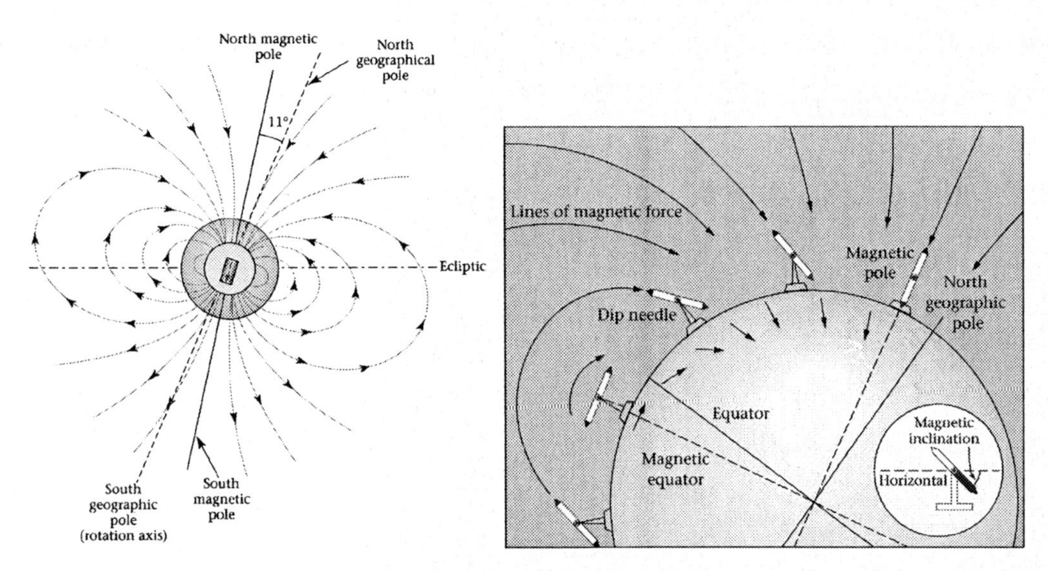

Figure 4.5 Left: Schematic diagram of Earth's magnetic dipole field. We can picture Earth's field by imagining that a bar magnet is located at its center. Presently, the magnetic north-south axis is tilted from the geographic north-south axis (i.e. the rotation axis) by 11°. Note that Earth's rotation axis is tilted 23.5° from the normal of the ecliptic, Earth's orbital plane. The magnetic field is therefore tilted 12.5° with respect to Earth's orbital plane. Right: An illustration of magnetic inclination. A magnetic needle that is free to rotate around a horizontal axis aligns with magnetic field lines. The compass needle is horizontal at the equator, vertical at the pole and tilted at latitudes in between. The angle the needle makes with the horizontal is the **inclination**. (source: ME)

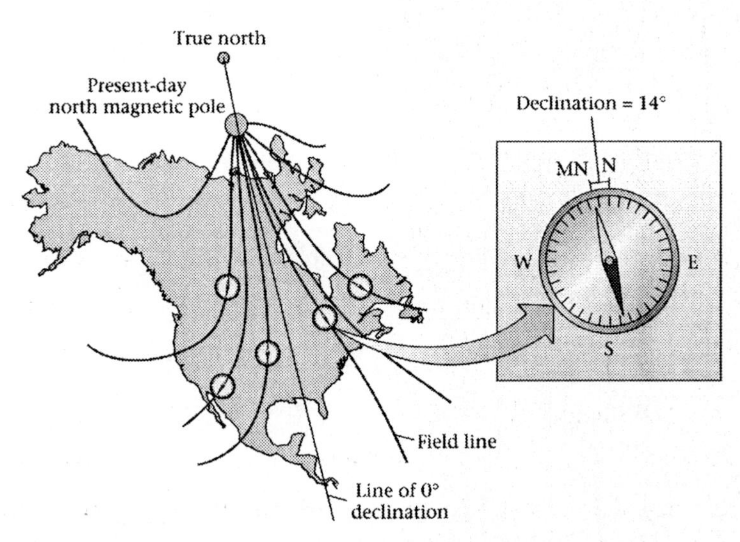

Figure 4.6 The horizontal projections of magnetic field lines in North America at present. As geographical longitudes align in a north-south direction but the magnetic field lines do not, in most places, a compass needle will have an angle with respect to geographic North, the **declination**. Only places along a line going through western New Orleans and the North Pole have 0° declination (in

2010). A compass needle in New York has a declination of -14° (i.e. 14° to the west). At UC San Diego, the declination currently is +12°12' (i.e. to the east; in 2010). (sources: ME; magnetic-declination.com)

4.2 Earth's Magnetic Field

A strong piece of instrument-based evidence for continental drift is the match or mismatch of apparent polar wander curves for different continents. Later, seafloor spreading was supported by the discovery of magnetic stripes on the ocean floor. To understand these concepts, we first need to recall the basic features of Earth's magnetic field.

Basics Features of Earth's Magnetic Field

As viewed from space, Earth's magnetic field approximately has the shape of that of a bar magnet that has a north (N) and a south (S) pole (Fig. 4.5). Such a field is called a dipole field. In space, Earth's magnetic field actually looks a bit like a tear drop pointing away from the Sun because it is affected by the solar wind. But we ignore this here. Unlike the field of a bar magnet in our laboratory, Earth's magnetic field does not dissipate (weaken) over time but is instead maintained by the **geodynamo**. Fluid motion in Earth's liquid outer core, that mainly consists of the metal iron, generates electric currents that in turn generate a magnetic field. The everlasting fluid motion and associated currents in the rotating planet sustain Earth's magnetic field.

Earth's magnetic field changes over time and magnetic North - which is physically the south pole of the equivalent bar magnet – currently does not align with geographic North but has an angle of 11° (Fig. 4.5). In 2000, magnetic North was at 81.5° N, 111.4° W and it is moving at about 15km per year. The wander of Earth's magnetic field changes as a result of changes in the geodynamo. Over time, Earth's field has changed from a mainly dipolar field to more complex fields. Sometimes, the field even reverses so that magnetic North aligns with geographic South. Such reversals have occurred several times in the last 2 million years and many more times throughout Earth's history. The cause of such reversals as well as the time it takes to revert the field is currently a hot topic in geomagnetism research. Recent best estimates for the duration of a reversal are a few thousand years (~2000 years), which is relatively rapid on the geological timescale.

At any location on Earth, the magnetic field is characterized by two angles, the **inclination** (Fig. 4.5) and the **declination** (Fig. 4.6), as well as the intensity of the field. The current **declination** in Southern California is about 13° E, which means that geographic North is 13° counterclockwise from what a compass indicates as North. At UCSD, the declination is currently 12°12' E. It is important to correct for declination in navigational operations as well as on field trips when a magnetic compass is used. For example, if the declination is not accounted for on a hike in San Diego's backcountry, an error of 0.21 mi results in the east direction for every mile going "north".

Global Positioning System (GPS) devices navigate with respect to geographic North and so are not affected by the mismatch between geographic and magnetic North. GPS devices receive a radio signal timed by atomic clocks from several of currently 30 satellites in orbit. The signal from at least four satellites are needed for trilateration, the determination of a geographic position as well as altitude.

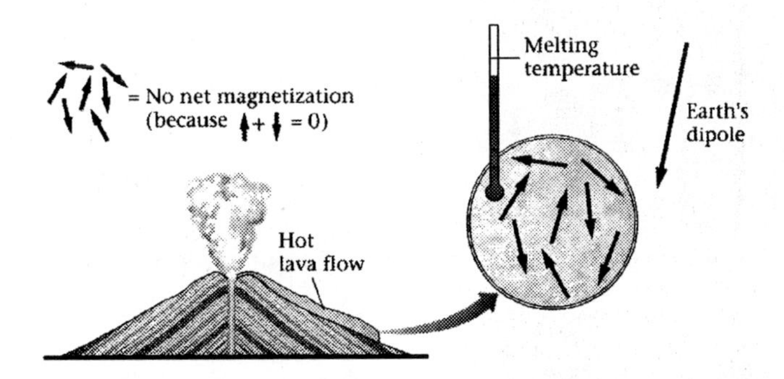

Figure 4.7a In hot, fresh lava the atom-scale magnetic domains are randomly oriented. Overall, the dipoles cancel each other out so that lava does not show a preferred magnetic orientation. (source: ME)

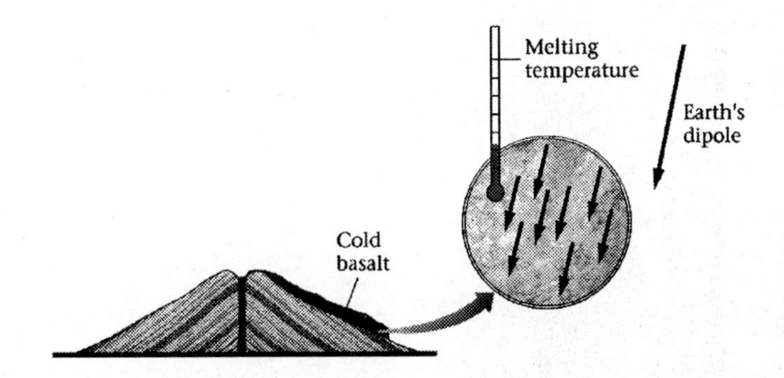

Figure 4.7b As lava cools to below the Curie temperature, the dipoles of the magnetic domains lock into alignment with the Earth's current ambient field. A cooled volcanic rock has a preferred magnetic orientation that can be observed by a magnetometer. (source: ME)

Recording and Preserving Earth's Ambient Magnetic Field in Rocks

In the field, a magnetic compass needle can actually deviate from magnetic North by angles that are different from the expected declination. Similarly, the inclination and the field intensity can also show anomalies. This phenomenon occurs particularly near iron ore deposits and some other sites such as volcanic dikes. The reason for this is that rocks that contain iron (Fe) or other substances such as nickel (Ni) and cobalt (Co) are magnetic and so locally add to Earth's magnetic field. In ferromagnetic minerals, electron spin directions can align with Earth's magnetic field within atomic-scale **magnetic domains**. Once aligned, this orientation remains when the external magnetic field is turned off or changes. The magnetic properties of an iron-bearing rock changes with ambient temperatures. When the rock is very hot or molten the magnetic domains remain at random orientations (Fig. 4.7a) and the rock exhibits no magnetism. When the rock cools below the **Curie temperature** the magnetic domains align with the ambient magnetic field and so enhance the current local magnetic field. The magnetic domains also preserve the information of this field (Fig. 4.7b) (**remnant magnetization**). When the field changes, then the remnant magnetization of the rock still points in the "old" direction and thereby influences the local magnetic field. The rock only loses magnetic orientations when it is remelted. Curie temperatures are relatively high, e.g. 770°C for pure iron, and a few 100°C lower for most iron-bearing minerals, but lower than the melting temperature of rock which typically ranges from 650°C to 1100°C. Over time, successive lava flows at a volcano can stack on top of each other to record and preserve ambient magnetic fields over time, with the oldest lava flow preserved at the bottom and the youngest on the top. This allows scientists to reconstruct a geologic record of the history of Earth's magnetic field at that location (Fig. 4.8a). Similar records can be reconstructed from a stack of lake sediments. These stack are also called **stratigraphic profiles**, leaning on the Latin word *stratum* for "cover" or "layer".

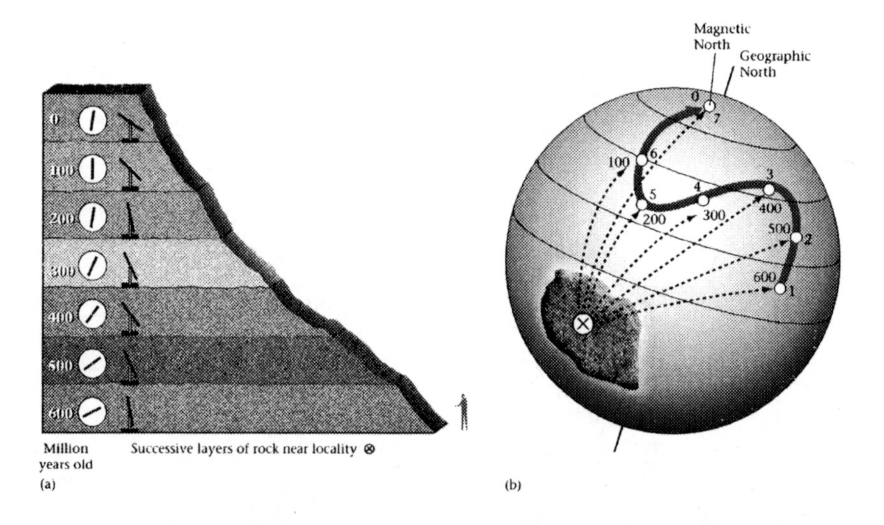

Figure 4.8 (a) Schematic view of a cliff at location X exposing a succession of dated lava flows or sediment layers (strata). A geoscientist can measure the preserved magnetic orientation of the past in each layer. The age of the layer is determined by other methods. (b) For each layer, the "paleo"-

location of the inferred magnetic pole is then drawn as a dot on a globe. The connecting line, the **apparent polar-wander curve**, then shows how the pole wandered over time, relative to the location of the rock cliff. (source: ME)

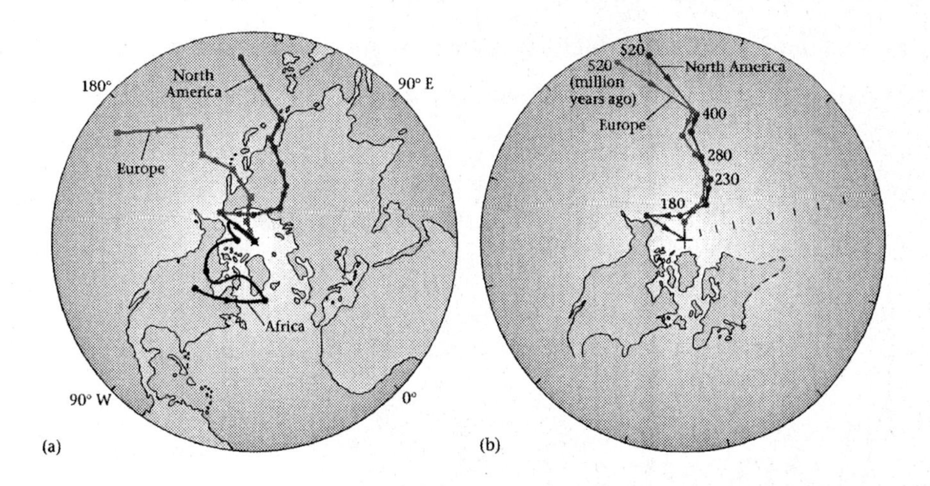

Figure 4.9 (a) Apparent polar-wander curves for North America and Europe for the last few hundred million year. (b) When continents are moved back, the polar-wander curves also move. During the time interval for which the curves match, the continents were joined together. North America and Europe therefore were joined in one larger continent between 280 and 180 million years ago, and also briefly about 400 million years ago. (source: ME)

Apparent Polar Wander Curves

At a given location (location X in Fig. 4.8a), the history of magnetic North over time can be reconstructed by measuring the inclination and declination in each layer in a stack of lava flows (or sediment layers). The age of each layer is thereby determined by other techniques such as by radiometric dating or, in the case of a sediment profile, by comparing fossils in the sediment layers. The curve connecting the reconstructed apparent magnetic North is the apparent polar wander curve (Fig. 4.8b). The curve shows the apparent location of magnetic North and not its absolute location because the curve shows only the relative location between North and the stack of lava flows at location X. An apparent movement of magnetic North results from either a true migration of magnetic North or the migration of the stack of lava flows, or usually a combination of both. So with only one curve, no definite conclusion can be drawn about either one. But the relative movement between two locations, such as those on two different continents, can be determined be comparing their two apparent polar wander curves (Fig. 4.9). Scientists rotate the curves to try and find a matching segment. When a match is found, then this indicates that the two continents were joined and moved together during the time corresponding to the matching segment. Any apparent polar wander in such a segment indicates again either true polar wander or the relative movement of the joined continents relative to the pole. From Fig. 4.9b, we can infer that North America and Europe were once joined between 280 and 180 million years ago, which overlaps with the era of Pangaea, and for a short time around 400 million years ago. At other times, the two continents moved with respect to each other, thereby supporting the continental drift hypothesis.

4.3 Seafloor Spreading

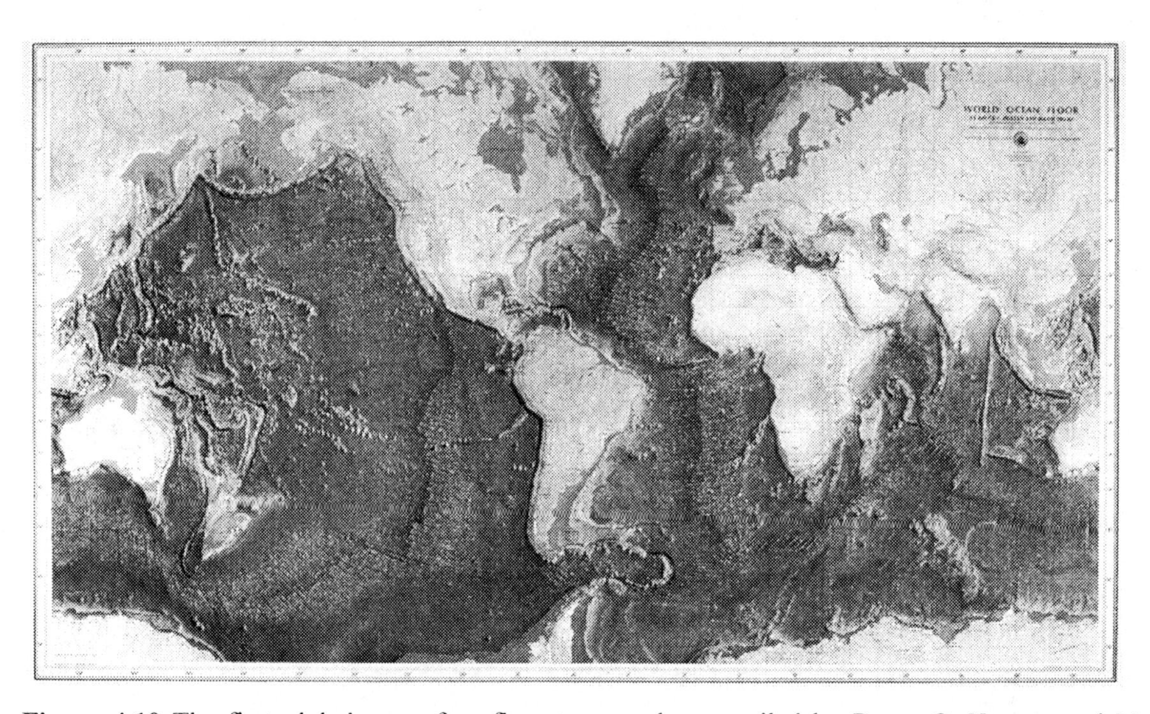

Figure 4.10 The first global map of seafloor topography compiled by Bruce C. Heezen and Marie Tharp in 1959. The mid-ocean ridges are exaggerated in this map. (source: Columbia University Archives).

Figure 4.11 Harry Hess (1906-1969) in his Navy uniform as Captain of the assault transport vessel USS Cape Johnson during World War II. (source: USGS and Princeton University Archives)

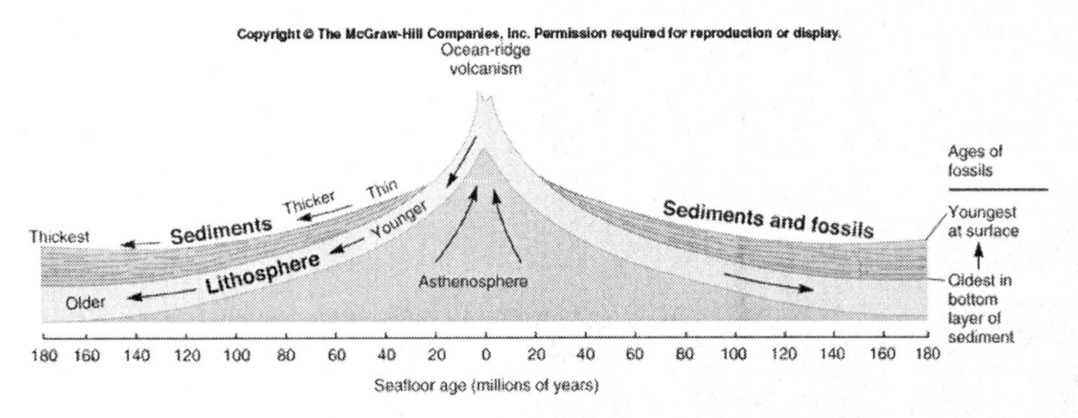

Figure 4.12 Schematic cross section through oceanic lithosphere perpendicular to the mid-ocean ridge. Moving away from the ridge, ocean age increases, the thickness of accumulated sediments increases as well as the ages of fossils in the sediments. The seafloor deepens with age as a result of the cooling, thickening and subsiding lithosphere. (source: AB).

The Proposal for Sea Floor Spreading

In the 1950s the U.S. Navy and scientists started mapping the ocean floor, and in 1959, Bruce Heezen and Marie Tharp compiled the first global seafloor map (Fig. 4.10). Early on, it became apparent that large ridges traverse the ocean basins. In his seafloor-spreading hypothesis, Harry Hess (Fig. 4.11) postulated that new lithosphere is formed along these mid-ocean ridges (MORs). Seafloor spreading would support Alfred Wegener's continental drift hypothesis because it delivers the mechanism that is needed to explain why continents move. As new seafloor is formed and spreads away from the MORs, continents that border these oceans also move apart. Hess based his theory not only on the new seafloor maps but also on other geological and geophysical observations. Hess observed that the seafloor deepens away from the MORs, so the seafloor appears to "settle" over time (Fig. 4.12). There are nearly no sediments along MORs but sedimentary layers thicken away from them, suggesting that MORs may be young features that have not yet experienced extensive erosion. The surface heat flux (the heat released through Earth's surface) is greater along MORs than away from them. This indicates the presences of some form of volcanism and therefore the production of young rocks. Dredging rocks from the ocean floor and deep sea drilling revealed that oceanic rocks are also different from continental rocks. As seismic instrumentation became more sophisticated, earthquakes were detected along MORs, perhaps indicating that young oceans crack as they cool. The conclusion of all these observations was that new oceanic rocks form along mid-ocean ridges and that they then move away as they cool. But even with the hypothesis of seafloor spreading not all scientists agreed with the idea of plate tectonics. In fact, Bruce Heezen, who started seafloor mapping by surveying the mid-Atlantic ridge, originally thought that the new maps support the idea of an expanding Earth before he eventually became a supporter for plate tectonics in the 1960s.

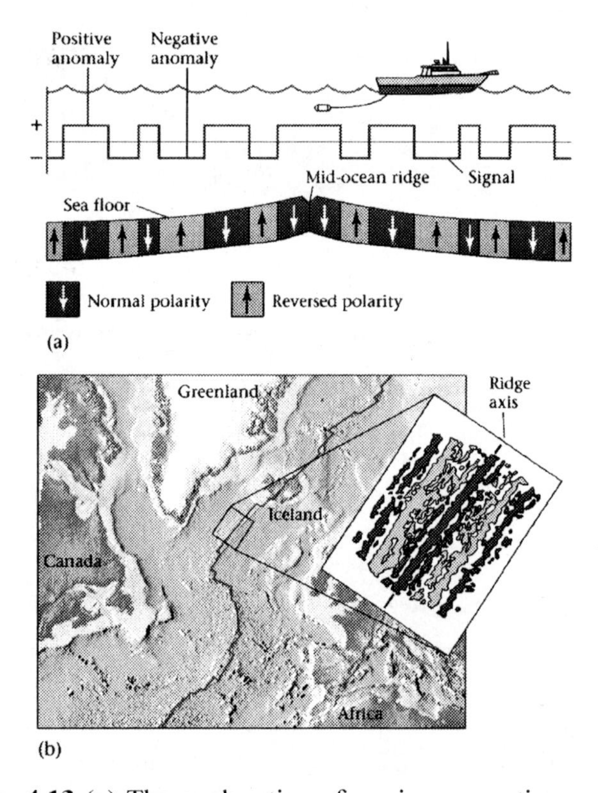

Figure 4.13 (a) The explanation of marine magnetic anomalies (magnetic "stripes") measured by a magnetometer that is pulled across the ocean above a MOR. The seafloor beneath positive anomalies has the same polarity in the remnant magnetization as Earth's current magnetic field and therefore enhances to it. The seafloor beneath negative anomalies has reversed polarity in the remnant magnetization and thus counteracts Earth's current field. (b) The symmetry of the magnetic anomalies

measured across the Mid-Atlantic Ridge south of Iceland. Individual anomalies are somewhat irregular, because the process of forming seafloor, in detail, happens in discontinuous pulses along the length of the ridge. (source: ME)

Magnetic Stripes – The Proof for Sea Floor Spreading

A breakthrough came when geomagnetic seafloor surveys provided another crucial piece of evidence. Using ship-towed magnetometers, surveys mapped stripe-like anomalies on the seafloor that were symmetric to mid-ocean ridges (Fig. 4.13). The stripes mark enhanced and diminished intensities of Earth's magnetic field. Recall that Earth's magnetic field reverses over time while, at the same time, volcanic rocks (and so also seafloor rock) preserve the ambient magnetic field when the rock cools below the Curie temperature. A rock magnetized in the direction of the current ambient field enhances the local magnetic field. A rock magnetized in the opposite direction (e.g. magnetized during a reversal) weakens the local magnetic field. Dredging and deep sea drilling work determined ages of seafloor rocks, and it was established that the magnetic stripes are records of the reversals of Earth's magnetic field over time. The symmetric arrangement of the stripes on both sides of the MORs documents how seafloor is formed and moved away from the MORs as time goes on.

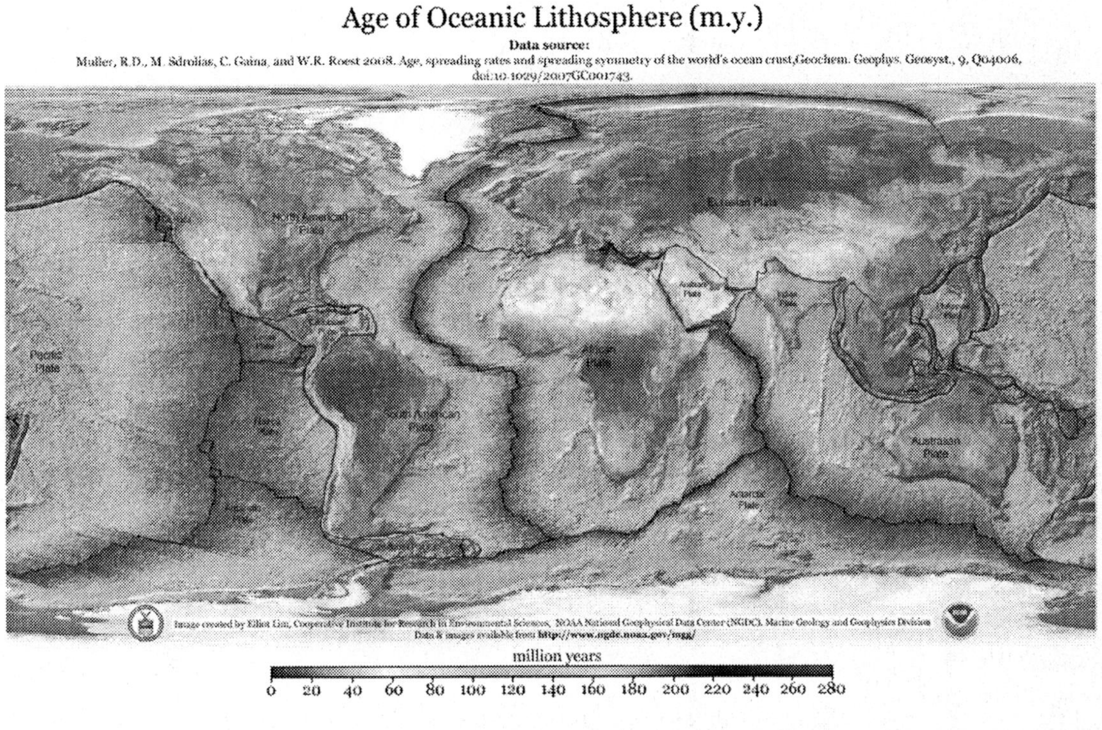

Figure 4.14 Age of the global ocean floor. The map was compiled by Dietmar Müller (then at SIO) and collaborators and is available at the National Oceanographic and Atmospheric Administration's (NOAA's) National Geographic Data Center (NGDC) website.

Deep Sea Drilling, Dredging and Subduction Zones

Though the hypothesis of seafloor spreading provided a mechanism to move continents, it left an important question unanswered. If new lithosphere is formed along MORs, does this imply that Earth is expanding? Evidence for an expanding Earth was not found as the length of day should increase on a growing planet. Where, then, does the growing lithosphere eventually go? The deep sea drilling project that operated between 1968 and 1983 helped solve this mystery. Scientists discovered that no oceanic rock exists that is older than 200 million years (Fig. 4.14). This is in stark contrast to the old ages of 4 billion years of some rocks in stable continental shields that are far away from plate boundaries. Together with the missing evidence for an expanding Earth, it was concluded that oceanic lithosphere must somehow be consumed. At the same time, some very deep trenches emerged from seafloor mapping along the oldest oceans. The Mariana Trench southeast of Japan is nearly 11 km (36,000 ft) deep. Very large earthquakes were found to occur along these trenches and some of the most violent volcanic eruptions occur not too far from these on land. Scientists identified these as the surface expression of **subduction zones** where oceanic lithosphere is consumed when an oceanic plate sinks below another oceanic plate or a continental plate. Further seismic imaging of subduction zones revealed complex blocks of material accumulating behind the trenches (Fig. 4.15). The shape and properties of these **accretionary prisms** is not unlike piles of dirt pushed together by a bulldozer. Accretionary prisms mostly consist of sediments that are scraped off a subducting oceanic plate as it sinks into the mantle. The prisms are tectonically weak in that they have many internal faults on which earthquakes can occur. Recent research provides mounting evidence that large subduction zone earthquakes can trigger submarine landslides on these prisms. The slides in turn can cause devastating tsunami. It is thought that a submarine landslide triggered by the 26 December 2004 Sumatra-Andaman earthquake locally enhanced the tsunami generated by the earthquake, and was ultimately responsible for the great destruction along some stretches of the Sumatran coast.

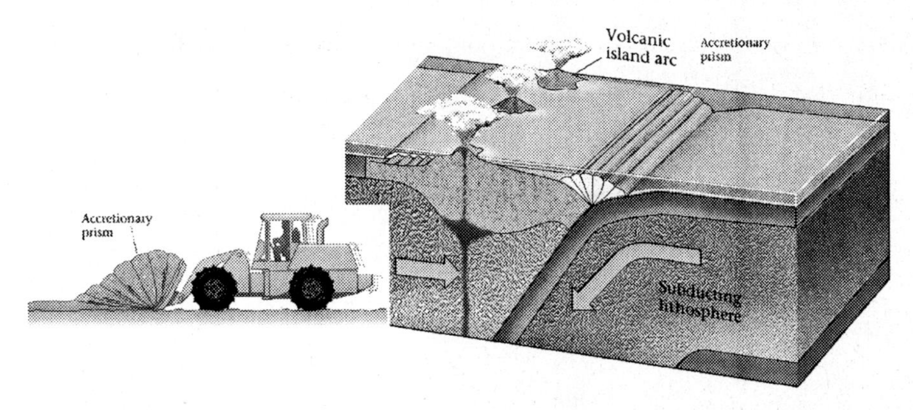

Figure 4.15 The accretionary prism in a subduction zone. (source: modified from ME).

4.4 Plate Tectonics

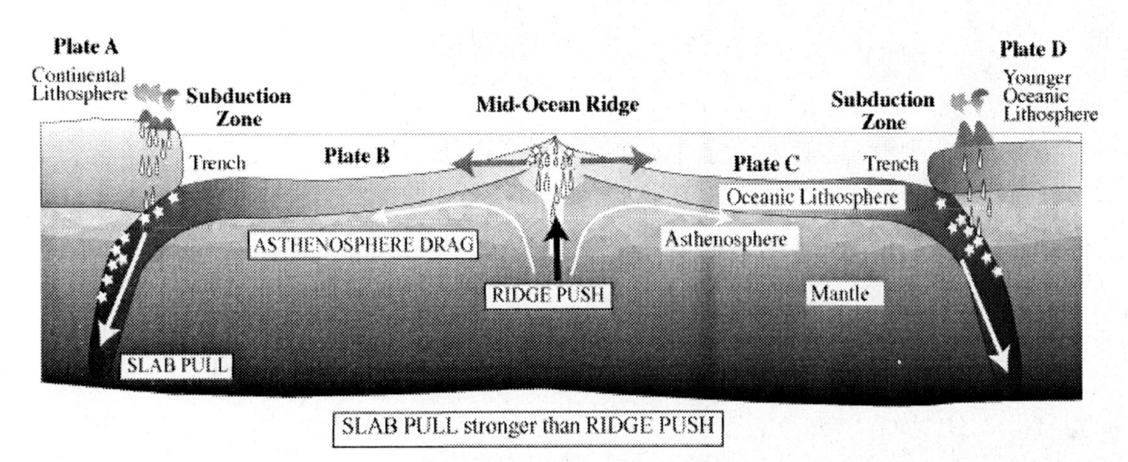

Figure 4.16 Schematic cross section of the tectonic cycle. Magma rises from the asthenosphere to the surface at the volcanic mid-ocean ridges, where it solidifies to form new lithosphere. Old, thick oceanic lithosphere subducts either beneath younger oceanic lithosphere or continental lithosphere to close the cycle. Earthquakes and volcanoes occur along MORs (divergent plate boundary) as well as along subduction zones (convergent plate boundary).

The Life Cycle of Oceanic Lithosphere

Close to a mid-ocean ridge, the new lithosphere is warm and has low density. It therefore is very buoyant and pushes up the seafloor (Fig. 4.16). New buoyant material from below pushes the two adjacent plates apart. This force is called the **ridge push**. As the lithosphere moves away from the MOR, the lithosphere cools, and its density increases. More mantle material also "freezes" to the bottom as the plate ages so that it thickens. As the plate ages, it becomes heavier and eventually is so dense that it is no longer able to float on the asthenosphere. It then starts to sink, or **subduct**, along a subduction zone beneath the **overriding plate**. The oceanic plate is recycled back into the mantle.

The sinking part of a subducting plate is called the **slab** and pulls the rest of the plate behind it. The **slab pull** is the driving force behind this process. Though only a minor contributor, a third force comes into play when the asthenosphere moves sideways. The overriding lithosphere is then dragged along by the so-called **asthenosphere drag**. Deep sea drilling and dredging work revealed that oceanic rock is much denser than continental rock. Continental lithosphere is too buoyant to sink into the mantle and therefore does not subduct. Only oceanic lithosphere subducts, as depicted in Fig. 4.16.

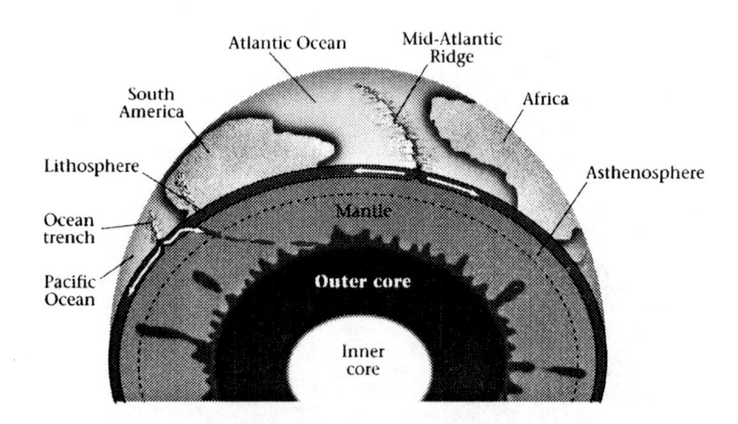

Figure 4.17 MOR activity, slab subduction, and ultimately plate tectonics involves the transfer of material from the mantle to the surface, and back down again. The insides and the surface of our dynamic planet are in constant motion through mantle convection. (source: ME)

Viscosity and Mantle Convection

As mantle material rises along mid-ocean ridges and slabs sink into the mantle, it becomes clear that Earth's mantle is not a rigid body but slowly turns over on a geological timescale (Fig. 4.17). Earth's mantle convects not unlike soup that is heated on the stove. This is possible because Earth materials have a certain viscosity that is not arbitrarily high. **Viscosity** is the resistance to flow. Highly viscose materials are difficult to deform while materials with very low viscosity tend to flow easily. The rock samples in our hands have a very high viscosity compared to low-viscosity materials such as cookie dough. But the high temperatures and pressures found inside Earth are capable of altering and deforming these rocks. Over geological timescales (i.e. many millions of years), the mantle behaves like a ductile material that moves at a rate of a few cm per year. Convection is driven by variations in heat and density where the speed of convection is controlled by viscosity. An interesting thought experiment is how long it would take to recycle subducting oceanic lithosphere. If we assume a velocity of 10cm/yr, it would take nearly 60 million years for a piece of lithosphere to travel from the surface to the core-mantle boundary and back. This is a rather short time in Earth's 4.5 billion year history.

Exactly what drives plate tectonics on Earth's surface is still a major research topic. It was postulated initially that the convecting mantle drags the plates along the surface and therefore asthenosphere drag was suggested as the major driving mechanism. However, recent research favors **slab pull as the major force**, with ridge push a close second.

Lithospheric Plates – A Closer Look

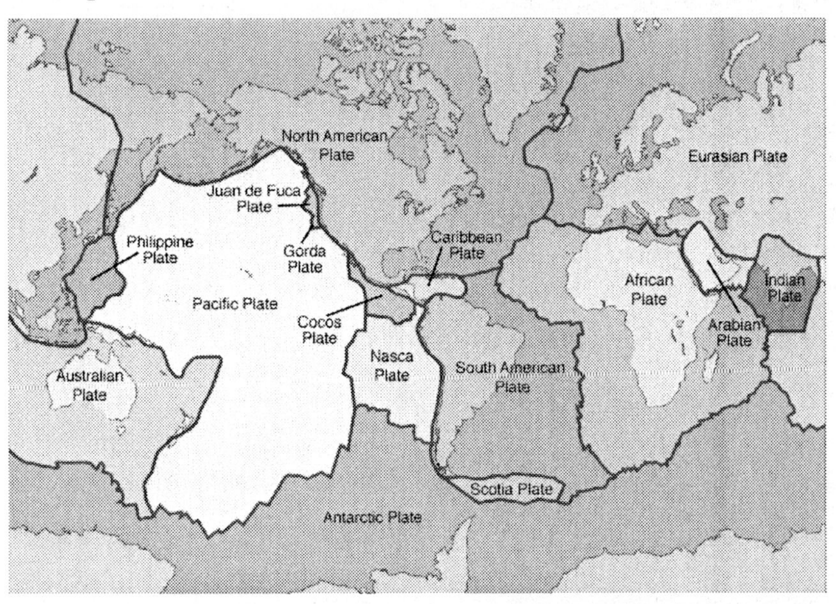

Figure 4.18 The 12 major tectonic plates. There are also 4 smaller plates (Caribbean, Scotia, Juan de Fuca and Gorda plates). Anatolia is sometimes shown as an additional plate. (source: modified from USGS[4]/wikipedia).

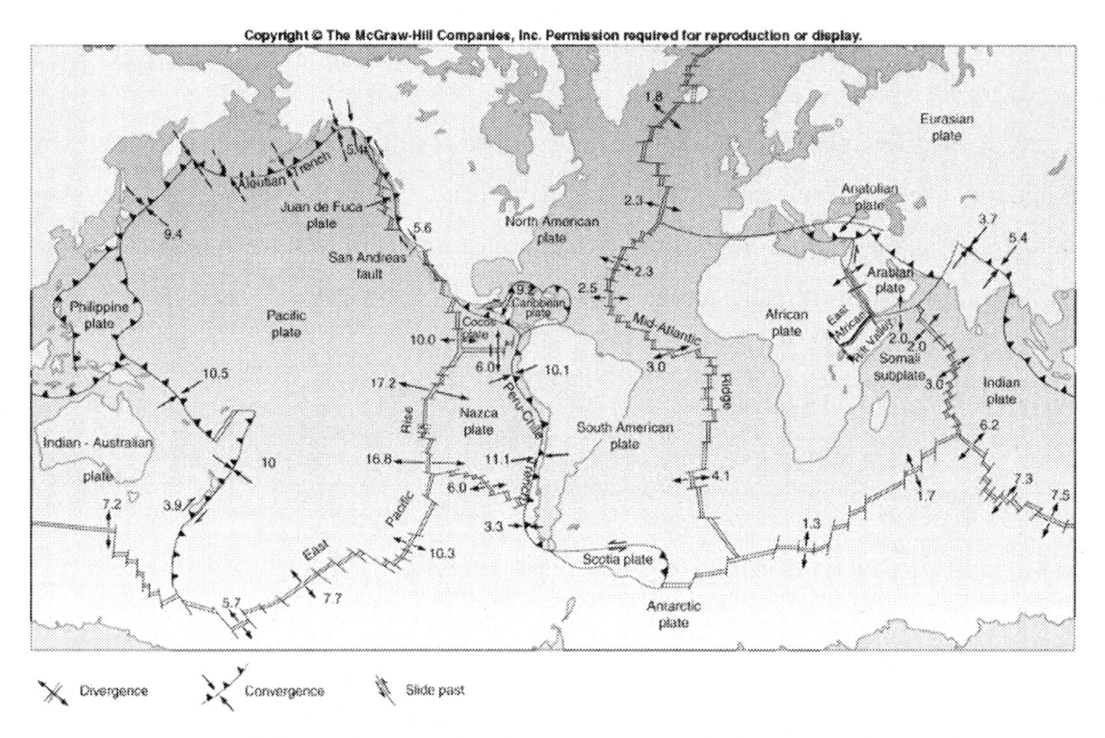

Figure 4.19 Map of the major tectonic plates together with the type and rate of motion along plate boundaries. Motion is labeled in cm per year. (source: AB)

Scientists identified 12 major tectonics plates (Fig. 4.18) that move about on Earth's surface at a rate of a few cm per year (Fig. 4.19). There are also a number of smaller plates – the Caribbean, the Scotia, the Juan de Fuca and the Gorda plates – and some references list yet smaller plate fragments as microplates.

Most plates consist of both continental and oceanic lithosphere, as shown in Fig. 4.18, and the depiction of one purely continental plate and three purely oceanic plate in Fig. 4.16 is somewhat idealized. When two plates converge, the denser one is forced to sink. In the case of a continent-ocean collision, the oceanic plate subducts, while in an ocean-ocean collision, the older, denser of the oceanic plates sinks. The subduction of a particular plate stops when all old oceanic lithosphere is consumed. The movement of the plates subsequently rearranges. For example, when the colliding plate remnants are both continental, then further convergence leads to mountain building as is found in the Himalayan mountain belt (Case Study 1). Mountain belts also form along the overriding continental plate in an ocean-continent collision as is found along the Andean mountains in South America. A subduction can even transform into a completely new regime in which plates move sideways past each other (Case Study 2). This regime is called a **transform plate boundary**. The different types of plate boundary are discussed in more detail in the next section. On a global scale and taking all plate movements into consideration, new lithosphere forms at the same rate as old lithosphere is consumed, so that Earth's overall size does not change.

Plate Tectonics – The Grand Unifying Theory

With hypotheses now in place on how new plate material forms along mid-ocean ridges and how old oceanic lithosphere is consumed along subduction zones, the concept of plate tectonics provides a mechanism for continental drift. It explains why continents move, why new oceanic lithosphere is found along mid-ocean ridges, why old oceanic lithosphere occurs along ocean trenches and why no oceanic lithosphere is older than 200 million years. It also explains many other features found on Earth's surface such as mountain belts.

The theory of plate tectonics also explains why some of the most devastating natural disasters, including earthquakes and volcanic eruptions occur at certain places but not at others. The vast majority of both earthquakes and volcanism occurs along mid-ocean ridges and subduction zones or, in more general terms, along plate boundaries. Mountain building, the formation of continental rifts and depressions, earthquakes and volcanism are the result of plates interacting with each other. Since the theory of plate tectonics explains all of these phenomena, it is often called the grand unifying theory of geosciences.

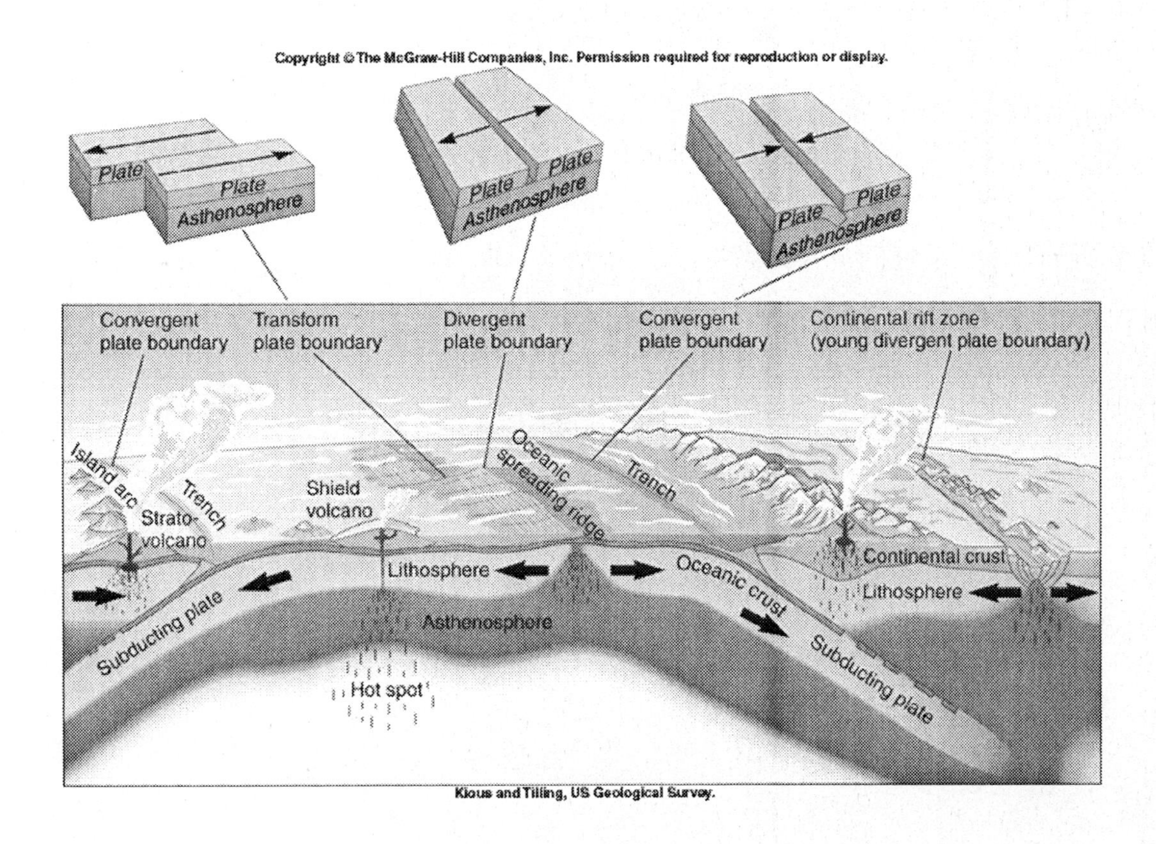

Figure 4.20 Schematic view of tectonic plates with divergent, convergent, and transform boundaries. Shield volcanoes form over oceanic hot spots while strato volcanoes form along subduction zones. (source: AB)

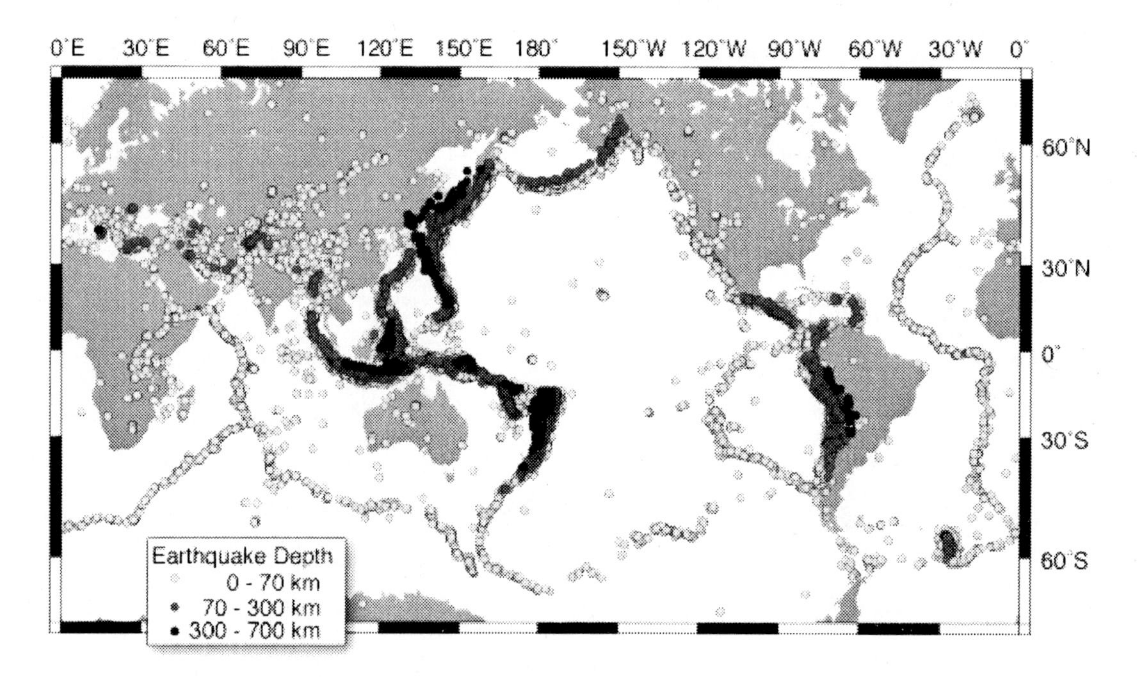

Figure 4.21 Earthquakes with magnitudes greater than 5.0, during a 15-year period. Most earthquakes occur along plate boundaries (all three types). Earthquakes occur in three depth ranges which are defined by the US Geological Survey (USGS) as follows(3): shallow events are located between 0 and 70 km depth; intermediate-depth events occur between 70 and 300 km depth; deep events occur below 300 km depth. Earthquakes occur to depths of about 700 km. Earthquakes deeper than 70 km occur only along convergent plate boundaries.

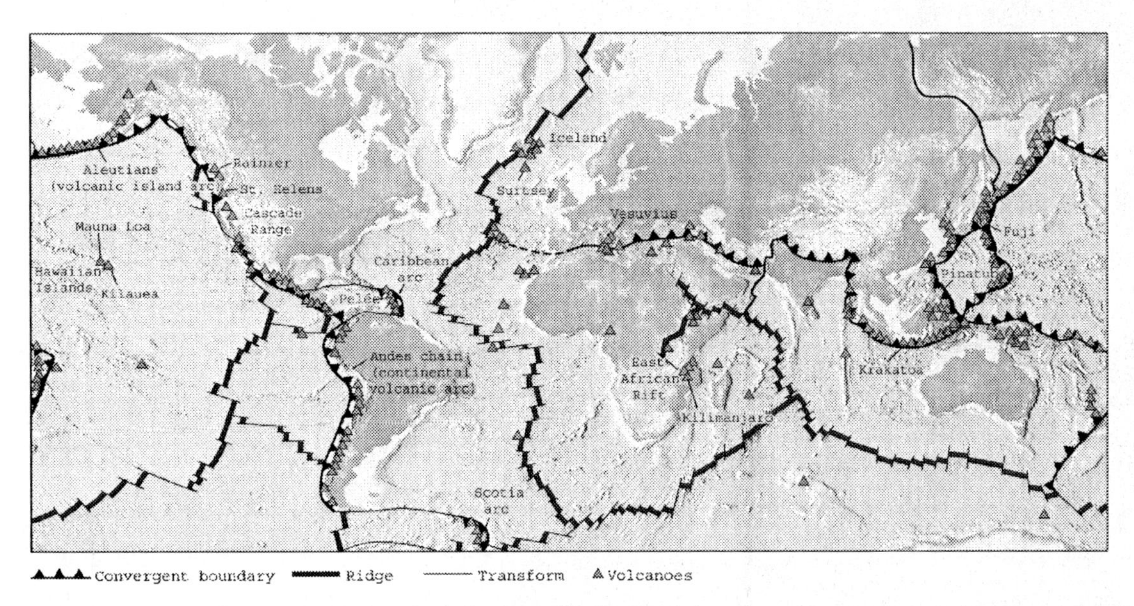

Figure 4.22 Volcanoes can be found along divergent plate boundaries, e.g. mid-ocean ridges and continental rifts. Volcanoes are also found along subduction zones and at hot spots, but not along continental collision zones and transform boundaries. (source: ME)

4.5 Plate Boundaries and Hot Spots

Plate Boundaries, Earthquakes and Volcanoes

Scientists distinguish between three different types of plate boundaries (Fig. 4.20). Plates move away from each other on divergent plate boundaries. Rifting occurs on these boundaries along so-called spreading centers. Mid-ocean ridges are the prime examples of this type of plate boundary. Continental rifts such as the East African rift are also divergent plate boundaries though they may not yet separate major plates. Some continental rifts also stopped after some time so that plate separation never occurred, and the rifts became “failed rifts”. Earthquakes occur along these boundaries because plates crack and move. Even ancient failed rifts can produce earthquakes because the rifting created faults along which these earthquakes now occur. This explains why some earthquakes seemingly occur far away from a plate boundary. Earthquakes occurring on divergent plate boundaries are typically small and shallow (Fig. 4.21). Rifting is usually associated with benign volcanism, especially along mid-ocean ridges, were lava oozes out of openings in the ocean floor. The rate at which two plates move apart (the absolute velocity of one plate plus that of the other) is the spreading rate. The highest spreading rates (about 17 cm per year) can be found along the East Pacific Rise (Fig. 4.19). In a map, a spreading center or rift is marked by a double line and double-sided arrows that mark the direction of spreading.

Plates collide at **convergent plate boundaries**. Scientists distinguish between ocean-ocean subduction zones (e.g. Aleutians, Japan, Tonga-Fiji, parts of Indonesia), ocean-continent subduction zones (e.g. Nazca/South America, Juan de Fuca/North America, Central America, New Zealand) and continent-continent collision zones where no subduction occurs (e.g. Himalayas, Alps, Zagros). Earthquakes along subduction zones occur at all depths down to a maximum depth of about 700 km (432 mi) (Fig. 4.21). Many of the largest observed earthquakes occur along subduction zones and cause tremendous damage. Many of these large events generate devastating tsunami that cross entire ocean basins to cause damage even far away from the earthquake. Such was the case after the 26 December 2004 Sumatra-Andaman earthquake when the tsunami crossed the Indian Ocean and killed people along Africa's east coast. Another example is the 22 May 1960 Valdivia, Chile earthquake. The largest recorded earthquake sent off a tsunami so powerful that it devastated Hilo on the Island of Hawaii, over 10,000 km (6200 mi) away. Often, violent volcanism can be found on the overriding plate along subduction zones (Fig. 4.22). As mentioned above, an exception is continent-continent collision were no volcanism is found. There is no mechanism for melting rock in these locations and the crust is too thick for magma to migrate to the surface. In a map, a subduction zone is marked by a barbed line. The arrows, which show the direction of subduction, are attached to the subducting plate (Fig. 4.19).

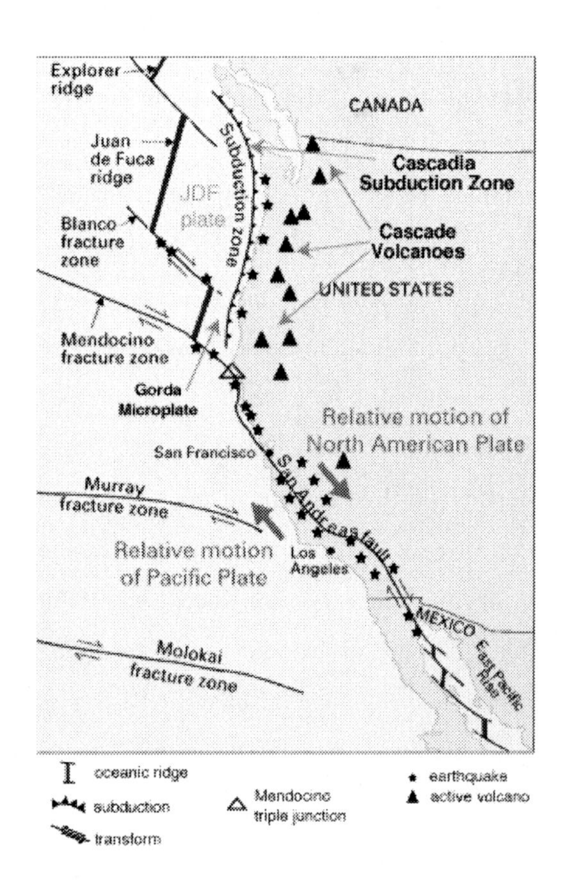

Figure 4.23 Schematic view of tectonic boundaries along the U.S. West Coast. (source: modified from USGS)

Transform boundaries form the third principal type of plate boundary. Plates scrape past each other and cause earthquakes, e.g. along the San Andreas Fault, Dead Sea Fault and North Anatolian Fault. Earthquakes occurring along these boundaries are typically small to large in size (Fig. 4.21). Since they are usually shallow, the larger earthquakes can cause significant damage. Volcanism is absent (Fig. 4.22) because transform boundaries are not associated with heat anomalies or other mechanisms that could melt rock. A transform boundary is called transform because it typically connects one type of boundary with another type. For example the San Andreas Fault (SAF) connects mid-ocean ridge segments in the Gulf of California in the south with the Cascadia subduction zone in the north where the Juan de Fuca plate subducts beneath the North American plate (Fig. 4.23; see also case study 2). In a map, a transform boundary is marked as a simple line. Arrows on each plate mark in which direction they move relative to each other (Fig. 4.19).

In concept figures, plate boundaries are often drawn with plate motions perpendicular to them. In the real world, this is often not the case. For example, plates can subduct obliquely so that in addition to a plate going down it can also slide sideways past the other plate. In another scenario, when a transform boundary is bent such as the southern segment of the San Andreas Fault, then plates sliding past each other exert compression across the bend (i.e. the Big Bend east of Los Angeles in Fig. 4.23). This can result in oblique motion and associated complex earthquakes that are otherwise typically found at subduction zones. We will revisit this topic in Chapter 5.

In some locations, three plates meet to form a triple junction. Three plate boundaries end at such a point. They may be of the same type but they are usually different. Some combinations are physically impossible such as three pure transform boundaries. At the Mendocino triple junction, the Pacific, the American and the tiny Gorda plates meet (Fig. 4.23). The San Andreas Fault approaches from the south, the Mendocino Fracture zone (which is also a transform boundary) from the west and the Cascadia subduction zone from the north.

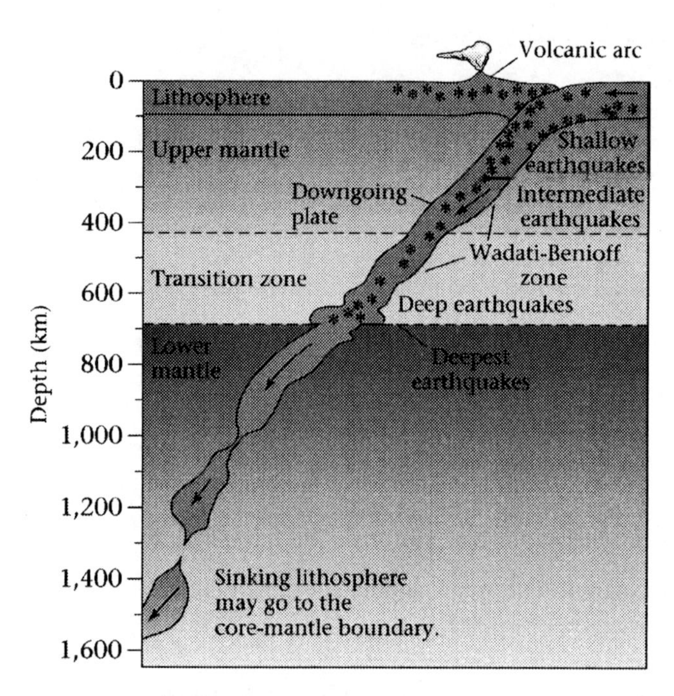

Figure 4.24 The Wadati-Benioff zone is the down-dipping band of earthquakes that occur in and next to a subducting oceanic plate. (source: ME)

Earthquakes in the Wadati-Benioff zone

Subduction zone earthquakes are unique in that they can occur in a wide depth range, as indicated in Fig. 4.21. Shallow earthquakes, with depths to about 70 km (43 mi), align with deep-sea trenches such as the Mariana Trench southeast of Japan, the Aleutian Trench in the North Pacific Ocean or the Peru-Chile Trench. But deeper earthquakes project behind the trench on the overriding plate where earthquakes project a farther away from the trench the deeper they are. A cross-section across the subduction zone shows that the earthquake locations trace the down-going slab (Fig. 4.24). The down-dipping band of seismicity is the **Wadati-Benioff zone** which was named after two seismologists who discovered this zone independently: Kiyoo Wadati of the Japan Meteorological Agency and Hugo Benioff of the California Institute of Technology.

Shallow earthquakes occur on both the subducting as well as the overriding plate in response to the compressional forces on the surface. Farther down, earthquakes occur within the subducting slab. Recent research shows that many intermediate-depth earthquakes (to a depth of about 300 km) occur near the top of the slab but deeper earthquakes happen in the middle of the slab. Even though Wadati-Benioff zones are strictly tied to subducting slabs, a similar pattern of seismicity (the distribution of earthquakes) can be observed on major continent-continent collision zones. In the case of the Himalayan collision, it turns out that subduction occurred prior to the collision, and the seismic activity now traces the last remnants of slabs that are still present in the upper mantle (case study 1).

Earthquakes occur only to a certain depth near 700 km (430 mi). The pressure and temperature at greater depth is so high that the slab is no longer brittle but ductile and so cannot break (also called brittle failure) or scrape along an existing fault (also called frictional sliding) against adjacent mantle to cause so-called tectonic earthquakes. In fact, the physical mechanism of very deep earthquakes below about 500 km (310 mi) is very different from shallower ones. At these depths, the crystal lattice of mantle minerals compresses in response to the surrounding pressure, and the mineral undergoes a phase transition. This phase transition occurs suddenly and causes earthquakes. Earthquakes occurring at such great depths usually do not cause any damage on the surface above but can nevertheless be felt far away, depending on how seismic energy is channeled through the Earth. Such a curious event was the magnitude-8.2 earthquake that occurred on 9 June 1994 at 631 km (390 mi) depth in Bolivia where the Nazca plate subducts beneath the South American plate. This earthquake caused little damage locally but was felt 7000 km (4350 mi) away in Canada.

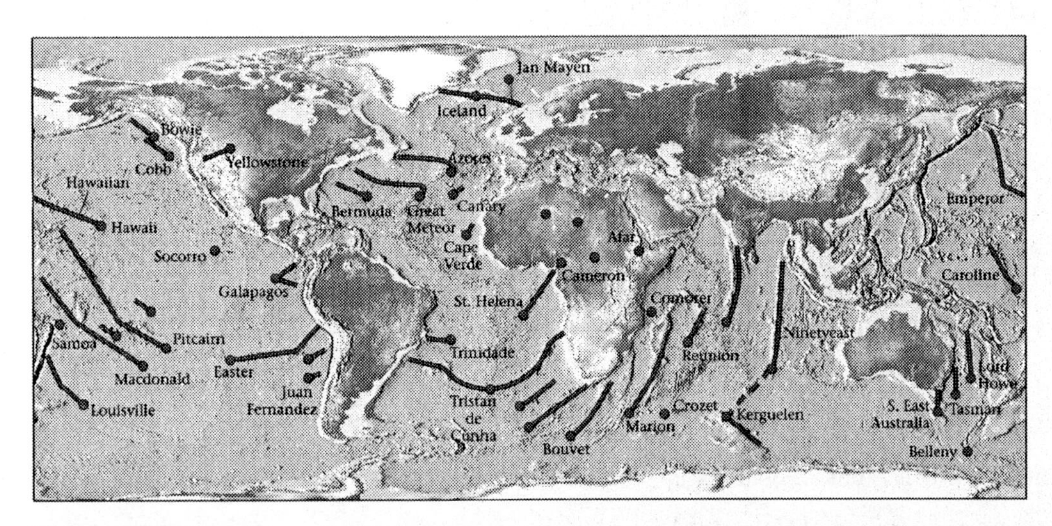

Figure 4.25 Locations of selected hot spot volcanoes. The tails represent hot spot tracks where the dots mark the youngest end. The older volcanoes along the tracks are extinct. Dashed tracks are places where the track was broken by seafloor spreading. (source: ME)

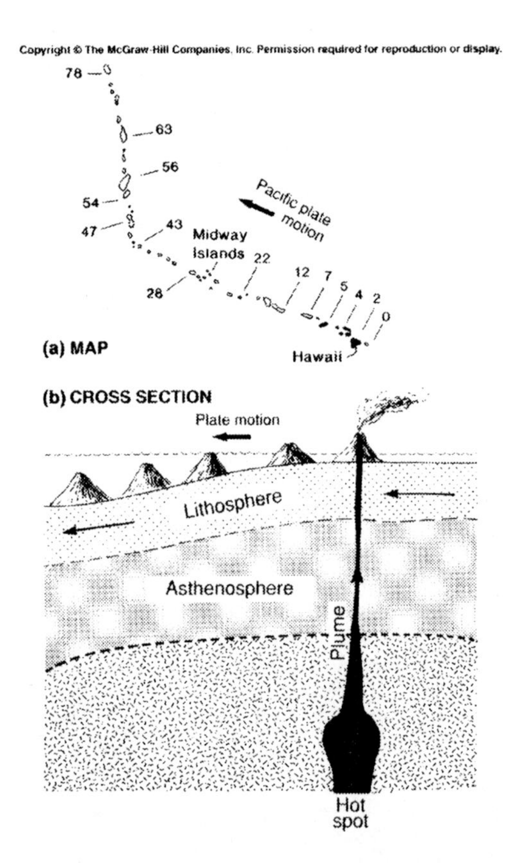

Figure 4.26 A deep mantle plume feeds hot spot volcanism on the surface. (a) map of progressing ages of volcanoes along the Hawaii-Emperor seamount chain. (b) the mechanism to explain this age-progressive volcanism. A plume feeds present volcanism. As the plate moves away from the plume, the volcano becomes extinct and a new one forms above the plume. Moving along with the subsiding oceanic plate, the extinct volcanic island submerges and becomes a seamount. (source: AB)

Hot Spots and Mantle Plumes

Not all volcanoes are located along plate boundaries. In fact, some of the best known isolated groups of volcanoes, such as those on the Hawaiian Islands or the Society Islands in the South Pacific are far away from any plate boundary (Fig. 4.22). The concept of plate tectonics therefore had to be augmented to explain this hot spot volcanism. Tuzo Wilson first described in 1963 that some ocean basins contain long chains of seamounts (extinct and submerged volcanic islands) that end at hot spots. The chains of active and extinct volcanoes are also called hot spot tracks. Many of these chains are parallel such as the Hawaiian, the Society, the Tuamotu and the Gulf of Alaska seamount chains in the Pacific Ocean. They are roughly aligned with the direction of plate motion of the Pacific plate (Fig. 4.25). Seafloor dredging revealed that volcanic rocks from the seamounts get progressively older along the seamount chain, away from the currently active volcanoes (Fig. 4.26a).

In 1971, Jason Morgan therefore postulated that this type of volcanism is fed by sources anchored deep in the mantle called mantle plumes. These plumes are relatively stationary over time, while the plates above move past them (Fig. 4.26b). As an active volcano sits above the plume, it is fed by magma transported upward through the plume conduit. But when the plate continues to move, that volcano moves away from the plume and becomes extinct. As it cools and ages with the lithosphere around it, the volcanic island then subsides below sea level to become a seamount. In the meantime, the plume "burned" a new pathway through the lithosphere to the surface where a new spot now moved above the plume, forming a new volcano. This process repeats itself to produce a whole chain of seamounts and volcanic islands. The progressively older volcanoes thereby point along the plate motion direction. The mantle plume theory explains intra-plate hot spot volcanism in the ocean, such as in Hawaii and the Kerguelen islands, but also on continents such as Yellowstone in North America and several others in Africa. Some of these very productive hot spots are associated with numerous but relatively benign volcanic eruptions that have become major tourist attractions such as Kilauea in Hawaii and Mt. Etna in Sicily.

As convincing as the plume theory appears, it also gives rise to much debate and ongoing research today. It turns out that it is quite difficult to find the 'smoking gun' scientific evidence that hot spot volcanism is in fact fed by sources located deep within the mantle. In 1973, Don Turcotte proposed an alternative explanation for island chain volcanism: plates moving from the equator toward the poles on a slightly oblate Earth encounter decreasing curvatures along the surface. The plates start to crack as they flatten, thereby allowing molten mantle rock to come to the surface to fed volcanoes. Progressing cracks in a plate would also explain age-progressing volcanism. Some scientists favor this explanation still today.

References and Websites

1. AB
2. ME
3. USGS at http://earthquake.usgs.gov/learn/topics/seismology/determining_depth.php
4. USGS Education and Outreach website on earthquakes: http://pubs.usgs.gov/gip/earthq1

Other Recommended Reading

Wegener, A., 2011. The origin of Continents and Oceans (translated by John Biram), Dover Publications, 272 pp.

Oreskes, N., 2003. Plate Tectonics: An Insider's History of the Modern Theory of the Earth, Westview Press, 448 pp.

Young, G. and Ed, M.S., 2008. Investigating Plate Tectonics: Earth and Space Science, Teacher Created Materials, 32 pp.

Cox, A. and Hart, R.B., 1986. Plate Tectonics: How it Works, Wiley-Blackwell, 416 pp. Somewhat pricey; for the mathematically/geometrically inclined who like to dig deeper into plate motion.

Chapter 5: Earthquake Seismology

In the last chapter, we learned that earthquakes do not occur everywhere on Earth's surface but primarily along plate boundaries. There are three principal types of earthquakes that can usually be associated with the three principal types of plate boundaries. In this chapter we learn how to describe the location, type and size of earthquakes, how often they occur along certain faults and what makes earthquakes go. This will help understand why some earthquakes are much more destructive than others despite their relatively small size. This chapter also includes a brief introduction to the four principal types of seismic waves. These waves can be categorized into two groups, body waves and surface waves. The latter waves are usually the most destructive waves and also the ones that can be felt over greater distances.

5.1 An Earthquake Along a Fault

The Fault Plane

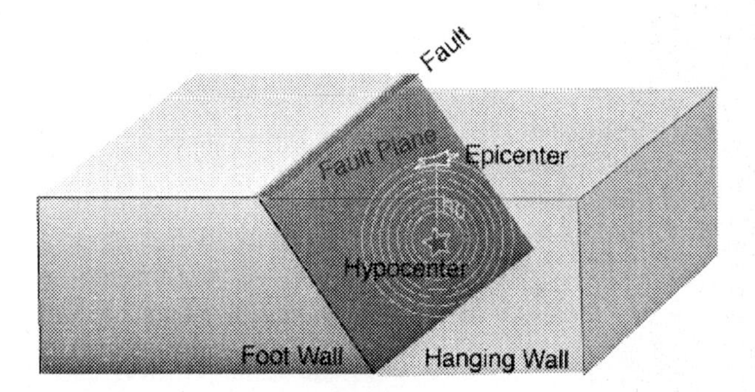

Figure 5.1a Some technical terms describing the area where an earthquake occurs.

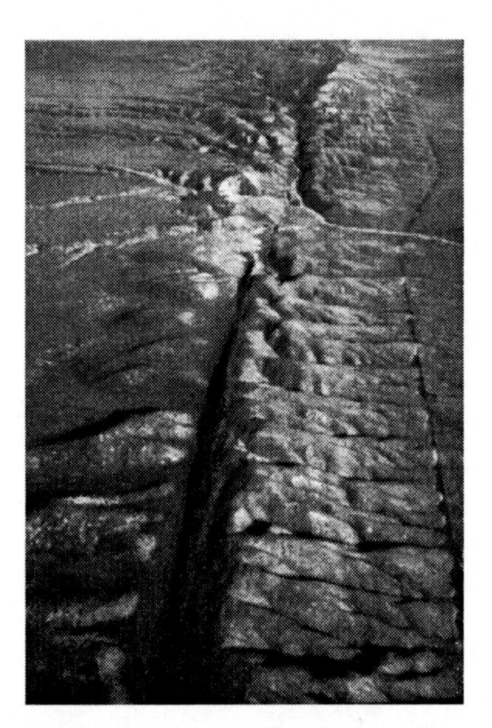

Figure 5.1b Arial view of the section of the San Andreas Fault extending through the Central Californian Carrizo Plain. (source: USGS [4]/Wikimedia)

Earthquakes are often described as the process occurring when a plate or a crustal block breaks during brittle failure, as shown in Fig. 2.1, but witnessing a new break is actually quite rare. Usually, earthquakes occur along pre-existing weak zones within Earth's crust in response to stress building up. In special cases, earthquakes can also occur in the mantle, and the unusual physics of very deep earthquakes was described in the last chapter. The fault plane describes the weak zone in which an earthquake occurs (Fig. 5.1a). It is the contact zone between two blocks (or plates) that are capable of moving relative to each other. The fault (or fault trace) is the surface expression of the fault plane. Geometrically, it is the line along which the fault plane intersects Earth's surface.

An expert can trace an earthquake fault in the field, e.g. when different rock types are found on each side of the fault, or their appearance is different, or structures like fences are offset. A change in vegetation often marks a fault through which water seeps. But to a non-expert, a fault can sometimes be hard to identify and define. To trace a fault, even experts sometimes need the help of imaging techniques. This is especially the case when no earthquake has occurred recently, and sediments cover the fault. Scientists dig trenches across faults to find out how and by how much the joining blocks moved over time. This is particularly important for the reconstruction of earthquakes that occurred before record-keeping and instrumental monitoring began. Sometimes, the sheer size and impact of a fault cannot be grasped by viewing it on the ground, but instead is best seen from above. For example, in Central California, the San Andreas Fault breaks through the Carrizo Plain leaving behind an impressive scar (Fig. 5.1b).

Adopting expressions from the mining industry, the two blocks on either side of the fault plane have different names. The block (or wall) above the fault plane is call the hanging wall while the one below the fault plane is the foot wall. The point where the earthquake initiates is the hypocenter, also called focus or source. Its projection on the surface is the epicenter. The epicenter therefore does not carry information on the depth of the earthquake, only its geographical position.

The initial rupture of an earthquake is called nucleation. The exact size of the nucleation zone (the area in that nucleation takes place) is currently unknown but an area of intense research. From there, the rupture propagates outward from the hypocenter within the fault plane until it eventually stops. The earthquake is then over. However, the seismic waves generated by this earthquake will propagate through the solid Earth and along its surface, and therefore the shaking will be felt some distance away and some time after the initial rupture took place. This is the reason why people in San Diego, for example, can feel an earthquake that occurred in the desert 100 km (60 mi) to the east but not in San Diego itself. Only part of the fault plane actually ruptures during an earthquake, not the whole plane. The rupture area describes the rupturing part of the fault plane. The rupture area is typically small for small earthquakes but involves larger sections of the fault plane in larger earthquakes. The very large magnitude-9.2 Sumatra-Andaman earthquake on 26 December 2004 ruptured over a nearly incredible length of 1600 km (1000 mi). Smaller, magnitude-8 earthquakes rupture for 100 km or so and smaller earthquakes for much less. Some sections of a fault or fault plane can also be locked and not move. These sections are often areas of intense monitoring because locked areas are likely places of increased stress buildup, thereby increasing the likelihood for being the locus of the next earthquake.

The source time (or event time) marks the time of the initial rupture of an earthquake while the source duration is a measure for how long the earthquake lasted. Smaller earthquakes have shorter source durations of up to a few seconds while very large earthquakes can shake for several minutes. The great 26 December 2004 Sumatra-Andaman tsunami earthquake lasted for an incredible 10 minutes. The source time is not to be confused with the duration of the shaking people feel farther away from the source. Waves sent out by an earthquake disperse or spread out over time so that the felt shaking can last longer, though diminished, than near the source.

The Movement on a Fault

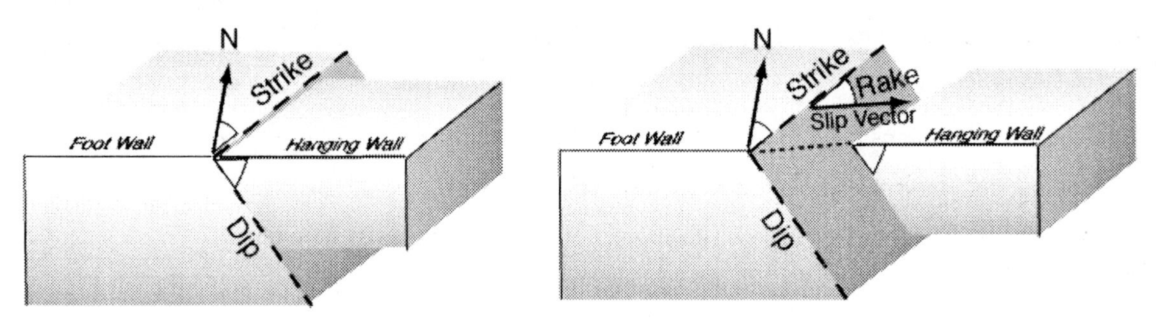

Figure 5.2 The three angles (strike, dip and rake) that define the orientation of a fault plane as well as the movement on the fault during an earthquake. The example shown here has a strike of about 30°, a dip of 45° and a rake of -25°.

Two angles define the orientation of a fault plane, the strike and the dip (Fig. 5.2). The strike is the direction of the fault relative to geographic North where the angle is counted clockwise, from 0° to 360°. To remove the 180°-ambiguity, e.g. distinguish between a strike of 10° and 190°, the hanging wall has to be to the right as one looks down the strike direction. The dip angle, between 0° and 90°, is measured from the horizontal and describes how steeply the fault plane inclines. A horizontal plane has a dip of 0° while a vertical fault plane has a dip of 90°. The slip vector describes how the hanging wall moves with respect to the foot wall. The slip vector lies in the fault plane and its orientation is measured clockwise with respect to the strike. The length of the slip vector defines the amount by which the hanging wall moved, the slip or displacement. Very large earthquakes can have slips of several meters. The 26 December 2004 Sumatra-Andaman earthquake had a horizontal slip of 15 m. Magnitude 8 earthquakes often have a horizontal slip of up to 5 m. The angle between the slip vector and the strike is called the rake. Measured counterclockwise, the rake goes from 0° to 180° if the hanging wall moves upward with respect to the footwall. Measured clockwise, the rake goes from 0° to -180° if the hanging wall moves downward. The rake typifies the type of earthquake (next section). The word slip is sometimes used loosely but falsely instead of rake.

5.2 The Three Earthquake Types and Plate-Boundary Equivalents

Scientists distinguish between three principal types of earthquakes: normal, reverse (and thrust) and strike-slip events. They can be likened to the movement along the three principal types of plate boundaries though, in detail, the occurrence of specific earthquakes on specific boundaries as well as the earthquake mechanisms themselves may be quite complex. Large earthquakes with magnitudes greater than 6.5 or so can generate tsunami if they occur in the oceans and rupture the seafloor through vertical movement.

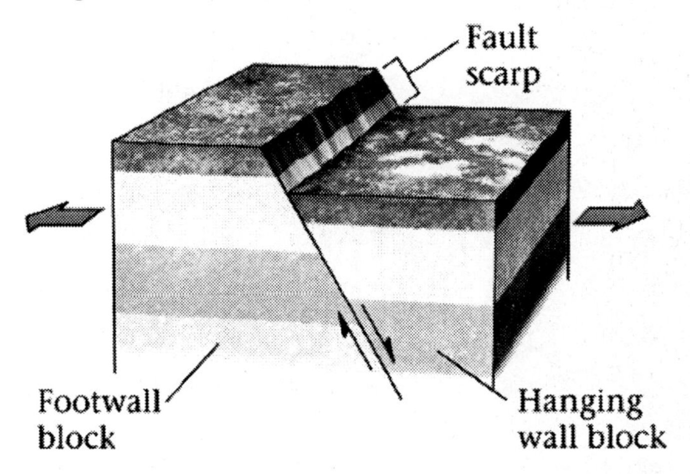

Figure 5.3 A normal fault. During an earthquake, the hanging wall moves down relative to the foot wall. An earthquake with a rake of -90° is a pure normal event. Fig. 5.2 shows an oblique-slip earthquake along a normal fault. Earthquakes on normal faults can be likened to the movement on divergent plate boundaries. (source: ME).

In normal faults and normal earthquakes, the hanging wall moves downward with respect to the foot wall (Fig. 5.3). In a pure normal event, also called dip-slip event, the rake is -90° and the hanging wall goes straight down. For all other rakes between 0 and -180° the general movement is an oblique normal slip. A normal earthquake can be likened to the movement along a divergent plate boundary because the two blocks involved move apart in an extensional stress regime. Normal events are typically small and occur at shallow depth. Normal events occur along mid-ocean ridges and continental rift zones. They can sometimes occur in the accretionary prism along subduction zones as well. If these events are large enough, they can generate a tsunami.

Table 5.1 The 16 Largest Earthquakes in the World since 1900[9]

Rank	Location	Date	Magnitude	Plate Boundary
1	Valdivia, Chile	22 May 1960	9.5	Nazca (plate) subduction
2	Prince William Sound, Alaska "Good Friday Earthquake"	28 Mar 1964	9.2	Pacific subduction
3	Sumatra-Andaman, Indonesia	26 Dec 2004	9.1	Indian subduction
4	Tohoku, Japan	11 Mar 2011	9.0	Pacific subduction
5	Kamchatka	04 Nov 1952	9.0	Pacific subduction

6	Maule, Chile	27 Feb 2010	8.8	Nazca subduction
7	Off the Coast of Ecuador	31 Jan 1906	8.8	Nazca subduction
8	Rat Islands, Alaska	04 Feb 1965	8.7	Pacific subduction
9	Northern Sumatra, Indonesia	28 Mar 2005	8.6	Indian subduction
10	Assam, India - Tibet	15 Aug 1950	8.6	India/Asia collision
11	Andreanof Islands, Alaska	09 Mar 1957	8.6	Pacific subduction
12	Southern Sumatra, Indonesia	12 Sep 2007	8.5	Pacific subduction
13	Banda Sea, Indonesia	01 Feb 1938	8.5	Pacific/Indian subduction
14	Kamchatka	03 Feb 1923	8.5	Pacific subduction
15	Chile-Argentina Border	11 Nov 1922	8.5	Nazca subduction
16	Kuril Islands	13 Oct 1963	8.5	Pacific subduction

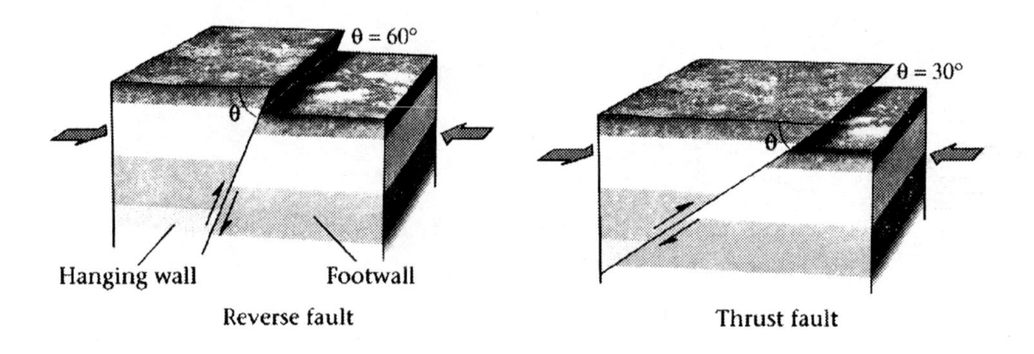

Figure 5.4 Left: A reverse fault. During an earthquake, the hanging wall moves up relative to the foot wall. An earthquake with a rake of 90° is a pure reverse event. Reverse faults typically occur along convergent plate boundaries. Right: A thrust fault. Same as a reverse fault but with dip angle of less than 45°. (source: ME).

In reverse faults and reverse earthquakes, the hanging wall moves upward with respect to the foot wall (Fig. 5.4). In a pure reverse-faulting event, the rake is 90° and the hanging wall goes straight up. For all other rakes between 0 and 180°, the general movement is an oblique reverse slip. A reverse earthquake can be likened to the movement along a convergent plate boundary because the two moving blocks approach each other in a compressional stress regime. Scientists speak of thrust faults and thrust earthquakes when the dip angle is less than 45°. Most earthquakes occurring along subduction zones are reverse or thrust events. They occur in a large range of depths in the Wadati-Benioff zone. Since some subduction zones are very long, earthquakes can grow to humongous events such as the 22 May 1960 Valdivia, Chile event, the 26 December 2004 Sumatra-Andaman event, or the 11 March 2011 Tohoku event in Japan. Such events are called mega-thrust events. With large amounts of vertical movement that rupture the seafloor, such earthquakes generate devastating tsunami. In fact, the 16 largest earthquakes in the last 90 years all occurred along convergent plate boundaries, and only one of them did not rupture along a subduction zone (Table 5.1).

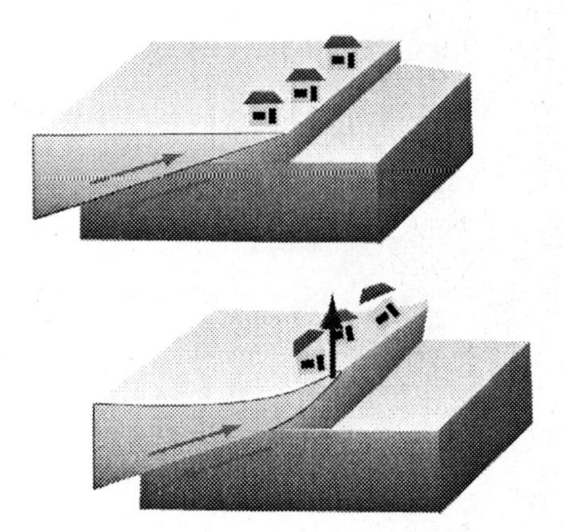

Figure 5.5 A shallow thrust fault is a thrust fault with a very small dip angle. The dip of the shown example is about 20°. At such faults, the edge of the hanging wall experiences the largest ground accelerations during an earthquake (arrow in the lower panel) and destruction there is most severe.

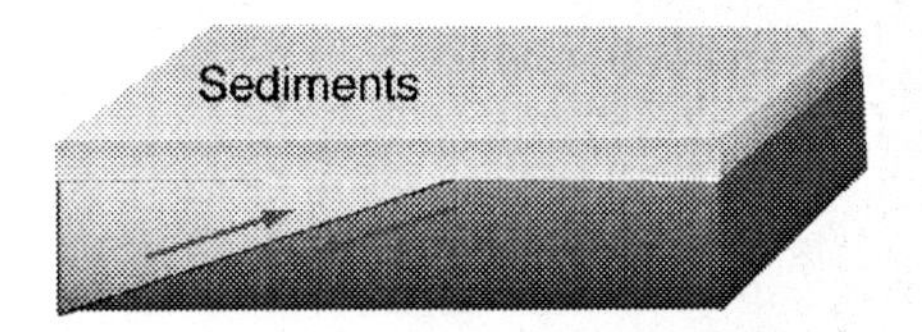

Figure 5.6 A blind shallow thrust fault. These are extremely dangerous faults because sediments cover the fault. It is difficult to map the fault and assess earthquake hazards using simple field methods alone. Expensive seismic imaging, however, helps reveal the crustal structure below the sediments. There are many such faults in the Los Angeles (sedimentary) Basin.

Some reverse faults and associated earthquakes can be particularly dangerous when the dip angle is very small (Fig. 5.5). During such **shallow thrust** events, the edge of the hanging wall can experience large upward accelerations which are often measured in terms of Earth's gravitational acceleration, g. It is not unheard of that accelerations during some shallow thrust events approach or even exceed g. During the 17 January 1994 Northridge, CA earthquake near the edge of the Los Angeles Basin, the maximum ground acceleration measured was $1.7g$, which suggests that things were literally thrown up in the air. This was one of the highest ever instrumentally recorded ground accelerations in an urban area in North America.

To scientists, the 1994 Northridge earthquake came as somewhat of a surprise because it occurred on a previously unknown fault. Seismic imaging later revealed this fault, the Pico fault, as a thrust fault covered by sediments. Such a fault is called a **blind thrust fault** (Fig. 5.6). They often remain unknown until they are either discovered by seismic imaging, or until an earthquake strikes. The seismic imaging in sedimentary basins can be quite expensive and usually involves either small explosions or, more commonly, the operation of several so-called vibrator trucks. The waves sent out by these "sources" are picked by countless geophones on long cables. One can imagine that such a program can be a logistical challenge in a big city such as Los Angeles.

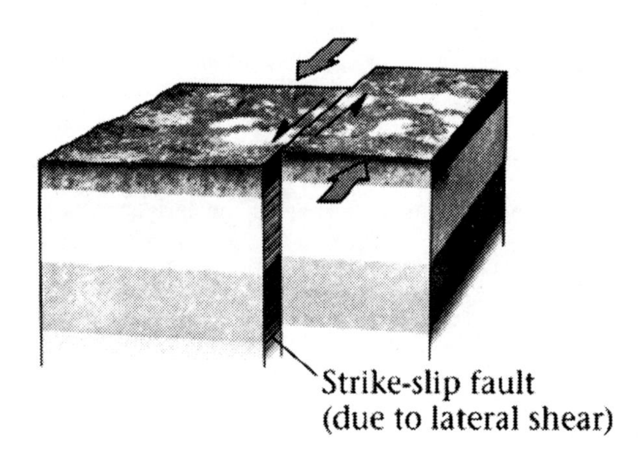

Figure 5.7a A strike-slip fault. During a strike-slip earthquake, the hanging and the footwall slide horizontally past each other. An earthquake with a rake of 0° or 180° is a pure strike-slip event. Strike-slip faults typically occur along transform plate boundaries. (source: ME).

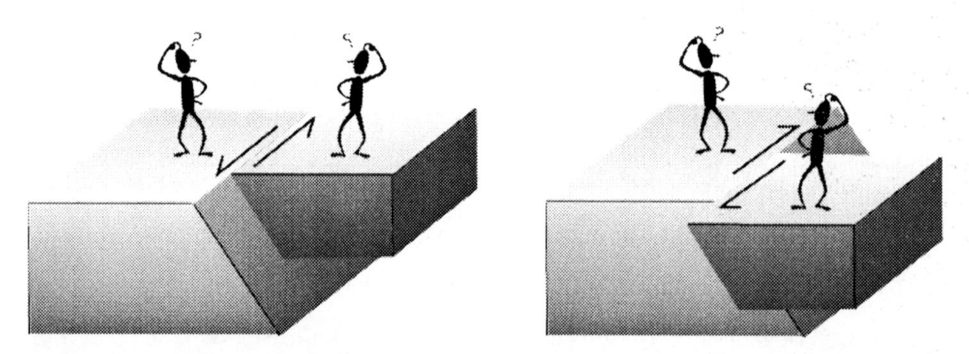

Figure 5.7b Right- and left-lateral strike-slip faults and earthquakes. Left: A person on a block facing the other one and watching it move to the left stands on a left-lateral fault. The rake is 0°. Right: A person watching the other block move to the right stands on a right-lateral fault. The rake is 180°. It does not matter on which side of a fault a person stands.

In transform boundaries, the dip of the associated fault is nearly vertical and the movement of adjacent blocks is horizontal. The fault is a strike-slip fault, and earthquakes with such movement are strike-slip events (Fig. 5.7a) in a lateral stress regime (or shear-stress regime). The rake is then 0° or 180°. A rake of 0° typifies a left-lateral strike-slip event (Fig. 5.7b) where the blocks move to the left relative to each other. A rake of 180° typifies a right-lateral strike-slip event. The San Andreas Fault is a right-lateral strike-slip fault where the Pacific Plate moves northward with respect to the North American plate. The North Anatolian Fault in Turkey is also a right-lateral strike-slip fault, while the Dead Sea Transform Fault in Israel and Jordan is a left-lateral strike-slip fault. In reality, the dip of the fault is rarely exactly 90° so that a hanging and a foot wall can be identified. If the motion is primarily horizontal, then the earthquake is still a strike-slip event. Concurrent small up or down movement will make it an oblique strike-slip earthquake, with a rake of within 20° or so to 0° or 180°.

Figure 5.8 Horizontal displacement from the 18 April 1906 San Francisco earthquake. About one kilometer northwest of Woodville on the E.R. Strain farm, this fence was offset by 3.3 m. On the Shafter ranch, a crevice momentarily opened wide enough to swallow a cow and then immediately

closed leaving only the cow's tail visible. In this area, the fault was a trench more than 2 m wide and filled with broken soil blocks. (source: NGDC/NOAA; photo: USGS).

Strike-slip earthquakes have a wide range in size but are typically not found among the largest earthquakes on Earth. They can nevertheless be very destructive as the 18 April 1906 San Francisco earthquake has shown. With a source depth of 10 km or even less this magnitude 7.9 and similar events are quite shallow and the horizontal offset can be several meters (Fig. 5.8). Most structures are not designed to withstand so much movement.

Strike-slip events occurring in the ocean generally do not generate tsunami because the horizontal motion on the seafloor does not displace any water above. However, locally, smaller tsunami may be generated by local submarine landslides or landslides entering the ocean that are triggered by the earthquake. Such was the case when the magnitude-7.0 earthquake on 12 January 2010 in Haiti produced mainly strike-slip motion but triggered a small submarine landslide and tsunami near the small fishing town Petit Paradis. The greater area of the earthquake has mainly strike-slip faults but later, more detailed research of this event suggested that the event may have occurred on a blind thrust fault associated with the main strike-slip fault system of the Enriquillo-Plantain Garden fault. This would make the general situation similar to the fault system of the 1994 Northridge earthquake where the earthquake did not occur on the San Andreas Fault but a blind thrust fault nearby, though the earthquake there was a thrust event and not a strike-slip.

In some convergent plate boundary systems, the curved geometry of subduction or collision zones generates complex stress regimes, thereby generating strike-slip faults and associated earthquakes. Such is the case in the Sumatra subduction zone, the Himalayas and the subduction arcs of the Caribbean plate (e.g. the Enriquillo-Pantain Garden fault).

5.3 Earthquake Sizes and Magnitude Scales

After the devastating 11 March 2011 Tohoku earthquake in Japan, news media in Germany, and perhaps elsewhere asked the question what if such a magnitude-9 earthquake happened in Germany. The short answer is that Germany does not have the plate boundary and associated fault system that could generate such an earthquake. There are earthquake faults however, and earthquakes with magnitudes 5.5 and above are not unheard of. In fact, an earthquake with an estimated magnitude of 6.7 destroyed much of the city of Basel, Switzerland on 18 October 1356, less than a decade after the plague ravaged much of Europe. To-date, this is the largest earthquake that has occurred in Europe north of the Alps. Of course, the discussion in the news media was not only about the earthquake but also about the associated destruction of nuclear power plants. Similar discussions also filled the news

COPYRIGHTED MATERIAL — DO NOT DUPLICATE, DISTRIBUTE, OR POST

in the San Diego area where some experts warn that earthquakes with magnitudes 7 or above are possible. Operators of the San Onofre nuclear power plant stress that they are prepared for such an event and the protecting wall is high enough to withstand a tsunami. A major concern that increases the likelihood for a large earthquake is that some of the off-shore faults may be connected. So what is it that makes earthquakes large?

Fault Geometry and the Size of an Earthquake

Instrument data from earthquakes currently indicate that very large earthquakes such as the mega-thrust events discussed above do not release their energy at once. They nucleate as a relatively small event somewhere along a fault and then grow as the rupture continues to move along this fault. For small earthquakes, the rupture typically lasts a few seconds or less but for very large earthquakes, the rupture can last several minutes. The longest-lasting recent earthquake was the 26 December 2004 Sumatra-Andaman earthquake that shook for an incredible 10 minutes! If the nucleation hypothesis applies to all earthquakes, the rupture has to stop at the end of a fault. In this case an earthquake can no longer grow, unless the rupture continues on a neighboring fault. The size of an earthquake therefore depends on the geometry of the fault system. More specifically it depends on the rupture area within a fault plane, if we assume that the earthquake is confined to a single fault. The size of an earthquake then scales with the length of the fault, or more precisely the size of the rupture area within the fault plane, and the slip. Unlike longer faults, short faults are therefore unlikely to generate large earthquakes.

The largest earthquakes are therefore only likely to occur along the mega thrust faults in long subduction zones, such as those along Chile, in Sumatra, the Aleutian islands and the western rim of the Pacific ocean. There is also potential for such earthquakes in the Cascadia subduction zone along the North American west coast. In the San Diego area, the recent discussion of the potential for larger earthquakes focuses on two faults. In San Diego, the Rose Canyon fault (RCF) is known to exist from Downtown to La Jolla. The southward continuation of the RCF is currently not well known. In the north, the RCF continues off-shore north of Mt. Soledad, but its exact continuation northward is not well-known. From the north, the Newport-Inglewood fault comes south. The destructive magnitude-6.4 Long Beach earthquake on 10 March 1933 occurred on this fault. The current question focuses on whether these two faults connect. If so, then the area should prepare for a magnitude-7 earthquake and possibly a tsunami.

Discussions so far mentioned earthquake magnitude without specifying what magnitude is. Scientists use the moment magnitude extensively to describe the "size" of an earthquake. Until recently, the news media have primarily used the Richter scale to report earthquake sizes but this scale has disadvantages. One is that it is strictly applicable only in California. Sometimes, reports on magnitudes mingle with those of intensities. Starting with the latter, these three magnitude scales are described in the following sections. Scientists also use other magnitudes such as the body wave magnitude, M_b, and the surface wave magnitude, M_S, to classify earthquakes. The comparison between M_b and M_S is of particular importance to distinguish between earthquakes and explosions in the effort to monitor compliance with the global Comprehensive Test Ban Treaty (CTBT; see section 5.7) on nuclear bomb testing.

Table 5.2 The Modified Mercalli Intensity Scale[7]

Intensity	Description
I	Not felt except by very few people under particularly favorable conditions
II	Felt only by a few people at rest, especially on upper floors of buildings
III	Felt quite noticeably by persons indoors, especially on upper floors of buildings. Many people do not recognize it as an earthquake. Standing cars may rock slightly. Vibrations similar to the passing of a truck.
IV	Felt indoors by many, outdoors by few during the day. At night, some awaken. Dishes, windows, doors disturbed; walls make cracking sound. Sensation like heavy truck striking building. Standing cars rock noticeably.
V	Felt by nearly everyone; many awaken at night. Some dishes, windows broken. Unstable objects overturned. Pendulum clocks may stop.
VI	Felt by all, many frightened. Some heavy furniture move; a few instances of fallen plaster. Damage slight.
VII	Damage negligible in well-designed and constructed buildings; slight to moderate damage in well-built ordinary structures; considerable damage in poorly designed and built structures; some chimneys broken.
VIII	Damage slight in specially designed structures; considerable damage in ordinary substantial buildings with partial collapse. Damage great in poorly built structures. Fall of chimneys, factory stacks, columns, monuments, walls. Heavy furniture overturned.
IX	Damage considerable in specially designed structures; well-designed frame structures thrown out of plumb. Damage great in substantial buildings, with partial collapse. Buildings shifted off foundations.
X	Some well-built wooden structures destroyed; most masonry and frame structures destroyed with foundations. Rails bent.

XI	Few, if any (masonry) structure remain standing. Bridges destroyed. Rails bent greatly.
XII	Damage total. Lines of sight and level are distorted. Objects thrown into the air.

Mercalli Intensity Scale

The twelve-level Mercalli Intensity scale (Table 5.2) is the oldest scale and describes the effects of an earthquake. The Italian volcanologist Giuseppe Mercalli based it on an older ten-level scale. It was introduced in 1902 and revised several times by Mercalli and other scientists. Charles Richter improved it in 1956. It is known today and the Modified Mercalli scale, or MM. On a scale from I to XII, the Mercalli intensity describes what people feel (lower levels) and the damage caused (upper levels) during the shaking. Nothing is felt at intensity I, while total destruction is observed at intensity XII when the ground acceleration can reach or exceed 1*g*. Various references attach attributes to the levels such as slight (III), strong (VI) and intense (X). But the USGS does not list them in their general interest publications, so they are omitted here. Smaller earthquakes tend to be felt less than larger earthquakes, so smaller earthquakes tend to have lower Mercalli intensities. But strictly speaking, the size of an earthquake cannot be derived from the Mercalli scale as other factors largely influence the degree of damage. Poorly built structures tend to be damaged by relatively small earthquakes. Local geology, such as weak sediments can enhance the shaking of an earthquake and the damage it causes.

Figure 5.9 1755 copper engraving depicting Lisbon, Portugal in ruins and in flames after the earthquake on 1 November the same year (also known as "The Great Lisbon Earthquake"). A tsunami

overwhelmed the ships in the harbor. An estimated 60,000 people in Lisbon died from the earthquake and tsunami. (source: Wikipedia/Museu da Cidade, Lisbon)

This scale has been used extensively to describe the damage of historical earthquakes before instrumental recording began. For example, the 1 November 1755 Lisbon, Portugal earthquake (Fig. 5.9) killed an estimated 60,000 people in Lisbon alone, a quarter of its population at the time, 10,000 in Morocco and many more elsewhere[25]. In Lisbon that experienced nearly total destruction by the shaking but also subsequent fires, the intensity was between XI and XII. At the time, Lisbon was one of the largest cities in Europe, and its destruction had a major traumatic impact on the whole continent. The earthquake was located along a mega-thrust fault in the Atlantic Ocean 200 km west of Portugal and lasted 10 minutes. An enormous tsunami, as high as 30 m (98 ft) in some places, arrived 30 min after the earthquake and devastated coastal areas. It crossed the Atlantic Ocean and arrived in the Lesser Antilles about 7 hours later. People there reported that the sea first rose by more than 1 m, followed by large waves a few meters high. Not lastly from estimates of the tsunami and its height along coasts in Europe as far as Finland and across the Atlantic Ocean the earthquake is now estimated to have had a moment magnitude of 8.5-9.0.

Figure 5.10 The 18 October 1356 Basel, Switzerland earthquake. This earthquake has been the largest in Europe north of the Alps. Its moment magnitude has been estimated at about 6.7. (source: NISEE)

Over four centuries earlier, on 18 October 1356 at 10 pm local time, the ground shook in Basel, Switzerland in the most destructive earthquake in central Europe in historical times (Fig. 5.10). The shaking and destruction reached intensity X and is often likened with that of the 1906 San

Francisco earthquake though scientists agree that the magnitude was only around M_W=6.7-7.1[22]. Much of Basel was destroyed by the shaking and subsequent fires, and the imposing cathedral was heavily damaged during the main shock or several aftershocks. Due to sparse and incomplete records, it has been a challenge to investigate the sequence of events in detail. At least 200-300 people were killed but estimates range from 100 to 2000. The number 2000 seems improbable given the fact that the total number of residents was only 6000 at the time. The population of Basel had already been decimated by famine in 1347 and the Black Death in 1349. The Black Death alone killed half of Basel's population. A reported 40 to 80 castles were destroyed within a radius of 40 km (25 mi). Intensity VIII shaking and damage was observed as far as 100 km southwest of Basel. The bell tower of the cathedral in Berne, 75 km (47 mi) away to the south, was destroyed where intensity was estimated at VII. The earthquake was also felt in Frankfurt, 290 km (180 mi) to the north. Basel's current population is 170,000. It is uncertain how often a significant earthquake strikes this region but evidence from the analysis of cave stalactites suggests that the spacing between large earthquakes is between 500 and 3000 years[23]. A shorter spacing would make an imminent event likely in an area that has prospered from chemical and pharmaceutical industry and has several nuclear power plants.

Figure 5.11 The water front of Messina in Sicily, Italy after the devastating 28 December 1908 earthquake and tsunami. An estimated 72,000 people were killed in Messina and tens of thousands elsewhere. (source: Wikimedia)

Reaching intensity XI, another deadly earthquake in Europe killed over 40% of the population of Messina, Sicily (Fig. 5.11). The ground shook for 30-40 s and destruction was reported within a 300 km radius. Messina is located on the island of Sicily where it meets the tip of the "boot" of Italy in Calabria. Sicily is located on the African plate along a convergent plate boundary system with the Eurasian plate on which the rest of Italy is located. Recent research indicates that the earthquake occurred along a local normal fault. The associated tsunami reached heights of 6-12 m on the coast of both Sicily and Calabria. At a magnitude of 7.1, as has been estimated for the Messina earthquake, it is capable of generating a tsunami but scientists have recently postulated that a submarine landslide triggered by the earthquake caused the tsunami [6].

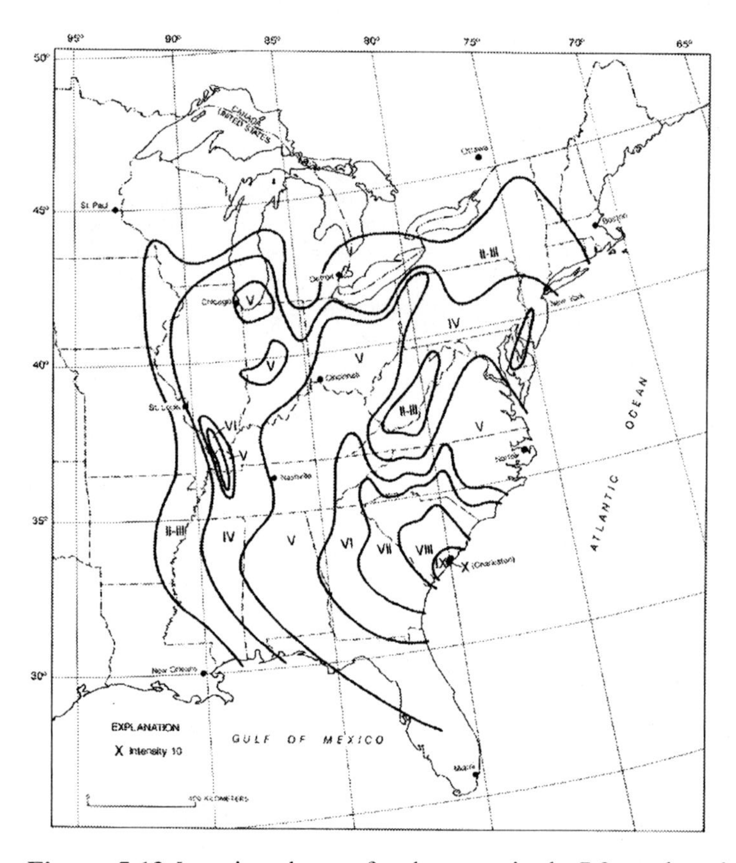

Figure 5.12 Isoseismal map for the magnitude 7.3 earthquake on 31 August 1886 (local time) near Charleston, South Carolina. Intensity X was registered near the epicenter. (source: USGS[4])

Destruction of an earthquake is not only registered near the earthquake but also farther away, though the damage reduces dramatically with the distance from the epicenter. To better understand the effects of earthquakes, scientists map the intensity of shaking in so called shake maps. In instrumented areas, the shaking can now be determined accurately. For older earthquakes, Mercalli intensity maps were assembled from reports of damage. An intensity X earthquake killing 60 people struck Charleston, South Carolina on 31 August 1886 (Fig. 5.12). It damaged or destroyed 2000 buildings in Charleston, but damage was also reported in Alabama, Ohio, Kentucky, Virginia and West Virginia several 100 km away and was felt even in Boston. In Chicago, the shaking and damage reached intensity V. Damage in Charleston was assessed at $6 million, a quarter of the value of all buildings in the city at the time. This earthquake is an example of an enigmatic and poorly understood intra-plate earthquake.

Even today, scientists gather information from the public to estimate the shaking and damage caused by an earthquake. For each noticeable earthquake mostly in the U.S. but increasingly also elsewhere around the globe, the USGS maintains a webpage where one can fill out a questionnaire(8). Data to be entered include what was felt, if others felt the quake, where the person was during the quake, reports of damages, description of the building a person was in and the zip code. After entering the data, an intensity value is returned based on the individual questionnaire, and the participant can check intensities of other zip codes and statistical data in the area.

Richter Magnitude Scale

The Richter magnitude, M_L, has long been used to describe the actual shaking from an earthquake, not the damage it causes. It was first introduced in 1935 by Charles Richter and Beno Gutenberg of the California Institute of Technology (CalTech) in Pasadena, CA. It was originally used to catalog earthquakes in California. The Richter magnitude is based on instrumental measurements and so is less subjective than the Mercalli scale. It is estimated from measuring the maximum amplitudes in a seismic record. The Richter scale formally has no lower or upper limit, and steps scale with the seismic amplitudes on a logarithmic scale. A ten-fold increase in shaking is associated with a one-step increase on the Richter scale. Until recently, the Richter scale was also widely used to compare earthquakes globally, but it has some drawbacks, which is why it was replaced by the moment magnitude scale for that purpose. The Richter scale is best used when seismic stations are within 500 km of the epicenter. Beyond that the signal from high-frequency waves are dissipated too much so that magnitude estimates are too low. The scale also saturates for large earthquakes (magnitude 6.5 and greater) because it does not account for the prolonged shaking during these events. Finally, the Richter scale depends on the geology of the region. A magnitude-6 event on the Richter scale in sediment-covered Southern California is difficult to compare with a magnitude-6 event in Hawaii that consists of solid volcanic bedrock. A Richter magnitude-6 event in Hawaii may release much more energy than a Richter 6 in Southern California where sediments enhance the shaking. However, the Richter magnitude is still used locally in California and elsewhere.

Table 5.3 The Relationship between Moment Magnitude and Radiated Seismic Energy[2]

Moment Magnitude (M_W)	**Radiated Seismic Energy** (Es in J)	**Examples/Comments**
9.5	3.6×10^{19}	world's largest recorded EQ (22 May 1960 Chile)
9.0	6.3×10^{18}	27 Mar 1964 Alaska "Good Friday", M_W=9.2 11 Mar 2011 Tohoku Japan 26 Dec 2004 Sumatra-Andaman 27 Feb 2010 Chile (M_W=8.8)
8.0	2.0×10^{17}	19 Sep 1985 Mexico City
7.0	6.3×10^{15}	17 Oct 1989 Loma Prieta, CA 17 Jan 1994 Northridge, CA (M_W=6.7) 12 Jan 2010 Haiti
6.0	2.0×10^{14}	significant structural damage may occur a M_W 5.5 can be felt as far as 500 km (eastern U.S.)
5.0	6.3×10^{12}	threshold for damage
4.0	2.0×10^{11}	can be felt as far as 100 km (eastern U.S.)
3.0	6.3×10^{9}	can be felt by many people in the area
2.0	2.0×10^{8}	approx. threshold to feel an EQ
1.0	6.3×10^{6}	recorded only by sensitive instruments
0.0	2.0×10^{5}	equivalent to 1000 lbs of TNT

Moment Magnitude Scale

To estimate the size of an earthquake, scientists now prefer to use the moment magnitude scale (Tables 5.2 and 5.3). It was introduced in 1979 by Tom C. Hanks and Hiroo Kanamori of CalTech. The **seismic moment**, M_0, describes the strength of an earthquake. It is measured in Nm (Newton meters) and equals the shear strength of the blocks along the fault times rupture area times average slip on the fault. It is the equivalent of the moment of force and results from the idea that the forces that play a role during an earthquake can be modeled by two pairs of forces (the double couple) that act on two sliding blocks without rotating them. Newton meters is also the unit of energy and so the seismic moment is often likened to the amount of energy released during an earthquake. However, not all of the potential energy that is stored in a loaded fault and released during an earthquake is actually transformed into radiated seismic energy. Some is lost to heat as well as cracking and deforming the rocks in the fault. The radiated seismic energy, Es, then scales with the seismic moment as $Es = (2 \times 10^{-4})\ M_0$[2]. The seismic moment is determined by examining the entire seismic record, not just the maximum amplitude, at many stations around the globe. The moment magnitude, M_W, then is $M_W = 2/3\ \log_{10}(M_0) - 6$. M_W scales with the radiated seismic energy, Es, on a logarithmic scale. For each step on the magnitude scale, an earthquake releases 30 times more energy (Table 5.3). On the moment magnitude scale, the 1960 Valdivia Chile earthquake has been the largest earthquake since 1900 (Table 5.2).

There is no direct relationship between the three magnitudc scales, but larger earthquakes with a larger moment magnitude typically cause more shaking and more damage, i.e. can produce a higher Mercalli Intensity. Approximate numbers for the relationship between magnitude, source duration, rupture length and intensity is summarized in Table 5.4.

Table 5.4 The Approximate Relationship between Moment Magnitude and other Parameters (modified from [1] and [3])

Approx. Moment Magnitude (Mw)	Approx. Mercalli Intensity at Epicenter	Duration of Shaking near Source (sec)	Slip or Offset (m)	Surface Rupture Length (km)
4	IV-V	0-2		0.1
5	VI-VII	2-5	0.0	11
6	VIII-IX	10-15	0.5-1.5	10
7	X-XI	20-50	2	50-100
8	XII	50-150	5	200-400
9	>XII	>200	20	1000

5.4 Earthquake Statistics and Recurrence

Table 5.5 Annual Worldwide Seismicity by Moment Magnitude (modified from [3])

Magnitude	# of EQs/year	Estimated Radiated Seismic Energy (× 10^{17} J)
8.5 and up	0.3	11.1
8.0-8.4	1	5.0
7.5-7.9	3	2.7
7.0-7.4	15	2.4
6.6-6.9	56	1.7
6.0-6.5	210	1.4

5.0-5.9	800	0.59
4.0-4.9	6200	0.15
3.0-3.9	49,000	0.04
2.0-2.9	350,000	0.008
0.0-1.9	3,000,000	0.002

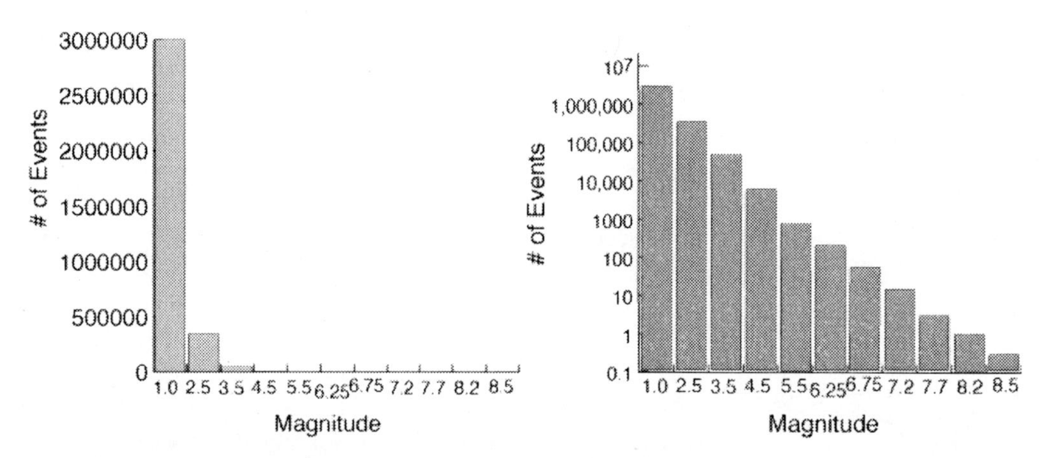

Figure 5.13 Annual worldwide seismicity using the data from Table 5.5. Left: on a linear scale. Right: on a logarithmic scale.

Seismic Activity and Earthquake Probability

Newsworthy earthquakes are terrible, destructive and fatal events. About 75 earthquakes with magnitude 6.6 and above occur each year (Table 5.5; Fig. 5.13). Many of them are located in remote areas and may not harm people or structures. But these events tell only a small part of the story on the overall seismic activity (seismicity). Each year, about 3.5 million earthquakes shake the globe. This is one earthquake every 9 seconds! The vast majority of these earthquakes go unnoticed in our every-day lives as three million events are registered only by sensitive seismic instruments (magnitudes below 2.0). The number in the next step up (2.0 – 2.9) is nearly ten times smaller. To compare the occurrence of small and large earthquakes in a diagram, one needs to use a logarithmic scale because on a linear scale, earthquakes with magnitude 4 and above do not show. Even though several very strong earthquakes in recent years have reminded us of their hazard, on average since global seismic monitoring began in the 1960s, a magnitude-8.5 and above earthquake occurs roughly only once in 3 years. Nevertheless, such earthquakes release nearly as much energy as all the smaller earthquakes taken together!

Table 5.6 Seismicity in Southern California in a Typical Week[11]

Magnitude	# of EQs
3.8-4.2	3
3.3-3.7	1
2.8-3.2	12
2.3-2.7	37
1.8-2.2	119
1.3-1.7	201
0.8-1.2	242
0.0-0.7	125

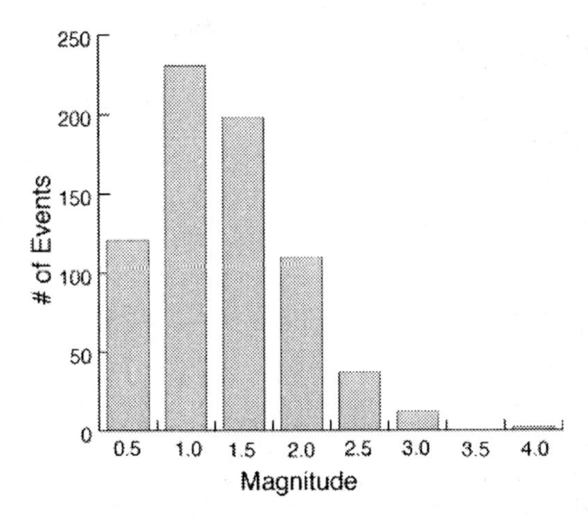

Figure 5.14 Seismicity in Southern California in a typical week. Data are taken from Table 5.6.

Zooming in on Southern California, Table 5.6 and Fig. 5.14 show the seismicity of a typical week as reported by the Southern California Earthquake Center (SCEC). The data are taken for the first week of October 2004 when no major event occurred previously. For times after major events, the seismicity is increased because movements along the involved fault system typically cause extensive aftershock seismicity. In a typical week with "normal" background seismicity, Southern California experiences upward of 750 earthquakes, probably more than 1000. The relatively small number of 125 events with magnitudes 0.7 and smaller, as shown in Fig. 5.14, is probably not real and rather shows the limitation of seismic monitoring in the area. At such low magnitudes, it becomes difficult to distinguish between the signals from earthquakes and those with anthropogenic origin or other, nonseismic natural origin (e.g. wind noise). Human activity such as traffic, industry and mining causes noise in the same frequency range as energy is radiated from small earthquakes (mostly 1-40 Hz). An increased noise level then raises the detection threshold. Most of the seismic energy radiated by an earthquake dissipates relatively quickly away from the epicenter. With a very small earthquake, the nearest stations may not have been close enough to detect it. As with global seismicity, the vast majority of Southern California seismicity is not felt by people. If we assume that only events with magnitude 2.3 and above are felt, then 53 earthquakes are felt per week, provided people are close enough. This sums to about 2750 events each year.

Seismicity catalogs and tables such as those just described form the basis of earthquake hazard assessment. An important question for governments, relief agencies as well as developers and construction companies is to know what the likelihood of an earthquake of a certain size is. A cost vs. risk analysis then often determines how the structure is built. The importance, or pitfalls, of this process has become painfully clear after the damage of the Fukushima, Japan nuclear power plant by the 11 March 2011 Tohoku earthquake and tsunami. A magnitude-9 earthquake was likely to occur in that region eventually. But it was felt that such a large earthquake would not happen during a human lifetime, or even beyond, and so many structures and the power plant were built to withstand a smaller event. The risk of an earthquake of a certain size is determined by its **recurrence time**. The recurrence time is the inverse of the number of events of that size in a given time period. For example, from Table 5.5 we see that 210 events per year occur in the magnitude range 6.0-6.5. The recurrence time of such an event is 0.00476 years or 1.74 days, and so such an earthquake occurs every 1.74 days worldwide, on average. An event with magnitude 0.0 to 1.9 occurs every 10.5 s, and the recurrence time for an event with magnitude 8.5 and above is 3.3 years. Of course, these times vary by region! The recurrence time of magnitude-8 earthquakes may be 1000s of years or much longer in seismically quiet regions but only a few decades in very active regions with long faults.

The recurrence time is not only determined by using catalog data. For pre-catalog events (typically before instrument-recording began), scientists rely on results from trench work to estimate the size and approximate time of earthquakes. This is particularly important for faults that are relatively quiet for long times, such as the southern segment of the San Andreas Fault east of the Salton Sea.

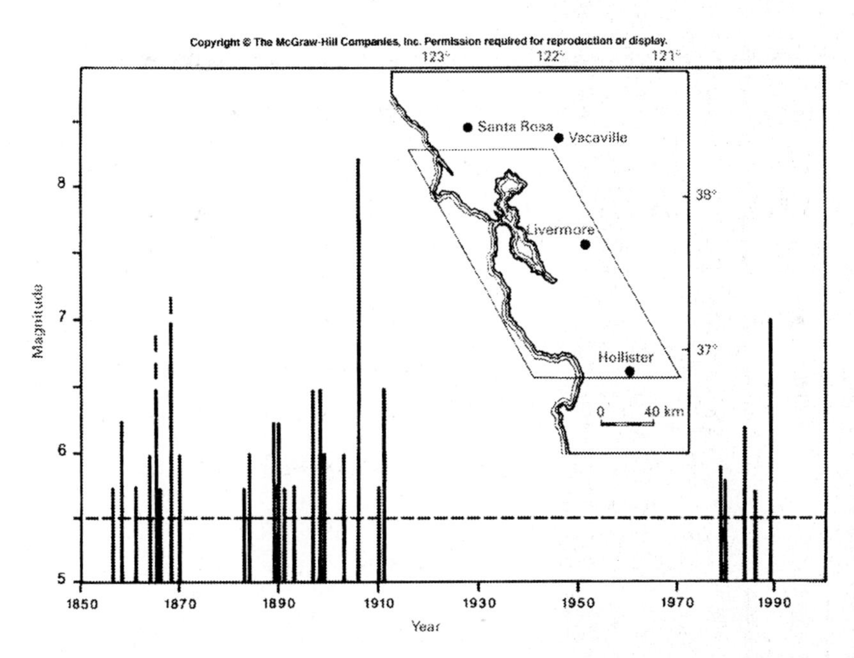

Figure 5.15 Distribution of earthquakes with magnitudes greater than 5.5 near San Francisco Bay, 1849-2007. The insert map identifies the area sampled. The earthquake catalog can be accessed online at http://www.ncedc.org/ncedc/catalog-search.html (source: AB)

Earthquakes also do not occur on a regular basis. Even in seismically active regions, long time intervals may pass before the next major earthquake strikes (Fig. 5.15). After the devastating 1906 earthquake, the San Francisco Bay area was seismically quiet for nearly seven decades before a magnitude M_L=5.8 earthquake occurred on 6 August 1979 near Morgan Hill south of San Jose. The recurrence time of an earthquake of a certain magnitude is therefore only a statistical estimate that has a certain **probability**.

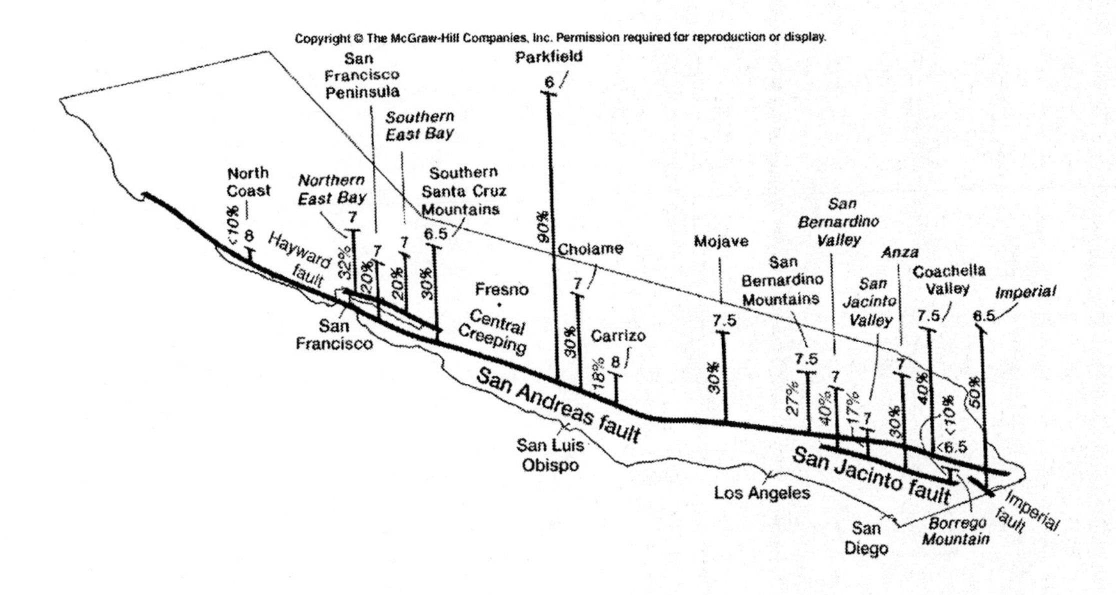

Figure 5.16a Working group analyses of expected earthquake magnitudes and their probabilities of occurring before the year 2032. Forecasts are based on paleoseismic research as well as current seismic monitoring (see Chapter 6). The prognosis was determined before the 2004 Parkfield earthquake. After this event, the probability for a magnitude-6 earthquake near Parkfield diminished. (source: AB)

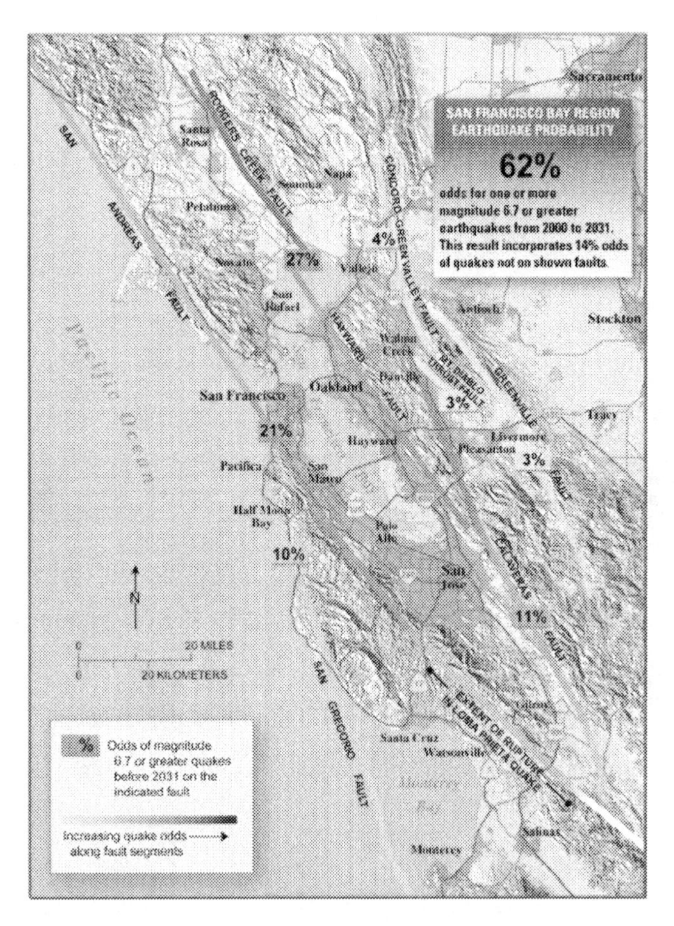

Figure 5.16b An updated map for the probability of a magnitude 6.7 earthquake before 2031 in the San Francisco Bay region, published by the USGS[10]. (source: USGS)

To be most useful, scientists provide probabilities for an earthquake of a certain size within a certain time frame (Fig. 5.16). The probability for a certain size earthquake is not the same everywhere along the San Andreas and adjacent faults. In many places along the SAF, the chance of a magnitude-7 earthquake to strike by 2032 is greater than 30%. For some people, this risk may be too high, and they prefer to live elsewhere. A magnitude-8 earthquake has only a 18% chance of occurring on the Carrizo section of the SAF, but there is a much higher, 50% chance of a magnitude-6.5 earthquake to occur on the Imperial Fault in the south. The USGS estimates a 67% chance for a magnitude-6.7 earthquake to strike the San Francisco Bay area before 2031 (Fig. 5.16b). Most likely, this event will occur on the Rodgers Creek, Hayward or San Andreas Fault, or along the northern segment of the Calaveras Fault. The probability of an earthquake along the SAF segment that ruptured during the Richter magnitude-6.9 Loma Prieta earthquake on 17 October 1989 is relatively low because it will take time to build up enough stress for the next big earthquake.

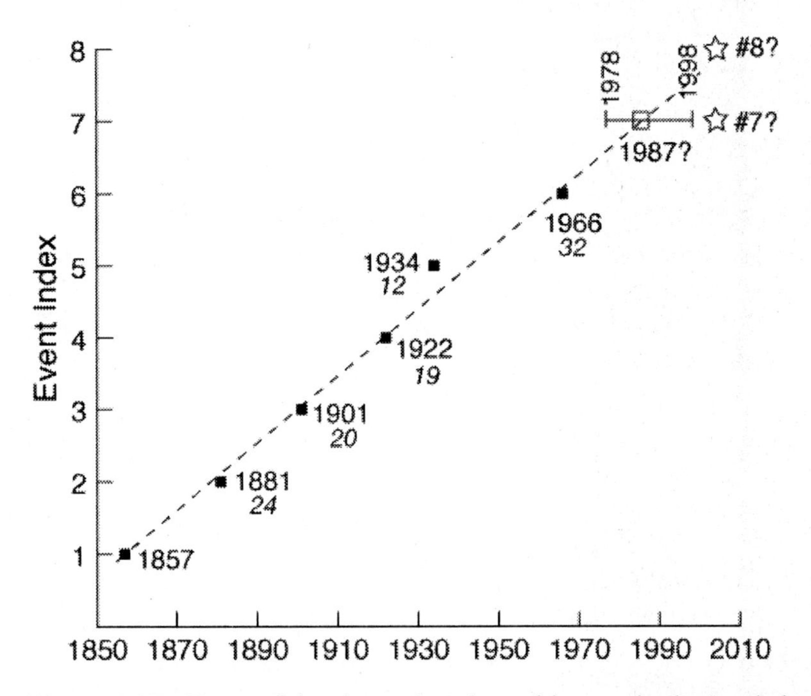

Figure 5.17a Years of the six earthquakes with magnitude 6 and above in the Parkfield, CA area prior to 2004[12]. The years of the events are labeled, together with the number of years elapsed since the last event (in italic). The time interval varies between 12 and 32 years and is 21.4 years on average. Taking this average, the next event after the 1966 earthquake was predicted for 1987 (square). The shortest time interval would place this event in 1978 and the longest in 1998. But no earthquake occurred until 2004 (star).

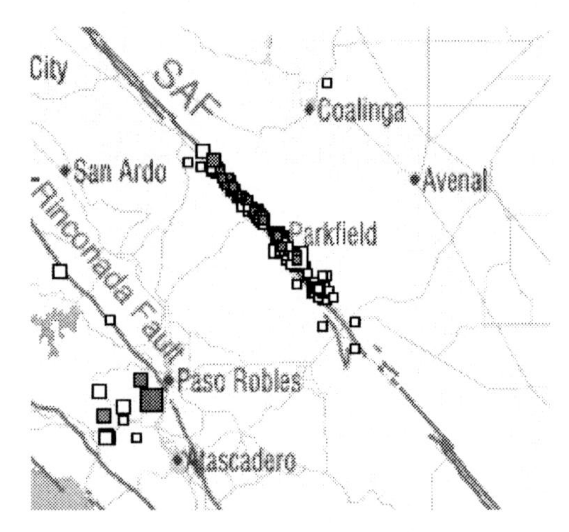

Figure 5.17b A one-week aftershock sequence of the 28 September 2004 Parkfield earthquake, published by the USGS on 11 October 2004. Aftershocks occurred mainly along the SAF, north of the main shock. The location of the main shock is near the southern end of the aftershock series. Also shown are "remote sympathetic events" on that day near Paso Robles. (source: modified from Wikipedia/USGS)

The Parkfield Prediction Experiment

An interesting case of seismicity and event probability occurred in the **Parkfield, CA** area. According to Fig. 5.16a, this area had a 90% chance for a magnitude 6 earthquake to occur before 2032. This segment of the SAF is located between the ruptured segments of the 1906 San Francisco earthquake to the north and the 9 January 1857 Fort Tejon earthquake to the south, the largest recorded earthquake in the lower 48 states. A unique and intriguing aspect of the seismicity near Parkfield was that six magnitude-6 and above events occurred between 1857 and 1966 with nearly regular intervals (Fig. 5.17a). Statistically, the next event was likely to occur in 1987 but, given data from previous events, most certainly between 1978 and 1998. Most scientists expected the earthquake between 1988 and 1993. Scientists were also excited that the seismograms of different earthquakes were nearly overlays of each other, i.e. the record of the 1922 event looked nearly exactly like that of the 1934 and 1966 events. This means that the same type of earthquake occurs in the same rupture area along the fault again and again. Some even speculated that the 1857 Parkfield event was a foreshock of the Fort Tejon event. The question then would be what makes a magnitude-6 earthquake near Parkfield grow into a great (very large) earthquake such as the Fort Tejon event. Given the high probability for a major earthquake to occur near Parkfield and the desire to observe all aspects of an earthquake while it is happening, scientists deployed an army of instruments in the area starting in 1985. A long-term forecast was issued for the year 1993 in the **Parkfield Prediction Experiment**.

But the earthquake did not occur until 28 September 2004, 38 years after the last event. The forecast clearly failed. And the seismic records provided perhaps more questions than answers. Extrapolating from previous events, the 2004 event was either a very late #7 or a nearly on-time #8. But then, why did #7 not happen? Scientists were also puzzled by the fact that the quake ruptured northward and not southward like most previous events. Most aftershocks concentrated toward the north of the main event, as expected, but an intriguing aspect of this series was that some "remote sympathetic events" occurred in Paso Robles 38 km (23 mi) to the west on the parallel Rinconada fault. Nine months earlier, the larger, magnitude 6.5 San Simeon earthquake occurred on 22 December 2003 on the Oceanic fault zone, 39 km (24 mi) west of Paso Robles (Fig. 5.17b). The most severe damage occurred in Paso Robles where intensity VIII was registered and century-old unreinforced masonry buildings collapsed in the city's historic downtown area, killing two people. Perhaps, faults "talk to each other", and scientists found that a transfer of stress likely happened from the Oceanic to the San Andreas fault thereby causing Parkfield's #8 earthquake to occur a few years early. Much was learnt from all the seismic data gathered. But it also became clear that statistical means are not adequate or at the very least not sufficient to forecast earthquakes. At this point, earthquake forecasting is far behind accurate severe weather forecasting.

5.5 Earthquake Physics – What makes an Earthquake go?

Asperities and Friction

In Chapter 2, we learned that strain, the deformation of blocks across a fault, is a result of stress buildup. Eventually, the stresses acting on the blocks are so large that the blocks can no longer hold together and an earthquake occurs. What causes the blocks to stick together so that they do not slide continuously? Simple laboratory experiments show that blocks of different materials slide past each other sooner or later depending on how rough their contact surfaces are. Smooth shiny metal blocks slide more easily than rougher wood blocks. Smooth metal surfaces have less friction than wood surfaces. This also applies to plates and blocks across an earthquake fault.

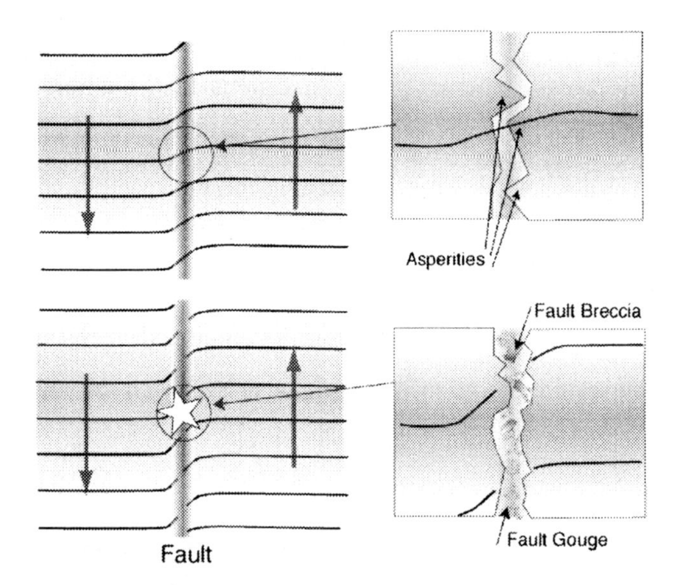

Figure 5.18 Stress buildup across an earthquake fault is possible as a result of asperities in the fault. The blocks hold together until the stresses are so large that the asperities start to break and an earthquake occurs (star in lower left panel). Fragments of asperities in the fault, also called fault breccia, are ground up by the movement within the fault to form the finer-grained fault gouge. The fault gouge zone can be several meters wide.

Rough surfaces have protrusions, so called asperities, that increase the friction (Fig. 5.18). Asperities prevent a fault from moving continuously and lock a fault. In a way, asperities act like anchors that hold a ship in place. Friction is a force that counteracts the ambient stresses. For an earthquake to occur the friction in the fault has to be overcome by these stresses. During an earthquake, or sometimes even between earthquakes, some of the asperities are crushed and ground up by the moving blocks to form fault breccia (coarser material) or fault gouge (finer, claylike material). Zones of fault breccia or fault gouge are hazardous areas to build tunnels and mines in because the loose, unconsolidated material form weak places were tunnels can collapse.

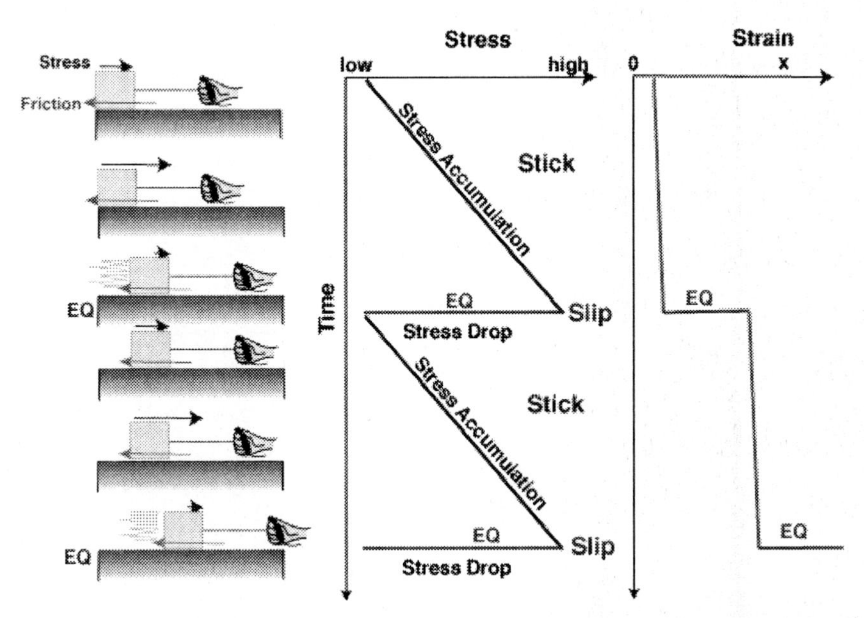

Figure 5.19 Illustration of the stick-slip model for earthquakes. Stick-slip behavior can be demonstrated on a block that is pulled over a rough surface. The middle panel traces the stress buildup before an earthquake and the stress drop during and earthquake. The right panel traces the strain accumulated before (nothing for a locked fault, little for a creeping fault) and during an earthquake (a lot).

The Stick-Slip Model

The occasional occurrence of an earthquake (seismic) and the stress buildup in between (interseismic) can be likened to what happens with a block pulled over a rough surface (Fig. 5.19). Even though we pull on the block, it does not slide (it **sticks**) until the friction is overcome (it **slips**). In the sticking phase, the force on the contact zone between the pulled block and the table builds up gradually until it drops instantaneously during the slip when the friction is overcome and the block gives way. Similarly, stresses on an earthquake fault increase during interseismic times (**stress buildup**) and drop instantaneously during and earthquake (**stress drop**). The repeated cycle of stick, slip and stress drop is called **stick-slip behavior**. During interseismic times, the block in the laboratory experiment does not move at all so that no strain accumulates. An equivalent fault in the field is then **locked**. In the field, it has also been observed that some faults **creep**, a very slow movement that is **aseismic**, i.e. it is not associated with earthquakes. In the old days, such movement was difficult to monitor. Nowadays, this is achieved by installing GPS (global positioning system) receivers on both sides of the fault and monitoring the relative movement between GPS receivers.

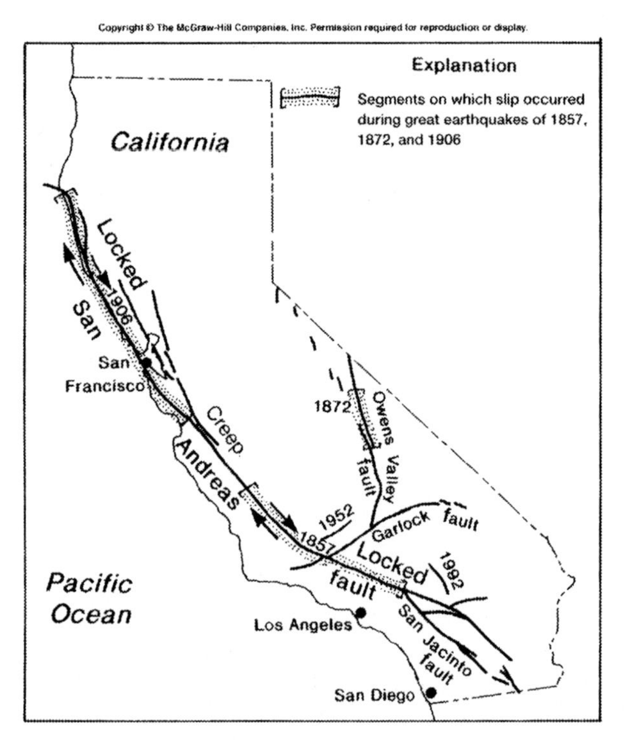

Figure 5.20 Historic behavior of some California faults. The northern "locked" section of the San Andreas fault ruptured for 400 km (250 mi) during the 1906 magnitude 7.8 San Francisco earthquake. The south-central "locked" section ruptured for 365 km (225 mi) in the 1857 magnitude 7.9 Fort Tejon earthquake. The central "creeping" section in between has frequent smaller magnitude 6 earthquakes. A major magnitude 7.5 or above earthquake in overdue on the "locked" southernmost section of the SAF. The Owens valley fault ruptured for 115 km (70 mi) in a 1872 magnitude 7.3 earthquake. The magnitude 7.3 Landers earthquake occurred on 28 June 1992 in the Mojave desert. (source: AB)

The stick-slip model is a simple model. It explains the repeated rupture along an earthquake fault, such as the 6 magnitude-6 or greater earthquakes in the Parkfield, CA area along the San Andreas Fault. But it does not quite explain why earthquakes occur at irregular intervals and with difference sizes. Hence, more complex processes must occur at a real earthquake fault. Different sections along the same fault may actually behave differently (Fig. 5.20). Large sections along the San Andreas Fault are locked so that large amounts of stress can build up and large earthquakes such as the 1906 San Francisco earthquake or the 1857 Fort Tejon earthquake can occur. The SAF also has a creeping section in between. This segment either moves continuously all the time or produces small earthquakes in short time intervals so that large amounts of stress capable of producing large earthquakes can not build up.

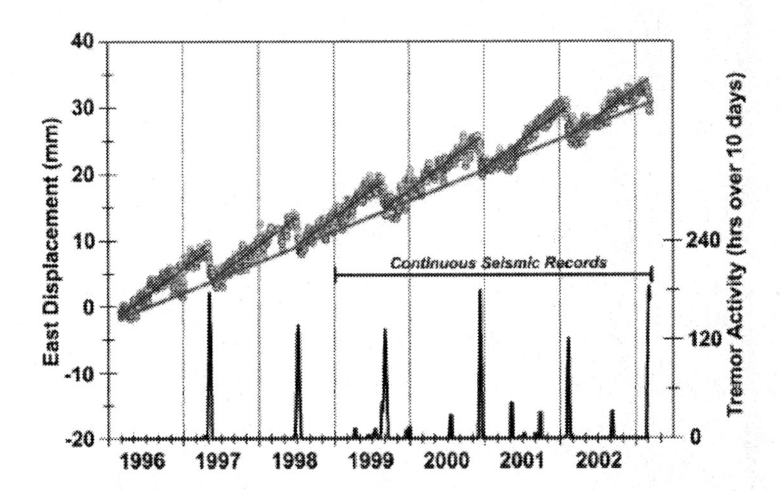

Figure 5.21a Comparison of slip in the east-west direction and tremor activity observed in the Victoria area on the Cascadia subduction zone[13]. The slip (in mm) was measured at a global positioning system (GPS) site on the Victoria peninsula with respect to a reference site in Canada's stable interior (upper dataset; left axis). The lower timeseries (right axis) shows tremor activity observed on a nearby seismic network. Episodic westward slip coinsides with tremor activity. The plot ends in early March 2003 with a predicted episodic slip. (source: Rogers and Hagerty, 2003 [13]).

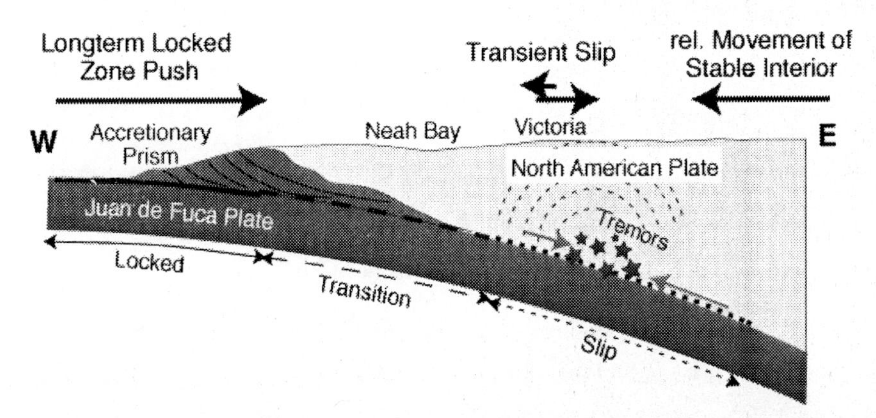

Figure 5.21b Concept figure depicting plate movements on Cascadia subduction zone. The upper section of the fault plane in the west, including the surface fault, is locked. The westernmost part of the North American plate moves east with the Juan de Fuca plate. Further east in Victoria, movement occurs eastward overall but episodic westward slip is associated with tremor activity.

Having the SAF in mind, many scientists thought that creep along a fault is a good sign because not much stress can accumulate and large earthquakes are less likely. But in 2003 scientists had interesting but also disturbing news for the Cascadia subduction zone. A GPS station in the Victoria, Canada had measured continuous movement with respect to a station in Canada's stable interior (Fig. 5.21). Over a time period of seven years starting in 1996, Victoria, that is located on the North American plate, had moved east by 3 cm (1.2 in). This means that the edge of the North American plate is being compressed, which is probably not too surprising. The fault at the edge of the plate is assumed locked because no earthquakes have occurred in recent times. An interesting aspect of the overall movement, however, that has puzzled scientists is that Victoria relapses every 14 months or so to slip back westward. At the same time, strange seismic tremor activity picks up, with 20-min long bursts of activity over a 2-week period that looks nothing like an earthquake. This movement is called **episodic tremor and slip** (ETS). Its high degree of periodicity led to the successful prediction of recurring ETSs in 2003, 2004, 2005 and 2007. Since the movement happens relatively slowly and no typical earthquake occurs - it would actually go undetected by the typical earthquake-monitoring seismometer networks - this movement is considered aseismic. Scientists have postulated that each ETS event is equivalent to a magnitude-7 earthquake, thereby loading the locked fault along the subduction zone for the next "big one" such as a mega-thrust event in 1700 that sent a large tsunami toward Japan (see Case Study 2 in Chapter 4). If such an event was to occur, the greater Victoria and Seattle area would experience major shaking and damage. Recent research shows that the ETSs are not unique to the Cascadia subduction zone, but that they occur along other subduction zones as well, e.g. off the coast of southern Japan, Alaska, Costa Rica and Mexico.

5.6 Seismic Waves

In the last section, we learned about seismic activity that is not considered earthquake activity. But what does a typical earthquake look like? A typical earthquake emits four different types of seismic waves. These waves are analyzed to locate an earthquake and determine its type. They are also used to image Earth's interior. Through seismic imaging, scientists initially learned about Earth's layers, that the outer core is fluid and the inner core is solid. Modern seismology is now capable of providing three-dimensional images of all major tectonic features including subducting slabs, mid-ocean ridges, stable continental cratons, volcanic magma chambers, complex mantle plumes and more.

The Four Types of Seismic Waves

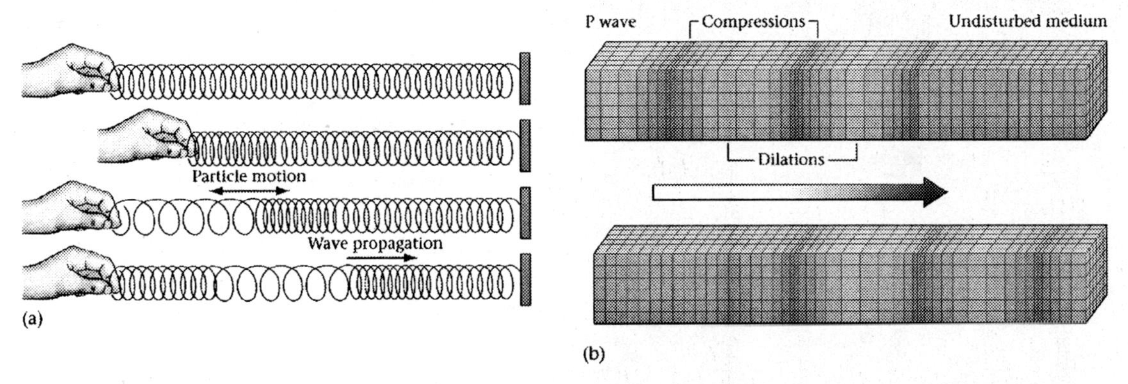

Figure 5.22 Two ways of picturing **compressional waves (P-waves)**. a) These waves can be generated by pushing on one end of a spring. The pulse of energy compresses in sequence down the length of the spring. Note that the back-and-forth motion of the coils occurs in the same direction the wave travels. b) P-waves propagating through a block of rock where a succession of compressions move through the rock. The **wavelength** of P-waves is defined as the distance between successive pulses of compression. (source: ME)

The first type of wave is the **compressional wave**. In a lab experiment, these waves can be reproduced by pushing on one end of a spring (Fig. 5.22). The push injects a pulse of energy into the spring that then progresses through the spring to compress neighboring areas in the spring in succession, in the direction in which the pulse propagates. Repeated pushes generate a succession of pulses. Pulses sent through rock behave the same way where the motion of any given point in the rock (the particle motion) is aligned with the propagation direction of the pulses. Sound waves in air and materials are a type of simplified type of compressional waves. At a given location along the spring, the time between two successive pulses is the **period**. Conversely, the number of pulses going through in a given amount of time gives the **frequency,** which is the inverse of the period. If we watch the spring as a whole, then the distance between two pulses is the **wavelength**. Compressional waves are also called **primary waves** or **P-waves** because they travel fastest and arrive at a seismic station as the first earthquake signal. From an earthquake, they travel through Earth's interior to a seismic station along a path that gives the shortest travel time. Such waves are called **body waves** (Fig. 5.23). Earth's body waves travel along arcs and not straight lines because the travel speed increases with depth. The waves are bent away from Earth structure with higher velocities. A P-wave emerging from an earthquake in the direction of a given station (e.g. station 2) therefore arrives at a closer station (i.e. station 1).

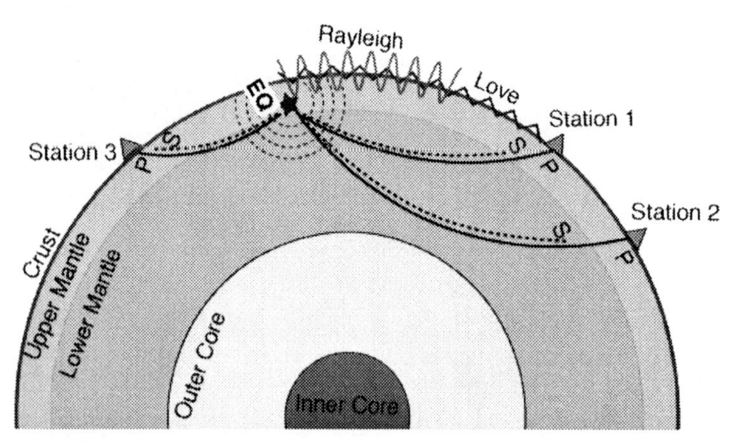

Figure 5.23a An earthquake (EQ) radiates seismic waves in all directions. P- and S-waves travel through Earth's interior to a seismic station directly. These are the **body waves**. S-waves travel at lower speeds than P-waves and arrive later. In addition, two types of waves travel along Earth's surface, the faster Love waves and the slower Rayleigh waves. These are **surface waves**. The time spent in Earth's interior or along the surface increases with distance from the source, the **epicentral distance**.

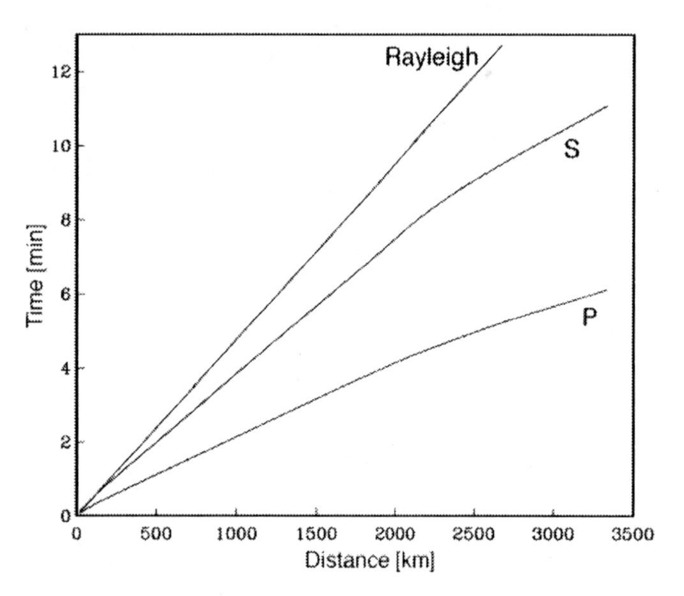

Figure 5.23b Travel time curves for P- and S-waves and for Rayleigh waves. For clarity, the curves are shown for epicentral distances only up to 3500 km. At such distances, it takes several minutes for the S-wave to arrive after the P-wave, and another several minutes for the Rayleigh to arrive. For distances less than 250 km, it becomes difficult to distinguish between the S-wave and the Rayleigh wave.

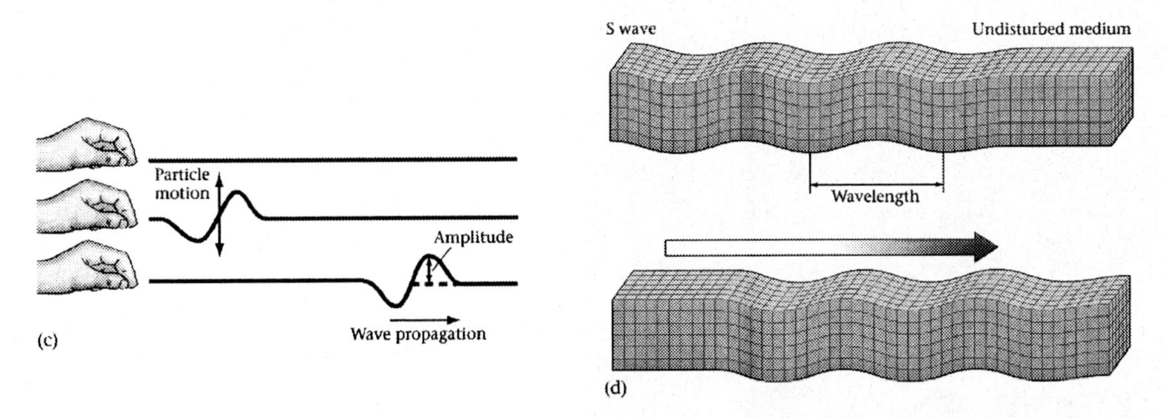

Figure 5.24 Two ways of picturing shear waves (S-waves). a) These waves resemble the waves along a rope after one end is jerked sideways. The particle motion occurs in a direction perpendicular to the direction the wave travels. b) Shear waves in a rock. The wavelength of S-waves is defined as the distance between successive peaks (or troughs). (source: ME)

Shear waves or **S-waves** travel at a lower speed than P-waves and arrive second at a recording station. They are also called **secondary waves**. With increasing distance from the earthquake, the S-waves arrives later and later after the P-waves (Fig. 5.23b). In a lab experiment, S-waves can be reproduced by jerking on one end of a rope (Fig. 5.24). Such a shearing jerk injects a pulse of energy into the rope that then progresses through the rope perpendicular to the direction in which the pulse propagates. Shearing jerks sent through rock behave the same way, and the particle motion is perpendicular to the propagation direction of the pulses. Such waves are called transversely polarized and can be decomposed into a wave that is only horizontally polarized and one that is vertically polarized. Light waves, or electromagnetic waves in general, are examples of such waves. The wavelength of S-waves is defined as the distance between two successive crests or two successive troughs. S-waves are the second type of body waves (Fig. 5.23a).

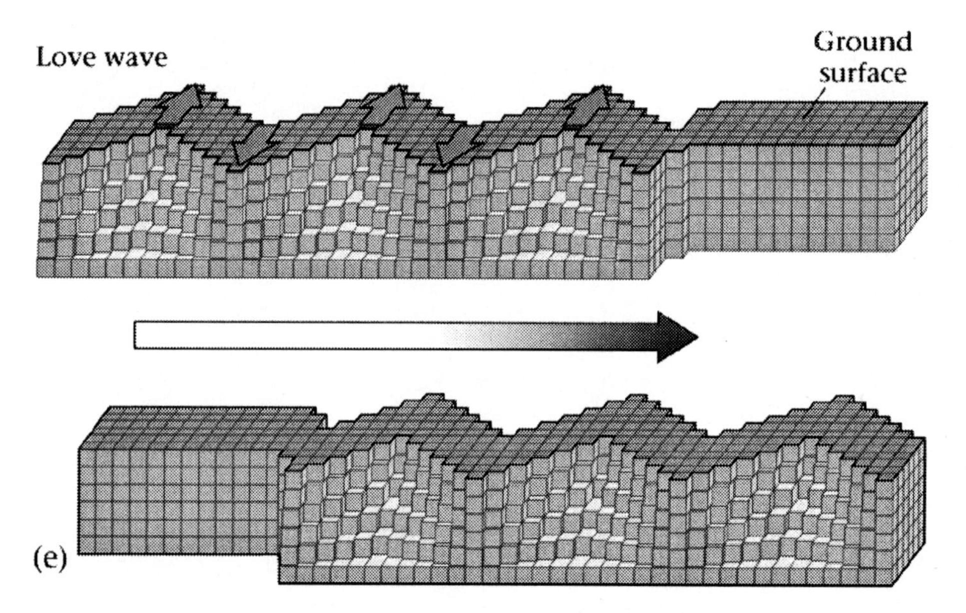

Figure 5.25 Love waves are slower than S-waves. They cause Earth's surface to shear sideways. The range of motion of individual particles decreases exponentially with depth. This motion can be compared to stalks in a wheat field that respond to a gentle breeze. (source: ME)

In 1911, Augustus Edward Hough Love predicted mathematically a type of **surface waves** in a layered medium. These **Love waves** (Fig. 5.25) are slightly slower than S-waves, and they are the third type of waves to arrive at a seismic station. Love waves can be understood as a superposition of horizontally polarized S-waves reflecting multiple times in Earth's uppermost layers. As with S-waves, the particle motion is perpendicular to the propagation direction but the amplitude of motion decreases exponentially with depth. Love waves are sometimes called Lg-waves when the earthquake was within a few 100 km. Love waves can be likened to the motion of stalks in a wheat field that respond to a gentle breeze.

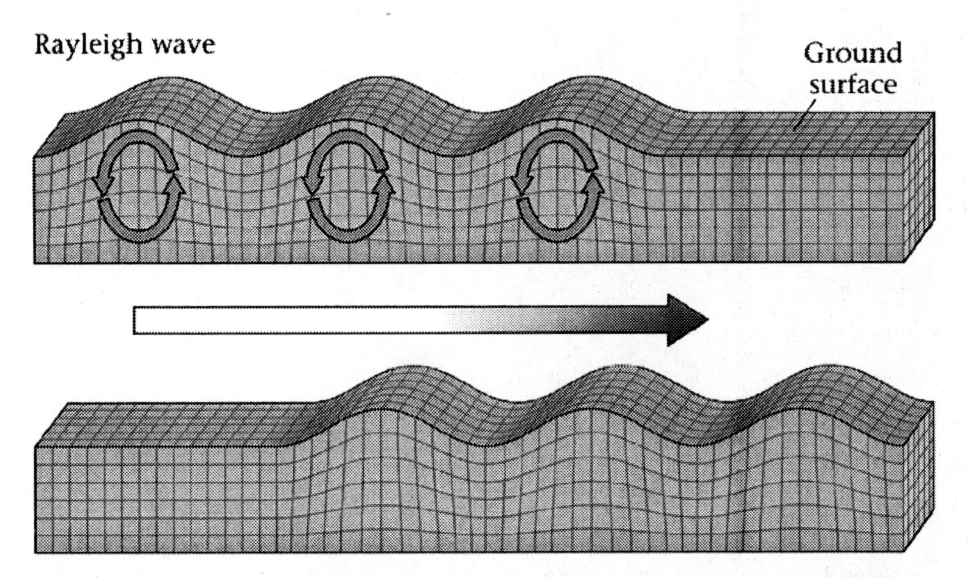

Figure 5.26 Rayleigh waves are the slowest but most destructive waves. They cause Earth's surface to roll in elliptical motion up and down as well as back and forth. The range of motion of individual particles decreases exponentially with depth. (source: ME)

Rayleigh waves, another type of surface waves (Fig. 5.26), were predicted in 1885 by Lord Rayleigh to exist in any solid medium. These waves, which arrive last at a seismic station, are a complex superposition of multiply reflected S- and P-waves. The particle motion of Rayleigh waves occurs in the direction of wave propagation but is elliptical, not on a straight line as for all the other waves. The particle motion is retrograde meaning that it orbits against the propagation direction. As with Love waves, the particle motion decreases exponentially with depth. Rayleigh waves can be likened to surface waves on the water, with a few exceptions. The particle motion of water waves generally is circular, except in shallow water, and is prograde. Rayleigh waves are sometimes called Rg-waves when the earthquake was within a few 100 km.

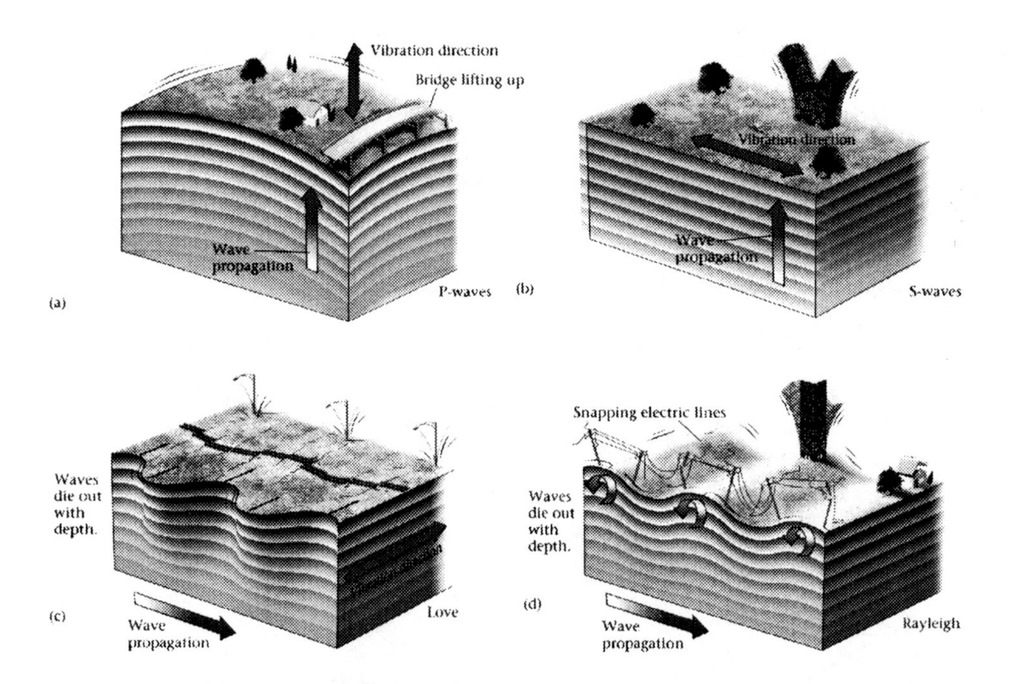

Figure 5.27 a) P-waves usually cause the ground to vibrate mostly up and down. b) S-waves cause mostly sideways shearing motion. The body waves shown here have vertical incidence but they can arrive at an angle so that S-waves cause some up-and-down movement. c) Love waves cause the ground to shake exclusively sideways. d) Rayleigh waves causes rolling motion like waves on a sea surface. (source: ME)

Surface waves are a reason for concern because they are the **most damaging seismic waves** (Fig. 5.27). They typically have large amplitudes, often larger than the S-wave, and the shaking from Rayleigh waves lasts a long time. Earthquakes emit body waves at relatively high frequencies between 0.1 and 20 Hz. Surface waves consist of waves at much lower frequencies between 0.001 and 0.2 Hz (corresponding to periods of 1000 to 5 s) though the very low frequencies are excited only by very large earthquakes. Some of the surface wave periods may be near the natural or resonance period of buildings and other structures. In this case, the shaking can build up to be violent and heavily damage or even destroy a building.

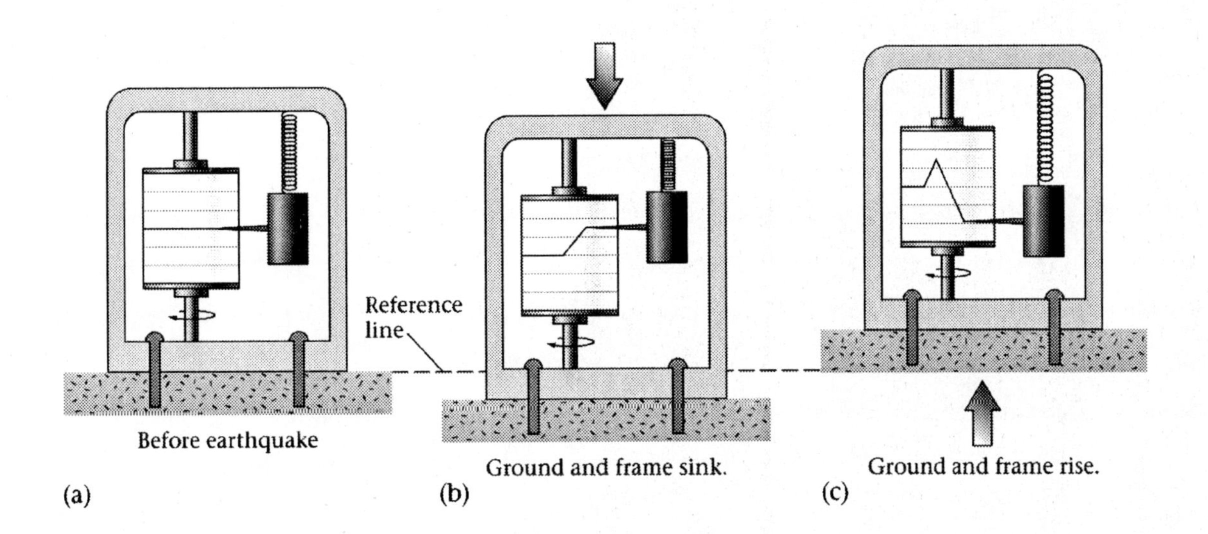

Figure 5.28a Concept figure of a vertical seismograph. a) A mass on a spring is attached to a housing that is bolted to the ground. Attached to the mass is a pen that records motion on a rotating drum. The recording, the seismogram, is flat when no motion occurs. b) If the ground moves down during an earthquake, the housing and the recording drum also move down. The spring compresses because the mass stays in place. The relative movement between the mass and the drum causes an upward wiggle on the rotating recording. c) If the ground moves up, the housing and the drum also move up. The spring expands and the drum records a downward wiggle. (source: ME)

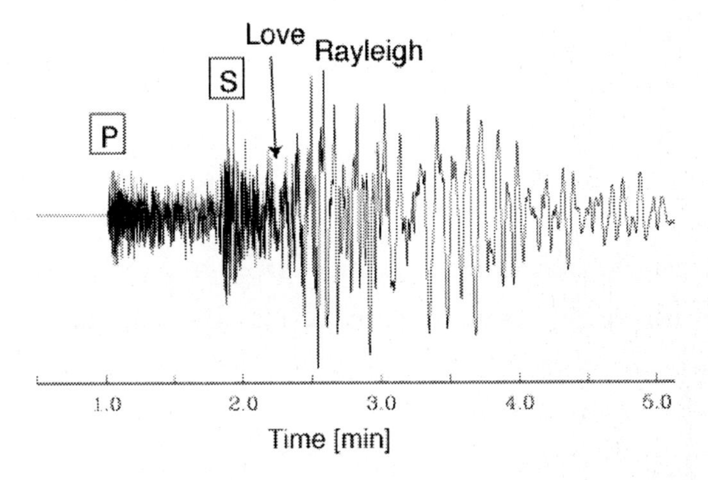

Figure 5.28b A seismic recording of the 22 December 2003 San Simeon earthquake at a station in Southern California, about 500 km (310 mi) away. Body waves (P and S) start abruptly and are relatively short-lived pulses while the surface wave phases are emergent and long-lasting. In the shown example, the Rayleigh wave has the largest amplitudes and is at least two minutes long. The Love wave, which is primarily horizontally polarized, is barely visible in this vertical recording.

The Phases in a Seismogram

Scientists record earthquakes at continuously running seismic stations (Fig. 5.28a). A simple seismograph, as it was used in the early days, essentially consists of a mass that is attached to a housing by a spring, and a rotating drum to record the movement. When the ground moves, the housing moves with the ground but the mass stays in place. A pen is attached to the mass and is capable of transcribing the relative motion onto the drum. The housing and the mass on the spring is the seismometer. The recording drum makes it a seismograph. Nowadays, seismologists use electronic recording systems instead of the drum. The mass and the pen are replaced by a magnet that is suspended on a spring and a coil within the magnet that is attached to the housing. When the housing and the coil move, an electric current is induced and its voltage is recorded. Knowing the relationship between ground motion and the voltage output, seismologists can then analyze the seismic recording, the seismogram. Similar to the vertical seismometer shown in Fig. 5.28a, seismologists also use two horizontal seismometers to record the full three-dimensional motion of the ground.

In a seismogram of an earthquake the waves arriving at different times are also called phases. The P-wave is the first phase recorded. It begins with a relatively sharp pulse and quickly decays as time goes on (Fig. 5.28b). Humans rarely feel the P-wave, except very close to the epicenter. The next distinct signal is the S-wave. It too rises relatively sharply and decays quickly. The S-wave is typically felt as the first sideways or back-and-forth jolt. The two surface waves arrive later and are felt as long-lasting, complicated ground motion in all directions. Surface waves consist of waves with widely different frequencies that travel at different speeds. In the seismogram, surface waves are dispersed, i.e. different periods are dominant at different times. A small earthquake lasts only a few seconds so that the different seismic phases can be identified quite easily. For large earthquakes, the shaking near the source can last up to a minute and phases become more emergent. For very large earthquakes, the shaking can last several minutes and the seismogram can become quite complicated, but seismologists can still analyze them.

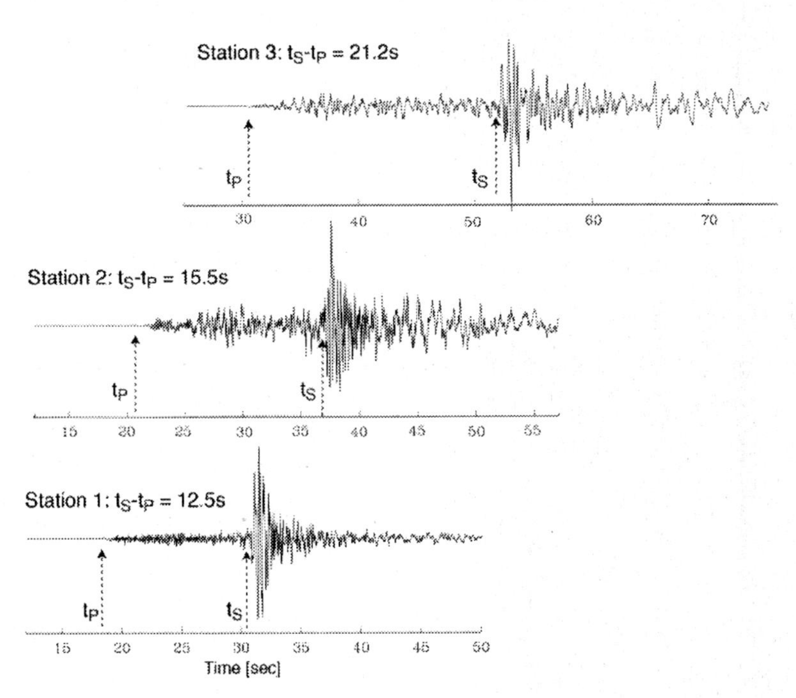

Figure 5.29a The first step in locating an earthquake. The time difference between the S-wave and the P-wave arrival is measured. At least three seismograms are needed. The travel-time difference increases with distance from the epicenter (see also Fig. 5.23). From this we infer that station 3 was farthest away from the source. Comparing the small travel time differences with the curves in Fig. 5.23, we can also conclude that the earthquake must have been at regional distances of less than 500 km.

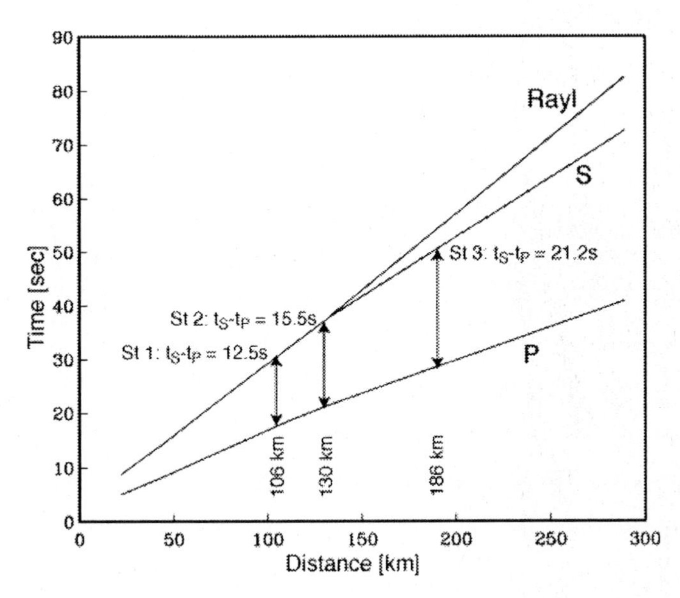

Figure 5.29b The second step in locating an earthquake. The time differences for each station from Fig. 5.29a are matched with predicted ones in a travel time chart (two-sides arrows). From this, the epicentral distance of each station can be determined by extrapolating the arrows to the x-axis.

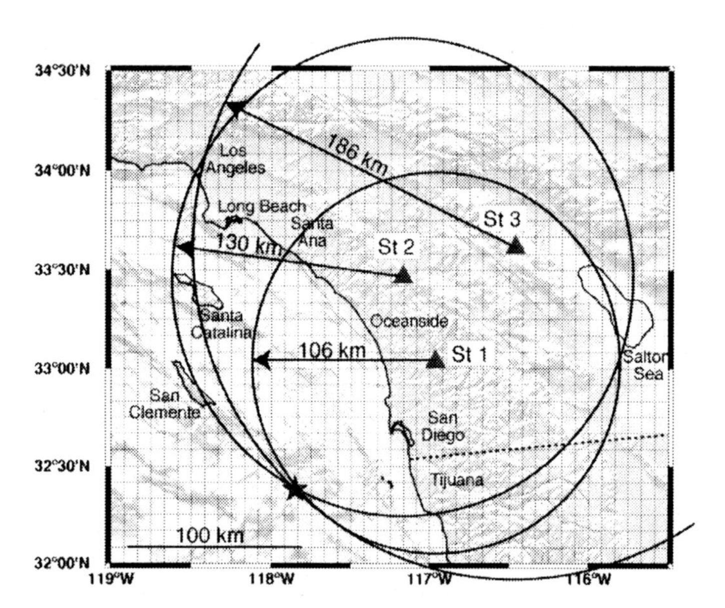

Figure 5.29c The third and final step in locating an earthquake. For each station, a circle is drawn in a map with the epicentral distance from Fig. 5.29b as radius. The epicenter is then at the point where all three circles intersect (star). The earthquake shown here was an off-shore earthquake that occurred on 15 June 2004 along the San Clemente Fault, one of the most active off-shore faults off the coast of Southern California. The magnitude M_L=5.2 event was felt throughout San Diego county and caused intensity IV shaking. It was a strike-slip event (rake=-20°) on a nearly vertical fault (dip=75°).

Locating an Earthquake

Seismologists pick the phases in a seismogram to find out where an earthquake struck. In principle, the three-component recording from a single station is sufficient to determine where the signals come from. By measuring the difference in the arrival time of P- and S-waves, the seismologist finds out the epicentral distance. Using the three-dimensional particle motion recorded for the phases, the seismologist finds out from with direction the waves arrived. The distance and the direction measured at a station uniquely defines the epicenter.

For earthquake monitoring and particularly as an early warning system, it is not a good idea to rely on a single station. If one component at a single station fails, the direction can no longer be determined. The station can also be "down" as a result of power failure, a part in the recording system could be broken, or the recording system could have lost accurate time. Seismologists therefore prefer to record on a network of stations. Using the vertical-component recordings of three stations allows the determination of the epicenter. A forth station is needed to also determine the source depth and so obtain the hypocenter. The epicenter determination is quite simple using triangulation. The arrival time difference between S- and P-waves is first determined at three stations (Fig. 5.29a). For each of the stations, this difference is compared with predicted time differences in a travel time chart (Fig. 5.29b). The travel time different can then be translated into an epicentral distance, for all three stations. At each station, we now know how far away the earthquake was but not in which direction. The earthquake lies on a circle with the epicentral distance as radius. In a map, circles can now be drawn for each of the three stations (Fig. 5.29c). The point at which the three circles intersect is the epicenter.

Ideally, a recording network is designed so that most of the seismicity occurs within this network. This ensures that the points at which the circles intersect can be determined most accurately. In the case of off-shore seismicity such as in Fig. 5.29, this cannot be achieved using land-based equipment alone. The circles then intersect at a low angle. A slight mistake in the time difference reading can then lead to a large shift of the intersecting point, or the circles not longer intersect at a single point. A 5% error in the epicentral distance estimates can then lead to a shift of the epicenter, a source mislocation, by nearly 40 km. Such errors are possible when seismograms are noisy (as a result of wind pounding on the sensor and human activity) so that the P- and S-phases cannot be read accurately, or if the phases are small or emergent. To reduce mislocation errors, seismologists use the seismograms of the whole network, not just three stations, and determine the area that is most likely to contain the epicenter.

5.7 Forensic Seismology and Induced Seismicity

Seismic monitoring does not exclusively apply to global earthquakes. In fact, the first comprehensive global network of seismometers was installed in the 1960s with the goal in mind to police clandestine nuclear testing. In more recent times, seismologists have detected more and more signals in their records that are not associated with earthquakes but other events such as landslides and even the recent breakup of large ice sheets and glaciers. Scientists also find signals from human-induced events such as blasts in large mining pits and also mine collapses but also other isolated events. Seismic records documented in shocking detail the initial hits and subsequent collapse of twin towers of New York's World Trade Center in the 9/11 attack in 2001. An interesting case of forensic seismology was made roughly a year earlier when the Russian submarine Kursk sank in the Russian Arctic in 2000. A legally charged case involves the 2007 collapse of the Crandall Canyon Mine in Utah. Human activity can also induce seismicity in previously aseismic areas and some examples follow.

The Monitoring of Nuclear Explosions

In the 1960s, the desire to monitor the future nuclear comprehensive test ban treaty (CTBT) led to the installation of a global network of seismic stations, the World-Wide Standardized Seismograph Network (WWSSN). This network could not only detect and locate earthquakes above a certain magnitude, even at remote locations, but also large explosions at nuclear test site, such as those at the Nevada test site and in the former Soviet Union. An important part of the treaty verification was to make sure that clandestine nuclear tests could not go undetected. Monitoring efforts with the WWSSN had to make sure that the catalog of events be complete, i.e. that as few as possible events above a certain energy threshold failed to be detected. One way to tell nuclear tests apart from earthquakes was their location. Events located near a known test site were likely explosions. But there was also concern that an explosion could be declared as an earthquake on a nearby fault. But scientists learned to distinguish between nuclear tests and earthquakes. A good way to detect an explosion is by comparing magnitudes derived from body waves, M_b, and from surface waves, M_S. Nuclear explosions set off significant energy that routinely exceeds that of a magnitude-5 earthquake. But the energy is primarily emitted at high frequencies and less so at low frequencies. Explosions generally also emit less shear energy relative to an equivalent-size earthquakes. Consequently, the ratio M_b:M_S is distinctly higher for explosions than for earthquakes(14). Seismic monitoring alone is not perfect however because it cannot distinguish between nuclear and chemical explosions. Nuclear explosions, even underground explosions, emit infrasound waves that travel around the globe in the atmosphere. Thus, modern monitoring of the CTBT therefore includes not only global seismic networks but also global networks of radionuclide, infrasound and hydroacoustic sensors.

The 2000 Sinking of the Russian Submarine Kursk

On 11 August 2000, the 154-m long Russian nuclear submarine Kursk sank in 100-m deep water in the Barents Sea, about 135 km (85 mi) from Severomorsk, Russia and about 300 km (185 mi) off the Norwegian north coast(15). The tragic event, that was likely caused by the accidental explosion of a torpedo, killed all 118 people on board. The submarine had sailed out to sea to practice and set off dummy torpedoes. At first, Russian officials blamed the sinking on a collision with a mystery foreign submarine that was not supposed to be in the area. Seismologists soon found out that two events occurred, an event of magnitude 2.2 was followed 135 s later by a more serious event of magnitude 4.2. Signals from so-called bubble pulses were found. The events therefore had clear signs of underwater explosions and not those of a collision or a breakup. By comparing the signals from the Kursk incident with that of other known explosions, scientists concluded that what they saw in the second event was the manifest of the explosion equivalent of several tons of TNT, or about 5 torpedo warheads. The first event was likely cause by a single torpedo exploding very near the submarine or even inside it. Prior to the first explosion, the sub had radioed for permission to launch its weapons. Film of the sunken sub showed that the periscope was up, indicating that it was very close to the surface when the first explosion occurred. Other evidence showed that the larger explosion likely occurred on the seafloor, perhaps after the sub sank and its impact set off the other torpedoes. Divers later found notes from submariners surviving the first explosion describing the situation as they tried to take stock and salvage the sub in the 2-min they had left. Some sources speculate that some submariners actually survived for several hours or even days after the explosion.

The 2007 Collapse of the Crandall Canyon Mine in Utah

The underground bituminous coal Crandall Canyon Mine in Emery County, Utah made headline news when six miners were trapped after a collapse on 6 August 2007. The event was recorded by a nearby seismic network[16]. Frantic rescue operations started immediately after the collapse. Holes were drilled at several locations of the mine to lower cameras and sensors to explore damage and air quality with the hope to eventually find and rescue surviving miners. Air quality was good enough to encourage rescue efforts to continue. Nine days into rescue efforts, new vibrations were picked up on 15 August though it was not clear whether the vibrations were cause by additional collapses, by (animal) activity on the surface or by the surviving miners. A day later, a second major collapse caused the death of three rescue workers. On 30 August the seventh and last hole was drilled but no sign of live was found and rescue efforts were suspended assuming the six miners were dead.

Immediately after the 6 August collapse, the operator had claimed that an earthquake on a nearby fault caused the mine collapse. In this case, the operator would not have been liable for the deaths of the miners. Scientists analyzed the signals recorded by the seismic network providing prove that the collapse was caused by an explosion and not an earthquake. Seismometers picked up the signal from an event of magnitude M_L=3.9 at the time and location of the collapse. More detailed analysis revealed that the event looked nothing like the typical double-couple earthquake. Firstly, the initial P-wave motion was downward at all stations. This is extremely unlikely in the case of an earthquake. A double-couple earthquake usually causes some initial motion to go upward at surrounding seismometers indicating if the slip occurred in the direction to or away from the seismometer. But all initial motion was away from all seismometers indicative of a collapse. A more detailed study of the event type also ruled out a typical earthquake of equivalent size. The scientists found that the collapse of a 920 x 220 m section of the mine with a height of 0.3 m best explains the seismic recordings, and the event began were the miners worked.

Scientists further noted in their final publication that the area was riddled with induced seismicity caused by the mining operations: explosions, collapses or actual earthquakes that would otherwise not have occurred. Prior to the 6 August collapse, there were 28 seismic events within a 3-km radius of the collapse. With magnitude $M_L \geq 1.8$, these earthquakes were large enough to be detected by the seismic network. Fifteen events occurred in late February and early March and 8 events in the 2.5 weeks before the collapse. The collapse also triggered an "aftershock" series of 37 events with ML $\leq$ 2.2. A M_L=1.6 event was associated with the collapse that killed the rescue workers. Forensic seismology provided crucial evidence that the mine collapse was not the result of an earthquake. The mine had a history of unsafe mining practices and was cited only a year prior to the disaster for lacking escape routes among other violations. But its 64 violations and $12,000 in fines then appeared to be in par with similar-sized mines through the country. Upon the release of the scientific report in May 2008, it became clear that this event was preventable, as experts agreed that the mine was unsafe. A section of the mine had collapsed previously. Nevertheless, the company requested a permit to continue work in another section, thereby concealing the earlier collapse. The earlier collapse was simply cleaned up and operations continued in the other section. In July 2008, the mine operator was fined $1.85 million in penalties by the U.S. Mining Safety and Health Administration for unsafe mining practices that led to the collapse. This has been the highest penalty for coal mine safety violations in U.S. history.

Wastewater Pumping and the 1962 Denver Earthquake Series

The following describes an example of how increasing pore pressure as a result of injecting fluids into the ground can induce seismicity. The case involved the pumping of wastewater into a deep hole in the Denver area[17]. Prior to 1962, there had been some historical earthquakes in the greater Denver area but at the time, seismicity was relatively low. In April 1962, there was a sudden increase, and by September 1963, about 700 events in the Richter magnitude range 0.7 - 4.3 occurred. Most of these events were within a 8-km-radius of the Rocky Mountain arsenal northeast of Denver were weapons were manufactured. The wastewater from this operation was initially evaporated but then pumped down a 3670-m (12,045 ft) well. The hole was cased to a depth of 3650 m (11,975 ft) leaving the rest of the well open so that the fluid could disperse. The wastewater was injected under pressure from 8 March 1962 through 30 September 1963. When the pumping ceased, so did the earthquakes. After pumping resumed in September 1964, new seismicity unsettled the population in the area until the pumping stopped. In 1969, the U.S. Geological Survey conducted a controlled study to find out how pumping water into a borehole induces seismicity. This was the first study to establish a connection between the two.

2006 Induced Seismicity in Basel, Switzerland by Geothermal Prospecting[18]

The Rhine Graben (German; a valley between two escarpments) is located between the Vosges mountains in France, the Black Forest in Germany and the Alps in Switzerland. It formed about 30 million years ago as part of an extensive rift system in western Europe. The Rhine Graben is bounded by active seismic faults. Most seismic events are small, with an occasional M_L=5 event. Hydrothermal circulation in the extensive fault systems brings warm and/or mineral-rich water to the surface as manifested by numerous health spas. Having examples to the north in mind, the greater Basel, Switzerland area has long been viewed as a potential site for geothermal power and spas but several shallow prospecting holes have been disappointing. The water was found not hot enough for profitable operations. Most such operations exploit natural water that circulates over heated rock and eventually surfaces as warm water. In 2006, a prospecting launched another attempt with the Deep Heat Mining Project. This time, the plan was to exploit a deep-seated fault in an area that was known to be seismically active. The new hot dry rock (HDR) technology promised independence from the natural water circulation of classical approaches. In HDR-enhanced geothermal systems, water is pumped into the ground, heated up by the hot rock and then pumped back from a production well. In many cases, the rock is not permeable enough to pump water through the rock for profitable production. So to prepare the production site, water is pumped under high pressure to fracture the rock (hydrofrac) and increase the permeability.

The Basel prospecting site was within less than 10 km from the fault that is thought to have ruptured during the 1356 earthquake that destroyed much of Basel. The site location was promising because it coincided with an area of anomalously high heat fluxes. The company drilled an over 4-km-deep hole to explore the possibility of geothermal power generation. The company had not done a hazard assessment study prior to their activity. On 2 December 2006, the company started to pump water at high pressure into the borehole. Six days later, people in the area felt shaking from a magnitude M_L=2.9 earthquake. The company stopped pumping but earthquakes continued, culminating in a magnitude 3.4 quake that day. More magnitude 3 or larger quakes occurred on 6 January 2007, 16 January and 2 February. The public protested strongly and 2700 claims of damage on homes both in Switzerland and Germany were processed. The city of Basel and the company had sponsored a six-station seismic monitoring network but already existing seismic networks also picked up about a tenth (only the largest events) of the unusual seismicity. The correlation between pumping and seismicity was very clear. After three years of study, the company canceled the project in December 2009. The study revealed that the area would have to endure more such earthquakes in the next 30 years, with multi-million dollar damages, if operations were to continue. The 2007 incident had already caused $7 million in damages. In the meantime, an already existing German geothermal power plant in the Rhine Graben received enhanced scrutiny.

In an interesting twist, this story has ties to the U.S.. A Seattle-based start-up company, partially sponsored by Google, planned similar operations in the Geysers area in Northern California and started an exploratory hole in 2009. But residents in the area protested after they were already concerned about seismicity induced by less invasive techniques, and a news article appeared in the New York Times[19]. The company had won a grant by the U.S. Department of Energy (DoE) to fracture bedrock on federal land. Questioned by the newspaper, however, government officials felt that the prospecting company was not forthcoming with information on the Basel Deep Heat Mining Project. The department launched its own investigation and announced that financing and issuing permits for fracturing rock at the drill site would depend on the outcome of their investigation. A day after the Basel project was shut down, the California company announced the termination of the Geysers project due to "technical difficulties". A month later, the New York Times reported that the DoE imposed new drilling safeguards that include requirements to monitor induced seismicity and to have an approved plan in hand to shut down if induced seismicity becomes too powerful. Companies must also file estimates of expected earthquake activity and subject their proposal to outside reviews. With the help of federal stimulus funding, the company launched a similar project near Newberry in Central Oregon, in a sparsely populated area. The company maintains on their website that the situation in California was much different from the situation in Basel (http://www.altarockenergy.com/nyt.html).

Large Reservoirs and Induced Seismicity

It is somewhat hard to imagine that the load of a reservoir could influence the behavior of hard rock. After all, the added weight of the water in a reservoir exerts a large amount of pressure on the bottom but this pressure is much smaller than the pressure exerted by natural processes within the Earth's crust. But water could leak from the reservoir and penetrate into existing faults. The water then reduces the friction along the faults and thereby increases the likelihood of earthquakes. Having the hydrofrac experiments in mind, one can imagine that the pressure of the water in the reservoir could cause the formation of new microcracks through which the water can migrate to reach existing faults. Another factor increasing the seismic hazard is a dramatically changing water table. This causes changes in ambient pressure and can weaken the underlying rock faster than if the pressure stayed content. Scientists expect that seismicity is induced some time after the increased pore pressure first overcomes any inhibiting effects of loading.

Dean S. Carder first established a relationship between changing water levels in **Lake Mead** and seismicity after the **Hoover Dam** was built and dedicated in 1935 (then called Boulder Dam)[20]. Within 3.5 years, the water level rose by 145 m (475 ft) and then fluctuated with seasonal river input. Within nine months, the first cluster of earthquakes occurred in the vicinity of Boulder City, Nevada. But the relationship was not a simple one. Sometimes, seismicity increased shortly after peak levels in the reservoir while in other years seismicity rose during sharply rising water levels from spring flooding on the Colorado River. Such correlations were also found elsewhere. At Lake Oroville, CA, seismicity decreases as the lake fills during winter and spring, and it increases as the water level drops during summer and winter. A magnitude-5.7 event occurred in 1975.

In Greece, the filling of **Lake Kremasta** caused a damaging earthquake. Similar events occurred at **Xinfengjiang**, China, **Nurek Reservoir** in Central Asia and **Kariba**, Central Africa. An induced magnitude-6.5 event at **Koyna, India** in 1967 caused 200 fatalities. It was established that not the size of the lake but its water depth has the most important impact where a depth of 100 m was established as an approximate threshold. This could explain why **Lake Nasser, Egypt** where the Aswan Dam controls the Nile River since 1965 was initially not known for induced seismicity. Even though it was the World's second most voluminous reservoir at the time, the maximum water depth of 72 m was not reached until 1975. Six years later a powerful magnitude-5.6 earthquake occurred on 14 November 1981, about 65 km from the lake's deepest point. It was felt in Khartoum, 900 km to the south and caused damage in Aswan though the dam itself was unharmed. Studies at Xinfengjiang and Nurek Reservoir indicate that induced seismicity has earthquake types different from what is expected along earthquake faults in the area. This fact can be used to discriminate between natural and induced seismicity.

The devastating M_W=7.9 **Wenchuan earthquake** in Sichuan Province, China on 12 May 2008 killed an estimated 80,000 people. This notorious event made headlines after it became known that 10,000 children were killed in collapsed classrooms and dormitories. China has a one-child-per-family policy to stem the exponential population increase, and many parents were in extreme emotional distress after they learned of the death of their only child. A few months after the earthquake, American and Chinese scientists suggested that the calamity may have been triggered by four-year-old Zipingpu Reservoir that is located less than a mile from a well-known major fault[21]. The earthquake occurred some 8 km (5 mi) from the reservoir. If the reservoir really triggered the earthquake, then this would be the largest known human-induced earthquake. Scientists were careful to emphasize, however, that the link between the reservoir and the event cannot be established conclusively. They speculated that even if the reservoir acted as a trigger, it would likely only have hastened an event that would otherwise have occurred naturally at some point.

The Three Gorges Dam on the Yangtze River has the World's largest hydroelectric power station (18,200 MW). The initial component was completed in 2008 and an update is planned for 2011 to increase the capacity to 22,500 MW. The dam reduces flooding downstream but its construction flooded numerous cultural and archeological sites. It displaced 1.3 million people and cause significant ecological changes. There is an increased risk for landslides as a result of erosion in the reservoir. During two incidents in 2009 20,000 and 50,000 m^3 (26,000 and 65,000 cubic yd) of material plunged into the flooded Wuxia Gorge of the Wu River. In the first four months of 2010, there were 97 significant landslides. Sedimentation from the Yangtze is also considered significant thereby reducing the lifetime of the reservoir. The reservoir also sits on a seismic fault and it is feared that earthquake-induced ground acceleration together with the immense weight of the water in the reservoir could breach the upstream face of the dam. The height of the dam is 181 m (590 ft). Even not considering the risk of earthquakes from the seismic fault, the height is far beyond the threshold for induced seismicity [(24)]. Perhaps, only time will tell if the concern of numerous opponents of the dam were right. But the fear is that the failure of the dam will claim numerous lives.

References and Websites

1. Hyndman, D. and Hyndman, D., 2006. Natural Hazards and Disasters. 2nd edition. Thompson Brooks/Cole, 510 pp.
2. Johnston, A.C., 1990. An Earthquake Strength Scale for the Media and the Public. Earthquakes and Volcanoes, **22**, 214-216.
3. Abbott, P.L., 2009. Natural Disasters. 7th edition. McGraw Hill, 526 pp.
4. USGS Education and Outreach website on earthquakes: http://pubs.usgs.gov/gip/earthq1
5. USGS webpage on the 1886 Charleston earthquake: http://earthquake.usgs.gov/earthquakes/states/events/1886_09_01.php
6. Billi, A., Funiciello, R., Minelli, L., Faccenna, C., Neri, G., Orecchio, B. and Presti, D., 2008. On the cause of the 1908 Messina tsunami, southern Italy. Geophys. Res. Lett., 35, L06301, doi: 10.1029/2008GL033251
7. USGS webpage on the Modified Mercalli Intensity Scale: http://earthquake.usgs.gov/learn/topics/mercalli.php
8. USGS "Did you feel it?" webpage on noticeable earthquakes:http://earthquake.usgs.gov/earthquakes/dyfi/
9. USGS "Ten largest earthquakes in the World" webpage: http://earthquake.usgs.gov/earthquakes/world/10_largest_world.php

10. USGS Fact Sheet 017-03 "The USGS Earthquake Hazard Program in NEHRP": http://pubs.usgs.gov/fs/2003/fs017-03/
11. assembled from data of the first week in October 2004; data from scec.org
12. dates available at the USGS website on Parkfield earthquake history: http://earthquake.usgs.gov/research/parkfield/hist.php
13. Rodgers, G. and Dragert, H., 2003. Episodic Tremor and Slip on the Cascadia Subduction Zone: The chatter of Silent Slip. Science, 300, 1942-1943. DOI: 10.1126/science.
14. Al-Eqabi, I., Koper, K.D. and Wysession, M.E., 2001. Source Characterization of Nevada Test Site Explosions and Western U.S. Earthquakes using Lg Waves: Implications for Regional Source Discrimination. Bull. Seismol. Soc. Am., 91, 140-153.and Stein, S. and Wysession, M., 2003. An Introduction to Seismology, Earthquakes, and Earth Structure, Blackwell Publishing, Oxford, UK, 498 pp.
15. Koper, K. et al., 2001. Forensic Seismology and the Sinking of the Kursk. EOS Trans. AGU, 82, 37. and Wikipedia site on the Kursk: http://en.wikipedia.org/wiki/Kursk_(submarine)
16. Berkeley seismolab webpage of the Utah Mine collapse at http://seismo.berkeley.edu/~peggy/Utah20070806.htm Pechmann, J.C., Arabasz, W.J., Pankow, K.L. and Burlacu, R., 2008. Seismological Report on the 6 August 2007 Crandall Canyon Mine Collapse in Utah. Seismol. Res. Lett., 79, 620-636. And Wikipedia site on the Crandall Canyon Mine: http://en.wikipedia.org/wiki/Utah_Mine_2007 Urbina, I., 9 May 2008. Utah Mine Disaster was preventable, report says, New York Times, available at http://www.nytimes.com/2008/05/09/us/08cnd-mine.html?pagewanted=all and Editorial, 12 May 2008. Crandall Canyon's Shame, New York Times, available at http://www.nytimes.com/2008/05/12/opinion/12mon3.html
17. Rocky Mountain Arsenal Remediation Venture Office, 2001. Deep Injection Well, Fact Sheet, http://www.rma.army.mil/cleanup/facts/deep-wel.html and USGS and Earthquake Information Bulletin November – December 1970. Colorado Earthquake History, USGS: tp://earthquake.usgs.gov/earthquakes/states/colorado/history.php
18. Wikipedia site on Induced Seismicity in Basel: http://en.wikipedia.org/wiki/Induced_seismicity_in_Basel and own and several local newspaper accounts in Germany and Switzerland
19. Glanz, J., 13 July 2009. Quake Fear Stall Energy Extraction Project. New York Times.and Glanz, J., 11 December 2009. Quake Threat Leads Swiss to Close Geothermal Project. New York Times. And Glanz, J., 12 December 2009. Geothermal Project in California is Shut

Down. New York Times. And Glanz, J., 16 January 2010. Geothermal Drilling Safeguards Imposed. New York Times.

20. Carder, D.S., 1945. Seismic investigation in the Boulder Dam Area, 1940-1944, and the influence of reservoir loading on local earthquake activity. Bull. Seismol. Soc. Am., 35, 175-192. And Adams, R.D., 1983. Incident at the Aswan Dam. Nature, 301, 14.
21. LaFraniere, S., 6 February 2009. Possible link Between Dam and China Quake. New York Times.
22. Fäh, D. and 11 co-authors, 2009. The 1356 Basel earthquake: an interdisciplinary revision. Geophys. J. Int., 178, 351-374. And Das Erdbeben von 1356. downloaded from http://www.altbasel.ch/dossier/erdbeben.html
23. Lemielle, F. and six co-authors, 1999. Co-seismic ruptures and deformations recorded by speleothems in the epicentral zone of the Basel earthquake. Geodinamica Acta, 12, 179-191.
24. Wikipedia webpage on the Three Gorges Dam
25. Wikipedia webpage on the 1755 Lisbon earthquake and Kozak, J.T., and James, C.D., Historical Depictions of the 1755 Lisbon Earthquake, National Information Service for Earthquake Engineering, UC Berkeley, online at http://nisee.berkeley.edu/lisbon/

Other Recommended Reading

Bolt, B.A., 2006. Earthquakes: 2006 Centennial Update. W.H. Freeman and Company, 320 pp.

Hough, S.E., 2002. Earthshaking science. Princeton University Press, 238 pp.

Chapter 6: Earthquake Hazards (draft)

Tropical cyclones and earthquakes are currently the deadliest natural disasters (Table 1.1). It is therefore important to understand why and how earthquakes occur, and to study their processes. This was described in the last chapter. It is also important to understand what effects earthquakes have locally and why, and what can be done to reduce the risk of damage. This will be discussed in this chapter.

Table 6.1 The 16 Deadliest Earthquakes in the World since 1900[6]

Rank	Location	Date	Magnitude	Fatalities
1	Tangshan, China	27 Jul 1976	7.5	255,000
2	Sumatra, Indonesia	26 Dec 2004	9.1	227,898
3	Haiti	12 Jan 2010	7.0	225,570
4	Haiyuan, Ningxia, China	16 Dec 1920	7.8	200,000
5	Kanto, Japan	01 Sep 1923	7.9	142,800
6	Ashgabat, Turkmenistan	05 Oct 1948	7.3	110,000
7	Wenchuan, Sichuan, China	12 May 2008	7.9	87,587
8	Kashmir, Pakistan	08 Oct 2005	7.6	86,000
9	Messina, Italy	28 Dec 1908	7.2	72,000
10	Chimbote, Peru	31 May 1970	7.9	70,000
11	Western Iran	20 Jun 1990	7.4	50,000
12	Gulang, Gansu, China	22 May 1927	7.6	40,900
13	Erzincan, Turkey	26 Dec 1939	7.8	32,700
14	Avezzano, Italy	13 Jan 1915	7.0	32,610
15	Bam, Southeastern Iran	26 Dec 2003	6.6	31,000
16	Quetta, Pakistan	30 May 1935	7.6	30,000

Table 6.2 Pre-1900 Deadly Earthquakes in the World[7]

Rank	Location	Date	est. Magn.	est. Fatalities
1	Shaanxi, China	23 Jan 1556	8.0	825,000
1a	Calcutta, India	11 Oct 1737	?	300,000°
2	Antioch, Turkey	21 May 525	8.0	250,000*
3	Aleppo, Syria	11 Oct 1138	?	230,000
4	Damghan, Iran	22 Dec 856	7.9	200,000

5	Ardabil, Iran	22 Mar 893	?	150,000
6	Genroku, Japan	31 Dec 1703	?	108,800
7	Lisbon, Portugal	1 Nov 1755	8.7	100,000
8	Chihli, China	27 Sep 1290	?	100,000**
9	Shemakha, Caucasia	Nov 1667	?	80,000**
10	Tabriz, Iran	18 Nov 1727	?	77,000**
11	Sicily, Italy	1 Nov 1693	7.5	60,000**
12	Silicia, Asia Minor	1268	?	60,000**
13	Calabria, Italy	4 Feb 1783	?	50,000**

° noted on the USGS website but not in their list;
fatalities may have been caused by a cyclone, not an earthquake

* not listed on USGS website

** not listed on Wikipedia website

6.1 Factors Controlling Damage and Fatality Rates

The damage and fatality rate at a certain location depends on both earthquake-specific parameters and but also parameters tied to local geology and human factors.

Figure 6.1a Damage from the 17 January 1994 Northridge, CA earthquake to an elevated highway section. (source: Wikipedia/FEMA)

Figure 6.1b Cliff collapse in Waipio Valley on the island of Hawaii. The collapse was triggered by the 15 October 2006 earthquake centered offshore just north of the island. (source: Wikipedia/FEMA)

Earthquake-Related Factors

Among earthquake-related factors that ultimately determine the damage is its size. Small earthquakes usually cause less damage than large earthquakes. The depth of an earthquake is also an important aspect. Deep earthquakes cause little damage, but shallow earthquakes can cause significant damage even though they may be relatively small. The location of an earthquake also plays a role. An oceanic earthquake that is large enough to cause a tsunami can kill more people that a same-size earthquake on land. An earthquake in a remote location causes insignificant damage but the same earthquake in a densely populated area can cause high fatality rates. The type of plate boundary on which an earthquake occurs is also relevant. Extremely large earthquakes are more likely to occur along subduction zones (Table 5.2). The type of an earthquake is also important. Normal and thrust earthquakes can cause devastating tsunami but strike-slip earthquake are much less likely to do so. The epicentral distance to an earthquake is also relevant. A building near the epicenter is much more likely to be damaged than that same building far away.

Site-specific Factors

A number of site-specific factors also determine the risk of damage. Figure 5.1 compares the consequences of two earthquakes of similar size, at similar distances. With a magnitude M_W=6.7 the **17 January 1994 Northridge earthquake** caused significant damages amounting to $18 billion and 65 fatalities. But an earthquake of the same size on **15 October 2006 off the north-coast of the island of Hawaii** caused relatively minor damage, with some landslides and cases of cliff collapse and power outages on Oahu. Though the type of earthquakes was different – the Northridge earthquake was a shallow thrust event while the Hawaii earthquake was an oblique strike-slip event - the local geology clearly also plays a role. The Northridge event occurred in an area that is covered by thick sediments that enhance the shaking from an earthquake (site amplification). The Hawaii earthquake occurred near volcanic islands that consist mainly of solid basement rock. Sediment-covered areas are also vulnerable to a particularly destructive phenomenon called liquefaction (see below) where structures temporarily lose the solid footing on which they were built. As seen in the Hawaii earthquake, even solid ground can harbor risks. Steep cliffs and hills are prone to cliff collapse and landslides, especially when internal faults already weakened the cliffs.

A Non-geological factor that determines the risk of damage and especially fatalities is population density. Large earthquakes in remote areas will hardly make the news but relatively small earthquakes that strike big cities with poorly constructed building can cause high fatality rates. The time of day of an event can also control fatality rates and they may be different in different areas. In areas with poorly built residential housing, the time when people are at home (i.e. at night) is probably the worst time for an earthquake to strike and will cause high fatality rates. In areas with well-built housing but extensive transit systems, the worst time for an earthquake is rush-hour time (see Case Study 1 in Chapter 1).

Figure 6.2a Damage from the 1994 Northridge, CA earthquake to an apartment building that collapsed. (source: Wikipedia)

Figure 6.2b The 25 January 1999 El Quindio, Colombia earthquake caused widespread damage in the city of Armenia. The damage was strongly affected by local site conditions (amplification of ground vibrations) and by local design criteria (dynamic response of buildings and foundations). (source: Nature)

Structural Designs and Local Building Codes

Building designs and local building codes clearly also affect the amount of damage (Figure 6.2). Many apartment buildings in California are essentially built on stilts where the first floor is used as garage space and the upper floors for dwellings. Such buildings are quite unstable. The first floor can collapse as photos from various earthquakes document, including those from the 1994 Northridge earthquake and from the **1989 Loma Prieta earthquake** in the San Francisco Bay area (see below and Case Study). The magnitude MW=6.4 **El Quindio, Colombia earthquake on 25 January 1999** caused nearly 1200 fatalities and the destruction of 8000 coffee farms in Colombia's coffee-growing region. The worst-hit areas were the cities of Armenia were 907 people died and 60% of the buildings were destroyed, including the police and fire stations, Calarca (60% of buildings destroyed) and Pereira (50% of buildings destroyed)[8]. The damage to older buildings not built to code was particularly disturbing as newer buildings next door survived the quake unharmed (Figure 6.2b). Brick-and-mortar buildings and adobe buildings are particularly vulnerable to the shaking of an earthquake (see also the 26 December 2003 Bam, Iran earthquake in chapter 1). In contrast, wood-frame structures can slightly bend and sway during an earthquake but are less likely to collapse.

Building codes do not and should not only apply to buildings but other structures as well such as roadways. Memorable images from the **17 January 1995 Kobe, Japan earthquake** included toppled freeways after the support pillars failed (Case Study 1). Economically, the most significant damage from the 1994 Northridge earthquake probably occurred because of collapsed freeways and overpasses (Figure 6.1 and Chapter 1). One of the affected freeways, the Santa Monica Freeway (Interstate 10) is the busiest freeway in the U.S., or perhaps even worldwide[9]. In an area that has relatively poorly developed public transportation infrastructure, the life of many millions was affected. The freeway reopened on April 12, less than three months after the earthquake and 74 days ahead of schedule, and the contractor received a $14.5-million bonus for this incredible feat. Repairs on the freeway were complete on July 23 where the last two ramps needing repair at La Cienega Blvd were reopened, four days ahead of schedule.

Other vital infrastructure, often called lifelines, that can be damaged during an earthquake include petroleum, gas and water lines. In the U.S., twelve of the twenty largest earthquakes have occurred in Alaska (Table 6.3). Vital petroleum pipelines in Alaska cross earthquake country. Many sections of the pipeline are therefore installed on rollers that can react to earthquake movements (case study 3). Ruptured gas lines can fuel explosive fires and the mix with ruptured water lines and lacking firefighting measures spells utter destruction that reminds of war bombings. The **1906 San Francisco** earthquake is probably one of the most compelling examples of this (chapter 1 and below under Fires). Much was done in the meantime to make gas lines safer. Shut-off valves should help prevent the next disaster, especially automated ones that react to sudden losses in pressure. But one should wonder how well we are prepared given California's never-ending budget crisis. Many water and gas lines are aging beyond their prime time and the 9 September 2010 gas explosion of San Bruno, CA[10], that was not even triggered by earth movements, may be a troubling sign that the San Francisco Bay area is ill-prepared for the next big one. It took rescue crews nearly an hour to determine that the cause of the fire was a gas pipeline explosion. How well will rescue teams be able to respond when several such explosions occur at the same time in different parts of San Francisco? Investigations of the San Bruno disaster revealed that older welds may break from higher gas pressures necessary for modern distribution. Ruptured water lines do not only hamper firefighting. They also disrupt the supply of drinking water which is why every resident in earthquake-prone areas are urged to stock bottled drinking water at all times. Ruptured water lines can also trigger landslides and other mass movements.

Figure 6.3 Large cliff collapse at Daly City triggered by the 17 October 1989 Loma Prieta Earthquake. Being 153 m (500 ft) wide at its base, and displacing 36,700 m^3 (48,000 cubic yard) of material, this was the largest slide encountered in San Mateo County. The slide has moved residences closer to the cliff edge. (source: USGS)

6.2 Earthquakes and Other Disasters

Landslides

Earthquakes can trigger landslides, cliff collapse (e.g. Figure 6.1) and other mass movements, especially on previously unstable hillsides. Nearly every major earthquake in California triggers landslides because much of near-coastal California is located on poorly consolidated sediments (Figure 6.3). According to the USGS, the 1994 Northridge earthquake triggered a staggering 11,000 landslides over and area of 10,000 km^2. Most of them occurred in the Santa Susana Mountains and in mountains north of the Santa Clara River Valley.[(17)]

Earthquake-triggered landslide can be particularly deadly when rivers or lakes are involved. The **8 October 2005 Kashmir earthquake** killed upward of 80,000 people[(12)]. Many of these were killed when a large landslide dammed the Neelum River destroying the village of Makhri on the northern outskirts of Muzzaffarabad (see also Chapter 8). The earthquake destroyed 50% of the city and killed 51,300 people there. Similar earthquakes rattled the region in 1555 and 1974.

Dam Bursts and Floods

Dam bursts can lead to catastrophic floods, and some are triggered by earthquakes. These include failures of temporary dams that form after landslides block the natural flow of rivers. Both the **16 December 1575 and the 22 May 1960 earthquakes in Valdivia, Chile** involved dam bursts. After a magnitude-8.5 earthquake in 1575, a landslide blocked a river flowing from Lake Rinihue, Chile. The dam lasted four months. European settlers evacuated on time, but 1200 native Indians perished downstream[11]. The 1960 Valdivia earthquake caused a similar landslide, damming the outflow from the lake and causing it to rise by more than 20 m (66 ft). The breach of the dam would have flooded many cities downstream, including Valdivia which prepared for evacuations. But a large team of workers prevented a disaster when they open a drainage channel through the landslide. The more recent magnitude **7.4 La Ligua, Chile earthquake on 28 March 1965** caused the failure of two tailings dams at the El Soldado copper mine. Tailings are the materials left over from mining and mineral extration processes. The flows resulting from the dam bursts destroyed the town of El Cobre and killed 400 people[14]. A tailings failure twice as large, also in Chile, was triggered by a large earthquake in 1928[14,15].

The 1994 Northridge earthquake ruptured the **Tapo Canyon tailings dam** which was one of the most striking failures of an earth structure to result from the quake. The dam burst involved a 60-m-wide breach of the 24-m high dam and 60 and 90-m downstream displacements of two sections of the dam. The failure was caused by liquefaction of the tailings. The resulting flow was several hundred meters long. The tailings dam was in a remote area and so did not cause any fatalities. Nevertheless, the failure of the dam considerable economic losses for the owner and affected a water-treatment facility downstream.[14,15]

Figure 6.4 The St. Francis Dam near Los Angeles before the 1928 failure. (source: USGS)

Earthquake-triggered dam bursts have also stimulated minds in the film industry. In 1974 the movie *Earthquake,* the Mulholland Dam in Los Angeles (named the Hollywood Reservoir in the movie) burst following a series of earthquakes to flood Hollywood. The "Hollywood Reservoir" demise was very similar of that of the **St. Francis Dam** on 12 March 1928. Though the St. Francis Dam breach was not triggered by an earthquake, it is nevertheless worth mentioning here. With more than 450 fatalities, the St. Francis Dam breach was California's second-deadliest disaster after the 1906 San Francisco earthquake and one of the worst civil engineering failures of the 20th century. The concrete dam was built between 1924 and 1926 about 65 km (40 mi) northwest of Los Angeles near present-day Santa Clarita to create a reservoir along the Los Angeles Aqueduct. While the reservoir filled during the following two years, cracks appeared in the dam. According to William Mulholland, the dam's builder, there were considered normal for a dam the size of the St. Francis. The reservoir reached full capacity only five days before it breached. On the morning of March 12, the dam keeper notice muddy water near the dam, a sign that its foundation was being eroded. After inspection, Mulholland declared the dam as safe. It failed catastrophically near midnight. The 38-m-high flood waters raced down San Francisquito Canyon into the Santa Clara River and eventually reached the Pacific Ocean 85 km (55 mi) away where the floodwaters were 5km (3 mi) wide. Bodies of victims were recovered as far south as the Mexican border. It is unclear whether landslides shortly before midnight may have contributed to the failure or were part of the failure but there was no earthquake.[(16)] According to the USGS, the dam failed because of a pre-existing landslide in the east abutment. The paleo-landslide became soaked by water from the reservoir and started to move thereby causing the dam failure.[(17)] After the failure of the St. Francis, citizens living below the Mulholland Dam, which was build in 1924, successfully petitioned the City of Los Angeles to reinforce it by piling large amounts of earth and rock on its face.

In recent years, the building of large reservoirs has met increasing resistance, maybe rightly so, especially in densely populated India and China. In late 2000, residents in the western state of **Gujarat, India** heavily protested the construction of the **Sardar Sarovar Dam**, the largest of over 3000 reservoirs to be built along the Narmada River. Residents are concerned that building the reservoirs would have a considerable environmental impact, displace millions of people and that its benefits were limited. The dam project is planned to be completed by 2025 to generate electricity and provide drinking water to millions across Gujarat and neighboring Madhya Pradesh. But the region has also been struck repeatedly by large earthquakes. On 26 January 2001, three months after the protests, a devastating magnitude M_W=7.7 intra-plate earthquake struck Gujarat. About 20,000 people were killed, 167,000 injured, and 600,000 were left homeless. Over 1 million structures were damaged or destroyed, including many historic buildings, tourist destinations, and 60% of usable food and water supplies in Kutch. Who can guarantee that the dams will withstand the earthquakes of the region?

In China, too, concerns about dam safety is on the rise after the **12 May 2008 Sichuan earthquake**. This quake damaged a total of 391 dams though most of them were small. Of particular concern was the integrity of the **Zipingpu Hydroelectric Power Plant** that was only 20 km (12 mi) away from the epicenter, after cracks had been found in the dam. The Zipingpu reservoir had supplied water to Sichuan's fertile for more than 2000 years. It has recently been suggested that this reservoir may have induced the earthquake (Chapter 5). The dam of the **Tulong Reservoir** upstream was in danger of collapse and 2000 troops were sent to ease the pressure by leading water through spillways. A suggestion was also to pour soil into the river leading into the reservoir in the hopes that some of the sediments would clog and cement the cracks in the dam[(13)]. More recently, on 14 April 2010, the magnitude M_W=6.9 Yushu earthquake struck remote and rugged terrain near the border of the Tibet Autonomous Region. The 12th century Thrangu Monastery and surrounding villages were severely damaged and 2700 people were killed. Numerous landslides destroyed infrastructure and made rescue missions a challenge. The earthquake damaged the **Changu Dam** on the Batang River. It was feared that it was close to breaching and endangering the lives of 100,000 people living downstream. There are plans to build 81 new large dams on the upper reaches of the Yangtze, the Mekong and the Salween rivers in Qinghai Province and the Tibetan Autonomous Region. According to internationalrivers.org, China's older dams have a bad safety record. Between 1954 and 2003, 3484 of the countries 85,300 dams collapsed[(18)]. Perhaps, new dams can be built safer than the older ones but there is also growing concern about induced seismicity by such dams (see Chapter 5).

Fires

Earthquakes can cause fire whenever open flames topple over and encounter flammable material, a candle for example. Fires can quickly grow out of control when highly flammable materials fuel the fire and/or the means of firefighting are insufficient or severely compromised. Two examples are mentioned in the following.

The **18 October 1356 Basel, Switzerland earthquake** was accompanied by fire that contributed to the destruction of much of the inner city. Historical records are too sparse to determine exact numbers. Buildings at the time consisted mainly of wood covered with wood shingle roofs though half-timbered houses, with a wood frame and filled with bricks or stones, and brick buildings also existed at the time. The quake destroyed most of the brick buildings and the more flexible half-timbered houses initially fared better. The quake happened late in the evening so many people probably used candles and oil lamps that toppled over during the quake. Fires broke out simultaneously in the city and there were not enough means to fight them. The quake killed relatively few people, perhaps a few hundred. Historians speculate that there were not many people to start with as over half of the population was killed in a famine and the Black Death a few years earlier. It is also possible that people were warned by foreshocks and had left the city before the main shock occurred. There simply may not have been enough people to fight the fires. The fires burned out of control and only stopped after they had consumed much of the inner city after eight days.

Figure 6.5 The burning of San Francisco after the 18 April 1906 earthquake, view from St. Francis Hotel. (source: Wikipedia/Library of Congress)

550 years later, the **18 April 1906 San Francisco earthquake** destroyed 80% of the city (see also Chapter 1). Most of the destruction (some sources cite 90%) was not caused by the shaking but by the subsequent fires that quickly raged out of control (Figure 6.5). Over 30 fires destroyed about 25,000 buildings on 490 city blocks. The fires burned for four days. Several factors led to this tragic disaster. Firstly, much of the city consisted of wood frame buildings and some were concerned before the earthquake that the city was a disaster waiting to happen. Secondly, many fires were fueled by ruptured gas lines. And even though no reservoir that stored the city's water supplied breached, many supply lines within the city broke so that no water was available to fight many fires. Many fire engines were also trapped in the collapsed buildings. Most disturbingly however seems the fact that many fires were fueled by uncoordinated disaster response and other human action. The city's fire chief died in the quake. Firefighters then lacked experienced leadership and coordination. In an attempt to stop the spread of fires, fire fighters and the army used dynamite and black powder to blast damaged buildings. Such attempts were unsuccessful as many ruins caught fire and accelerated the spread of the already burning fires. Lastly and perhaps most disturbingly, many people purposely set their homes on fire. The reason was that their insurance would not pay for earthquake-damaged homes but would pay for burned homes.(19)

Both events described here happened a long time ago so one could argue that modern disaster relief efforts are much more efficient than they were 100 years ago. The German tourist guide book mentions that San Francisco's shiny fire engines are now always parked on the street to prevent a repeat of the 1906 debacle. But with the modernization and improvement of relief efforts also came an explosive development of structures and lifelines that could break as they age (see the San Bruno, CA gas explosion). And large earthquakes along the San Andreas Fault will most certainly happen again. What if ... ?

Volcanoes

On a volcano that is shortly before an eruption, there is a fine balance between the pressure exerted from the magma chamber and the ambient pressure. The pressure in the magma chamber drives the ejection of material while the ambient pressure tries to prevent it. An eruption occurs when the ambient pressure can no longer counteract the pressure from the magma chamber. Rupturing earthquakes change the ambient stress field (force) and therefore ambient pressure (force per area). Earthquakes therefore influence the likelihood of imminent eruptions. The most prominent example of this is the **18 May 1980 eruption of Mt. St. Helens, WA** in the Cascade volcano chain. Before the eruption, the volcano was closely monitored after an earthquake in March indicated increased activity. A magnitude-5 earthquake on the morning of 18 May triggered a massive landslide on the northern flank of St. Helens. There is some debate whether the landslide itself produced the earthquake signal or the earthquake triggered the landslide.

Within seconds, the sudden pressure relief allowed the volcano to explode in one of the most impressive volcanic eruptions in the last 35 years.

The Mt. St. Helens earthquake and eruption occurred on the same volcano. But even earthquakes farther away can influence volcanic activity. The great 22 May 1960 Valdivia, Chile earthquake triggered the eruption of **Puyehue-Cordon Caulle volcano** 38 hours later, 135 km (85 mi) inland from the Pacific coast. This eruption received little attention because the volcano is located in a sparsely populated area. The eruption occurred through a 5.5-km-long (3.4 mi) fissure (a crack) from 21 vents (volcano openings). The eruption ended two months later on 22 July.

Earthquakes can also affect volcanic activity very far away. The 9 November 2002 Denali earthquake (case study 3) was large enough to trigger small earthquakes in **Yellowstone National Park** 3250 km (2000 mi) away. Scientists speculate that the strong Rayleigh waves temporarily changed the stress field in Yellowstone enough to trigger these quakes. More recent research by scientist at the University if Utah also found that the Denali earthquake changed the timing and behavior of some of Yellowstone's geysers and hot springs. Some small hot springs suddenly started to boil with eruptions as high as 1 m. The temperature of some springs changed from 42°C to 93°C (108°F to 199°F) and became less acidic. Some geysers erupted more frequently while other erupted less frequently(20).

Earthquakes can also trigger the formation and eruption of volcano-like features, including Sand blows, sand volcanoes and mud volcanoes. These involve the liquefaction of sediments as discussed in the next section.

Earthquakes

Earthquakes change the stress field in the greater area around a fault; this change can trigger earthquakes further down on the same fault or neighboring faults; e.g. the magnitude 7.2 28 June 1992 Landers Earthquake in the Mojave desert triggered a magnitude 6.5 earthquake in Big Bear 3 hours later naturally occurring fluctuations in ground water table can also reduce friction and trigger earthquakes and landslides

Earthquake triggering

Earthquakes can trigger other earthquakes on nearby faults;

- e.g. the M_W7.3 28 June 1992 Landers EQ in the Mojave desert triggered the M_W6.5 Big Bear earthquake 3 hours later. The Big Bear earthquake is not an aftershock of the Landers quake!
- e.g. the M_W6.5 22 December 2003 San Simeon EQ that occurred on a neighboring fault of the SAF and killed 2 people in Paso Robles has likely increased the chance of an EQ to occur on the SAF in the Parkfield area. This quake also triggered hot springs activity. A M_W 6.0 EQ occurred near Parkfield several months later on 28 September 2004.

Diseases

- diseases: e.g. when access to clean drinking water is disrupted;
- cholera, Haiti.
- e.g. an unusual effect of the 1994 Northridge EQ was an outbreak of Valley Fever, a respiratory disease. Spores that cause valley fever were released from the soil by the landslides and blown toward coastal communities were people died from the disease [17]

Pollution

- pollution: e.g. when chemical, bio-hazardous, radioactive materials from damaged human structures enter environment

6.3 Liquefaction

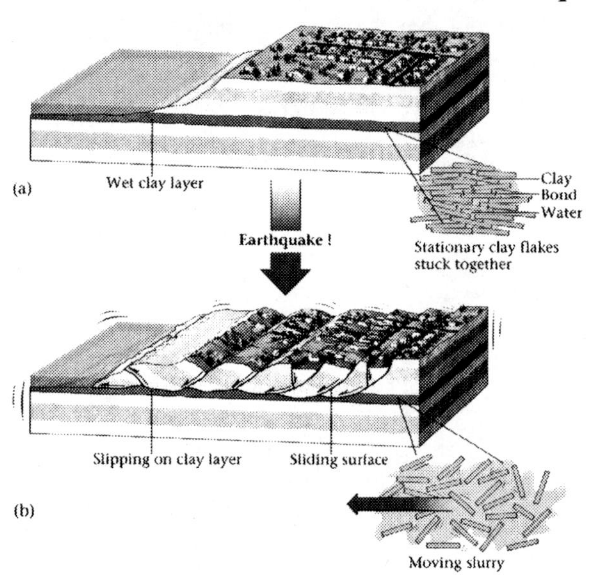

Figure 6.6 The Turnagain Heights, Alaska disaster. (a) Before the earthquake, wet clay packed together in a subsurface layer of compacted but wet mud. (b) Ground shaking caused liquefaction of the wet clay layer in the sediment beneath a housing development. As a consequence, the land slumped, blocks slid seawards, carrying the homes with them. (source: ME)

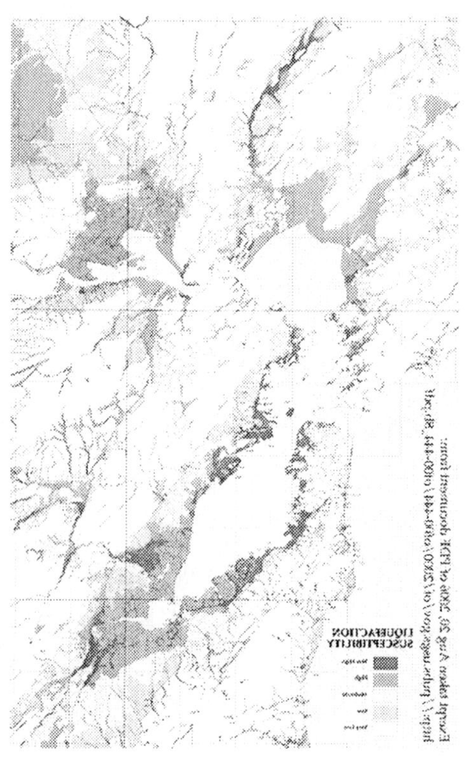

Figure 6.7 A liquefaction susceptibility map – excerpts of USGS map for the San Francisco Bay Area. Many areas of concern in this region are densely urbanized. (source: Wikipedia/USGS)

Figure 6.8 Liquefaction in San Diego could occur in Mission Bay and Lindbergh Field which are located on sediments near sea level (high groundwater level). Being located on old river sediments, La Jolla Shores further north is also susceptible to liquefaction. (source: google maps)

the shaking from earthquakes can make certain types of ground behave temporarily like a liquid

- o significantly enhances effects from even relatively small EQs
- o ground may seem solid/stable under normal conditions
- o structures build on such ground lose support, may topple over as complete structure

examples for geology prone to liquefaction:

- o unconsolidated sediments
- o clay
- o weak sediments underlying mechanically strong layers particularly treacherous as risk often not recognizable on the surface

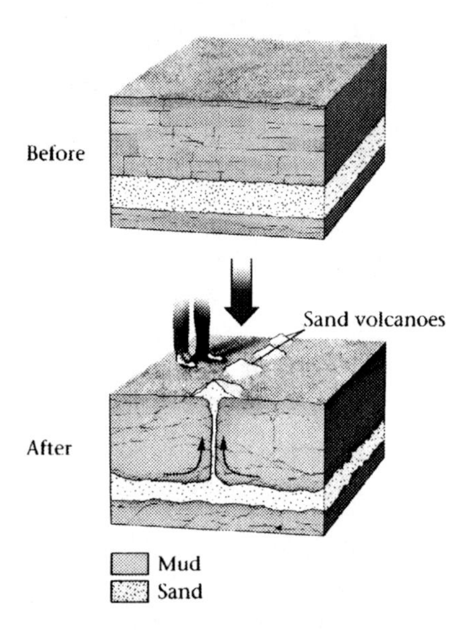

Figure 6.9 Sand blows, sand volcanoes and mud volcanoes can form during ground shaking when shaken wet sand from below the surface squirts up between cracks. Generally, these features are small but large ones, tens of meters across, formed after the 1811 New Madrid earthquake. (source: ME)

Sand blows were observed after the magnitude 7.7 Charleston, SC earthquake on 31 August 1886 and the 1811-1812 series of earthquakes in the New Madrid seismic zone (see section 6.5). More recently, the magnitude M_W=7.1 Canterbury earthquake in New Zealand on 4 September 2010 and the 11 March 2011 Tohoku, Japan earthquake caused sand blows. Sand blows and **sand volcanoes** are formed when sand from below is ejected onto the surface from a central point. **Mud volcanoes** form in more vigorous processes when gases such as methane are involved. Triggered by an earthquake, all these phenomena result from soil that is liquefied underground and escapes through an opening such as a newly formed crack. Mud volcanoes are naturally occurring features and are discussed in chapter 7.

Sand and mud volcanoes: form when the pressure in a weak layer below a strong layer get so great that material gets pushed upward through an edifice or crack. The eruption of the **Lusi mud volcano** is a still ongoing human-induced mud volcano.

Figure 6.10a The May 2006 eruption of the Lusi mud volcano in Indonesia shortly after an earthquake 280 km (174 mi) away. It was later established that drilling for gas and oil caused the volcano. (source: Xinhua/Reuters and Xi'an Center of Geological Survey)

Figure 6.10b July 2011 Google Earth image of the Lusi mud volcano. Names of destroyed and nearby villages were added.

The **Lusi mud volcano**, about 10 km (6 mi) south of the city of Sidoarjo in East Java, Indonesia was blamed on earthquake triggering but it was likely caused by human activity.[21] Lusi is a contraction of Lumpur (Indonesian for mud) Sidoarjo. The mud volcano is a clearly visible feature in Google Earth where it is called the Lapindo mudflow. On 29 May 2006 at 5:00 am local time, a mud volcano started to form. The production of mud peaked at 180,000 m^3 (6.36 million cubic ft) per day and still averaged 30,000 m^3 (1 million cubic ft) as of early 2011. The mud flooded a large area and initially destroyed more than 10,000 homes, schools and other structures. It displaced 30,000 people in Ketapang, Siring and Kalisampumo (Figure 6.7). It is within a few 100 m of the Kali Porong river and some fear that it will have an impact on coastal and marine ecosystems. Since 2008, the mud volcano is contained by levees but the resultant flow still disrupts highways and villages. Scientists estimate that the mud volcano will likely continue to erupt for two to three decades.

So why did the mud volcano form? The answer lies in the local geology. The mud volcano is located in the East Java Basin that contains large amounts of oil and gas. There are three major oil and gas fields near the mud volcano. At the site of the mud volcano, a prospecting company drilled 2835 m (9300 ft) through clay and other impermeable rock into a permeable layer that promised to contain the desired gas. At about 5:00 am on 29 May, water, steam and a small amount of gas erupted at a location about 200 m (655 ft) away from the well. The exact sequence of events seems somewhat unclear but consistent between several references is the fact that on the night before, the prospector drilled into limestone that contained pressurized water. The upper part of the borehole was protected by steel casing but the lower part was not. The prospector noticed a 'kick', a massive influx of water but then a drop in pressure at 9:30 pm, a sign that the water escaped elsewhere out of the borehole. Scientists hypothesize that hydro-fracturing produced cracks through which the water could reach the surface. There were earlier unusual signs that something is amiss. At about 6:00 am on 27 May, the operator observed a loss of circulation. This occurs when the mud pumped into the borehole during drilling is not circulating back to the surface. Some time after that, the borehole suffered a complete loss of mud. This indicates that the mud penetrated into the surrounding rock thereby setting the stage for the formation of the mud volcano.

But the prospector blamed the calamity on the magnitude M_W=6.3 earthquake that struck Java on 5:54 am local time on 27 May, about 250 km (155 mi) from the borehole. At an average travel speed of 3.5 km/s Rayleigh waves (the strongest wave) would take 71 s to reach the borehole but the operator emphasized that mud loss was noted 7 minutes after the quake. The quake killed 5400 people in the Bantul-Yogyakarta area. Shaking near the epicenter reached MMI intensity IX at Bantul but only II to III at the borehole site. There were several strong aftershocks, including a M_b=4.4 on 27 May at 8:07 am local time, a M_b=4.8 on 27 May at 10:10 am and a M_b=4.6 at 11:20 am. None of these aftershocks coincided with any unusual event at the borehole. The Center of Volcanology and Geological Hazard Mitigation (CVGHM) of the Smithsonian Global Volcanism Program (http://www.volcano.si.edu) states that a magnitude 5.9 earthquake coincided with volcanic activity (pyroclastic flows) on Mt. Merapi. Since the National Earthquake Information Center of the USGS (http://neic.ucsd.gov) does not list an event of such a magnitude, the M_W=6.3 may have been meant here. Activity at Merapi, Indonesia's currently most active volcano, had increased long before the earthquake so an immediate relationship is not clear. But Merapi was only 47 km (29 mi) from the epicenter so it is possible that the quake influenced Merapi's activity. In late 2008, scientists discussed the possible causes of the Lusi mud volcano at a meeting of the American Association of Petroleum Geology. The vast majority of scientists agreed that the drilling was at least partially responsible for this disaster.

6.4 Tsunami

Figure 6.11 The 2004 Sumatra tsunami strikes Ao Nang, Thailand, 2.5 hours after the earthquake. The earthquake was felt in Thailand, 2000 km (1242 mi) from the epicenter. (source: Wikipedia)

Figure 6.12 A town near the coast of Sumatra lies in ruin on 2 January 2005. This picture was taken by a U.S. military helicopter crew from the USS Abraham Lincoln that was conducting humanitarian operations. (source: Wikipedia)

Figure 6.13 The flooding and destruction of the Sendai airport by the 11 March 2011 Tohoku earthquake tsunami. (source: wikipedia)

Figure 6.14 People in Hilo, HI running from the tsunami caused by the 1 April 1946 Aleutian Islands earthquake. The tsunami killed 159 people in Hawaii) and prompted the creation of the Seismic Sea Wave Warning System. (source: Wikipedia/USGS)

Figure 6.15 This iconic photo shows parking meters in Hilo, HI bent by the tsunami from the 22 May 1960 Valdivia, Chile earthquake, the largest recorded earthquake to-date. Despite the then functioning tsunami warning system 61 people in Hawaii were killed and 282 seriously injured. (source: NOAA/USGS)

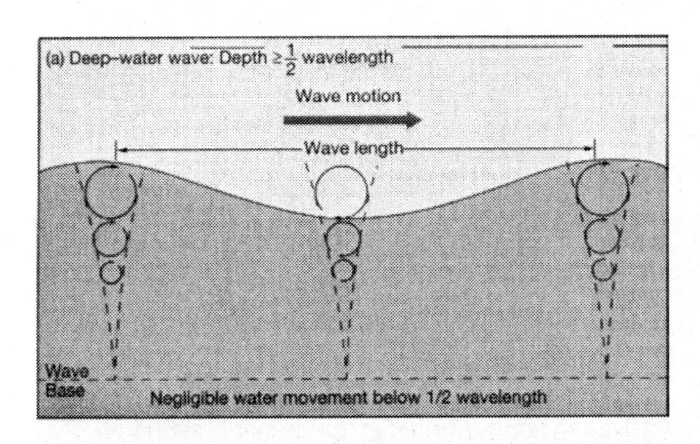

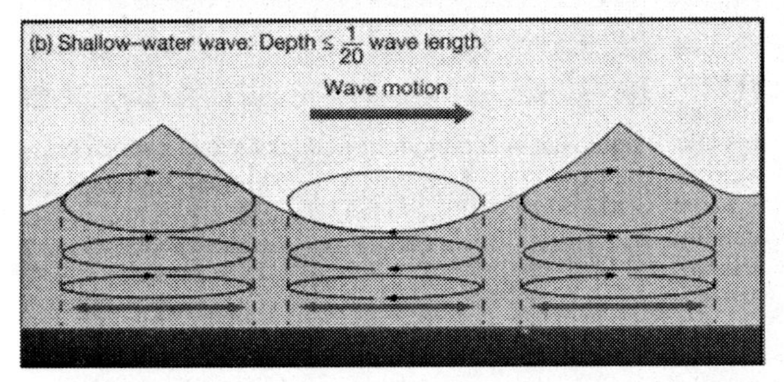

Figure 6.16 Characteristics of deep- and shallow-water waves. (a) Deep-water waves exhibit a diminishing size of the circular orbits of water particles with increasing depth. (b) Shallow-water waves, where the ocean floor interferes with circular orbital motion causes to orbit to become flattened. (source: TT)

Ocean Wind Waves

Tsunami are very difference from regular ocean waves that we observe every day on the beach. We therefore need to review some basic features of ocean waves first. Regular ocean waves are generated by wind that blows across the sea surface. They are call ***gravity waves*** because the propagation speed of these waves is controlled mainly by gravitational forces.

- unlike tsunami, ordinary ocean waves (ocean gravity waves) are generated by winds blowing across the sea surface
- wave height depends on
    - wind speed
    - duration of winds
    - fetch: size of area over which winds blow consistently
- wave height usually less than 3m
- wavelength: usually between 60 and 150 m
- deep-water waves: Waves don't sense the bottom. When water depth is greater than 1/2 wavelength **l**. Wave speed depends on wavelength: v=l/T, where **T** is the wave period. It turns out that if **T** is known, v can be obtained through v=g x T/(2π), where **g** is the gravitational acceleration (9.81 m/s^2) (the dependence on g gives ocean waves the term ***gravity waves***.)
- shallow-water waves: Waves sense the bottom. When water depth is less than 1/20 wavelength **l**. Wave speed depends on water depth **D**: v=sqrt(D x g) (see tsunami above)

Largest wave heights are typically less than 18m but can approach this height in the Southern Oceans (around Antarctica); some rogue waves have

Tsunami

Tsunami have the potential to be one of the great killers during an earthquake, particularly in areas that lack a functioning tsunami warning system (see 26 December 2004 Sumatra-Andaman earthquake). The term tsunami comes the Japanese word from Japanese "tsu" for harbor and "nami" for wave (singular/plural both tsunami). This name may come from the fact that tsunami are not noticeable in the open ocean but only when they approach the coast, or a harbor. Tsunami are **not** generated by wind, but by any process that involves **vertical movement** is ocean-floor topography: 1) submarine volcanic eruptions; 2) submarine landslides or sudden collapse; 3) submarine normal and reverse earthquakes. Tsunami are falsely called "tidal waves" even though they have nothing to do with tides.

- generated by initial displacement of incompressible water
- tsunami are **shallow-water waves**: the wavelength **l** is so large (200km) that they sense the bottom even in the deepest oceans; the speed **v** then depends only on the water depth **D** and the gravitational acceleration **g** (9.81m/s^2).

$$v = \text{sqrt}\,(D \times g)$$

- in the open ocean, tsunami are extremely fast; e.g. at water depths of around 5500m, tsunami travel at 230m/s (835km/h; 515mph)
- due to low dissipation (slow decay in wave amplitude), tsunami can cross large ocean basins
- using the formula: tsunami take about 6.5h from Alaska to Hawaii; about 14h from Chile to Hawaii; about 5h from Hawaii to California
- in the open ocean, tsunami are hard to detect:
    1. the amplitude of tsunami is less than a meter
    2. the wavelength is so large that a ship does not notice a passing tsunami
- upon approaching the shore tsunami "built up": trailing parts, traveling in deeper water, are faster and catch up with leading parts traveling in shallower parts (see equation above)

- drawdown

    - the largest tsunami can be 10m high or even higher after crossing an ocean basin (e.g. Chile 1960 tsunami arriving in Hawaii was nearly 18m in some bays)
    - due to the long wavelength, the time between a wave crest and the next crest is 15 min (time=l/v), in the open ocean; approaching the coast, the waves slow down and this interval becomes longer

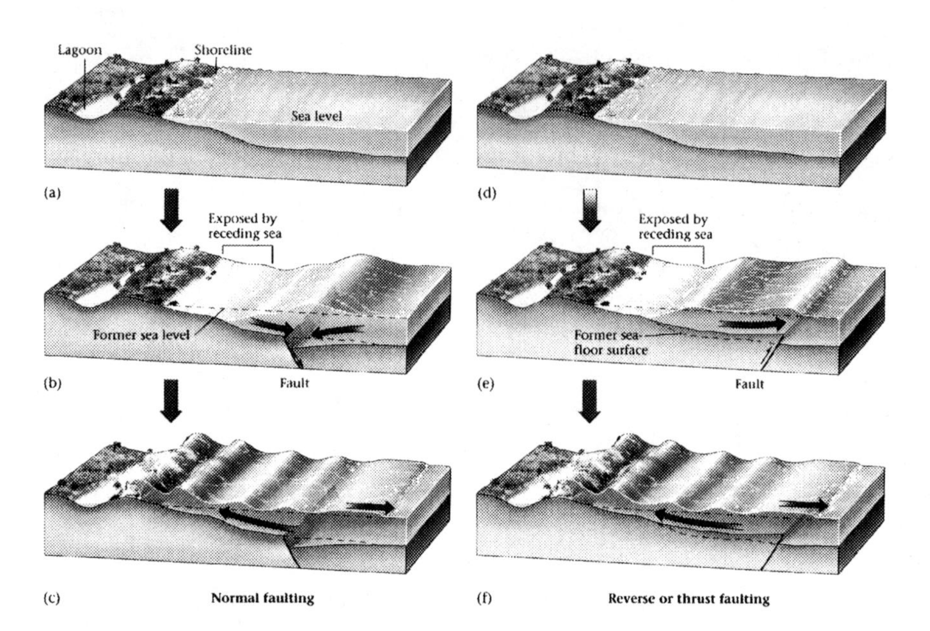

Figure 6.17 Diagrams illustrating how sea level behaves after a tsunamigenic earthquake. (a) and (d) before a normal and a reverse earthquake. (b) and (e) shows the initial recession of sea level after both types of earthquakes. (c) and (f) the run-up for the first tsunami wave. (source: ME)

Figure 6.18 Maximum recession of the ocean at Kata Noi Beach, Thailand, before the third and strongest tsunami wave strikes at 10:25 am local time. The ocean water is visible in the upper right corner and the beach on the extreme left. (source:Wikipedia)

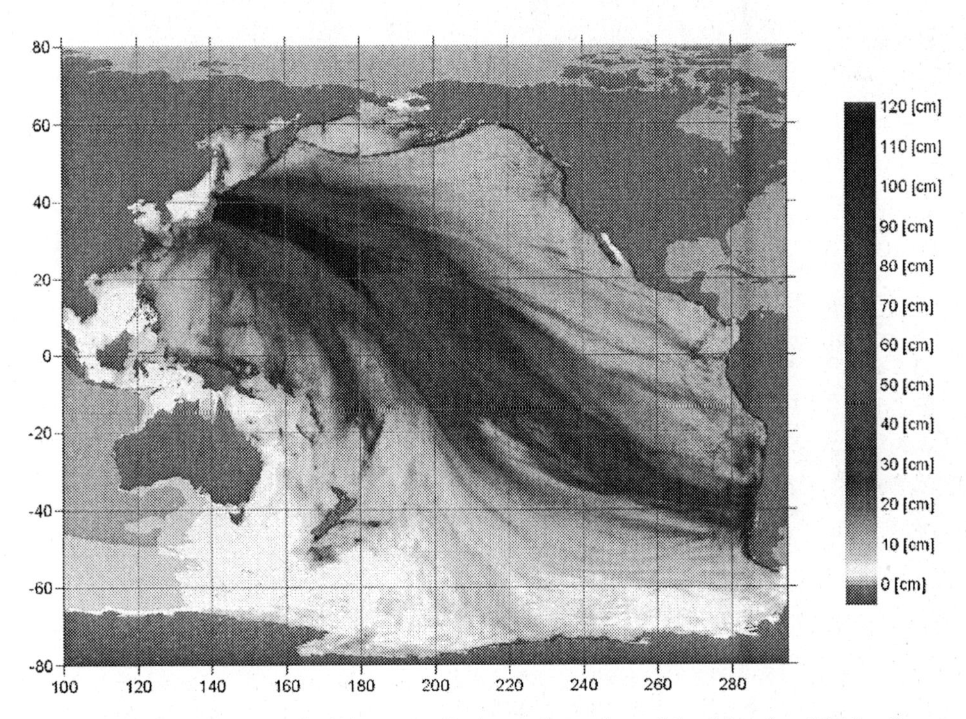

Figure 6.19 Wave height prediction for the 11 March 2011 Japan tsunami. (source: Wikipedia/NOAA)

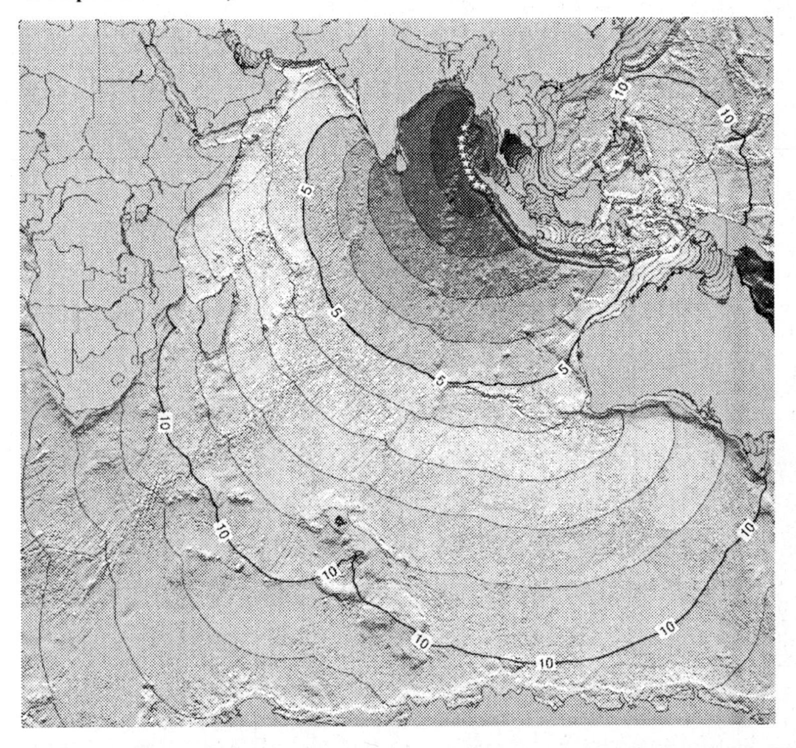

Figure 6.20 Tsunami travel time map (in hours) for the 26 December 2004 Sumatra-Andaman tsunami. (source: NGDC/NOAA)

Crescent City and the 1964 EQ!

Figure 6.21 Tsunami damage in Crescent City from the 1964 Good Friday Alaska earthquake. (source: npr and [(2)])

NB: depending on the geometry of the beach, a tell-tale sign of a coming tsunami is often that the water recedes in the space of a few minutes; a person knowing the tide calendar would therefore know that something unusual is about to happen

0. water recedes faster than the usual low tide
1. timing of receding water is wrong

People not informed about tsunami may miss these tell-tale signs. Tragically, this was the case during the tsunami of the great December 26, 2004 Sumatra earthquake. Many people misread the sign of fish flopping on the beach sand. Instead of getting to higher grounds, people collected the fish ... and drowned in the tsunami a few minutes later.

- a tsunami warning system was put in place in the Pacific Ocean after a fatal tsunami generated by the 1946 Aleutian Islands earthquake killed people in Hawaii
- a tsunami warning system was incomplete/not functioning in the Indian Ocean during the 2004 Sumatra-Andaman Earthquake. Tsunami in the Indian Ocean are less common than in the Pacific Ocean. As a consequence of this, there is less collective memory/education about tsunami risks.

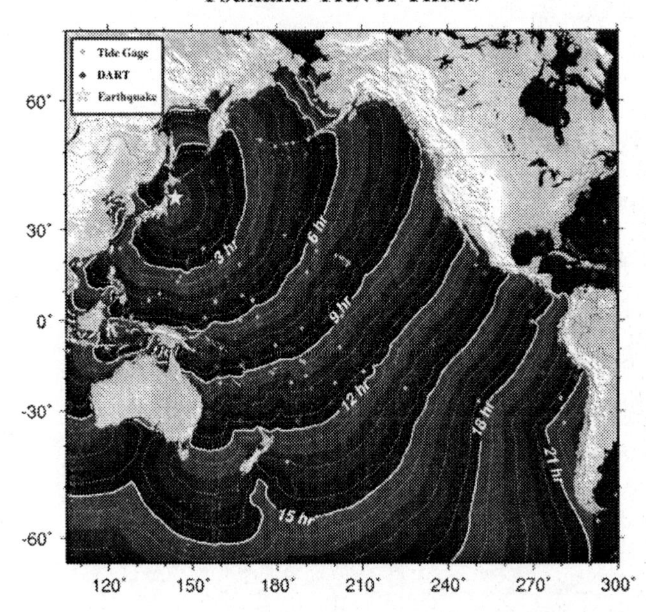

Figure 6.22 A widely circulated tsunami travel time map for the 11 March 2011 Tohoku, Japan tsunami. This map was modified from the original NOAA map. (source: NOAA)

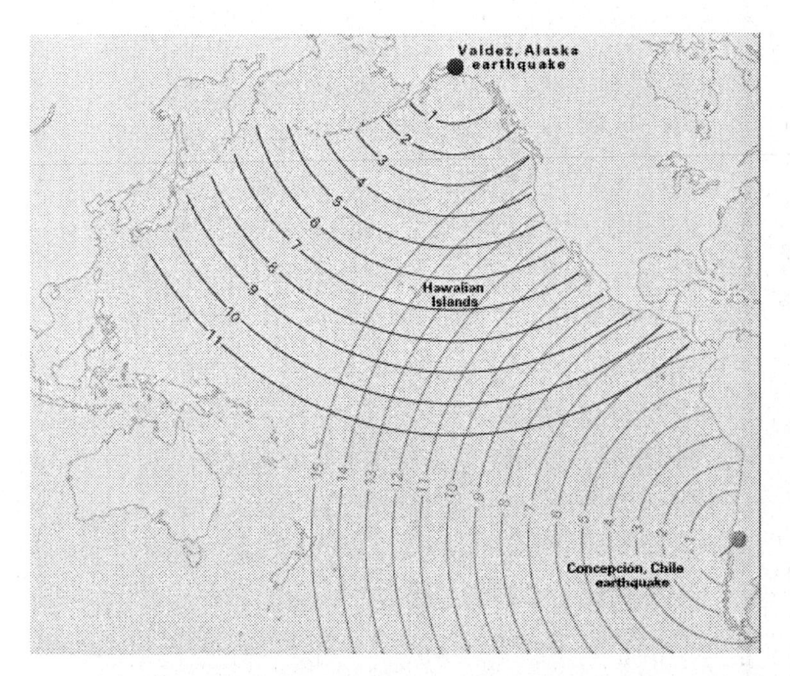

Figure 6.23 Tsunami travel time map (in hours) for the 1964 Alaska Good Friday earthquake and the 22 May 1960 Chile earthquake. (source: Wikimedia/USGS)

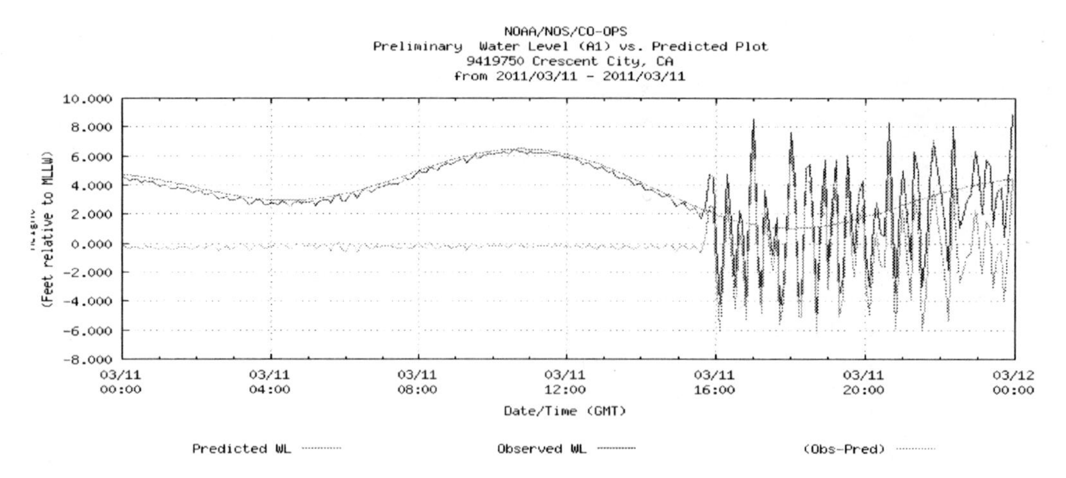

Figure 6.24 Tide gauge record at station 9419750 Crescent City, CA of the 11 March 2011 Japan tsunami. Tide gauge plots can be created interactively at http://tidesandcurrents.noaa.gov (source: NOAA)

Figure 6.25 Anomalous colliding currents from the 11 March 2011 Japan tsunami observed in San Diego Bay (just behind the boat) as they leave a little inlet. The waves are not from wind (no wind that day), ebbing tides (do not occur as turbulent currents) nor boats. The height of these waves is nearly insignificant, yet the powerful currents can cause damage to docks and other structure. (photo: Eric Buck, SIO)

Table 6.3 The 20 Largest Earthquakes in North America[5]

Rank	Location	Date	Magnitude
1	Prince William Sound, Alaska	28 Mar 1964	9.2
2	Cascadia subduction zone	26 Jan 1700	~9
3	Rat Islands, Alaska	04 Feb 1965	8.7
4	Andreanof Islands, Alaska	09 Mar 1957	8.6
5	East of Shumagin Islands, Alaska	10 Nov 1938	8.2
6	Unimak Islands, Alaska	01 Apr 1946	8.1
7	Yakutat Bay, Alaska	10 Sep 1899	8.0
8	Denali Fault, Alaska	03 Nov 2002	7.9
9	Gulf of Alaska, Alaska	30 Nov 1987	7.9
10	Andreanov Islands, Alaska	07 May 1986	7.9
11	Near Cape Yakataga, Alaska	04 Sep 1899	7.9
12	Ka-u District, Island of Hawaii	03 Apr 1868	7.9
13	Fort Tejon, California	09 Jan 1857	7.9
14	Rat Islands, Alaska	17 Nov 2003	7.8
15	Andreanof Islands, Alaska	10 Jun 1996	7.8
16	San Francisco, California	18 Apr 1906	7.8
17	Imperial Valley, California	24 Feb 1892	7.8
18	New Madrid, Missouri	16 Dec 1811	7.7
19	New Madrid, Missouri	07 Feb 1812	7.7
20	New Madrid, Missouri	23 Jan 1812	7.5

6.5 Earthquakes and Risk in North America

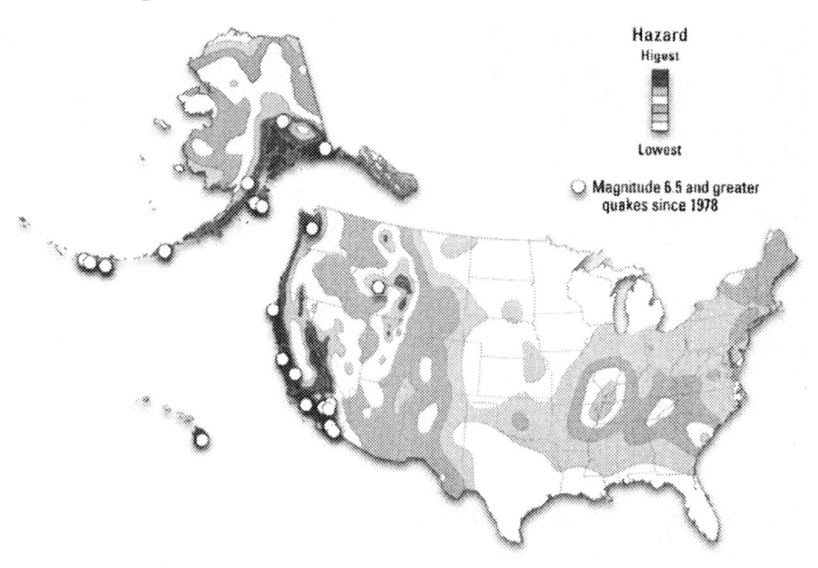

Figure 6.26 Seismic hazard map for the U.S.. (source: USGS)

Features of the North American/Pacific Transform Boundary System

- complicated; SAF dominant/longest fault but not the only one (Fig. 5.22/4th 4.24)
- faults mostly parallel to SAF, but some are not
- SAF not only fault on which earthquakes occur (Fig. 5.29/4th 4.31); e.g. in San Francisco area, Hayward and Calaveras faults also quite active
- structure along entire length of SAF complicated (Fig. 5.12/4th 4.13)
- Big Bend north of L.A. makes SAF go east before continuing south, east of Salton Sea
- many earthquakes in Southern California occur in Mojave desert, east of SAF; and on faults west of SAF
- San Jacinto fault most active fault in San Diego county

CASCADIA!

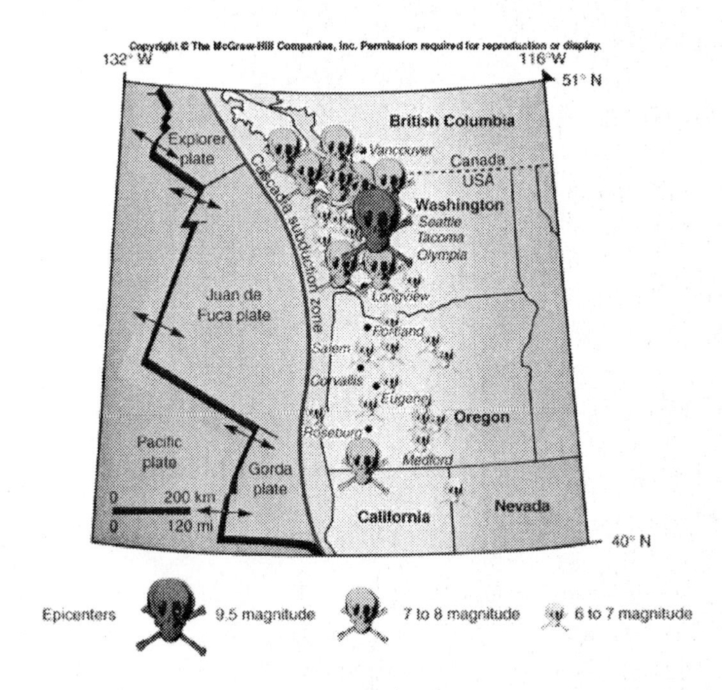

Figure 6.27 Epicenters for the 1960 Chile earthquake sequence are plotted over the Cascadia subduction zone. Earthquake magnitudes were one 9.5; nine between 7 and 8; and 28 between 6 and 7. (source: AB)

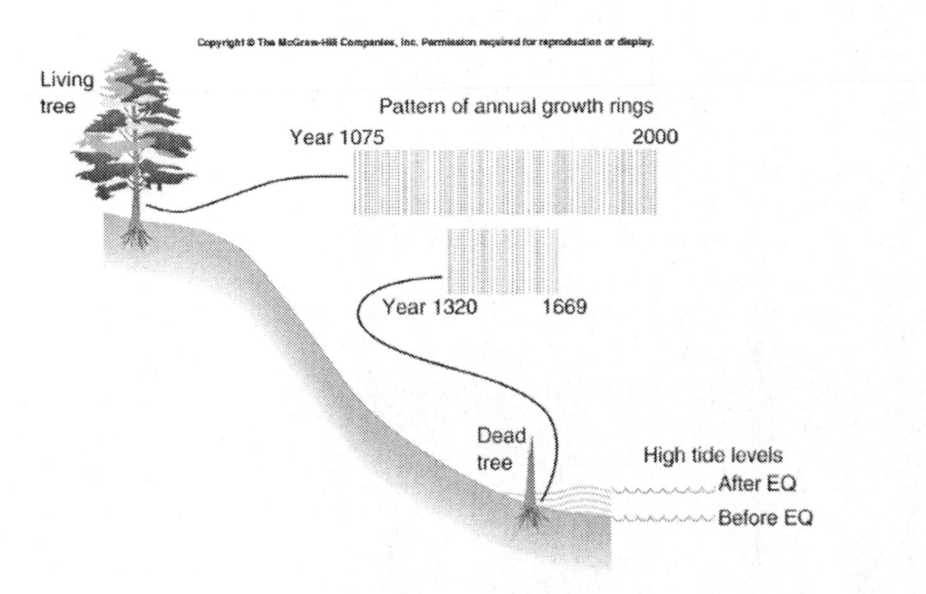

Figure 6.28 Annual growth rings in drowned trees along the Oregon-Washington-British Columbia coast tell of their deaths after the 1699 growing season. Seawater flooding occurred as land dropped during a magnitude 9 earthquake. (source: AB)

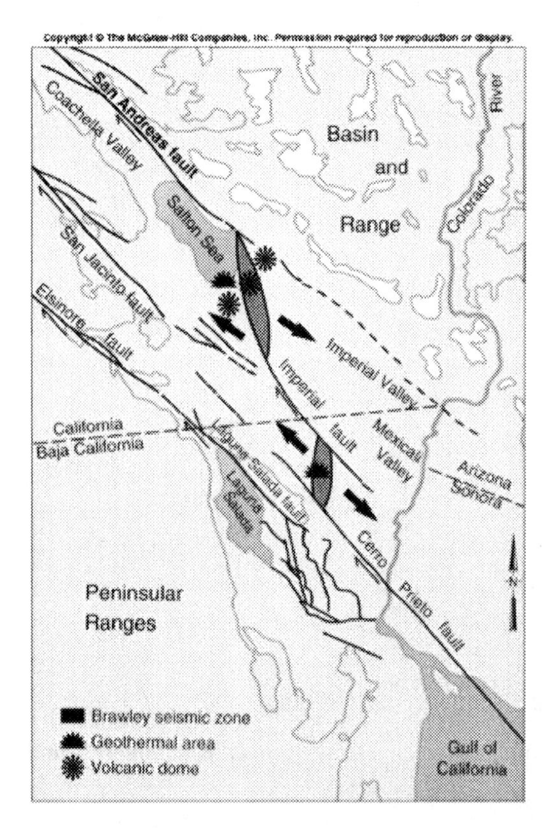

Figure 6.29 Map of the northernmost section of the Gulf of California together with active seismic faults. (source: AB)

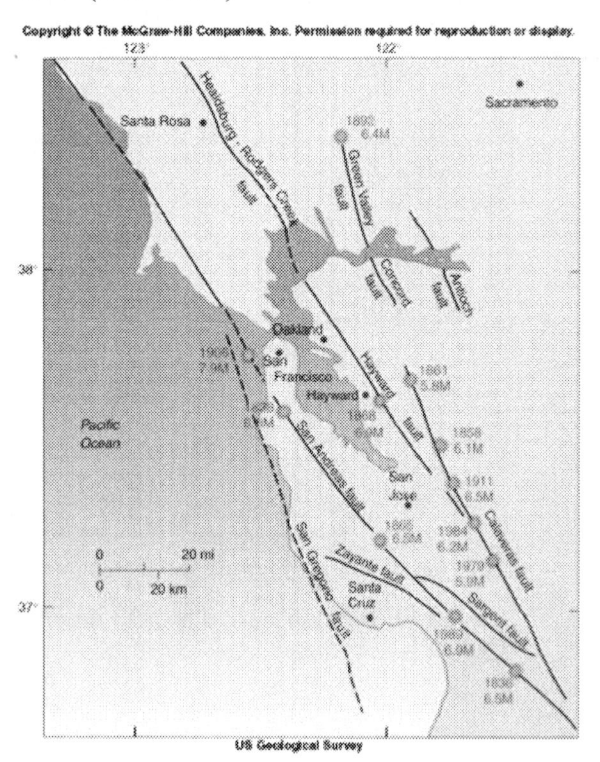

Figure 6.30 Locations and approximate sizes of some larger Bay Area earthquakes. (source: AB)

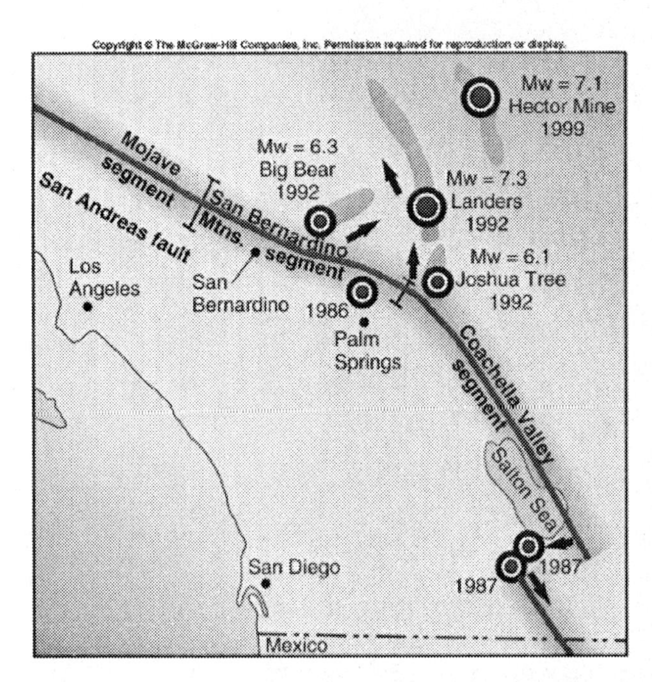

Figure 6.31 Map of major earthquakes near the northern and southern ends of the Coachella Valley segment of the San Andreas Fault. The triangular block of crust near the northern end has moved northward. (source: AB)

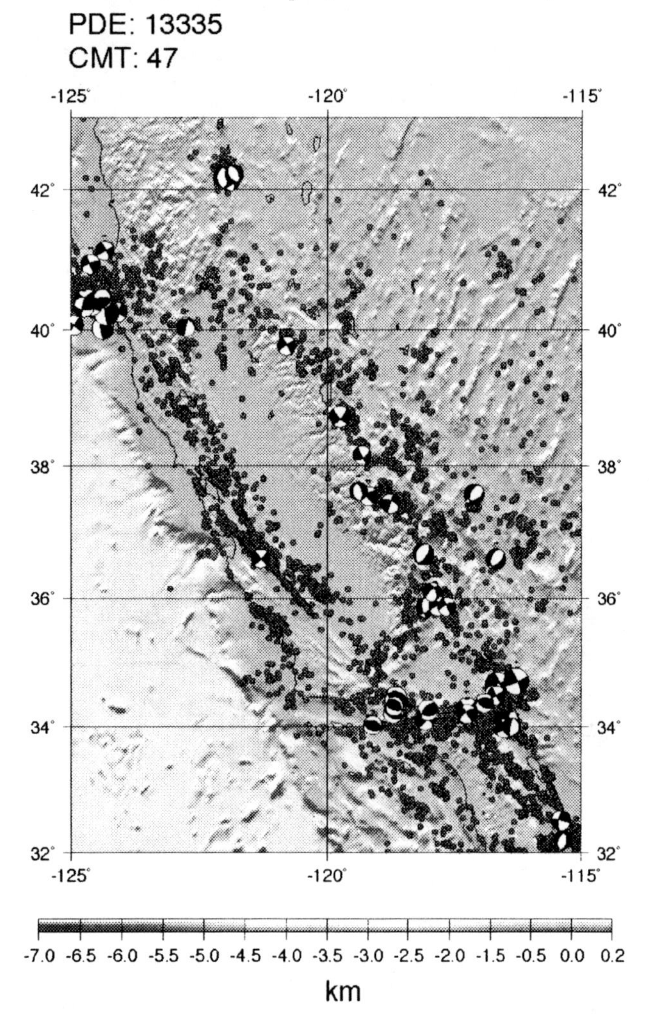

Figure 6.32 Earthquakes in California 1990 through 2003. The events marked by a beach ball have magnitudes of about 5 and greater. Earthquakes do not only occur on the San Andreas Fault. In fact, increased seismicity occurs at the northern end of the SAF and along the Big Bend in the south.

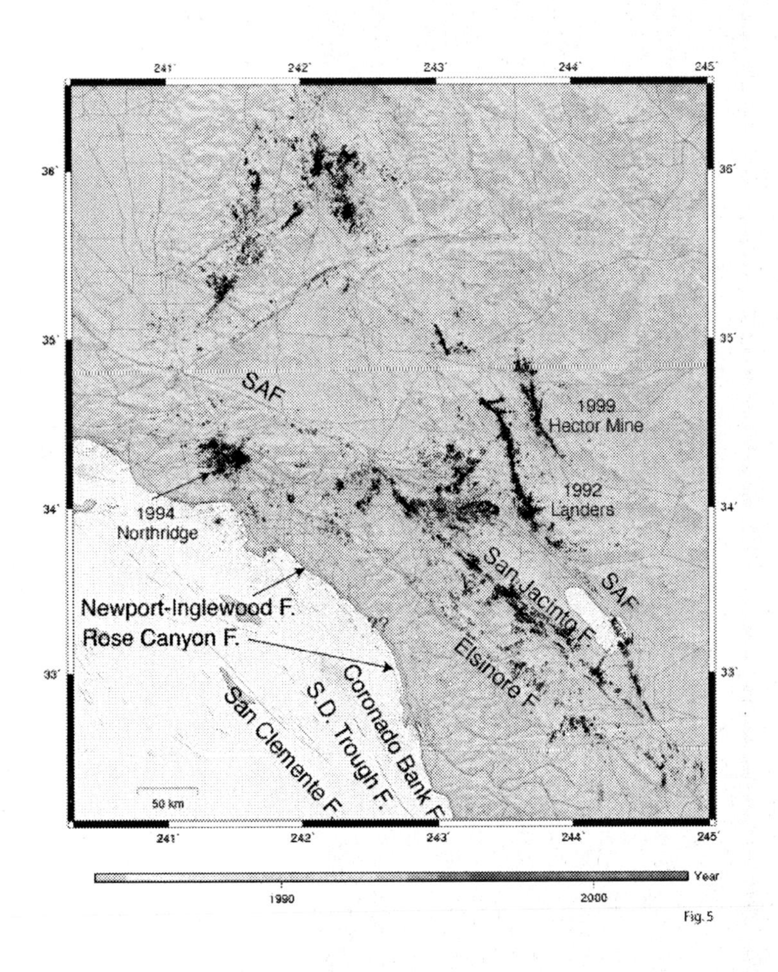

Figure 6.33 Earthquakes in Southern California through 2003. Major faults and earthquake sequences are marked. (courtesy Peter Shearer)

Where does San Diego come in?

- compared to other southern CA areas, San Diego county has sparse seismicity
- but active faults do exist, even within the City of S.D.
- seismic records (instruments and historic) poor (no records prior to 1800)
- currently, San Jacinto fault most active fault in S.D. county
- recurrence time for mag. 6 EQ along San Jacinto 8-12 years
- last mag. 6 EQ along San Jacinto: 1987!!!!! (we are overdue)
- Rose Canyon fault (RCF) along I-5 thought to be one of S.D. city's most hazardous
- RCF continues off-shore north of La Jolla for at least 25km
- Right-lateral slip has occurred along the RFC within the last 8000 years at an average rate of about 1-2mm/yr

- if slip occurred in EQs (and not aseismically), then RCF might have been fault of two mag 6 EQ (1800/1862)
- epicenter of two intensity VII EQ within city limits in last 30 years (1964 San Diego Bay; 1986 S.E. San Diego)
- Oct. 7, 2003, intensity III EQ in San Ysidro largest within city limits since 1986
- many active faults off-shore
- connectivity of off-shore faults uncertain (e.g. Does RCF connect north to Inglewood fault, making a very large EQ more likely)
- Jun. 15, 2004, mag. 5.2 EQ along San Clemente fault (100km off-shore) caused intensity IV shaking in San Diego; closing down Amtrak and Sea World for damage inspection.

Particular Seismic Hazards in San Diego

- Mission Bay, Lindberg Field, North Island build on unconsolidated sediments (liquefaction) (Figure 6.8)
- vertical movement on off-shore faults can cause tsunami

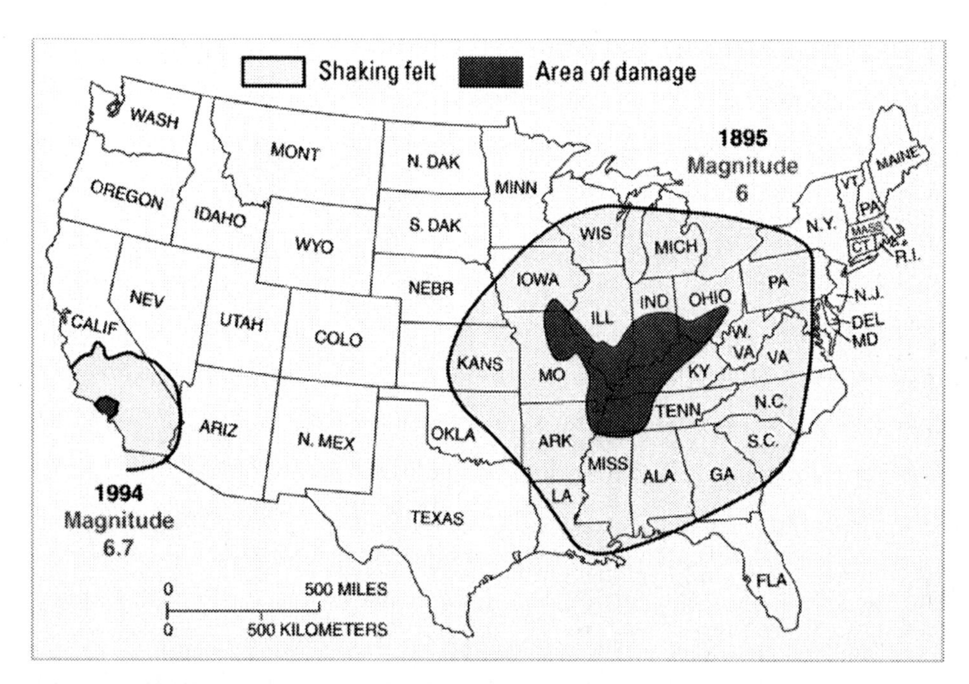

Figure 6.34 The USGS warns that, due to the different geologies, even a moderately large earthquake in the Eastern U.S., such as the magnitude 6 event of 1895 that occurred near St. Louis, will affect a much larger area than the 1994 Northridge earthquake that killed 67 people. (source: USGS)

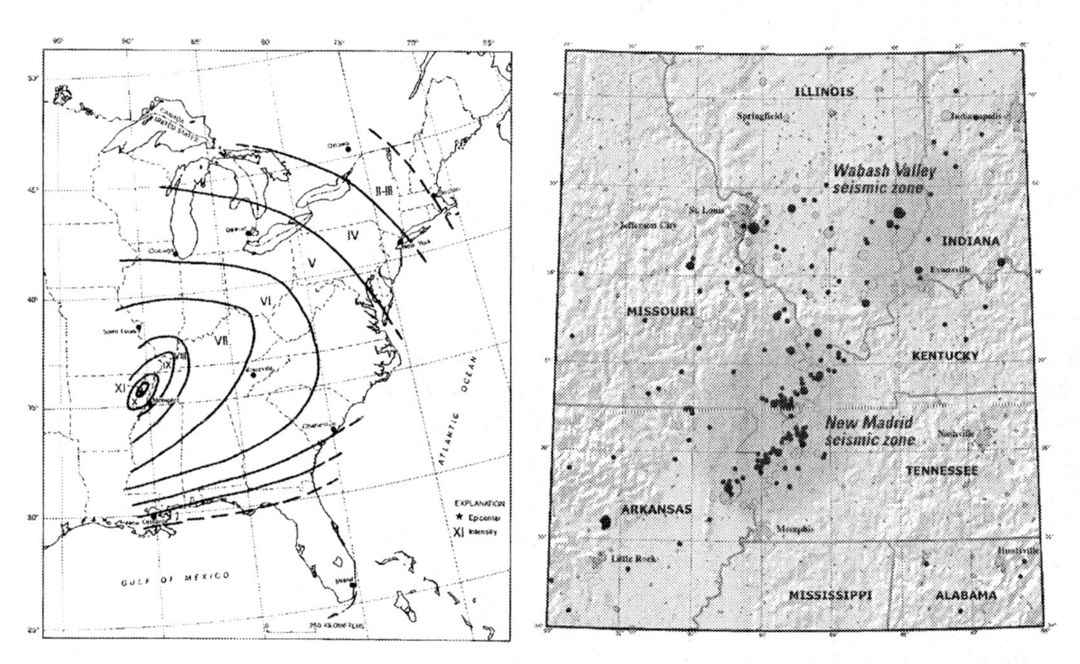

Figure 6.35 Left: Map showing estimated Mercalli intensities for the magnitude 7.7 16 December 1811 Arkansas earthquake, the first of the three magnitude 7.5 or greater 1811-1812 "New Madrid" earthquakes. The other two events occurred on 23 January 1812 (magnitude 7.5) and on 7 February 1812 (mag. 7.7). Right: recent seismicity (source: Wikipedia/USGS)

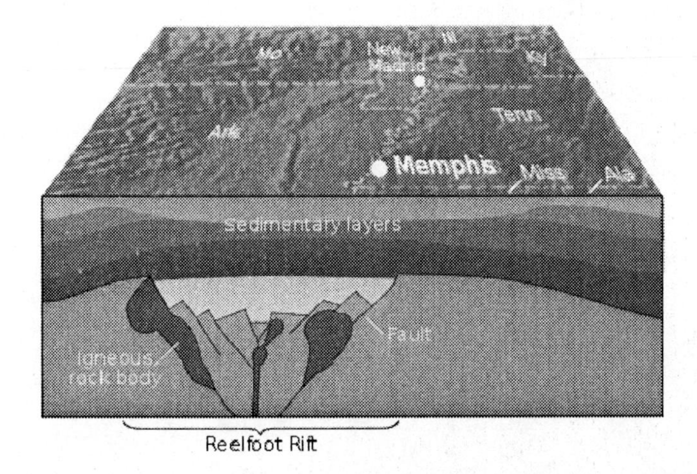

Figure 6.36 Geology of the Reelfoot Rift. It formed 500 million years ago but the stopped progressing was subsequently covered by sediments. Today, earthquakes occur along the associated ancient faults. (source: Wikipedia/USGS)

COPYRIGHTED MATERIAL — DO NOT DUPLICATE, DISTRIBUTE, OR POST

Intra-Plate EQs or "Some Odd Ball Earthquakes in North America"

In the U.S. some of the most destructive or intense Earthquakes occurred in the interior of the North American plate and appear to defy class plate tectonics theory. A closer look however, reveals that the EQs are manifests of long ago plate tectonic processes.

New Madrid (Figs. 6.34-6.36)

- far away from any plate boundary
- Winter 1811/1812, 3 mag. 8.1+ caused intensity X shaking; largest quakes in the continental US in recorded history
- area still seismically very active
- New Madrid Fault zone ancient failed rift (>500Mio years old) with thick wedge of sediments above (enhances ground shaking) (Fig. 6.36)

Figure 6.37 Damage from the 31 August 1886 Charleston, SC earthquake. Fissure and a wrecked brick house on Tradd Street. (source: USGS)

Charleston, S. Carolina (Fig. 6.37): edge of continent but NOT near plate boundary

- large 1886 mag. 7.7 caused up to intensity X shaking
- no surface rupture; fault hidden beneath sediments
- thought to be onshore extension of oceanic fracture zones
- Oceanic fracture zones are the extension of oceanic transform faults, where no relative movement between ridge segments occurs. Thought to be aseismic though seismic monitoring on ocean floor extremely poor. Still puzzling!

Figure 6.38 left: tsunami sirens installed along the coastline in Hawaii. right: a tsunami evacuation sign in La Jolla, CA. (photos: Gabi Laske)

travel time maps can be downloaded at http://wcatwc.arh.noaa.gov/travel.times/ttt.htm and http://www.ngdc.noaa.gov/hazard

6.6 Earthquake Hazard Mitigation

Figure 6.39 Structural damage caused by the 17 January 1994 Northridge, CA earthquake. A parking structure of the Northridge California State University collapsed. The bowed columns are of reinforced concrete. The structure has precast moment- resisting-concrete frames on the exterior and a

precast concrete interior designed for vertical loads. The inside of the structure failed, and with each aftershock the outside collapsed slowly toward the inside until finally the west side failed totally. The reinforced concrete columns were extremely bent. (source: NGDC/NOAA; photo: M. Celebi, USGS)

Figure 6.40 Damage from the 19 September 1985 Mexico City earthquake. Nuevo Leon fifteen-story reinforced concrete structure. Part of the building was only slightly damaged while another part of it collapsed. (source: M. Celebi, USGS)

Figure 6.41 Most people died in the 17 October 1989 Loma Prieta Earthquake when support column failure led to the collapse of the upper deck of the Cyprus Viaduct (I-880). (source: USGS)

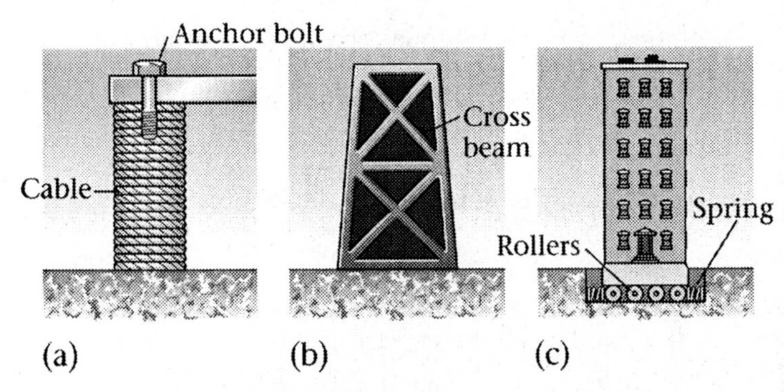

Figure 6.42 How to prevent damage and injury during an earthquake. (a) By wrapping a bridge's support columns in cable (preventing bucking of the columns) and bolting the span to the columns (preventing the span from separating from the columns), the bridge will not collapse as easily. (b) Buildings will be stronger if they are wider at the base and if cross beams are added inside. Sometimes braces are also added outside. (c) Placing buildings on rollers (or shock absorbers) will lessen the severity of the vibrations. (source: ME)

Seismic Monitoring

Unlike volcanic eruptions, scientists currently cannot forecast earthquakes and no 'tell-tale signs' have led to evacuations. Scientists, however, have carefully recorded seismic events to better understand EQ processes which perhaps eventually may lead to limited forecasting capabilities.

- traditional monitoring:
  - seismometers: record seismic activity; locate events; locations of aftershocks trace active fault; increase in seismicity can potentially be foreshocks to a larger event
  - trenches: paleoseismologists dig trenches across faults that had EQs in the past; by comparing the relative offset between layers and their ages, sizes of past EQs and recurrence times can be estimated
  - reflection seismic imaging to image lower reaches and geometry of EQ faults: emission of small seismic signals (explosions or vibrator trucks) and recording on seismic sensors; reflections off EQs faults show up as waveforms with increased amplitudes
- additional recent tools
  - GPS and laser strainmeters: measure the relative movement between two blocks; this has led to surprising findings that plates can creep aseismically (i.e. move without causing EQs)

    - satellite-based interferometry; measure altimetry with satellite and see how this changes over time; use is very limited to areas with no changes otherwise, e.g. areas with no vegetation (e.g. Southern California deserts)
    - computer simulations of physically consistent EQ fault models
    - statistical models: trying to understand the repeatability of EQs by hindcasting past EQ catalogs using statistical theories; if hindcasts are consistent with catalogs, try to forecast next EQ; no physics involved
- tsunami warning
    - traditionally with tide gauges and comparison of suspicious signal with tide calendars
    - may use a variety of new sensors after the Dec 26, 2004 Sumatra earthquake (e.g. hydrophone arrays)

A fully functional tsunami warning system is in place around the Pacific Ocean but not elsewhere. Most earthquake-caused tsunami are generated in the Pacific Ocean. Danger: last EQ that generated large tsunami that crossed the Pacific Ocean was the Good Friday 1964 EQ in Alaska. People may 'forget' what to do when tsunami sirens go off.

Long-term Earthquake prediction

- extremely difficult (see Parkfield)
- each fault system is unique, has different characteristics
- use past seismic records (historic and instrumental) to obtain recurrence time of certain magnitude
- use paleoseismology/trench work where seismic records are sparse (e.g. San Diego)
- some faults are hidden beneath unperturbed sediments (blind fault) and only discovered after an EQ (e.g. the fault that ruptured in the 1994 Northridge in L.A. was previously unknown)

Short-term Earthquake prediction (seismic)

- also very difficult
- seismometers, which are designed to record motion during an earthquake, cannot record what happens between earthquakes

- use GPS and laser strain meter to determine continuous movement along fault, also at aseismic times (time when no EQ happens); advantage: continuous data; disadvantage: only spot measurements where instruments are deployed
- determine stress pattern after EQ: rupture on one fault can change stress level on nearby fault (e.g. San Simeon Dec 22, 2003 EQ may have made Parkfield Sep 28, 2004 EQ more likely)
- How much is aseismic (creep, no EQ)?
- Is there significant increase in strain rate just before EQ?
- How long before an EQ?
- studies still in its infancy

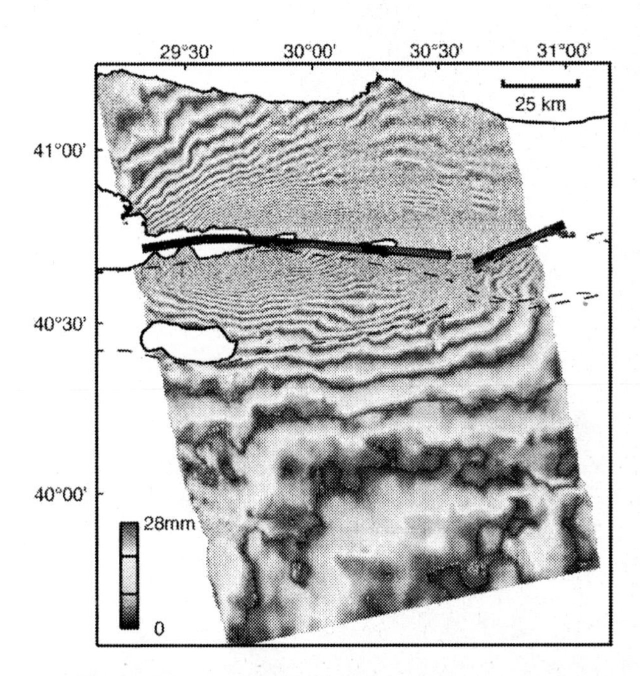

Figure 6.43 An InSAR image of the 17 August 1999 Izmit, Turkey earthquake. The Mw=6.7 quake claimed over 17,000 lives. Interferogram produce by comparing satellite altimeter data from 13 August and 17 September 1999. Large fringes indicate little movement while very densely spaced fringes indicate large movement, i.e. along the fault that is also marked. (source: Wikipedia/NASA/JPL)

INSAR

A relatively new field that was originally used to image displacement caused by an earthquake. Makes use of data collected by a satellite that passes an area every few days. Complete images are composed of swath data measuring the distance between Earth's surface and the satellite. Images are compared before and one after an EQ and are overlaid . At places where displacement occurred, the images are out of sync. Relatively small displacements can be measured by counting the number of fringes by which the images are out of phase.

This technique can also be used at times when no EQ occurs, e.g. along the southern SAF. Recent studies have found that the far field (area away from the fault) has moved as much as it would take to generate a magnitude 8 EQ while no movement was imaged close to the SAF. This implies that the southern segment of the SAF is about ready for a magnitude 8 EQ (see previous lectures on loading a fault)

Unlikely Prediction Tools (Extremely questionable!!)

- comets (unless they actually hit the ground)
- alignment of planets (if Earth tides don't work, planets don't work!!!)
- weather (hurricanes may influence volcanic eruptions but no influence found for EQ!)
- stock market (yeah, right!)
- grandma's arthritis (probably more weather related)

What Scientists can do

- (seismic) monitoring (also GPS, possibly EM, ground water table in wells)
- Tsunami warning system
- educate residents how to prepare and react; see EQ Safety Booklet published by Southern California Earthquake Center (SCEC), USGS and others
- recommend building codes
- develop effective ways to build EQ-safe structures and retrofit existing structures (e.g. external braces for high-rise buildings; ball bearings for footing of bridges and freeway overpasses)

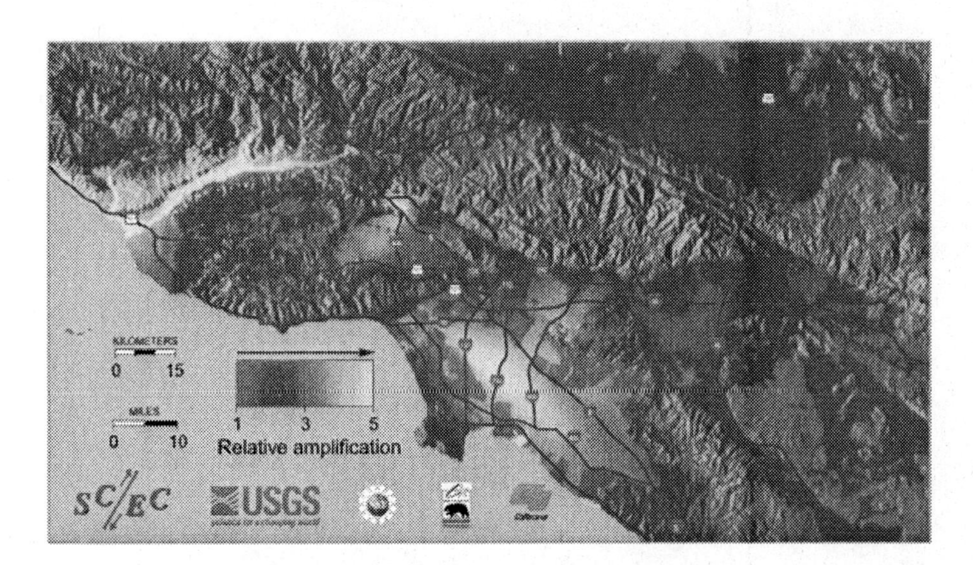

Figure 6.44 Site amplification map for the Los Angeles Basin. Resonating sediments enhance the shaking. (source: USGS and SCEC)

site amplification maps: get info on local geology and predict how local (thick) sediments enhance shaking (e.g. L.A. Sedimentary Basin vs. hard-lava-rock volcanic islands of Hawaii)

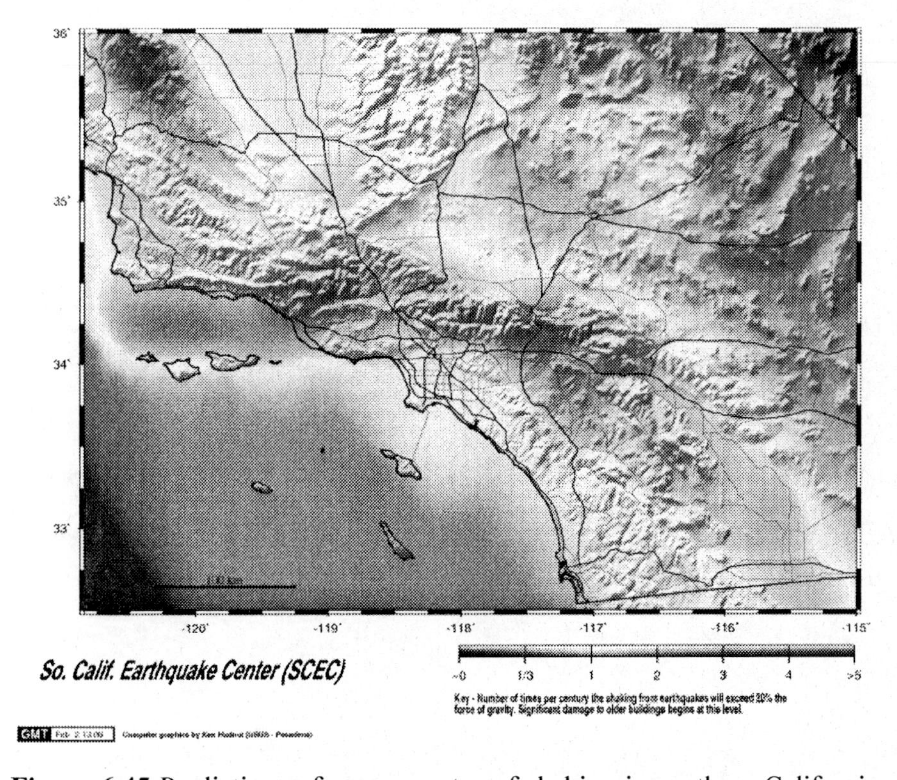

Figure 6.45 Predictions of average rates of shaking in southern California, expressed as number of times per century the shaking exceeds 20% of the gravitational acceleration g. Significant damage

occurs to older buildings at this level. SCEC predicts that people in this area should experience a magnitude 7.0 or greater earthquake about 7 times each century. About half of these will be on the San Andreas "systems" (incl. San Jacinto, Imperial and Elsinore Faults) [4]. (source: SCEC)

make shake probability maps: get info on fault type and geometry, seismicity patterns and calculate probability of of shaking of certain strength

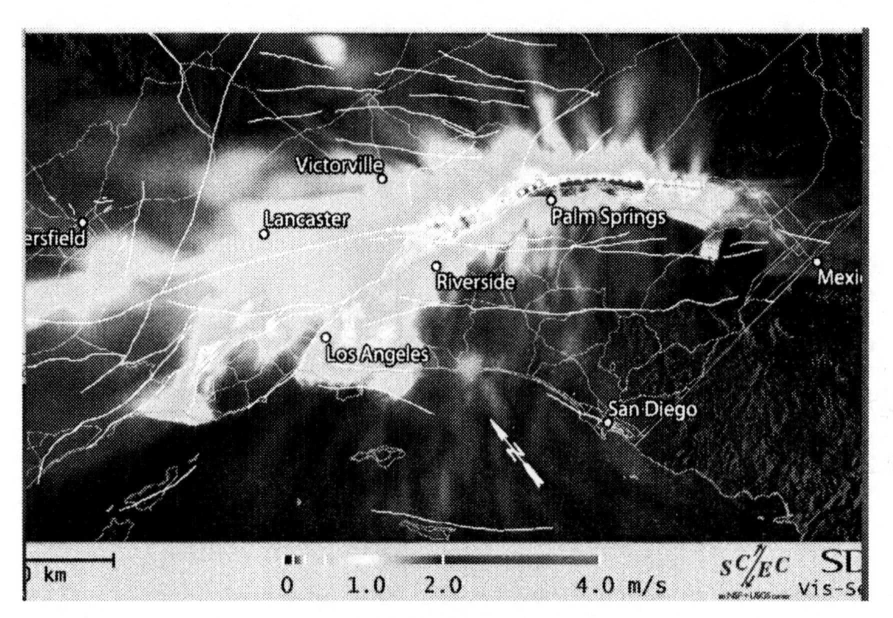

Figure 6.46 Screenshot of the Terashake animation of a magnitude 8 earthquake that originates on southern San Andreas Fault rupturing north. In the animation, the energy does not stay along the SAF but branches into smaller faults leading into Los Angeles. (source: SCEC)

TERASHAKE: A large group of scientists affiliated with SCEC have recently collected all know information on earthquakes in southern California, such as faults, seismicity and local geology. Together with colleagues at the San Diego Supercomputer Center, they created a simulation of what would happen if the "Big One" on the southern end of the SAF would occur. The simulation predicts that, if the rupture of a magnitude 8 earthquake would start at the southern end and then propagate northward, then the seismic energy would be handed over onto neighboring faults that reach into the Los Angeles basin and cause major shaking there.

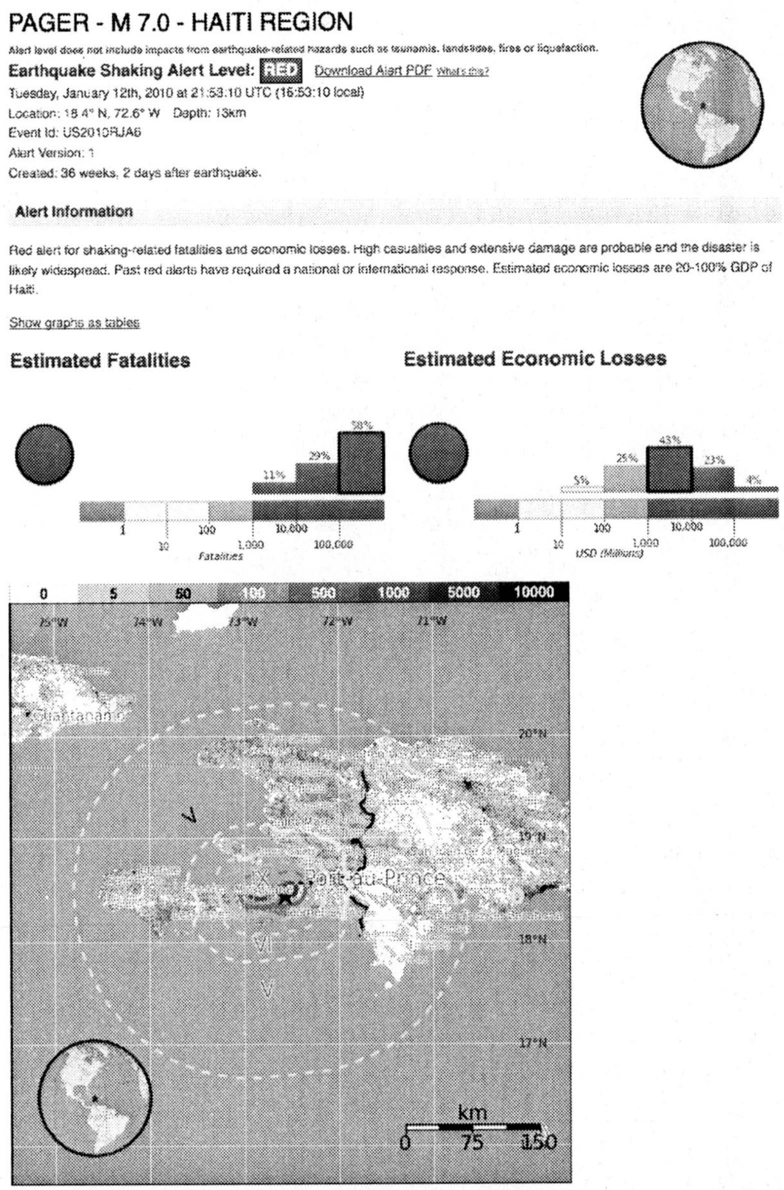

Figure 6.47 PAGER first response webpage for the 12 Jan 2010 Haiti earthquake. Such pages are issued by the USGS. (source: NEIC)

immediately after an earthquake, determine (preliminary) fast-response shake maps, depending on population densities and direct relief teams to likely worst-hit areas; this is a relatively new program by the U.S. Geological Service called PAGER (Prompt Assessment of Global Earthquakes for Response).

What Governments and we can do

- dont' build in hazardous area, build according to building code (e.g. use wood, not bricks)
- construct lifelines to earthquake building codes (both problematic in less developed countries)
-
- grading according to code (landslide risk!)
- secure life lines/make disaster plans
- have water, flashlight, canned food at home

- DID YOU FEEL IT?

References and Recommended Reading

1. adapted and adjusted from "Natural Hazards and Disasters" by D. Hyndman and D. Hyndman; Thompson Brooks/Cole, 2006 ISBN: 0-534-99760-0
2. "The Raging Sea - The Powerful Account of the Worst Tsunami in U.S. History", by Dennis M. Powers; Citadel Press, 2005 ISBN: 0-8065-2682-3 (paperback); describes the tsunami of the "Good Friday" 1964 Alaska Earthquake and how it damaged and affected Crescent City, California
3. "A Crack in the Edge of the World", by Simon Winchester; Harper Collins, 2005 ISBN:0-06057199-3 (hardcover); 2006 ISBN: 0-06057200-0 (paperback); describes the 1906 San Francisco earthquake
4. http://www.data.scec.org/general/PhaseII.html
5. USGS webpage on the largest earthquakes in the U.S.:http://earthquake.usgs.gov/earthquakes/states/10_largest_us.php
6. USGS webpage on the deadliest earthquakes in the world: http://earthquake.usgs.gov/earthquakes/world/world_deaths.php
7. Wikipedia webpage on the deadliest earthquakes on record: http://en.wikipedia.org/wiki/Lists_of_earthquakes#Deadliest_earthquakes_on_record and USGS webpage on earthquake with 50,000 and more fatalities http://earthquake.usgs.gov/earthquakes/world/most_destructive.php
8. USGS website on the 1999 significant earthquakes http://earthquake.usgs.gov/earthquakes/eqarchives/significant/sig_1999.php and Wikipedia page on the 1999 Armenia, Colombia earthquake

9. Zamichow, N. and Ellis, V., 6 April 1994. Satan Monica Freeway to Reopen on Tuesday: Recovery: The contractor will get a $14.5-million bonus for finishing earthquake repairs 74 days early. Los Angeles Times. And 23 July 1994. Santa Monica Freeway Ramps Reopen 4 Days early. Los Angeles Times.
10. Lagos, M. and 3 co-authors, 10 September 2010. San Bruno fire levels neighborhood – gas explosion. SFGate.com and Wikipedia webpage on 2010 San Bruno pipeline explosion.
11. NGDC website on significant earthquakes. And Wikipedia webpages on the 1575 and 1960 Valdivia earthquakes and the Rinihue Lake.
12. NASA Earth Observatory webpage on the 8 October 2005 Pakistan earthquake: http://earthobservatory.nasa.gov/IOTD/view.php?id=5952
13. 15 May 2008. Earthquake dams pose floods risk. BBC news. http://news.bbc.co.uk/2/hi/asia-pacific/7402489.stm and Wikipedia webpage in the 2008 Sichuan earthquake
14. USGS webpage in the La Ligua, Chile earthquake: http://earthquake.usgs.gov/earthquakes/world/events/1965_03_28.php And Rudolph, T. and Coldewey, W.G., Implications of Earthquakes on the Stability of Tailings Dams. Retrieved from pebblescience.org
15. Harder, Jr., L.F. and Stewart, J.P., 1996. Failure of Tapo Canyon Tailings Dam, J. Performance of Constructed Facilities, 10(3), 109-114.
16. Wikipedia webpage on the St. Francis Dam
17. Highland, L.M. and Schuster, R.L., Significant Landslide Events in the United States. http://landslides.usgs.gov/docs/faq/significantls_508.pdf
18. website of International Rivers pages on recent Chinese earthquakes http://www.internationalrivers.org/en/node/5285 http://www.internationalrivers.org/china/dam-safety-concerns-amid-qinghai-earthquake
19. Wikipedia webpage on the 1906 San Francisco Earthquake and Timeline of the San Francisco Earthquake, April 18-23, 1906, Virtual Museum of the City of San Francisco at http://www.sfmuseum.org/hist10/06timeline.html
20. Quake in Alaska changed Yellowstone geysers. May 27, 2004. News Center at the University of Utah. http://www.unews.utah.edu/old/p/030306-10.html
21. December 2008. Mud Volcano Cause Discussed. Explorer Magazine of the Am. Assoc. Petrol. Geol. Webpage at: http://www.aapg.org/explorer/2008/12dec/mud.cfm and Morgan, J., 10 November 2008. The Eruption of the Lusi mud volcano in Indonesia was caused by drilling for oil and gas. China Xi'an Center of Geological Survey.

COPYRIGHTED MATERIAL — DO NOT DUPLICATE, DISTRIBUTE, OR POST

http://www.xian.cgs.gov.cn/english/2008/1110/article_176.html and Nasa Earth Observatory webpage on the Lusi mud volcano: http://earthobservatory.nasa.gov/IOTD/view.php?id=42526&src=eoa-iotd and Wikipedia webpage on the Sidarjo mud flow: http://en.wikipedia.org/wiki/Lusi_mud_volcano

A few EQ web sites

- The US Geological Survey National Earthquake Information Center
- Recent EQs in California and Nevada at the Southern California Earthquake Center (SCEC)
- Recent Bay Area EQs at the Southern California Earthquake Center (SCEC)
- EQ info at the Berkeley Seismological Laboratory
- Important public input after an EQ: shake maps did you feel it? questionnaires at the USGS

Chapter: 7 Volcanoes (draft)

It is important to understand how volcanoes erupt because volcanoes have become an increasing cause of deaths over time. It is important to recognize the hazards associated with certain types of volcanoes and try to mitigate hazards (see also previous lecture) and prevent losses from eruptions

7.1 Volcanic Activity and Human Impact

The Increasing Fatality Rate from Volcanic Eruptions

Table 7.1 Number of Eruptions and Fatalities over Time[1]

Century	# of Fatal Eruptions
14th	14
15th	6
16th	20
17th	32
18th	45
19th	105
20th	215

This does not mean that there are more volcanic eruptions now than there were a few hundred years ago. But there are two main reasons why the fatality rate is increasing:

1. the world's population has grown exponentially over the centuries
2. people tend to move to volcanoes because volcanic soil is very fertile

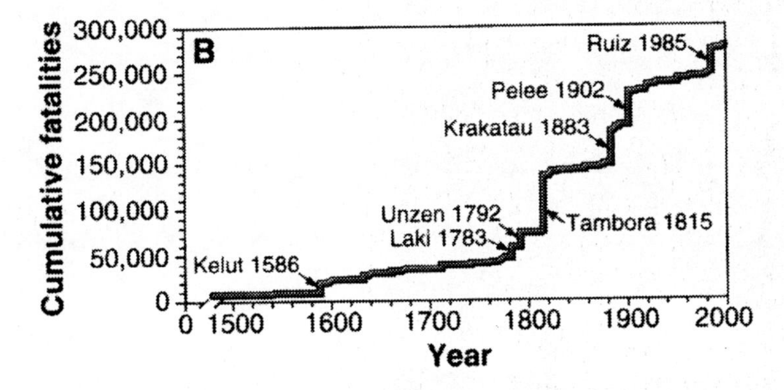

Figure 7.1 Cumulative volcano fatalities since AD 1. The seven eruptions that dominate the record, all claiming 10,000 or more victims, are named. These account for two-thirds of the

total and heavily influence studies based on number of fatalities alone. (source: Simkin et al., 2001(1))

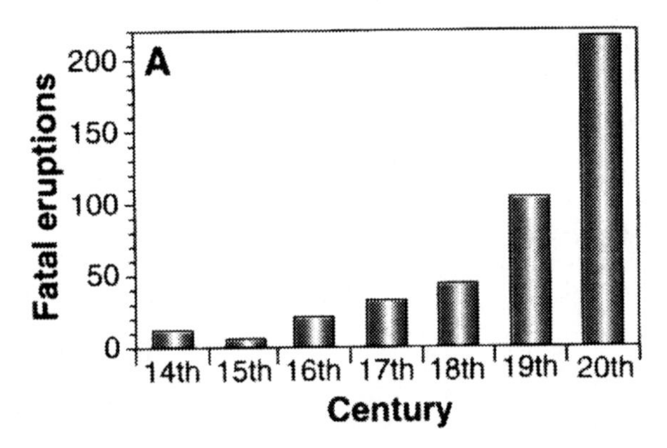

Figure 7.2 Fatal volcano eruptions per century. (source: Simkin et al., 2001[(1)])

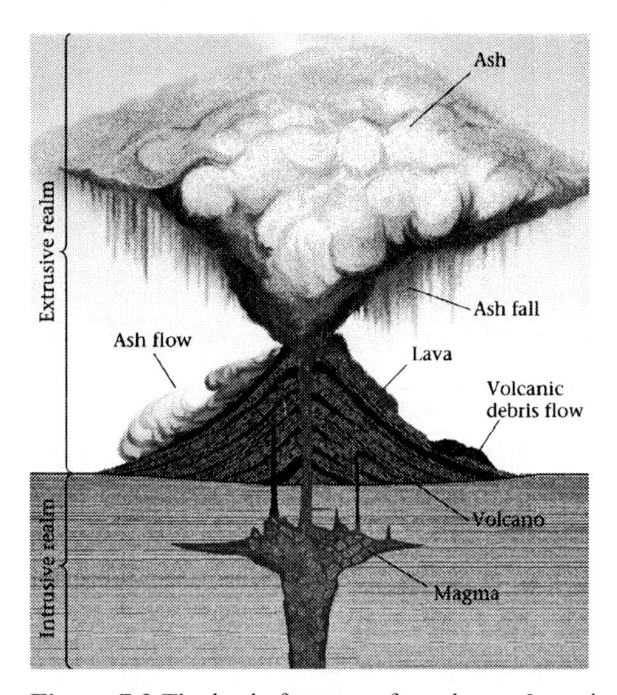

Figure 7.3 The basic features of a volcano. Lava is produced on the surface in the extrusive realm, while magma resides below in the intrusive realm. Ash is ejected into the atmosphere. When an ash cloud cools, it collapses and an ash flow races down the flanks of the volcano. The material ejected from a volcano is also called tephra. (source: ME)

What is a Volcano?

A typical volcano is a structure at the surface that is formed during volcanism, the process that transforms magma (molten rock at depth) into lava (molten rock at the surface).

A typical volcano has a crater (typically < 500m across), a vent and a magma chamber. For a typical eruption rock has to melt (through decompression or volatiles) and escape through an opening (crater or crack/fissure). The eruption is driven by the release of dissolved gases during further decompression on the way up to the surface. A volcanic eruption is much like opening a coke can.

Examples of volcanoes: Kilauea (Hawaii), Grimsvötn (Iceland), Mt. Etna (Sicily), Mt. Vesuvius (Italy), Mt. St. Helens (Cascades, Washington), Mt. Pinatubo (Philippines), Nevado del Ruiz (Colombia), Mt. Fujiyama (Japan), Mt. Klyuchevskoi (Kamchatka), Merapi (Indonesia), Krakatoa (Indonesia), Kilimanjaro (extinct; East Africa).

What is Magma and Lava?

both are molten or partially molten rock, but while lava cools at the surface to make volcanic rock, magma cools below ground to make plutonic rock. Above ground, rocks form in an extrusive environment while below the surface rocks form in an intrusive environment.

What are Lava and Magma made of?

1. molten rock
2. + dissolved grains
3. + up to 15% dissolved volatiles (e.g. water - 50%; CO_2 - 20%; N_2; H_2; SO_2; H_2S)

Active, Dormant and Extinct Volcanoes

There is no exact definition of an active vs. dormant volcano but the Smithsonian Global Volcano Program defines a volcano that has erupted in the last 10,000 years as an active volcano. Others (such as the U.S. Geologic Survey) classify as ***active*** a volcano that currently shows any sign of activity, including seismic (tremors and earthquakes) and gas emissions. A ***dormant*** volcano has erupted in historic times (last 10,000 years) but currently shows no sign of activity. E.g., the USGS classifies Mt. Shasta, CA, that had 11 eruptions in the last 3400 years, as "currently not active" (dormant) because is currently shows no signs of activity. Mt. St. Helens, WA can be dormant for 1000s of years between eruptions. It was dormant since 1857 before it erupted on 18 May, 1980. An ***extinct*** volcano is one that is currently not active nor dormant and is unlikely to erupt again (e.g. Mt. Kilimajaro in Africa, old volcanoes on the Hawaiian island of Oahu).

How many are there and how often do they erupt?

There are about 1500 volcanoes that are known to have erupted within the last 10,000 years. Volcanic eruptions can be quite regular (e.g. Stromboli/Italy every 15 min or so, for the last 1000 years;

Arenal/Costa Rica is currently continuous), but eruptions at volcanoes are usually irregular. Some volcanoes may be dormant for hundreds of years lulling residents in a false sense of safety. The number of eruptions and the number of active distinct volcanoes at any given time is sometimes difficult to pin down because some blasts may belong to the same eruption or come out of the same volcanic field but may "look" distinct (e.g. two different cinder cones). Here is a summary from the Smithsonian volcano web site

Table 7.2 Typical Number of Eruptions within a given time span[(2)]

Time span	# of Eruptions
any time	20
each year	50-70
each decade	160
total historical eruptions	550
total known in last 10,000 years	1300
total known and possible in last 10,000 years	1500

7.2 Volcanic Hazards

Depending on the eruption style volcanoes pose threats to property and lives through a variety of volcanic characteristics (see also Lecture 10).

- lava flows
- ash/tephra fall (ground and air)
- pyroclastic flows (mixture of tephra and volcanic gases)
- lahars (mud flows, mixture of tephra and water)
- gas exhalations (e.g. Lake Nyos in Cameroon)
- erosion
- tsunami (typically when island volcanoes collapse and form calderas; e.g. Krakatoa, Santorini)
- submarine landslides, only recently discovered as hazard (e.g. Hawaii)
- earthquakes
- climate change (e.g. Mt. Pinatubo, 1991; Tambora, 1815 "the year without a summer)"

Volcanic Material

the three major groups of volcanic products are: Lava flows, pyroclastic debris and volcanic gases.

lava flows

- basaltic: pahoehoe (low-viscosity; flows easily; ropy structure); colder a'a' (low-to-medium viscosity; somewhat stagnant; blocky structure)
- andesitic: high viscosity; usually short; sometimes gets stuck in the vent and forms a lava dome clogging the vent; gas pressure can build up underground and lead to an explosion
- pillow lava: when lava gets in contact with water; the outer surface solidifies instantaneously; cracks force the lava to ooze out into another blob

pyroclastic debris

- ash: powder size; < 2mm; sharp glassy particles
- lapilli or cinder: marble-to plum-size
- bombs: basketball-to house-size

volcanic gases

most magma contains dissolved gases, incl. water, CO_2, SO_2, H_2S (up to 9%). Generally rhyolitic lava contains more gas than mafic lava. Volcanic gases can still escape long after an eruption and may be the only sign of volcanic activity (e.g. dormant volcanoes). Volcanic gases escape in fumaroles.

Hazards from Volcanic Material

lava flows: cause significant structural damage but usually too slow to kill

ash fall:

- covers ground, sometime many feet to yards deep, smothering living organisms
- airborne ash hazardous to air traffic: i.e. to engines of large planes flying through an ash cloud get clogged up; e.g. 1989 Mt. Redoubt eruption caused engine failure of a Jumbo Jet; $80 Mio damage; similar problems arise on other volcanoes (see case studies for Lecture 10)

- large eruptions can lead to temporary global climate change (e.g. Mt. Pinatubo, 1991; Tambora, 1815, "the year without a summer")

lahars: ash mixing with water form dangerous mudflows at speeds of 50km/h (a car in fast city traffic!)

Generation of Lahars:

- rain caused by eruption (e.g. Vesuvius, A.D. 79)
- rivers (e.g. Mt. St. Helens, 1980)
- drainage of crater lake (e.g. Mt. Kelut)
- melting ice cover (ice cap or glacier) (e.g. Nevado del Ruiz, 1985)
- post-eruption storms (e.g. typhoon after Pinatubo, 1991)

pyroclastic flows: ash mixing with air and hot gases compose extremely destructive fast-moving (300km/h; 3 times as much as freeway traffic) pyroclastic flows (also called "nuee ardente", French for glowing cloud). (e.g. Mt. Pelee on Martinique in Apr. 1902 that killed 29,000 people leaving 2 alive; Pompeii 79 A.D.).

Generation of Pyroclastic Flows:

- dome collapse (e.g. Mt. Unzen, 1991)
- overspilling of crater rim (e.g. Mt. Pelee, 1902-1903)
- directed blast (Mt. St. Helens, 1980; Mt. Pinatubo, 1991)
- collapse of eruption column (Mt. Unzen, 1991; Vesuvius 79; Mt. Mayon, 1968

Factors that kill people:

- physical impact
- inhaling superhot and toxic gas
- burns

gas exhalations: emission of massive amounts of toxic volcanic gases leads to death; e.g. CO_2Lake Nyos, Cameroon killed most people and livestock in a valley but plants survived (see case studies). Another example of volcanic gases as hazard is Popocatepetl/Mexico, a volcano that is about 100km from the world's largest city, Mexico City. Popocatepetl can emit massive amounts of CO_2 (greenhouse gas, also results from traffic) and SO_2 (same pollutant that results from burning coal) that further degrades Mexico City's air quality.

erosion and mudflows caused by rainstorms (e.g. Lake Atitlan area, Guatemala, after Hurricane Stan, October 8, 2005)

Factors that increase risk:

- steep slopes (close to angle of repose)
- unconsolidated material
- little vegetation after recent eruptions

submarine land slides, only recently discovered as hazard; if slides don't creep but happen suddenly, tsunami generation of global proportions (e.g. Canaries, Hawaii)

The Top Three Killers

- proclastic flows
- Indirect (Famine)
- Tsunami

Lahars are #4 when it comes to counting fatalities. Perhaps surprisingly, lava flows are not the principle killers of all volcanic processes. In fact, they are responsible for extensive property damage but less than 1% of fatalities. The reason is that lava flows usually travel slowly enough for people to get to a safe place. Pyroclastic flows and lahars are so deadly because their fast velocities. The indirect impact comes to play after extremely large eruptions that inject enough volcanic material to temporarily change global climate. These changes can be so severe that they lead to crop failure elsewhere, and to widespread famine.

Table 7.3 Deaths from Volcanic Hazards, according to historical records[1]

Volcanic Hazard	# of Fatalities in %
Pyroclastic Flows	29
Indirect (Famine)	23
Tsunami	21
Lahars	15
Pyroclastic Fall (bombs)	2
Debris Avalanches	2
Volcanic Gas	1
Floods	1

Lava Flows	<1
Earthquakes	<1
Lightning	<1
Unknown	7

# of fatalities are percent of total (275,000)

7.3 Volcanic Eruption Styles and Types

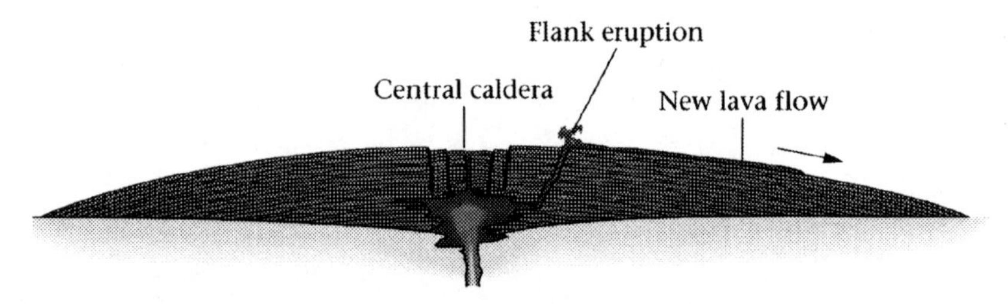

Figure 7.4 A shield volcano, made from basaltic lavas with low viscosity, has very gently slopes. Eruptions at shield volcanoes are typically effusive. (source: ME)

Figure 7.5 Mauna Loa, HI is a good example of a shield volcano. (source: wikipedia)

COPYRIGHTED MATERIAL — DO NOT DUPLICATE, DISTRIBUTE, OR POST

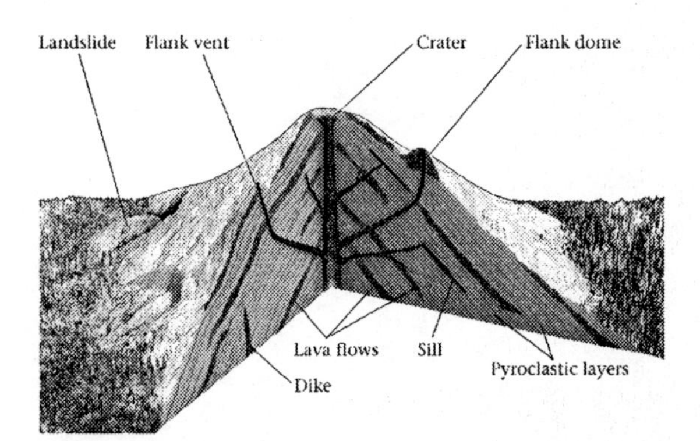

Figure 7.6 A strato volcano, or composite volcano consists of alternating layers of tephra and lava. Eruptions at strato volcanoes are typically explosive. (source: ME)

Figure 7.7 Mt. Fujiyama, Japan, also called Mt. Fuji, is a prominent example of a strato volcano. (source: ME)

Figure 7.8 Mt. Mayon, Philippines is a composite volcano with probably the steepest flanks of any volcano. The flanks of this beautiful volcano are subject to severe erosion. (source: GVP, Smithsonian)

The Two Principal Types of Volcanoes and Eruption Styles

We typically distinguish between two main types of volcanoes: **shield volcanoes** (e.g. Iceland, Hawaii, Mt. Etna): are typically much wider than high; slope angle is often less than 5°; shield volcanoes are formed when erupting lava has extremely low viscosity and typically for during **effusive** volcanism; occur at only few places on Earth, typically near hot spots.

Strato (or composite) volcanoes (e.g. Mt. St. Helens, Mt. Shasta, Stromboli Mt. Vesuvio, Mt. Fujiyama): are typically only a few times wider than high; the slope angle is often much larger than 5°, sometimes reaching 35°; strato volcanoes are typically composed of alternating layers of lava and tephra (ejecta) indicating alternating effusive and **explosive** activity; vast majority of volcanoes, mostly along subduction zones.

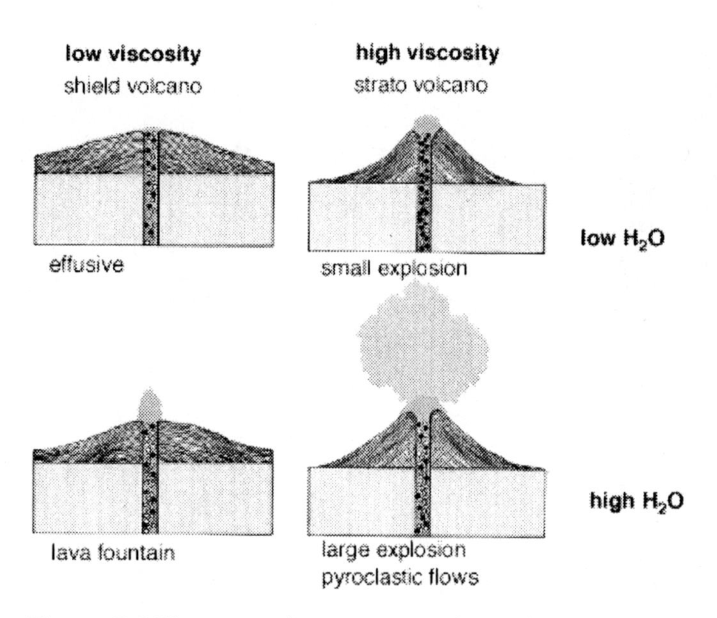

Figure 7.9 The two volcano types and eruption styles. (source: modified from RO)

Eruption Styles and the Three Vs

the eruption style and vigor of a volcano depends on the three Vs:

1) **viscosity** of magma (hence SiO_2-content, T)

2) amount of dissolved gases (**volatiles**); e.g. the water content determines whether a magma at Hawaii will erupt as lava flow oozing out of the vent or as lava fountain

3) in addition to these, the **volume** of erupted material determines the significance of an eruption (e.g. a volcano oozing out gigantic amounts of low-viscosity lava erupting in an icelandic type eruption

can become a VEI 4, due to the large volume of erupted lava; a tiny amount of high-viscosity lava may erupt in an explosion small enough to be watched safely relatively close to a volcano)

The following table helps relate the amount of viscosity and volatiles with the two principal eruptions styles.

Table 7.4 Eruption Styles, Viscosity and Volatiles

Eruption Style	Viscosity of lava	low in volatiles	high in volatiles
effusive	low	lava flows	lava fountains
explosive	high	small explosions	dramatic explosions

Figure 7.10 Pu'u ka Pele on the southeast flank of Mauna Kea, HI is a cinder cone. It is 95 m high and the diameter of the crater at the top is 400 m. (source: USGS; photo: J.P. Lockwood)

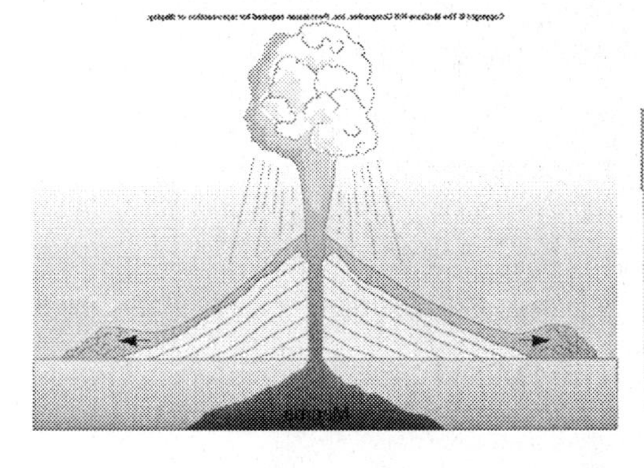

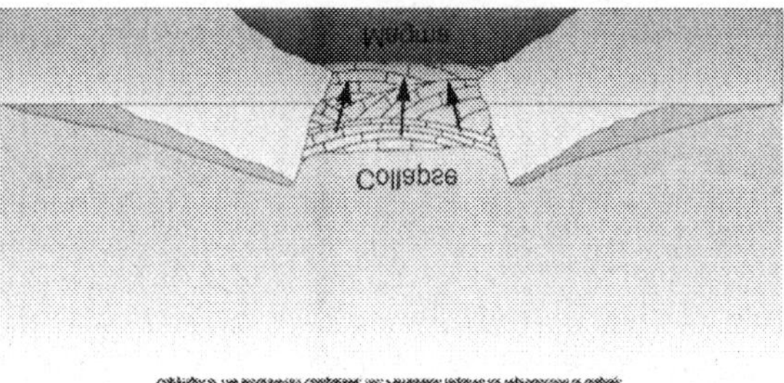

COPYRIGHTED MATERIAL — DO NOT DUPLICATE, DISTRIBUTE, OR POST

Figure 7.11 Left: an explosive volcano erupting. Right: a volcano may erupt to a point when the emptied magma chamber can no longer support the volcano above. The volcano collapses and forms a caldera which is larger than a crater. (source: AB)

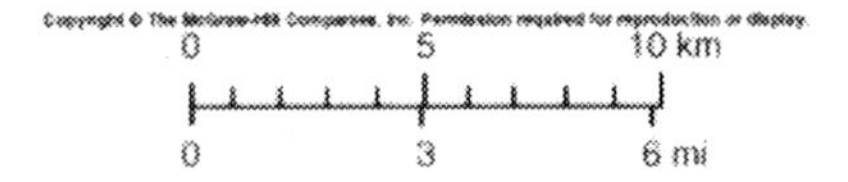

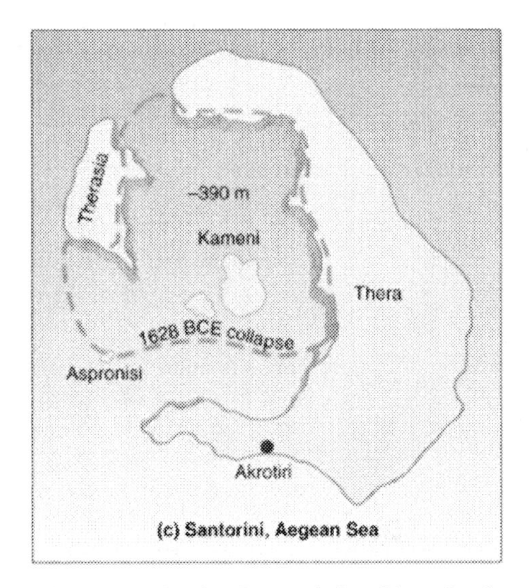

Figure 7.12 The Santorini caldera in the Aegean Sea formed about 1650 B.C.. (source: AB)

Some Special Features

Cinder Cones (or Scoria cones): form when the volume of ejected lava is relatively small (and volatiles relatively high)

Mud volcanoes, e.g. Azerbaijan

Geysers, e.g. Iceland, Yellowstone

Caldera: forms after large amounts of magma escaped from the magma chamber; roof collapse due to loss of support (according to the course book, calderas are associated with high viscosity, high volatiles, very large volume eruptions. This is not always the case!)

Phreatic: (not mentioned in the book but worth knowing about); phreato-magmatic, also hydro-volcanic; from Greek "phrear" for well. Water interacts with lava to form vigorous eruptions.

- underwater volcanic eruptions (e.g. Surtsey/Iceland)
- volcano on dry land intersects aquifer

- lava or pyroclastic flow moves over water saturated sediments
- water runs over hot rock to form steam explosions
- subglacial eruptions

something curious: A yet unknown source of pumice rafts in the Fiji Islands: according to the Smithsonian GVP website, reports have been received on September 16 when observers aboard the M/V National Geographic Endeavour noted almost continuous rows of pumice that day as they traveled about 90km east-southcast to Vatoa Island, where the pumice was present on the beaches. Large rafts of pumice were also passing through the northern Lau Group around Naitauba Island on September 19. The source of the pumice is unknown at this time.

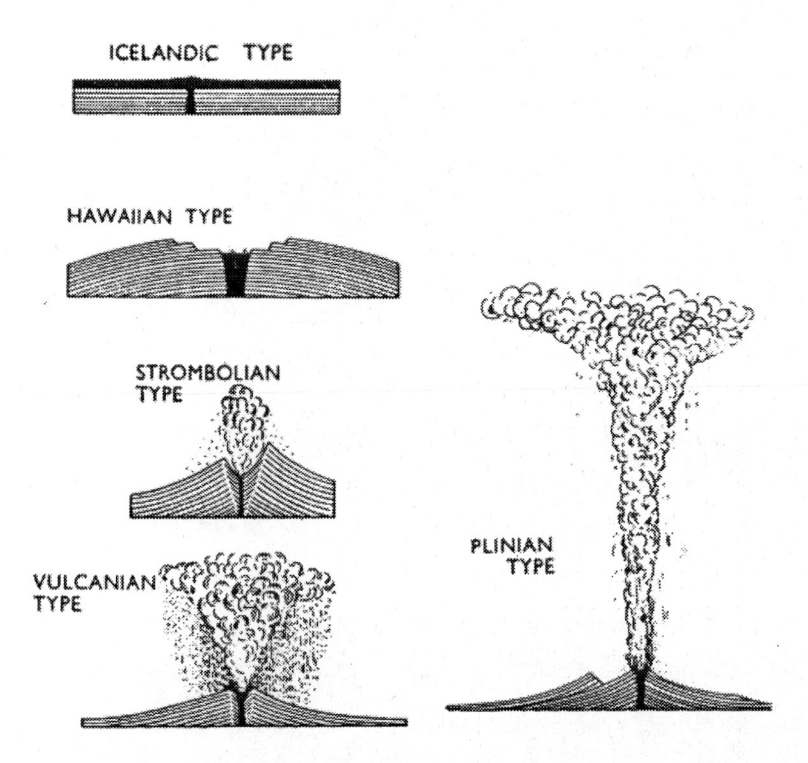

Figure 7.13 The principal types of volcanic eruptions to classify its severity. (source: AB)

Eruption Types: A More Detailed Classification by the Three Vs

Table 7.5 Characteristics of Eruption Types

Eruption Type	Characteristics
Icelandic	large amounts of very low-viscosity lava; non-explosive; forming plateaus

Hawaiian low-viscosity lava; non-explosive; forming shield volcanoes

Strombolian relatively small amounts of moderately high-viscosity lava; usually peaceful; forming scoria or cinder cones

Vulcanian high-viscosity lava; moderately violent eruptions; moderately high-volatile content; moderately large eruption cloud

Plinian very high-viscosity lava; violent eruptions; very high-volatile content shoots tephra high into atmosphere; pyroclastic flows; tephra alternating with lava flows form strato volcanoes

Volcanic Explosivity Index (VEI)

	VEI								
	0	1	2	3	4	5	6	7	8
Volume of ejecta (m^3)	$<10^4$	10^4–10^6	10^6–10^7	10^7–10^8	10^8–10^9	10^9–10^{10}	10^{10}–10^{11}	10^{11}–10^{12}	$>10^{12}$
Eruption column height (km)	<0.1	0.1–1	1–5	3–15	10–25	>25			
Eruptive style		<-----Hawaiian----->		<-----Vulcanian----->					
			<-----Strombolian----->				<------------Plinian------------>		
Duration of continuous		<-----<1----->		<-----1–6----->			<--------->12--------->		
blast (hours)					<-----6–12----->				

Figure 7.14 The volcanic explosivity index to categorize the size and severity of a volcanic eruption. (source: AB)

The Volcanic Explosivity Index (VEI)

A scale from 0-8 that describes the violence of a volcanic eruption (see table 6.8 in the course book). Effusive eruptions typically have VEI values of 0-1. Very explosive eruptions with lots of tephra/pyroclastic material being injected into the stratosphere typically have VEI values of 4 and greater. Cataclysmic (very catastrophic) eruptions are 6-8. Icelandic and Hawaiian eruptions typically fall into the 0-1 category (though there are exceptions!), while Plinian eruptions are VEI 5 or greater.

Table 7.6 Volcano Examples for VEIs

VEI index	Volcano
0	Lake Nyos (Cameroon) 1986; Kilauea (Hawaii) 1974
1	Mt. Unzen (Japan), 1991; Kilauea (Hawaii) 1983-2003; Nyiragongo (Dem.Rep.Congo) 2002
2	Stromboli (Italy) 2003
3	Surtsey (Iceland), 1963; Heimaey (Iceland), 1973; Nevado del Ruiz (Colombia) 1985; Soufriere Hills (Montserrat) 1995

4	Laki (Iceland) 1783; Pelee (Martinique) 1902; Soufriere (St. Vincent) 1902; Paricutin (Mexico) 1943
5	Mount St. Helens (Washington) 1980
6	Santorini (Greece) B.C. 1650; Vesuvius (Italy) 79; Krakatau (Indonesia) 1883; Pinatubo (Philippines) 1991
7	Crate Lake/Mt. Mazama (Oregon) B.C. 5000; Tambora (Indonesia) 1815
8	Yellowstone (Wyoming) B.C. 650,000

Figure 7.15 The 1783 eruption of the Laki fissure in Iceland was effusive but had a VEI of 4 because it erupted immense volumes of volcanic gases. (source: AB)

Figure 7.16 The Laki fissure is part of the Grimsvötn volcano whose eruption disrupted air travel in May 2011. The photo here was taken in November 2004. (source: GVP, Smithsonian)

7.4 Why are Volcanoes Different? – The Link to Plate Tectonics

Factors Controlling the Type of Volcanism

Crustal Properties

Review from Lecture 04: continental crust is different from oceanic crust.

- continental crust is thicker, less dense, has light colored rock
- oceanic crust is thinner, denser, has dark colored rock
- continental rock is silica-rich (silicic) while oceanic crust is silica-poor (mafic)
- NB: silica is SiO_2

Composition of Volcanic Rocks

Controlled by

- crustal thickness
- silica-content

Viscosity

Review from Lecture 03: viscosity is the resistance of a material to flow. More viscous magma causes more violent eruptions.

Factors controlling the viscosity of magma:

- silica-content (silica increases viscosity)
- temperature (temperature decreases viscosity)

Where do we find volcanoes and what do we find?

~90% of activity near plate boundaries; the other 10% near hot spots and/or intra-plate rift zones

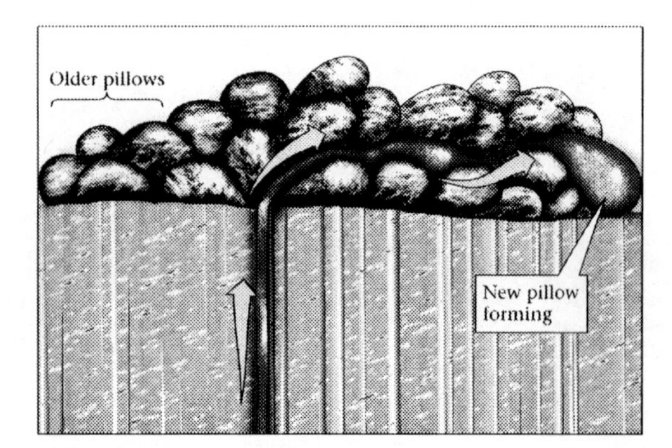

Figure 7.17 The formation of pillow basalt on the ocean floor. (source: ME)

Figure 7.18 Cross section through a lava pillow that was uplifted after its formation and is now on land. (source: ME)

Figure 7.19 A black smoker is a mineral-rich hydrothermal vent often found near a mid-ocean ridge. (source: Wikipedia/NOAA)

Mid-ocean ridges: majority of igneous activity (80% of total magma production); as sea-floor spreads new magma from asthenosphere gets to surface and drifts apart (like two divergent conveyor belts); *decompression melting* leads to partial melting of mafic magma; eruption at ocean floor leads to *pillow lava* as water chills lava instantaneously, forming a thin skin around lava blobs; hot mineral-rich water escapes in "black smokers"; *eruption is submarine, remote and relatively peaceful*

Subduction zones: subducting oceanic lithosphere *releases volatiles* that melt the surrounding material in the asthenosphere; as the silicic minerals have the lowest melting temperatures the rising magma is more silicic than the source it came from comes to the surface in *continental arcs* in continent-ocean subduction zones and *island arcs* in ocean-ocean subduction zones.

In **ocean-ocean subduction zones**, basaltic magma can rise directly to the surface leading to *more effusive and less explosive* volcanism.

In **continent-ocean subduction zones**, rising magma melts a lot of continental crust (low melting T!) and magma becomes more silicic resulting in *more explosive* volcanism.

Continental rifts: thinning and rifting leads to breaking up of a continent, enabling hot magma to rise and penetrate to the surface, causing volcanism; typically very silicic (e.g. East Africa); 1 billion years ago, the Mid-Continent Rift visible from Lake Superior to Kansas almost split North America.

Hot spots: Hot spots are fixed relative to moving plates and plumes are thought to rise from the core-mantle boundary (plumes are solid rock, not molten magma!) E.g. Hawaii; magma rises from a stationary mantle plume giving rise to volcanism at the surface; a chain of volcanoes is created when plate moves over the hot spot; many hot spots are in the oceans, creating basaltic volcanism similar to that along mid-ocean ridges (though slight differences in mineral content); the volcanism at continental hot spots is typically more silicic (e.g. Yellowstone 600,000 years ago).

Large igneous provinces: Some hot spots can be very hot producing more partial melt (10% instead of the usual 4-6%) and large amounts of lava leading to flood basalts (e.g. Columbia River, 17 Mio. yrs ago; Deccan Traps 65 Mio yrs ago; Parana Plateau with Iguacu Falls).

generally NO volcanic activity near transform boundaries and continent-continent collisions

The Types of Volcanic and Plutonic Rocks

Volcanic rocks freeze/crystalize from melt above ground, while plutonic rocks crystalize below ground. Plutonic rocks have more time to cool and form larger crystals than volcanic rocks.

Table 7.7 Igneous Rock Types

Type	Silica content	Volcanic	Plutonic
silicic	>65%	Rhyolite	Granite
intermediate	55-65%	Andesite	Diorite
mafic	< 55%	Basalt	Gabbro
(or basaltic)	rich in Fe and Mg		

- o The silica content defines types of magma and igneous rocks
- o Si-rich mineral weigh less than mafic minerals
- o The rate of cooling determines the grain size, i.e. the texture (the faster the cooling, the finer the grain).
  - ▪

therefore

- o mafic rocks are typically darker than silicic rocks
- o mafic rocks are denser than silicic rocks
- o plutonic rocks are typically coarser than volcanic rocks

Minerals, Silica and Melting

- o different rocks are composed of different minerals
- o a rock usually has several different minerals
- o Si-rich minerals (e.g. Quartz) are usually found in silicic rocks
- o Si-poor minerals (e.g. Olivine) are usually found in mafic rocks
- o Si-rich minerals (e.g. Quartz) melt at lower T (650°C;1200°F)
- o mafic minerals (e.g. Olivine) melt at very high T (>1000°C;1850°F).
- o Therefore certain types of magma or solid rock can exist only at certain temperatures.
- o Si-rich magma is relatively cool while mafic magma is hot
- o Partial melting results in a more silicic magma and more mafic residual rock as the more silicic minerals escape into the magma.

Melting of Rock/Freezing of Melt

- o when melting a rock, the silicic minerals are first to melt
- o when melting a rock, the mafic minerals are last to melt
- o when freezing a melt, the mafic minerals are first to crystallize
- o when freezing a melt, the silicic minerals are last to crystallize

Partial Melt: Where does Magma come from?

from partially molten rock; the molten grains migrate into the magma while the solid grains stay behind. Different types of grains have different melting temperature so when mantle rock heats to a certain temperature only the ones whose melting temperature is low will melt. Partially molten mantle rock has typically only 2-10% melt. The molten minerals escape into the rising magma while solid minerals stay behind.

What causes a rock to melt?

- heating (e.g. heat transfer from adjacent hot magma body; e.g. geysers)
- decompression melting
- injection of volatiles

Earth's Two Major Melting Processes

decompression melting: the geotherm traces the temperature within the Earth as function of depth (or pressure). Upper mantle temperatures reach the melting temperature of rock at about 100km depth. The rocks are solid however due to increased pressure (e.g. water does not boil in a pressure cooker at 100°C). If rock ascends from a certain depth pressure decreases, i.e. eventually melting the rock.

injection of volatile: volatile (e.g. water brought into the mantle in subduction zones) break the bonds between atoms at the mineral grain boundaries and the rock begins to melt.

7.5 Volcanism in Western North America

Figure 7.20 An eruption of Castle Geyser in Yellowstone National Park. (photo: Gabi Laske)

Yellowstone Hot Spot/Columbia River LIP

- The Yellowstone hot spot is thought to be stationary while the American plate moves to the southwest, creating a hot spot track (oldest calderas found at the southwestern end of the hot spot track).
- Activity 17 Mio years ago was that of a very hot plume causing rifting in the crust and producing vast amounts of flood basalts (up to 30m thick lava sheets flowed as far as 550km).
- Several lava flows now pile up as a "large igneous province" in an up to 500m thick and 220,000km^2 large plateau. Noteworthy: the Deccan Traps that formed 65Mio years ago and are thought to be one of the mechanisms (the other one is an asteroid impact) responsible for the extinction of the dinosaurs at the K-T boundary was much larger.

The volcanic activity (that continues today in the Yellowstone National Park in form of geyser activity) started a hot spot track across the Snake River Plain. A humongous eruption took place ca. 600,000 years ago. Pyroclasts were carried as far as the Mississippi River. The last volcanic eruption happened 70,000 years ago.

Currently, the only "volcanic activity" includes geysers, hot springs, mud pools and fumaroles. Yellowstone National Park currently harbors more than 200 active geysers. Some are quite irregular but Old Faithful (which is located in the more active Upper Basin) is known to erupt every 75-95 min, depending on the weather, tides, rainy season and other factors. The mineral-rich Mammoth Hot springs are located near the northern end of the park.

In the middle sits Norris Basin that exhibited increased heat and steam emission at the end of July 2003. Changes noticed near the Back Basin Trail include:

- new mud pots
- changes in geyser activity (eruption intervals)
- a significant increase in ground temperature (which is now at 95°C; 200°F)
- dying vegetation
- a heightened rate of steam discharge

The Basin was closed on July 23, 2003 to tourists and was carefully monitored by the Yellowstone Volcano Observatory with seismometers, GPS (Global Positioning System) and temperature loggers.

Scientists DID NOT expect a volcanic eruption but did not exclude violent hydrothermal (phreatic) explosions. Parts of the Basin were reopened in October. A year later, scientists noticed increased seismic activity, called a swarm, which has happened repeatedly in the last 50 years. Seismicity has decreased since then. The caldera floor is currently rising at 5 cm/year, which indicates the "normal" uplift and subsidence cycle of an active caldera and is no sign of an imminent eruption.

The Cascade Mtn. Range

The Cascade volcanoes stretch from Northern California through Washington into Canada. They are the surface expressions of large intruding batholiths, just like the long-gone volcanoes related to the batholiths of the Sierra Nevada (intrusions more than 50 Mio years ago). The volcanism is typical continental volcanic arc volcanism along the ocean-continent plate boundary. The volcanic rocks are mostly intermediate to silicic (andesitic to rhyolitic).

The majority of volcanoes are considered dormant but can erupt again (i.e. they have been active during the past 2000 years). Three of the most hazardous volcanoes, Mt. Rainier, Mt. Shasta and Lassen Peak, are described in the case studies.

Mount St. Helens

Mt. St. Helens is located in the northern Cascades in Washington and is the most active of the Cascade volcanoes. For decades, Mt. St. Helens showed no activity. The last eruption was 1857. Nevertheless, the explosive eruption on May 18th 1980 did not come unexpectedly. Mt. St. Helens is the most active of the Cascade volcanoes and usually produces rocks with a high silica content (sometimes up to 67%). Activity restarted with a small earthquake on March 20, 1980. The USGS started monitoring the volcano after a small crater opened at the summit emitting gas and pyroclastic debris and seismic activity increased within a week of the earthquakes. Within the next month however, seismic activity decreased, roadblock were removed and there was some dispute as to what precautions should be taken and how to proceed with an evacuation. Then the northern end started to bulge. The growth of the bulge was monitored and well documented with geodetic surveys (theodolites and tilt meters*) and on May 18th Mt. St. Helens erupted. The eruption was more violent than expected because the hillside started to slide and give way to a huge explosion. The eruption killed 57 people . Four of these people were inside the restricted evacuation area: David Johnson, a USGS scientist who was supposed to observe and document the eruption from the north side, Harry Truman (84), who lived in his lodge at Spirit Lake all his life and had special permission to stay and two amateur volcanologists, Bob Kaseweter and Beverly Wetherald who had permission to take readings at Spirit Lake. Some of the other victims were as far as 13 mi from the mountain that were considered safe.

NB: At the time of the 1980 eruption, GPS (Global Positioning System) that is now used to monitor small displacements, was not yet available (started in mid 1980ies).

Activity at Mt. St. Helens has significantly increased in late September 2004 that included the growth of a new dome. Some scientists likened many signs of an imminent eruption to those found prior to the 1980 eruption. In subsequent years, the dome growth slowed and stopped completely in January 2008, and the volcano has so far been quiet.

Mammoth and the Long Valley Caldera

The Long Valley caldera formed after a cataclysmic eruption about 760,000 years ago that was not quite as large as the eruption at Yellowstone, 160,000 years later. No classic hot spot seems to be associated with its activity. Seismic activity along the Sierra Nevada and White Mountain fault systems and volcanic activity in the Long Valley Caldera area started about 4 Mio years ago. Volcanic activity near the Mono Lake area to the north started only about 400,000 years ago. There were eruptions in the Long Valley 600 years ago and in the Mono Lake area just 150-250 years ago. On May 25-26, 1980, shortly after the Mt. St. Helens eruption, the Long Valley area experienced numerous earthquakes, incl. four magnitude 6 events. The resort town of Mammoth Lakes reported damage to structures. On May 27, 1980 the U.S. Geological Survey issued a "Earthquake Hazard Watch". In October 1980, a survey showed that the caldera floor was resurging by more than 25cm (10in). By early 1982, some magma found its way from 10km depth to only 3km. New fumarole and seismic activity prompted the U.S. Geological Survey on May 25, 1982 to issue a "Notice of Potential Volcanic Hazard" (the lowest level of alert). Real estate prices fell 40% over night and tourism dropped dramatically but there was no serious volcanic activity. This incident was highly publicized and geologists were accused of misleading the public.

In the early 1990ies, trees started to die on Mammoth Mountain, due to increased emission of subsurface CO_2. Seismicity picked up and the ground surface rose. To this date, there has not been an eruption. The 1982 events are considered "false alarm" but volcanologist continue to have a close eye on the area.

In the 1990s, seismicity picked up again near the south rim of the caldera but this time geologists did not issue a warning. Subsequent analyses of the seismic data have revealed that magma migrated upwards and came very close to the surface. Some scientists take this as indication that an eruption might have been imminent, and call their findings 'a close call'. But lessons learnt in the 1982 debacle made them more cautious not to issue warnings prematurely.

7.6 Volcanic Hazard Mitigation

Volcanic eruptions cannot be prevented but some measures can reduce the damage and number of fatalities.

- monitoring (see Lecture 11)
- diversion of lava flows (moderately successful); e.g. the residents of Heimaey/Iceland saved their harbor in a 1973 eruption when they slowed down a lava flow by spraying it with water
- draining crater lakes before eruptions: e.g. Mt. Kelut (moderately successful)
- education
- evacuation (often a political rather than scientific problem)

  Case Study:

  Mt. Mayon: an eruption in 1993 killed 75 farmers in a ravine that had been declared off limits. 50,000 residents were evacuated within 10km of the volcano in anticipation of a possible larger eruption. Many residents had to be forcibly removed but returned during the day to tend their fields (without the crop people would starve)

Are there any Benefits from Volcanic Activity?

- ejecta are extremely rich in minerals so supply fertile soil
- volcanic rock is used as building material
- pyroclastic ejecta used for many purposes (e.g. ash and lapilli on icy streets)
- some activity (e.g. black smokers) create significant mineral resources
- volcanism forms new islands (e.g. Iceland, Japan)
- geothermal activity is used to heat spas, provide electricity (e.g. Iceland, New Zealand)
- hot springs

Monitoring Volcanism

- heat flow and surface temperature
- shape of the volcano by geodetic survey (theodolites, GPS, tiltmeters, laser strainmeters)
- uplift of caldera floor (in cases there is a caldera)
- seismic activity; tremors (magmatic) and earthquakes (tectonic); changes in source depth

- chemical analysis of ground water (acidity and chemical content)
- fumarole activity (gas emission); chemistry; occurrence
- above/below ground CO_2 emission (basement of houses/tree kill)
- remote sensing and chemical cloud analysis using satellite systems

References and Websites

1. Simkin, T., Siebert, L. and Blong, R., 2001. Volcano Fatalities – Lessons from the Historical Records. Science, 291, 255.

Visit the USGS Yellowstone Observatory website.
Visit the University of Utah Yellowstone website.
Visit the USGS Cascade Volcano Observatory website
story on people during Mt. St. Helens eruption: click here .
link to USGS.
Visit the USGS Long Valley Caldera Observatory website

new activity/unrest and ongoing activity can be checked out at the Smithsonian/USGS Weekly Volcanic Activity Report

Other Recommended Reading

"Pompeii" by Robert Harris, Random House, 2003 ISBN: 0-67942889-5 (hardcover); 2005 ISBN: 0-81297461-1 (paperback); describes the life of people when Vesuvius erupted in 79 AD, using Pliny the Younger's letters

"Krakatoa, Harper Collins, 2003 ISBN: 0-06621285-5 (hardcover); 2005 ISBN:0-06083859-0 (paperback); describes the eruption of Krakatau in 1883

"Volcanoes, Crucibles of Change", by Richard V. Fisher, Grant Heiken, Jaffrey B. Hulen; Princeton Univ. Press, 1998, ISBN: 0-691-01213-X (paperback); describes the some volcanic eruptions and related hazards, including the 19889 eruption Mt. Redoubt in Alaska

Chapter 8: Mass Movements (draft)

Figure 8.1 The 18 June 1972 Po Shan Road and Conduit Road, Hong Kong landslide killed 67 people and destroyed 2 buildings. The two roads were built in 1925 and in 1910, respectively. At the time, work was done on a construction site above Conduit Road where the major part of the landslide was located. In 1971, two landslides occurred on the construction site. Cracks in Po Shan Road were found 11 months prior to the main slide, after the passage of Typhoon Rose. Attempts to seal the cracks on 16 June 1972 failed. A day later a slip occurred on the construction site and continued to creek until the main detachment happened on 18 June. Heavy rain and building on steep, unstable slopes is the perfect recipe for disaster. (source: ME)

Figure 8.2 Yungay, Peru, before (left) and after (right) the devastating 31 May 1970 mudflow from the 6400-m-tall Nevado Huascaran. The mudlside buried the towns of Yungay and Ranrahirca and with it 18,000 people (some sources report even 25,000 people). Only 92 people survived at the

highest spots in town, a cemetery and the stadium. A magnitude 7.9 earthquake off-shore Peru broke off an 800-m-wide slab of Huascaran's permanent ice cap. The slab disintegrated and mixed with rocks and soil on the way down to produce the 910-m-wide 1.6-km-long mudflow that traveled 18 km down the slope. Speeds of the mudflow were estimated at near 300 km/h (185 mph). The disaster unfolded in only a few minutes, not allowing residents to get to a safe place. Sadly, this is a recurrent hazard. On 10 January 1962, a landslide and avalanches killed up to 2000 people in Ranrahirca. These also were triggered by an earthquake. Could this disaster have been prevented? A few years before the Yungay disaster, climbers had recognized the instability of the ice cap and published a warning but no one took notice. After the 1970 event, geologists determined that Yungay was built on debris from ancient landslides. Unbelievably, despite this knowledge, Yungay was rebuilt only 1500 m away from the old town that was buried in 1970. (source: ME)

Figure 8.3 On 8 October 2005, a magnitude 7.6 earthquake in Kashmir triggered a massive landslide along the Neelum river and buried the northern part of the village of Makhri on the northern edge of the city of Muzaffarabad. This landslide and several others forced the Neelum river to leave its bed. The destruction of the town, only 10 km from the epicenter, and extensive flooding killed 51,300 people. The quake caused a total of at least 75,000 fatalities in Pakistan and several thousand more elsewhere. Many fatalities were attributed to the landslide but also due to poorly constructed buildings. It has been the deadliest earthquake on the Indian subcontinent. At the time, it was feared that 15,000 more people could succumbed in the aftermath as a harsh winter was in the forecast. (source: NASA)

8.1 Mass Movements as a Growing Threat

The Urgency of Understanding Mass Movements

mass movements become an increasing threat because of

- population grows
- expansion of cities
- building on unstable ground

Mass Movements

- a mass movement is the downward movement of material pulled by gravity
- in the U.S., mass movements cause $1.5 billion in damage and 25 deaths annually
- most mass movements are associated with
    - slope failure
    - collapse of structures beneath

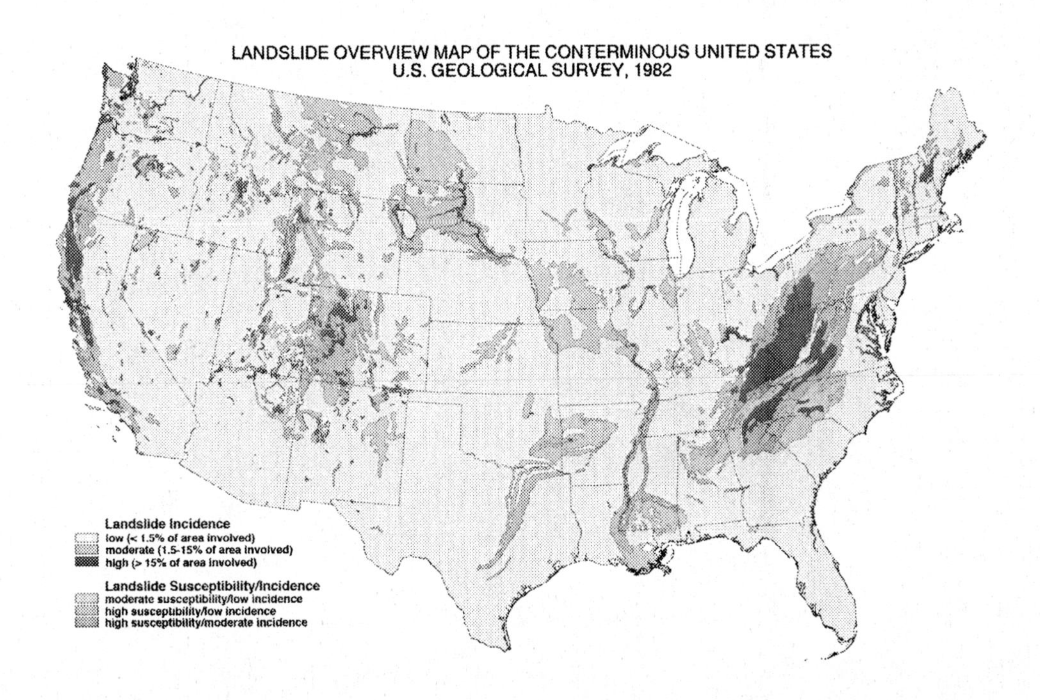

Figure 8.4 Areas of landslide incidence and susceptibility in the lower 48 U.S. states. (source: USGS)

Mass Movements in the U.S.

- areas of high precipitation
- areas of steep slopes
- areas of high run-off rates
- large meandering rivers (see also lectures on floods)

According to the USGS, mass movements are a serious geologic hazard in almost every U.S. state. They case in excess of $ 1 billion in damages and 25 – 50 fatalities every year.

8.2 The Causes and Processes of Mass Movements

Factors Controlling Landscape Development

- erosion: agents are water, ice and wind
- elevation: slope steepness determines the speed of erosion
- climate: determines the predominance of eroding agents
- life activity: conserving (tree roots) and weakening (bacteria) organisms
- plate tectonics: uplift, earthquakes
- substrate composition: controls response to erosion
- human activity: mining pits, housing and urban developments, dams, agriculture, deforestation and others

The Role of Gravity

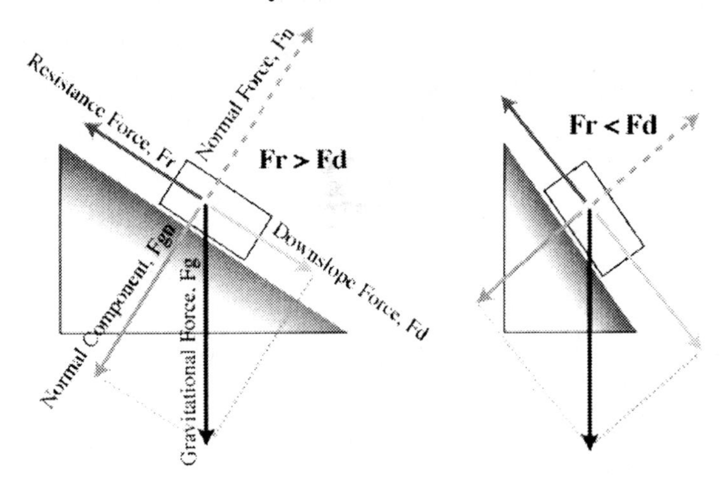

Figure 8.5 The interplay of forces that cause movement of a mass on a slope and hold it in place.

- pull of gravity is constant and continuous by pulling down vertically
- on a tilted slope, pull of gravity is decomposed into components perpendicular (**normal force**) and parallel (**downslope force**) to the slope
- friction holds mass in place; friction causes a slope-parallel **resistance force** with opposite direction of the downslope force
- the balance between resistance force and downslope force holds a mass into place or lets it move downward

- if resistance force is smaller than downslope force, mass moves down the slope
- some events can lower friction and trigger mass movements: rainstorms, irrigation, earthquakes, volcanic eruptions

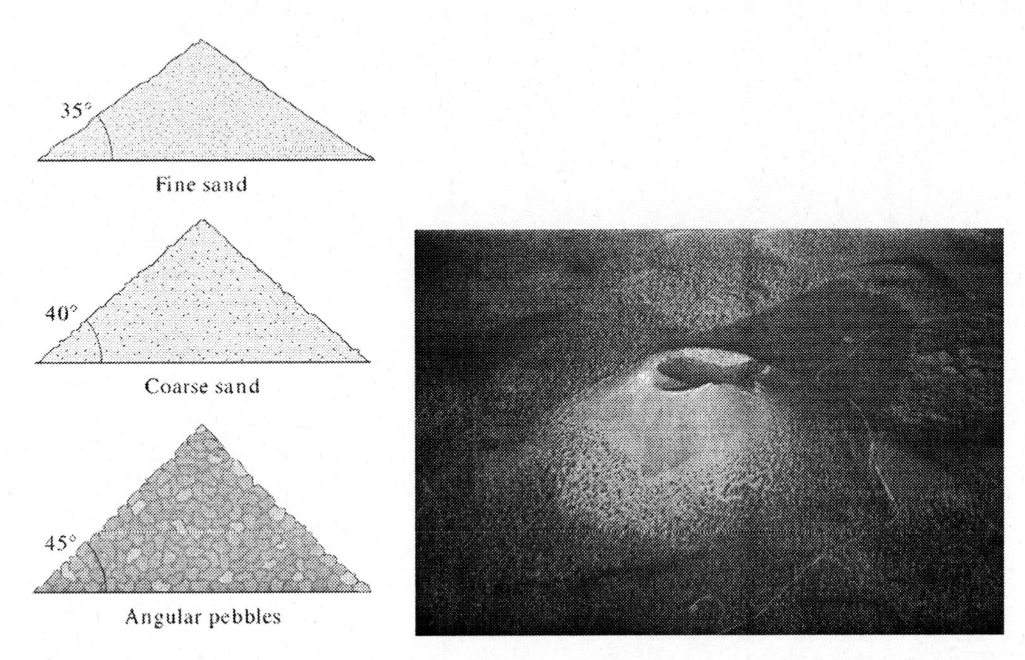

Figure 8.6 Left: The angle of repose increases with particle size. Right: a cinder cone. (source: ME; photo: Wendell Duffield, USGS)

The Angle of Repose

- maximum angle a pile of debris can retain without disintegrating
- up to 45°
- depends on properties of particles with the pile:
    - large, rough particles have large angles
    - small, smooth particles have small angles

NB: a substance that acts is cohesive increases the angle of repose; example: when building a sand castle, people use water to make the sand stick together. When the water dries, the castle starts to collapse because the walls are steeper than the angle of repose.

External Causes of Slope Failure

- steepening of slope (e.g. by fault movement)
- removal from low end (toe) of a slope (e.g. river cutting into its banks)

- o adding mass on the high end (head) of a slope (e.g. sediment accumulation)

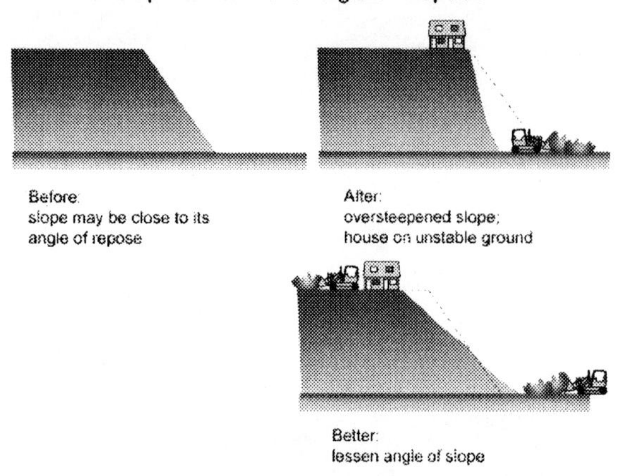

Figure 8.7 Housing development can raise or lower the slope angle, thereby increasing or reducing the likelihood of mass movements.

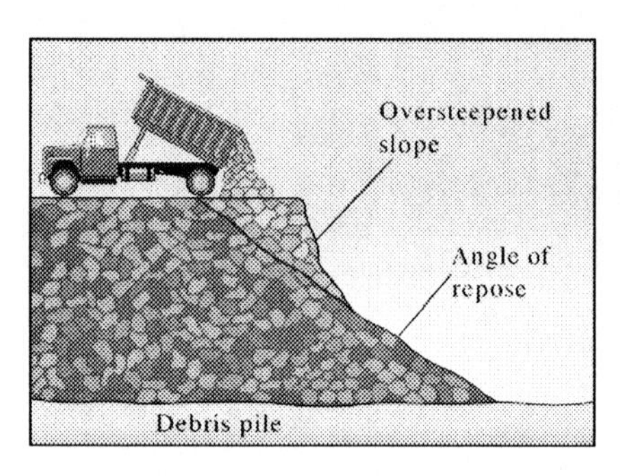

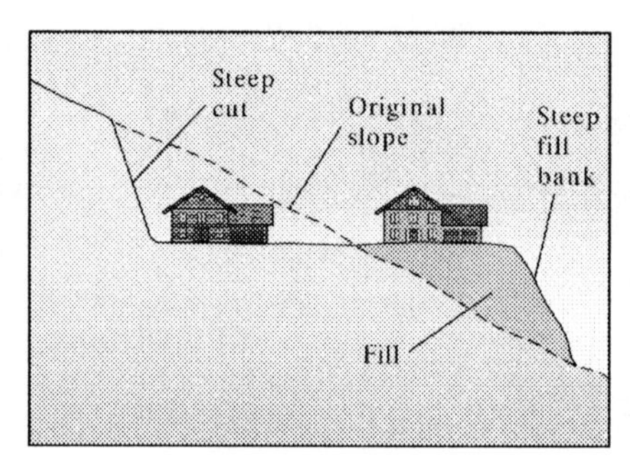

Figure 8.8 The two ways how regarding a hillside for housing or other development can oversteepen a slope beyond the angle of repose: 1) adding mass to the top of a slope; 2) removing mass

Basic Features of Slope Failure

- o scarp or tear-away zone upslope (head scarp)
- o pile-up zone or zone of accumulation downslope (toe)
- o glide horizon/surface of weakness
- o transverse cracks where tear-away zone goes into zone of accumulation

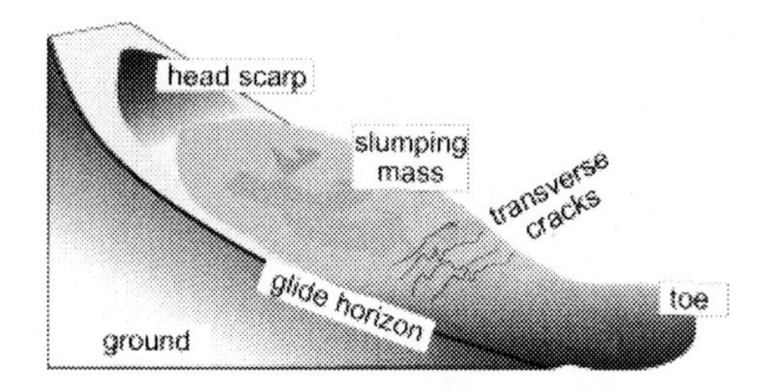

Figure 8.9 The features of a slump.

Figure 8.10 Head scarp of a slump caused the magnitude 6.8 28 February 2001 Nisqually, WA earthquake. The quake caused damage to property and highways in the greater Seattle area. (source: USGS and Wash. Dept. of Transportation)

Figure 8.11: Toe of the 3 October 2007 slump on Mount Soledad, La Jolla, CA.

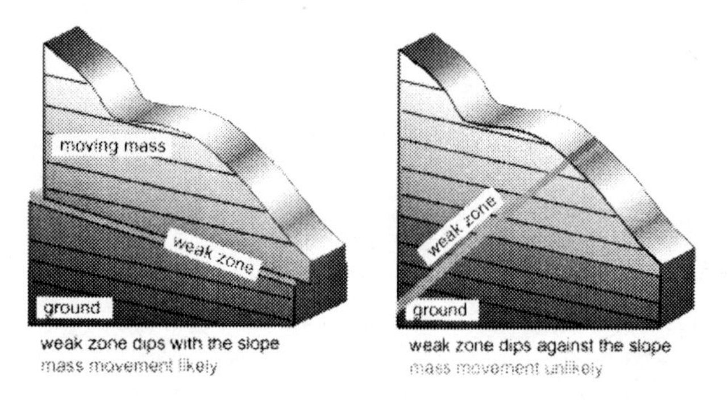

Figure 8.12: The risk of slumping and the orientation of the weak zone.

Internal Causes of Slope Failure

- inherently weak materials (e.g. clay-rich layers)
- water in different forms (see below)
- decreasing cohesion (e.g. when buried layers get exposed)
- adverse geologic structures (e.g. preexisting ancient slide surface, weak zones that dip with the slope (daylight bedding), structures within the rock (rock not cemented together; clay lenses; joints)

Surfaces of Weakness

The contact zone between the sliding mass and the ground below.

- faults
- joints (extensive cracks in layers of rocks)
- clay-rich layers
- soft rocks slipping on hard rock

- o hard rock being broken apart by movement in underlying soft rock

NB: clay is a very fine sediment that results from the weathering of silica rich rocks. Clay particles have a book-like structure and it can absorb large amount of water. The clay expands and shrinks, depending on how much water it absorbs and releases, thereby weakening the ground.

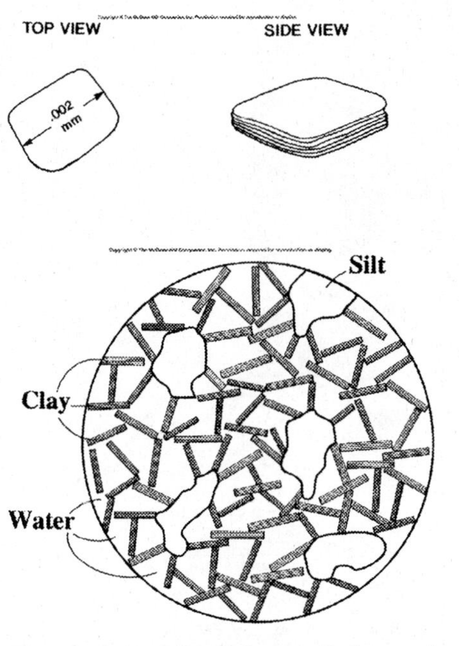

Figure 8.13: The makeup of clay and quick clay.

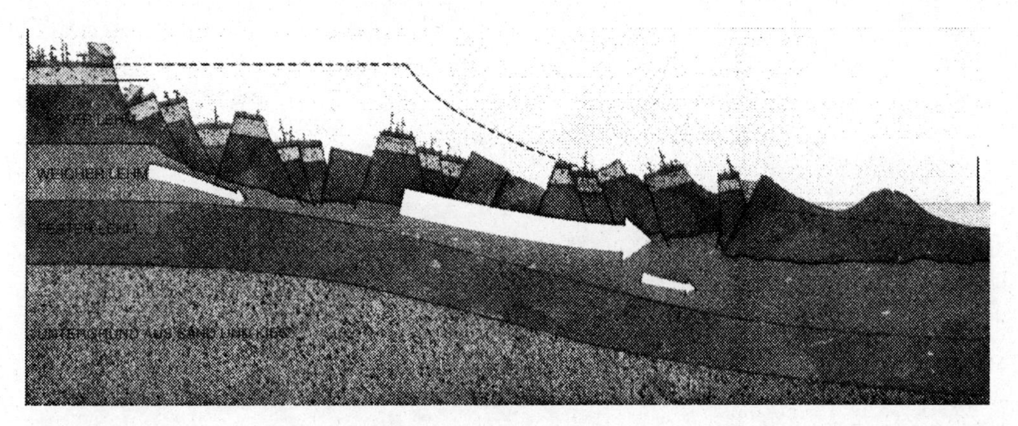

Figure 8.14a: Layering of clay at Turnagain Heights, AK. A mix of sand and gravel (light brown) is overlaid by a sandwich of hard clay, soft clay, hard clay. The soft clay can liquefy and initiate mass movements.

Figure 8.14b: The 1964 Alaska earthquake-triggered widespread liquefaction and damaged homes in Turnagain Heights area in Anchorage. (source: USGS)

The Role of Water in Slumping

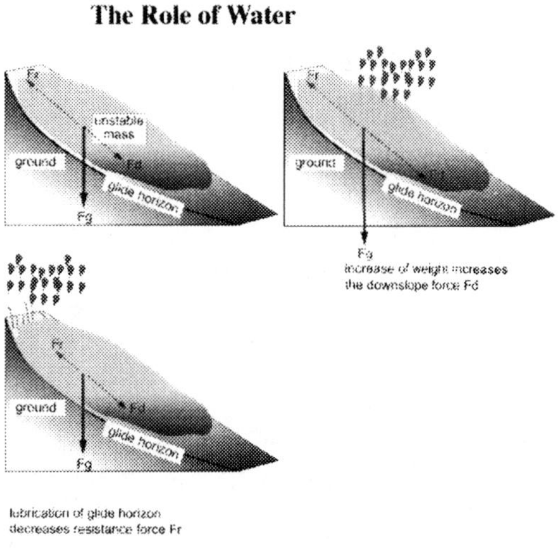

The Role of Water

increase of weight increases the downslope force Fd

lubrication of glide horizon decreases resistance force Fr

Figure 8.15 Water increases the likelihood of a mass movement in two ways: 1) the added weight increases the downslope force; 2) water that penetrates into the glide horizon reduced the friction and therefore the resistance force

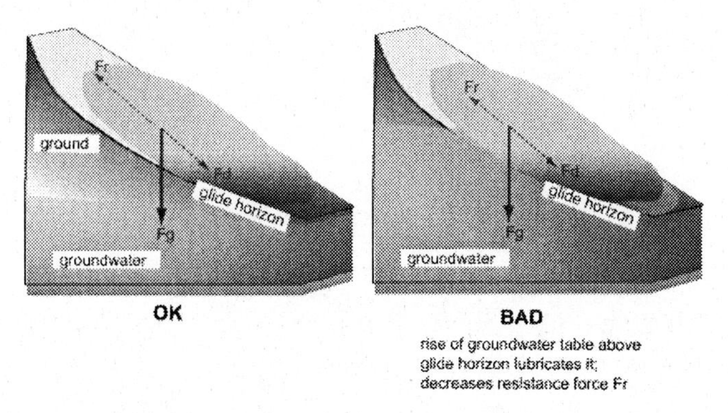

Figure 8.16: The risk of slumping and the groundwater table.

Water in general, especial when it penetrates the weak zones/glide horizon:

- interplay with clay minerals
- reducing of cohesion
- change in pore pressure
- reduction of friction and lubrication along glide horizon

Rain and Irrigation

- increase in weight

acceleration of surface and subsurface erosion

Groundwater Table

- can rise after prolonged rain or by irrigation
- can also rise when pools, septic tanks or sewer lines leak

increases slumping risk when water table rises **into or above** glide horizon

Summary: How Can Humans cause Slope Failure?

- mining, grading, streets, housing development
    - steepening slopes beyond angle of repose
    - adding mass to head (increasing potential energy)
    - removing mass at toe (removing support)
    - removing vegetation that used to anchor mass into ground
- dams
    - change of pore pressure
    - change of groundwater table
    - increase in weight may open faults beneath or cause new faults

- inappropriate drainage; excess irrigation
- deforestation

Deforestation

Deforestation is particularly devastating in the tropics and causes *mass wasting*. Contrary to what a lush tropical rain forest would suggest, the fertile humus layer is very thin so is easy to erode. In fact, the soil is good only for a few seasons so farmers have to move on to new sections in the forest. High precipitation rates accelerate erosion leaving behind huge gullies that cannot be used for farming etc(badlands). New tropical soil accumulates extremely slowly. A tropical rainforest takes thousands of years to come back.

Fig. 8.17 Deforestation in Madagascar has led to one of the world's most severe cases of soil erosion. (source: wildmadagascar.org).

8.3 Classification of Mass Movements

Slow and Rapid Mass Movements

slow movement can cause extensive property damage but fast movements are the big killers

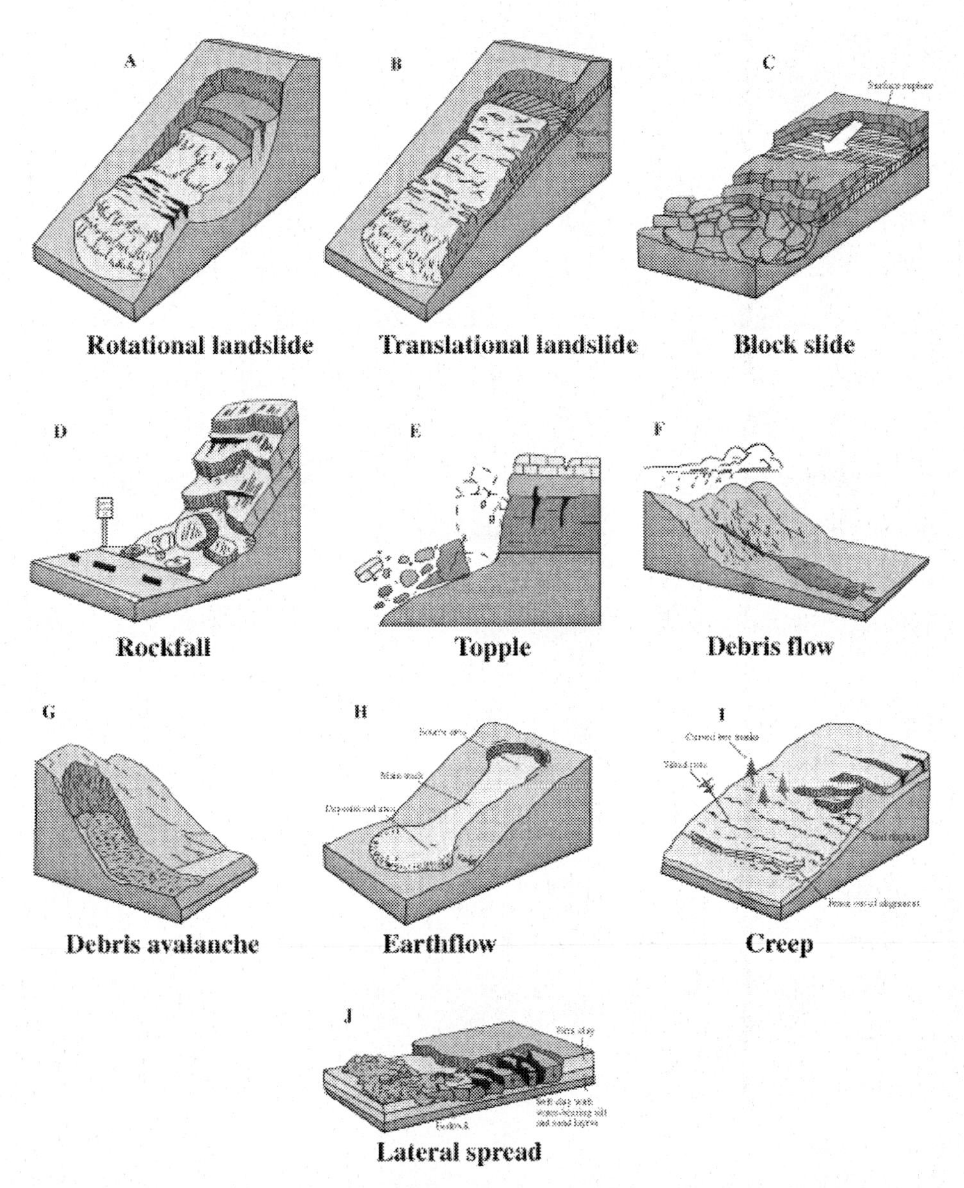

Figure 8.18 Types of mass movements. (source: USGS).

Classification by How a Mass Moves

- typically by speed and water content (see below)
- Abbott's book classifies by how a mass moves
    -
        - falls
        - slides
        - flows
        - subsidence or subsides

Falls

- free fall of mass
- detachment of mass along near-vertical cliff
- dominantly vertical downward movement of mass
- mass moves in separate blocks
- possible triggers: rain, frost wedging, earthquakes

Flows

- flow over landscape
- mass movements that behave like viscous fluids
- internal movement (turbulence) dominates
- can reach speeds > 320km/h(200mph) but some are slow (e.g. Portuguese Bend; Case Study 3)
- no or only short-lived slip surfaces

Slides

- movements above a failure surface (curved or planar)
- move as semisolid mass (no internal turbulence)
- some pre-slide coherence of mass is maintained during movements (i.e. can break apart somewhat)

- rotational slides/**slumps**:
    - movement on curved failure surface
    - short distance
    - a resisting force at the toe works against the driving force (Swedish Circle)
    - head scarp can cause more instability when water accumulates near top
- translational slides:
    - slide on surface of weakness
    - can move long distances
    - type 1: remains coherent as one mass
    - type 2: mass deforms and disintegrates into debris slide
    - type 3: can spread laterally, when hard material at top breaks apart

Subsidence

ground moves down

- slow:
    - sags gently
    - compaction of loose fluid saturated sediments (e.g. New Orleans 3m in last 50yrs)
    - removal of ground water or oil (e.g. San Joaquin Valley 8.5m since 1925, Las Vegas 1-2m since 1935)
- fast:
    - drops catastrophically
    - compaction of loose, water saturated sediments or rapid collapse of caves (e.g. sinkholes); cave collapses (e.g. Winter Park, FL; May 10, 1981) or (e.g. San Diego due to failing infrastructure)

Figure 8.19 Aerial view of a sinkhole that formed in May 1981 in Winter Park, FL. The sinkhole was 100 m (325 ft) wide and 34 m (110 ft) deep. The white structure to the left of the sinkhole was a municipal swimming pool. (source: AB)

Figure 8.20 A sinkhole the formed on 19 October 2005 in a front yard in the Point Loma neighborhood of San Diego, CA. The sinkhole was the results of a water main break. (source: San Diego Union Tribune)

in the U.S. 10-30% of water is unaccounted for. Near 30% in cites.

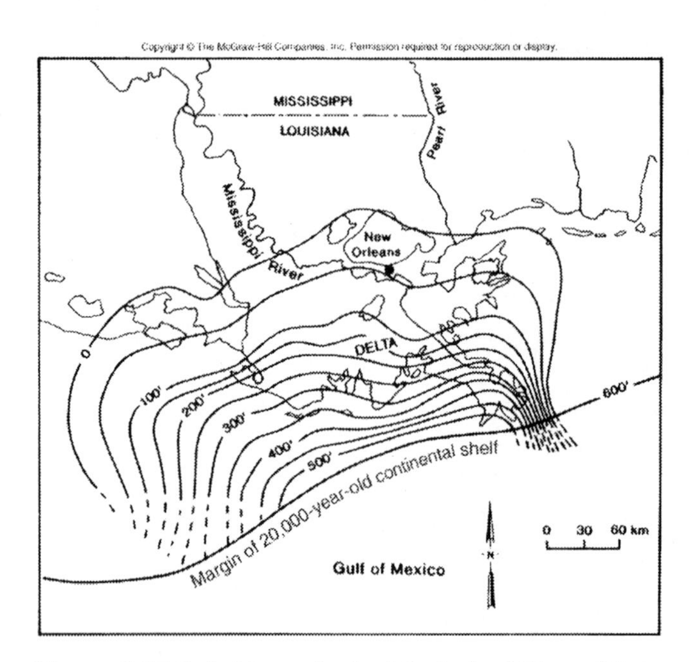

Figure 8.21 Subsidence in the Mississippi River delta area during the last 20,000 years. Contour lines are in feet. (source: AB)

Classification by Speed and Water Content

- solifluction
- creep

- slumps
- slides
- earth and mudflow
- debris and snow avalanches
- rock fall

Solifluction: occurs in permafrost areas were only the top 1-3cm thaw in summers. Meltwater cannot penetrate the ground building wet soggy unstable ground that moves down a hillside. rock glaciers: mixtures of rock fragments and ice.

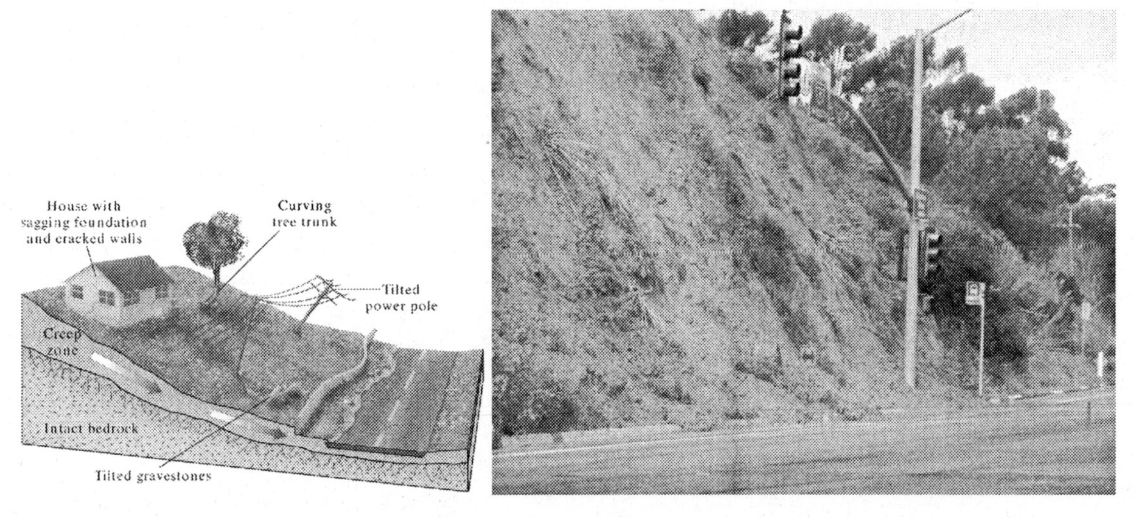

Figure 8.22 Left: Telltale signs of creep. (source: ME) right: signs of creep along Gilman Drive, La Jolla near UCSD (Photo: Gabi Laske)

Creep: slowest but most widespread form of slope failure; thawing and freezing of water in soil changes its volume and makes mass move down slope; movement is so slow that it is imperceptible; can be observed through deformed structures such as fences, bent tree trunks, tilted telephone poles and grave stones; can be cause by swelling and shrinking of ground;

- swelling:
    - freezing water in rock pores expands by 9%
    - expandable material (e.g. clay) is soil
    - heating by sun
- shrinking:
    - soil thaws

COPYRIGHTED MATERIAL — DO NOT DUPLICATE, DISTRIBUTE, OR POST

- soil dries
- soil cools

Debris Flows

- water may aid as lubricant
  but some flows are dry!
- generation of steam as lubricant
  considered unlikely because some flows contain ice
- frictional melting of material within moving mass
  considered unlikely because some flows contain ice
- air cushions
  popular but unlikely because similar sturzstroms are found on ocean floor (no air), Moon (no atmosphere) and Mars (thin atmosphere)
- trapped vibrational (acoustic) energy/acoustic fluidization
  somewhat provocative; each particle remains in its relative position with respect to each other (as found in many flows) (i.e. laminar, not turbulent flow)

Snow Avalanches

- similar to earthen mass movements though usually smaller
- heavy snow fall/end of season
- travel up to a few km at 370km/h (230mph)
- typically 0.6-0.9m (2-3ft) thick, 30-60m (100-300ft) wide; drop 90-150m (300-500ft)
- dry avalanches at 65-100km/h (40-60mph)
- wet avalanches at 30-65km/h (20-40mph)
- downhill skier can trigger snow avalanches
  two basic types:
- type 1:
  - loose powdery snow; 95% pore space
  - start at one point and grow
  - start at steepest points of slopes (30-45°)
  - then move down less steep slopes (>20°)
  - come to end in runout zone (<20°)
- type 2:
  - large avalanches from breaking free of slabs of cohesive snow

- analogous to translational slides
- ice slab contains many layers of packed snow
- melt surfaces are potenital failure surfaces
- mass breaks free and slides down as a whole on a glide horizon
- typically turn into flows during downhill movement

Submarine Mass Movements

- rotational slumps within river deltas;
- distorted rock masses at subduction zones
- submarine volcanoes

Hawaii: largest mass movement on Earth; slumps and debris avalanches; some debris avalanches over 200km (125mi) long; volumes > 5,000km^3 (1120mi^3) (10,000 large sub-arial flows! Blackhawk Event); catastrophic volcano flank collapses; 70 in last 20 Mio years; tremendous tsunami hazard; some corals and shells found 365m (1120ft) above sea level on Lanai and 60m (200ft) above water and 2km inland on Molokai; Molokai has no craters anymore (i.e. slid down into ocean); next event quite likely to happen southeast of Hawaii where rift zone weakens structure of island; right now movement is 25cm/yr; early November 2000 as fast as 6cm/day (silent earthquake); 10,000 people live on it but more are threatens through tsunami; there is 1 such flank collapse per 10,000 years worldwide;

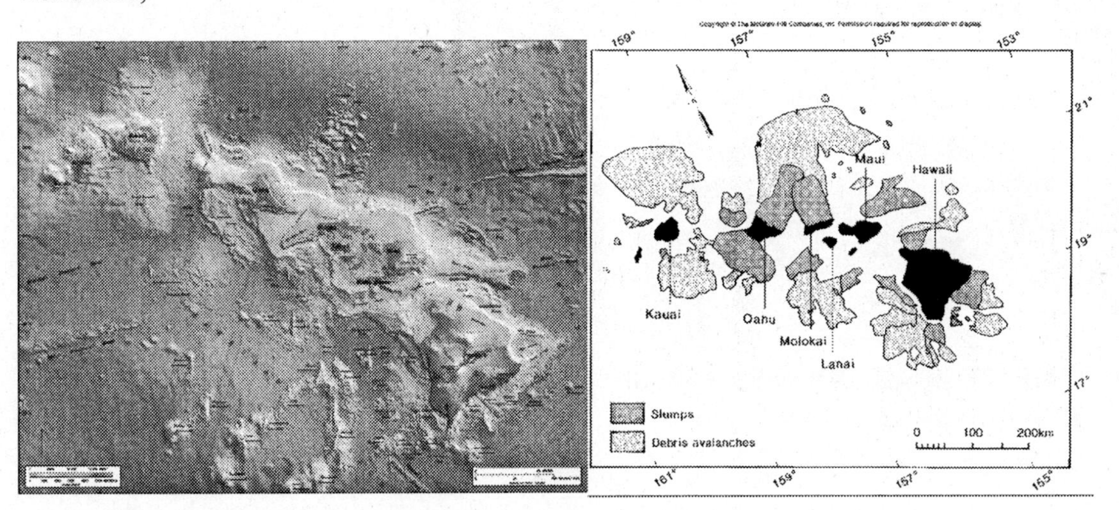

Figure 8.23 Left: seafloor bathymetry around the Hawaiian islands. (courtesy Barry Eakins) Right: How a geologists classifies seafloor features. (source: AB)

8.4 Triggers of Mass Movements

General Triggers

- preexisting instability
- rain
- earthquakes
- thawing of frozen ground
- human construction (roads, houses, grading)
- irrigation
- failing infrastructure (water mains, etc.)

The Role of Plate Tectonics

- causes uplift, therefore relief/topography
- causes faulting which weakens the crust
- causes volcanism, weathering and erosion
- causes earthquakes that trigger landslides/slumps

8.5 Mass Movements in Southern California

Figure 8.24 Fallen house after the June 2005 Laguna Beach, CA landslide. This landslide occurred due to record heavy rainfall in Southern California in the preceding months. It occurred next to the 1978 bluebird Canyon Landslide. (source: USGS)

Figure 8.25a March 1978 oblique aerial view of the U.S. (source: Kuhn and Shepard)

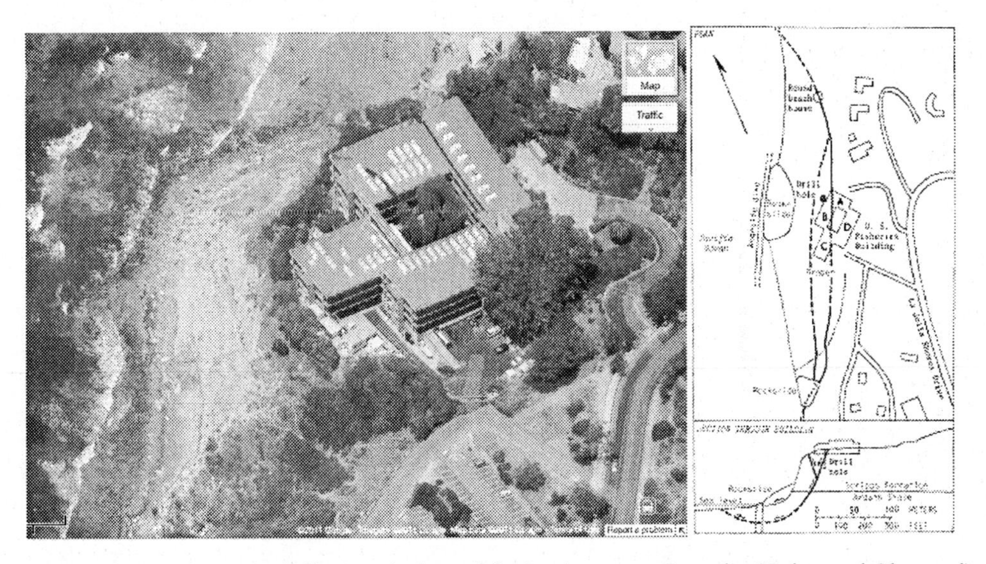

Figure 8.25b same building and plan of faults (source: Google; Kuhn and Shepard)

Plate Tectonic Setting

- southern CA has been an active plate boundary for at least the last 50 Mio yrs (see updated lecture 4)
- weak substrate in general as part of mesozoic (250 - 67 Mio years ago) accretionary prism that was later uplifted and now forms much of southern California (recall that whole North American west coast used to be subduction zone of old and now gone Farallon plate!, see Lecture 4 for details)
- faulting during past 15 Mio years (fractures are planes of weakness; steep cliffs)
- compression across SAF (oblique strike-slip!) causes uplift (slopes!)
- earthquakes shake the ground loose/reduce friction of potentially sliding masses

Environmental Effects

- o chemical weathering in faults (e.g. through salt water)
- o physical weathering in faults (e.g. through heating and freezing in the high desert)
- o wave erosion cut caves near bottom of cliffs (**undercuts**); upper portions of cliffs lose support below
- o semiarid climate cause sparse vegetation
- o brush fires destroy vegetation; subsequent rains cause mudflows
- o west coast climate brings occasional (but) heavy rains
- o coastal erosion

Human Effects

- o irrigation on unstable cliffs
- o planting of non-native vegetation that requires increase in irrigation
- o development on unstable ground, e.g. on old landslides (Mt. Soledad) and too close to cliff edge (some La Jolla Farms homes; western edge of Palos Verdes, CA)
- o landscape alteration by aggressive development (over-steepening, overloading)
- o changes to **groundwater table** (see Lecture 12)
  - excess withdrawal can lead to subsidence. Some extreme examples outside of Southern California:
    - groundwater withdrawal for farming and cities in San Joaquin valley resulted in drop of groundwater table by 120m since 1925, surface subsided by 8.5m
    - groundwater withdrawal for city of Las Vegas has resulted in subsidence by up to 3m since 1935
  - irrigation/leaky systems can raise the ground water table above glide horizon and lubricate it (reduce friction/resistance force) (Point Fermin, CA; see below)
  - OK: ground water table below glide horizon
  - NOT OK: ground water table above glide horizon
- o failing infrastructure (e.g. leaky pools, leaky irrigation, storm drains, water pipes, septic tanks)

8.6 Protection against Mass Movements

- some areas are subject to repeated hazards
- avoidance (don't build in hazardous areas)...yeah right, tell a southern Californian!
- geologists look for movements in the past (slump head scarps, swaths of flattened or tilted trees, piles of loose debris at base of hills, hummocky land surfaces) to assess potential hazard risk

IDENTIFYING REGIONAL RISKS

- roads, buildings, pipes crack
- power lines too tight or loose
- cracks of ground (potential head of slumps)
- bulge of ground (toe of slump) ("pressure ridge")
- anomalies in vegetation (sudden surplus or lack of water)
- bent tree trunks
- instrumental surveys (tilts, distances)

ITEMS FOR POTENTIAL HAZARD AND LANDSLIDE MAPS

- slope steepness
- strength of substrate
- degree of water saturation
- dip of bedding, joints, foliation relative to slope
- vegetation cover
- climate
- undercutting
- seismicity

PREVENTING MASS MOVEMENTS

- re-vegetations
- re-grading (Fig 8.7)
- reducing subsurface water (improve drainage, reduce irrigation; repair leaky pools)
- preventing undercutting (offshore breakwater or pile riprap)
- constructing safety structure (stabilize slopes; retaining walls; spray road cuts with shotcrete; protect highways with avalanche sheds)

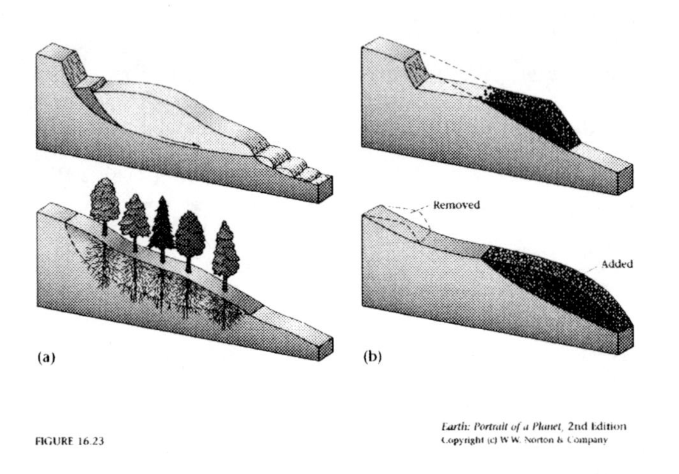

Figure 8.26a Means to "fix" an inherently unstable slump. (source: ME)

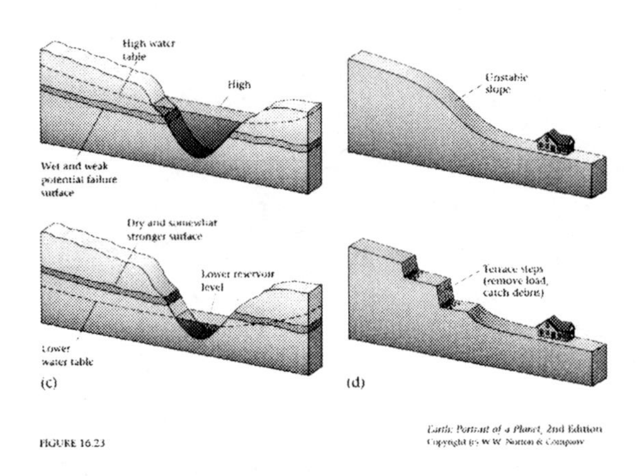

Figure 8.26b Means to "fix" an inherently unstable slump. (source: ME)

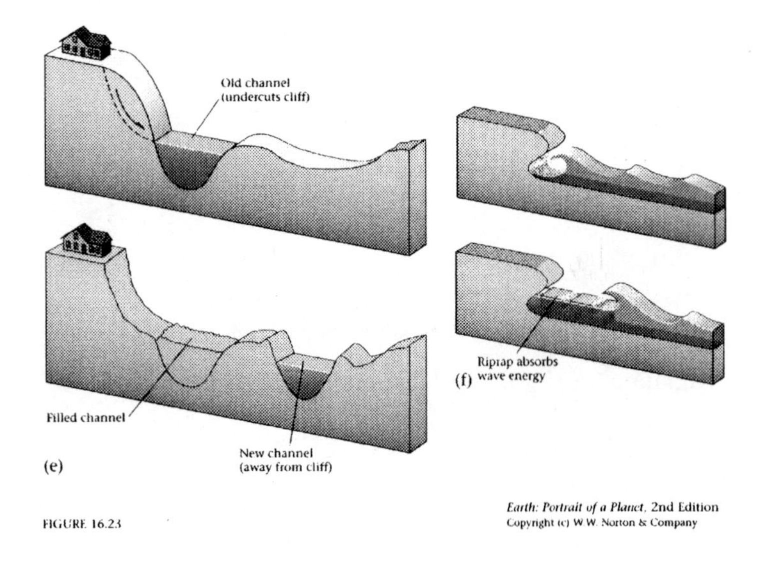

Figure 8.26c Means to "fix" an inherently unstable slump. (source: ME)

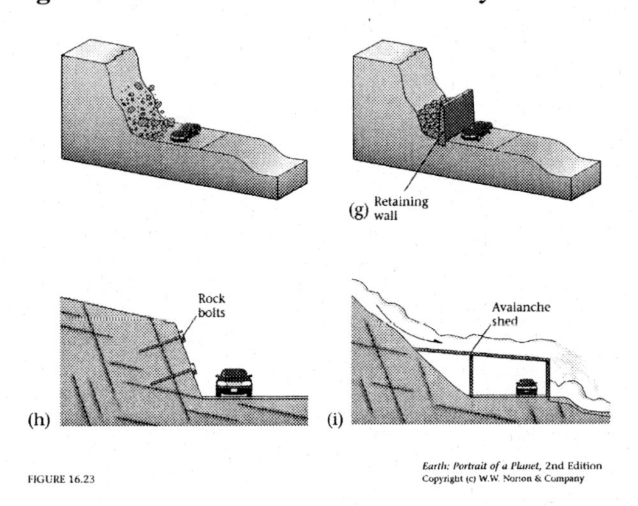

Figure 8.26d Means to "fix" an inherently unstable slump. (source: ME)

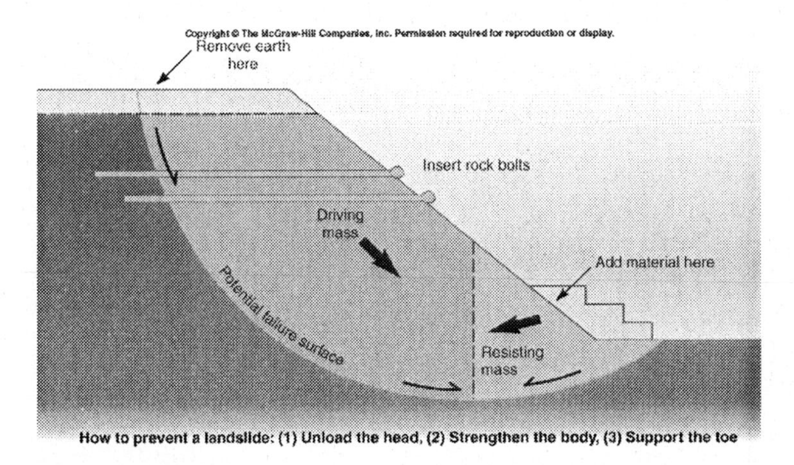

Figure 8.26e Means to "fix" an inherently unstable slump. (source: AB)

Figure 8.26f Means to "fix" an inherently unstable slump. (source: ME)

8.7 Coastal Erosion

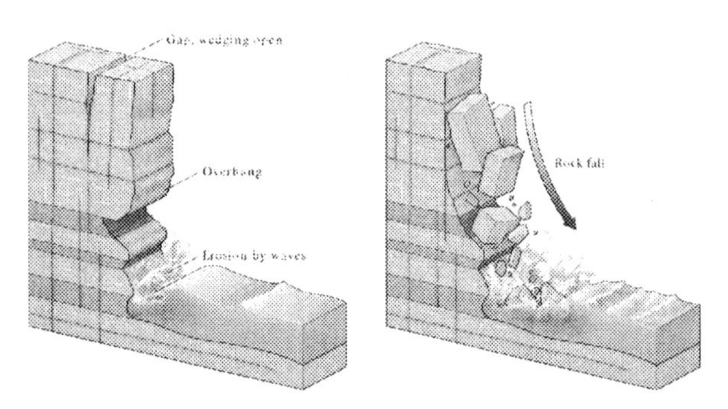

Figure 8.27 Undercut in cliffs and subsequent cliff collapse caused by ocean waves. (source: ME)

Figure 8.28 Precarious housing on unstable cliff tops in Rancho Palos Verdes (source: Google)

Natural Hazards to Ocean Coasts and River Banks

- wave action
- ocean tides
- storm surges (hurricane surges in particular)
- tsunami
- landslide, cliff collapse

Hurricanes and other great storms cause extensive damage to coastline. Stretches that are particularly vulnerable:

- barrier islands along Gulf of Mexico and East Coast

- o exposed areas along northeaster part of East Coast
- o northern part of West Coast (winter storms)
- o other extremely vulnerable coasts include those below sea level (e.g. New Orleans and Bangladesh), that need to be protected by levees that can fail

global warming may increase frequency and severity of severe weather, coastal erosion

Barrier Islands

Barrier Islands:

- o natural sand bars that are deposited by longshore current
- o lagoons form on landward side of bars
- o coast is protected by barrier islands
- o barrier islands often only a few meters high
- o barrier islands are particularly vulnerable to storms and sea level rise
- o sea level rises at about 1ft/century
- o barrier islands are often developed for tourism
- o the deadliest natural disaster to strike the U.S. was in 1900, when a hurricane made landfall near Galveston, TX, which is built on a barrier island; 6000 (some sources cite 8000) people got killed, when hurricane forecast did not yet exist;
- o Galveston now protected by seawall, but threats still exist

Figure 8.29 Housing near the head scarp of a slump along Black's Beach, La Jolla. (photo: Gabi Laske)

Coastal Protection

Several measures exist to protect costal areas and structures at the beach:

sea walls: built to ward off waves; e.g. at Scripps Beach to protect Hubbs Hall

rip-rap and boulders: structures made of concrete or natural rock placed either in undercuts to protect cliffs or around structures, e.g. piers. An example is found on Scripps Beach to protect the IGPP building that sits on the cliffs above. Rip-rap and boulders take up some of the wave energy that would otherwise be used by the waves to erode that targets.

jetties: structures in the form of dams to protect passages into a harbor, e.g. at Mission Bay.

breakwaters: built often parallel to the coastline to divert currents or approaching waves

example Santa Barbara: do not build housing on or near cliff tops. Instead, use area as public space if form of a parks. This kills two birds with one stone: 1) prevent precarious housing; 2) give public access to beach front

Is Protection a Good Thing?

Some measures, such as seawalls, are only temporary fixes:

- Often, the waves erode the beach at the base of the seawalls much faster than they would without the seawall; eventually the seawalls lose support and collapse
- sand in front of the seawalls is often transported further out into the ocean and deposited offshore instead of on the beach; the beach then loses its supply of sand
- at places where the beach sand comes from local cliff erosion and not from long-shore transport of sand from somewhere else, the beach is eroded quickly to never come back
- much of the U.S. East Coast is losing its beach sand which has to be replenished by expensive measures

NB: as many of the San Diego beaches, Scripps Beach, loses its beach sand in winter. During summer, sand is deposited as the longshore drift (driven by the California Current) brings sand from further north (e.g. San Dieguito River). In winter, storms that come from the North along the coast bring winds and strong surf from the west that pull the beach sand offshore. In years with particularly strong storms, the beach can be stripped completely of its beach sand, exposing the rocks beneath. The last time this happened at Scripps Beach was during the El Niño winter 97/98.

References and Websites:

1. **Natural Disasters in General:** Earthshock by Andrew Robinson, Thames and Hudson Publ., 1993, ISBN: 0-500-27738-9

Kuhn, G.G. and Shepard, F.P.,1984. Sea Cliffs, Beaches, and Coastal Valleys of San Diego County. Univ. California Press, Berkeley.

Chapter 9: Floods (draft)

9.1 The Causes of Floods

Why Do We Need to Study Floods?

- people like to live near rivers thereby increasing the risk of being a flood victim
- rivers provide an effective transport system and water for industry thereby attracting people
- floodplain provide fertile ground for agriculture thereby attracting people
- hazards of flood seem to be increasing despite our efforts of protection costs of floods are increasing
- we control river flow at will
- dams and levees have limited lifetime thereby posing a flood risk

General Causes of Floods

- local thunderstorms can cause flashfloods
- extended rainfall can cause weeklong regional floods
- tropical cyclone causes storm surges
- snow melt
- break-up of winter ice clogs water flow, then fails and floods
- dam break of short-lived natural dams
- failing of human-built dam and levees
- (excessive) sedimentation in lakes and reservoirs
- tsunami

Table 9.1 U.S. Flood Deaths and Damages by Decade, 1960 – 2010 (data from NOAA)

Decade	Deaths	Damages [$ billion 1995]	Yearly Ave. [$ billion 1995]
1960-1969	1297	14.418	1.44
1970-1979	1819	38.862	3.89
1980-1989	1097	21.560 (excl. 1980-1982)	3.08
1990-1999	992	49.840	4.98
2000-2010	647	10.759 (only 2000-2003)	2.69

Totals 50 years 5852 135.439

Average Fatalities: 117/year

Average Damages: $3.30 billion/year

Weather-Related Floods

Flash Floods

- typically the type of flood with high fatalities
- local short thunderstorms in areas of steep topography; most flood-related deaths by flash-floods; about 50% car-related
- flash floods occur in deserts many miles away from storm; narrow canyons (slot canyons) obstruct view
  (e.g. 12 hikers died in Antelope Canyon, 1997)
  NB: more people drown in deserts than die of thirst!
- unusually strong thunderstorms can lead to cloudbursts
- habitation of floodplain increases risk of death and loss

Regional Floods

- typically the type of flood with few deaths but extensive damage
- high water inundating extensive region for weeks
- large river valleys with low topography
- prolonged (heavy) rain from extended cyclonic systems
- 2.5% of land and 6.5% of population in floodplains
- can also be caused by snow melt and ice blockage

9.2 How Streams and Rivers Work

Run-off and the Hydrologic Cycle

- water stored in atmosphere (water vapor): 3,100mi^3
- water stored as groundwater: 1,920,000mi^3
- water stored on land: lakes and rivers (48,000mi^3); ice and snow (6,960,000mi^3)
- precipitation in balance with evaporation
- evaporation: sea (103,200mi^3); land (16,800mi^3)
- precipitation: sea (93,600mi^3); land (26,400mi^3)

- run-off from land that potentially cause floods: (9,600mi^3)

Flow Velocity

- the flow velocity is not constant throughout the stream
- the highest kinetic energy is associated with the highest flow velocity (E_{KIN}=0.5 * m * v^2, where v=velocity, m=mass of water)
- this energy can be used to carry a load (i.e. sediments) or to erode the stream bed
- high friction (causing erosion!) at the bottom and the sides slow the flow down
- in a cross-section of a symmetric stream, the highest velocities are found near the top center

Discharge

- the amount of water discharged through the stream's cross-section per second
- discharge=average flow velocity * height of water column*width of channel
- latter two multiply to be the area of the cross-section
- Amazon: 200,000mi^3/s (15% of world's run-off!)
- Congo: 40,000mi^3/s
- Mississippi: 17,000mi^3/s

The Stream Profile (Longitudinal Profile)

- longitudinal cross-section of rivers and streams are concave with steeper slope/higher gradient near the top;
- **gradient**: height vs. horizontal distance travelled; determines how fast a river flows and cuts into its bed;
- near top of stream profile: origin/spring; highest potential energy, high gradient, quickest gain in kinetic energy, high erosion rate (energy is spent on erosion)
- near bottom of stream profile: end of stream (flowing into other stream/river, lake or ocean) lowest potential energy, low gradient, lowest gain in kinetic energy, high sedimentation rate (energy is gained by dumping sediment load)
- **base level**: level below which a stream cannot erode (i.e.. ocean, lake or river the stream flows into)

- after scaling, all rivers have the same profile, independent of environment and size

Stream Gradient and Stream Length

- fast flowing streams are down a steeper slope and straight (symmetric V-shaped river bed)
- slow flowing streams are down a more gentle slope and start to meander (asymmetric river bed with an erosion and a deposition bank)

Equilibrium Stream

- equilibrium: state of balance where a change causes compensating actions (negative feedback mechanism); a stream tries to maintain equilibrium state
- controlling factors: 1) discharge of water; 2) amount of sediments to be carried away (stream load); 3) slope gradient; 4) channel pattern

- **Feedback Systems**
  - positive feedback: enhances changes; could end in runaway process; e.g. snow falls in cold climate; snow reflects more sunlight making climate colder
  - negative feedback: system strives to obtain equilibrium state; counteracts changes
- **Graded Stream Theory**
  - every stream adjusts stream bottom to changes in condition, thus sustaining graded stream (equilibrium state); a graded stream has a concave longitudinal profile and it tries to maintain this profile; a stream tries to use up all the energy that it has at the top by the time it reaches the bottom of the longitudinal profile (incl. sedimentation at the end); it can use only the energy it had at the beginning and not more; any disturbance to this balance results in compensating actions;
- **Case 1: Too much discharge**
  - more water causes stream to flow more rapidly (has more energy)
  - erodes stream bed (near top) (energy consumed)
  - carries more sediments away (stream load increases; energy consumed)
  - sediments deposited near bottom
  - process eventually decreases overall slope steepness
  - could eventually lead to meandering
- **Case 2: Too much load**

- more sediments choke stream
- sediments are soon deposited (near top) increasing stream gradient
- water then flows faster
- thus having more load carrying capacity
- could eventually lead to braided streams when load changes significantly

- **Case 3: Stream flows into lake**
  - carries sediments into the lake until water overflows dam
  - stream then flows down the steeper dam gradient being more erosive

Drainage Basins

- within small drainage basins, short heavy downpour can cause short-lasting flash floods
  (e.g. Guadalupe River in central TX rose from 0.9m to 9m in just 2h)
- within large drainage basins, widespread rains lasting for weeks can cause long-lasting floods
  (e.g. Mississippi River, Missouri; 1993 Jefferson Missouri)

Floodplains

- floor of streams during floods
- subject to seasonal change
- usually flat but have same gradient as river in normal condition
- can to lead terraces when river cut deeper into its bed and its ability to flood the entire floodplain

Meanders

- form on gentle slopes, plains with low gradients
- river may change course seemingly randomly, depending on small perturbations to river bed, discharge, load and other conditions
- unlike rivers on steeper slopes, the cross section of a meandering river is not symmetric; it has an outer and an inner bank
- outer bank: high flow velocities, high kinetic energy -> erosion; undercutting can cause cliff collapse
- inner bank: low flow velocities, low kinetic energy, river unable to carry its load -> deposition of sediments
- over time growing curvature of meanders as a result of prolonged erosion on outer banks

- in a full loop of a meander, there are three places with outer banks, one near the outer top and one on each side
- over time, the loop is enhanced (positive feedback enhances erosion)
- eventually, river cuts off its loops forming an **oxbow lake**
- new path is shorter and has a steeper gradient, increasing the flow speed and discharge at the downstream end of the meander, increasing the risk of flooding

9.3 Flood Frequency Curves and Hydrographs

Flood Frequency Curves

Flood frequency curves are used by FEMA (Federal Emergency Management Agency) to issue recommendations in disaster management and prevention. Issues addressed by flood frequency curves:

- small floods happen more often than large floods that have longer recurrence interval
- different rivers have different flood recurrence intervals, hence different flood frequency curves
- flood frequency curves help to calculate the probability of a flood of a certain size to occur within a given time span;
- the 100-year flood is a large flood (with a large discharge) that has a recurrence time of 100 years; i.e. it occurs approximately once every 100 years
- the probability of a certain-size flood is not a cumulative process, i.e. a 100-year flood has 1% probability to occur within any given year; but only 63% within 100 years.

  The physical reasoning behind this statistical number: Since the recurrence time is only an average time, the time to the next flood of the same size could be slightly shorter or longer. Therefore, there is a chance that a 100-year flood does not occur within the next 100 years but a little later. Consequently there is only a 63% probability.
- need to know the flood frequency curve of a stream, river or channel in construction projects for canals, flood channels and levees!!

How to construct a flood frequency curve?

like earthquakes, floods can't be predicted; use statistical measures to estimate likelihood of a flood of a certain size

- determine peak discharge for each rainfall in each year
- rank all floods from largest to smallest (ignore chronologic order)
- recurrence interval for each year's maximum flood R = (N+1)/M; N=number of years of flood records; M numerical rank of the individual flood
- plot each value in diagram
- probability of a flood in any given year is the reciprocal of the recurrence interval
- the longer the records the better the flood-frequency curve is

Hydrographs of Streams

- plots amount of rainfall and flood runoff over time during and after a storm
- runoff lags behind rain fall
- maximum runoff smaller than maximum rainfall
- runoff curve has steep rising limb and gentle falling limb (rise of flood quicker than recovery)
- recovery prolonged because infiltration feeds stream from below ground

Runoff and Infiltration

- **runoff**: flowing water from precipitation above ground
- **infiltration**: rain water than penetrates the ground
- natural setting: 80-100% infiltration; 0-20% runoff
- urban setting: 0-10% infiltration; 90-100% runoff
- runoff in urban setting increased due to paving and roofing, prohibiting water from penetration ground
- maximum runoff/discharge much larger in urban than natural setting
- maximum runoff/discharge reached much earlier in urban than natural setting

9.4 Levees

Levees are supposed to

- keep a river in its bed
- keep an ocean from flooding land near or below sea level
- avoid spill into floodplains
- serve as flood gates for controlled flooding

- Levees fail due to

- wave attack
- overtopping
- slumping
- piping

PROS:

- flood damage would have been greater without levees

CONS:

- building cost of levees may exceed value of structures to protect
- levees create false sense of security, encouraging more development
- should tax payer provide funds for disaster relief (people live there against better knowledge)
- higher levees eventually raise river above floodplain
  e.g. floodwater height increased 2-4m in engineered upper Mississippi basin, while no change in not-engineered sections

PERHAPS:

- remove some levees to restore wetland habitat and ease flood risk
- move towns
- provide disaster relief funds only for moving to higher grounds or raise floor level of existing houses

many levees near the end of their lifetime!

9.5 The Mississippi River

- Geography
  - world's third largest drainage basin
  - length: 2,340 mi
  - length of Missouri-Mississippi River system: 3,710mi
  - world's 4th longest river system
  - one of world's largest river deltas
  - delta arms change on 100-1000 year time-scales
  - large accumulations of oil-bearing sediments (up to 10km)

- one of world's largest sedimentary basins

- The fight of Human vs. Nature:
  - building levees shortens but steepens path
  - increase of flow speed
  - potential floods in bottleneck downstream
  - Mississippi flooded several times in the last 100 years
  - flood severity seems to increase
  - costs of floods increase
- Example 1: The fight against the Atchafalaya River
  - 1831: dig channel to by-pass one of Mississippi's "irritating" meanders
  - Mississippi accepted the new channel but Old River arm of meander remained active (Red River upstream!)
  - 1839: removed 30mi log jam down Atchafalaya River to ease flow (of Red River/Old River)
  - Mississippi now drained into Old River, ultimately into Atchafalaya River
  - took away too much water downstream (Baton Rouge/industry; New Orleans/tourism)
  - 1939: Atchafalaya took 30% of Mississippi's flow
  - 1963: decision to keep Atchafalaya from draining more of Mississippi's water; USACE built old River control structure
  - 1973: barely withstood flood; greatly damaged!
  - 1980: auxiliary structure that wasn't ready for 1983 flood; first structure barely survived
- Example 2: The 1927 flood

  The 1927 flood of the Mississippi River has probably been the most troublesome flood of this River, with the most controversial measures taken to protect a certain area, on the expense of another.

  - according to Abbott: flood breached levees in 225 places; inundating 19,300mi ^{2} and drowning 183 people
  - according to Wikipedia: flood breeched levees in 145 places and flooded 27,000mi ^{2}, killing 246 people
  - according to *Earthshock* by Robinson (and many other sources): the flood threatened New Orleans; therefore, levees downstream were dynamited to allow the water to flood other areas, including plantations; this flooding

COPYRIGHTED MATERIAL — DO NOT DUPLICATE, DISTRIBUTE, OR POST

created a sea covering 26,000 mi^2 in 7 states; it was 18ft deep and 80mi across; this flood cost 246 lives, maybe 500 and made 650,000 people homeless, half of them African-American; according to Wikipedia, many blacks were detained and forced to labor at gunpoint during flood relief efforts.

- A flood gage in Vicksburg, MS measured a flood level of 17.1m (56.2ft) above normal. Had the levees held, the Mississippi River would have risen to a staggering level of 19.0m (62.2ft).
- according to *Rising Tide* by Barry: the flood profoundly changed race relations, government, and society in the Mississippi River valley; many blacks left the area and migrated to northern cities. The flood also had profound consequences for plantations in the deep South.
- check out Great Flood of 1927 at Wikipedia
- the causes and consequences of the 2005 Hurricane Katrina flood has been likened in its impact to the 1927 flood

- Example 3: The 1993 flood
  - at $ 12 billion 4th costliest natural disaster in U.S. history
  - upper Mississippi flooded for 160 days
  - greatest inundation flood in 140 years
  - Mississippi alone flooded 12,500 mi^2
  - 50 fatalities
  - adverse weather situation: wet winter, spring and summer due to south-shifting jet stream
  - largest discharge of any Mississippi flood
  - check out Great Flood of 1993 at Wikipedia
- Example 4: New Orleans and Hurricane Katrina 2005
  - Katrina was a category 3 hurricane when she roared ashore August 29, 2005
  - New Orleans levees broke at dozens of locations, flooding large portions of city, especially the poor 9th Ward
  - three key locations where levees broke: London Ave (canal to Lake Pontchartrain); 17th Street Canal (Lake Pontchartrain); Industrial Canal
  - investigations revealed that levees were not overtopped but failed due to shifting soil beneath

- the Inner Harbor Navigational Canal meant to direct and protect shipping traffic into Lake Pontchartrain acted as funnel to enhance effects of storm surge into Industrial Canal
- New Orleans is below sea level and subsiding and protection requires update
- canal system prevents flooding of wetlands thereby contribution to subsidence
- according to Tidwell in an interview on CSPAN-2 in October 2005: restoration of wetlands would help preventing such floods; estimated cost: $18 Billion

Figure 9.1 Trash collecting near Smuggler's Gulch flood control channel along. (source: San Diego Union Tribune, 2 Dec 2008; John R. McCutchen)

9.6 Southern California's Flood Control Channels

The Case San Diego

San Diego adopted the L.A. style channel system: most runoff channels are lined with concrete, even the bottom. Since San Diego practically has a desert climate (only 10.5in/year rainfall), these channels are dry most of the year, but can become a death trap during the few heavy winter storms. Also, a growing problem in recent years - with decreased maintenance resources due to budget woes: the sedimentation in some channels can get quite heavy. The channels lose their effectiveness if these sediments and other debris is not removed. There have been reports in recent years of some cases of extensive littering and growth (reed and even trees).

Two channel projects have repeatedly made the flooding news as a result of poor planning:

- San Diego/Tijuana joint Project to redirect the Tijuana River
    - 80% of river in Baja; 20% in California; last few miles in San Diego before entering the Pacific Ocean
    - Agreement to adopt L.A. style channel system
    - Tijuana realized its part south of the border
    - San Diego part stalled due to environmental concerns
    - a flood followed in Jan. 1978 that submerged farmland south of Imperial Beach
- Mission Valley Project (San Diego River)
    - centuries worst flood was in Jan 1916 with a discharge of 72,000ft^3/s
    - U.S. Army Corps of Engineers (USACE) built a 245m wide channel with natural floor in 1940ies to hold maximum discharges of 115,000ft^3/s
    - in 1950ies, central part of Mission Valley was developed (incl. Fashion Valley at the eastern end); channel joined existing one to the west; 7.5m wide holding maximum discharge of 8000ft^3/s (seems ok though not designed to hold the "100-year flood")
    - in 1980ies, eastern part of Mission Valley was developed; channel joint small channel to the west; 110m wide holding maximum discharge of 49,000ft^3/s
    - the current situation is that the easternmost channel (upstream) can hold larger discharge than the one it is directing the water into (smaller one to the west)
    - devastating floods in the central part (incl. Old Town and Hotel Circle) an inevitable!
    - nearly every winter, storms cause some degree of flooding in Fashion Valley. In early 2010, the mall parking lot was completely flooded, and closed to the

public. The I-8/163 connector can also be closed as a result of flooding after heavy winter storm rainfall.

References and Websites

1. AB
2. Earthshock by Andrew Robinson, Norton and Comp., 2002, ISBN: 0-500-28304-4

The 1927 Mississippi Flood: Rising Tide by John M. Barry, Simon&Schuster Publ., 1998, ISBN: 0-684-84002-2

The loss of the Mississippi delta wetlands: Bayou Farewell by Mike Tidwell, Vintage Publ., 2004, ISBN: 0-375-72517-2

USACE: http://www.usace.army.mil

Other Recommended Reading:

Earthshock by Andrew Robinson, Thames and Hudson Publ., 1993, ISBN: 0-500-27738-9

Johnstown Flood: "The Johnstown Flood" by David McCullough; Touchstone Book by Simon & Schuster Publ., 2. ed, 1987, ISBN: 0-671-20714-8

A possible location of the biblical/Gilgamesh flood "Noah's Flood" by W. Ryan and W. Pitman, Touchstone Book by Simon & Schuster Publ., 2000, ISBN: 0-684-85920-3

COPYRIGHTED MATERIAL — DO NOT DUPLICATE, DISTRIBUTE, OR POST

Chapter 10: Wildfires

Ever since plant life developed on land, Earth has experienced uncounted wildfires. These fires ranged from small, very local and short-lived fires to very large, widespread fires lasting for months, maybe even a year. Since the time when humans appeared on Earth, the nature and geometry of wildfires has completely changed. Very often, wildfires are suppressed immediately after they started to protect human structures. It turn out that this strategy may help in the short term but could increase the risk of very large fires in the long term. Urban sprawl has pushed communities farther into the surrounding **wildland** (a natural area that is not used or modified for human activity). But at the same time, firefighting costs have increased tremendously in recent decades. Already, we have witnessed in San Diego that potentially life-saving but costly equipment is grounded because of budget constraints. In a warming global climate, some areas, for example much of California, will experience drier climates that are more likely to allow larger wildfires.

Figure 10.1 A wildfire in California, 5 September 2008. (source: Wikipedia/Bureau of Land Management)

10.1 Fires and the Natural Environment

The Types of Fuel and Wildfires

The terms wildfires and wildland fires are used in this chapter in general terms and include forest fires, bushfires, brush fires and grass(land)fires. In detail, the different terms describe the type of environment or the type of fuel involved in the fire.

A **grassland fire**, grass fire or prairie fire burns mostly grass. Flames are typically 1 – 2.5 m (3 – 7.5 ft) tall. The fire moves at about 6.5 km/h (4 mph) though can spread significantly faster when pushed by winds. Grass fires may burn with less intensity and often are not as spectacular as forest fires. Nevertheless, they can get very large, especially during droughts. In the wet seasons, the sawgrass marshes of the Everglades, FL experience numerous grass fires that are initiated by lightning. These fires are superficial and only burn the parts of sawgrass above water, dead plants and tips of trees, while leaving the roots intact to allow re-grow of the grass. During a drought, fire can ignite in the peat. Such fires penetrate deeper and destroy the roots (see below). Grass fires are also common in Texas, particularly when grass thrives during a wet season but then dries out during a long dry, windy period. On 15 March 2006 the second largest fire in U.S. history burned in grassland near Amarillo, TX (Table 10.1). Whipped at times by 55 mph (35 km/h) winds, the fire consumed 850,000 acres (240,000 ha or 3,400 km^2) within only 4 days. Eleven people lost their lives. Apart from its size, this fire made the news because 10,000 cattle were killed in the fire. The fire spread so fast that the cattle had no chance to run away. Five years later, the Bastrop County Complex fire set a new record as Texas' most destructive wildfire, burning nearly 15 times as many homes as the previous record-holder of 2005 (Case Study 6).

A **brush fire** burning brush (or shrubs) typically is more intense than grass fires. Shrubs are typically 0.6 – 3.5 m (2 – 12 ft) high. Flames can reach a height of 15 m (50 ft), and fires can spread at speeds of 15 km/h (8 mph). In Mediterranean climates, where shrubs often contain large amount of highly flammable, aromatic oils, fires can burn more intensely and spread faster. The term **scrub fire** (scrub meaning thicket or jungle) may be a stand-in for such fires.

Flames in a **forest fire**, **tree fire** or **woodland fire** can become very tall (Fig. 10.1) but the intensity of a forest fire depends on the makeup of the vegetation. Fires become most intense and fast-moving when they can spread into the treetops to become crown fires (see below, Case Study 7).

The term **bushfire** is a general stand-in for large wildland fires in Africa and Australia. Bushfires involve the entire vegetation, including grasses, scrubs (low shrubs), bushes and trees. In Australia, this includes eucalyptus trees and banksia shrubs. In Africa, bushfires mainly burn in the savanna, a grassland ecosystem in which trees (e.g. acacias) are sufficiently widely spaced that they do not form a closed canopy.

Slash is a type of fuel consisting of woody debris after a forest had been clear-cut. Slash can also accumulate in a natural environment such as in a eucalyptus grove when trees frequently discard leaves and branches. **Duff** often stands in for the forest floor in general. But in the discussion of wildfires, duff usually consists of a 1-ft (30 cm) thick layer of highly flammable pine needles. Its potential danger is easily underestimated, even by professionals. After fighting a wildfire, duff can look extinguished but it continues to smolder inside. When the temperature becomes sufficiently high, the duff can re-ignite and start a new fire. Such was the case during the 20 October 1991 Oakland, CA Firestorm (Case Study 3).

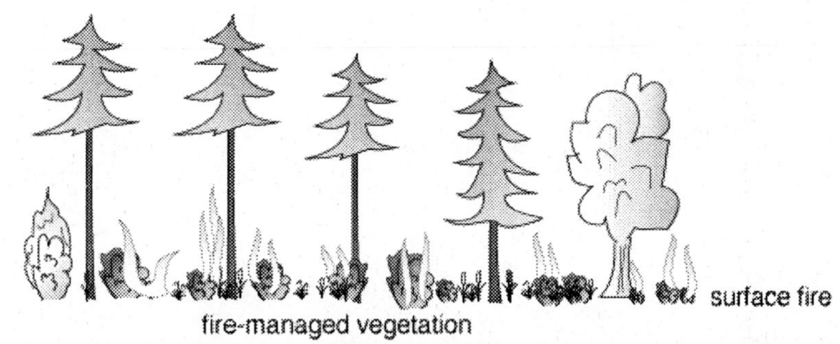

Figure 10.2 Ladder fuels. Top: Vegetation of varying heights allows fires to climb from grasses to tree tops, similar to moving up the rungs of a ladder. Bottom: in a forest lacking vegetation at intermediate heights, a surface fire is less likely to develop into an intense crown fire.

Fire administrations often classify wildfires into surface, crown and ground fires. **Surface fires** typically burn rapidly at a low intensity and consume light, low-lying fuels while presenting little danger to mature trees and root systems (Fig. 10.2). Light fuels include leafs, duff, slash, debris, grass and low shrubs. Surface fires could occur in grassland but also in the understory of forests.

Crown fires (or **canopy fires**) literally burn material high up in the trees, at the canopy level (Fig. 10.2; Case Study 7). Such fires may develop from a surface fire, but lightning can also ignite crown fires. Such fires release embers and falling burning branches, which aids in the extensive spread of crown fires as well as surface fires. Crown fires therefore often consume the entire vegetation. Whether a surface fire can potentially become a crown fire depends on the geometry of the vegetation. So-called **ladder fuels** occur in multi-story forests, i.e. vegetation exists at all heights. Fires then easily spread from the surface level up to higher levels until it spreads in the canopy. Slash-and-burn clearing in the Amazon rain forest that has a multi-story vegetation or other tropical rain forests often develop into devastating crown fires that destroy ecosystems that are not adapted to wildfires. In forests managed to reduce wildfire risks, the intermediate stories in the vegetation should be removed. The downside of this is that such "skeleton" forests are not capable of supporting the same biodiversity as an uncontrolled forest.

Ground fires burn at least partially below the surface. Such fires are relatively rare but very intense as they destroy all vegetation and organic matter, leaving only bare earth. Particularly dangerous are **peat fires**. Peat is an accumulation of partially decayed vegetation in wetland bogs, moors and peat swamp forests. Peat is also the first stages in a long process to form coal (geological timescales apply!). Peat can ignite underground and smolder for months, burning surface vegetation as they progress. Since the source of such fires is inaccessible, the fires are particularly difficult to fight.

During droughts, **peat fires in the Everglades, FL** can causes extensive damage to the marshes. Since peat fires penetrate deeply, the root systems of the sawgrass is also destroyed, in contrast to the frequent grassland fires from which the marshes recover quickly (see above). Charcoal found in the peat in portions of the Everglades indicate that the region has endured severe fires for years at a time, although this trend seems to have abated since the last major occurrence in 940 B.C..

The devastation of peat fires of recent years, with a significant human-impact component in both effect but also cause, is dramatically illustrated by the **2010 wildfires in Russia** (Case Study 1). The heat wave and poor air quality led to a high fatality rate of 56,000 but previous draining of the peat bogs dramatically increased to risk of peat fires.

The Need for Wildfires in Natural Environments

In the natural environment, wildfires have occurred ever since organic materials have been available for burning. Before humans arrived, the vast majority of wildfires were sparked by lightning, but volcanic eruptions, spontaneous combustion, sparks from rock falls or, relatively rarely, impacts of extraterrestrial bodies also started fires. Earth's flora has adapted well to wildfires and some plants developed protective measures. The Giant Redwood (sequoia sempervirens), also called California Redwood lives along much of costal California. Growing to a height of 60 - 100 m (200 - 330 ft) and 3 – 4.5 m (10 - 15 ft) in diameter, it is the tallest tree on Earth. It can grow 1800 years old though

COPYRIGHTED MATERIAL — DO NOT DUPLICATE, DISTRIBUTE, OR POST

older trees are known. Its tannin-rich bark is soft and up to 30-cm (1 ft) thick. The bark protects the tree from wildfires as does the fact that the foliage starts high above the ground. In addition, the wood is intrinsically fire-resistant, one reason why it is popular in home-construction.

Figure 10.3 Two probably 2000-year old Giant Sequoia in the Giant Sequoia National Monument, CA. The tree on the right experienced burns in a previous wildfire but survived thanks to the thick bark. (source: Wikipedia)

The Giant Sequoia (Sequoiadendron giganteum) lives on the western slopes of the Sierra Nevada Mountains that are frequently by wildfires. Sequoias are the largest trees on Earth in terms of volume. They grow to an average height of 50 – 85 m (160 – 280 ft) and 6 – 8 m (20 – 25 ft) in diameter, but single specimen can be larger. Some sequoias are 3,500 years old, and have likely survived many fires that other plants have succumbed to. These trees have an astoundingly thick, fibrous, furrowed bark, sometimes reaching 90 cm (3 ft) in thickness at the base of the tree (Fig. 10.3). This bark gives the tree the protection needed in wildfires. The seed cones remain green and closed for up to 20 years before either fire or damage by insects opens them. Several species of California Oak also have an unusually thick bark, giving them protection from wildfires. While young saplings die in fires, the mature trees usually survive and grow shoots from the base. In fact, some natural environments depend on wildfires to thrive.

Figure 10.4 Left: A group of eucalyptus trees at SIO; Right: the multi-layered bark of a eucalyptus tree at SIO.

Eucalyptus trees also have a soft, multi-layered protective bark (Fig. 10.4) and hard, leathery leaves that resist heat and drought. After a fire, regenerative shoots can grow from so-called epicormic buds located deep within the bark or from so-called lignotubers, a starchy swelling of the root crown near the base of the tree.

Fire is needs for a variety of reasons: recycling of organic materials in otherwise unfavorable climates; nutrient supply; germination; removal of competitors; pest control; influence insect behavior.

1) Only three climates do not need fires for the **recycling of organic materials**: the tropical climate is the only climate with speedy recycling as temperatures and moisture are both high enough to effectively decompose organic matter in a short time; deserts where plant growth is so slow that recycling is not needed; polar regions where plant growth is not abundant. Antarctica is therefore the only continent on which no wildfires occur. In a natural setting, all other climates and environments depend on the help of fires. This includes grasslands, seasonal tropical forests, some temperate-climate forests an Mediterranean shrublands. In Mediterranean climates (found around the Mediterranean Sea, in Australia and in California), winters are mild but nevertheless too cold and summers are too hot and dry to effectively decompose organic materials. The burning of organic matter produces ash that ultimately provides **nutrients** for new growth.

2) Some plants need fire for **germination** through the impact of either heat or smoke, or through the removal of taller vegetation that deprive smaller plants of sunlight. Lodgepole Pines (Pinus contorta) occur along the coast and inland areas of western North America. It is the only

conifer native to both Alaska and Mexico. It can grow 6 – 24 m (20 – 80 ft) tall and 0.3 – 0.9 m (1 – 3 ft) in diameter. The tall trees typically grow densely packed, allowing fires to spread into the crowns. Fires in such forests are relatively rare, due to the lack of understory vegetation, but they are intense. The pine cones usually remain closed until the heat of a fire opens them, and the forest regenerates in the post-burn areas. The Knobcone Pine (Pinus attenuata) is another tree that needs fires for germination. This tree occurs along the coast and inland of western North America from Oregon to Baja California. The 9 - 24 m (10 – 80 ft) tall trees form almost pure stands on poor, coarse, rocky soil. The cones may become embedded in the trunk when the tree grows. The cones may survive for 30 years before a fire opens them to release the seeds.

3) Fires are needed to **eliminate competitors**. In a forest of tall trees, the understory vegetation often looses out because the tall trees deprive it of light. The removal of the tall trees by fire gives smaller plants the chance to return. It is curious that one of the tall trees actually needs these same fires for this same reason. The seeds from giant sequoias can thrive in moist needle humus (duff) in spring, but the seedlings only survive in full sunlight and will die during summer as the duff dries. Sequoias need wildfires to clear the duff and the vegetation that competes for sunlight to give young seedlings a chance to survive. Redwoods also benefit from the removal of competitors. Some plants, such as Eucalyptus plants, contain flammable oils that encourage fires that may eliminate competitors. On hot days, eucalyptus trees release flammable aromatic volatiles (gases), causing produce a blue haze above extensive forests in Australia, their native continent. In addition the trees often dispose of branches, contributing to a rapid build-up of wood litter on the forest floor, thereby increasing the risk of fires. The tree's hard leaves and thick bark ensure its dominance over less fire-tolerant species.

4) Fires are also needed for **pest control**. Insects including beetles are an integral part of a healthy ecosystem. Many insects help the reproduction and spread of plants through pollination. Others help decomposing and recycling organic matter. However, an unusual increase in parasites can devastate even the largest of landscapes. In climates where winters get very cold, with temperatures well below freezing, pests such as the bark beetle die off and may survive only in balanced quantities. But with a warming climate, winters in some locations may no longer become cold enough to kill off the pests. Droughts also contribute to an imbalance between host trees and insects that otherwise are vital for deadwood recycling. In German and other European forests, a drought in 2003 and extensive bark beetle infestations therefore contributed to extensive tree kills. In Mediterranean climates, winters are not cold enough to kill parasites and natural environments need fires for pest control. Since 2009, the Goldspotted Oak Borer has been decimating three species of California Oak in San Diego County. The larvae remain in cut oak logs and firewood, and human can potentially spread the

beetle by taking firewood home from a camping trip. In a natural environment, a fire would reduce the pests to a level that allows oak groves to survive. One would think that beetle infestation and the resulting so-called "deadkill" lead to more devastating forest fires but this long-held belief has recently been challenged. Ongoing NASA studies suggest that infestation of Lodgepole pine stands in the Rocky Mountains by the Mountain pine beetle may actually reduce available small fuels and so limit the effect and reach of fires[15].

5) Wildfires also influence **insect behavior**. Migrating insects such as some butterflies likely change their migration routes in the quest to find food. Some insects increase populations after a fire to make up for losses but others decrease. Some insects react to heat or smoke. This is perhaps most obviously demonstrated by beekeepers who calms bees down by smoke. Some beetles are also sensitive to the smell of smoke and are capable to detect it from up to 50 mi (80 km) away. A variety of jewel beetle (Melanophila acuminata) has heat-sensitive organs that help the beetle find freshly burned conifers into which the beetle lays its eggs.

10.2 The Global Distribution of Wildfires

Since the moment humans discovered how to ignite a fire, it has been used to their many advantages. Above all, fire is used to prepare food, heat the home, and also to direct and control game animals and predators. But fire has also played an important role in the development and advance of civilizations. In the slash-and-burn clearing for forests, valuable land is gain for agriculture, if only for a short time in tropical regions. The ash left behind often serves as cheap fertilizer. Land-clearing and burning is also used for logging, mining and road-building. Fire is also used to harden materials of all kinds, including pottery but also weapons. Fire ultimately facilitated the industrial revolution that started in the second half of the 18th century, and life as we know it today. Of course, fire is also used as a means of controlling people and of warfare. The vast majority of wildfires today, some 85%, are caused by human action. On any given day, the number of fires worldwide can vary between a few thousand and many thousands. A rather instructive Quicktime movie of worldwide fires from 2000 through 2012 is available at NASA's Earth Observatory[3] (Fig. 10.5). As a result of seasonal variations, February is the month with the fewest fires globally, with an estimated 3000 per day. During July through September, on the other hand, an estimated 6000 fires can burn each day. With the seasons, the location of the fires varies throughout the year. In the tropics, the time from November through February is the time for slash-and-burn clearing, with December and January having the most intense activity. Slash-and-burn fires are particularly evident for the Sahel region in Africa, but they occur elsewhere as well. In Venezuela, most fires occur on the plains between the

Guiana Highlands in the east and the Cordillera in the west. In Central America, the fire season usual lasts from April through May. A large number of fires also burn in Southeast Asia during that time. In temperate climates, "fire seasons" shift either to the dry season or to a time when farmland is prepared for planting. Fires return every August and September to huge areas in South America (mainly Brazil and Argentina), the southern part of Africa from the equator to the Tropic of Capricorn, including Madagascar, and the Ukraine and neighboring areas north of the Black Sea. In August 2008, a significant fire was ignited by warfare and destroyed 250 ha (620 acres) in the Borjomi-Kharagauli National Park in Georgia, one of the largest national parks in Europe. An estimated 150,000 m^3 of standing trees were lost and endemic species almost completely destroyed[(4)].

In southern Europe along the Mediterranean Sea, wildfires are common during the summer months. Many fires may be sparked by lightning but a large number is also deliberately set. In well-developed Europe, fires are *not* set for agricultural land clearance. Plain arson may be a cause, but fires are also often set by greedy land speculators. The devastating **2009 fires in Attica, Greece** dramatically illustrate the threats of urban sprawl into the surrounding wildland. The exact source of these fires will likely never be known but heat wave conditions, strong winds and arson have been cited. The fires started on 21 August about 25 km northeast of Athens, the Greek capital and spread quickly, engulfing 14 towns in suburban Athens within only 3 days. The 1000 firefighters and soldiers fought with 19 planes and helicopters but where soon overwhelmed. Help was called from Italy, France and Cyprus that supplied 14 more helicopters, for a cost of 30 million Euros ($39 million). The fires burned 21,000 ha (52,000 acres) of pine forest and olive groves, destroying 60 homes and damaging 150 more. The World Wildlife Fund (WWF) had criticized the country for its inability to effectively manage its forests. Thousands of illegal housing projects had been pushing into the wildland ahead of government elections, while the WWF lamented that the fire service had been weakened since 1981 by political interference and the abolishment of specialized firefighting units. Greece's fire and forest services are severely depleted, with 3,000 permanent vacancies in the fire service alone. Without doubt, ignorance, negligence and greed contributed to this dilemma. Greece's current financial crises does not bode well for a speedy solution. The 2009 fires luckily did not cause any fatalities but it is only a matter of time until the next inferno strikes.

Australia is well-known for devastating bushfires. They usually occur during the dry season from November through January though the northern part of the Northern Territory may be struck throughout the year[(3)]. Particularly deadly fires occurred in Victoria in South Australia from 7 February through 14 March 2009. The Black Saturday bushfires, the deadliest bushfires in Australian recorded history, caused 173 fatalities, destroyed over 2000 homes and 2000 other structures, and burned more than 450,000 ha (1.1 million acres). Two years earlier, the Great Divides Fire, also in

Victoria, raged for over three months starting in 1 December 2006 and burned over 1 million ha (nearly 2.5 million acres), killing one and destroying 51 homes.

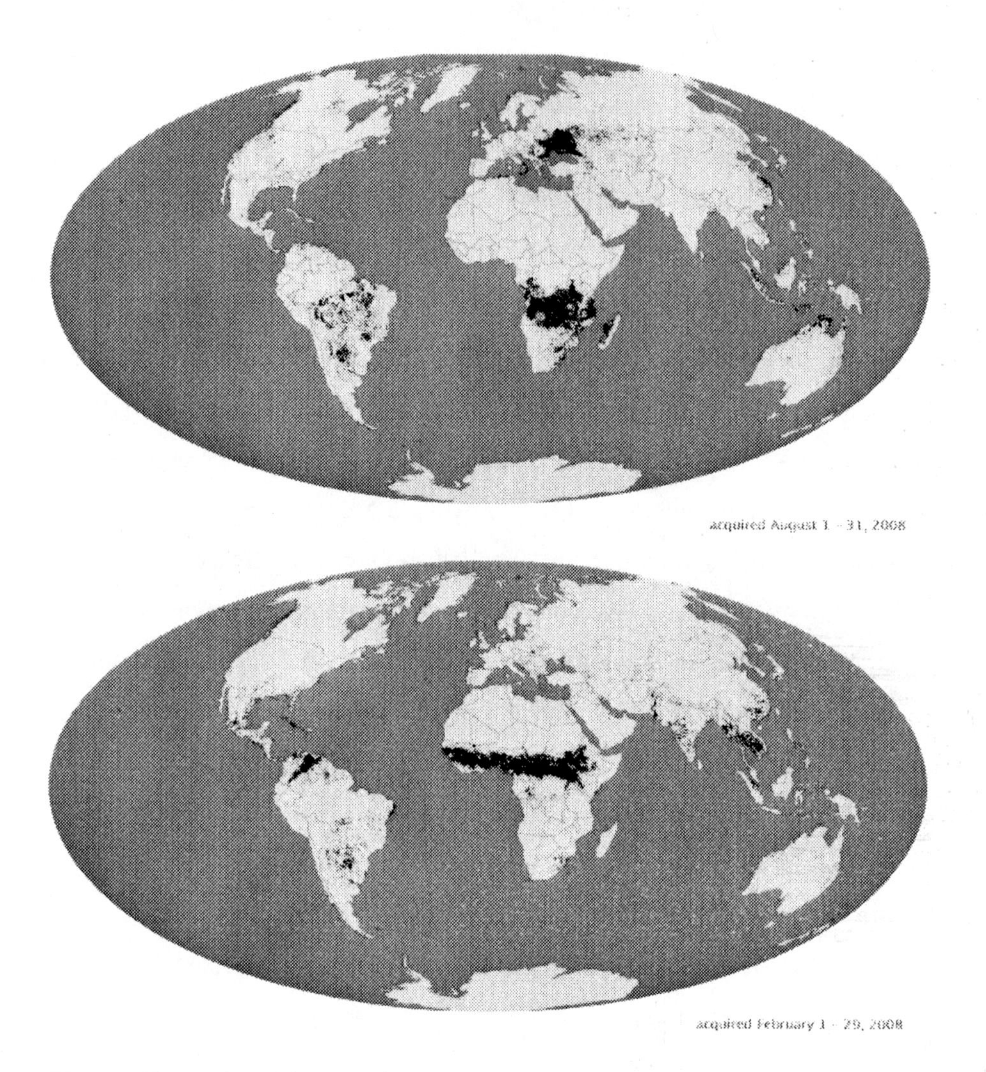

Figure 10.5 MODIS images from NASA's Terra Satellite showing the global distribution of wildfires for two time intervals. bottom: 1 – 29 Feb 2008; top: 1 – 31 Aug 2008. (source: NASA Earth Observatory)

10.2 Wildfires in North America

Wildfires are common in North America. In Canada alone, nearly 9000 wildfires occur each year, burning a total of 2 million ha (nearly 5 million acres) or 20,000 km^2 (7,700 sq mi) of land. These fires usually burn in remote areas and rarely threaten town. An exception was the **2011 Slave Lake wildfire** that started through arson on 14 May and burned 4,700 ha (12,000 acres) of wildland in only three days[(5)]. Pushed by strong, 100 km/h (60 mph) winds, it destroyed 433 buildings and damaged 89 more, and one firefighter died in a helicopter crash. The complete population of Slave Lake (7,000 residents) evacuated before the fire destroyed 1/3 of the town, including the town hall, the library and the radio station. The total damage was estimated at $1.8 billion. With about $700 million in insured losses, this has been Canada's second costliest disaster, after the North American ice storm of 1998 (that caused $6 billion in damages). The fire was one of over 100 wildfires that had burned over 105,000 ha (260,000 acres) across the province of Alberta in an unusually dry season, with about a quarter of the fires burning out of control.

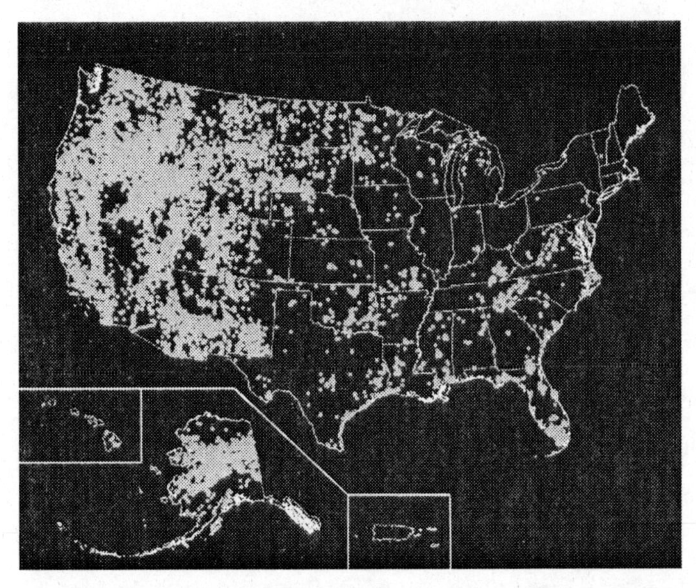

Figure 10.6 Locations in the U.S. with wildfires burning more than 250 acres, from 1980 to 2003. (source: NASA/Wikipedia[11])

In the U.S. about 100,000 wildfires burn on the order of 5 million acres (2 million ha) each year, the same number of acreage burned as in Canada, but by ten times as many fires. Hence, fires in Canada are typically ten times larger than in the U.S.. Presumably, many of the Canadian fires burn in remote areas, not threatening any human structures, and so are not suppressed by firefighting measures. According to the U.S. Fire Administration, 115,000 wildfires occurred annually from 1988 through 1997, burning 4 million acres[6]. For the years 2001 through 2011, the National Interagency Fire Center estimated that 76,000 fires burned 6.7 million acres each year[8], where single large fires burned hundreds of thousands of acres (Tables 10.1, 10.2). The large majority of fires occur in the western half of the country, including Alaska (Fig. 10.6). Relatively to the rest of the country, 2% or less of the fires occur in Alaska but the fires can be very large in some years. In 2005 and 2009, more than half the acreage burned in the U.S. occurred in Alaska. The 2004 fire season was Alaska's worst on record. 270 fires ignited by lightning burned a total of 6.5 million acres, making up 82% of the acreage burned in the entire country. The Taylor Complex fire of that season was the largest fire in the 8 years from 2000 through 2007, and the 2004 fire season in Alaska produced half of the 10 largest fires in the country from 2000 through 2007. In fact, the **Taylor Complex Fire** was the largest in the U.S. since 1997. Some fires threatened the Trans-Alaska Oil Pipeline. With some 430 human-caused fires, that season also experienced about 50% more human-caused fires than usual for Alaska but the acreage burned was relatively small. The largest wildfire in recorded U.S. history was probably the 1910 Bitterroot Blowup when 3,000,000 acres (12,000 km^2) land burned (Case Study 7). The highest loss of lives occurred during the **1871 Peshtigo Fire, WI** when 1,500 people perished.

Table 10.1 Large Wildfires greater than 150,000 Acres in the Lower 48 States 2000 – 2007[(7)] (with one 2011 update)

Year	Fire Name	Location	Acres burned
2006	East Amarillo Complex	Texas Forest Service, TX	907,245
2007	Murphy Complex	BLM Twin Falls District, ID	652,016
2011	Wallow	White Mountains, AZ/NM	538,049
2002	Biscuit	Siskiyou National Forest, OR	499,570
2002	Rodeo/Chediski	BIA Fort Apache Agency, AZ	468,638
2007	Big Turnaround Complex	Okefenokee Nat. Wildlife Refuge, GA	388,017
2007	Milford Flat Fire	SW Area, State Division of Forestry, UT	363,052
2007	Cascade Complex	Boise National Forest, ID	302,376
2007	East Zone Complex	Payette National Forest, ID	300,022
2000	Valley Complex	Bitterroot National Forest, MT	292,070
2003	Cedar	Cleveland National Forest, CA	279,246
2005	Cave Creek Complex	Tonto National Forest, AZ	248,310
2006	Winters	BLM Winnemucca and Elko Districts, NV	238,458
2006	Derby Fire	Gallatin National Forest, MT	223,570
2006	Crystal	BLM Idaho Falls District, ID	220,042
2000	Clear Creek	Salmon-Challis National Forest, ID	216,961
2005	Clover	BLM Twin Falls District, ID	192,846
2000	Eastern Idaho Complex	BLM Idaho Falls District, ID	192,450
2006	Charleston Complex	BLM Elko District, NV	190,421
2000	SCF Wilderness	Salmon-Challis National Forest, ID	182,600
2005	Delamar	BLM Ely District, NV	170,046

2006	Day	Los Padres National Forest, CA	162,702
2000	Command 24	BLM Spokane District, WA	162,500
2002	McNally	Sequoia National Forest, CA	150,696
2006	Sheep	Elko Field Office, BLM, NV	150,270

Table 10.2 Large Wildfires greater than 150,000 Acres in Alaska 2000 – 2007[7]

Year	Fire Name	Location	Acres burned
2004	Taylor Complex	Alaska Division of Forestry	1,305,592
2004	Eagle Complex	BLM Upper Yukon Zone	614,974
2004	Solstice Complex	BLM Upper Yukon Zone	547,505
2004	Boundary	Alaska Division of Forestry	537,098
2004	Central Complex	BLM Upper Yukon Zone	451,162
2004	Pingo	BLM Upper Yukon Zone	285,885
2004	Winter Trail	BLM Upper Yukon Zone	279,865
2004	Chicken	BLM Tanana Zone	257,720
2004	Wolf Creek	BLM Upper Yukon Zone	197,067
2004	Camp Creek	Alaska Division of Forestry	175,815
2000	Zitziana	BLM Tanana Zone	164,387
2005	Chapman Creek	BLM Alaska Tanana Zone	162,670

Natural Causes vs. Human Causes of Wildfires

The number one natural cause of wildfires is lightning strike. Every hour, about 1800 thunderstorms occur around the globe. Six million lightning strikes occur each day, or about 70 each second. Recalling that during the "busy" fire season, 6000 fires can burn globally each day, only a very small fraction of lightning strikes actually cause a fire: only 1 in 1000 lightning strikes, if we assume that all fires are actually caused by lighting. But they are not, and so the number is actually much smaller.

In the U.S., lightning strikes start about 15% of all wildfires, while the rest are caused by human action (Fig. 10.7)! This means that human activity is 7 times more likely to ignite a wildfire than lightning does. These numbers include only wildfires (so a house fire is not included here) and the fraction of human-caused wildfires is fairly uniform from year to year. In California, the fraction of wildfires with natural causes usually stays below 10% (Table 10.3) but can vary from year to year and depends on drought severity and occurrence of thunderstorms.

In developing countries, the number one human cause of wildfires is land clearing for farming. Even though land clearing by means of fire is not a common practice in Europe, 95% of forest fires are started by humans[12]. In California, about ¼ of wildfires are started by debris burning (burning of waste on private land and in landfills) and equipment use (e.g. sparking equipment in dry conditions). The **2007 Zaca Fire** (Table 10.4) started on 4 July. Sparks from a grinding machine ignited the fire during repairs to a water pipe on private property. The fire had burned for almost four months in rugged areas in Santa Barbara and destroyed over 240,000 acres of land (97,000 ha or 970 km^2). Firefighting costs amounted to $117 Mio by 2 September when the fire was declared contained. About 2 in every 100 wildfires are started by vehicles, e.g. when the vehicle itself burns or a faulty catalytic converter ejects hot debris. Another 2 wildfires are started by carelessly discarded cigarettes. Arson, the deliberate start of a fire to do harm, has been attributed to 4% of the wildfires. The **2006 Esperanza Fire** west of Palm Springs claimed the lives of five firefighters who tried in vain to save a home (Case Study 4). It was allegedly set by Raymond Lee Oiler who was thrilled by setting fires. A signal fire started by a lost hunter on 26 October 2003 caused the **Cedar Fire** in San Diego County, the largest wildfire in California history (Case Study 5; Table 10.4). Almost exactly four years later, sparking power lines and TV cables ignited the **Witch Creek and Guejito Fires**, burning the second-largest area in San Diego history. In California, naturally-caused wildfires tend to be much larger than human-caused wildfires, but variations are large from one year to the next (Fig. 10.8). For California, Cal Fire reports that seven of the twenty largest wildfires since 1932 were ignited by lightning (Table 10.4). During the 20 – 21 June 2008 Northern California Lightning Series, 25,000 – 26,000 dry lightning strikes (lightning from a thunderstorm that does not produce precipitation on the ground) ignited more than 2,000 fires. Over 1 million acres (400,000 ha or 4,000 km^2) had burned by 29 August. The 2003 Cedar Fire, the 2007 Zaca Fire and the 2007 Witch (Creek) Fire were exceptionally large human-caused wildfires in California.

Table 10.3 Causes of 2009 Wildfires in California[7]

Cause	% Fraction of Wildfires
Natural	
lightning	7
Human	
debris burning	12
equipment use	12
arson	4
smoking	2
vehicle	2
campfires	1

power lines	0.7
miscellaneous	30.5
undetermined	28.5
total (4736 fires)	100

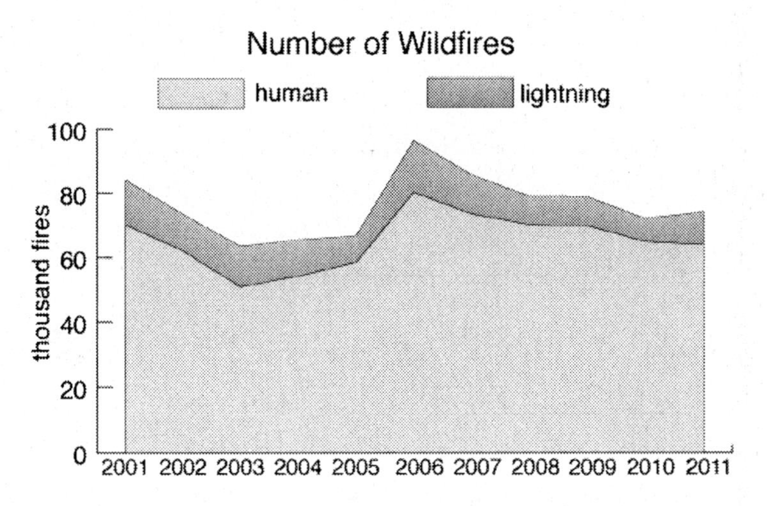

Figure 10.7 Human causes vs. natural causes of wildfires in U.S. during the last decade. Most wildfires (typically 15%) are consistently caused by humans. (source: NIFC[8])

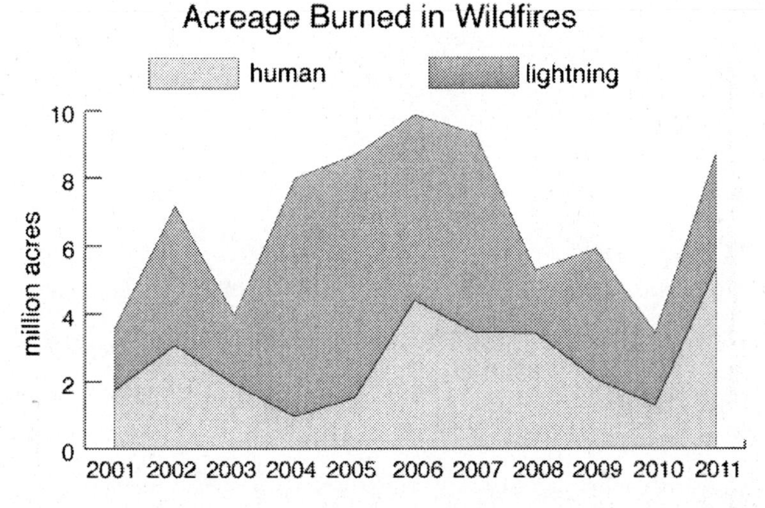

Figure 10.8 Acreage burned during wildfires in the U.S. that have a human cause vs. a natural cause, in the last decade. Naturally-caused wildfires tend to be larger than human-caused wildfires, but variations are large from one year to the next. (source: NIFC[8])

Table 10.4 The 20 Largest Wildfires in California 1932 – 2009*[(2)]

Rank	Date	Fire Name/Cause	County	Acres burned
1	10/2003	Cedar/Human	San Diego	273,246
2	07/2007	Zaca/Human	Santa Barbara	240,207
3	09/1932	Matilija/undetermined	Ventura	220,000
4	10/2007	Witch/Powerlines	San Diego	197,990
5	06/2008	Klamath Theater Complex/Lightning	Siskiyou	192,028
6	07/1977	Marble Cone/Lightning	Monterey	177,866
7	09/1970	Laguna/Powerlines	San Diego	175,425
8	06/2008	Basin Complex/Lightning	Monterey	162,818
9	06/2006	Day/Human	Ventura	162,702
10	08/2009	Station/Human	Los Angeles	160,557
11	07/2002	McNally/Human	Tulare	150,696
12	08/1987	Stanislaus Complex/Lightning	Tuolumne	145,980
13	08/1999	Big Bear Complex/Lightning	Trinity	140,948
14	08/1990	Campbell Complex/Powerlines	Tehama	125,892
15	07/1985	Wheeler/Arson	Ventura	118,000
16	10/2003	Simi/under investigation	Ventura	108,204
17	08/1996	HWY 58/vehicle	San Luis Obispo	106,668
18	06/2008	Iron Alps Complex/Lightning	Trinity	105,805
19	09/1970	Clampitt/Powerlines	Los Angeles	105,212
20	07/2006	Bar Complex/Lightning	Trinity	100,414

* the largest wildfire in California history was probably the September 1889 Santiago Canyon Fire (formerly known as the Great Fire of 1899). It burned up to 300,000 acres (125,000 ha) in Orange and San Diego Counties.

10.2 The Science of Fire

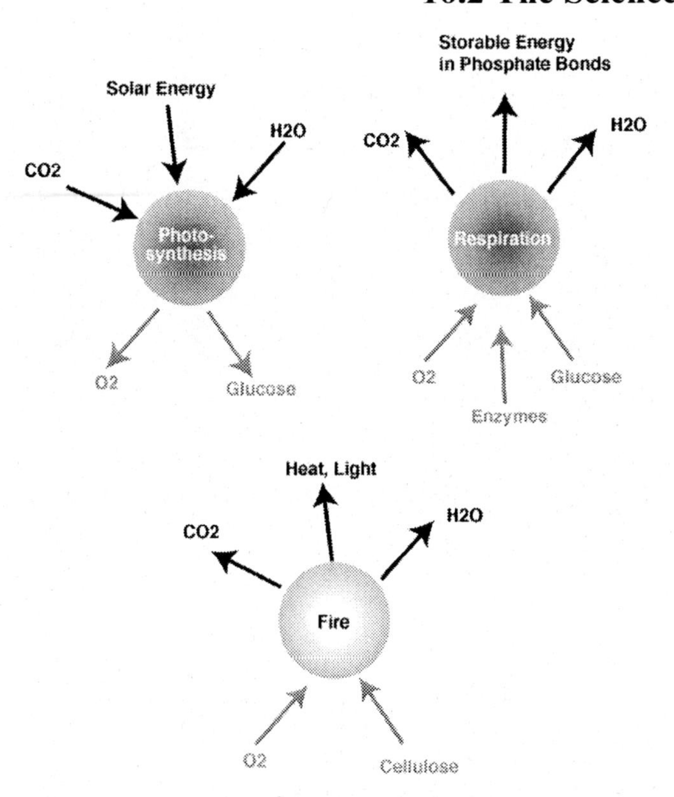

Figure 10.9 Photosynthesis, cell respiration and fire. Chemically, respiration and fire are the opposite of photosynthesis. In the latter, solar energy is used to convert carbon dioxide, CO_2 and H_2O to sugars, e.g. in form of glucose, $C_6H_{12}O_6$. The chemical reaction during photosynthesis then is: $6\ CO_2 + 6\ H_2O + \text{solar energy} \rightarrow C_6H_{12}O_6 + 6\ O_2$ With the help of enzymes, respiration converts O_2 and glucose to storable energy (adenosine triphosphate, ATP), and CO_2 and H_2O. In fire, O_2 and cellulose ($C_6H_{10}O_5$) are converted to CO_2 and H_2O, and heat is released. The chemical equation describing this process is: $C_6H_{10}O_5 + 6\ O_2 \rightarrow 5\ H_2O + 6\ CO_2 + \text{energy (heat + light)}$.

The Chemical Process and the Stages of Fire

During a wildfire, the organic material undergoes **combustion** where fuel reacts with an oxidant (oxygen in our case) in an exothermic chemical reaction (heat is released). Since oxygen is involved, this chemical process is an oxidation of the fuel. We generalize for now that the term fuel primarily stands for plant material. In a chemical and simplified sense, a wildfire can then be understood as a reversed photosynthesis (Fig. 10.9). Fire is the rapid combination of oxygen with organic material (e.g., cellulose) in a reaction that produces water, CO_2, flames, heat and light. Cellulose makes up about 1/3 of all plant matter, but this fraction can be much higher in particular plants (e.g., 90% in cotton). In wood, about ½ of the matter is cellulose. Another compound in wood is lignin that strengthens the cell walls and makes up about ¼ to 1/3 of dry wood. Heat and light are energy that is released during the chemical process, while the flames are the visible (light-emitting), gaseous part of the fire.

Perhaps surprisingly, fire involves the release of water, which comprises the first of several stages of a fire. During the **preheating phase**, water is driven out from potential fuel and eventually evaporates. Since water has a high heat capacity, large amounts of heat energy is needed in this phase to prepare the fuel (i.e. wood) for the next stage. Heat is also needed to raise the temperature of the fuel high enough to start the next stage of a fire. For example, cellulose has to be heated to about 300°C (575°F)[9]. This is just beyond the capabilities of a regular kitchen oven! The preheating phase applies to living as well as dead organic materials. The need for the preheating phase explains why it is often hard to get a campfire started. Drought conditions that removed much of the water in plants often aid or even stand in for this first fire stage. Dry plants are much more likely to catch fire than well-watered plants. On the other hand, keeping potential fuel wet dramatically reduces the risk of a fire.

Once temperatures pass the threshold for the second stage, the **pyrolysis**, the heat starts to decompose the chemical structure of the wood. A product thereby is tar that further decomposes and leads to the release of flammable hydrocarbon gases that react with each other and with oxygen. During this reaction, more heat is released feeding a chain reaction. Combustion can then begin.

If oxygen is present, the flammable gases ignite, producing flames. This is the third stage of the fire in which **flaming combustion** takes place. In a strict sense, this is the actual process taking place in combustion. In this stage, the pyrolized surface of the wood (the surface on which pyrolysis takes place) burns fast and hot, and large amounts for heat is released that is now the driving force of the fire. The larger the amount of heat released, the hotter the gases become and the faster the fire spreads. We will deal with the external spread of a fire in a later section. Internally, the fire spreads through conduction (see Chapter 2). The heat moves toward the interior, starting the preheating phase and moving the pyrolysis surface toward the center. This is also called **charring**. The conduction of heat in woods is relatively slow (wood is a good insolator!) so the interior of the wood may remain below the 300°C threshold for a long time while the wood burns on the outside. The time it takes to consume a whole wood log depends on its size. With the same speed at which heat is conducted, smaller logs burn much faster than larger logs. The **charring rate**, the speed at which the pyrolysis surface moved toward the interior, depends on the type of wood. Wood panels and plywood have charring rates about twice as high as those of soft- and hardwood, i.e. they burn faster. Even laminate veneer lumber burns more slowly than plywood and wood panels. The extensive use of plywood in the construction of a home in a fire-prone area is therefore not a good idea. The Forest Products Laboratory of the U.S. Department of Agriculture publishes charring rates for various woods and combinations thereof at http://www.fpl.fs.fed.us.

As a wood log heats, it expands in volume and forms cracks through which the gases can escape to feed the flames. In cases when cracks do not form, the wood "pops", and sparks and embers are thrown out. Picked up by winds, the embers can start new fires well ahead of the main fire. This was the case during the **2008 Freeway Complex fire** near Corona, CA when many burning palm trees ejected embers like sparklers (Fig. 10.10a). This was also a particular problem in the 2007 San Diego Witch (Creek) fire where embers picked up and carried by strong Santa Ana winds, ignited homes 5 miles away from the fire front! Burning palm trees have been a problem in other fires as well, particularly when the trees are not maintained and dead palm fronds hang in clusters from the trees.

After flaming combustion passed through the wood, it goes into the final phase of a fire, the **glowing combustion**. The active flames have disappeared but the wood still glows. Oxygen now combines with the carbon compounds in the charred solid wood rather than with the gases.

Figure 10.10a A storm of embers rains down from palm trees that caught fire in the Hollydale mobile home park in Brea, CA during the 15 November 2008 Freeway Complex Fire, also called the Triangle Complex Fire, or Corona Fire. Flying embers from the palm trees are evident through the light traces coming off the trees. The fire burned 30,000 acres (12,000 ha or 120 km^2), destroying or damaging over 300 residences. Firefighting costs were estimated at $16 million. It is thought that this

fire was ignited by hot particles ejected from a car exhaust along State Route 91. (photo: Don Bartletti, Los Angeles Times).

Figure 10.10b Firefighters watch a monstrous cloud of smoke approach as a 10-mile-wide wall of flames climbs the mountains during the 26 October 2003 Cedar Fire in San Diego county. The photo was taken on 28 October after the winds shifted from easterly Santa Ana winds (blowing from the east) to westerly winds, pushing the fire eastward. (source: Lenny Ignelzi, Associated Press).

Smoke also forms during a fire and consists of small, mainly carbon-containing particles, liquid particulates and gases, including air. It forms during both pyrolysis and combustion. Smoke inhalation is the primary cause of death in victims of indoor fires. Depending on particle size, smoke can be visible or invisible to the naked eye. In wildland fires, smoke can greatly decrease the visibility, sometimes to the point where airborne firefighting has to cease (Fig. 10.10b). Some particulates in smoke can be toxic and highly corrosive, e.g. when hydrochloric acid forms from burning fire retardants or when other materials burn in homes engulfed by a wildfire. Depending on the extent of the fire and wind directions, the smoke can also be carried a long distance. In Southeast Asia, smog is a regular problem resulting from land and forest fires in Indonesia, especially Sumatra and Kalimantan.

According to Fig. 10.9, CO_2 is released during a fire. As discussed in more detail in Chapters 11 and 18, CO_2 is a greenhouse gas, and the recent global increase in CO_2 enhances global warming. Much of the increase is caused by the burning of fossil fuels: oil, gas and coal. But wildfires also contribute to this increase. Forest fires in Indonesia in 1997 were estimated to have released between 13% and 40% of the annual CO_2 emission from fossil fuel burning. It is tempting to make the case that, because 85% of wildfires are caused by humans, then the human impact on the global increase of CO_2 and global warming becomes an even more pressing issue. But critics argue that the burning of live plants (in contrast to fossil fuels) is just the recycling of CO_2 that was recently in the atmosphere anyway before the plants grew, and some wildfires actually promote new growth and so lead to an accelerated uptake of CO_2. On the other hand, research also suggests that the smoke and soot from (human-caused) wildfires enhances the absorption of solar radiation during the winter months at high latitudes (e.g. in the Arctic) by as much as 15%, thereby enhancing the warming effect.

The Spread of a Fire

The spread of a fire depends on several factors such as the 1) topography, the 2) current weather conditions, the 3) type of fuel involved and the 4) behavior of the fire itself.

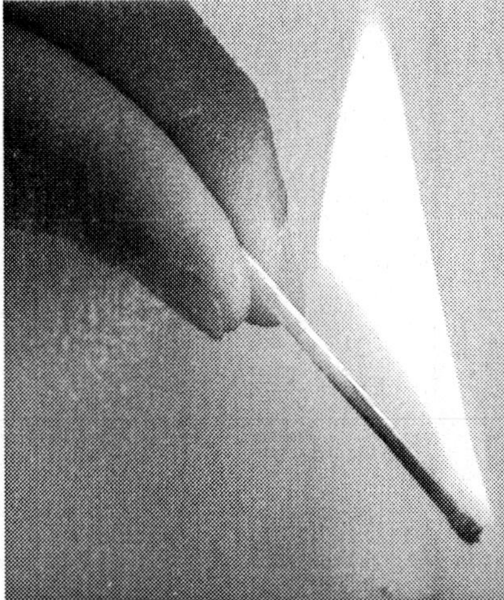

Figure 10.11 Concept of how a wildfire burns and spreads on a slope. Left: Since heat rises, a match held upright burns only slowly downward, similar to a fire initiated on the top of a hill. Right: A sloping match, with the initially burning end at the bottom, burns rapidly upward, similar to a fire that initiated on the bottom of a hill.

Figure 10.12 The 26 October 2003 Cedar Fire spreading uphill near San Diego Country Estates in San Diego's backcountry. (source: John Gastaldo, San Diego Union Tribune[13])

1) Inside wood, fire spreads by conduction. In air, the heat supply for the preheating phase spreads by means of the much faster radiation, diffusion of air particles and convection. Hot air rises so fires tend to spread much faster upslope than downslope. A simple experiment illustrates this point. An ignited match held upright with the flame at the top burns very slowly. In our experiment, the match burns itself out about halfway down, after more than 30 s (Fig. 10.11). A match from the same box completely burns within only seconds when the burning end is held downward, allowing the flames to climb up the match. The heat released from the flame advances along the sloped match and starts the preheating phase. In an upright match, with the flame on the top, only little heat reaches the match below the flame, allowing the match to survive much longer. Similarly, in calm winds, a wildfire that starts on top of a mountain will burn very slowly down a hill, or may even burn itself out. A fire starting at the bottom of this hill will climb up, quickly engulfing the entire hillside (Fig. 10.12). Hence, wildfires spread more effectively upslope than downslope.

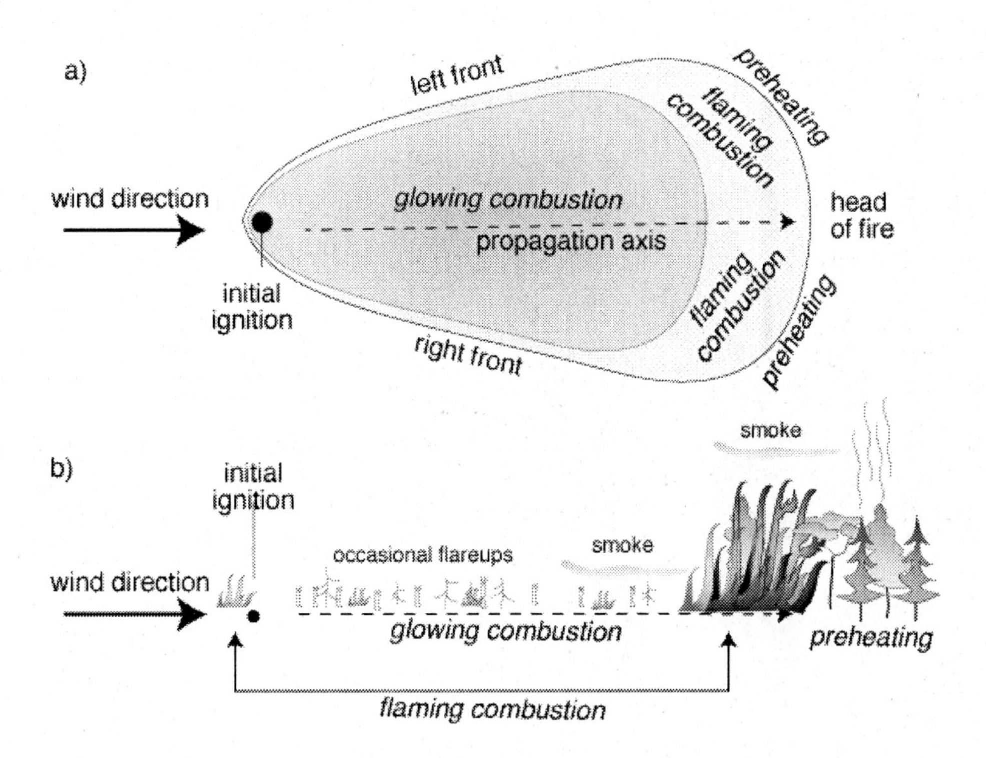

Figure 10.13 The spread of a wildland fire and the stages of fire in map view (a) and along a vertical-profile view along the propagation axis (b).

2) Weather conditions in a certain environment greatly influences the likelihood of a fire. Prolonged droughts and unusually dry conditions reduce the water content in living plants and dead material alike, conditioning them for the first fire stage, the preheating. Moist, water-rich plants and parts thereof are much less likely to burn. A blowing wind controls how a fire spreads. The fire will spread in all directions, but it spreads fastest in the direction the wind blows. The wind blows the heat from the fire ahead of it, thereby starting the preheating phase and preparing the fuel for pyrolysis. Numerical modeling of fire propagation predicts that the burned area assumes a squeezed-tear-drop shape (Fig. 10.13). The widest burn area is found near the head of the fire, while some flaming combustion can occur along the left and right fronts of the fire, and to a small extend also at the rear end beyond the initial point of ignition. In strong wind conditions, fires can be pushed downslope even though they would not spread otherwise. This fact makes firefighting more difficult and dangerous because fires can become less predictable than in calm wind conditions.

3) The spread of a fire depends on the type of fuel available. This was already touched in the discussion of fuel and the types of fires. Plants thriving in Mediterranean climates typically contain aromatic oils that aid the spread of fires. Plants with a high water content, on the other hand, are much

less likely to burn up in a huge fire. The abundance of ladder fuel allows a grass fire to burn into a much more intense crown fire that is also more difficult to fight. As mentioned above, trees whose wood does not tend to form cracks to allow the gases to escape during pyrolysis tend to explode or send off ambers that in turn can ignite new fires well ahead of the head of the fire.

Figure 10.14 A photo taken in East San Diego County during the 26 October 2003 Cedar Fire. The palm trees indicate the direction and intensity of the winds during the fire. (source: Dan Trevan, San Diego Union Tribune[13])

Figure 10.15 A fire tornado captured near Langdon, ND on 24 October 2011. The tornado formed in a grassland burn. (source: storm spotter Kelly Schwartz, downloaded from NOAA[14])

Figure 10.16 A pyrocumulus cloud from the 26 August 2009 Station Fire seen from downtown Los Angeles, CA. The largest wildfire fire in L.A. County's modern history was started by arson in steep terrain in the Angeles National Forest northwest of Pasadena and northeast of Los Angeles. The blaze threatened 12,000 structure in the National Forest and nearby communities, La Cañada Flintridge, Glendale, Acton, La Crescenta, Littlerock and Altadena as well as the L.A. neighborhoods of Sunland and Tunjunga. The fire burned on the slopes of Mt. Wilson and threatened numerous TV, radio and cell phone antennas and well as the historic astronomical observatory. A 40-mi (64-km) stretch of State Highway 2 (Angeles Crest Highway) was closed indefinitely due to guardrail and sign damage. Two firefighters lost their lives, 89 homes and 120 other structures were destroyed, and over 160,000 acres (65,000 ha or 650 km^2) before the fire was 100% contained on 16 October. The fire cost nearly $100 Mio to fight. (source: Wikipedia)

4) The behavior of a fire itself also determines how it spreads. Large fires can form huge convection clouds (Figs. 10.10b; 10.16) that produce their own weather. They resemble cumulus clouds (cauliflower clouds, see Chapter 12) and are also called pyrocumulus clouds, in reference to the Latin word *pyro* for "fire". The strong updrafts within these clouds cause intense surface winds that can whip a fire in otherwise calm wind conditions. At this point, a fire has become a **firestorm**. The photo in Fig. 10.14 shows evidence of strong winds at the time when the palm trees burned. These may have been Santa Ana winds but may also have been winds generated by the fire. The 2007 Guejito Fire burned around the San Diego Wild Animal Park but not through it. Nevertheless, the fire generated intense winds down the valley within the park, which was later evidenced by numerous wind-tilted trees.

A fire racing up a slope burns particularly hot near the top. This causes an updraft on the other side of the slope. A fire burning on that side can thereby be pulled up and accelerated. When the wind shear in a fire is just right, fire whirls and fire tornadoes can form (Fig. 10.15). The intense updraft from a fire whirl draws fire and hot burning debris to high altitudes. The fire intensifies and the burning debris can be dropped far away to start a new fire. A fire whirl can last for several minutes. A fire tornado forms when the fire whirl extends from the surface all the way to the base of the pyrocumulus cloud.

The Fire Triangle, Firefighting and Fire Prevention

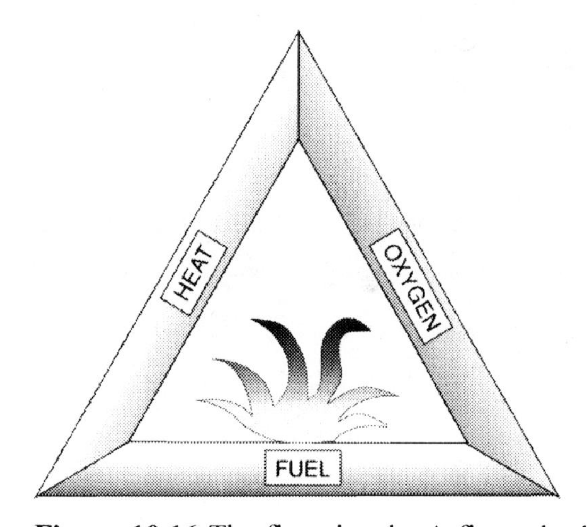

Figure 10.16 The fire triangle. A fire only thrives when all three sides are present. The removal of one side during firefighting extinguishes the fire.

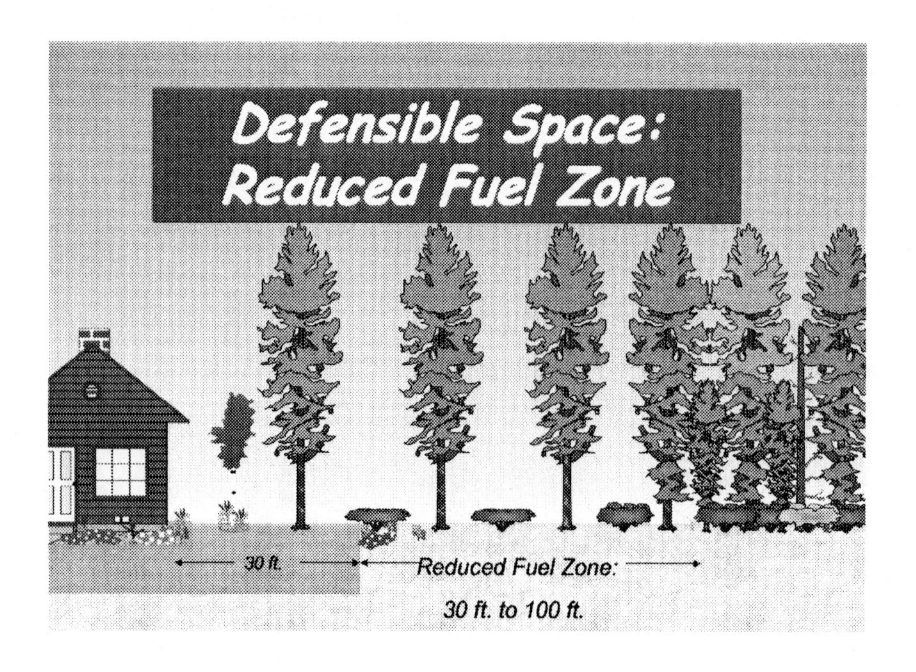

Figure 10.17 Illustration of the implementation of California State Law to clear the defensible space around a home in a rural setting. Vegetation has to be cleared most comprehensively within 30 ft (10 m) of a home, with certain exceptions. Large trees and dead vegetation may not be located within 10 ft (3 m) of a chimney or stovepipe. Dead vegetation has to be cleared in the Reduced Fuel Zone from 30 to 100 feet (10 to 30 m) from a home and large trees have to have a certain distance. According to Public Resources Code (PRC) 4291, homeowners are generally not responsible for the space beyond 100 ft (30 m) from a home (source: State Board of Forestry and Fire Protection[10])

Fire thrives on the three sides of the **fire triangle** (Fig. 10.16): fuel, heat and oxygen. A fire needs fuel to start and maintain itself, it needs heat for the preheating phase and pyrolysis, and it needs oxygen for the processes of combustion. This concept is important as it can be used to understand how to effectively fight a fire. Since a fire needs all three sides, firefighters try to remove at least one side.

The **removal of the fuel** from the fire triangle involves the clearing of vegetation. As a precaution before the upcoming fire season, dead vegetation should be cleared around a home or structure. In California, state law requires homeowners adjoining a mountainous area, forest, brush-covered land, grass-covered land or land covered with flammable material to clear the defensible space within 100 ft (30 m) around homes and structures. Public Resources Code (PRC) 4291 distinguishes between a zone within 30 ft (10 m) of a home as the **most intense zone** and the **reduced fuel zone** beyond that (i.e., 30 to 100 ft from a home). The State Board of Forestry and Fire Protection published practical, illustrated guidelines accordingly which are available on the Cal Fire website[10] (Fig. 10.17). Homeowners are required to maintain a **firebreak** (essentially a gap in the vegetation to prevent a fire from spreading) in the most intense zone, clearing away all flammable vegetation and other combustible growth, with certain exceptions. For example, single well-pruned trees are allowed, but none of their branches shall reach within 10 ft (3 m) of a chimney or stovepipe. Dead parts of vegetation that overhang a building and material on a roof need to be removed. In the reduced fuel zone, dead and dying vegetation is to be removed and ladder fuel is to be reduced. The allowable spacing between shrubs depends on their height. If the home is located near a steep slope, then the allowed distance of a home to vegetation and the spacing between vegetation is larger. The law also includes directives about the dimensions of access roads (i.e. a fire truck has to be able to reach the home).

Firefighters often start **prescribed fires** (also termed **controlled burns**) to reduce the fuel ahead of a fire season. While this is a very good idea in general, such precautions have also gone awfully wrong when such fires burn out of control. Several years of abnormally wet years in New

Mexico in the early-to-mid 1990s followed by several years of severe drought set the stage for the **2000 Cerro Grande Fire**, the inferno that burned 400 homes in Los Alamos and structures in the Los Alamos National Laboratory (Case Study 2).

Figure 10.18a Fighting a fire by removing a side of the fire triangle: fuel. A bulldozer cuts a firebreak in advance of a forest fire in South Carolina. (source: Wikipedia)

Figure 10.18b An aerial view of the 26 October 2003 Cedar Fire in San Diego advancing on Interstate 15 after it burned Scripps Ranch and jumped State Route 52 (running horizontally in the

photo). North-south running Interstate 15 cuts diagonally through the left half of the photo. (source: John Gibbins, San Diego Union Tribune[13])

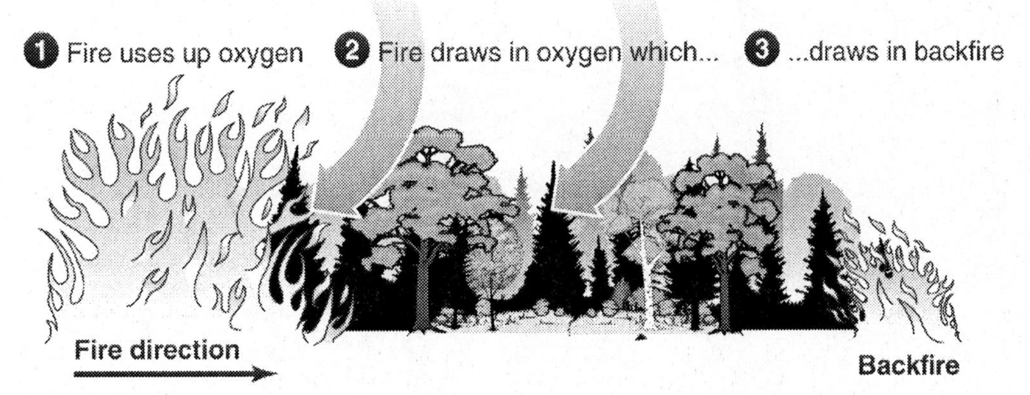

Figure 10.18c Firefighting through the ignition of a backfire. As the advancing wildfire draws in oxygen to continue burning, it also draws in the backfire. The backfire, in turn, eliminates the fuel in front of the wildfire. (source: AB)

During a fire, firefighters clear vegetation to build firebreaks (Fig. 10.18a). And advancing fire may then be slowed, or even stopped. In fact, any clear space may act as a firebreak. Such was the hope during the 26 October 2003 Cedar Fire in San Diego that the westward advancing fire would not cross the 10+ lanes of Interstate 15. But pushed by intense Santa Ana winds, the fire this large firebreak (Fig. 10.18b). Firefighters also start backfires to deprive an advancing fire of fuel (Fig. 10.18c). As a large wildfire advances, it draws in oxygen from all sides, including in front of it. The draft of the big fire forces the backfire to spread in the direction of the big fire not away from it. By the time the fires meet, the backfire produces a large firebreak because the fuel ahead of the big fire is now consumed.

Figure 10.19 U.S. Air Force Reserve aircrews drop fire retardant from a C-130 Hercules. The aircraft can carry up to 3000 gallons (11,500 l) of retardant covering an area one-quarter of a mile long and 60 ft wide (400 x 20 m). (source: Wikipedia/U.S. Air Force)

The **removal of oxygen** from the fire triangle is also an important way to fight a fire. A campfire is often extinguished by covering it with dirt. A blanket is used to cover and stop a small fire. Many fire extinguishers have are filled with carbon dioxide or dry powder. All of these smother a fire by depriving it of its oxygen source. Large wildfires are often fought by releasing a reddish fire retardant from a plane (Fig. 10.19). The retardant interrupts the use of oxygen in the chemical chain reactions of pyrolysis. It is died so that the treated area can be seen from the air. The retardants for wildfire applications are usually mixtures of water and chemicals. Newer gel retardants have a different color to distinguish them from traditional retardants. Though fire retardants may contain potentially harmful chemicals, forest fire retardants are generally considered non-toxic. But even these can be harmful when applied in great quantities. Also, some retardants release dioxins and furans when they burn, and so become toxic to fish, wildlife and firefighters. Drops within 300 ft (100 m) of bodies of water are therefore prohibited unless human lives our property is in direct danger.

Figure 10.20a Fire engine ANF 13, used for firefighting in the Angeles National Forest (ANF). This engine is staffed by a captain and 4 crew members. (source: Wikipedia/Charles P. White)

Figure 10.20b An Italian firefighting helicopter fills its 125-gallon bucket (a Bambi bucket) in a public swimming pool in Naples, Italy, to help fight a fire in September 2004. (source: Wikipedia/U.S. Navy)

Lastly, the **removal of heat** from the fire triangle is an effective firefighting measure. Dropping water on a campfire is an effective way of extinguishing it. Since water has a high heat capacity, the addition of water tremendously slows down the preheating phase of the fire. It also takes the heat from an already burning fire. Wildfires are therefore fought and contained by hosing down the perimeter of the fire from a fire truck (Fig. 10.20a). A typical fire truck used to fight wildfires is smaller than fire engines seen in cities but they are better at negotiating difficult terrain. Typical wildland fire engines can carry up to 800 gallons (3,000 l) of water. But even these versatile fire engines do not reach everywhere in rugged mountainous areas. Firefighting helicopters and air tankers then come into play (Fig. 10.20b). Firefighting helicopters cannot carry large amounts of water but they can recharge in a matter of minutes. A few early-response drops by a helicopter in otherwise inaccessible terrain can mean all the difference in the potential growth of the fire.

10.3 Wildfires in California

Fuel in Southern California

Much of Southern California, and even the rest of California, experiences a Mediterranean climate with mild, wet winters and hot, dry summers. The green phase of the landscape is near the end of the winter, from February through May, depending on elevation, before much of the vegetation dries out as rainfall starts to cease. By May, the first serious wildfires occur. Matching the climate, Southern California has native as well as non-native Mediterranean-style vegetation. The eucalyptus tree falls into the latter category and its relationship with wildfires has already been discussed. The tree does extremely well from Southern California to Oregon, and even parts of British Columbia. It was introduced in California during the Gold Rush in the 1850s as a fast-growing renewable source of timber for construction, furniture making and railroad ties. But it soon turned out that the dry wood was too tough to drive rail spikes into. The wood also twisted while it dried. But the tree still has played a major role as windbreaks for orchards, highways and farms, and as a pleasant shade tree in gardens. Thirty years before the 1991 Oakland Fire (Case Study 3), critics emphasized the increased fire hazard of eucalyptus trees and the rapid build-up of wood litter beneath. But the trees remained ubiquitous in California landscapes, contributing to the 1991 Oakland Firestorm, and more to come. The fire-resistance of eucalyptus over anything else became shockingly clear during the 2003 Cedar Fire when San Diego's Scripps Ranch neighborhood burned (Case Study 5). Thousands of homes were incinerated, but eucalyptus trees standing between the homes remained green. Lately, palm trees have been added to the discussion of introduced plant species that enhance the spread of fires in an urban setting.

A signature tree lining many streets in Southern California is the Mexican Fan Palm (Washingtonia robusta) that is also found in many parks and yards. It can grow 25 – 30 m (80 – 100 ft) tall. It is native to western Sonora and Baja California Sur, but most trees found in Southern California are introduced. The tall trees have a bare stem and some 20 – 30 large palmate fans of 1-m long leaflets. The fans rapidly die but hang on to the stem to accumulate a cone-shaped clump of palm fronds. Because the clump serves not only as nest for birds but also rats, city landscaping regularly clears the palm trees of these "nests". But many remain in private yards. The stem and the nest give off large amounts of embers when they burn (Fig. 10.10) and so contributed significantly to the spread of the 2008 Freeway Complex Fire in Corona, CA and the 2007 Witch (Creek) Fire in San

Diego's Rancho Bernardo neighborhood (Case Study 5). In the Witch Fire, embers picked up and carried by strong Santa Ana winds, ignited homes 5 miles away from the fire front!

Figure 10.21 Chaparral, Santa Ynez Mountains, near Santa Barbara, CA. Mature chaparral, that has not burned for a long time, is characterized by nearly impenetrable, dense thickets. It covers about 5% of California. An additional 3.5% is covered by coastal sage scrub. (source: Wikipedia)

Figure 10.22 A Manzanita in the San Diego River Valley in Santee, CA growing new shoots a year after the 2003 Cedar Fire (Case Study 5).

Of course, the eucalyptus tree is not to blame (alone) for California's fire problems. California (or the North American west coast from Baja California to British Columbia) has its own native vegetation that is destined to burn. A hiker in the San Diego backcountry can enjoy a multitude of fragrances released from the aromatic oils in herbs and shrubs, including various kinds of sage. All of these are highly flammable. Though it may not appear so, Mediterranean scrubland (incl. Coastal sage scrub and chaparral) contain more than 20% of the world's plant diversity. Accordingly, the scrubland habitats support a large number of animal species, including many endemic birds. The signature vegetation of Southern California's hillsides is the **chaparral**, a collection of perennial evergreen scrubs with small, hard and drought-tolerant leaves (Fig. 10.21). The term comes from the Spanish word *chaparro* for "small" in reference to the size of the brush. In the natural environment, chaparral fires occur infrequently, with intervals between 10 to 100 years. A typical plant of the mature chaparral vegetation is the manzanita, an evergreen brush with distinct a red bark. The wood burns very hot and gives excellent firewood. It is thought that manzanitas (arctostaphylos) need fire for germination but no one has ever found a Manzanita seedling. Instead, and in most cases, manzanitas rejuvenate after a fire by re-sprouting from underground root growth (Fig. 10.22). It is thought that the stressed plant responds to the fire by first growing a burl underground from which new shoots grow.

Weather Conditions that Increase the Risk of Fire Infernos

In earlier sections, we learned that the removal of water from the vegetation prepares it for a fire. Prolonged periods of dry weather, such as droughts, set stage for wildfires. Exactly what constitutes a drought depends on the climate of an area, as a dry period in a more moderate, temperate climate may be a drought but that same condition in a hotter, dry climate may actually be normal. But droughts alone do not cause the ultimate inferno. Before the dry spells, there also have to be periods with excessive growth in the vegetation. The 25 March 2006 Texas grassland fire described in section 10.1 would not have been so severe had there not been a wet season that makes the grass grow. In principle, this is part of the normal wet-dry-season cycle. Southern California has the same cycle.

Some winters are much wetter than normal so that vegetation can grow much more. The dry season following immediately does not necessarily have to be the season of infernos. In fact, quite often, the fire problem is enhances a season or two later. Without going into too much detail until Chapter 12, very strong El Niño winters appear to be particularly wet winters in Southern California. The **1997/1998 El Niño** was one of the strongest in the century. The rain arrived unusually late, with some of the strongest rainfalls occurring in March and April when the plants in the mountains started to grow. Usually the rain arrives two months earlier, on time for vegetation at lower elevations but too early for the mountains to induce growth. An unusually lush vegetation followed in the year 1998 that subsequently died as rainfall declined below normal during a subsequent La Niña. But it was not until 2003 when Southern California experienced the worst fires in its history (Case Study 5). But the strong El Niño five years earlier, with its delayed but strong rainfall set the stage.

Similarly, the region experienced its third-wettest winter on record during the 2004/2005 winter, with a total rainfall of 22.5 inches in San Diego where 10 inches are closer to normal. The rain in that winter started earlier, with devastation rains associated with a Pineapple Express in January triggering the deadly landslide in La Conchita. Following this wet season, the 2007 firestorm came earlier to San Diego than the last one, only three years after wet season.

Dangerous Wind Conditions

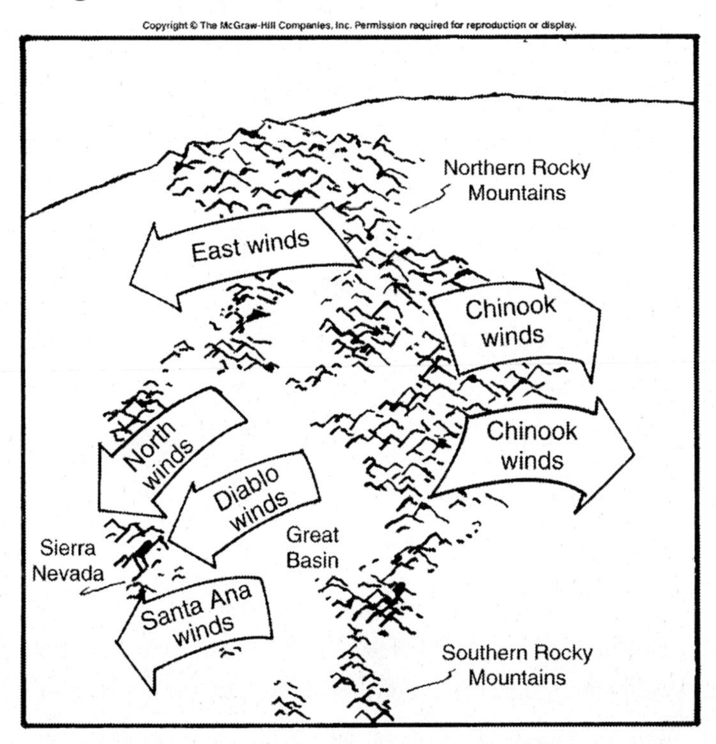

Figure 10.23 The naming convention of Foehn winds spilling from the Great Basin and Rocky Mountains. Foehn winds are warm, dry, high-speed winds. Westerly Chinook winds blow eastward into the Great Plains. East and North winds blow from the Northern Rockies and the northern Great Basin western toward the Pacific. In Northern California, the easterly Diablo winds blow from the Great Basin, and in Southern California, these winds are called Santa Ana winds. Unlike ocean currents, winds are named for the direction they come from. (source: AB)

Current weather conditions and, above all, strong winds often make the difference of whether a fire is manageable or becomes uncontrollable (e.g. Case Studies 2 – 5). The significance of the wind factor depends on the setting of each fire and there is no defined threshold. But winds at 30 km/h (20 mph) can already spell disaster. Increased winds occur in a variety of atmospheric situations, which is discussed in detail in Chapter 12. Winds occurring near a cold front, so-called, **cold-front winds** enhance the fire risk, especially when no precipitations are associated with the cold front during the warm season. **Local winds** depend in the local geography and topography. Sea and land breezes fall under this category as well as winds associated with hillsides (slope winds) and valleys (valley winds). **Foehn winds** increase fire hazard in the entire western half of the U.S.. Named after winds in the European Alps, these warm, dry winds occur in the U.S. when air spills from the western Great Basin and the Rocky Mountains to the surrounding lower-lying areas during a period of persistent high air pressure (Fig. 10.23). Sustained wind speeds of 30 km (20 mph) and higher are common and gusts can top 80 km/h (50 mph). The winds can be enhanced if a low-pressure system is nearby and increases the pressure gradient (the change in pressure over distance). Since Foehn winds descend from higher elevations, they heat during the process, and the air becomes drier (adiabatic heating, see Chapter 12), where the relative humidity can fall below 20%, sometimes into the single digits. Foehn winds typically occur from September through April but can occur throughout the year.

In Southern California, a full-blown "**Santa Ana**" is associated with strong winds, extremely low relative humidity and anomalously high temperatures. Santa Ana winds blow for several days and typically occur from October through February though they can be as late as April and develop at other times as well. Sometimes, depending on the air temperature within the high-pressure system, Santa winds are actually not anomalously hot. The news media then refers to these as cold Santa Anas. Along San Diego's coast, a thick layer of brownish smog builds up over the horizon as the smog is pushed into the Pacific Ocean by the northeasterly winds. The National Weather Service issues a **red flag warning** when sustained winds are 25 km/h (15 mph) or higher and the air humidity is low, implying extreme fire danger. Santa Anas are particularly dangerous when they occur early in the fall, before the first serious winter storms in October or November drench the backcountry wilderness. Pushed by the winds, and due to the low humidity, even the smallest of fires can easily explode into uncontrollable firestorms.

Fires during a Santa Ana typically come in **fire clusters** and, as the usual weather patterns also do, Santa Anas "migrate" from the north toward the south. This often results in a north-south timing of fire outbreaks, with fires burning out of control last in San Diego. An agreement exists between fire agencies to exchange fire-fighting resources where they are needed most. During the Santa Ana in late October 2003, a large fraction of San Diego's fire-fighters and equipment was therefore sent north on 25 October to help fight the Old, the Grand Prix and the Padua Fires that ultimately burned more than 90,000 acres (36,500 ha) in San Bernardino and Los Angeles Counties. That evening, the Cedar Fire started in San Diego's eastern backcountry to become California's largest wildfire since 1889 (Case Study 5). Similar miscommunications also occurred during the 1991 Oakland fire also occurred between San Diego county and San Diego city fire fighters. And military fire fighting crews idled on their bases. Rules were subsequently changes and communication improved, but during the Witch Creek Fire four years later, military fire fighting crews still could not be utilized to their fullest capacity because of a lack of civil fire observers to accompany military fire fighters. Very often, fire fighters do not stand a chance to control before Santa Ana winds weaken. Usually, winds calm during the night, but this hiatus is often too short, and success is not guaranteed even after a Santa Ana completely dies. During the 2003 Cedar Fire, winds shifted after a few days to become westerly winds. Most homes were destroyed before winds shifted, but more acreage burned after the shift than before.

Dangerous wind conditions that whip wildfires out of control are not limited to the western half of the U.S.! In 2011, the winds from Tropical Storm Lee pushed the Bastrop County Complex Fire out of control to become the most destructive wildfire in Texas' history (Case Study 6).

10.4 Fire Fighting and Fire Suppression

Figure 10.24 The development of vegetation in the Bitterroot National Forest in Montana by 1909 (left), 1948 (middle) and 1989 (right). The increase in vegetation density and the development of ladder fuel is attributed to fire prevention efforts since 1895. (source: Wikipedia)

The controversy over whether and how to fight and prevent wildfires, especially near the wildland-urban interface, has continued since the turn of the 20th century, when the federal government suggested fire suppression as a primary goal of managing the nation's forests. At a time before extensive and aggressive urban sprawl, the major efforts concentrated on protecting valuable timber resources. In 1908, the government first allowed deficit spending for fire prevention and suppression. Wildfire prevention programs include **wildland fire use** where wildland fires ignited by natural causes are monitored but allowed to burn. **Controlled burns** are strategically ignited by government agencies under less dangerous weather conditions to reduce the overall fire risk.

The **1910 Bitterroot Big Blowup** in which 3 million acres (12,000 km^2) of land burned (Case Study 7) fundamentally shaped the U.S. Forest Service and fire suppression measures. Before the epic event, debates flared on whether to let wildland fires burn because they were part of the natural cycle and were expensive to fight. Others argued for fighting fires at all costs to protect the forests. After the 1910 Blowup, the U.S. Forest Service was charged with the task to do the latter. By 1935, U.S. Forest Service policy stipulated that all wildfires were to be suppressed by 10 a.m. the morning after they were first spotted. By the late 1930s, over 8,000 fire lookout towers were erected and by 1940, so-called smokejumpers would parachute out of planes to fight fires in remote locations. In 1944, the Forest Service developed an add campaign to educate the public on fire suppression using a cartoon black bear named Smokey Bear. As a result, the acreage burned in wildland fires has decreased from over 26 million acres per year in the 1920s to only a few million acres per year in the last 5 decades. On the other hand, forests have become much denser than before fire suppression measures (Fig. 10.24). Before, about 30 big trees stood on each acre of a mature forest, while now this number is between 300 and 3,000. In addition, considerable growth in the understory vegetation provides ladder fuel that enhances the severity of a potential fire.

The debate renewed after the catastrophic 1988 wildfire in the Yellowstone National Park (Case Study 8). Decades of aggressive fire suppression allowed the significant growth of ladder fuel vegetation. Nearly since fire fighting policies took effect, it was also felt that fires are part of the natural cycle. And so, starting in 1972, naturally caused fires were therefore allowed to burn as **prescribed natural fires**. This strategy seemed to work as in the following years until 1987, only 15 of the total number of 235 fires burned larger than 100 acres (40 ha) and only one reach 7,400 acres (3,000 ha). Problems started with 5 wet years which allowed understory vegetation to thrive, followed by a dry year in 1988. Fires starting in June, and despite massive fire fighting efforts, the fires burned for over 5 months until snowfall brought relief in November. The fire cost $120 million to fight. Contrary to initial news reports, surprisingly few large animals perished and all but moose recovered and the park's forests are slowing growing back. Responding to the lessons learned, a new, stricter fire policy for Yellowstone was implemented in 1992, with additional amendments in 2004. All natural wildfires are now allowed to burn, as long as potential danger is not exceeded (incl. size and weather conditions). The rest of the fires are to be suppressed. And as of 2007, fuel is reduced within 400 ft (120 m) of structures and other high-priority locations. The biggest lesson learned though was that ecosystems such as that of Yellowstone are well adapted to large and intense fires. Of course, large fires that may burn out of control are unacceptable at the wildland/urban interface.

The failure or success of fire suppression clearly depends on the type of environment. In the Okefenokee National Wildlife Refuge in Southern Georgia, the vegetation consists of scrub and brush in a swamp environment. Wildfires there restore the prairie ecosystem, allowing prairie grass to grow back. Firefighters therefore allow the fires to burn and concentrate work on protecting structures and keeping the fire within the refuge. Nevertheless, fires can become large. Lightning started the **Honey Prairie Fire** on 28 April 2011. It burned over 60,000 acres within a little over a week, spread to nearly 270,000 within 2 months and still burned 5 months later in September. By this time it had consumed 310,000 acres (125,000 ha or 1,250 km^2) and became Georgia's largest fire in its history. Four years earlier, several fires merged into the **Bugaboo Scrub Fire** that raged from April to June to become the largest fire in the history of both Georgia and Florida at the time. The fire burned a total of 600,000 acres (2,400 km^2). The causes of the fires during drought conditions varied. A tree falling on a power line during high winds from a nor'easter storm caused the initial fire in Georgia. Lightning started a fire in May in the refuge that quickly crept into Florida. Strong winds from Subtropical Storm Andrea initially spread the fire in early May.

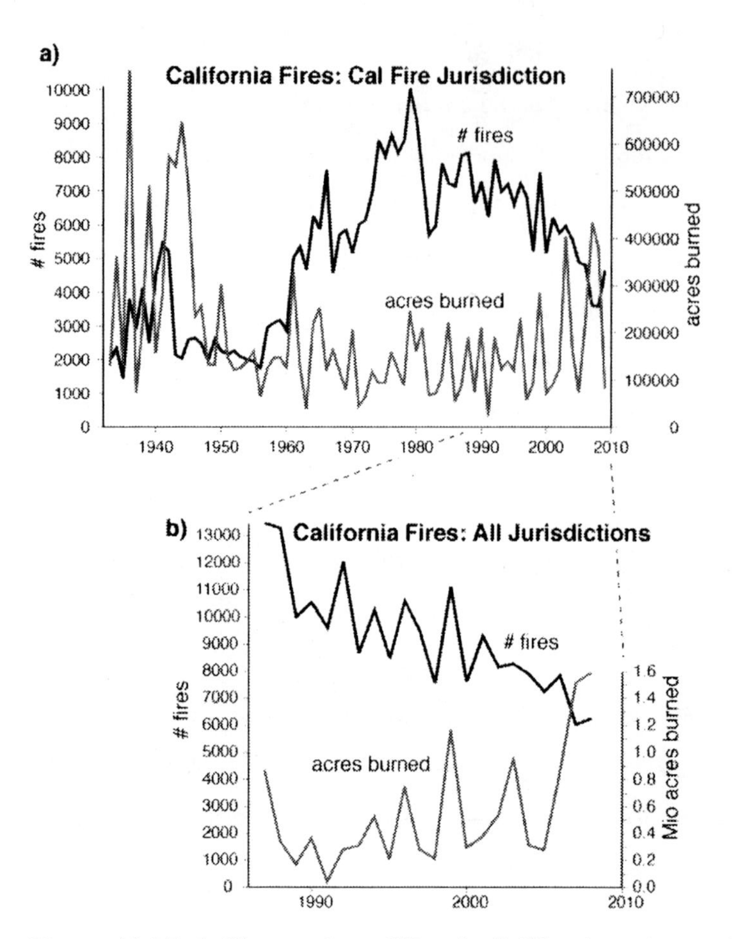

Figure 10.25a,b The number of fires in California and acres burned in California within the Cal Fire jurisdiction only (a) and all jurisdictions (incl. federal) (b). (data from Cal Fire)

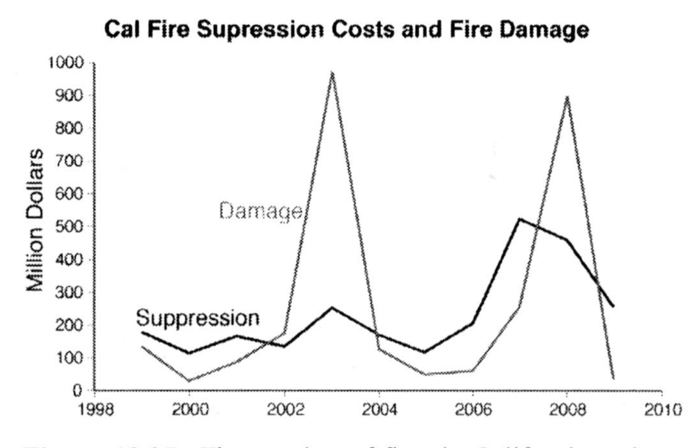

Figure 10.25c The number of fires in California and acres burned in California within the Cal Fire jurisdiction only (top) and all jurisdictions (incl. federal) (bottom). Damages in 2003 result mainly from the Cedar Fire, while damages in 2008 result from a large cluster of lightning-induced fires in Central and Northern California in the summer and several Santa Ana-driven fires in November. (data from Cal Fire)

Zooming in on California

Since the 1988 Yellowstone fires, cooperation between federal and state agencies has been coordinated on a national level by the National Interagency Fire Center. But this seems to be an area that still leaves room for improvement, as subsequent fires in California showed (e.g., Case Studies 2 – 5). Summary statistics on individual fires as well as overall fire seasons may be confusing when fires burn land covered by different jurisdictions. Individual jurisdictions often publish acreage burned and fire fighting costs only as covered by their jurisdiction, not to the entire fire(s) (Fig. 10.25a). For example, Cal Fire reports for the 1987 California fire season that a total of 86,945 acres burned, and 34,488 acres of these burned in lightning fires. But including other jurisdictions, a total of 13,476 fires actually burned 873,000 acres in California. Most of this was due to dry lightning (lightning without precipitation) between 29 August and 11 September 1987. The majority of the fires and acreage were located within U.S. Forest Service jurisdiction explaining Cal Fire's rather low numbers (of only 1/10). In a typical wildfire in California, 1/3 of the burned land falls under Cal Fire jurisdiction, while 2/3 falls under other jurisdictions and is also their responsibility (compare Figs. 10.25a,b). Excellent coordination in firefighting efforts is therefore crucial. However, a disturbing trend is evident: while the number of wildfires has gone down since 1980, the acreage burned has nearly tripled since the mid-1990s. It is also estimated that the value of damaged structures has tripled over the last 20 years (while the number of deaths by fire has declined by a factor 2). Firefighting costs published by Cal Fire alone have nearly tripled since 2000 (Fig. 10.25c), and this is likely the case for federal expenses as well.

Mexico has different fire fighting policies than the U.S.. It is therefore interesting to peek across the border and compare numbers between densely populated Southern California (SoCal) and sparsely populated Baja California (Baja) (Table 10.6). While SoCal conducts aggressive fire prevention and fighting measures, wildfires in Baja often remain unattended until they burn themselves out. During the years from 1972 – 1980, the same fraction of chaparral (about 8%) burned on both sides of the border, despite firefighting measures in SoCal. More than twice as many fires in Baja burn half as much land than in SoCal, contrasting small fires, with 2.5 acres burned in each fire in Baja, with larger fires, with 10 acres burned in SoCal. On average, then, fires in Baja occur twice as often but are 4 times smaller. The question then arises what is it about SoCal firefighting measures that allows fires to be larger, despite firefighting. It turns out that the timing of the fires is crucial to answer this question. Fires start to occur as soon as the vegetation dries out after the wet season, in June or earlier. Such fires burn in Baja, while they are extinguished in SoCal. By September, 80% of chaparral fires have burned in Baja while most of the fires in SoCal are yet to come, during the season of severe Santa Ana winds. This could imply that firefighting in SoCal prevents wildfires earlier in the season but increases the hazard of uncontrollable fires during a Santa Ana.

Table 10.6 Chaparral Areas Burned, 1972 – 1980[18]

	Southern California	Baja California
# Residents	37,700,000	3,200,000
Population Density	242/sq mi	120/sq mi
Total Area (1000 ha)	2,019	1,202
Area Burned (1000 ha)	166	95
% Area Burned	8.2	7.9
% Burned after 1 September	72	20
# Fires	203	488

The suggestion of prescribed fires early in the season has therefore been made for SoCal but debates, and sometimes controversy and confusion, exist over this topic, yet again, between land management, ecologists and not lastly the public. Many arguments speak against prescribed fires in densely populated SoCal, including health impacts from loading the already ozone-burdened summer air in SoCal with smoke. An alternative to prescribed burns would be the mechanical removal of understory vegetation, but this process is labor-intensive and may be invasive on natural habitats (see Box 1 for pros and cons of fire suppression, prescribe fires and mechanical fuel removal). Given California's prolonged budget problems, the big question is who is going to pay for all these precautions. On the other hand, with wildfires becoming larger, more expensive fire fighting equipment, such as JumboJet-size air tankers, have to be deployed (Case Study 9). Urban sprawl further presses deeper into the wildland, requiring and increase in fire prevention. Costs are therefore heading up in the near future. In the current situation, states can request federal disaster relief funds after a catastrophic wildfire to help pay for the fire fighting costs and help fire victims that would otherwise not be available, setting the fact aside that human lives are wiped out. Cash-strapped cities, such as San Diego, often wait with maintenance work until something breaks (e.g. the water main and sewer system) to then deal with the problem. Wildfires may be a size too large for this kind of lifestyle.

In 2011, the California governor and lawmakers approved the collection of a firefighting fee from its rural residents that rely on state-funded firefighting measures to offset revenue losses to Cal Fire[19]. The argument was made that "rural landowners within state responsibility areas receive disproportionately larger benefits from fire-prevention activities". For San Diego's 73,000 backcountry dwellers scattered across 1 million acres administered by Cal Fire, this means the payment of a minimum $150 annual fee. The controversial measure is expected to generate $100 million statewide and $10 million in San Diego. Cal Fire's total expenditures were estimated at $256 million for the year 2009-2010, with an estimated $15 million of tax revenue coming from San Diego residents[19]. Residents who already pay to a fire agency, nine out of ten, would get a $35 credit (this reduces San Diego's fee contribution to $8.4 million). More than $6 million would go to the state Board of Equalization to identify and bill those who rely on Cal Fire within "state responsibility areas". Critics contest that the new fee will not increase Cal Fire's firefighting capacity and residents vehemently oppose the new fee arguing that they already pay firefighting fees. Moreover, so the argument, many landowners already pay for fire protection through their property tax and special assessments. In the 2007 fire season, the July Zaca Fire in Santa Barbara county cost $118 million to fight though this was outside of the Cal Fire jurisdiction. Three months later, the Witch, Poomacha, Rice and Harris Fires in San Diego county cost 18, 20.6, 6.5 and 21 million dollar to fight.

The firefighting costs do not reflect the losses that occurred during a wildfire which can reach or even exceed $1 billion, as was the case in the 2003 Cedar Fire. As much as it is a fire agency's responsibility to protect the wildland and the public from fires, reducing the fire hazard is also in the residents' hands. To start with, aggressive urban sprawl into the wildland contributes to the problem, be it for making a quick buck as developer or for moving there to get into touch with what is perceived as nature. At the very least, it is in residents' hands to make it harder for a fire to consume a home by building it fire-safe (Table 10.7). Even if up-front costs may be higher for a tile roof, a wood-shingle roof in an environment that is destined to burn just is not a good idea.

Table 10.7 Building a Fire-safe Home in California [1]

measure increases fire risk	**measure reduces fire risk**
house built above a canyon edge	set house back by 100ft
wooden deck overhanging edge of canyon	see above
lots of dead brush on the slope below the house	clean up brush; plant ice plants
trees and brush touching/overhanging the house	clear brush 30 ft around the house
Pine trees and eucalyptus trees near house	plant less burnable, native trees
fire wood stored next to house	store wood away from house
wood shingle roofing	tile or composite roofing
shingle or other wood siding	spray stucco
far-reaching eaves	boxed eaves that do not extend far beyond

	house
large, single-pane windows facing the canyon	smaller windows
no spark arrester on chimney	install spark arrester
narrow driveway	wide driveway to allow access by fire trucks

BOX 1: Pros and Cons of Fire Prevention Methods

	Pros	Cons
Fire Suppression	- reduces human health impacts - protects forest and agricultural resources - saves residential and commercial buildings	- labor intensive - requires high level of planning - expensive - particular strategies can be inefficient (e.g. retardant drops) - increases intensity and likelihood of future fires - inhibits natural ecological processes
Prescribed Fires	- provides habitat for wildlife - improves forest and agricultural resources - reduces hazardous fuel loading - mimics natural processes but in controlled manner	- expensive - requires skilled workforce - requires high level of planning - impacts human health (e.g. smoke affects people with asthma and allergies)
Mechanical Fuels Reduction	- provides habitat for wildlife - improves forest and agricultural resources - reduces hazardous fuel loading - no smoke buildup from fires	- requires use of heavy machinery/use of fossil fuels and its impact/ soil compaction - expensive - does not mimic natural processes

References and Websites

1. AB
2. BH: Best, D.M. and Hacker, D.B., 2010. Earth's Natural Hazards: Understanding Natural Disasters and Catastrophes. Kendall/Hunt Publishing, pp. 310.
3. Earth Observatory Image of the day, December 31, 2008, Global Fires. Accessed May 2012.
4. http://earthobservatory.nasa.gov/IOTD/view.php?id=36220
5. Wikipedia webpage on 2008 Borjomi wildfire, accessed May 2012.
6. Wikipedia webpage on 2011 Slave Lake wildfire, accessed May 2012.
7. Wildland Fires: A Historical Perspective. U.S. Fire Administration Topical Fire Research Series, 1(3), October 2000 (Rev. December 2001). and Lightning Fires. U.S. Fire Administration Topical Fire Research Series, 2(6), August 2001 (Rev. March 2002).
8. Accessed at the website of the U.S. Fire Administration in May 12: http://www.usfa.fema.gov/statistics/reports/wildland.shtm
9. Cal Fire: http://www.fire.ca.gov
10. Fire Statistics at the National Interagency Fire Center: http://www.nifc.gov/fireInfo/fireInfo_statistics.html and http://www.nifc.gov/fireInfo/fireInfo_stats_lightng.html
11. http://www.nifc.gov/fireInfo/fireInfo_stats_totalFires.html
12. InnoFireWood, Burning of wood, Innovative eco-efficient high fire performance wood products for demanding applications. Accessed in May 2012 at
13. http://virtual.vtt.fi/virtual/innofirewood/stateoftheart/database/burning/burning.html
14. State Board of Forestry and Fire Protection, California Department of Forestry and Fire Protection, 8 February 2006, General Guidelines for Creating Defensible Space, available at the Calfire Website at www.fire.ca.gov/cdfbofdb/pdfs/4291finalguidelines2_23_06.pdf
15. NASA Data Helps Pinpoint Wildfire Threats; online at http://www.nasa.gov/centers/goddard/news/topstory/2006/wildfire_threat.html
16. credits for the image include: Bureau of Land Management/U.S. Forest Service/U.S. Fish and Wildlife Service/Bureau of Indian Affairs/National Park Service/USGS
17. Wikipedia webpage of wildfires, accessed May 2012.
18. The San Diego Wildfires Education Project. Accessed May 2012 at http://interwork.sdsu.edu/fire/

19. Cal Fire/California Department of Forestry and Fire Protection, Pest Management Program, accessed May 2012 at http://www.fire.ca.gov/resource_mgt/resource_mgt_pestmanagement.php
20. Jennifer Shoemaker, NASA Satellites Reveal Surprising Connection Between Beetle Attacks, Wildfire, NASA's Goddard Space Flight Center, accessed May 2012 at http://www.nasa.gov/topics/earth/features/beetles-fire.html
21. Giant Texas Fire Complex Claims 11 Lives, 15 March 2006. Environment News Service, accessed May 2012 at http://www.ens-newswire.com/ens/mar2006/2006-03-15-01.html
22. NASA/Goddard Space Flight Center, Wildfire Growth around Yellowstone National Park in 1988 (WMS); accessed May 2012 at http://svs.gsfc.nasa.gov/vis/a000000/a002900/a002909/index.html
23. modified from AB, based on data by Minnich, R.A., 1983. Fire Mosaics in Southern California and Northern Baja California, Science, 219, 1287 – 1294.
24. Board OKs $150 fee for rural state fire protection, 10 November 2011, San Diego Union Tribune; and Rural Dwellers want out of fire fee, 17 May 2012, San Diego Union Tribune. And Fire EMS Firefighter Blog, First Read: 2010-2011 Cal Fire "Emergency fire suppression" budget sliced by $32 mil. accessed May 2012 at http://firefighterblog.com/2010/05/calfirebudget/
25. Wikipedia webpages on the 2011 Texas wildfires And Wildfires in Parched Texas Kill 2 and Destroy Homes, New York Times, 5 September 2011, accessed September 2011, http://www.nytimes.com/2011/09/06/us/06wildfire.html And Texas Wildfire Destroys More Than 1,500 Homes, New York Times, 11 September 2011, accessed September 2011, http://www.nytimes.com/2011/09/12/us/12wildfire.html

U.S. Fire Administration: http://www.usfa.fema.gov

Cal Fire: http://www.fire.ca.gov

California Board of Forestry and Fire Protection: http://www.bof.fire.ca.gov

Interfire: resources for fire services, fire insurers and law enforcement training:
http://www.interfire.org/

Chapter 11: The Atmosphere, Weather and Climate

Some of the most devastating natural disasters occur as a result of processes in the atmosphere. Hurricanes come to mind, but also severe thunderstorms that spawn tornadoes, and even simply inclement, or severe weather in general. Severe weather often results in flooding (discussed in Chapter 9), and it can trigger mass movements (Chapter 8). Certain (extreme) weather conditions set the stage for devastating, uncontrollable wildfires. Apart from the fact that severe weather causes other types of natural disasters, it is also the type of natural hazard that most likely affects all of us one way or the other. But before understanding why and how extreme and severe weather forms, we need to understand the processes in the atmosphere responsible for "regular" weather and why "regular" weather at one location may spell "extreme" in another. Why would 38°C/100°F-weather not raise eyebrows in Cairo, Egypt but make people moan and run for air-conditioned rooms in London, U.K.? Why does Fargo, ND have extreme temperature swings between -44.4°C/-48°F in the winter and 45.6°C/114°F in the summer, but Seattle, WA which is located at nearly the same latitude has never experienced temperatures below -17.8°C/0°F? Why are some regions dry but others humid and wet? In fact, why do we have seasons? The first part of this chapter discussed the basic concepts of how solar energy, that ultimately drives air circulation, winds and weather (Chapter 12) is received and distributed over the planet. The concepts of seasons, climate and weather are introduced. Oceans play an important role as climate moderators.

The second part of this chapter summarizes the basic characteristics of Earth's atmosphere: its composition and structure, the greenhouse effect that is vital to life on Earth. The moisture content in air is related to the stored latent heat that can ultimately drive severe weather.

Figure 11.1a Climate in San Diego: Beach weather on the beach by the Scripps Institution of Oceanography, La Jolla. At 32.5° latitude, the inland San Diego area experiences Mediterranean dry climate, with mild, wet winters and hot, dry summers. Along the coast, summers are cool compared to those along the Mediterranean Sea because of the proximity to the cold California Current. Coastal areas therefore have a cool semi-arid climate (see Box 1).

Figure 11.1b Weather in San Diego: Rain in San Diego on 27 October 2004. (source: San Diego Union Tribune)

Climate and Weather

When discussing processes in the atmosphere, we need to distinguish between weather and climate. Climate typically describes an average long-term condition that goes beyond the time scale of one or several years. For example, overall, the climate in the San Diego, CA overall is similar to a Mediterranean climate, with mild, wet winters and dry summers. In detail though San Diego experiences marked differences in small-scale climates (micro climate). While summers are hot inland, they can be quite chilly along the coast that is more characteristic of a cool semi-arid climate (Fig. 11.1a). In contrast to San Diego, the climate in Chicago, IL is humid continental with cold, windy, snowy winters and hot, humid summers.

In contrast to "typical" long-term characteristics of climate, weather can change on a time scale of days, hours or, in extreme cases, even minutes, e.g. when a passing front produces violent weather. An example of weather is shown in Fig.11.1b, when San Diego experienced a rainstorm on 27 October 2004. Climate is commonly defined as the weather averaged over a long period. The standard averaging period is 30 years. Climate is typically controlled by location while weather is controlled by anomalies in the atmosphere. Climate zones are defined on a regional to continental scale while weather occurs on a local to regional scale. To quote Lazarus Long in Robert Heinlein's *Time Enough for Love* "Climate is what you expect, weather is what you get."

11.1 The Physics of Earth's Climate

The Two Principal Controlling Factors of Climate

Many factors determine local variations in climate, such as the proximity to a large body of water. But two main factors determine the principal climate category at a certain location: 1) how much solar radiation is received at a particular location (insolation) and 2) how much of this radiation is retained at that location (albedo).

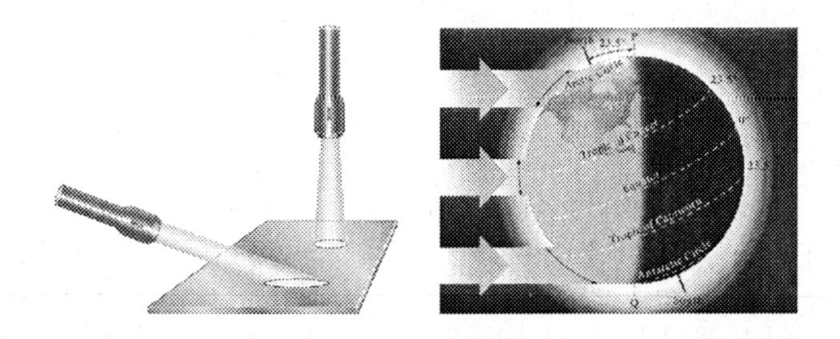

Figure 11.2a The concept of insolation: the amount of sunlight received in a given area. Left: a narrow flashlight beam illuminates a sheet of paper from straight above. This situation is equivalent to sunlight received at noon on the summer solstice along the Tropic of Cancer in the Northern Hemisphere (right), or on the spring and autumn equinoxes at the equator, or on the southern summer solstice (northern winter solstice) along the Tropic of Capricorn in the Southern Hemisphere. Insolation anywhere else is oblique, i.e. a narrow flashlight beam illuminates a sheet of paper from an angle. This angle is smallest near the poles, or if the Sun is observed near the horizon. The illuminated area is largest and so insolation is smallest. (source: ME)

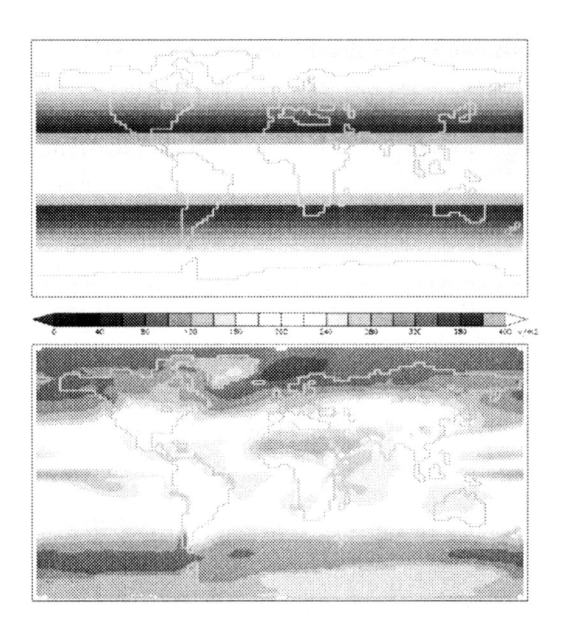

Figure 11.2b Insolation determined from a climate model. Top: at the top of the atmosphere; bottom: on Earth's surface. The highest insolation can be found in equatorial regions at both the top of the atmosphere as well as Earth's surface. But due to differences in cloud cover and absorption and reflection of sunlight by the atmosphere, Earth's surface receives varying amounts of sunlight along a line of latitude. Note, for example, that the insolation on ice-covered but cloudless regions in Greenland and Antarctica is relatively high compared to the surrounding oceans. (source: wikipedia)

Table 11.1 The Albedo of Earth Materials

Material	Albedo (in percent)
fresh snow	80
thick clouds	60-90
thin clouds	30-50
old snow	50
sand	40
planet Earth	31.3
desert	25-30
grass	25
farmland	15-25
savanna	18-20
meadows	10-20
forest	8-15
water/oceans	7-10

The amount of sunlight received at a location on Earth per area (insolation) depends on its angular position relative to the Sun (Fig. 11.2), i.e. how high the Sun appears in the sky. Since the Sun is so far away from Earth, the rays in a beam of light (a ray bundle) are parallel and the ray density or the sunlight per given area is the same at every location on Earth that faces the Sun. If Earth were a disk facing the Sun, then the insolation would be the same everywhere. But due to Earth's curvature, the area illuminated by a ray bundle is different at different locations. If the Sun is directly overhead (e.g. at the equator at noon during the spring and fall equinoxes), the area hit by a ray bundle is smaller, so the insolation is higher. But if the Sun is near the horizon or an observer is near the poles, the illuminated area is larger so the insolation is smaller. There are seasonal variations which will be discussed below. Overall however, beyond the Tropics of Cancer and Capricorn (23.5° N and 23.5° S) the insolation always declines with increasing latitude. As a result of this, over the course of a year, equatorial areas receive 2.4 times more solar energy than the polar regions. We discuss Earth's energy budget in more detail in the next section.

Depending on the material properties of a body that receives insolation, some sunlight is absorbed which warms the body. But some sunlight is also reflected back to space. The ratio between reflected and absorbed solar energy is the body's albedo (Table 11.1). Fresh snow has a high albedo so snow-covered areas remain relatively cool. Oceans which absorb large amounts of solar energy, on the other hand, have a low albedo. Compared to landmasses, which can heat up relatively quickly, oceans still stay cool as a result of the high heat capacity of water.

The actual insolation on Earth's surface can be quite complex (see Fig. 11.2b). This results from marked differences in water vapor content in the atmosphere and cloud cover at different locations on Earth which both contribute to variations in albedo. Note however, that the overall amount of insolation at Earth's surface is still largely controlled by latitude.

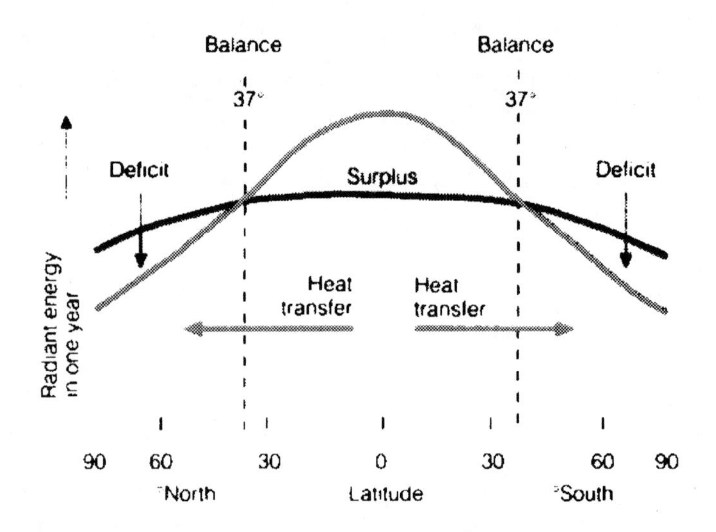

Figure 11.3a The average annual incoming solar radiation absorbed (grey line) and infrared radiation emitted (black line) by Earth and the atmosphere. Areas within 37° latitude of the equator receive more energy than is lost to space (surplus), while higher latitudes receive less (deficit). The low-latitude heat surplus is eventually transported toward the poles within both the atmosphere and oceans. (source: MT)

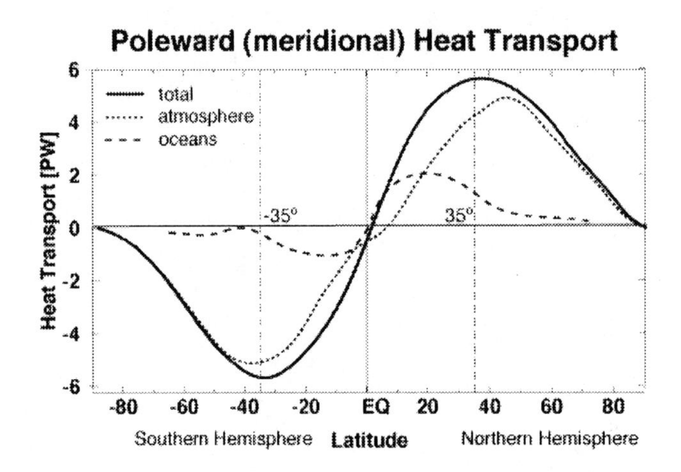

Figure 11.3b Estimated meridional (along a meridian) transport of heat. The total heat transport was inferred from satellite data measuring the net radiation at the top of the atmosphere. Transport in the oceans and atmosphere is obtained from surface instruments and climate modeling. Poleward heat transport in the oceans dominates at low latitudes, between 0 and 17°, especially in the Northern Hemisphere. At higher latitudes, transport in the atmosphere dominates. Poleward transport peaks at 35° latitude. At 35° S, 92% of the heat is transported in the atmosphere, while at 35° N, the role of the oceans is greater and only 78% of the heat is transported in the atmosphere. Heat transport is given in petawatts (10^{15} W). (modified from Trenberth and Caron, 2001[(10)])

Earth's Energy Budget

Because insolation and albedo vary with latitude, the amount of solar energy that Earth's surface and atmosphere absorb also varies with latitude (Fig. 11.3a). Some of the absorbed energy is re-radiated as heat (infrared energy). In regions within 37° of the equator, more energy is absorbed than lost, while the opposite applies to near-polar regions. A balance is reached at 37° latitude where both absorbed and emitted energy are the same. Since Earth tries to reach energy equilibrium, the surplus heat from low latitudes is transported to higher latitudes thereby driving global circulation in both the atmosphere and oceans. The energy transport occurs through 3-dimensional circulation in the atmosphere and oceans (see next chapter). In the atmosphere, for example, hot air rises near-vertically along the equator before it is transported nearly horizontally toward the poles. Scientists have measured this horizontal poleward flow as function of latitude, within both the atmosphere but also the oceans (Fig. 11.3b). Perhaps somewhat surprisingly, it turns out that heat transport in the oceans is a major contributor at lower latitudes, particularly in the Northern Hemisphere (e.g. through the Gulf Stream in the North Atlantic or the Kuroshio Current in the Western Pacific). But at higher latitudes, heat is transported poleward mainly in the atmosphere.

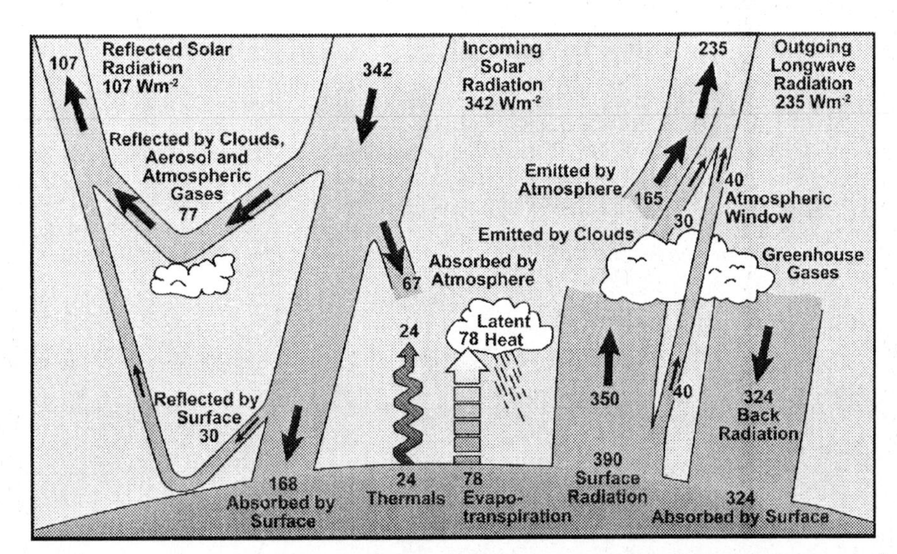

Figure 11.4 Earth's annual global mean energy balance, measured in Wm^{-2} (which is actually the unit for power per area). At the top of the atmosphere, Earth receives as much energy as it loses (342 Wm^{-2} = 107+235 Wm^{-2}). The left-hand side shows the shows the amount of incoming solar energy (342 Wm^{-2}) as well as the amounts that are reflected back to space (77 Wm^{-2} reflected by clouds and 30 Wm^{-2} reflected by Earth's surface for a total of 107 Wm^{-2}). Solar energy is absorbed by the atmosphere (67 Wm^{-2}) and by Earth's surface (168 Wm^{-2}). Of the energy absorbed at the surface, 24 Wm^{-2} are re- radiated through thermals (warming the air in contact with the surface), 78 Wm^{-2}

through evapotranspiration (transfer of water from Earth's surface to the atmosphere by evaporation, sublimation and transpiration). The rest is spent on heating the planet and, ultimately, re-radiation of IR. The right-hand side shows the balance of IR radiation. The surface receives 324 Wm^{-2} in IR energy from the atmosphere through back radiation from greenhouse gases. The surface radiates an additional 66 Wm^{-2} of IR energy for a total of 390 Wm^{-2}. Some of this energy is recycled in the greenhouse but, at the top of the atmosphere, 235 Wm^{-2} is lost to space so that received and lost energy are in balance (342 Wm^{-2}). Overall, Earth's surface emits as much energy ($24 + 78 + 390\ Wm^{-2} = 492\ Wm^{-2}$) as it receives ($168 + 324\ Wm^{-2} = 492\ Wm^{-2}$). (source: IPCC, 2007)

How is solar energy absorbed and reflected in detail? Recall from Chapter 2 that the amount of Earth's internal heat reaching the surface is about 5000 times smaller than that from the Sun so internal heat is nearly insignificant when it comes to Earth's total energy budget (Fig. 11.4). Of the 342 Wm^{-2} energy received from the Sun, nearly one third ($107\ Wm^{-2}/342\ Wm^{-2} = 31.3\%$) is reflected back to space, giving planet Earth an albedo of 31.3%. About one half of solar energy ($168\ Wm^{-2}/342\ Wm^{-2} = 49.1\%$) is absorbed at Earth's surface. This leaves 19.6% that is absorbed by the atmosphere. Solar energy is received in form of **electromagnetic waves** (light) of different wavelengths. These include the visible light with wavelength between about 380 nm (blue) and 740 nm (red). At shorter wavelengths, sunlight consists of ultra-violet (UV) light, X and gamma rays. At longer wavelengths, it consists of infrared (IR) light, microwaves and radio waves. Light at shorter wavelength has more energy. Different Earth materials are transparent to some wavelengths but absorb and re-emit others. For example, the atmosphere absorbs large amounts of IR light but is transparent to visible light. Earth's surface warms when it receives and absorbs sunlight and it re-emits IR energy. Some of this energy is lost to space but some is trapped by Earth's greenhouse (discussed later in this chapter). The energy absorbed by the atmosphere ultimately fuels all processes that lead to extreme and severe weather.

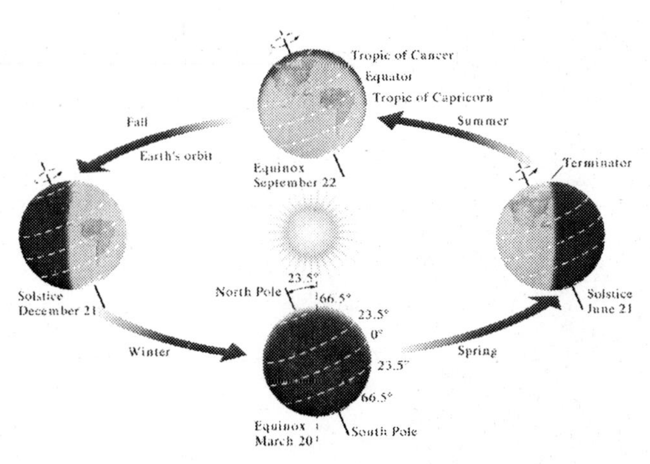

Figure 11.5 Earth has seasons because its rotation axis (or spin axis) is tilted 23.5° with respect to Earth's orbit around the Sun. During summer in the Northern Hemisphere (right side), the northern hemisphere is tilted toward the Sun. It is winter in the Southern Hemisphere. As Earth moves into the September 22 equinox (top), day and night are equally long everywhere. It is fall in the Northern Hemisphere and spring in the Southern Hemisphere. As Earth moves into the Northern Hemisphere winter solstice (left), days are shortest in the Northern Hemisphere. It is summer in the Southern Hemisphere. Finally, at the March 20 equinox (bottom), day and night are equally long again. It is spring in the Northern Hemisphere and fall in the Southern Hemisphere. (source: ME)

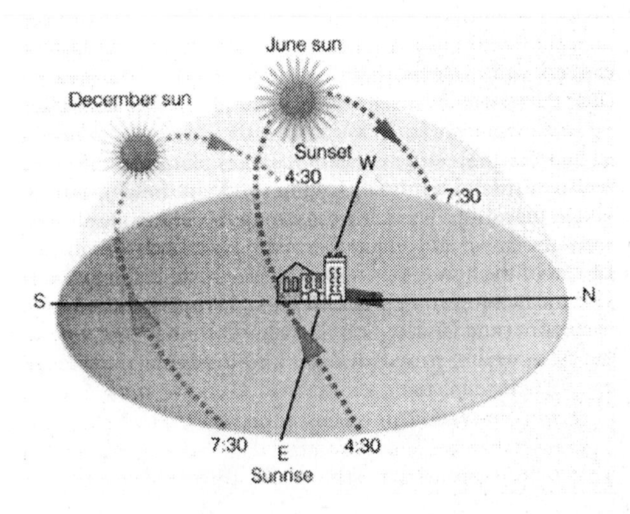

Figure 11.6a The changing position of the Sun throughout the year, as observed at mid-latitudes in the Northern Hemisphere. In summer, the Sun rises high into the sky and days are longer. In winter, the Sun stays closer to the horizon and days are shorter. (source: MT)

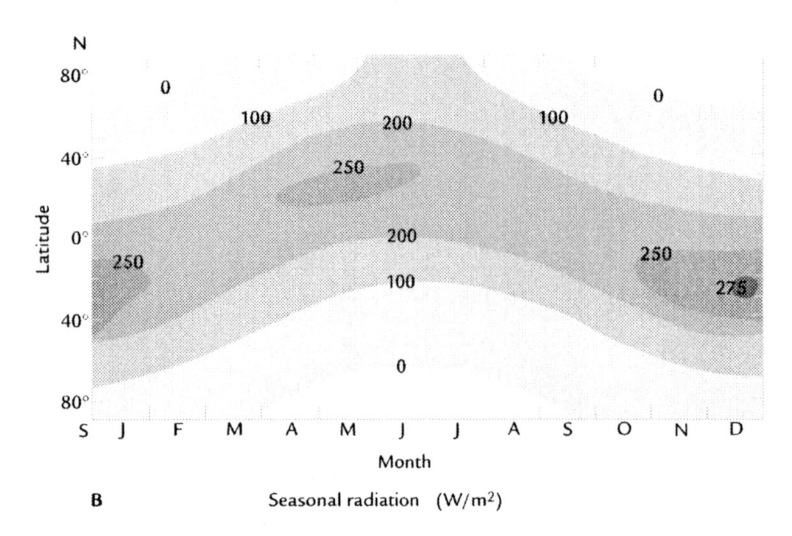

Figure 11.6b Seasonal changes in solar radiation received at Earth's surface. Solar radiation is given in Wm^{-2}. While seasonal changes are not dramatic near the equator, higher latitudes receive much less sunlight in winter than in summer. (source: RU)

Earth's Tilted Rotation Axis: Why Does Earth Have Seasons?

Before elaborating on different climates on Earth, we explore the reasons for daily and seasonal variations in the climate at one location. Changes between night and day occur because Earth rotates about its rotation axis (or spin axis) at a rate of once a day. Earth orbits around the Sun at a rate of once a year. Earth's spin axis is tilted with respect to its orbit around the Sun by 23.5° (Fig. 11.5a). The consequence of this is that the local insolation at a certain point on Earth now changes throughout the year. Earth goes through 4 phases throughout the year:

1. During the (astronomical) northern summer, Earth's spin axis is tilted toward the Sun. Locations in the Northern Hemisphere receive more insolation than locations with the same latitude in the Southern Hemisphere, where it is winter. In the Northern Hemisphere, the Sun rises and sets farther north and rises higher in the sky throughout the day (Fig. 11.6a). Days are longer than nights. On the summer solstice (usually 21 June), days are longest in the Northern Hemisphere, the Sun is overhead at noon at the Tropic of Cancer (23.5° N). On this day, locations along the equator actually receive less sunlight than locations along the Tropic of Cancer. The North Pole receives little insolation but has daylight for 24 hours a day, while the South Pole receives no sunlight. Note here that the astronomical seasons do not quite align with the seasons in our current calendar. The astronomical summer centers on the summer solstice, 21 June, and lasts from early May through early August, while the calendar summer lasts from the summer solstice through the autumn equinox.

2. During the (astronomical) **northern autumn**, or southern spring, Earth's rotation axis moves toward a location along the orbit where it is not tilted with respect to the Sun but only with

respect to the orbital plane. This location coincides with the autumn equinox (22 or 23 September) when night and day are equally long (12h) everywhere on the planet. The area with the highest insolation is along the equator where the Sun is overhead at noon.

3. During the (astronomical) **northern winter**, it is summer in the southern hemisphere. The rotation axis is now tilted away from the Sun. In the Northern Hemisphere, days are shorter than nights. The Sun rises and sets farther south and it remains closer to the horizon throughout the day. On the northern winter solstice (21 or 22 December), the area with the highest insolation is now along the Tropic of Capricorn (23.5° S) where the Sun is overhead at noon. On this day, locations in the Northern Hemisphere experience the shortest day of the year, the North Pole receives no sunlight and the South Pole has daylight for 24 hours a day.

4. Finally, during the (astronomical) **northern spring**, it is autumn in the southern hemisphere. Again, Earth's rotation axis moves toward a location along its orbit where it is not tilted with respect to the Sun. This location coincides with the spring equinox (usually 20 March) and the area with the highest insolation is along the equator where the Sun is overhead at noon. Night and day are equally long everywhere.

While the amount of sunlight received at a location near the equator does not change dramatically throughout the year, summer-winter differences can be quite dramatic at higher latitudes (Fig. 11.6b). At 21° N, Honolulu, HI receives 150 Wm^{-2} in December and about 230 Wm^{-2} in June, which causes very mild seasonal variations throughout the year. At 32.5° N, San Diego, CA receives about 100 Wm^{-2} in December, but more than 250 Wm^{-2} in June, so there is a larger summer-winter difference in energy received, causing a more pronounced summer-winter difference in climate. Seattle, WA at 47.5° N, on the other hand, receives about 75 Wm^{-2} in December and 220 Wm^{-2} in June. The summer-winter difference is similar to that in San Diego but the overall insolation received throughout the year is less which is why Seattle's climate is overall cooler than that in San Diego. At nearly 65° N, Fairbanks, AK receives 3h and 43 min of sunlight during the winter solstice. This appears as barely measurable solar energy (> 0 Wm^{-2}) in Fig. 11.6b. On the summer solstice, the sun is up for 21h 49 min and Fairbanks receives about 170 Wm^{-2} of solar energy. This explains the surprisingly mild summers (10-21°C/50-70°F, with 37°C/99°F as the highest recorded temperature) but merciless, cold winters (with average low temperatures at -26°C to -32°C/-15°F to -25°F and -54°C/-66°F as the lowest recorded temperature).

Earth's orbit around the Sun is slightly elliptical so the insolation Earth receives changes slightly throughout the year. Because of this, the Northern and Southern Hemispheres receive slightly different isolation during their respective summers and winters. Earth is a little closer to the Sun during the southern summer (Earth in perihelion) than during the northern summer (Earth in aphelion). Figure 11.6b shows that the maximum of energy received reaches 275 Wm^{-2} in southern summers but only just above 250 Wm^{-2} in the northern summer. In Principle, southern summers can therefore get hotter than northern summers at the same latitude. Accordingly, northern winters can be warmer than southern winters. In the northern winter, the latitude receiving only 100 Wm^{-2} in December is 28° N, but the equivalent in the southern winter in June is at only 23° S, and so closer to the equator.

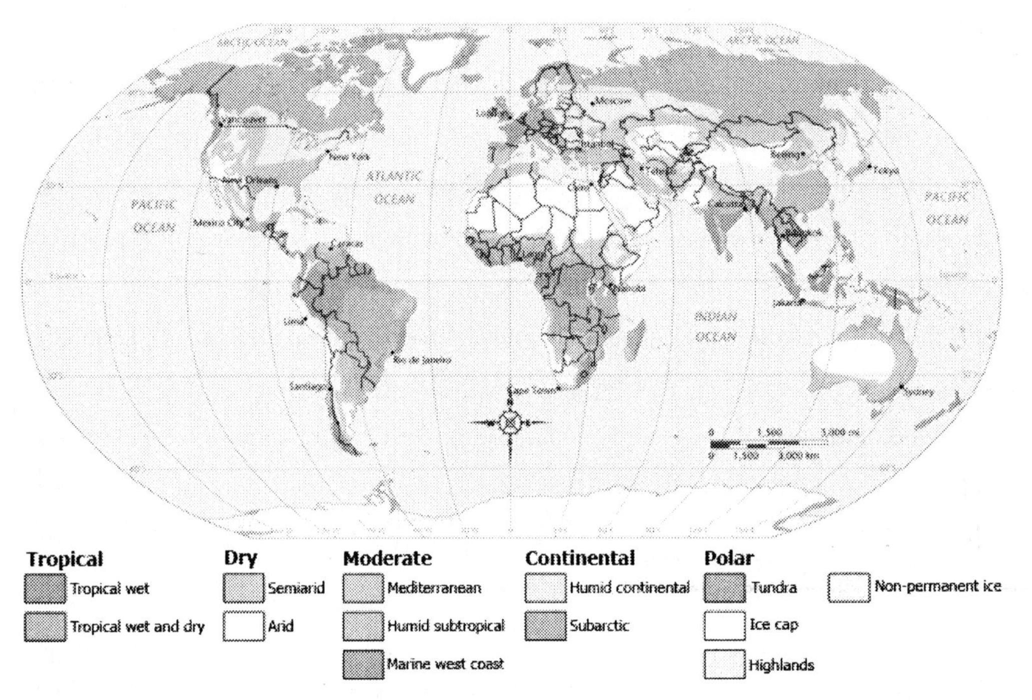

Figure 11.7 Simplified map of the World's climate zones. The classification of climate zones does not only depend on latitude (and the amount of insolation) but also on other factors such as altitude and proximity to an ocean. A more detailed and one of the most widely used classifications is the Köppen climate classification. See Box 1 for details. (source: wikipedia)

Table 11.2 Average High and Low Temperatures in January and July

City	Latitude	January T (°C/°F)	July T (°C/°F)	Köppen Climate
San Diego, CA	32.75° N	9.7 – 18.7 49.5 – 65.7	18.8 – 24.0 65.9 – 75.2	arid-Mediterranean
Seattle, WA	47.5° N	2.8 – 8.3 37 – 47	13.3 – 24.4 56 - 76	cool, dry-summer subtropical
Des Moines, IA	41.5° N	-11 to -3.3 12 – 26	18.6 – 30 65.5 – 86	humid continental

Boston, MA	42.5° N	-5.6 to 2.2	18.6 – 27.8	humid continental with
		22 – 36	65.5 – 82	maritime influence
New York, NY	40.75° N	-2.8 to 3.9	20.5 – 29.4	humid subtropical
		26.9 – 39.1	68.9 – 84.9	
Honolulu, HI	21.25° N	19.0 – 26.9	23.6 – 31.2	tropical savanna
		66.2 – 80.5	74.4 – 88.2	
London, U.K.	51.5° N	2.3 – 8.1	13.9 – 23.5	temperate oceanic
		36.1 – 46.6	57 – 74.3	
Berlin, Germany	52.5° N	-1.5 to 2.9	14.7 – 24.0	humid continental
		29.3 – 37.2	58.5 – 75.2	
Kiev, Ukraine	50.3° N	-8.4 to -2.9	15.0 – 25.3	humid continental
		16.9 – 26.8	59.0 – 77.5	
Madrid, Spain	40.5° N	2.6 – 9.7	18.4 – 31.2	continental Mediterranean
		36.7 – 49.5	65.1 – 88.2	
Sydney, Australia	33.75° S	18.7 – 25.9	8.0 – 16.3	temperate to
		65.7 – 78.6	46.4 – 61.3	humid subtropical

11.2 Earth's Climate Zones: The Geography of Earth's Climate

If insolation were the only factor controlling Earth's climate, there would be only 3 major climate zones. These would be classified by latitude: a frigid zone near the poles, a torrid zone near the equator and a temperate zone in between.

The **actual climate zones** are much more complicated because they depend on many geographical factors. Simplified climate zone maps such as that in Fig. 11.7 distinguish between five basic climate zones: tropical, dry, moderate, continental and polar. But other maps categorize climate into eight climates: polar, sub polar, tundra, temperate, subtropical, continental, tropical, dry land (including desert, steppe, savanna). A more detailed and widely used classification of climates is the Köppen-Geiger climate classification. It is based on the concept that native vegetation is the best expression of climate. It combines average annual and monthly temperatures and precipitation, and the seasonality of precipitation.

Here, we discuss the six most dominant geographical factors. As we already know, the **geographic latitude** of an area controls the amount of insolation and hence the temperature. The cold polar regions receive little insolation while the warm tropics receive much more. Honolulu, HI therefore has a much warmer climate than cities at higher latitudes (Table. 11.2). The **altitude** of an area also controls its overall temperature. Temperature generally decreases with increasing elevation, so mountain tops are typically much colder than the valleys below. One of the most impressive examples contrast the chilly mountain tops of the Sierra Nevada with the hot desert floor of Death Valley that lies below sea level just east of the Sierras, at the same latitude. Mt. Kilimanjaro in Kenya and Tanzania in Africa is located near the equator but has an ice cap near its summit at nearly 6000 m. But climate variations can be much more subtle. Madrid, Spain has a relatively cool climate for its latitude in the Mediterranean because, at 650 m (2,133 ft) altitude, it is situated high in the country, sometimes even receiving snow and below-freezing temperatures. The **proximity to water** also influences climate. Because of the large heat capacity of water, warming and cooling of near-coastal areas occurs more slowly than farther land. Because this effect is so profound, it will be discussed in its own section below. **Proximity to ocean currents** also plays a role. Ocean currents can have cooling effects, such as the California Current cools coastal areas along the California coast. San Diego coastal areas are much cooler than inland areas, particularly in June when low coastal clouds before noon keep temperatures low. Warming ocean currents such as the Gulf Stream affects coastal areas on the North American east coast. The geography of cold and warm currents will be in more detail in Chapter 12. The **proximity to orographic barriers** such as mountain ranges also influences climate. The Sierra Nevada along the California/Nevada border, for example, blocks moist air from the Pacific ocean from moving into Nevada thereby causing a dry and hot climate. Similar rain shadow effects can be found in many regions elsewhere. In the Hawaiian islands, trade winds bring moist air from the northeast. On the larger islands (Kauai, Oahu, Maui and Big Island), the high mountains cause the formation of rain-producing clouds. Much of this rain falls on the eastern side of the islands and in the mountains, leaving the western side considerably drier. In fact, Big Island has a desert in the southwest. Finally, **the proximity to high- or low-pressure zones** also controls climate. Many deserts are found near 30° latitude on both sides of the equator. Stable high-pressure zones exist at these latitudes. The reason for this results from global air circulation patterns and will become clear in the next chapter. On the other hand, low-pressure zones along the equator harbor daily thunderstorms thereby providing plenty of rainfall to nurture the equatorial topical rainforests.

Due to its high albedo, the presence of ice (**glaciation**) and snow can profoundly cool the climate locally along glaciers or on a larger scale on and near ice sheets such as Greeenland and Antarctica. Oddly, and perhaps counterintuitively, a climate can be extreme (hot summers, cold winters) yet a more moderate climate may lead to more glaciation when summers are not warm enough to melt ice that is accumulated during the winter. This explains why some glaciers can be found even at lower latitudes (though altitude is an important factor). In addition, the amount of precipitation in the cold period determines whether ice can grow or not. If the cold period is too cold to produce snow, glaciers will not advance (grow).

The Role of Water in the Atmosphere

Water in the atmosphere plays a dual role when it comes to Earth's climate. Depending on its phase, in liquid or gaseous form, H_2O in the atmosphere can have both a cooling or a warming effect. For example, after water vapor condenses, H_2O exists in the liquid phase. It forms clouds that reflect incoming sunlight back to space, during daytime. Earth's surface below the clouds is thereby ***cooler*** than it would be without clouds, during the day. At night, clouds trap the heat released from Earth's surface, and so the area stays ***warmer***. H_2O can influence atmosphere temperatures in another way. After water evaporates, H_2O exists in the gaseous phase. The molecules absorb and release IR sunlight (see also greenhouse effect below). This has a ***warming*** effect on the atmosphere. On the other hand, the process of evaporation is endothermic and so the air immediately above the evaporating body of water is ***cooled***. The various impacts of water content in the atmosphere need to be taken into account carefully when studying details in climate research, including global warming.

11.3 The Oceans as Climate Moderators

As a result of the high heat capacity of water (Table 2.1), oceans store vast amounts more heat than the atmosphere. During spring and summer heating, the rise in ocean temperatures therefore is much slower than that on land while during fall and winter cooling, ocean temperatures decline more slowly. This delay in warming and cooling affects coastal areas. Oceans therefore have a buffering or moderating effect on the climate of near-coastal areas, making winters milder and summers cooler than inland areas. The interior of continents has more extreme and often drier climates than coastal areas.

For example, Des Moines, IA, is centrally located in North America at 41.5° N, away from any major body of water. It has a humid continental climate with hot, humid summers and cold, snowy winters (Table 11.2). Boston, MA on the east coast is located just a degree farther north. It also has a humid continental climate but with maritime influence, with warm, rainy and humid summers and cold, windy and snowy winters. Summer temperatures may be comparable to those in Des Moines, but winters do not get as cold as a result of the maritime influence.

A similar comparison holds for Kiev, Ukraine and London, U.K.. Both cities are located at similar latitudes but, with a temperate oceanic climate, London has much milder winters than Kiev in the middle of the continent. Severe winters with fatal cold snaps are therefore much more likely to occur in Kiev than in London. With temperatures dropping below -30°C (-22°F), the severe cold wave in late January 2012 killed over 300 people in Ukraine and Russia[5]. On the other hand, London is not very well prepared for snow and relatively light snowfall can bring down daily activity. In December 2010, Heathrow, Europe's busiest airport had to shut down because they did not have enough

COPYRIGHTED MATERIAL — DO NOT DUPLICATE, DISTRIBUTE, OR POST

equipment to remove light snow from the runways. This would not have been a problem at the Denver, CO International Airport.

Even relatively close to the coast, the weakening of the ocean moderating effect as one moves away from the beach is quite profound. For example, on a sunny spring or summer day in San Diego, CA, temperatures in the inland valleys can be more than 6°C (11°F) higher than along the coast. The impact of the coastal cooling effect can be felt within only a few miles when driving away or toward the beaches. Cooling is also strongly felt on a 30-min drive on westbound freeways, for example when driving west along I-8 toward the beaches. An early whiff of cool air can sometimes be felt as far inland as Alpine, 50 km (31 mi) from the coast. And the air feels definitely cooler after passing El Cajon, 30 km (19 mi) from the coast. The cooling effect on this particular 30-min drive during the summer is actually often enhanced, or even caused by the so-called inversion that San Diego often experiences in summer. Warm air then sits on top of cool air in a stably stratified atmosphere. The cooling effect is then felt upon descending from the mountains.

Ocean vs. Atmospheric Processes

There are fundamental differences in processes between oceans and the atmosphere. To start with, we learned earlier that air above and near the oceans warms and cools much more slowly than the air on land, because the heat capacity of water is more than 3000 times larger than that of air (Table 2.1). Other major processes in the oceans are slower than in the atmosphere. For example, the movement of ocean currents is much slower than that of atmospheric winds. The speed of a typical ocean current is tens to hundreds of kilometers per day (see Chapter 12) while winds in the lower atmosphere are tens of kilometers per hour (fair weather) to hundreds of kilometers per hour (severe storms).

Ocean currents play a role in long-term processes while atmospheric changes typically involve short-term processes. Much more than atmospheric winds, ocean currents are diverted by land masses leading to complex circulation patterns (see Chapter 12). Ocean currents control climate in profound ways. For example the warm Gulf Stream in the Atlantic Ocean influences climate along some of the east coast of North America. It is a popular belief that even climate in Western Europe is influenced by the warming **Gulf Stream**. A common comparison is the Mediterranean climate in Madrid, Spain with that in New York City, NY at the same latitude. New York City experiences more extreme seasonal temperature swings with much colder winters (Table 11.2). But some atmospheric scientists contest the level of impact of the Gulf Stream on European climate as oceanographers often suggest. They argue that Seattle, WA, which is located along the south-flowing cold **California Current** and lies further north than both Madrid and New York City, also experiences a relatively mild climate, although summers are cooler in Seattle. The jury may still be out on this issue but it turns out that the California Current may start farther south and may have less of an impact on Seattle's climate than some assume. Seattle and British Columbia may also benefit from the fact that the north-flowing warm **Alaska Current** passes the area, thereby inhibiting the upwelling of colder deep water (see Chapter 12).

The **California Current** actually serves another purpose. On the U.S. West Coast, this cold current prevents the northward migration of many east Pacific hurricanes and so has protected Southern California from the landfall of a major hurricane. This is an interesting and also disturbing aspect in the current climate debate because some scientists suggest that global warming could weaken ocean currents such as the California Current, thereby making Southern California more vulnerable to frequent hurricanes in the East Pacific. This is discussed in Chapter 15 in more detail.

Oceans do not just absorb solar heat and for that reason play an important role in the current global warming debate. Oceans also **absorb atmospheric CO_2**, a greenhouse gas, and so have had a double-role at slowing recent global warming. CO_2 dissolves in water to form carbonic acid and even though the oceans are huge, the CO_2 uptake has already led to measurable **ocean acidification**. If this process affects organisms at the bottom of the food chain, as laboratory experiments suggest it could, then Earth could face a **major mass extinction**. These issues will be discussed in Chapters 18 and 21.

Table 11.3 The Composition of Dry Air

Component	Symbol	Fraction
Nitrogen	N_2	78.08%
Oxygen	O_2	20.95%
trace gases		1%
Argon	Ar	0.93%
Carbon Dioxide	CO_2	280 ppm[5] pre-industrial, 390 ppm (2011); greenhouse gas
Neon	Ne	18 ppm
Helium	He	5 ppm
Methane	CH_4	2 ppm; greenhouse gas
Krypton	Kr	1 ppm

(5) parts per million

11.4 The Composition and Properties of Earth's Atmosphere

Earth's atmosphere is unique in the solar system and consists mostly of nitrogen (78%) and oxygen (21%) (Table 11.3; Box 2). The remaining 1% is made up of trace gases, with argon being the most abundant one at 0.93%. Compared to the atmospheres of other planets such as Venus (96.5%) and Mars (95.3%), the fraction of carbon dioxide (CO_2) in Earth's atmosphere is extremely small. Nevertheless, despite their small fractions, the recent exponential increase of CO_2 and methane (CH_4) has fueled discussion on the causes and consequences of global warming because these gases are greenhouse gases. Water vapor, the most important greenhouse gas, is present in the atmosphere at varying concentrations, e.g. from 0 to 4% at the surface. But it is usually excluded from discussions on atmospheric composition because these focus on dry air. Apart from these, other minor contributors to Earth's atmosphere are neon, helium and krypton.

Contributors at less than 1 ppm include hydrogen (0.55 ppm), nitrous oxide (N_2O at 0.3 ppm), carbon monoxide (CO at 0.1 ppm) and xenon (at 0.09 ppm). At even smaller amounts, ozone (O_3) is a naturally occurring constituent of Earth's atmosphere at 0.0 to 0.07 ppm. Radon (Rn) results from the natural decay of Uranium (U) but concentrations in the atmosphere are usually insignificant. At a diameter of less than 1μm, aerosols are also found in the atmosphere. Possible sources for aerosols are water droplets, sea salt, dust, volcanic ash, clay, soot and pollen. Nitrous oxide, carbon monoxide, ozone (depending on where it occurs) and aerosols are considered atmospheric pollutants and will be discussed in Chapter 18.

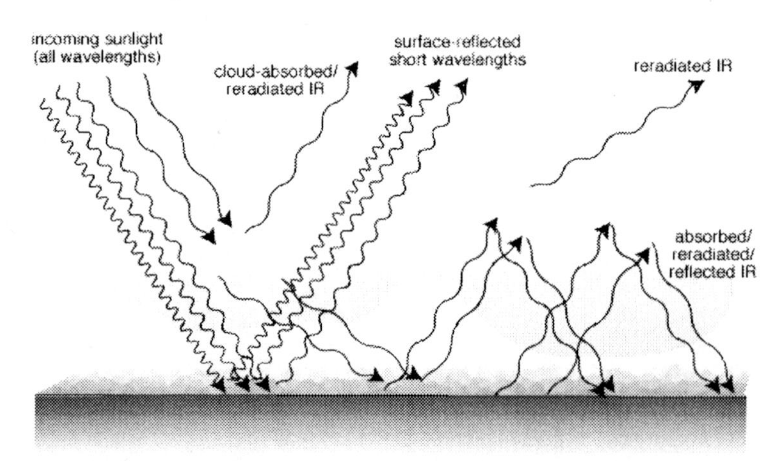

Figure 11.8 Concept of the greenhouse effect. Sunlight of many wavelengths arrives at the top of the atmosphere. The atmosphere is transparent to visible light (wavelength between .38 and 0.74 μm) and shorter-wavelength radiation (e.g. UV), so that the light can reach the surface. Greenhouse gases absorb (trap) the IR light and reradiate it in all directions. Some of the IR is radiated back to the surface. Short-wavelength light reflected at the surface can travel back through the atmosphere to escape into space but the IR light is trapped. The greenhouse gases thus heat the atmosphere and the

ground. The term greenhouse was adopted for the atmospheric warming effect because it was thought initially that this effect was very similar to that of a greenhouse where the glass is transparent to visible light but traps IR light. Recent experiments revealed however that the warming effect of the glasshouse is so strong because it inhibits convection and therefore the efficient removal of heat (see Chapter 2). So the actual greenhouse paradoxically does not quite have the same greenhouse effect as Earth's atmosphere.

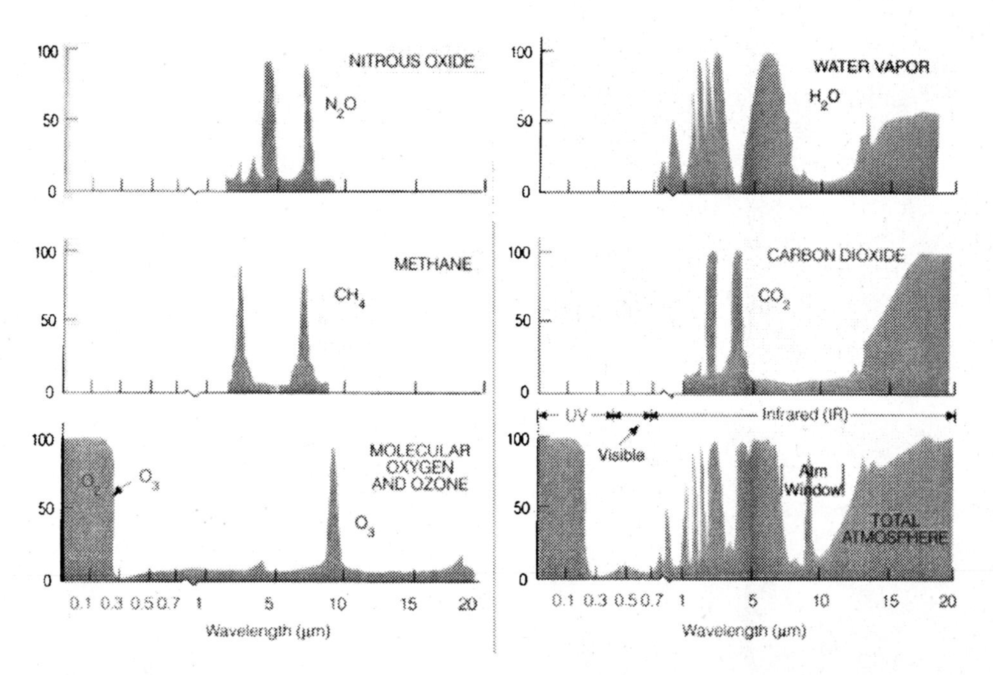

Figure 11.9 Absorption and re-radiation by greenhouse gases in the atmosphere. Shaded areas represent the percentage amount absorbed. Gases are considered greenhouse gases if the absorption occurs at wavelengths longer than about 0.7 μm. The strongest absorbers of infrared (IR) radiation are water vapor and carbon dioxide. IR radiation at wavelength between 8 and 11 μm is not absorbed very well and so can escape in a cloudless sky back into space more easily. This wavelength range is called the atmospheric window. Clouds, however, can absorb IR at these wavelengths. (source: MT)

The Greenhouse Effect

One reason why Earth is a habitable planet to life as we know it is because its atmosphere acts like a greenhouse in that it is transparent to visible light but traps IR light (Fig. 11.8). Earth's surface is on average a balmy 16°C (61°F). Without the greenhouse effect, it would be 34°C (61°F) colder, i.e. a rather chilling -18°C (0°F)! The latter is Earth's radiative equilibrium temperature. At this temperature Earth is absorbing solar radiation and emitting infrared radiation at equal rates and its average temperature does not change. But how does Earth's greenhouse work? Some atmospheric gases are transparent to the higher-energy visible and UV light. But the energy amount of IR sunlight is just right to cause the greenhouse gas molecules to resonate, i.e. to absorb this energy and oscillate. The exact amount of energy depends on the chemical composition of the gas (Fig. 11.9). The molecules absorb and re-emit this energy within the atmosphere. At a fraction of anywhere between 0% in deserts to 4% in the tropics, water vapor (H_2O) is the most important greenhouse gas, followed by CO_2. Even though only a tiny amount of CO_2 is abundant in the atmosphere, it is a much more potent greenhouse gas than H_2O is, i.e. fewer molecules are necessary to obtain the same effect. The relative contribution to the greenhouse effect is also called radiative forcing. Methane (CH_4) is an even more potent greenhouse gas than CO_2 but luckily its current abundance is nearly 200 times less than that of CO_2. After the industrial revolution in the late 1700s, atmospheric CO_2 has risen sharply (exponentially) by nearly 40%, as a result of the burning of fossil fuels. CH_4, which is released by anthropogenic (human-caused) processes such as agriculture, has also risen exponentially. Most climate scientists agree that this increase of anthropogenic greenhouse gases dramatically accelerates the greenhouse effect and global warming. This is discussed in more detail in Chapter 18.

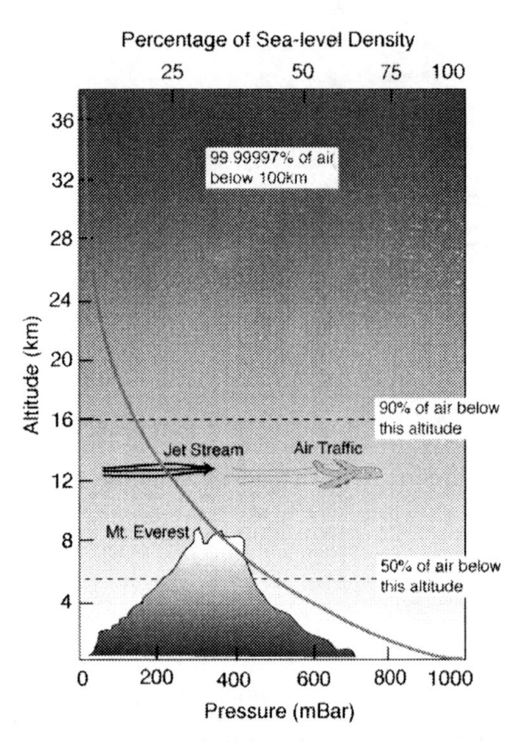

Figure 11.10 Air pressure (bottom scale) and density (top scale) in Earth's atmosphere as function of altitude. Both have a simple, exponential decline with increasing altitude. Most of Earth's air (90%) is found below an altitude of 16 km. People who climb the high summits of the Himalayas breathe air that has a molecule concentration of only 1/3 of that at sea level. Many therefore need the help of bottled oxygen above about 8000 m. The cruising altitude of commercial airline planes has only about

20% of the sea level air concentration and pressure. Planes therefore pressurize passenger cabins to mimic conditions of about 2100 m in altitude. Newer planes can mimic conditions as low as 1500 m, the altitude at which some people start to experience physiological problems such as hypoxia or altitude sickness.

Atmospheric Pressure and Density

Anybody hiking in the high Sierra Nevada or other high mountains may experience an unusual shortness of breath. In fact, most people climbing Mt. Everest or other high summits on the Himalayan Mountains typically need to bring along oxygen supplies. Because helium is lighter than air, we can enjoy helium balloons rising into the sky. But they eventually pop before they ever return to the surface. Airplanes have to pressurize passenger cabins shortly after take-off. All this is related to the fact that the air becomes dramatically thinner with increasing altitude. In fact, the density (the number of molecules per volume) of air declines exponentially with increasing altitude (Fig. 11.10). One half of all air molecules are found below an altitude of 5.6 km, 90% of air molecules are found below an altitude of 16 km and a tiny amount of just 0.00003% of air molecules exist at altitudes above 100 km. The air molecules exert a push on its surroundings. Hence, due to the weight of the overlying atmosphere, air pressure at sea level is highest with 1013.25 mbar (or 1 atm or 14.7 PSI). The air pressure of 1 atm is also referred to as the standard atmosphere. The most commonly used unit nowadays is mbar (in 10^3 dyn·cm^{-2} = 10^3 g·cm^{-1}s^{-2}), or in SI units, 1 hectopascal (hPa) (= 100 Pa = 100 kg·m^{-1}s^{-2}). In the U.S., some barometers measure air pressure in mercury inches, in Hg, (or mm Hg, in SI units). The reason for this will become obvious in Chapter 13. In this unit, the standard atmosphere measures 29.92 in Hg (or 760 mm Hg). Like density, air pressure declines exponentially with altitude (Fig 11.10). The exact exponential decline of air pressure with altitude depends on the sea level standard temperature (15°C/58.97°F), Earth's gravitational acceleration (9.81 ms^{-2}) and the molecular mass of air (which depends on humidity). The relatively simple formula can be found on the Wikipedia webpage on air pressure[(6)].

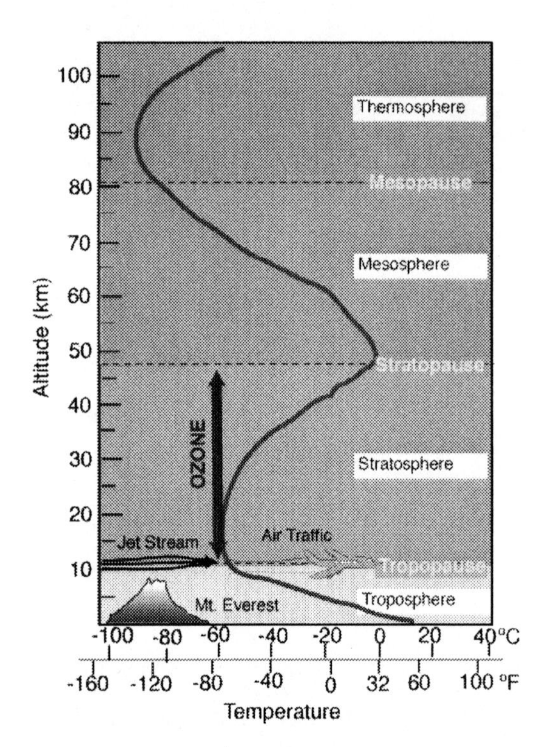

Figure 11.11 Air temperature in Earth's atmosphere as function of altitude. Also shown are the four principal atmospheric layers whose definition follows the principal changes in temperature.

Atmospheric Temperature

Unlike the simple change of air density and pressure, the change of temperature (T) with altitude is non-uniform as a result of the complex interplay of the density of air molecules and their interaction with the incoming solar radiation (Fig. 11.11). Starting at sea level with a balmy average 16°C, air temperature declines linearly until reaching about -60°C at an altitude of about 10 km. This gives a temperature gradient of -6.3°C/km, where the "-"-sign indicates that temperature decreases with increasing altitude. In other units, the gradient is about -11.4°F/km or about -3.5°F/1000 ft. This rate is also called the lapse rate (or also environmental lapse rate). The temperature declines as a result of declining air density. Air molecules have to travel farther to bump into the next one, the principal mechanism by which temperature is kept high. If molecules are packed together closely, they interact more often than if density is lower, so temperature declines. Above an altitude of 10 km, the temperature stays nearly constant for about 15 km but then rises until peaking at about 0°C at about 50 km altitude. The heating in this altitude range occurs because ozone molecules absorb solar radiation (ozone is a greenhouse gas and also absorbs UV radiation, see Fig. 11.9). Above an altitude of 50 km, the temperature declines from 0°C to less than -80°C at about 90 km altitude. The low concentration of air molecules at this altitude and the lack of absorption of solar energy cause this region to be the coldest place on Earth. Above this altitude, the temperature rises again until an altitude of at least 100 km. The air is so thin that an air molecule can travel 1 km before hitting another. The temperature increases there because air molecules are heated by the Sun's short-wavelength radiation. Temperature in the usual sense in not well defined there. And even though the temperature can rise to 1,500°C (2,700°F), the atmosphere does not contain much heat.

The Atmospheric Layers

The change in temperature in the atmosphere is approximately aligned with the four principal atmospheric layers.

Troposphere: This layer received its name from the Greek word *tropos* for "turning". The term expresses the fact that the air in this layer constantly undergoes vertical and horizontal mixing. It convects. Starting at sea level, this layer rises to about 10 km (9 km near the poles and 12 km at the equator). Near the top of the troposphere, which is typically the cruising altitude of passenger jets, T can drop to less than -60°C. Some airlines broadcast this information in the passenger cabin on long-hall flights. The weather-controlling jet streams (see Chapter 12) can also be found at this altitude. Since most weather-related processes occur in this layer (an exception would be severe thunderstorms that reach higher), the troposphere is also called the weather layer. The troposphere also experiences seasonal variations.

Stratosphere: This layer received its name from the Latin word *stratum* for "cover" or "layer". The term expresses the fact the stratosphere usually has no vertical mixing, which means that no air is exchanged with higher or lower layers. The stratosphere layer extends from about 10 km to about 50 km in altitude. It is very dry and stably stratified (layered). However, strong horizontal winds can occur. The stratosphere is perhaps best known as the location of the 'ozone layer' that protects life on Earth. Ozone absorbs much of the UV radiation arriving from the Sun. Since the air then heats, a warm layer lies above a cold layer. This **temperature inversion** high in the stratosphere is stable in contrast to the case when warm air below a cold layer wants to rise and explains why there is no vertical mixing within the stratosphere itself and with the troposphere below. In the 1980s, scientists discovered the ozone hole, a thinning of the ozone layer in the Antarctic spring. A similar thinning was later found for the Arctic region as well and even for lower latitudes. The thinning of the ozone layer means that more harmful UV radiation can reach Earth's surface. The thinning is caused by the production of anthropogenic ozone-depleting compounds. Some of these have been phased out successfully, but the long lifetime of these compounds and the continued production of other ozone killers mean that the ozone hole will be a continuing problem for at least the next few decades. The ozone hole is discussed in Chapter 18.

Mesosphere: This layer extends from roughly 50 km to 85 km in altitude. This layer has less ozone and does not absorb much solar energy and, due to the lack of air molecules interaction with each other, the temperature can become very low. Meteoroids and space junk entering the atmosphere burn up in this layer. Pieces large enough create meteors, the light phenomenon of shooting stars. Due to the low temperature, the little water vapor present up here is frozen and forms ice clouds (or noctilucent clouds) that are visible in deep twilight, and most commonly observed between 50° and 70° latitude. Noctilucent clouds are not fully understood and a relatively recent discovery[7]. There are no records of their observation before 1885 when they were first observed after the eruption of Krakatau volcano in 1883. The generation of these clouds requires very specific conditions and their occurrence can be used as a sensitive indicator of the state of the upper atmosphere. Some scientists speculate that the recent increase in observations of noctilucent clouds is related to global warming.

Thermosphere: This layer starts at an altitude of about 85 km and extends a few hundred kilometers above. The thermosphere contains less than 1% of Earth's air. The temperature increases with altitude but the layer contains little heat because of the small number of air molecules. The space station travels in this layer at an altitude between 320 and 380 km. The thermosphere is poorly mixed and compositionally stratified. The height to which the thermosphere extends varies because it depends on the level of solar activity. It reaches to altitudes between 350 and 800 km.

Exosphere: Above the thermosphere is the exosphere. It consists mainly of hydrogen and helium, the lightest elements. The particles are so far apart that they can travel hundreds of kilometers without colliding with each other. Some of these free-moving particles can actually escape Earth's gravitational pull and may migrate into and out of the magnetosphere or the solar wind.

Below the thermosphere, the atmosphere is typically well-mixed. This lower zone is also called the **homosphere**. The region above is called the **heterosphere**.

The boundaries between the atmospheric layers are called **pauses,** except for the boundary between the thermosphere and the exosphere. The tropopause separates the troposphere from the stratosphere above, the stratopause is above the stratosphere and the mesopause is above the mesosphere. The boundary between the thermosphere and the exosphere is the exobase.

At altitudes above 60 km, oxygen is ionized during the absorption of short-wavelength, high-energy solar radiation. The region between 60 and 400 km altitude therefore is the **ionosphere**. It hosts the spectacular **aurora borealis** (or the Northern Lights) in the Northern Hemisphere and **aurora australis** in the Southern Hemisphere. These light phenomena appear when charged particles (protons and electron) ejected from the Sun (especially from solar flares) interact with ions in the ionosphere and prompt them to release energy in the form of visible light. This usually happens at high latitudes where the particles get trapped by Earth's magnetic field. The ionosphere acts like a mirror for radio waves sent from Earth. Hence it enables radio transmission over great distances. During strong solar activity, long-range radio communication can be disrupted.

11.5 Moist Air, Dry Air and Latent Heat

Dry air during a Santa Ana contributes to one of San Diego's most hazardous weather conditions. It removes water from dead and living vegetation and increases the danger of devastating wildfires to extremely high levels. How do we know when the air is dry? The best indicator for this is the relative humidity but other indicators are also in use.

To get us started and as discussed above, air has a water vapor content between 0% in areas of no rainfall, such as the dry deserts and polar deserts, and 4% in the rain forest. This is the absolute water vapor content by volume, the **absolute humidity**. Air with a certain temperature and pressure can absorb and "hold" only a certain amount of water vapor before the vapor starts to condense. Air in this condition is called **saturated air**. The use of the term hold here is somewhat unscientific. Without going into too much detail, it expresses the ability to retain a balance between water vapor and liquid water. In the tropical rain forest, where it rains much of the time, air is often close to being saturated, even though at only 4% the total water vapor content in air is relatively small. On the other hand, air on cold rainy winter day in temperate climate may contain much less water but yet it rains all the time. So the absolute humidity may not be the best way to describe weather conditions.

The **relative humidity** compares the actual water content at any given time with the maximum water content the air at the current condition (temperature and pressure) could potentially "hold", i.e. the saturated state. So during rain or fog, the relative humidity is close to 100%. With the same absolute amount of water vapor, cold air is closer to saturation than warmer air. With no change in water vapor content and pressure, an increase in air temperature therefore lowers the relative humidity, while a decrease in air temperature raises the relative humidity. Air that we feel as comfortable typically has 40 – 70% relative humidity, though the "comfort zone" depends on air temperature.

Table 11.4 Relative Humidity, Dew Point and How the Air Feels on a Hot Day

Rel. Humidity at 32°C/90°F	Dew point [°C]	Dew point [°F]	Human perception
> 65% and higher	> 26	> 80	severe, may be deadly for asthma-related illnesses
62%	24-26	75-80	extremely uncomfortable, fairly oppressive
52-60%	21-24	70-74	very humid, quite uncomfortable

44-52% 18-21 65-69 somewhat uncomfortable for many people 37-46% 16-18 60-64 OK for most, but all notice the humidity

38-41% 13-16 55-59 comfortable

31-37% 10-12 50-54 very comfortable

30% < 10 < 49 a bit dry for some

Hot air with 70% rel. humidity feels quite muggy. Summer days in Louisiana are quite a challenge when the relative humidity is typically 80%. Muggy days are unpleasant to many people and can actually turn into a health hazard, especially for the elderly or ill people (Table 11.4). Because of the high relative humidity on muggy days, sweat does not evaporate easily from the skin. But it is this evaporation that cools the skin and ultimately the human body. Recall that evaporation is endothermic and the heat necessary for this process is taken from the air surrounding the skin. In high humidity, the body temperature can then rise and people can suffer heat exhaustion or even heat strokes. The hazard of humid air during heat waves will be discussed in more detail in Chapter 13.

On the other hand, in natural deserts, the relative humidity is very low, 10 - 20% though in irrigated areas, such as the Imperial Valley, CA, the relative humidity is considerably higher. In these dry deserts, people do not feel just how much they sweat because the sweat immediately evaporates. They do not feel hot and so there is less awareness to drink in order to replenish lost fluids and electrolytes. People then get dehydrated easily. Symptoms include headaches, decreasing blood pressure, dizziness and even fainting, and severe dehydration can be fatal if untreated.

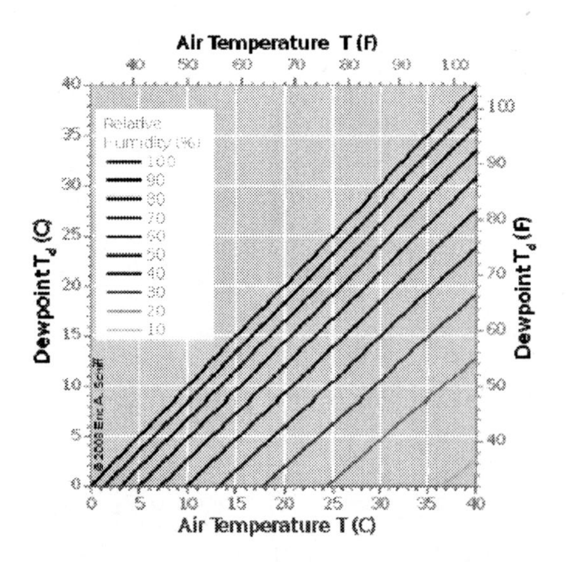

Figure 11.12a The relationship between air temperature, relative humidity and dew point, at air temperatures above freezing. The dew point depends on the relative humidity of air. For a given

temperature and relative humidity, the dew point gives the limit to with the air temperature can decline. High relative humidity results in high dew points, while low relative humidity results in a low dew point. The dew point can never decline below the air temperature. (source: Wikipedia[8])

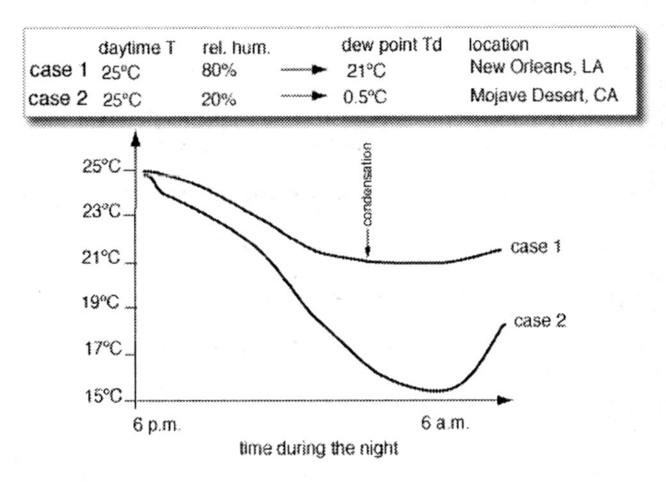

Figure 11.12b Two possible scenarios of how low temperatures could decline during the night on a summer day. In muggy New Orleans, the relative humidity is so high that the dew point remains above 20°C. During the night, the temperature then falls to 21°C at which point condensation occurs and T no longer declines. In the Mojave Desert, on the other hand, the relative humidity is very low and so the temperature can drop significantly during the night. The dew point is only 0.5°C and therefore just above freezing. On a summer day, it is unlikely that this point is ever reached before sunrise, and the atmosphere warms up before reaching the dew point. The cases were constructed using the chart in Fig. 11.12a.

Figure 11.13 Citrus in a grove in Lakeland, FL, covered with icicles to avoid frost damage. (source: denverpost.com)

The relative humidity is measured with a hygrometer. Different instruments are based on different concepts but the one first used is the hair tension hygrometer. It uses a human or animal hair under tension. The length of the hair changes with humidity. This change is magnified by a mechanism and indicated on a dial or scale. The traditional folk art device known as "weather house" works on this principle. Electronic versions nowadays replace the traditional instruments.

A very useful parameter that farmers are often interested in is the **dew point** (Fig. 11.12). It expresses at which point during cooling the air becomes saturated. Knowing the dew point is also useful in estimating the likelihood of fog setting in later in the day or at night. Fog or clouds form in air that cools to the dew point. The dew point therefore gives us an idea about the minimum temperature air can achieve under current conditions (i.e. no change in pressure and moisture content). Recall that during condensation, **latent heat** is released to warm the surrounding air. Air cooling to the dew point therefore begins to be heated by condensation and so cannot become colder. Of course, whether the dew point is actually reached or not depends on the current weather conditions. Two examples are given in Fig. 11.12b. In humid New Orleans, the relative humidity is very high so the dew point is also very high. It is very likely that the dew point is reached during nighttime cooling. Condensation occurs and the air cannot cool further, which is quite stressful to the human body. In the Mojave Desert on the other hand, the relative humidity is very low and so the dew point is also very low. Nighttime temperatures can therefore plummet before the morning sunshine warms the air again.

Viewed from a different angle, the difference in temperature and dew point gives us an idea about relative humidity. If the dew point is nearly as high as the current temperature, the relative humidity is high. Precipitation may be imminent. A large gap between temperature and the dew point indicates low relative humidity. Above freezing, a simple mathematical relationship exists between air temperature, relative humidity and the dew point, and the interested reader can find it on the Wikipedia webpage on the dew point[(7)].

Frost forms when the dew point is below freezing (the dew point is the called the **frost point**). Watching the dew point in winter is very important for California and Florida farmers who may need to protect their frost-sensitive crops such as citrus and avocado fruit[(9)]. If the air is very dry on winter nights, the dew point is very low and the temperature can drop below freezing. If air cools down to this point, the crop may become frost-damaged and ruined. To avoid this from happening, farmers then turn on the irrigation sprinklers in their orchards either to raise the relative humidity and therefore the dew point. Or, if it is already freezing cold, the freezing spray builds an insulating ice crust around the fruit to protect it (Figure 11.13). If, on the other hand, the dew point stays above freezing during the day, farmers know that this precautionary measure is not necessary and save costs that they would otherwise spend on unnecessary irrigation. A particularly severe freeze in California occurred in January 2007 (see Case Study).

The dew point is also a good indicator of the actual water vapor content air. High dew points indicate high water vapor content, while low dew points indicate low water vapor content. For example, compare **polar air**, with an air temperature of -2°C (28.5°F) and a dew point of -2°C (28.5°F) to **desert air**, with an air temperature of 35°C (95°F) and a dew point of 10°C (50°F). The relative humidity of the polar air is 100%, while that of desert air is only 21%. Nevertheless, the desert air holds more water vapor than the polar air, as indicated by the dew point. Note here that the dew point in moist, tropical air, which holds even more moisture is likely above 20°C (68°F).

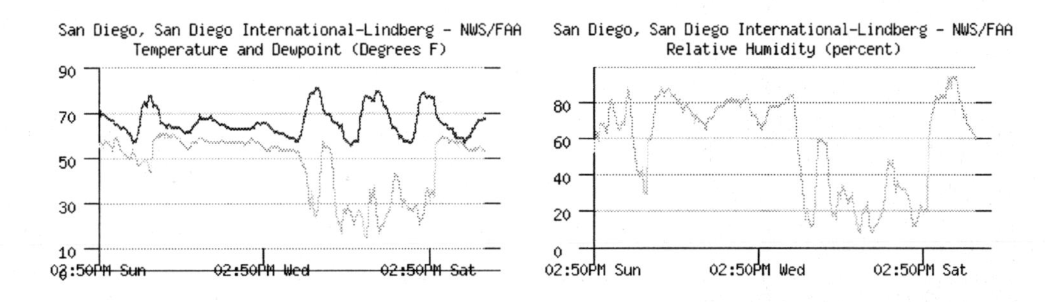

Figure 11.14 Temperature (top left), dewpoint (bottom left) and relative humidity (right) at the National Weather Service (NWS) station San Diego International Lindbergh (San Diego Lindberg Field international airport), for the week of 22 – 28 October 2006 (Sunday through Saturday). (source: NWS)

It is interesting to explore the relationship between temperature, relative humidity and dew point during a Santa Ana in San Diego (Fig. 11.14). On a typical day in autumn in San Diego, the temperature is about 65°F (18.5°C) and the relative humidity ranges between 65 and 80% (Sunday through Wednesday). Day-night temperatures do not vary much because the water vapor in the air stores and releases latent heat to moderate temperatures. The dew point is at around 60°F (15.6°C). Since the temperature and the dew point are so close together, it does not take much cooling for fog to form, a hazard that drivers at night should be aware of. In fact, patchy fog may have formed along I-5 Monday night when relative humidity was near 85%. During the Santa Ana on Thursday through Saturday, the relative humidity dropped to below 30% and the dew point dropped to 30°F (-1.1°C), which is below freezing. So the air would have had to cool below freezing before any precipitation (incl. fog) could form. Since the air was so dry, little latent heat was stored in the little water vapor available to moderate temperatures. Less water vapor also implies a weaker greenhouse effect. Air therefore heated quickly during the day and cooled quickly at night, and there was a pronounced difference in day–night temperatures, with nighttime temperature dropping below 60°F (17.5°C) but daytime temperatures rising above 80°F (26.5°C).

Box 1: Simplified Köppen Climate Zones

The Köppen climate classification is one of the most widely used climate classification systems though others exist, such as the Trewartha climate classification. The following is a brief description of the quite complex system that was first introduced by German-Russian climatologist Wladimir Köppen in 1884 and later modified. The system uses a two-, three- or four-letter code. The first letter denotes the five principal climate groups (Table 11.5). The terms mega-, meso- and microthermal broadly express how warm it is during the winter, i.e. mesothermal climates are hot and microthermal winters are cold. Subsequent letters specify characteristics within these group and divide climates into types and subtypes. For example, in group B the second letter describes whether precipitations surpass a certain threshold or not and the third letter characterizes typical temperatures.

Table 11.5 The Five Principal Köppen Climate Groups

Letter	Climate Group	Main Characteristic
A	tropical/megathermal	all months have average T > 18°C (64°F)
B	dry/arid and semiarid	deficient precipitation most of the year
C	temperate/mesothermal	moist, mid-latitude climates with moist winters
D	continental/microthermal	moist, mid-latitude climates with cold winters
E	polar	extremely cold winters and summers

Group A: tropical

Climates in group A have constant high temperatures where all 12 months have average temperatures of 18°C (64°F) or higher (at sea level).

Tropical rainforest (*Af*): typically within 5-10° of equator; 12 months of rainfall ≥ 60 mm (2.4 in); e.g. Belém, Brazil; Hilo, Hawaii; Singapore. No natural seasons though some locations have distinctly wet months, e.g. Palembang, Indonesia; Sitiawan, Malaysia

Tropical monsoon (*Am*): mostly in S. America; controlled by changing monsoon winds; driest month with < 60 mm (2.4 in) after winter solstice but more than a certain threshold (Pr_{th} = 100 – Pr_{ann}/25, where Pr_{ann}=total annual precipitation in mm); most precipitation falls in 7-9 hotter months; e.g. Cairns, Australia; Chittagong, Bangladesh; Miami, FL

Tropical wet, dry or savanna (*Aw*): pronounced dry season with driest months < 60 mm and also < Pr_{th}; precipitation during the wet season typically < 1000 mm (39.4 in); e.g. Jakarta, Indonesia; Rio de Janeiro, Brazil; Honolulu, Hawaii

Group B: dry (arid and semi-arid)

Group B climates are typically at 20 – 35° latitude and in large continental regions of mid-latitudes, often surrounded by mountains. Precipitation is less than the potential evapotranspiration threshold which depends on the annual temperature and when in the year, summer or winter, the main precipitation occurs. With T_{av} being the average annual temperature in °C, the precipitation threshold (in mm) is defined as $Pr_{th} = T_{av} \cdot 20 + 280$, if 70% or more of the total precipitation is in the summer half (high-sun half) of the year. $Pr_{th} = T_{av} \cdot 20 + 140$, if 30-70% falls during this period. $Pr_{th} = T_{av} \cdot 20$, if less that 30% falls during this period. The second letter in the code signifies the amount of precipitation relative to the Pr_{th}. A third letter distinguishes between temperatures. The letter *h* signifies that the average temperature in the coldest month is above 0°C (32°F) and the letter *k* stands for a colder coldest month.

Desert/dry arid (*BW*): precipitation < ½ of threshold; covers 12% of global land surface; e.g. Yuma, AZ (*BWh*); Dubai, UAE (*BWh*); Turpan, China (*BWk*); Isfahan, Iran (*BWk*). Desert areas along west coasts near cold ocean currents; e.g. Lima, Peru (*BWn*); Walvis Bay, Namibia (*BWn*).

Steppe/dry semiarid (*BS*): ½ of threshold < precipitation < threshold; grassland climate that covers 14% of global land surface; low latitude steppes include Murcia, Spain (*BSh*); Porto Santo, Portugal (*BSh*).; foggy coastal steppes along west coasts near cold ocean currents (*BSn)* includes some of coastal San Diego, CA; colder middle latitudes steppes include Boise, ID (*BSk*); Denver, CO (*BSk*).

Group C: temperate/mesothermal

Group C includes moist subtropical mid-latitude climates with humid summers and mild winters. The average T > 10°C (50°F) in the warmest month and between 0°C (32°F) and 18°C (64°F) in the coldest month. The second letter distinguishes between dry winters (*w*) and dry summers (*s*) with < 30 mm in precipitation, and precipitations in all seasons (*f*). The condition of the dryness for type *w* and *s* climates is determined through the average precipitation of the driest month. It has to be less than

1/10 of the average precipitation of the wettest month in the wet season. A third letter distinguishes the intensity of summer heat. The letter *a* signifies an average T > 22ºC (72ºF) in the warmest month, with at least 4 months averaging above 10ºC (50ºF). The letter *b* signifies an average T < 22ºC (72ºF) in the warmest month, with at least 4 months averaging above 10ºC (50ºF). The letter *c* signifies less than 4 months averaging above 10ºC (50ºF).

Dry-summer subtropical/Mediterranean (*Csa, Csb*): usually on western sides of continents between 30º and 45º latitude; hot, dry summers, except for coastal regions with milder summers with fog but no rain. Examples of *Csa*: Athens, Greece; Antalya, Turkey; San Remo, Italy; Cape Town, S. Africa; Adelaide, S. Australia; Los Angeles, CA. Examples of *Csb*: Coimbra and Porto, Portugal; Valladolin, Spain; Risan, Montenegro; Concepción and Santiago, Chile; San Francisco and San Jose, CA.

Humid subtropical (*Cfa, Cwa*): usually in interiors of continents or on east coasts; 20º - 30º latitude (sometimes to 45º in Eurasia); summers generally humid (in contrast to Mediterranean climate). Examples of *Cfa* include Atlanta, GA; Houston, TX; New York, NY; Philadelphia, PA; Edirne, Turkey; Brisbane, Australia; Milan, Italy; Corvo, Azores; Tbilisi, Georgia; Buenos Aires, Argentina. Examples of *Cwa* include Guadalajara, Mexico; Lahore, Pakistan; Zhengzhou, China; Hong Kong; Hanoi, Vietnam.

Dry-summer subtropical but non-Mediterranean (*Cfb*): meets some of Mediterranean climate but not usually associated with it; typically marine, close to oceans; e.g. Pacific Northwest; Portland, OR; Seattle, WA; Victoria, B.C., Canada; Vancouver, B.C.

Maritime temperate/oceanic (*Cfb, Cwb*): usually on western sides of continents, 45º - 55º latitude (up to 63º in Europe); immediately poleward of Mediterranean climates; changeable, often overcast weather; cool summers; mild, cloudy winters. Examples of *Cfb* include Bilbao, Spain; Limoges, France; Liverpool, England; Hamburg, Germany; Hobart, Tasmania; Valdivia, Chile (though similar to *Csb*); Valdivia, Chile. Examples of *Cfb* at high elevation in subtropical and tropical areas include Boone, NC; Pico, Azores; Crkvice, Montenegro (*Cfsb* holds Europe's annual precipitation record with 4927 mm/m^2).

Temperate with dry winters (*Cwb*): typically in highlands inside the tropics but also found outside of tropics; very dry winters, very rainy summers; e.g. Cuzco, Peru; Mexico City, Mexico; Bogotá, Colombia; La Paz, Bolivia; Johannesburg, S. Africa; Gangtok, India.

Maritime subarctic/subpolar oceanic (*Cfc*): poleward of maritime temperate; typically confined to narrow coastal strips on western polward margins of continents; e.g. Reykjavik, Iceland; Faroe Islands; Monte Dinero, Argentina; Bodo, Norway.

Group D: continental/microthermal

Average T > 10°C (50°F) in warmest month; average T < 0°C (32°F) in coldest month; usually in continental interiors or east coasts at latitudes > 40°; the second and third letter in the code is used as in group C. A third letter *d* indicates 3 or less months with T > 10°C *and* coldest month with T < -38°C (-36°F).

Hot summer continental (*Dfa, Dwa, Dsa*): *Dfa* usual in high 30s° and low 40s° latitude; much drier in Europe than in North America; *Dwa* extends farther south in Asia than elsewhere because of influence of Siberian High. Examples of *Dfa* include Bucharest, Romania; Dnipropetrovsk, Ukraine; Rostov-on-Don, Russia; Toronto, Canada; Cleveland, OH; Santaquin, UT. Examples of *Dwa* include Beijing, China; Seoul, South Korea. *Dsa* exists only at higher elevation adjacent to hot-aummer Mediterranean climates. Examples include Cambridge, ID; Saqqez, Iranian Kurdistan.

Warm summer continental/hemiboreal (*Dfb, Dwb, Dsb*): *Dfb* and *Dwb* immediately poleward of hot summer continental climates; generally high 40s° to low 50° latitude in North America and Asia, and high 50s° to lowest 60° latitude in Europe and Russia. *Dsb* is found at even higher elevations than *Dsa,* or at higher latitudes though the latter is chiefly restricted to North America where the Mediterranean climates extend farther north than in Eurasia. Examples of *Dfb* include Kars, Turkey; Helsinki, Finland; Växjö, Sweden; Minsk, Belarus; Moncton, Canada; Saskatoon, Canada. Examples of *Dwb* include Vladivostok, Russia. Examples of *Dsb* include Mazama, WA.

Continental subarctic or broeal (taiga) (*Dfc, Dwc, Dsc*): *Dfc* and *Dwc* poleward of other group D climates, mostly in 50s° and low 60s°, sometimes into low 70s° latitude. Examples of *Dfc* include Kirkenes, Norway; Luleå, Sweden; Murmansk, Russia; Mount Robson, B.C., Canada; Anchorage, AK. Examples of *Dwc* include Irkutsk, Russia. Examples of *Dsc* include Zubacki kabao, Montenegro; higher elevations of Massif Central, France; Galena Summit, ID.

Continental subarctic, with extremely severe winters (*Dfd, Dwd*): temperature in the coldest months lower than -38°C (-36°F). Only in eastern Siberia.

Group E: polar

Average temperatures remain below 10°C (50°F) in all twelve months.

Tundra (*ET*): warmest month with average T between 0°C (32°F) and 10°C (50°F); along northern edges of North America and Eurasia and on some islands near Antarctic Circle, and at high elevations outside of polar regions, above tree line. Examples include Barrow, AK; Iqaluit, Canada; Nuuk, Greenland; Provideniya, Russia; Vardø, Norway; Longyearbyen, Svalbard; Grytviken, South

Georgia. Examples of high-elevation *ET* include Mt. Washington, NH; Mt. Pico, Azores; Jotunheimen, Norway.

Ice cap (*EF*): all twelve months have average temperatures below 0°C (32°F). Examples include Antarctica and inner Greenland. A third letter is sometimes added to characterize dry seasons, with similar attributes as for groups C and D, e.g. *ETs* for dry summers on Pic du Midi de Bigorre in the French Pyrenees; *ETw* for dry winters on Herschel Island off the coast of Canada's Yukon Territory; *ETf* for uniformly spread precipitation in Hebron, Labrador.

Box 2: Some Relevant Elements and Compounds

Symbol	Name	Comment
H	Hydrogen	most basic element
He	Helium	second most simple element
C	Carbon	
O	Oxygen	
S	Sulfur	
N	Nitrogen	
Ar	Argon	noble gas
Ne	Neon	noble gas
Kr	Krypton	noble gas
Rn	Radon	radioactive; decay element of U
U	Uranium	last naturally occurring element in periodic
CO	Carbon monoxide	toxin; from volcanoes, fires and burning of fossil fuels
O_3	Ozone	in higher atmosphere: protects against UV radiation in lower atmosphere: aggressive, unhealthy pollutant
CO_2	Carbon dioxide	greenhouse gas; from volcanic eruptions and fossil fuel burning
CH_4	Methane	greenhouse gas from decay of organic matter and some geologic processes
CFC	chloro-fluoro-carbons	greenhouse gas; destructor of stratospheric ozone; not naturally occurring

References and Websites

1. MT
2. AB
3. ME
4. IPCC and webpage http://www.ipcc.ch/publications_and_data/ar4/wg1/en/faq-1-1.html
5. Wikipedia webpage on the 2012 European cold wave: http://en.wikipedia.org/wiki/2012_European_cold_wave
6. Wikipedia webpage on air pressure: http://en.wikipedia.org/wiki/Air_pressure
7. Wikipedia webpage on noctilucent clouds: http://en.wikipedia.org/wiki/Noctilucent_cloud
8. Wikipedia webpage on dew point: http://en.wikipedia.org/wiki/Dew_point
9. Ferrer, A. Freeze Damage Protection for Citrus Trees. Website contribution at seminolecountyfl.gov
10. Trenberth, K.E. and Caron, J.M., 2001. Estimates of Meridional Atmosphere and Ocean Heat Transport. J. Climate, 14, 3433-3443.

National Weather Service Severe Weather Forecast:
http://www.weather.gov/

NOAA's Annual State of the Climate Global Analysis (since 1997):
http://www.ncdc.noaa.gov/sotc/global/

various Wikipedia pages, including the one on climate:
http://en.wikipedia.org/wiki/Climate

Other Recommended Reading

Introductory books on oceanography:

Garrison, T., 1999. Oceanography. 3rd edition. Brooks/Cole Thomson Publishing, Pacific Grove, CA, pp. 541.

Trujillo, A.P. and Thurman, H.V., 2005. Essentials of Oceanography. 8th edition. Pearson Prentice Hall, Upper Saddle River, NJ, pp.518.

COPYRIGHTED MATERIAL — DO NOT DUPLICATE, DISTRIBUTE, OR POST

Chapter 12: Ocean Currents, Winds and Weather

In the last chapter, we learned that processes in both the ocean and the atmosphere influence climate and weather. One of the most basic ocean-atmosphere interactions is the generation of wind-driven waves. These were described briefly in Chapter 6. This chapter lays out the large-scale processes, namely the main circulation patterns in the oceans and the atmosphere. Ocean circulation is largely driven by differences in temperature and salinity, the thermo-haline circulation, also called the ocean heat conveyor. Air flow is largely driven by differences in temperature and pressure. Large-scale air circulation results in the formation of low- and high-pressure systems that in turn control where stormy or fair weather is happening. Extremely fast high altitude winds, the jet streams, control the path of some of the large low- and high-pressure systems. High-pressure systems are typically associated with fair weather while low-pressure systems are typically associated with storms. So, typically, low-pressure systems are associated with severe weather. However, high-pressure systems in the North American Southwest can set the stage for strong winds that fuel devastating, uncontrollable wildfires. Seasonal variations in the near-equatorial air circulation patterns result in the arrival of monsoons that bring much-needed rain, but monsoons can also cause devastating floods. Severe droughts and famine can occur when monsoons fail to arrive.

Clouds form when water vapor condenses. Their shape and ability to produce precipitation, and which type, depends on where and how the clouds form and develop. Some clouds are harbingers of fair weather, but some are clear signs of imminent severe weather. This is discussed at the end of this chapter.

12.1 Ocean Circulation

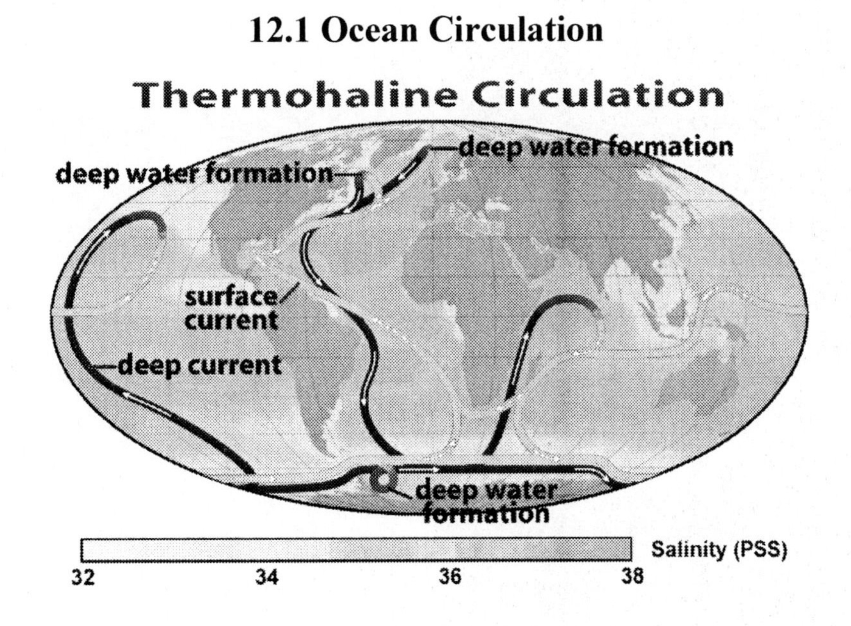

COPYRIGHTED MATERIAL — DO NOT DUPLICATE, DISTRIBUTE, OR POST

Figure 12.1 Schematic drawing of the global ocean heat conveyor that is driven by thermo-haline circulation. Black paths represent deep-water currents, while light grey paths represent surface currents. (source: wikipedia)

The Ocean Heat Conveyor

An often-used concept of large-scale ocean circulation is the **global ocean heat conveyor** (Fig. 12.1). It explains how oceans transport heat between the equator and the poles. The ocean heat conveyor is driven by density variations in the water and the concept makes use of the fact that cold, salty water is denser than warm, fresh water. Colder and saltier water therefore sinks, as found in the high-latitude North Atlantic Ocean and in the South Atlantic off Antarctica, while warmer or fresher water floats. Circulation within the oceanic heat conveyor is therefore called **thermo-haline circulation.** Many oceanographers currently think that deep-ocean circulation is the driving force of the heat conveyor mainly through the formation of dense deep-water masses in the North Atlantic and in the Southern Oceans. But unlike surface currents, deep ocean currents are difficult to study. In fact, recent studies by atmospheric scientists contest that the heat transfer from the equator to the poles is much larger in the atmosphere than in the oceans and that the thermo-haline circulation is only relevant at shallow ocean depths. However, recall that Fig. 11.3b documents that the relative importance of ocean and atmosphere heat transport is a strong function of latitude.

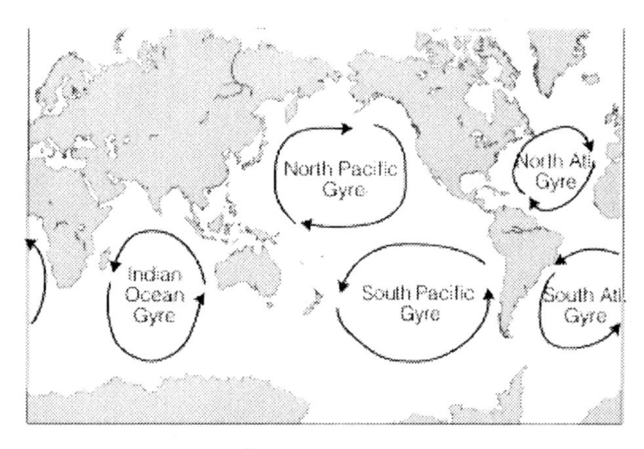

Figure 12.2a The five mayor ocean gyres.

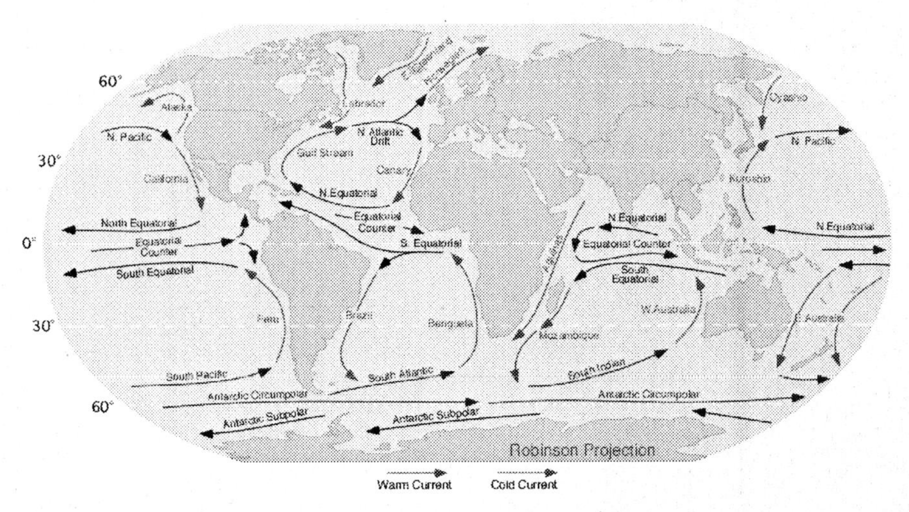

Figure 12.2b Major ocean surface currents. Western boundary currents are typically warm (e.g. Gulf Stream), while eastern boundary currents are cold (California Current). (source: wikipedia)

Ocean Currents

Ocean currents affect climate in important ways, on both local and on larger scales. The way ocean currents move is not only controlled by the coastline geometry but in a profound way by the Coriolis effect (see Chapter 2). As a result of the Coriolis effect, wind-driven surface currents are deflected to the right in the Northern Hemisphere and to the left in the Southern Hemisphere. Clockwise (Northern Hemisphere) and counter-clockwise (Southern Hemisphere) rotations result within ocean basins to form loops, the so-called **gyres** (Fig. 12.2a). There are five major subtropical gyres in the world's oceans: the North Atlantic and the South Atlantic Gyres, the North Pacific and the South Pacific Gyres, and the Indian Ocean Gyre. There are also a number of smaller gyres such as the subpolar gyres at high latitudes (Fig. 12.2b). In subtropical gyres, currents along the western edge of ocean basins, the **western boundary currents**, are intensified compared to **eastern boundary currents**, in both hemispheres (Table 12.1). They are narrow but deep and fast-moving, while the eastern boundary currents are wide but shallow and slow-moving. Western boundary currents, such as the Kuroshio Current off Japan, are typically warm currents while eastern boundary currents, such as the California Current, are relatively cold (Fig. 12.2b), in both hemispheres. Currents are named for the direction they are moving in, i.e. a southward current flows from north to south.

Table 12.1 Characteristics of western and eastern boundary currents

Current Type	Example	Width	Depth	Speed	Other Properties
western boundary current	Gulf Stream Brazil Current	narrow (< 100 km)	deep (< 2 km)	fast (100s km/day)	warm little/no upwelling

COPYRIGHTED MATERIAL — DO NOT DUPLICATE, DISTRIBUTE, OR POST

	Kuroshio Current					
eastern boundary current	Canary Current Benguela Current California Current	wide (< 1000 km)	shallow (< 0.5 km)	slow (10s km/day)	cool	coastal upwelling

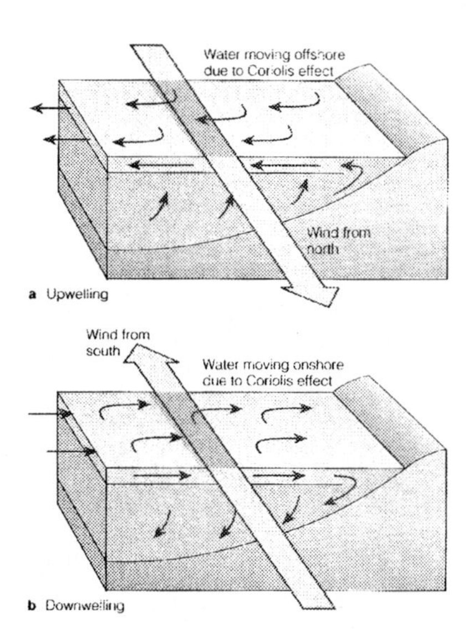

Figure 12.3 Coastal upwelling (top) and downwelling (bottom). Where winds from the north blow parallel to a west coast in the Northern Hemisphere, Ekman transport carries surface waters away from the continent. Coastal upwelling of deeper water occurs to replace the surface water that has moved away. The opposite is the case for winds from the south. They cause coastal downwelling. In the Southern Hemisphere the situation is reversed, with winds from the south causing coastal upwelling along west coasts. (source: TG)

An interesting aspect thereby is that the eastern boundary currents are not necessarily cold because they transport cold Artic water toward lower latitudes. **Coastal downwellings** along western boundary current and **coastal upwellings** along eastern boundary currents also play a role, if not the major role. Why do these currents have up- or downwellings? Due to the Coriolis effect, the average movement of water below the sea surface is deflected from the surface-wind direction by 90°, to the right in the Northern Hemisphere and to the left in the Southern Hemisphere. This is one of the main features of **Ekman Transport** that describes in more detail than given here how water at a particular depth is deflected from the surface wind direction. Along near-coastal currents, the Ekman Transport explains how surface water sinks (downwelling) or deep water rises to the surface (upwelling) (Fig. 12.3). A prolonged wind blowing southward along the North American west coast causes surface water to move westward away from the coast. This allows upwelling along the coast as water from below tries to fill the space vacated by the westward moving surface water. A prolonged wind blowing northward does the opposite: water is pushed toward the coast causing coastal downwelling. The reverse occurs in the Southern Hemisphere were northward blowing winds cause upwelling. As a consequence of this, subtropical eastern boundary currents are typically associated with costal upwelling while western boundary currents are not (Table 12.1). The upwelling of deeper, colder water along eastern boundary currents brings nutrient-rich waters to the surface. It causes near-coastal cooling but is also important to sustain large fish populations (see case study 1).

12.2 Moving Air

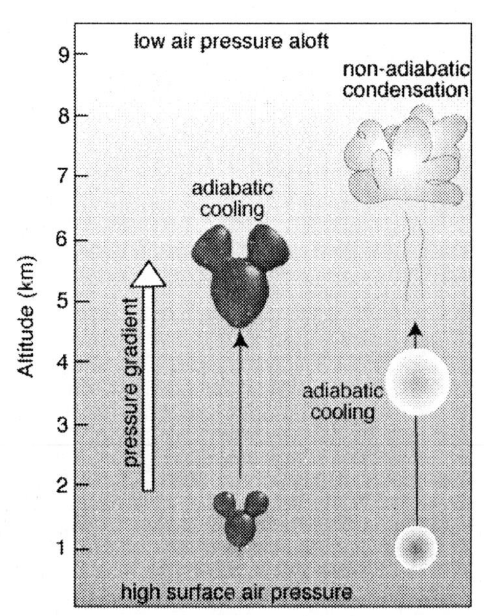

Figure 12.4a The concept of **adiabatic cooling**. A pocket of air rises and expands (e.g. air in a balloon). The relative humidity increases. When condensation occurs, the process is no longer adiabatic. The **pressure gradient** is the difference in pressure divided by the distance.

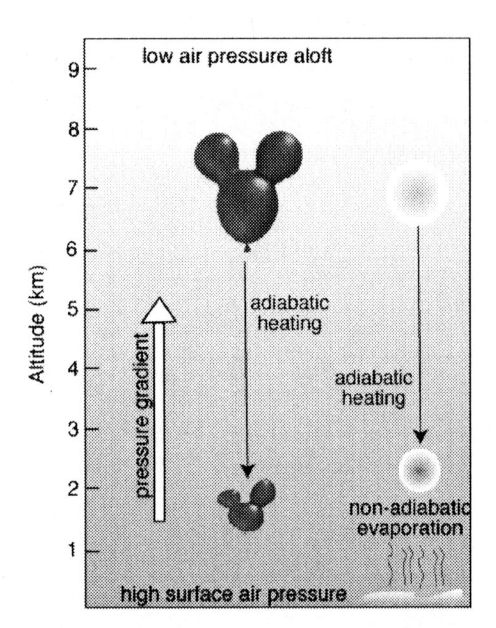

Figure 12.4b The concept of **adiabatic heating**. A pocket of air sinks against the pressure gradient and compresses. The relative humidity declines. When evaporation occurs, the process is no longer adiabatic.

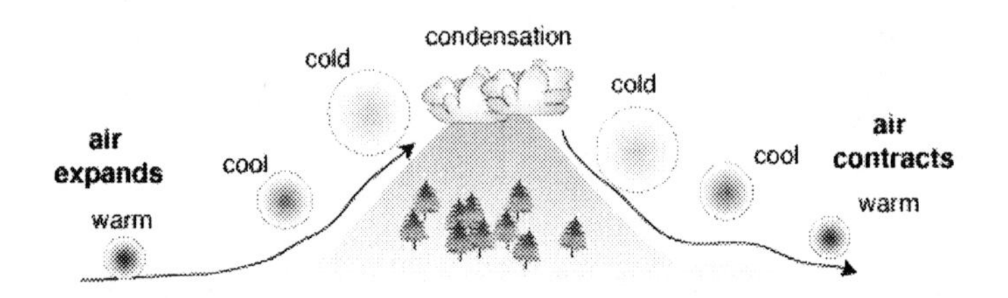

Figure 12.4c Orographic lifting. When horizontally flowing air encounters a mountain, it is forced to rise. It experiences adiabatic cooling. When it reaches the mountain top, condensation may take place and clouds form. Adiabatic heating occurs when air flows down a mountain slope. The air contracts and heats.

Adiabatic Changes

A parcel of air can move without exchanging energy in form of heat with its surroundings. This process is called an **adiabatic process**. The parcel of can expand or contract in response to changing pressure. In adiabatic processes, compression always results in warming, and expansion results in cooling.

During **adiabatic cooling**, air moves from high air pressure to low air pressure. At lower air pressure, molecules can travel farther before colliding with other molecules. The collision of molecules is what determines air temperature. An example of adiabatic cooling is air that moves from high surface pressure to low pressure aloft (Fig. 12.4a). When air rises and expands, molecules now have more space, and air cools at a rate of 6-10°C per km. With the same amount of water vapor, cooler air is more likely to experience condensation than warmer air. The relative humidity increases until the dew point is reached, condensation occurs and clouds form. Latent heat is released during condensation at which point the process is no longer adiabatic. Rising air leaves a **low surface pressure** behind relative to the surrounding area (Fig. 12.5). An example of adiabatic cooling is rising air at the equator or in a developing storm system or during orographic lifting (Fig. 12.4c).

During **adiabatic heating**, air moves from low air pressure to high air pressure against the **pressure gradient**. An example of adiabatic heating is air that moves from low pressure aloft to high surface pressure (Fig. 12.4b). When air sinks, it contracts and molecules are pushed together. Collision is more likely and the air temperature rises. Warmer air can hold more water vapor so the relative humidity decreases increasing the likelihood to take up more water vapor. Evaporation leads to the absorption of latent heat at which point the process is no longer adiabatic. Sinking air causes **high surface pressure** relative to the surroundings (Fig. 12.5). An example of adiabatic heating is sinking air over deserts at subtropical latitudes or in stable air masses.

Adiabatic temperature changes are quite significant. As long as a rising or sinking air parcel remains unsaturated (i.e. the relative humidity is well below 100%) it cools and heats at the **dry adiabatic rate**: 10°C/1000 m. For example, during a Santa Ana the dry air from the desert has to pass the local 2000-m-high mountains east of San Diego. Because of the low relative humidity, condensation does not take place in the mountains. After this passage, adiabatic heating can increase the temperature by a whopping 20°C (36°F) before the air reaches the coast. Note that this applies to dry air only. The temperature in nearly saturated air rises more slowly during adiabatic heating because cooling from evaporation would offset some of the temperature increase.

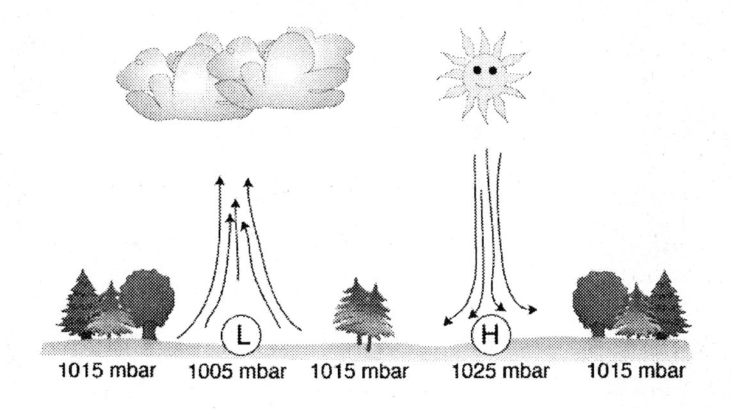

Figure 12.5 Rising air leaves low pressure behind at the surface (L). Sinking air increases to pressure on the surface (H). Low-pressure systems are typically "bad weather" areas, while high-pressure systems are typically associated with fair weather.

During sunny days in San Diego, particularly in summer, we can experience a daily cycle of local surface high and low pressure (Figs. 12.6 and 12.7). During daytime, winds typically come from the west (westerly winds). This on-shore flow is part of a loop formed by low surface pressure inland and high pressure on the ocean. This reverses at night when low pressure over the oceans causes an off-shore flow. To people living west of freeways and other main thoroughfares traffic then appears much louder.

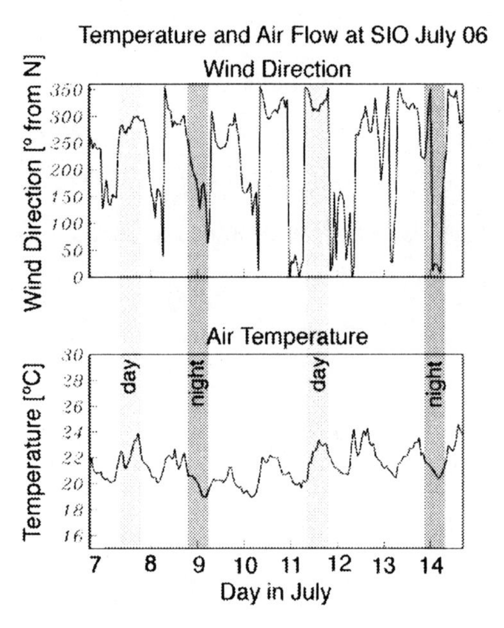

Figure 12.6 Temperature (bottom) and wind direction (top) measured at the La Jolla SIO pier during the eight days of 7-14 July 2006. Day times are characterized by rising temperatures and on-shore flow. The **wind direction** (from which the wind comes) is about 280° (westerly) and so comes from the ocean (**on-shore flow**). At night, when temperatures decline, the wind direction changes to 120° (easterly) and so winds come from land (**off-shore flow**).

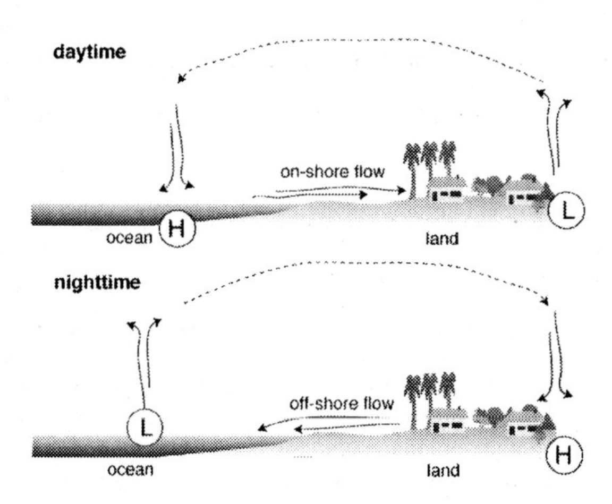

Figure 12.7 Air flow pattern across San Diego's coast on a summer day and during nighttime. **During daytime**, air above land heats much faster than above the ocean. Air on land rises, causing low surface pressure. Air from the ocean replaces the surface air to form **on-shore flow**. **At night**, air above the oceans is warmer than on land. Low surface pressure forms over the ocean. Air moving from land to replace ocean surface air forms **off-shore flow**.

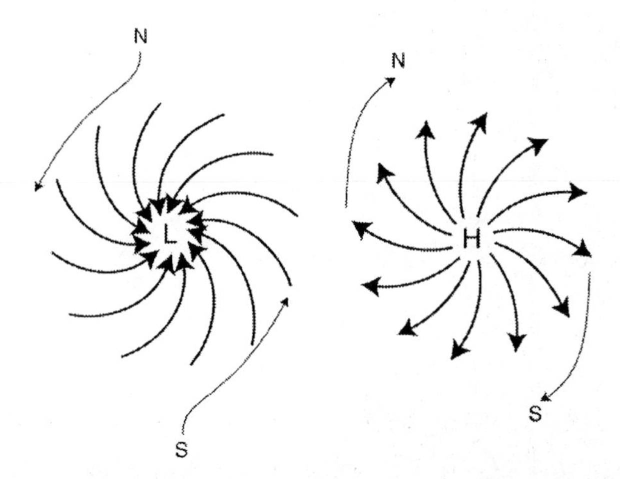

Figure 12.8 Surface air flow into a low pressure center (left) and from a high pressure center (right), in the Northern Hemisphere. The Coriolis effect causes air to flow counter-clockwise into the low pressure, while air flows out of a high pressure in a clockwise manner. In the Southern Hemisphere, the sense of rotation is reversed for both low and high-pressure systems.

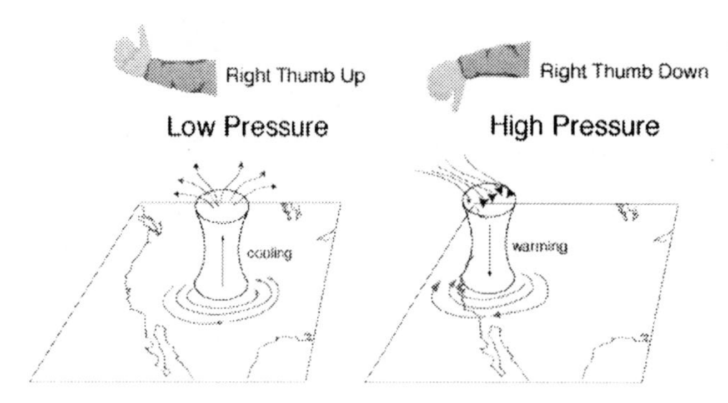

Figure 12.9 In the Northern Hemisphere, air spirals upward and counterclockwise (creating a cyclone) in a low-pressure system. Air spirals downward and clockwise (creating an anti-cyclone) in a high-pressure system. Use the **right-hand rule** to determine air flow in the Northern Heminsphere.

Table 12.2 Naming Convention of Air Flow in Low and High Pressure Systems

	Northern Hemisphere		Southern Hemisphere	
Coriolis Effect in general	to the right		to the left	
	Low (Northern)	High (Northern)	Low (Southern)	High (Southern)
Air Flow	counter-clockwise	clockwise	clockwise	counter-clockwise
Naming	cyclonic	anti-cyclonic	cyclonic	anti-cyclonic

12.3 Low- and High Pressure Systems

So far, we have concentrated on vertical air flow in high and low pressure and horizontal flow between high and low pressure. But what happens to air when it leaves a low pressure or enters a high-pressure area? If Earth did not rotate and there was no Coriolis effect, air would flow radially into a low-pressure center and radially out of a high-pressure center along lines like that resemble the spokes on a wheel. The Coriolis effect forces air to rotate around the pressure center in a cyclonic or anti-cyclonic flow. More details on the forces involved and wind directions are discussed in Chapter 13.

Cyclonic and Anti-Cyclonic Air Flow

First, we examine air flow in a **low-pressure system** which is also called **cyclonic** (Table 12.2). Adiabatically rising and cooling air in the center of a low-pressure system forces outside surface air to follow and flow toward the low-pressure center. In the Northern Hemisphere, air thereby gets deflected to the right. It pushed air in front of it toward and around the center in a **counter-clockwise** manner (Figs. 12.8 and 12.9). In the southern hemisphere, air moves toward and around the center in a clockwise manner. Sometimes, the center of a low-pressure system can be elongated to form a **trough**.

Air flow in a **high pressure system** is labeled **anti-cyclonic**. Adiabatically sinking and heating air in the center of a high-pressure system forces surface air to move away from high-pressure center. In the Northern Hemisphere, air thereby gets deflected to the right to rotate around the high pressure in a **clockwise** manner (Figs. 12.8 and 12.9). Sometimes, the center of a high-pressure system can be elongated to form a **ridge**.

In weather maps, the center of low and high-pressure systems form the centers of concentric lines of equal air pressure, the **isobars** (see Chapter 13 for an example).

It is useful to use the right hand to determine how air flows around low and high-pressure centers. This is known **right-hand rule.** One raised the right hand to form a makes a fist. The thumb sticks out and points in the direction of air flow in the center of the system: upward in the case of a low-pressure system. For a high-pressure system, the hand is rotated so that the thumb points downward. The other four fingers then indicate the rotation of winds around the system.

12.4 Global Air Circulation

As pointed out in Chapter 11, the imbalance in solar radiation leads to energy transport by circulation in both oceans and the atmosphere where excess heat is transported toward the poles.

In a **simplified single-cell circulation model**, we assume that Earth is not rotating and covered uniformly by water. We also ignore seasons and assume that the Sun is always above the equator. Air rises near the hot equator leaving a low-pressure system at the surface. At high altitude, the air no longer rises and starts to move toward the poles. Near the poles air sinks to form a high-pressure system at the surface. Air then flows back toward the equator along the surface. This type of air flow forms a single circulation cell, the **Hadley cell**, one in Northern and one in the Southern Hemisphere.

The fact that Earth rotates breaks up the single-cell air circulation to form three cells in each hemisphere (Fig. 12.10). In **the three-cell circulation model,** the Hadley cell reaches from the equator to 30° latitude, the Ferrel cell is between 30° and 60°, and the Polar cell between 60° and the pole, in each hemisphere. Near the equator, air still rises to form low surface pressure. Near the equator air can rise adiabatically to high altitudes to set the stage for thunderstorms and heavy rain to sustain lush tropical rain forests. The high surface pressure near 30° is associated with very dry conditions to cause many deserts such as the Sahara Desert in Africa. High surface pressure is also found at the poles and is responsible for extremely dry climate. In fact, with only 8 in of precipitation per year along the coast and far less inland, the high interior receives 2 in per year, Antarctica is home to the driest place on Earth. The Dry Valley has seen no precipitation in 2 million years. In the low-pressure belts surface air flow converges while in the high pressure belts surface air flow diverges.

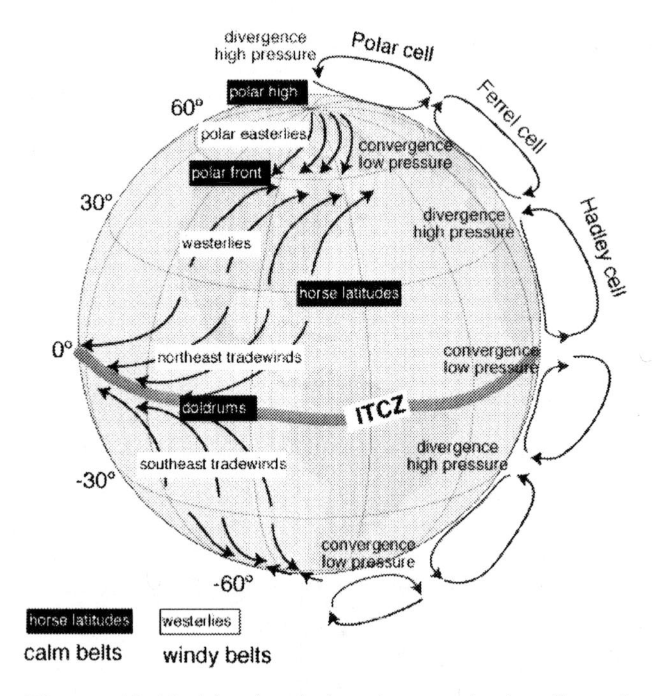

Figure 12.10 Air circulation (convection) cells and prevailing surface winds in both hemispheres. Two belts of convergent surface air flow near the equator and 60° latitude are regions of low surface pressure. The one along the equator is the inter-tropical convergence zone. Its seasonal migration controls monsoons. A divergent belt near 30° latitude and the pole are regions of high surface pressure. All four belts are calm regions with only weak surface winds.

Prevailing Surface Winds

If Earth was not rotating, winds would blow directly from high to low-pressure zones along the pressure gradient. However, the Coriolis effect deflects the winds. In general, winds can then blow around the pressure gradient, almost parallel to the isobars (lines of equal air pressure). Unlike ocean currents, winds are named after the direction they come from. An easterly wind therefore blows from east to west.

There are **three windy belts** in the three-cell circulation model (Fig. 12.10). Between 30° latitude and the equator in the Northern Hemisphere, **northeast trade winds** blow in a southwesterly direction. In the Southern Hemisphere the corresponding counterparts are the southeast trade winds that blow in a northwesterly direction. The trade winds received their name because historical trading ships used these winds to get from Europe to the Americas. Prevailing **westerlies** exist between 30° and 60°. They blow in a northeasterly direction in the Northern Hemisphere and in a southeasterly direction in the Southern Hemisphere. Finally, **polar easterlies** exist between 60° and 90°. They blow toward the southwest in the Northern Hemisphere and toward the northwest in the Southern Hemisphere.

There are also four calm belts that are found along the low and high surface pressure belts. The doldrums are found near the equator. At 30° latitude are the horse latitudes. According to some accounts, the name comes from the fact that many horses died on early sail boats crossing the Atlantic. Some blame heat exhaustion but others cite lack of water on the severely prolonged journey when caught in the "no-wind" zone. Dead or dying horses were then thrown overboard. At 60° and the pole, we find the polar front and the polar high.

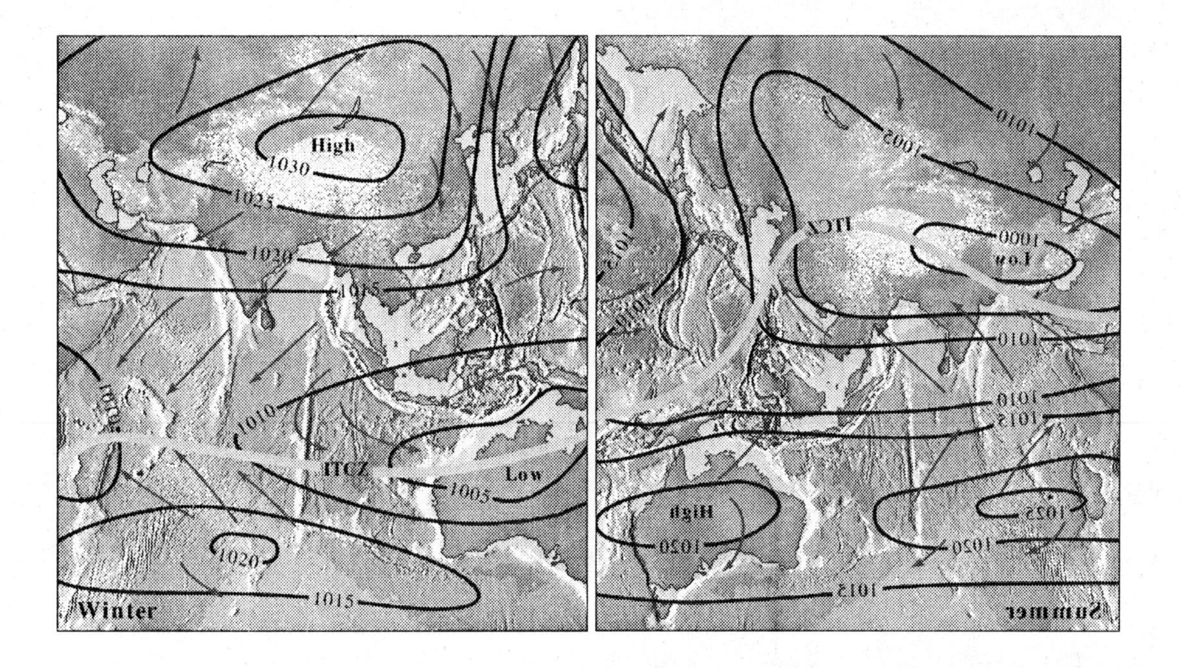

Figure 12.11a In the monsoonal climate of Asia, each year can be divided into a dry and a wet season. Left: During the dry winter, a large high-pressure cell develops over central Asia and the ITCZ lies well south of Asia. Right: During the wet summer, a low-pressure cell develops over Asia allowing warm, moisture-laden water from the Indian Ocean to flow landward. Orographic lifting along the Himalayas leads to cloud formation and intense rainfall. (source: ME)

12.5 The Intertropical Convergence Zone and Monsoons

The low-pressure zone near the equator is also called the inter-tropical convergence zone (ITCZ). Its geographical position changes throughout the seasons. The northward or southward shift of the ITCZ controls the start and end of rainy and dry seasons in the subtropical latitudes. **Monsoons** are seasonal rainy periods that are linked to major reversals in wind directions as a results of the shift of the ITCZ (Fig. 12.11). The arrival of monsoons is vital to agriculture as monsoons often bring the majority of the annual rainfall in certain areas. Monsoons exist throughout the world but the perhaps best-known monsoon is the **South Asian Monsoon** for which the term was first used in India. In southern Asia the monsoonal wind reversal causes a shift from a very dry to a very wet season. In winter, a stable high-pressure cell forms over central Asia due to prevalent cold, dry air. The ITCZ gets pushed out south into the Indian Ocean. As a result of this the Indian subcontinent remains dry. In summer, central Asia heats up dramatically thereby creating a pronounced low-pressure area. As a result, the ITCZ moves north, thereby allowing moisture-rich marine air to follow and bring sometimes heavy and prolonged rain to India. Traditional scientific thinking has been that the South Asian monsoon was driven by the strong differences in heating between continents and ocean, in particular the rapid heating of the high plateau in Central Asia. There is geological evidence that the Asian monsoons became stronger over millions of years with the uplift of the Tibetan Plateau. But modern global circulation modeling reveals that the monsoon would also exist if Earth were covered completely by water. Hence, the monsoon can be driven by seasonal changes in surface heating alone. Details on how monsoons really work can be quite complex and surprising. For example, recent research indicates that the mountains along East Africa play an important role at steering the South Asian monsoon.

The ITCZ does not always shift in the same way so that in some years, monsoons are weakened dramatically or they do not arrive at all, resulting in severe crop failure and even extended famine. Some of the most catastrophic droughts and related famines occur in the **sub-Sahara Sahel**, such as the drought in the 1960s to early 1980s that affected most of the Sahel's 50 million people. Apart from the fact that the increased influx of people into the Sahel has exacerbated the problem and clearly adds a human impact component to it, there is mounting scientific evidence that the shift in the African Monsoon is influenced by human activity through local deforestation and the local and global burning of fossil fuels. This will be revisited in Chapter 19.

On the other hand, monsoons can also cause catastrophic flooding that through the increase in human globalization can affect people worldwide[4,5]. A recent example is the **flooding in Thailand** in summer and fall of 2011 that lingered into 2012. Epic monsoon rains and typhoons starting in late July battered a vast swath of Asia, killing hundreds of people from the Philippines to India. Thailand was among the hardest hit where the worst floods in half a century swamped more than two thirds of the country. Bangkok, Thailand's capital was affected when the Chao Phraya River overflowed its banks and massive efforts concentrated on protecting the capital's inner district. 20,000 km^2 (7700 sq mi) of farmland was submerged and damaged, and more than 1000 factories shut down, including those of Toyota Motor Corp. and Western Digital Corp., a major manufacturer of computer hard drives. Thailand accounts for up to 45% of worldwide hard drive production and by November, the flooding caused serious hard drive shortages and price spikes that could linger well into 2012. After the devastating 11 March 2011 Tohoku, Japan earthquake, the flooding was a repeat reminder of the vulnerability of global supply chains to natural disasters. Thailand became a hub for Japanese car manufacturers in the 1980s and 1990s but the flooding in Thailand forced Toyota and Honda to slow production in factories worldwide from the Philippines over Pakistan to North America. As of February 2012, the flooding resulted in 815 deaths and directly affected 13.6 million people. As of December 2011, the World Bank has estimated nearly $46 billion in economic damages and losses due to the flooding, which brings the ranking of this event as the fourth-costliest natural disasters after the Tohoku, Japan earthquake (> $200 billion), the 1995 Kobe, Japan earthquake (~ $100 billion) and 2005 Hurricane Katrina (> $80 billion), though numbers on this may vary by source.

A year earlier, a perhaps less-well publicized monsoon **flooding in Pakistan** caused economic damages of similar magnitude ($43 billion). Flooding of the Indus River basin on the order of nearly 800 km^2 (310 sq mi) caused one-fifth of Pakistan to be under water, killing 2000 and affecting 20 million people. The flooding that started in late July was caused by the worst monsoon rains in 80 years. A record-breaking 274 mm (10.8 in, approximately the amount of San Diego's annual rainfall) of rain fell in Peshawar during a 24-h period. The previous record was logged in April 2009 at 187 mm (7.4 in). By mid-August, within only few weeks, the death toll stood at over 1500. Nearly 560,000 homes had been destroyed and over 6 million people displaced. By mid-September, when the floods started to recede, more than 1.9 million homes were destroyed. International disaster relief was requested and initial pledges were prompt. The US and EU each pledged on the order of $500 million for humanitarian support and to repair infrastructure and the contributions from a large number of countries and private entities brought the total support close the $1.8 billion. But the government was criticized for sluggish and disorganized relief efforts.

Figure 12.11b Thunderstorms during the North American Monsoon as seen from El Cajon in San Diego County, CA. The monsoon pushes up against the Peninsular Ranges but only rarely manages to push over the mountains to the coastal strip. During the monsoon season, southern Californians have a spectacular view of the thunderstorms only a half-hour's drive away. (source: wikipedia)

There is also a **North American monsoon**[6]. It occurs from late June into September and originates in Mexico. It usually spreads into the southwestern United States by mid-July, stretching from West Texas into California and as far north as Nevada. Sometimes, it pushes as far west as the Peninsular Ranges of Southern California and is responsible for severe, hail-producing thunderstorms in the Mojave and Sonora deserts and in the mountains. Sometimes the humid monsoon air even reaches the Pacific coast. As it lingers and the influence of the cooling ocean breeze is reduced, the monsoon brings unusually ''muggy'' days to San Diego, and clouds in the skies. For San Diego country weather, this means that low-pressure systems approach from the south unlike the winter storms that push in from the north, guided by the jet stream. Rainfalls or periods of moisture-rich air during the monsoon are not continuous but occur during burst periods. In fact, monsoonal episodes in San Diego rarely last longer than a few days but can recur several times over the summer.

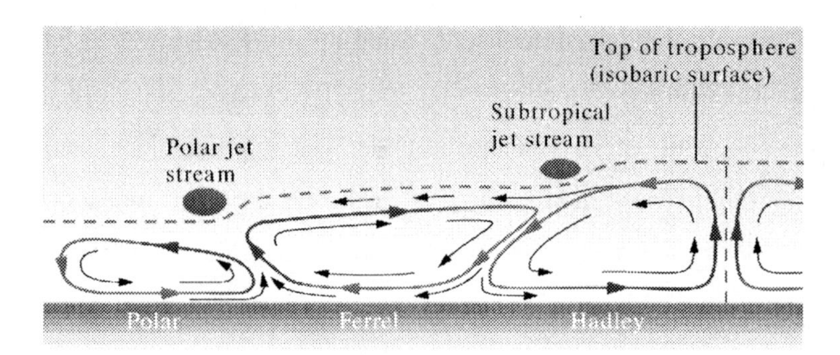

Figure 12.12 In the equator-to-pole cross section of the atmosphere, we see that the troposphere varies in thickness. The atmosphere above the troposphere, however, is essentially uniform in thickness from the equator to the pole. The tropopause is therefore an isobaric surface. At a given altitude within the troposphere, there is a pressure gradient from the equator to the pole, causing a northward flow of air at the top of the troposphere. Because of the Coriolis effect, high-altitude westerly winds develop in the Northern Hemisphere. At places where the pressure gradient is particularly steep (at the boundary between convection cells), the air moves very fast, creating the jet stream. (source: ME)

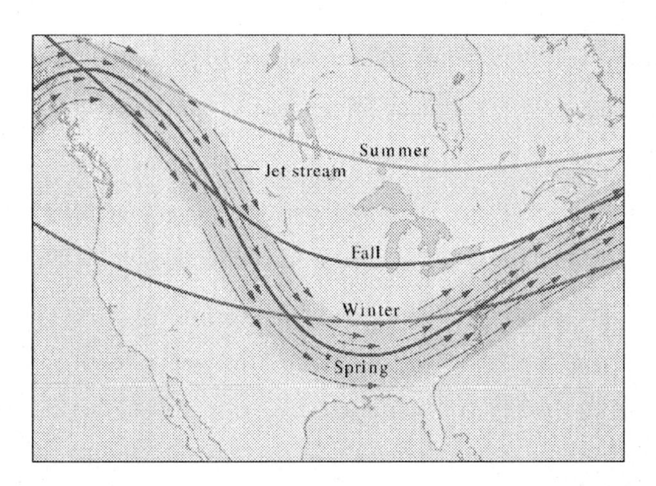

Figure 12.13a The polar jet stream meanders with time, and changes its overall position with the season. It tends to reach farther south in the winter than in the summer (source: ME). Jet stream maps for North America can be downloaded from the archive at San Francisco State University [9].

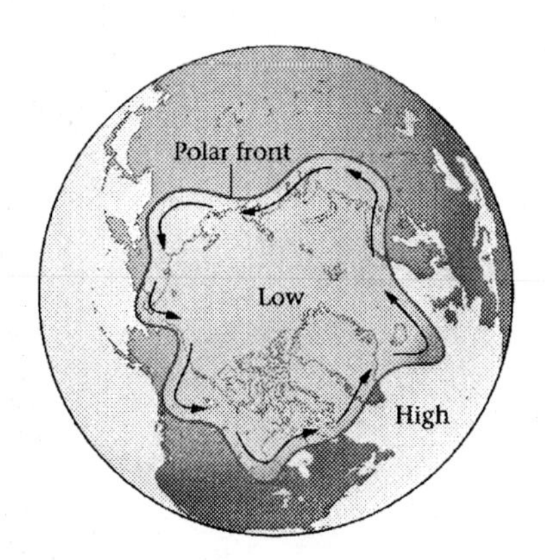

Figure 12.13b Global of the summer-time northern polar jet stream viewed from above the North Pole (source: ME).

12.6 Jet Streams

Jet streams, though a global feature, control local weather in profound ways. It is largely up to the jet stream whether San Diego has a wet or a dry winter season, and when and where the midcontinent experiences severe arctic chills. The polar jet stream pushes large storm systems across a continent and even contributes to the huge tornado risk in the U.S. (this case will be discussed in Chapter 14). Not lastly, jet streams also have a "hand" in whether or not airplane travel is on time.

Jet streams are **high-altitude winds** near the top of the troposphere. They essentially occur because the 3 basic convection cells do not have the same shape (Fig. 12.12) and the height of the cells decreases for those closer to the poles. In the figure, the isobar tracing the top of the convection cells is at higher altitude near the equator than at the pole. This means that, at the same altitude, air pressure is lower at the pole than at the equator. Winds near the top of the troposphere start to move along the pressure gradient northward away from the equator toward pole (southward in the Southern Hemisphere). But due to the Coriolis effect, these winds become westerlies. In each hemisphere, the boundaries between the three convection cells are special places where the pressure gradient is particularly steep and winds are therefore particularly strong. These winds are the **polar front jet stream** (between the Ferrel and the Polar cell; also called the polar jet stream) and **the subtropical jet stream** (between the Hadley and the Ferrel cell). Jet streams have extremely high wind speeds, between 200 and 400 km/h (125 and 250 mph). Such high wind speeds are found near the surface only in storms with hurricane-force winds and tornadoes. Airline pilots flying from west to east try to "ride the jet stream" because the tailwind pushes planes to arrive earlier and saves fuel. Accordingly, flights from east to west are longer either because of headwinds or because the plane has to fly along a longer route to avoid the jet stream.

Jet streams meander (change their path) throughout the year (Fig. 12.13). In summer, the polar front jet stream is far north. But in winter and spring, it typically reaches far south. It then also undulates in large loops around high and low pressure systems allowing for very cold Arctic air to move deep into the south. Most of the short but heavy rain that Southern California receives from **winter storms** comes from storm systems that earlier developed in the North Pacific and were then pushed by the polar jet stream south along the North American west coast. If the jet stream does not reach south enough, Southern California remains dry even though the storms dump large amounts of rain further north. A special, hazardous case of the alignment of the jet streams and winter storms is the **Pineapple express** (see case study 2). A southern branch of the polar jet stream or a far north reaching subtropical jet stream provides moisture-laden tropical air to fuel severe rain storms that cause flooding and landslides along the North American west coast.

12.7 Clouds and Precipitation

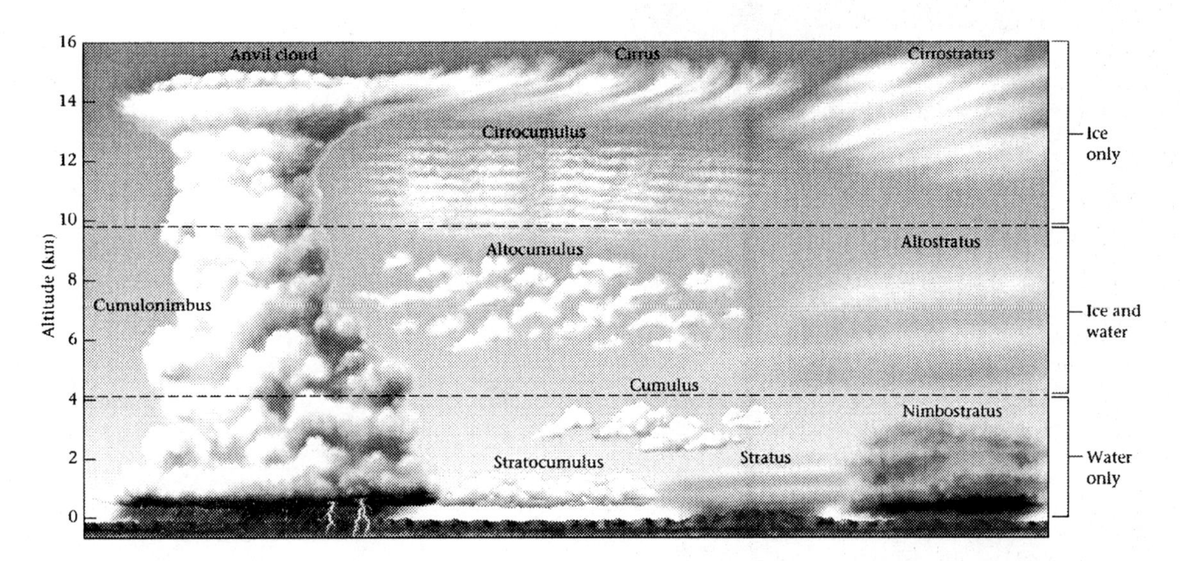

Figure 12.14 The type of cloud that forms in the sky depends on the stability of the air, the altitude at which moisture condenses, and the wind speed. The temperature determines if condensed H_2O exists in the form of water (potentially producing rain) or ice (potentially producing hail). In the lower troposphere, up to 4 km altitude, warm temperatures cause the existence of only water. Between 4 and 10 km altitude, both water and ice exist. At altitudes above 10 km, only ice exists. (source: ME).

Figure 12.15a A tropical cumulus cloud near Mazatlan, Mexico. (source: UCAR)

Figure 12.15b A cumulonimbus cloud from a severe June 2004 thunderstorm near Denver, CO. (source: UCAR)

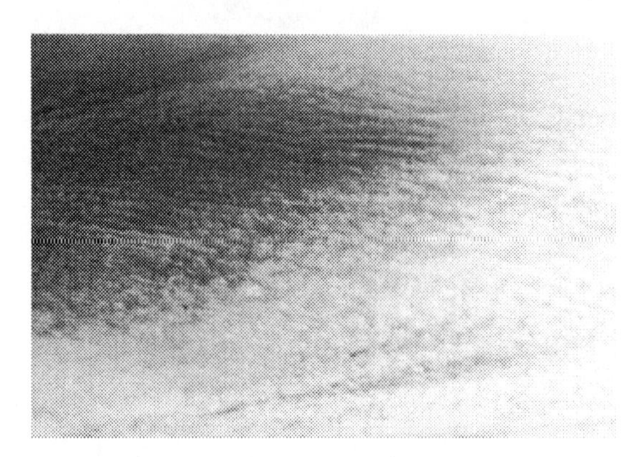

Figure 12.15c Cirrocumulus clouds are usually short-lived. (source: UCAR)

Figure 12.15d Cirrus clouds consist entirely of ice crystals. These clouds are frequently blown about into feathery strands called mares' tails. (source: UCAR)

Figure 12.15e Cirrostratus clouds form at slightly lower altitude than Cirrus clouds. They are very thin white sheets composed of ice crystals. (source: UCAR)

Figure 12.15f A supercell thunderstorm Cumulonimbus Anvil. This cloud has grown into a severe thunderstorm that could produce destructive hail and spawn tornadoes. (source: UCAR)

Figure 12.15g Mammatus clouds that are often found near severe thunderstorms. (source: UCAR)

Clouds

As rising moist air cools adiabatically, the water vapor contained in the air eventually condenses to tiny droplets of water to form clouds. The droplets are typically 20 μm in diameter, less that 1/3 of the diameter of a human hair. **Fog** forms when cloud formation occurs near the surface. In regard to Earth's climate, clouds actually have a dual function. Firstly, clouds reflect incoming sunlight hence cool Earth's surface during the day. But they also trap heat coming from the Earth's surface, hence keeping it warm at night. On the other hand, the water in the clouds is a greenhouse gas and so traps IR radiation, thereby keeping the atmosphere warm.

Several mechanisms can force air to lift and eventually condense and form clouds. On land masses surrounded by large bodies of water, such as islands or Florida, **convective lifting** pushes the air upward. **Frontal lifting** occurs when air masses collide and cold air pushes beneath warm air thereby forcing the warm air to rise (see next chapter). **Convergent lifting** occurs near the equator where air has nowhere else to go but up. Covergent lifting is also relevant in Florida, where the sea breezes from the Atlantic Ocean and the Gulf of Mexico collide. Finally, **orographic lifting** in mountainous areas also forces air to move upward (Fig. 12.4c).

Clouds are named by their shape and the altitude they occur at (Figs. 12.14, 12.15). **Cumulus clouds** are puffy and are often likened to cauliflower. As the name suggests, **stratus clouds** (stratus is Latin for layer) are thin, sheet-like clouds. And **cirrus clouds** are wispy, featherlike clouds that are sometimes curly. Cirrus clouds typically occur at high altitudes. Classifying clouds by altitude, the **prefix cirro** indicates high clouds (altitude above 7 km), e.g. cirrocumulus and cirrostratus, and the prefix alto indicate mid-altitude clouds (2 to 7 km), e.g. altostratus and altocumulus. Low clouds do not have a prefix. Finally, the name **nimbus** or the prefix **nimbo** is attached when the cloud produces precipitation (e.g. cumulonimbus and nimbostratus).

When thunderstorms develop, cumulonimbus clouds can transform into cloud-monsters. Fueled by immense amounts of latent heat, strong updrafts exist and the developing cloud can reach into the stratosphere. At that point, the cloud starts to spread sideways to form an **anvil cloud** near the top (Fig. 12.15f). Severe thunderstorms can develop very large anvils. When the updrafts start to rotate, the thunderstorms become **supercell thunderstorms**. **Mammatus clouds** (Fig. 12.15g) (from the Latin word for breast, "mamma") look like smooth, rounded puffs hanging from the underside of anvil clouds from a severe thunderstorm that may even spawn tornadoes. Mammatus clouds are formed by downdrafts in which the air is cooler than the surrounding air. Though occurring with severe weather clouds, Mammatus clouds themselves do not produce severe weather.

Figure 12.16a Freezing rain coating the branches of a tree. (source: David Wert, NOAA)

Figure 12.16b Ice-covered car in Versoix near Geneva, Switzerland, after the epic January 2005 ice storm. Very cold conditions (-12 to -8ºC, 10ºF) combined with winds over 100 km/h (62 mph) resulted in a spray formed by water from Lake Geneva that froze on everything it touched. A similar freeze occurred in late January in 2012. (source: photosfan.com)

Figure 12.16c Sleet, also referred to as ice pellets by the U.S. National Weather Service. (source: NOAA)

Figure 12.16d Snowfall on Christmas Eve 2010 in the author's hometown Weil am Rhein in Southern Germany, near Basel, Switzerland. A "White Christmas" in the sheltered Upper Rhine Graben is quite rare but the winter of 2010 brought early and frequent snowfall. A few days earlier, Winter Storm Petra dumped ice and snow on much of Europe. It caused the closure of major airports (London, U.K.; Amsterdam, Netherlands; Frankfurt, Germany) and brought rail and road traffic to a standstill in the midst of one of the busiest travel periods of the year. Ten days earlier, a similar winter storm had already closed the Eiffel tower and Charles de Gaulle airport in Paris, France.

Figure 12.16e Graupel forms when supercooled water droplets condense on snowflakes. (source: sitkanature.org)

Figure 12.16f This hailstone fell on 14 May 2004 in Harper, Kansas. With a diameter of nearly 13.5 cm (5.25 inches), it is one of the largest hailstones found in the U.S.. (source: NOAA)

Precipitation

In clouds and fog, water vapor condenses on a pre-existing liquid or on solid aerosols that serve as **condensation nucleus**. The tiny droplets are buoyant enough to stay suspended in the cloud. The circulating air around the droplet thereby provides lift, which counteracts gravity. In warm clouds, the droplets grow in size by collision and coalescence, i.e. the droplet bump into each other and stick together to form a larger drop. This process continues until the drop is too large and heavy to stay suspended. It starts falling as **rain**. A typical raindrop is 2 mm across and falls at a speed of 20 km/h (12 mph). Some raindrops can grow to be 5 mm across but they tend to be broken up into smaller drops during their fall. If the droplets in the falling rain are smaller than 0.5 mm in diameter, the rain falls as **drizzle**. Most drizzle falls from stratocumulus clouds. It may also form near the surface after rain fell through unsaturated air aloft and partially evaporated on its way down. **Virga** (from Latin "virga" for rod or branch) is rain that never hits the ground. It is visible as streaks or "rain streamers" emerging below clouds. Virgae are the result of rapid evaporation and are a common sight in San Diego skies.

If rain falls through colder air near the surface, it freezes to **sleet**. The rain then falls as small ice pellets. **Freezing rain** (Fig. 12.16a) forms when the below-freezing air is just along the surface. Freezing rain then covers everything in continuous layers. A curious but spectacular case of "freezing rain" that actually did not come from rain but from spray from a storm in 2005 on Lake Geneva, Switzerland is shown in Fig. 12.15b. Photos from this storm circulate widely on the internet, sometime falsely associating them with other locations.

Snow (Fig.12.16d) can fall when cold clouds contain a mix of very cold droplets and tiny ice crystals. The water droplets evaporate faster than ice because water molecules are less tightly bound to water droplets than to ice. The moisture then condenses around the ice crystals, leading to the growth of hexagonal snowflakes. If the air below the snow-producing cloud is very cold, the snow falls in powder-like flakes. If the air is close to melting T (0°C/32°F), then large, wet clumps of snowflakes fall. And if the air is above 0°C, then snow transforms to rain. **Graupel,** or **soft hail,** falls when supercooled water droplets condense on snowflakes. Graupel has a diameter of 2 to 5 mm. Graupel commonly forms in high altitude climates and is both denser and more granular than ordinary snow.

Hail (Fig. 12.16f) develops in clouds with strong updrafts (e.g. thunderstorms). Ice crystallizes to ice balls at high levels where the temperature is below freezing. Hail falls over a few minutes, in a typically 2-10 km long streak in the direction a storm moves. The diameter of hailstones typically ranges from 5 mm (0.2 in) to 15 cm (5.9 in). Most hailstones are pea-sized though the largest recorded hailstone in the U.S. reached a diameter of 20 cm (8 inches) and weighed 880 g (1.9 lb). The hailstone fell in Vivian, South Dakota on 23 July 2010. Large hailstones are usually produced in severe thunderstorms. In the U.S., most hail-producing storms occur in Tornado Alley from Texas through Oklahoma into Kansas and north (see Chapter 14).

References and Websites

1. Ahrens, C.D., 2003. Meteorology Today, 7th edition. Brooks/Cole-Thomson Learning, Pacific Grove, CA, pp. 544.
2. Marshak, S., 2005. Earth: Portrait of a Planet. 2nd edition. Norton & Company, New York, pp. 748.
3. Abbott, P.L., 2009. Natural Disasters. 7th edition. McGraw-Hill, New York, pp. 504.
4. New York Times News Service and Associated Press, 12 November 2011. Thai floods hit global supply chains. San Diego Union Tribune.
5. Wikipedia page on the Thailand flood: http://en.wikipedia.org/wiki/2011_Thailand_floods
6. Wikipedia page on the North American Monsoon: http://en.wikipedia.org/wiki/North_American_Monsoon
7. Wikipedia page on forage fish: http://en.wikipedia.org/wiki/Forage_fish
8. Orlove, B.S., Chiang, C.H. and Cane, M.A., 2000. Forecasting Andean rainfall and crop yield from the influence of El Niño on Pleiades visibility. Nature, 403, 68-71.
9. Archive of jet stream maps for North America at San Francisco State University at http://virga.sfsu.edu/crws/jetstream_fcsts.html

10. San Diego May Grey/June Gloom: meteora.ucsd.edu/cap/gloom.html

Government Sites:

National Weather Service (NWS): nws.noaa.gov

NOAAs State of the Global Climate Analysis (since 1997): http://www.ncdc.noaa.gov/sotc/global/2011/13

NWS Climate Prediction Center: spc.ncep.noaa.gov/products/

NWS Storm Prediction Center: spc.ncep.noaa.gov/products/

National Hurricane Center: www.nhc.noaa.gov

San Diego weather at NWS: wrh.noaa.gov/sgx/

Commercial Sites:

Accu Weather: accuweather.com

The Weather Channel: weather.com

Other Recommended Reading

wikibook on air movement:

http://en.wikibooks.org/wiki/High_School_Earth_Science/Air_Movement

Chapter 13 Severe Weather: From Heat Waves to Great Storms

To establish the context for extreme single weather events, it is useful to reflect the usual ranges and extremes in some of the parameters that describe daily weather. The most obvious one is air temperature, but this chapter also explores air pressure and wind speeds. Scientists suggest that global warming will increase the occurrence of extreme weather as more latent heat is stored in the atmosphere that can drive storms. Heat waves and episodes of extreme cold are often associated with stable, long-lasting high-pressure systems and so are usually not the first thing that comes to mind when thinking about severe weather. It is feared that the frequency of heat wave and droughts, and also cold episodes will increase on a warming Earth. In the second part of this chapter, we will move on to the more classical storms and how they form. Again, the garden-variety storm may lose out to hurricanes and tornadoes when it comes to natural disasters. But most people will never experience a tornado or hurricane. Instead, the perils of storms are very real to a great many people, and some storms are killer-storms. Tropical cyclones (such as hurricanes) are the most powerful storms on Earth and will be discussed in a dedicated chapter (Chapter 15). But even some of the most powerful extratropical storms can pack wind speeds comparable to those of hurricanes. The formation of storms is ultimately caused by the movement and collision of air masses that have difference characteristics.

Table 13.1 Temperatures in our Solar System

Planet	**daytime**	**nighttime**
Mercury	427°C	-173°C
Venus	480°C	
Mars	20°C	-140°C

COPYRIGHTED MATERIAL — DO NOT DUPLICATE, DISTRIBUTE, OR POST

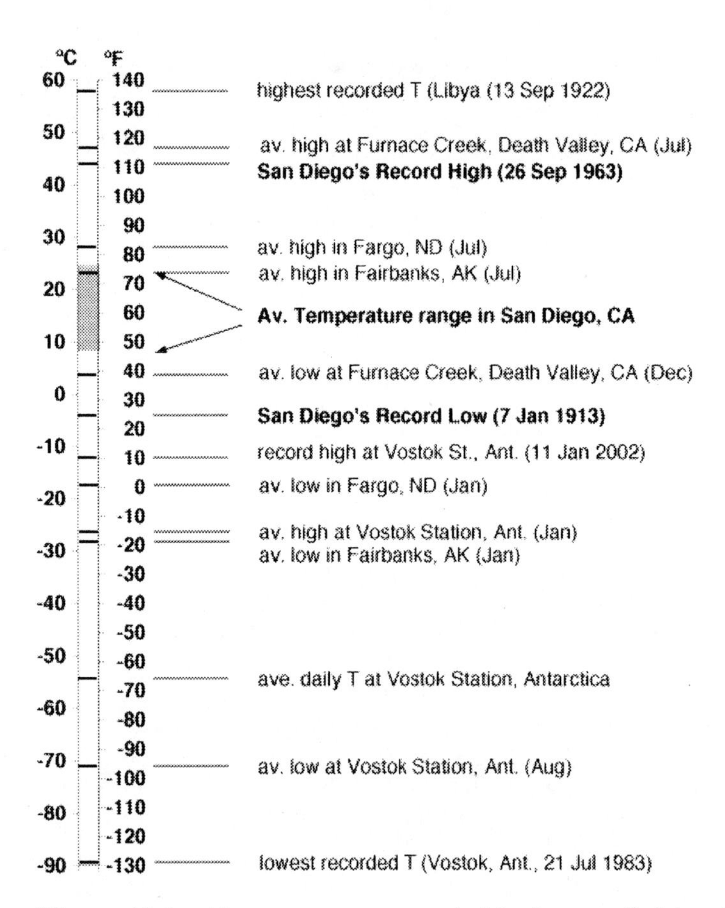

Figure 13.1a Air temperature recorded in degrees Celsius (°C) and degrees Fahrenheit (°F), at some locations in the U.S. and the world. For comparison, the temperature at the top of the troposphere is about -60°C/-76°F. The conversion between temperatures in °C (T_C) and in °F (T_F) is $T_F = T_C*9/5 + 32$ and $T_C = (T_F - 32)*5/9$.

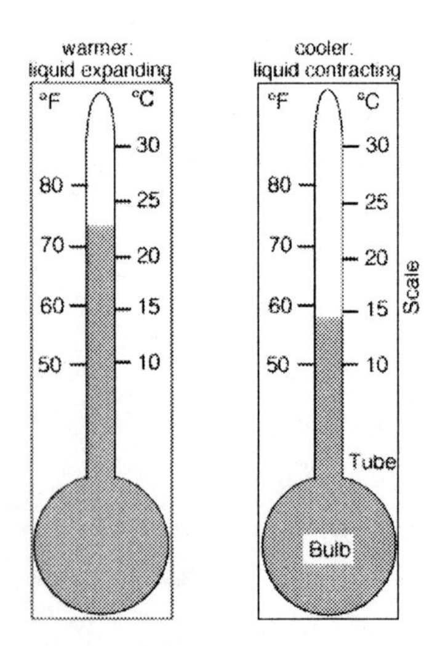

Figure 13.1b The concept of a liquid-in-glass thermometer. As the liquid heats, it expands into the tube. As the liquid cools, it contracts back into the bulb. In 1724, Daniel Gabriel Fahrenheit proposed

the Fahrenheit temperature scale that has 180 points between the freezing and boiling points of water. In 1742, Anders Celsius introduced his scale that has only 100 points. Degrees Celsius are also known as degrees centigrade. Nowadays, this scale is used in most countries though Fahrenheit remains the official scale of the U.S., Cayman Islands and Belize.

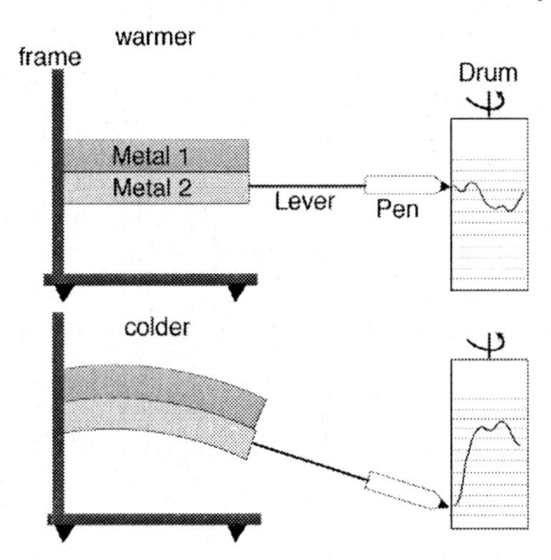

Figure 13.1c The concept of a bimetallic thermometer. Two metal strips with different thermal expansion coefficients are glued together, with one end fixed to a frame. When the temperature changes, the combined strip bends. A long lever at the bending end enhances this movement, and a pen attached to the end records the temperature onto a rotating paper drum. The concept here is exaggerated. In a real thermometer, the metal strips are very thin.

13.1 Temperature, Pressure and Wind Speed

Air Temperature

As established in Chapter 11, Earth's surface enjoys a balmy average temperature of 16°C (61°F). However, temperatures can fluctuate immensely (Fig. 13.1a and Box 1). The highest temperature ever recorded was a most oppressive 57.8°C (136°F) in Al Aziziyah, Libya. With the highest recorded temperature in the western hemisphere, Death Valley, CA was not far below this. In hot deserts such as Death Valley and the Libyan Desert temperatures in the summer usually climb above 38°C (100°F) from later spring through early fall. At the other extreme, the lowest temperature ever recorded was -89.2°C (-128.6°F) at Vostok Station, Antarctica. The average daily temperature at Vostok is -55.2°C (-67.4°F). Vostok is located at an elevation of 3488 m (11,444 ft) which also makes the air very thin. Consider now that the freezing point of typical jet fuel is -45°C (-49°C) though some kerosenes reach freezing points of -60°C (-76°F) which is just barely below the average daily temperature at Vostok.

Planes can therefore reach Vostok only during the Antarctic summer. Despite these temperature extremes, Earth still has much less extreme temperatures that our neighbor planets in the solar system (Table 13.1). On Mercury, the range in temperatures is much larger. At 480ºC (896ºF), the high temperature on Venus causes the rocks to glow red. Mars has a balmy 20ºC (68ºF) as high temperature but the temperature low is much lower than that on Earth.

It is important to place average temperatures and temperature extremes into the proper context. A temperature of 38ºC (100ºC) Fahrenheit may seem normal as summer temperature in the Egyptian desert. But if the thermometer climbs to this point in London, U.K., we would deal with an oppressive, deadly heat wave, as occurred in July and August 2003. As we saw in Chapter 11, the important point here is that the Egyptian Desert and London are located in difference climate zones. Temperature ranges also vary with different locations. In San Diego, CA, average monthly temperature highs and lows remain within 15ºC (27ºF). Death Valley, CA has a much larger range of over 40ºC (72ºF). The missing water vapor in the much drier desert air is responsible for the lack of temperature moderation.

Temperature (T) characteristics at a location are expressed by average daily T, average daily low T, average daily high T. Monthly averages, highs and lows characterize seasonal changes (see Box 1). Typically, averages are determined over the most recent 30 years. This is important to realize, particularly in a warming world, when warmer averages become the "new normal". However, it is also important to remember that some natural climate cycles are on the order of decades and so a recent warming (or cooling) over the last 5-10 years by itself does not indicate a long-term warming (or cooling) trend. Single extreme temperature swings are recorded as record highs or record lows. They usually do not affect temperature averages.

Air temperature (in the context of weather observations on Earth) is measured with a thermometer. The simplest thermometer involves a liquid that expands and contracts with changing temperature. In the classical liquid-in-glass thermometer, a larger bulb that holds most of the liquid is connected with a very thin vertical tube (Fig. 13.1b). When the liquid heats it wants to expand but there is no place to go in the bulb. It therefore pushes into the tube. Since the cross section of the tube is very small the liquid rises high, thereby allowing a precise reading on a scale. In principle, one could use dyed water as a liquid. But such a thermometer would be useful only in the temperature range between 4ºC (39.2ºF), the point when water is densest, and 100ºC (212ºF), the boiling point of water. This is quite a bit higher than the highest temperatures measured in the desert, so a water thermometer could theoretically be used on hot days in the desert. But water cannot be used below 4ºC because it expands again. It also freezes at 0ºC (32ºF), and freezing would likely destroy the thermometer.

In 1654, Ferdinando II de Medici, Grand Duke of Tuscany used alcohol in the first modern-style thermometer. The low boiling point of alcohol limits the use of such thermometers to temperatures below 78°C (172°F), which is still ways about temperatures we find in hot deserts. In 1724, Daniel Gabriel Fahrenheit used mercury as liquid. Mercury expands more easily when warming (higher thermal expansion coefficient), so it allows more precise measurements. The boiling point of mercury is 356.7°C (674.1°F), which is even higher than that of water. But mercury has a melting point of -38.8°C (-37.9°F), and so is not suitable for use in very cold climates. Ethanol-filled thermometers can be used for temperatures as low as -70°C (-94°F) and are therefore the preferred thermometer for meteorological measurements. This is still far above the average low temperature measured in winter at Vostok Station. It takes a special type of thermometer to measure air temperature in such extreme environments. At the German Neumayer Research Station in Antarctica, scientists use a platinum resistance thermometer[(5)]. It basically consists of a platinum wire through which a small electrical current flows to measure electrical resistance. The electrical resistance of platinum depends on temperature in a very defined and repeatable way, and so recording it can be used to measure temperature very accurately, to temperatures as low as -272.5°C (-458.5°F), which is nearly absolute zero (-273.15°C). Other types of thermometers exist and the interested reader is referred to Wikipedia. One other important type is the bimetallic thermometer (Fig. 13.1c) because it readily and autonomously records temperature readings. The principle works on the fact that two metal strips that are glued together have different thermal expansion coefficients and behave differently when the temperature changes. A lever-pen system records the temperature onto a rotating paper drum. Nowadays, the drum-recording system is replaced by electronic dataloggers.

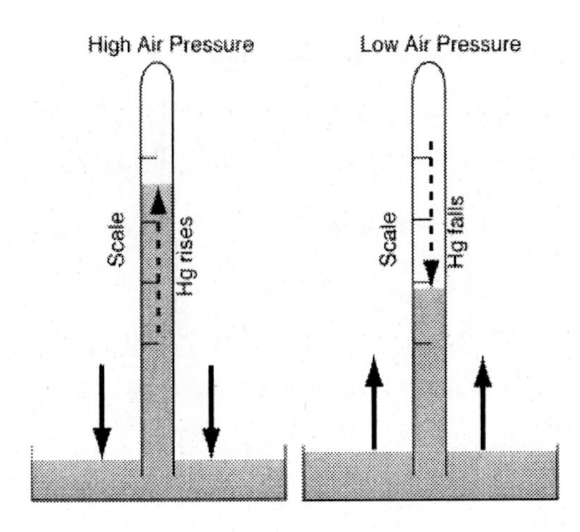

Figure 13.2 The barometer is an upside-down tube filled with mercury (Hg) with the top end closed. The lower end sits in a larger dish of Hg. The Hg in the tube rises and falls, depending on air

pressure. When the air pressure is rising, the air presses increasingly on the Hg in the dish. As a consequence, the Hg in the tube is rising. When air pressure is falling, it releases pressure on the Hg in the dish. As a result, the Hg in the tube is falling.

Air Pressure

Air pressure is measured with a barometer (Fig. 13.2). In principle, this works with any device that has an inverted fluid, with a large dish as a base and a narrow feeder tube. A high-quality barometer, of course, measures pressure most accurately. Since air pressure declines exponentially with altitude, a barometer has to be calibrated when it is used at a location that is above sea level.

To get a feel for how important this is, consider the range of air pressure that is caused by changes in weather (Fig. 13.3). The lowest sea level pressure ever recorded was during 1979 Typhoon Tip in the Western Pacific Ocean (870 mbar; see Box 1), while the highest pressure was recorded in 2001 in Tosontsengel, Zavkhan Province, Mongolia (1085.6 mbar). This 216-mbar range in air pressure is the same as one would experience when traveling from sea level to an elevation of 2160 m. This is nearly the air pressure equivalent to which the cabin pressure is set on a typical airplane (it is actually 2100 m). To get a feel for this pressure difference, consider being in a plane at cruising altitude, or in the mountains at this altitude. You just emptied your water bottle and put the cap back on. By the time you get down to sea level, this empty bottle is crushed because the ambient pressure (outside the bottle) presses on the fewer trapped air molecules that resembles the atmosphere at 2100 m. Weather changes occur relatively slowly so we usually do not notice the related changes in pressure. But our ears would pop if this happened more rapidly.

It turns out that the Tosontsengel weather station is at an elevation of 1723 m[(6)]. The pressure actually experienced during the record-breaking December day in 2001 was much lower than the 1085.6 mbar reported for sea level. To be able to compare air pressure at different weather stations around the globe, meteorologists "correct" the actual air pressure for elevation. The relationship to recalibrate the exponentially changing atmospheric air pressure with altitude is mathematically simple but not easy to remember. However, we can apply a trick to make this easier (Fig. 13.3). In the first 3500 m above sea level, the exponential change in air pressure closely resembles a linear change with slope 100 mbar/km, or 100 mbar/1000 m (dashed line). This is a number easy to remember. Using this slope, we have to subtract 172.3 mbar from the reported sea level pressure to obtain the actual air pressure at the Tosontsengel station. This gives an actual pressure of 913.3 mbar. We can also determine this graphically by moving the dashed line in Fig. 13.3 until it intersects the sea level pressure at 1085.6 mbar. Then we read the corresponding pressure along this line for an elevation of 1723 m.

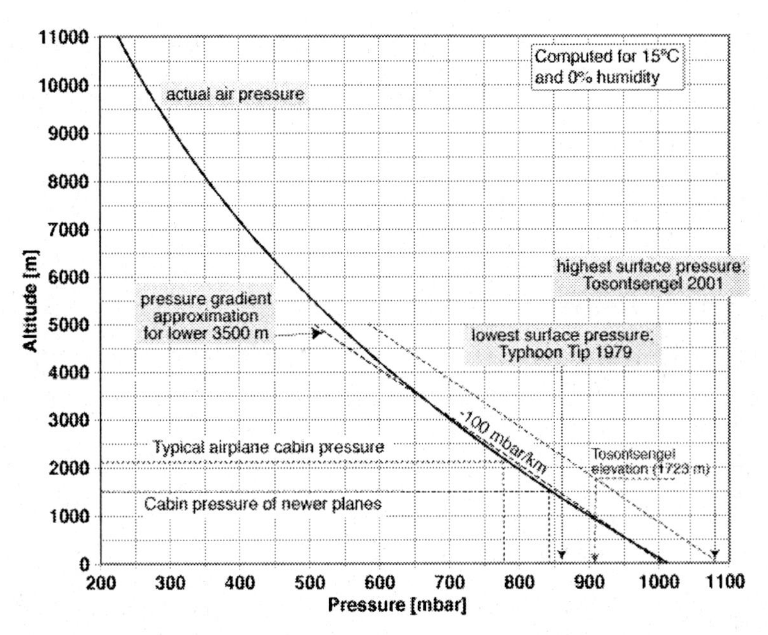

Figure 13.3 Air pressure recorded in a barometer as mbar and as mercury height (in inches) during various weather events. The surface pressure during Typhoon Tip (1979) was equivalent to the air pressure at about 1400 m altitude. In the lower 3500 m, the exponential decline of pressure with altitude can be approximated by a straight line with slope -100 mbar/km, where the "-"-sign indicates a decline in pressure as the altitude increases (see also Ahrens, 2003).

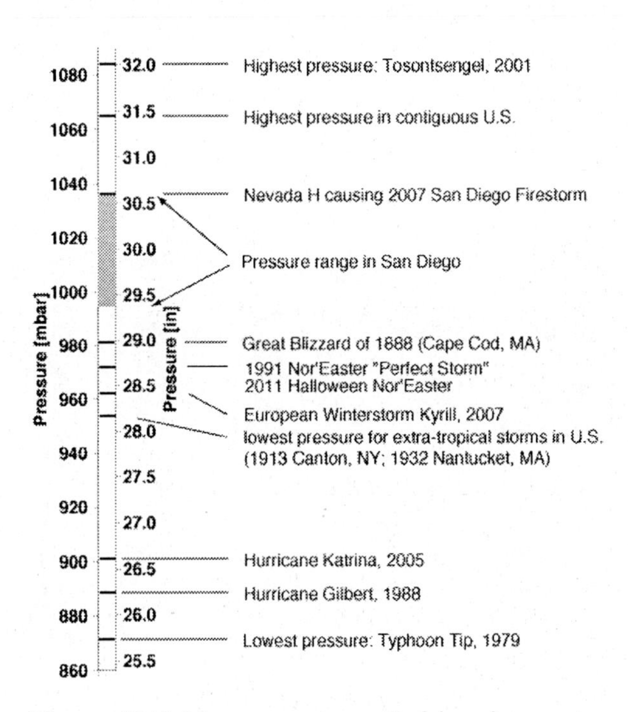

Figure 13.4 Air pressure recorded in a barometer as mbar and as mercury height (in inches), during various atmospheric conditions and weather events.

As with air temperature, the range in air pressure experienced in San Diego, CA is relatively small (Fig. 13.4) compared to the pressure observed worldwide. The lowest measured pressures are associated with the center of tropical cyclones, with the pressure of 870 mbar during Typhoon Tip being the lowest recorded to date. The lowest pressure in even the most powerful extratropical storms stays markedly higher[2]. The lowest pressure is such a storm was recorded at 913 mbar on 1 Nov 1993 in the Atlantic Ocean near the Shetland Islands. The lowest pressure in the U.S. was recorded at 955 mbar during two storms in 1913 and 1932.

Figure 13.5 The Robinson anemometer invented in 1846 by John Thomas Romney Robinson. Designs used today still resemble this early model. (source: Wikipedia)

Table 13.2 The Beaufort Scale

Beaufort number	Description	Wind Speed km/h	mph	kn (knots)	Land Conditions
0	calm	0-2	0-1	0-1	smoke rises vertically
1	light air	2-6	1-3	1-3	drifting smoke indicates wind direction
2	slight breeze	7-11	4-7	4-6	wind felt on face; leaves rustle; wind vanes move; flags stir
3	gentle breeze	12-19	8-12	7-10	leaves and small twigs move; light flags extend
4	moderate breeze	20-29	13-18	11-16	dust and loose paper raise; small branches move; flags flap

5	fresh breeze	30-39	19-24	17-21	small leave trees begin to sway; flags ripple
6	strong breeze	40-50	25-31	22-27	large tree branches in motion; telegraph wires whistle; umbrella use is difficult
7	high wind	51-61	32-38	28-33	whole trees in motion; effort needed to walk against wind; flags extend
8	gale	62-74	39-46	34-40	wind breaks twigs off trees; walking difficult
9	strong gale	75-87	47-54	41-47	slight structural damage (signs and antennas blown down); some branches break off trees
10	whole gale	88-101	55-63	48-55	trees uprooted; considerable damage; poorly attached shingles peels off roofs
11	storm	102-119	64-74	56-64	widespread damage to vegetation; roofs damaged;
12	hurricane force	≥ 120	≥ 75	≥ 65	extensive damage; some windows break; poorly constructed sheds and barns damaged

Wind Speed

Winds speeds are measured with anemometers and reported in m/s, km/h, mph or kn (knots). The knot, 1 nautical mile per hour, (1 kn = 1 nm/h = 1.852 km/h) is a unit often used in meteorology and navigation. The proper SI unit of wind speeds is m/s but the use of km/h is very common. Widely used **windmill anemometers** look just like miniature wind mills, while modern **cup anemometers** still resemble the original model invented in 1846 (Fig. 13.5). A large variety of anemometers exist today and the reader is referred to the Wikipedia page on anemometers for details. Nevertheless, it is still not possible today to measure directly the strongest winds on Earth, the ones occurring in a tornado. Instead, they are estimated by indirect methods using Doppler Radar. The Beaufort Scale (Table 13.2) categorizes wind speeds and was devised in 1805 by Francis Beaufort. The scale relates wind speeds to observed conditions at sea and on land. For simplicity, the sea conditions are omitted here and the interested reader is referred to the Wikipedia page on the Beaufort scale. Hurricane-force winds occur for wind speeds above 120 km/h. Such wind speeds are reached in severe winter storms as well. Hurricanes have their own scale, the Saffir-Simpson scale, where the strongest hurricane winds have so far clocked in at an incredible 407 km/h (253 mph) on 10 Apr 1996 in Barrow Island, Australia, during a 3-s wind gust in Cyclone Olivia (Box 1). The strongest winds not related to tornadoes or tropical cyclones have been measured at the weather station on Mt. Washington, NH. At 372 km/h (231 mph) on 12 Apr 1934, these winds correspond to those of a category 5 hurricane.

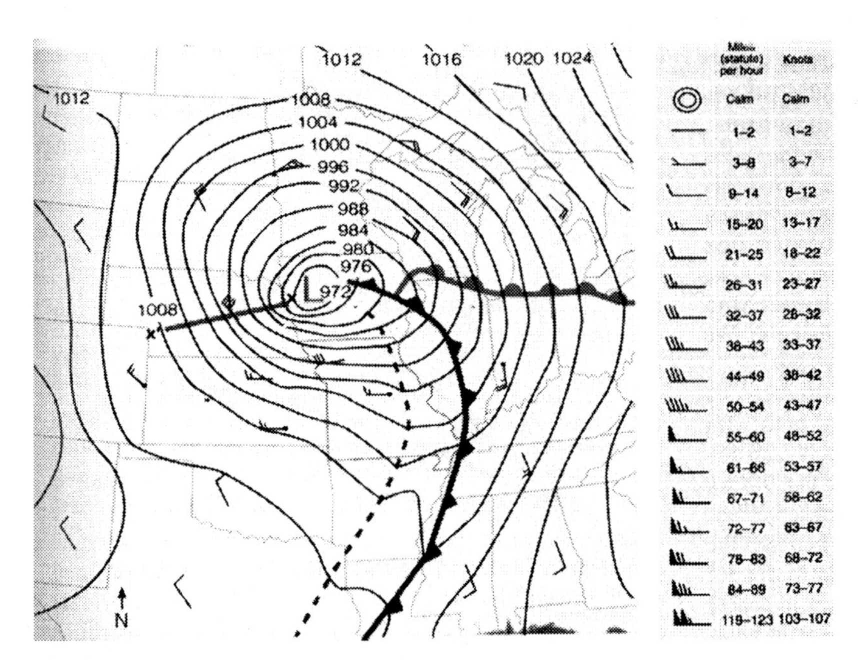

Figure 13.6 Surface weather map or 6 a.m. (CST) on Tuesday, 10 Nov 1998. Dark concentric lines are the isobars (in mbar). A deep low, with a central air pressure of 972 mbar (28.70 in) is moving over northwestern Iowa. Winds are marked by barbed and flagged sticks where the barbed end of the stick indicates the direction from which the winds blow. Winds speeds are indicated by the number of barbs or flags. The distance along the x'-x line in the west is 500 km and the difference in air pressure is 32 mbar. This gives a pressure gradient of 32 mbar/500 km. The tightly packed isobars in the west are associated with strong northwesterly winds of 40 kn. The barbed black line is a cold front and the lobed line is a warm front. The dashed line marks a trough. (source: MT)

In a weather map, winds are marked by barbed and flagged sticks (Fig. 13.6) where the barbed end of the sticks indicates the direction from which the winds blow. The pressure gradient controls the wind speed. In a map, the pressure gradient can be estimated through the density of the isobars (lines of equal air pressure). The closer the lines are together, the larger the pressure gradient is and the stronger the winds blow. As already noted in Chapter 12, the winds blow counter-clockwise around the low-pressure system. On the other hand, we learned that winds blow from high pressure to low pressure, along the pressure gradient. So why do winds tend to blow along the isobars and not perpendicular to them? As stated in Chapter 12, it has something to do with the Coriolis Effect.

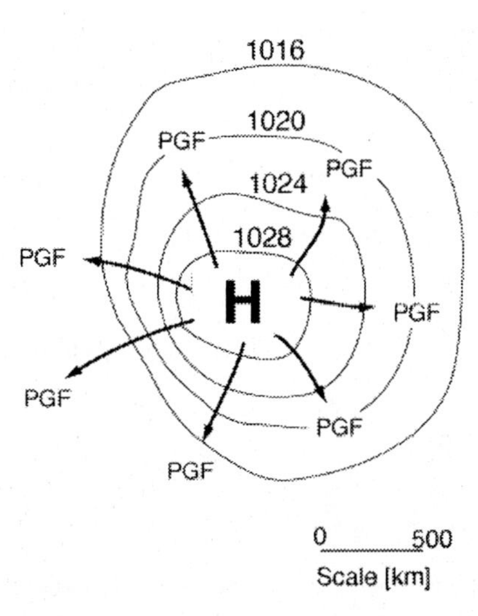

Figure 13.7 Concept figure of the pressure gradient force (PGF). The PGF follows the pressure gradient, which is always perpendicular to the isobars. The pressure gradient and the PGF are larger for closely spaced isobars than for widely spaced isobars. The wind speed increases with the PGF. Without the Coriolis Effect, winds would blow parallel to the PGF.

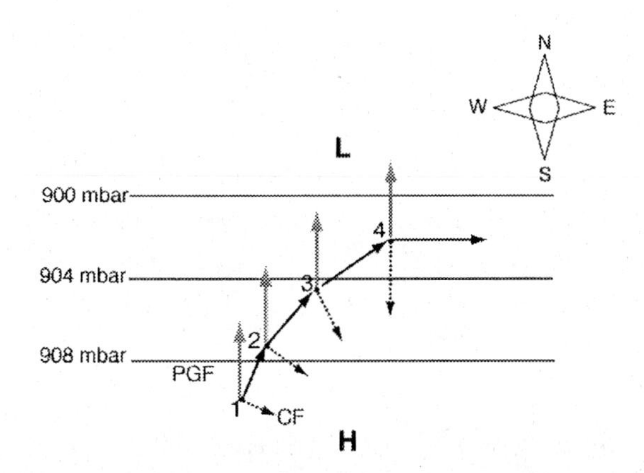

Figure 13.8 Concept figure of the interplay between pressure gradient force (PGF), Coriolis force (CF) and the resulting wind direction. Air initially at rest will accelerate until it flows parallel to the isobars at a steady speed, with the PGF balanced by the CF. Winds blowing under these conditions are called **geostrophic**.

At altitudes above 1000 m where friction along Earth's surface plays no longer a role, winds are controlled by two forces: the **pressure gradient force** (PGF) and the **Coriolis force** (CF). The pressure gradient force is determined by the pressure gradient as illustrated in Figs. 13.6 and 13.7. The PGF is proportional to the pressure gradient and like the gradient is always perpendicular to the isobars. Wind speeds increase with the PGF. The Coriolis force is perpendicular to the direction an object moves and increases with latitude, □: CF = m·Ω·v·sin □□□where m is the mass of the moving object, Ω Earth's rotation rate (1/day) and v the speed of the moving object. It is strongest near the pole and zero at the equator. When air starts to follow the PGF, it will also start to feel the CF (Fig. 13.8) that acts perpendicular to the direction of motion. The wind speed will increase as a result of the acting PGF, and the increasing CF will bend the path until the wind blows parallel to the isobars at which point the PGR and the CF are in balance. The wind speed then remains the same. If the isobars bend around a high- or low-pressure system, the winds will also follow the bent isobars (see Fig. 13.6). Winds in this balanced condition are called **geostrophic winds**.

Many locations on Earth have reoccurring winds that are characteristic of that area, be it because of special processes in the atmosphere or geographical features. These winds often have names. For example the **Föhn winds** is a warm dry southerly wind off the northern side of the Alps. In general, Föhn winds are a type of dry down-slope wind that occurs on the lee (downwind) side of a mountain range. It is also a rain shadow wind because the ascending air on the windward side of the mountains lost at least some of its moisture during orographic lifting and condensation and relative humidity decreases further on the lee side during adiabatic heating. The **Santa Ana** is a type of Föhn wind that is characteristic of Southern California. Similar winds are called **Diablo winds** in the San Francisco Bay area (northeasterly winds from the Sierra Nevada) and **Chinook winds** east of the Rocky Mountains. The **Sirocco** is a hot southerly wind in Italy that originates in North Africa. As the wind originates from the desert, it brings along large amounts of dust. They can cause health problems. The winds can pick up speeds of gale-force 100 km/h (62 mph), and the dust causes abrasion in mechanical devices and penetrates buildings. The **Mistral**, on the other hand, is a northerly cold wind occurring in summer in the Provence in southern France. It originates in the mountains in Central France and in the Alps. Wind speeds in excess of > 50 km/h are common around Marseille and visibility is often reduced. At the French Mediterranean beaches, the wind pushes warm surface waters out to sea, allowing colder, deeper water to well up, surprising many unsuspecting summer visitors. In Central Asia, the **Karaburan** is also called black storm. The violent northeasterly wind occurs in spring and summer. It is a **katabatic wind** (a strong downslope wind that is often enhanced by funneling through mountain valleys) and picks up large amounts of dust and soil on its way, thereby darkening the skies[7]. There are many more local winds and the interested reader finds a list at Wikipedia[8].

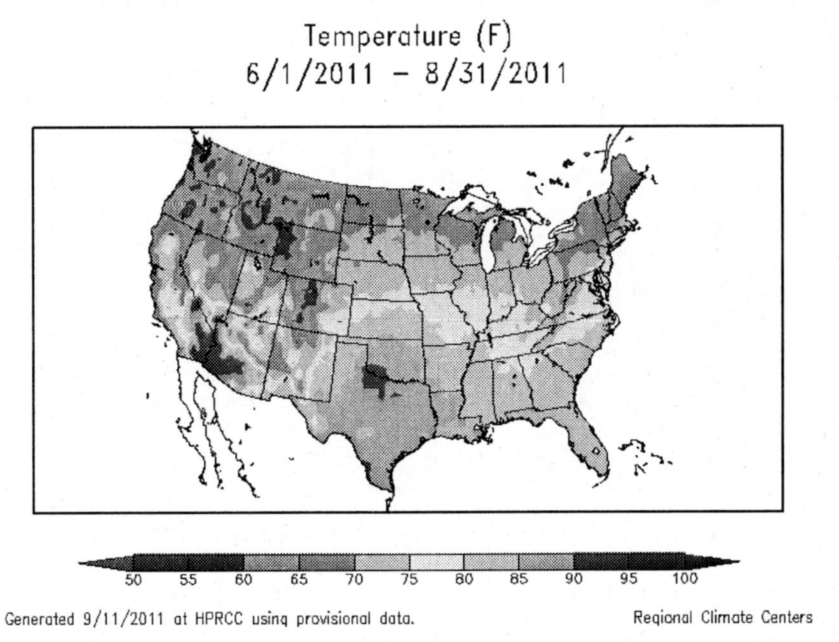

Figure 13.9a Average temperature (in °F) in the contiguous U.S. for the period of June through August 2011. The Southwest stands out as the hottest region. During the summer of 2011, northern Texas also experienced high temperatures. (source: NOAA[9])

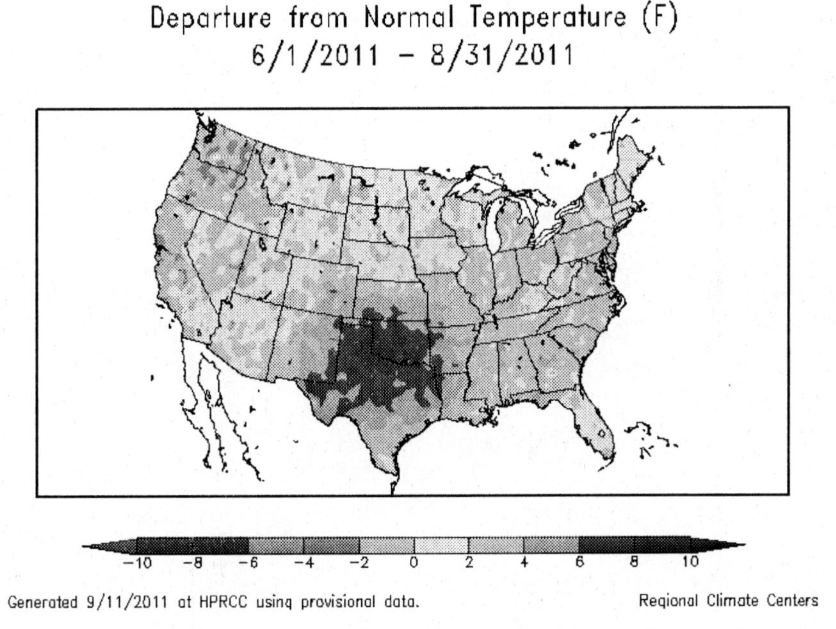

Figure 13.9b Average temperature (in °F) in the contiguous U.S. for the period of June through August 2011, but now as deviation from "normal". Normal is usually the 30-year average. (source: NOAA[9])

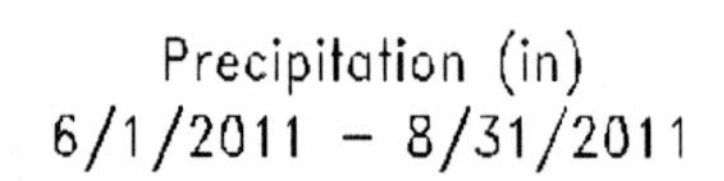

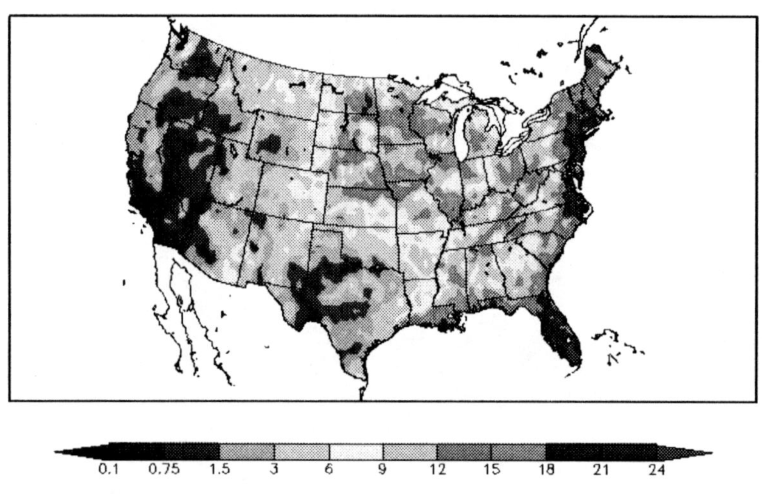

0.1 0.75 1.5 3 6 9 12 15 18 21 24

Generated 9/11/2011 at HPRCC using provisional data. Regional Climate Centers

Figure 13.9c Precipitation (in inches) in the contiguous U.S. for the period of June through August 2011. (source: NOAA[9])

Percent of Normal Precipitation (%)
6/1/2011 - 8/31/2011

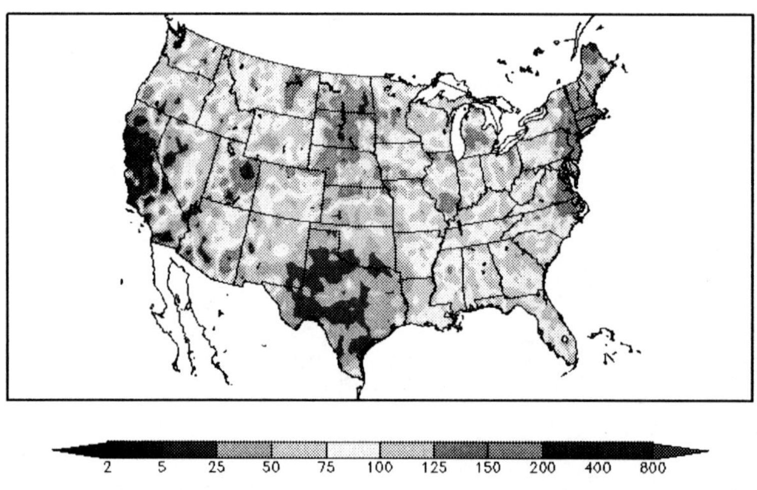

2 5 25 50 75 100 125 150 200 400 800

Generated 9/11/2011 at HPRCC using provisional data. Regional Climate Centers

Figure 13.9d Precipitation (in inches) in the contiguous U.S. for the period of June through August 2011, now as percentage fraction of "normal". Normal is usually the 30-year average. (source: NOAA[9])

13.2 Heat Waves and Droughts

With the regions of Arizona, California and Nevada along the Colorado and Gila Rivers and the depression of Death Valley, the U.S. Southwest is the hottest area in the United States and among the hottest in the world (e.g. Fig. 13.9). Many criteria can be used to compare the hotness of an area with that of others. The first is the annual average temperature. In this case, places that are warm throughout the entire year are most likely to make the list of hottest places, and so Key West, FL and Honolulu, HI lead the list of hottest major cities in the U.S. (Box 3). Another criterion is the average temperature in the hottest month (usually July in the Northern Hemisphere, but it could also be August). By this standard, Palm Springs, CA and Yuma, AZ lead the list of hottest U.S. cities (Box 3). Another way of defining hotness is by the average maximum temperature in the hottest month. The maximum and minimum average temperature for each month is now available in the climate section of Wikipedia pages for every major city though the source data can also be obtained at NOAA directly[(11)]. Yet another way to identify hot places is by the number of days in a year on which temperatures are above a certain threshold, say 90°F. In the U.S. Southwest and in the southern tip of Texas, the annual number of days above 90°F is greater than 150[(2)]. This heat is not considered a heat wave because it is normally occurring, according to climate zone. Fig. 13.9a shows the average temperature between June and August in 2011. The Southwest shows up as the hottest area but the entire state of Texas also was pretty warm, the north in particular. The heat is normal for the Southwest. For Texas, on the other hand, Fig. 13.9b, which shows the deviation from normal temperatures, documents clearly that the heat was unusual. Texas had a prolonged heat wave. Moreover, Texas also experienced an epic drought[(10)], as a comparison of precipitation Figs. 13.9c and 13.9d underlines. Ultimately, this drought led to dust storms and devastating wildfires. More than 47% of the acreage burned in the U.S. that year was burned in Texas.

Memorable Heat Waves

With 750 in Chicago alone, the 1995 Chicago Heat wave (Case study 1) stands out as one of the most fatal single-event natural disasters in U.S. history, being surpassed only by 2005 Hurricane Katrina (1833 fatalities), the 1889 Johnstown Flood (2200) and the 1900 Galveston Hurricane and Flood (> 6000). More recently, the 2003 European heat wave (Case study 2) left over 35,000 dead after weeks of oppressive heat and severe drought conditions. Drought conditions elsewhere in the world, particularly in Africa, put millions of people at risk of starvation. This is discussed in more detail in Chapter 19.

In the U.S. Midwest, the most intense and widespread heat wave ever to occur took place during July and August of 1936, during the Dust Bowl years (see Chapter 19). This heat wave

COPYRIGHTED MATERIAL — DO NOT DUPLICATE, DISTRIBUTE, OR POST

followed one of the coldest winters on record. Fifteen state absolute-maximum temperature records measured during this heat wave still stand today. Readings of 120°F (48.9°C) or higher were recorded in Arizona and California, where they are considered normal, but also in Arkansas, Kansas, Oklahoma, South Dakota and Texas. Incredibly temperatures even in North Dakota peaked at 121°F (49.4°C). In North and South Dakota, as well as in northern Nebraska and eastern Minnesota and eastern Iowa, temperatures were up to 12°F (6.7°C) above normal. Perhaps the hottest night ever recorded in the U.S. outside of the desert Southwest occurred on 15 July 1936 at Lincoln, NE, when the minimum temperature fell to only 91°F (32.8°C).

Perhaps less well known but more deadly than the 1995 Chicago heat wave, drought and a heat wave during the summer of 1980 in central and the eastern U.S. caused an estimated 10,000 heat-related deaths and an estimated $20 billion (1980 dollars) in damages and losses to agriculture and industry[(15)] though other sources report heat-related fatalities of no more than 2,000[(17)]. Temperatures were highest in the southern plains. From June through September, temperatures remained above 90°F (32.2°C) on all but two days in Kansas City, MO. The Dallas/Forth Worth area, TX experienced 42 consecutive days with high temperatures above 100°F (37.8°C), with temperatures reaching 117°F (47.2°C) at Wichita Falls, TX on 28 June.

In 1988, again, during intense heat spells in combination with one of the worst multi-year droughts in U.S. history caused a combined 5,000 to 10,000 deaths though some sources put this number even higher. This drought caused $60 billion in damages and losses. This was the same year when the Yellowstone National Park fell victim to the worst recent wildfire disasters.

The American Southwest, already the hottest region of the United States, experienced its highest temperatures ever during late June and early July 1994. From West Texas to Arizona, most weather stations attained their all-time maximum temperatures, including Midland, Texas (116°F/46.7°C); El Paso (114°F/45.6°C); Albuquerque, New Mexico (107°F/41.7°C); and Roswell, New Mexico (114°F/45.6°C). Santa Fe, normally cool even in summer because of its 2,100-m elevation (7000 ft), reached 100°F (37.8°C) for the first time in recorded history. Five state records were broken or tied in Nevada (125°F/51.7°C), Arizona (128°F/53.3°C), New Mexico (112°F/44.4°C), Texas (120°F/48.9°C), and Oklahoma (120°/48.9°C).

The 2006 North American heat wave and drought affected a wide area of the U.S. and parts of Canada from July through August. Over 220 deaths were related to this heat. Temperatures in the parts of South Dakota exceeded 115°F (46.1°C). California experienced unusually high temperatures, with records ranging from 100 to 130°F (37.8 – 54.4°C). On 22 July, the County of Los Angeles recorded its highest temperature ever at 199°F (48.3°C). Humidity levels also were unusually high for

California climate. In that year, the western half on the country also experienced numerous wildfires, burning nearly 10 million acres, a new record for the period since 1960.

The Impact of Heat Waves

Heat waves are prolonged periods of excessively hot weather for the climate zone this heat occurs in. Heat waves typically occur when strong upper atmospheric high-pressure systems become locked over a region for an extended time. This can occur over weeks, but also months. The exact recognition of a heat wave as such is somewhat vague and clearly depends on the climate zone. Prolonged heat normal for Phoenix, AZ and Las Vegas, NV in the hot desert may spell killer heat wave in more temperate climates. On the other hand, the same summer heat usually experienced in the Midwest can spell heat wave for coastal San Diego that is used to cooler summers. If present, humidity vastly enhances the already oppressive situation during a heat wave.

Heat waves pose a serious hazard for the elderly, very young children and the sick as well as overweight as their bodies have difficulties to adjust to the heat. Chronically ill and elderly often take medication (e.g. diuretics, blood pressure medication, antipsychotics and anticholinergics) that interferes with the body's ability to sweat. The inability to cool the body by sweating leads to heat exhaustion and heat stroke (hyperthermia), and a dangerous increase in body temperature. It is for this reason that two of the perhaps most memorably heat waves, the 1995 Chicago heat wave and the 2003 European heat wave killed disproportionally many elderly.

Severe heat waves have caused thousands of deaths from hyperthermia and widespread power outages. The power outages in turn add to the agony caused by the heat wave. In the U.S., power outages likely increase the death rate because life-saving air conditioning in homes and most public places no longer work (they likely still work in hospitals because they are expected to have emergency backup power). In the U.S., power outages are caused mainly because the increased usage of residential air conditioning during heat waves overloads the power grid. In Europe, where air conditioning in homes still is the exception, power outages occur during droughts when hydroelectric power generation breaks down. As pointed out in Chapter 1, heat waves have been the #1 killer in the U.S. in at least the last 20 years. Skeptics sometimes argue that heat waves only kill people who would have died soon anyway, pointing to evidence that mortality rates after heat waves are lower than normal. On the other hand, it is feared that there is many unrecorded deaths are in fact caused by heat waves.

If scientists' predictions are correct that the number and severity of heat waves will increase, the fatality rate from heat waves will likely also increase, unless there will be a change in awareness in the population. Heat waves are usually not perceived as a "natural disaster", not only because they occur over a longer time and so do not compare with tornadoes in terms of newsworthiness. But people also often do not perceive heat as an immediate personal threat. As a result, people also do not adhere to emergency recommendations to drink increased amounts of water. Typically, elderly may be concerned that their elderly neighbors are affected but not they themselves.

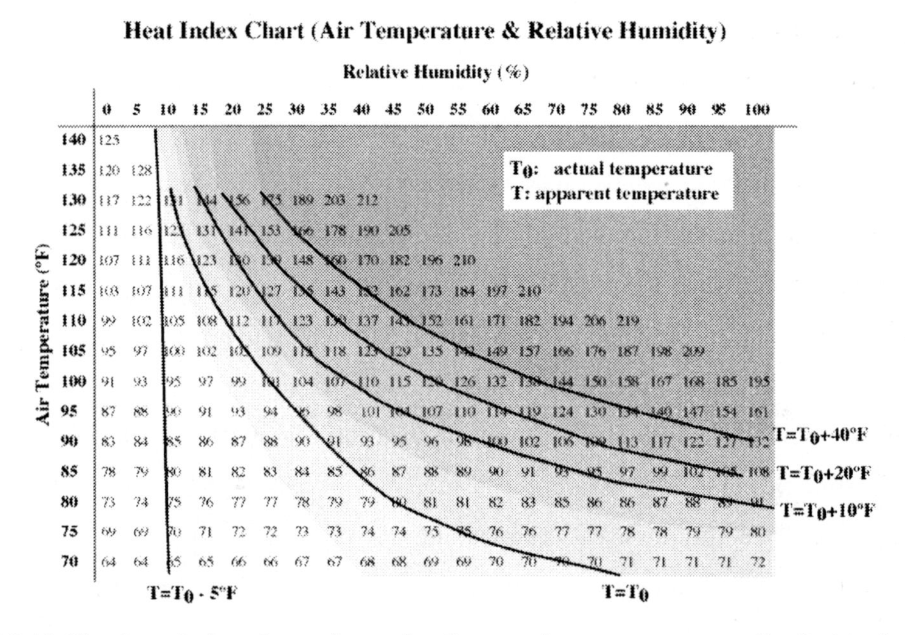

Figure 13.10 The heat index shows how air of a certain temperature really feels when the relative humidity is taken into account. This felt hotness is the apparent temperature. The numbers in the chart provide the apparent temperature for a given temperature (left axis) and relative humidity (top axis). For example, if the temperature is 90°F and the relative humidity is 50%, then the air feels like it is 96°F hot, i.e. the apparent temperature is 96°F. The curved lines mark apparent temperature that is higher or lower than the air temperature by a certain amount (e.g. 20°F). The formula to generate the curved lines can be found on the Wikipedia page on the heat index.

Table 13.3 The Heat Index and How Humans are Affected[12]

Category	Apparent Temp. [°F]	Apparent Temp. [°C]	Heat-related Symptoms
caution	80 – 90	27 – 32	Heat fatigue possible
extreme caution	90 – 105	32 – 41	sunstroke, heat cramps, heat exhaustion possible
danger	105 – 130	41 – 54	sunstroke, heat cramps, heat exhaustion likely
extreme danger	> 130	> 54	heatstroke and sunstroke imminent

The Heat Index

The high July temperatures in the U.S. Southwest appear forbidding but, setting air conditioning aside, people can usually cope with the heat because it is dry heat. On the other hand, agricultural irrigation in Southern California's Imperial Valley south of the Salton Sea increases humidity, and so the heat is perceived as being much more oppressive. The summer heat in southern Louisiana is routinely oppressive although one would not infer this from the map in Fig. 13.9a. This is because the high humidity adds to the misery. The **heat index,** which takes relative humidity into account, is an important tool to judge when hot weather becomes an extreme health hazard (Table 13.3). The heat index (Fig. 13.10), also referred to as the temperature humidity index, was developed in 1978 by George Winterling and adopted by the National Weather Service a year later. When air warms beyond a certain threshold (about 21°C/70°F), humid air feels warmer than dry air. This is defined as the apparent temperature. The effect becomes more severe with increasing air temperature, and with increasing humidity. The apparent temperature has something to do with the human body's ability to sweat. With increasing humidity, this ability is impeded and the body is less able to cool down. Why? Recall that evaporation costs energy that will be stored as latent heat in the water vapor. When this happens, the location where evaporation takes place cools down. Hence evaporation of sweat cools the air around the skin and so ultimately cools the body. In hot, humid air, sweat cannot evaporate easily and so the cooling of the skin does not take place.

In certain situations, the human body is actually quite effective at cooling down. In Fig. 13.10, these situations remain left and below the $T=T_0$ line when the apparent temperature is lower than the actual temperature. This is the case when temperatures are moderate (below 80°F/26.7°C), even for relatively high humidity (50% or larger). Even at high temperatures above 100°F (37.8°C), the apparent temperature is lower than the actual one, if relative humidity is less than 15%.

Starting at about 85°F (29.4°C) even a moderate humidity of 60% causes the air to feel 5°F (2.8°C) warmer than it really is. And at 95°F (35°C), this humidity makes the air feel nearly 20°F (11.1°C) hotter. If we now consider that the average high temperature in New Orleans, LA is over 90°F (32.2°C) in July but the humidity is nearly 65% or larger on virtually every day of the month, then the felt temperature average high is over 100°F (37.8°C), for the whole of July! Taking the heat index into account, coastal and southern Texas are to hottest summer locations in the U.S.. With an apparent July and August afternoon average temperature of 113°F (45°C), Del Rio, TX tops the list and Corpus Christi, TX ranks as the hottest major city, with an apparent July-August temperature of around 110°F (43.3°C). Even Death Valley's real July average afternoon temperature of 115°F (46.1°C) would feel cooler than Del Rio because the relative humidity is less than 15%. Such a low humidity poses its own health risks, however, such as dehydration when we are not aware how much we sweat and therefore do not drink enough to replenish the lost fluids and electrolytes.

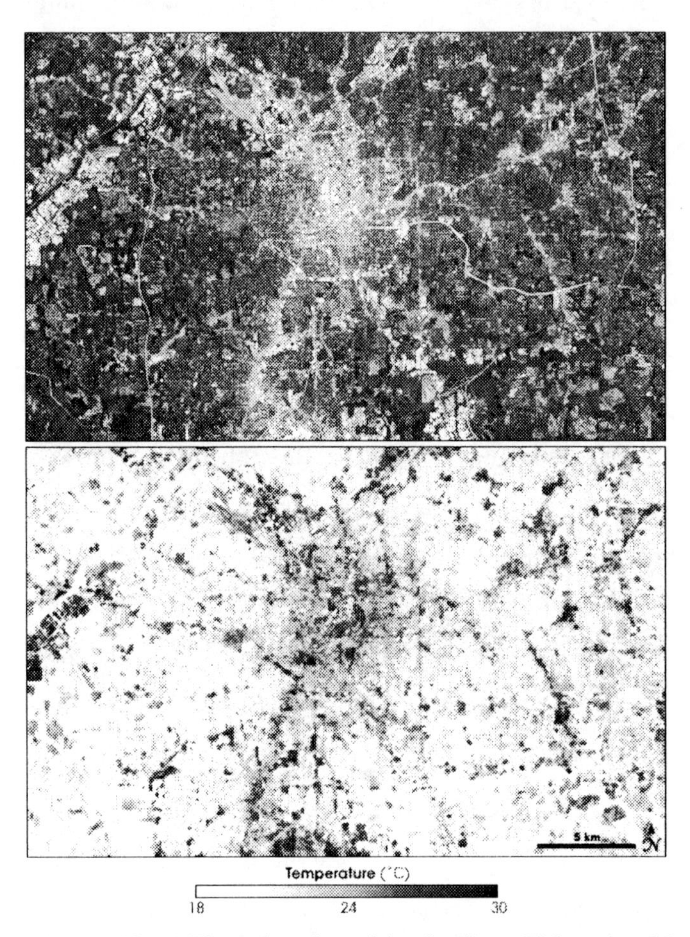

Figure 13.11 The urban heat island effect. This pair of Landsat satellite images provides a true-color view (top) and a satellite temperature image (bottom) of urban Atlanta, GA on 28 September 2000. The urban core is near the upper center of the images. Developed and paved urban areas can be up to 8°C (14.5°F) warmer than surrounding suburban or natural landscapes. This urban heat island affects not only the amount of energy a city needs to keep its residents cool and comfortable, but it also appears to influence where and how much it rains in the vicinity. (source: NASA Earth Observatory)

The Urban Heat Island Effect

Humans can cope with heat during the day if nighttime temperatures fall below a certain threshold (80°F/26.7°C). During heat waves, this may not occur, particularly in big cities where the **urban heat island effect** prevents nighttime cooling. In essence, the urban heat island causes a temperature difference that hardly noticeable during the day but pronounced during the night. Paved ground, concrete structures and building retain heat during the night and prevent the ground from cooling as much as soil would. For example, black surfaces absorb more radiation during the day than lighter surfaces that is then reradiated as infrared radiation during the night. In addition, the lack of vegetation that would release cooling water at night during respiration (in essence evapotranspiration similar to the water vapor released when people sweat) is also missing. Thirdly, the use of air conditioners to cool houses inside heats the air outside further. Buildings also block winds that may otherwise blow and cool an area. Pollution from traffic, industry and other source may also heat the air. The urban heat island is present all the time, and it is the reason why cities may have less snow in winter than the surrounding rural areas. But during heat waves, the urban heat island is enhancing an already dire situation by raising the nighttime temperatures in the city by a few degrees Celsius over that in the surrounding rural areas. In Barcelona, Spain, the urban heat island is estimates to raise temperatures by 0.2°C (0.4°F) during the day, but by 2.9°C (5.2°F) at night. An early account reported the excess nighttime temperature of London, U.K. to be 2.1°C (3.7°F) in 1810[(13)]. This is likely higher nowadays.

The urban heat island effect can profoundly affect long-term temperature records at single weather stations[(14)]. Temperatures can jump by 2°F (1.1°C) if a court next to the weather station is paved over. On the other hand, vegetation encroachment can lower measured temperatures. Many weather stations are near airport runways, and so measured temperatures may represent an upper limit of true temperatures in the city. Satellite imagery, on the other hand, delivers very helpful overviews and demonstrates quite graphically the impact and extent of the urban heat island (Fig. 13.11) that can raise temperature in the city by up to 8°C (14.5°F). Because of large amounts of rising heat, large cities generate their own atmospheric convection cells. Scientists suspect that the urban heat island effect may influence rainfall in their surrounding area and maybe even the frequency and severity of thunderstorms. In the U.S. Southeast, the rainfall downwind of major urban areas can be as much as 20% greater than in its upwind areas.

Heat waves can be hazardous to **air traffic**. A continental tropical air mass, stretching from southern California to the heart of Texas brought record warmth to the desert Southwest during the last week of June 1990. On 26 June 1990, a heat wave in Phoenix, AZ drove temperatures to 50°C (122°F). The extreme heat lowered the air density to the point where aircraft lift was reduced. Officials had to suspend aircraft takeoffs at Sky Harbor Airport.

13.3 Air Masses and Fronts

Like heat waves, many cold waves are associated with high-pressure systems. But the line between cold waves and blizzards, and ultimately cold winter storms, may sometimes seem blurry. To fully appreciate their differences, it is there useful to step back and review the movement and collision of air masses and the formation of fronts, the lines along which "bad weather" often occurs.

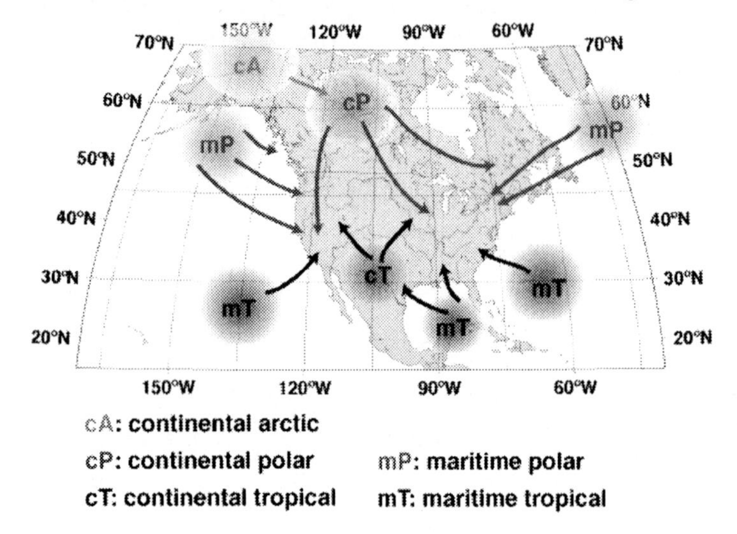

Figure 13.12 Air masses of North America, their source regions and their paths.

Air Masses

An air mass is a large body of air with similar temperature and humidity in every horizontal direction (at the same altitude). Air masses can be larger than 1500 km (932 mi) across. They typically form over a large flat area, the **source region**, where surface winds are calm or light. Ideal source regions are areas with high surface air pressure. These include ice- and snow-covered arctic plains in the winter and the subtropical oceans in the summer. The middle latitudes, on the other hand, where surface temperatures and moisture characteristics vary considerably over short distances, are not good source regions. They are instead transitional zones where air masses of different physical properties move in, collide and produce weather activities, and at times severe weather. Air masses are moved about by winds aloft.

Air masses are classified by temperature and humidity. There are cold and warm air masses and well as dry and humid ones. They are grouped into four general categories according to their source region. A capital letter denotes the general latitude of the source region: P for polar, and T for tropical. Furthermore, a lowercase letter denotes the type of surface area over which the air masses form: c for continental, and m for maritime. In winter, a continental Arctic air mass is a fifth type of air mass relevant to weather in the contiguous U.S. (Fig. 13.12; Table 13.4).

Table 13.4 Air Masses Relevant to Weather in the U.S.

Label	Full Name	Properties
cA	continental Arctic	very cold, dry, stable

cP	continental Polar	cold, dry, stable
mP	maritime Polar	cool, moist, unstable
cT	continental Tropical	hot, dry, stable air aloft; unstable surface air
mT	maritime Tropical	warm, moist; usually unstable

The bitterly cold weather that invades southern Canada and the U.S. in winter is associated with cP and cA air masses. In fact, the character of a cP air mass is different between winter and summer. In winter, over very cold areas (T= -20°C/-4°F), the cooling effect from low surface temperatures is stronger than the adiabatic heating effect from sinking air in a high-pressure system. Cold air is then trapped near the surface below warmer air aloft. This causes the formation of an **inversion layer**. Over large cities, pollutants can accumulate within a few days, thereby lowering visibility and air quality.

Along the U.S. west coast, mP air usually gets transported southward to feed winter storms. However, during a Pineapple Express, mT air can reach the west coast. This air typically traveled a long distance (over 1500 km/932 mi) and is very warm and moist. It produces heavy precipitation along the west coast that can lead to damaging and fatal mass movements.

The characteristic of an air mass changes when it leaves the source regions and moves over new areas. For example, an unstable mP air mass loses moisture when it moves east and crosses the Rocky Mountains. It then travels over a cold, elevated plateau that chills the surface air and slowly transforms the lower level into dry, stable cP.

In spring, a particularly dangerous situation unique to the U.S. can occur over the southern Midwest when warm mT air moves in from the Gulf of Mexico and particularly cold cP air still exists and moves in from the north. If the jet stream aloft flows nearly parallel to the line of collision between these air masses (the front), violent thunderstorms can form that spawn tornadoes.

Fronts

Figure 13.13 Arial photo of an approaching storm front in Mexico. (source: UCAR)

The transitional boundaries between two interacting air masses are called fronts. Weather can change drastically when one air mass replaces another. The term front was therefore first used by Norwegian meteorologists during WW I who likened the often severe weather with military actions on the battlefield[(26)]. Depending on which air mass occupies and area and which air mass is moving in, different types of fronts can form with each having its characteristic weather pattern.

With the study of weather maps and with the help of weather forecasts one usually gets a good idea about impending changes in weather. The exact timing of the passage of a front at a certain location as well as and the type of the passing front can usually be determined by studying the changes in the following parameters:

- temperature change over distance and over time
- changes in air moisture content/dew point
- shift in wind direction
- changes in air pressure
- clouds and precipitation patterns
- the occurrence of severe weather

A nice summary of the different types of fronts and other features, and their signature in a weather map can be found on the NOAA/NWS website[(26)].

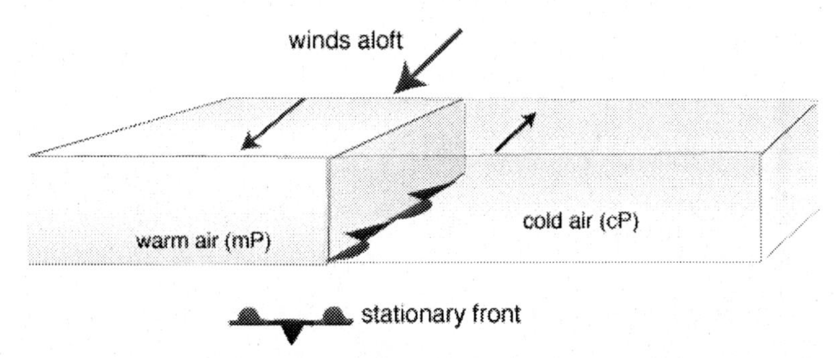

Figure 13.14 Principal features of a stationary front between a warm and a cold air mass. In a weather map, the front is marked with alternating barbs and lobes pointing in opposite directions. In a colored map, the lobes are red and the barbs are blue.

A **stationary front** forms when two air masses butt up against each other but nearly no movement of air occurs toward the front (Fig. 13.14). Surface winds blow parallel to the front but in opposite directions on each side of the front. Winds aloft usually blow parallel to the front. The temperature and moisture content in the air may change drastically across the front but the weather is usually clear to partly cloudy. No severe weather involving high winds, large amounts of precipitation or thunderstorms occur. Sometimes, light rain can fall if warm, moist air starts to ride up and over the cold air. However, in this case, the front is no longer a stationary front but starts to become a warm front.

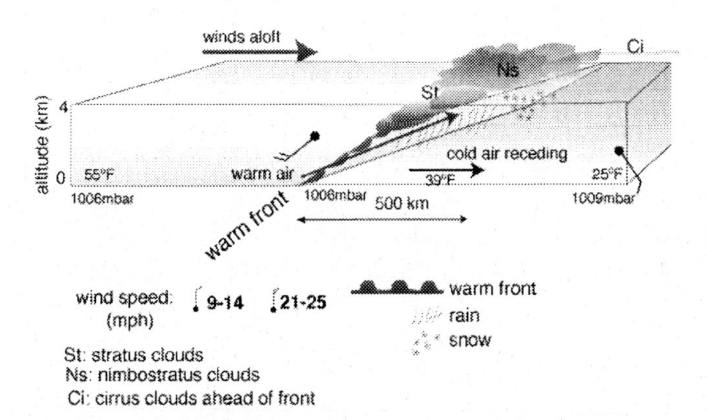

Figure 13.16 Principal features of a warm front, where a warm air mass rides up and over a cold air mass. Extensive rainfalls occur in a large area leading the front. In a weather map, the front is marked with a lobed line where the lobes point in the direction in which the warm front advances. In a colored map, this line is red.

A **warm front** forms when a warm air mass moves over a cold air mass. Since the humid air is forced to climb, it cools and eventually saturates, leading to condensation and precipitation. The slopes on which the warm air mass advances over the cold air mass is rather shallow, therefore providing a large area on which widely spread, long-lasting rainfall can occur well ahead of the front. Winds aloft blow in the direction the warm air mass is moving. These winds push cirrus clouds far ahead of the front, and the clouds are harbingers of the impending change in weather. Since the cold air mass is replaced by the advancing warm air mass, the surface temperature increases steadily during the passage of a warm front and stays steadily high after the passage. The air pressure falls when the front approaches. It levels off while the front passes, then slightly rises before it falls again. In the Northern Hemisphere, winds change during the passage of a warm front from southeasterly to southwesterly.

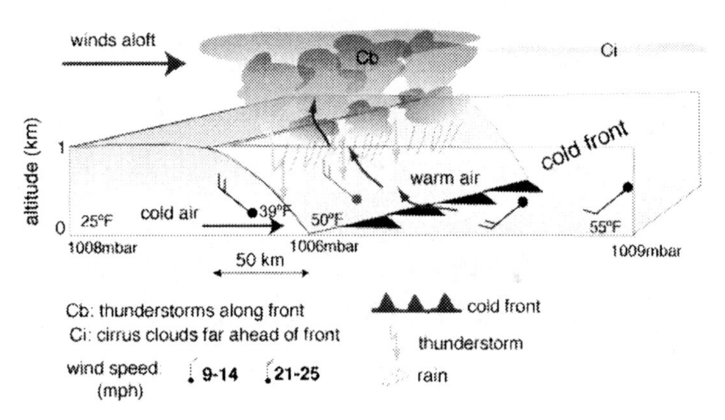

Figure 13.16 Principal features of a cold front, where a cold air mass pushed beneath a warm air mass. Severe weather with heavy showers of rain or snow forms behind the front. Hail-producing thunderstorms can also occur. In a weather map, the front is marked with a barbed line where the barbs point in the direction in which the cold front advances. In a colored map, this line is blue.

A **cold front** forms when a cold air mass pushes beneath a warm air mass. A cold front advances about twice as fast as a warm front. The slope of the contact zone between the two air masses is steep. This forces the warm air mass to ascent quickly, thereby leading to the rapid development of heavy precipitation. The air can also rise very high which can lead to the formation of thunderstorms and severe weather with hail. Precipitation concentrates on a relatively small area around the front, mostly behind it, but thunderstorms can form even ahead of the front. The line of thunderstorms associated with the cold front is also called a **squall line**. As with warm fronts, winds aloft push cirrus clouds far ahead of the front to warn of an impending change in weather. The surface temperatures drops rapidly during the passage of a cold front and may decline further but more slowly thereafter. There is also a marked drop in air pressure during the passage, followed by a steady rise thereafter. In the Northern Hemisphere, winds change from southwesterly to northwesterly.

The passage of a cold front can be quite spectacular. Though no temperature records are available, it is reported that temperatures dropped nearly instantaneously from a balmy 40s°F (to 0°F when a cold front moved through Illinois on 21 December 1936.

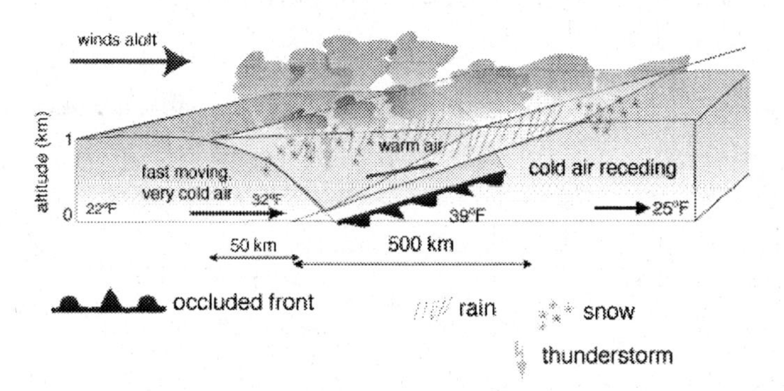

Figure 13.17a Principal features of a cold-type occluded front where a fast-moving very cold air mass pushed beneath a warm front. This lifts the warm air mass, leaving only cold air near the surface. Extended rainfall occurs leading the front and severe weather occurs along and behind the front. In a weather map, the front is marked with a alternating barbs and lobes that point in the direction in which the front advances. In a colored map, this line is purple.

An **occluded front** forms when a cold front catches up with or even overtakes a warm front. Occluded means "closed off". In the first scenario, a fast-moving very cold air mass pushed beneath the warm front. The warm air mass gets lifted up and loses contact with the surface. This is called a **cold occlusion** and the front is a **cold-type occluded front**. The weather along a cold occlusion is a combination of the weather from both the warm and the cold front, with extensive precipitation leading the front and severe weather along and behind the front. The most violent weather occurs when the cold front is just overtaking the warm front. During the passage of an occluded front in North America, already low temperatures drop further and are lower afterward. The air pressure is falling before the passage, is at a low point during passage and then rises. Winds are usually southeasterly before the passage and northwesterly afterward. Cold occlusions are the most prevalent type of front that moves into the Pacific coastal states and into interior North America. Occluded fronts frequently form over the North Pacific and North Atlantic, as well as in the vicinity of the Great Lakes.

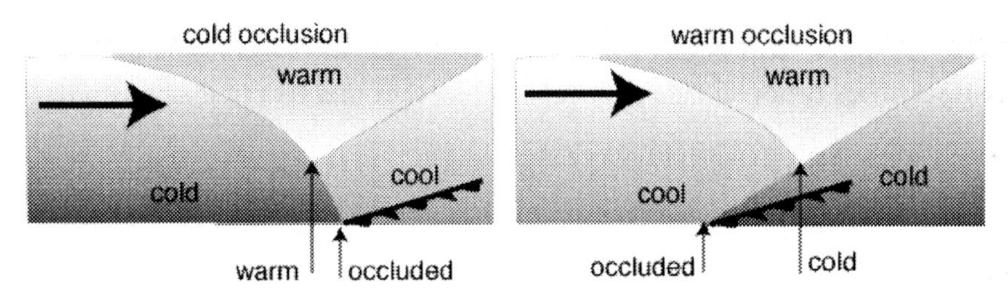

Figure 13.17b The relative position of the upper-level front with respect to the surface occluded front for a cold occlusion (left) and a warm occlusion (right).

Sometimes the cold air mass behind the cold front moving in is warmer than the cold air mass leading the warm front. This occurs when cool mP air moves from the Pacific Ocean into eastern Washington and Oregon, where a cold cP air mass has been sitting on the continent. In this case, the cold front catches up with the warm front but the cold air mass cannot lift the warm front and push beneath the cP air mass. The cool mP air then rides up on the cold cP air. A **warm occlusion** forms with a **warm-type occluded front**. The surface weather associated with a warm occlusion is often similar to that of a warm front. In western European winters, many of the occlusions are warm because the maritime air moving in from the North Atlantic is warmer than the continental air mass sitting on the continent. In a warm occlusion, the upper-level cold front precedes the surface occluded front, while in a cold occlusion, the upper-level warm front follows the surface occluded front (Fig. 13.71b).

Figure 13.18 Weather forecast maps issued daily by NOAA and the NWS, as published on the website of the Hydrometeorological Prediction Center[27]. The map shown here was issued and downloaded on 14 November 2007. NOAA and the HPC provide a nice archive of daily weather charts at http://www.hpc.ncep.noaa.gov/dailywxmap. Weather charts and data in Europe can be accessed through the German Weather Service (Deutscher Wetterdienst) at http://wetterzentrale.de/topkarten/fsreaeur.html (charts go back as far as 1871). Weather data and other information on global climate can be obtained from UK's Met Office at http://www.metoffice.gov.uk. Current and historical data at global meteorological stations are readily available at http://www.wunderground.com.

In a weather forecast map (Fig. 13.18), we are now able to identify not only high- and low-pressure systems but also fronts. On 14 November 2007 much of the northwestern U.S. experienced high-pressure weather while a cold front stretched from Southern California over the Rockies and across Texas (interrupted by a small warm front near the New Mexico/Colorado border). Weather in the eastern U.S. was influenced by two low-pressure systems, where the advancing cold front was likely to bring rain. The close proximity of a High in the Gulf of Mexico likely caused the warm air mass to be particularly warm in Louisiana and Mississippi, which increased the chance of thunderstorms in the Deep South. Occluded fronts formed in two Lows near Hudson Bay and the Pacific Ocean off-shore the Pacific Northwest.

13.4 Cold Waves and Cold Storms

With the knowledge on air masses, their movement and the formation and passage of fronts, we can now examine some properties of cold waves and cold storms. Locally, cold waves are defined as events with a rapid fall in temperature within a 24-hour period that requires a substantial effort of protection for agriculture but also anything else that is sensitive to severe "below-normal" cold. Usually, cold waves occur during the wintertime when cold Arctic air spreads unusually far south. In late January 2012, a combination of high- and low-pressure systems over and around Europe together with extreme excursions of the jet stream set the stage for one of the most deadly cold waves in recent history. It lasted nearly a month and killed upward of 650 people, with nearly 550 in areas of the former Soviet Union (Case Study 3).

In the U.S., the greatest cold wave in modern history occurred during the first two weeks of February 1899. This weather event also carries the name of **Great Blizzard of 1899** or **Snow King.** With record-breaking low temperatures, the South was affected in particular severely. For the first and only time, temperatures fell below zero in every state of the Union, including Florida, where

Tallahassee dropped to an unbelievable -2°F (-18.9°C). In Mobile, AL the reading of -1° (-18.3°C) was a full 13°F (7.2°C) colder than the previous records low of 12°F (-11.1°C). Even tropical Havana, Cuba, was affected of 14 Feb, when the temperature did not rise above 54°F (12.2°C). There are reports of hard frost killing or damaging many crops. In New Orleans, LA the lowest reading was 6.8°F (-14°C), and ice floes choked the Mississippi River and even drifted into the Gulf of Mexico. Three inches (7.6 cm) of snow fell in New Orleans. The arctic outbreak spread south out of Canada on 7 Feb, dropping temperatures to -61°F (-51.7°C) in Montana and -59°F (-50.6°C) in Minnesota.

The greatest nationwide cold wave of the 20th century occurred in January 1977. The core of the cold air extended from Florida to New Hampshire and west to Iowa and Missouri. At the center of the cold wave was Ohio, where every weather station measured its coldest month on record. At -25°F (-31.7°C), Cincinnati experienced its coldest temperature since records began in 1820. Snow was observed as far south as Miami (a first). From eastern Iowa to western Pennsylvania and northward, temperatures failed to rise above freezing during the entire month. In Minneapolis, the wind chill dropped to the lowest on record: -78°F (-61.1°C). Near hurricane-force winds brought the worst blizzard ever to Buffalo, NY, with white-out conditions and snowdrifts up to 30 ft deep. At temperatures around 0°F (-17.8°C), the winds contributed to a wind chill of -60°F (-51.1°C).

During a cold wave in January 1994, many parts of the northeast and north-central U.S., as well as southern Canada logged the coldest January since at least the late 1970s. Many overnight record lows were set. Detroit, MI saw its coldest temperature since 1985. The cold extended into Texas farther south than usual, bringing snowfall and temperatures below -20°F (-28.9°C). Florida also experienced extreme cold for its climate and snowfall, with flurries reported in Miami. The damage to the citrus crop was extensive. A similar cold wave affected much of the U.S. Deep South and Florida in early 2010 and set snowfall records elsewhere. In Burlington, VT 33 in (84 cm) of snow fell over the first January weekend, topping the 1969 record of 30 in (76 cm). The weather caused hundreds of schools to close in Arkansas, Ohio, Pennsylvania, West Virginia and the North Carolina mountains.

The **1983/1984 Siberian Express** was responsible for one of the coldest winters on record across North America since reliable records began in 1931, with the coldest Christmas ever. The term was introduced by the news media and describes the unusually far south-reaching arrival of an extremely cold air mass of Arctic origin in the U.S.. Winds gusting to 45 kn (83 km/h, 52 mph) add the term "express". To use technical terms that become clear in the next section, a cold cP air mass covered most of the northern and central plains in early December. Far to the north, a huge bitter cold cA air mass formed over the Canadian Northwest Territories. An upper-level ridge (high pressure) formed over Alaska, pushing strong northerly winds (associated with the polar jet stream) along its eastern side. This allowed the cA air to travel south over the prairie provinces of Canada. A portion of this cold air broke away and swirled as an anticyclone into the U.S. Temperatures dropped to some of the lowest readings ever recorded for December. On 22 December, Elk Park, MA recorded an unofficial -53°C (-64°F), only 4°C higher than the all-time low of -57°C (-70°F) for the contiguous U.S. (recorded at Rogers Pass, MA on 20 January 1954). The high-pressure system eventually stretched across North America into the Gulf of Mexico. By 24 December, the sea-level pressure at Miles City, MA reached an incredible 1064 mbar (a new record high that topped the old mark of 1063 mbar set in Helena, MA on 10 January 1962). By this day, Arctic air covered almost 90% of the U.S.. Hard freeze caused hundreds of millions of dollar in damage to the fruit and vegetable crops in Texas, Louisiana, and Florida. On 25 December, 125 record low temperature readings were set in 24 states. That afternoon, it was colder in Atlanta, GA (-13°C/8.6°F) than in Fairbanks, AK (-12°C/10.4°F). By 1 January 1984, the extreme cold eased, only to return 10 days later. On 18 January, an all-time record low of -54°C (-65°F) was recorded at Middle Sinks, UT. On 19 January, temperatures at the airports in Philadelphia and Baltimore plummeted to a new low of -22°C (-7°F). The Siberia Express was then followed by one of the warmest Februarys on record. Most states east of the Rockies were affected by a heat wave fueled by warm, humid air from the Gulf of Mexico.

The Wind Chill Factor

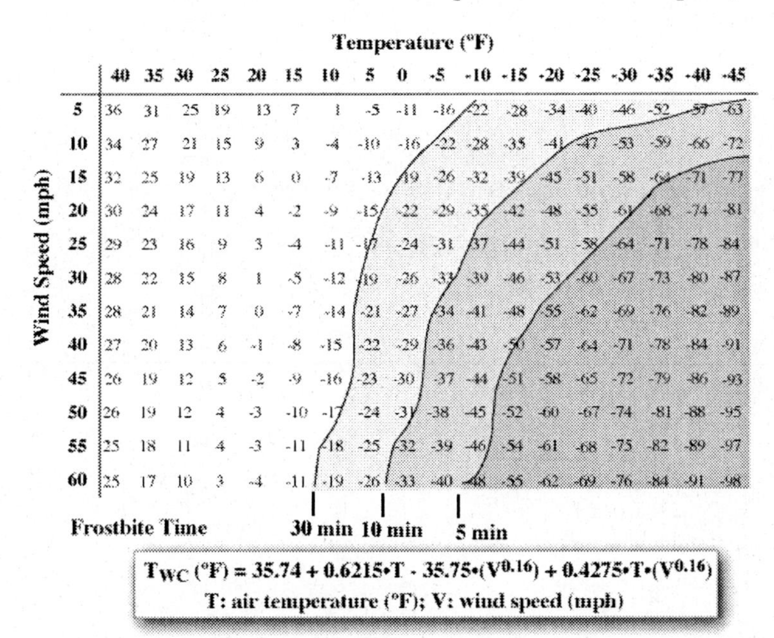

Wind Chill Chart (Air Temperature & Wind Speed)

Wind Speed (mph) \ Temperature (°F)	40	35	30	25	20	15	10	5	0	-5	-10	-15	-20	-25	-30	-35	-40	-45
5	36	31	25	19	13	7	1	-5	-11	-16	-22	-28	-34	-40	-46	-52	-57	-63
10	34	27	21	15	9	3	-4	-10	-16	-22	-28	-35	-41	-47	-53	-59	-66	-72
15	32	25	19	13	6	0	-7	-13	-19	-26	-32	-39	-45	-51	-58	-64	-71	-77
20	30	24	17	11	4	-2	-9	-15	-22	-29	-35	-42	-48	-55	-61	-68	-74	-81
25	29	23	16	9	3	-4	-11	-17	-24	-31	-37	-44	-51	-58	-64	-71	-78	-84
30	28	22	15	8	1	-5	-12	-19	-26	-33	-39	-46	-53	-60	-67	-73	-80	-87
35	28	21	14	7	0	-7	-14	-21	-27	-34	-41	-48	-55	-62	-69	-76	-82	-89
40	27	20	13	6	-1	-8	-15	-22	-29	-36	-43	-50	-57	-64	-71	-78	-84	-91
45	26	19	12	5	-2	-9	-16	-23	-30	-37	-44	-51	-58	-65	-72	-79	-86	-93
50	26	19	12	4	-3	-10	-17	-24	-31	-38	-45	-52	-60	-67	-74	-81	-88	-95
55	25	18	11	4	-3	-11	-18	-25	-32	-39	-46	-54	-61	-68	-75	-82	-89	-97
60	25	17	10	3	-4	-11	-19	-26	-33	-40	-48	-55	-62	-69	-76	-84	-91	-98

COPYRIGHTED MATERIAL — DO NOT DUPLICATE, DISTRIBUTE, OR POST

Figure 13.19 Wind chill chart for dry bulb temperature (temperature measured on a thermometer freely exposed to air but shielded from radiation and moisture). This is the felt temperature on exposed skin due to wind. The wind chill temperature is always lower than the measured air temperature. The formula to generate the table is also given and can be found on the Wikipedia page on the wind chill. The chart can be found on Wikipedia as well as on the NOAA and NWS websites [23].

The effects of cold waves and cold storms are enhanced when strong winds occur. Normally on a cold day, a thin layer of warm air close to the skin protects humans from the surrounding cooler air and from rapid loss of heat. However, once winds start to blow, the insulating layer is swept away, and heat is rapidly removed from the skin. The heat loss is greater as winds speed up. Similar to heat index that describes how hot humid air feels, the **wind chill factor** describes how cold the air feels with increasing wind speeds (Fig. 13.19). Tables for the wind chill factor were first developed before WWII and were available at the NWS by the 1970s. These tables were based on physical properties of air in plastic bottle suspended in the wind. The formula currently applied in North America was introduced in 2001. It was developed by scientists and medical experts and models the skin temperature. At temperatures below 40°F (4.5°C) even light winds reduce the apparent temperature by 5°F (2.8°C), and by more than 10°F (5.6°C) at temperatures below 10°F (-12°C). A strong breeze of 30 mph (48 km/h) at 10°F results in an apparent temperature of -12°F (-24.5°C). Frostbites start to occur when skin is exposed to temperatures of -20°F (-28°C) for more than 30 min.

Blizzards

By definition, and as described on the NOAA website, the difference between a winter storm and a blizzard is that the latter is a particularly severe winter storm with strong winds and low visibility[24]. While blizzards are commonly associated with snowfall and severe cold, they are not required for a winter storm to be classified as a blizzard. Blowing snow, on the other hand, is a criterion for a blizzard. If snow is not falling during a blizzard, then strong winds pick up already fallen snow from the ground. Canada and the U.S. have different criteria to call a snowstorm a blizzard. In the U.S., a blizzard has to meet the following four criteria:

- sustained wind speeds 56 km/h (35 mph) or larger
- blowing snow
- visibility of 0.25 mi or less ($\leq$ 0.4 km)
- lasting for at least 3 h

Canada, on the other hand, also makes a wind chill of less that -25°C (-13°F) a criterion but is less restrictive with regard to sustained wind speeds and visibility. In Canada, a snow storm is considered a blizzard if wind speeds are at least 40 km/h (24.8 mph) and the visibility is less than 1 km (0.6 mi). As with the Pineapple Express and the Siberian Express, the term blizzard was introduced by the news media. In the 1870s, an Iowa newspaper first used the word blizzard to liken fierce snowstorms to a canon shot or a volley of musket fire. Because of strong winds at low temperatures, the wind chill plays a major role for blizzards to become deadly.

Blizzards can bring “whiteout” conditions, and can paralyze regions for days at a time, particularly where snowfall is unusual or rare. With at least 4,000 fatalities, the 1972 Iran blizzard was the deadliest in recorded history. A week of snowstorms 3 – 9 February dumped more than 3 m (9.8 ft) of snow across rural areas in northwestern, central and southern Iran. In the South were most deaths were reported snow piled as high as 8 m (26 ft). Reportedly, the city of Ardakan and surrounding areas were hardest hit, with no survivors in Kakkan and Kumar. In the northwest, near the border with Turkey, the village of Sheklab and its 100 residents were buried.

The second-deadliest blizzard worldwide occurred in February 2008 in Afghanistan and is responsible for 926 fatalities. The weather also claimed more than 100,000 sheep and goats, and nearly 315,000 cattle. Temperatures dropped as low as -30ºC (-22ºF) and snow piled up to 1.8 m (6 ft). But sadly, and perhaps illustrating deteriorating living conditions in that country, this blizzard was not a record breaker in the meteorological sense. After walking barefoot in the freezing cold mud and snow, at least 100 people suffered frostbites so severe that amputations had to be performed.

The third-deadliest blizzard was the **Great Blizzard of 1888** in the U.S. that occurred on 11 – 13 March of that year. It claimed 400 lives. Meteorologically interesting is that this event was a Nor’Easter. It is therefore discussed in more detail in that section. On the website of the NWS Weather Forecast office, this storm is also listed as one of the biggest snowstorms in the U.S. from 1888 to present(24). In fact, there were actually two blizzards within three months. The first one occurred on 12 – 14 January in the Great Plains and is also referred to as **Great Plains Howler**(2). It affected the Midwest from Texas to the Dakotas before roaring east into Wisconsin. The storm caused 237 fatalities, an unusually high number considering the low population density at the time. The storm also killed so much livestock that many historians refer to this blizzard as seminal event in the downfall of the Plains free-range cattle industry.

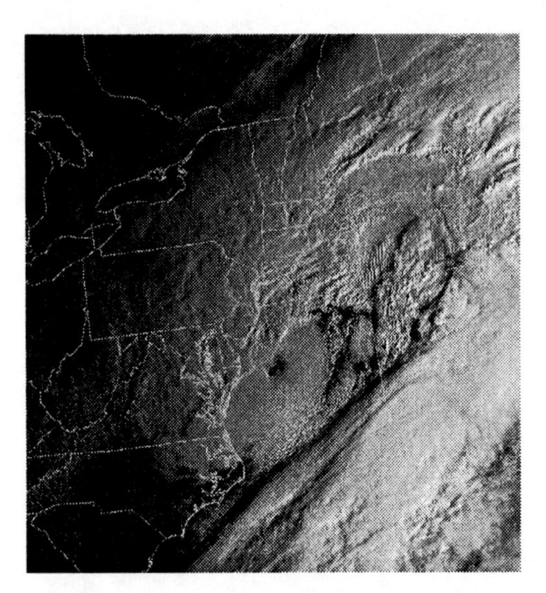

Figure 13.20 GOES satellite image of the 11 February 2006 Blizzard. A calm eye resembling that of a hurricane is clearly discernible. This is unusual feature for an extratropical cyclone. (source: NASA, Wikipedia)

On 11 – 13 February 2006, the **North American Blizzard of 2006** affected the northeastern U.S. from Virginia to Maine. The storm caused $5 million (2006 dollars) in damages, and 3 people died from indirect effects. In this respect, this storm did not break any records. However, it is unique because it formed a calm eye in the center similar to that of a hurricane (Fig. 13.20). The central low pressure reached 971 mbar, the central pressure of a typical category 2 hurricane. The storm brought heavy snow, upward of 32 in (81 cm) in some places, from Virginia to Maine and into Canada. Major northeast cities from Baltimore to Boston received upward of 1 ft of snow (30 cm), with an all-time largest amount in New York City (26.9 in; 68 cm) since at least 1869 when records began. Whipped by 95 km/h (60 mph) winds a 1-m high (3 ft) storm surge caused coastal flooding, particularly in Massachusetts. Offshore, waves reached a height of 8 m (25 ft). The storm also caused the closure of LaGuardia, John F. Kennedy International and Newark Liberty International Airports in New York, for the first time since 9/11 in 2001. Many flights were cancelled elsewhere, e.g. 90% of flights were cancelled at Logan International Airport, Boston, MA. Despite the large societal and economic impact, some scientists contend that at least in New York, the use of the term blizzard is not appropriate because the conditions for blizzards were not met (high sustained wind speeds and low visibility).

Figure 13.21 Chicago's Lake Shore Drive during the 2011 Groundhog Day Blizzard. The lakeside boulevard was closed on February 2 as 900 cars and busses were stranded. Tow trucks started pulling the vehicles to temporary lots and the road was reopened before rush hour on the next day. Many drivers had difficulties to retrieve their cars from the lots as the city did not record license plates and the lots were filled bumper-to-bumper.

More recently, the 2011 North American Blizzard made the News[25]. Also called the **Groundhog Day Blizzard of 2011**, it lasted from 31 January through 2 February. The storm brought cold air, heavy snowfall, blowing snow and mixed precipitations on a path from New Mexico and northern Texas to New England and Eastern Canada. An ice storm traveling ahead of the storm's warm front brought hazardous conditions to much of the American Midwest and New England. Many areas experienced ice accumulations of over 1 in (2.5 cm). In the warm sector of the storm ahead of a cold front, several tornado touchdowns were reported in Texas. The blizzard is perhaps most memorable as it brought live in Chicago to a standstill. Whiteout conditions were reported in the city's North Side, with winds exceeding 35 mph (56 km/h). 1,300 flights were canceled at O'Hara and Midway airports. Lake Shore Drive was shut down because it became impassable, stranding 900 cars and buses (Fig. 13.21). Drivers were trapped in their vehicles for as long as 12 h. Schools were closed for the first time since the 1999 Blizzard. Heavy snow and high sustained winds gusting at more than 50 mph (80 km/h) caused rail switched on the CTA's Red Line to freeze, A portion of the roof of Wrigley Field was blown off. The major universities in the city and the area canceled classes for the first time in over a decade and 39,000 state worker were asked to not come to work, the largest number since a blizzard in 1978. Mail services stopped for a day, as was Amtrak train service. 21.2 in (54 cm) of snow fell at O'Hare Airport, making this the third largest total snowfall in Chicago history (after the 1967 and 1999 blizzards). In all, the storm claimed at least 36 lives and caused $1.8 billion in damages.

Also in 2011, the **2011 Halloween Nor'Easter** brought unusually early snowfall to the northeaster U.S. and the Canadian Maritimes. The storm reached a central low pressure of 971 mbar. As it moved up the East Coast, the snowfall broke records in at least 20 cities and some communities considered to cancel or at least postpone Halloween festivities. In Peru, MA, 32 in (81 cm) accumulated. The snow caused significant damage to power infrastructure and the associated widespread power outage of nearly a week affected 12 states and 3 Canadian provinces. More than 3 million people lost power in the middle of this cold storm that killed 39 people. Traffic accidents killed at least 6 and 2 were electrocuted by falling power lines. Two rail services were closed in the New York area, and Amtrak service across the region was either delayed or canceled altogether. Flights were canceled at Newark International Airport after 4 p.m. on 29 October, and flights out of New York's major airport were delayed by up to 5 hours. Several JetBlue flights departing from Bradley International Airport in Hartford, CT sat on the tarmac for up to 7 hours due to hazardous conditions. In some areas, the number of downed trees and the length of blackouts broke records set just two months earlier by Hurricane Irene. Delays in restoring power led to the resignation of the CEO of Connecticut Light & Power amid widespread criticism that the company mishandled both Irene and the nor'easter.

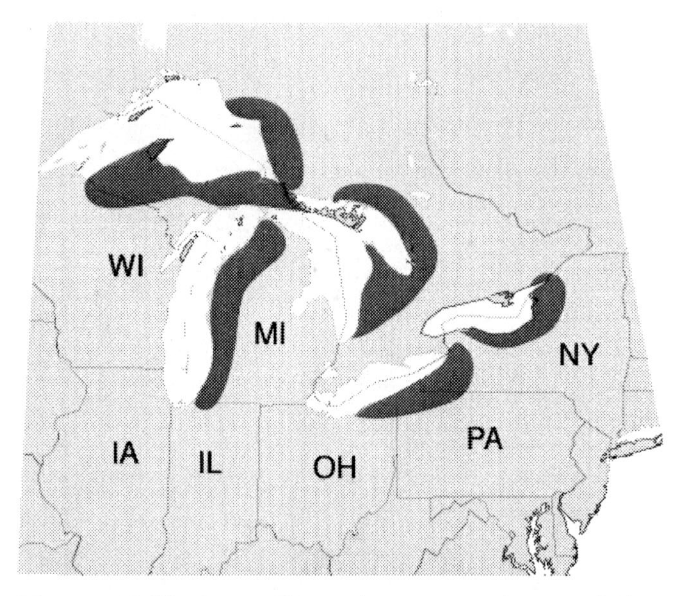

Figure 13.22 Areas along the eastern shores of the Great Lakes in Canada and the U.S. that experience heavy lake-effect shows (dark grey areas). (modified from MT and Wikipedia)

Lake Effect Snow

Under certain conditions, the eastern shores of the Great Lakes and areas up to 50 km (31 mi) inland, receive large amounts of snow. This is the result of the **lake effect** (Fig. 13.22). The phenomenon often occurs from November through January. During this time, the Midwest is often dominated by cold and dry cP air. If this air moves over a relatively warm body, such as the Great Lakes, the still warmer lakes warm the air from below, which then becomes unstable. Through evaporation, the air picks up moisture that eventually condenses to fog after the air rises adiabatically. The air can rise further to form clouds that bring heavy snowfalls as they move eastward. Snowstorms that form on the downwind side of one of the lakes are known as lake-effect snows. The contrast between temperatures of the water and the cold air above can be as much as 25°C (45°F), and studies show that the chances of heave lake-effect snow increase with an increasing temperature difference. As the cold air moves farther east, the heavy snow showers usually taper off, but the western slopes of the Appalachian Mountains give the air lift and increases the possibility of more and heavier snowfalls.

East of the Rockies, Mt. Washington in New Hampshire (elev. 6,288 ft/ 1917 m) holds the seasonal snowfall record with 566 in (14.4 m) in the 1968 – 1969 winter. In an average season, 300 in (7.6 m) of snow falls. At lower elevations, Old Forge NY (pop. 1,061) is the snowiest location, where the lake effect over Lake Ontario contributes to a total of 227 in (5.8 m) of annual snowfall. Some of the most intense snowfalls recorded anywhere in the world occur in this region of the state. The village of Adams holds New York state's official 24-hour record, when an impressive 68 in (1.7 m) fell on 9 January 1976. The snow belts around the Great Lakes that receive the heaviest snowfalls are actually not along the shore but about 30 km (19 mi) inland. Here, hills provide the orographic lift necessary to squeeze out the moisture of the wind-pushed air. The duration of the winds and the distance over which they blow (the fetch) across the lakes also increase the likelihood of heavy snowfalls.

Figure 13.23 High-voltage utility towers in St. Bruno, Quebec collapsed 10 January 1998 under the weight of ice following 80 hours of freezing rain. (source:AP photo/Jacques Boissinet at vpr.net)

Ice Storms

Rain falls when the air temperature stays above freezing for about 2500 m (8,200 ft) above the surface. Snow falls when the temperature stays below freezing. In winter, the maximum temperature occurs somewhere between the surface and 2500 m. If this maximum is at about 1500 m (4,900 ft) and the temperature is above freezing but below freezing at higher altitudes and near the surface, other forms of precipitation can fall. If the altitude range of above-freezing temperatures is only narrow and is high above the surface (called a deep freezing layer), then sleet falls. But if the range is larger and is closer to the surface (e.g. 300 – 2500 m in altitude; called a shallow freezing layer), then the falling rain freezes after it hits the surface, thereby coating everything from tree branches to roads with crusts of ice that can be several cm thick. Driving conditions on ice-covered roads are disastrous, much worse than in snow. Coated tree branches break under the load of the ice. Ice-coated power lines sag until utility poles collapse.

Freezing rain is common in Canada and New England. But the January 1998 North American ice storm was the most expensive natural disaster in Canadian history. Starting on 4 January 80 hours of freezing rain fell on an already wet region and formed 5 cm (2 in) thick coats of ice that downed power lines, bent trees and caused over $5-7 billion (2005 U.S. dollars) in damages in eastern Canada and northeastern U.S. (Fig. 13.23). The highest accumulations of > 10 cm (> 4 in) of freezing rain by 10 January occurred along a corridor on the U.S.-Canadian border from Lake Ontario to Montreal and southern Quebec. Prior to 1998, the last major ice storm to hit Montreal occurred in 1961, depositing 3 – 6 cm (1.2 –2.4 in) of ice. But the 1998 storm left deposits twice as thick. Power was lost to millions of people, with thousands of customers lacking power for nearly a month. The ice storm also was responsible for 7 deaths in New England and 28 in Canada. Parts of Vermont, New Hampshire, Maine and New York were declared disaster areas.

13.5 Mid-latitude Cyclones

So far, we have used the term "storm" loosely but have not yet examined what it means and how storms form. The word storm may refer to different things in different languages and popular usages. In the glossary of the NWS, a **storm** is defined as "any disturbed state of the atmosphere, especially affecting the Earth's surface, and strongly implying destructive and otherwise unpleasant weather. Storms range in scale from tornadoes and thunderstorms to tropical cyclones to synoptic-scale extratropical cyclones." The term "unpleasant weather" may be subjective but, in plain words, this includes severe winds, rainfall, snowfall and lightning when they become strong enough to be bothersome or dangerous. A sea breeze, on the other hand, is not a storm even though it may be unpleasantly chilly. Nor is dreadful drizzle coming out of SoCal's marine layer in June.

Storms then fall into two major categories, the more localized storms and the larger cyclones that are associated with a large-scale low-pressure system. Some localized storms form where large pressure gradients develop. An example would be a thunderstorm developing near a cold front (recall that the air pressure changes quickly across the front). Storms also form locally where warm, moist surface air is unstable and rises. Examples of such storms are thunderstorms developing in summer in the mountains were orographic lifting forces the air to rise. Thunderstorms developing through local heating and convection in tropical areas such as those in Florida also fall in this category. These localized storms develop when rising air leads to condensation and the release of latent heat. This latent heat, in turn, heats the surrounding air, thereby causing it to rise further still until the air becomes cool enough to form cumulonimbus clouds. The warm, moist bottom air that follows from near the surface and rises upward keeps feeding the process until the clouds become thick enough for precipitation to fall. The falling rain causes a downdraft within the storm that eventually overcomes the updraft. Deprived of its fuel, the storm then starts to dissipate.

Wave Cyclones

The other type of storm is the very large-scale (low-pressure) cyclone that always involves rotational movement of air. Recall from Chapter 12 that a low-pressure system is associated with cyclonic air movement around it. Tropical cyclones are also known as hurricanes in certain regions of the world and are discussed in Chapter 15. Here, we concentrate on the extratropical cyclones or mid-latitude cyclones. These are usually not as powerful as tropical cyclones but they can locally still pack hurricane-force winds and be similarly destructive.

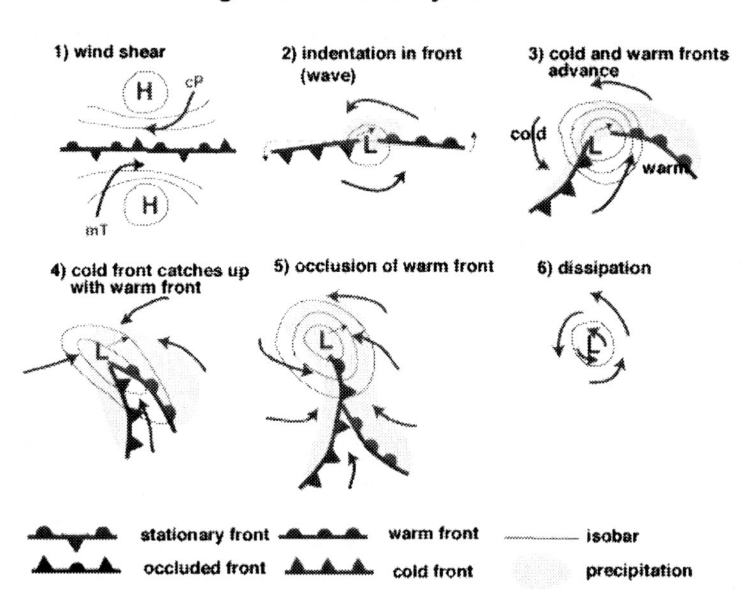

Figure 13.24 The idealized life cycle of a wave cyclone in the Northern Hemisphere based on the polar front theory. As its life progresses, the cyclone moves eastward. The small arrow next to each L shows the direction of storm movement.

The **polar front theory** is a very simplified model of how mid-latitude cyclones develop from two colliding air masses along a stationary front and then evolve as they move eastward, steered by the winds aloft. In our example, a cP air mass approaches from the north and a mT air mass approaches from the south (Fig. 13.24; stage 1). Recall from Chapter 12 that a low-pressure belt and a convergence of surface air exists near 60° latitude. The cP air mass belongs to the Polar circulation cell, while the mT air mass comes from the Ferrel circulation cell. Due to the Coriolis effect, air from the north gets deflected westward, while air from the south gets deflected eastward. This type of flow sets up a cyclonic wind shear, with low air pressure now in the center, not along a line anymore. Under the right conditions, a wave-like indentation develops along the front. A warm front develops on the leading side and a faster-moving cold front develops on the trailing side of the developing low-pressure center (stage 2). The wave that forms is known as a frontal wave and watching its life cycle on a weather map is like watching a water wave from its side first develop, then break and finally dissipate on the beach. This association is why a cyclonic storm is also called a **wave cyclone**.

Leading into stage 3, the indentation becomes more pronounced, and the low pressure near the center intensifies. The storm is now a fully developed "open wave", about 12 – 24 hours after initially forming. Precipitation falls well ahead of the warm front and in a narrow swath behind the cold front. The shrinking region of warm air between the cold front and the warm front is called the warm sector. The weather is usually partly cloudy but some precipitation may fall here as well in the air is unstable. The cyclone is now fueled through several sources of energy: 1) as cold and warm air try to attain equilibrium, warm air rises and cold air sinks, thereby transforming potential energy into kinetic energy that drives thc motion of air; 2) the release of latent heat during condensation provides energy to drive air convection; 3) as surface air converges into the low pressure center, wind speeds and thereby the kinetic energy increase.

Moving into stage 4, the central air pressure decreases and the storm intensifies. The faster cold front catches up with the warm front, until it starts to occlude the warm front in stage 5. The occlusion progresses from the low-pressure center outward. The weather is most severe where the occlusion occurs. The point where all fronts come together is called the triple point. The triple point moves outward from the low-pressure center and the occlusion progresses. Since cold air lies on both sides of the occluded front, the triple point eventually is so far away from the center that the supply of energy from warm rising air to feed to storm is cut off. The storm is now is stage 6 where it dissipates.

Some lows receive their name from were they form. The **Hatteras Low** develops in the Atlantic off the U.S. eastern seaboard. The **Gulf Low** forms in the Gulf of Mexico and the **Colorado Low** on the eastern side of the Rockies near the eastern state boundary. The **Alberta Clipper** forms on the eastern side of the Rockies in Alberta, Canada, then rapidly skirts across the northern tier states.

Wave cyclones can develop to large powerful storms that destroy large stretches of forests. For example, a storm with 130-kn winds (240 km/h; 150 mph) in January 1921 flattened 8 times as many Douglas firs in Washington and Oregon (a process called **wind throw**) as Mt. St. Helens destroyed on its flanks during its 18 May 1980 eruption through its powerful pyroclastic flows.

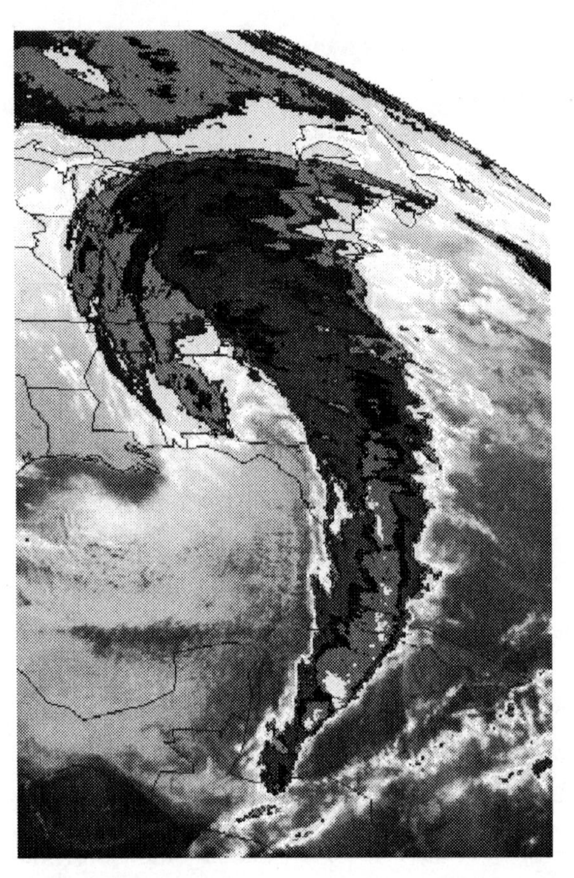

Figure 13.25 NASA satellite image of the 1993 North American Storm on 13 March at 10:01 UTC. The cloud band is in the shape of a comma that covers the entire eastern seaboard. Such **comma clouds** indicate that the storm is still developing and intensifying. (source: Wikipedia)

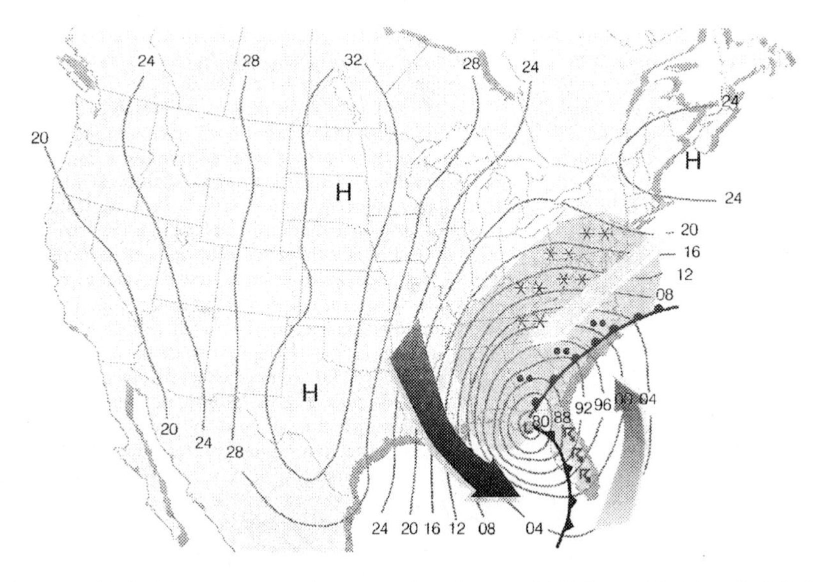

Figure 13.26 Surface weather map for 4 a.m. (EST) on 13 March 1993. Lines on the map are isobars (add 900 mbar for readings of 50 - 99 and 1000 mbar for readings 00 to 49). The air pressure near the

center of the storm was less 980 mbar (960 mbar at the time when it was most intense). Shaded areas mark precipitation. Thunderstorms occurred ahead of the cold front. Three air masses came to play: a warm, humid mT air mass from the Gulf of Mexico rode over a cold, moist mP air mass from the northeast. On the western side, the jet stream pushed a cold cP air mass beneath the mT air mass from the Gulf. High-pressure systems surrounded thc storm. This increased the pressure gradient, and therefore enhanced wind speeds. (source: MT)

The 12 – 15 March 1993 White Hurricane was an immense cyclone that covered the eastern U.S. from Florida to Maine and from the Appalachian Mountains to the Atlantic coast, affecting 26 U.S. states and much of eastern Canada (Fig. 13.25). It is also referred to as the "Storm of the Century" or the '93 Superstorm, and many consider it the most intense and extensive extratropical storm ever to rake the Atlantic seaboard[2]. Hurricane-force winds gusted at over 160 km/h (99 mph) driving a blizzard with snow and sleet into the eastern seaboard and sending a storm surge into coastal areas. The storm caused 310 fatalities from Cuba to Canada, including 48 sailors that were lost at sea. At one point, 10 million people were without electric power. Damages exceeded $6 billion (2008 U.S. dollars). The hardest hit area was Florida where on the order of 15 tornadoes killed 44 people, more people than Hurricane Andrew killed in 1992. The storm brought record-low temperatures on its wake to parts of the U.S. South and East. This storm rivaled the legendary Great Blizzard of 1888 that brought wind gusts of up to 135 km/h (85 mph) and snow drifts of up to 6 m (20 ft) deep. The 1993 White Hurricane occurred during an unusually wet winter and spring that set the stage for the great 1993 Mississippi flood in August (the costliest regional flood and one of the costliest natural disasters in U.S. history). The unusually large and late winter wave cyclone formed on the early morning of 12 March over the western Gulf of Mexico. It then traveled across the Gulf, passed northern Florida, traveled along the eastern seaboard. The center of the storm reach Maine within only 48 hours. The lowest pressure was measured at 4 p.m. on 13 March, when the storm's center was passing the Washington, D.C. area. The storm was fueled by three colliding air masses. Toward the east, a warm, moist mT air mass from the Gulf of Mexico advanced on a cold mP air mass from the northeast; 2) on the western side, the southward looping jet stream pushed a cold cP air mass from the north under the warm air mass from the Gulf (Fig.12.26).

Extratropical Cyclones in Europe

While winter storms are common in Europe, three violent extratropical cyclones of the "storm of the century" category have hit much of Europe within only 17 years, Daria in 1990, Lothar and Martin in

1999, Kyrill in 2007. In the most recent winter in 2011, another strong storm hit a few weeks prior to the February deep freeze, with a minimum low pressure less than that during Lothar.

Figure 13.27 Wind damaged forest near Mössingen, Germany from Extratropical Cyclone Lothar in 1999. Trees were uprooted but many tree trunk broke in half. (source: wikipedia)

In 1999, a pair of violent extratropical cyclones storms wreaked havoc in Europe[(28)]. On 25 – 27 December 1999, **Winter Storms Lothar and Martin** killed 110 and 30 people. The storms caused more than $6 billion in insured damages and 11.5 billion Euros in economic losses. With winds gusting at 272 km/h (169 mph) on the Hohentwiel near Singen, southern Germany on 26 December, Lothar flattened in excess of 100 millions trees with many tree trunks breaking like matches (Fig. 13.27). The author's parents lost all 3 fir trees in their backyard. Lothar traveled in excess of 100 km/h, with hurricane-force winds. On Feldberg, the highest mountain in the Black Forest, Germany, the anemometer broke at 212 km/h (132 mph). In Switzerland, 10 million trees fell, when the storm crossed the northern part of the country from west to east within only 2.5 hours. France was worst hit, recording most fatalities and losing 140 million m^3 (4.9 billion cubic ft) worth of wood. In southern Germany 13 people lost their lives and trees were destroyed worth 35 million m^3 (1.2 billion cubic ft) of wood. In Switzerland, 14 people died and the equivalent of 13 million m^3 (459 million cubic ft) of wood was lost. Fifteen more people died in that country in the aftermath as untrained personnel attempted to clear out the forests. The total loss of wood in Europe amounted to 200 million m^3 (7 billion cubic ft). The German Weather Service was criticized for failing to issue an early warning, apparently as a results of a software bug. In France, Storm Martin arrived on 27

December on the tail end of Lothar, with nearly equal force. The central low pressure in the storms reached 962 and 960 mbar. Wind speeds on the Eiffel Tower in Paris clocked at 216 km/h (134 mph). Over 200 utility pylons toppled knocking out power to more than 3 million homes for several days. Flooding at the Blayais Nuclear Power Plan knocked out power and safety systems and resulted in a "level 2" event in the International Nuclear Event Scale. Nine years earlier, **Extratropical Cyclone Daria** devastated northern and central Europe on 25 – 26 January 1990 after a then rather mild winter. Daria formed over Scotland and reached a central low pressure of 950 mbar. It caused at least 94 fatalities in the UK, Belgium, France, Netherlands and Germany and caused 4-6 billion Euros in insured damages.

Figure 13.28 Wind damaged utility tower near Magdeburg, Germany from Extratropical Cyclone Kyrill in 2007. (source: wikipedia)

In 2007, Europe was again hit by a "storm of the century"[(28)]. **Extratropical Cyclone Kyrill** wreaked havoc on 18 January 2007 across central and western Europe (Fig. 13.28). Winds gusted at 225 km/h (149 mph) and the central low pressure dropped below 960 mbar. Forty-seven people lost their lives, with 13 in Germany and 11 in the UK. Kyrill caused damages estimated at $10 billion from Ireland to the Czech Republic. In anticipation of the storm, hundreds of flights were canceled at London's Heathrow Airport and at the Frankfurt am Main Airport. Ferry traffic between France and the UK and Ireland across the Channel stopped, as did the Eurostar train through the tunnel from Paris to London. Power outages across the U.K. brought rail traffic to a stand still. As a precaution, high-speed trains in Germany first traveled at lower speeds, causing widespread delays. In the evening of 18 January, long-distance rail traffic stopped altogether, for the first time in its history. Thousands of passengers were stranded in railway stations and many slept in the trains. The storm

knocked down major power pylons and caused structural damage to Berlin's newly opened central railway station. At least 150,000 homes lost power across the country. A tornado damaged 20 apartment buildings and a church (a UNESCO world heritage site) in Wittenberg, Sachsen-Anhalt.
As Lothar did in 1999, Kyrill caused significant damage to forests. In Germany alone, 37 million m^3 (1.3 billion cubic ft) of wood was lost. The storm initially developed on 15 January over Newfoundland and crossed the Atlantic Ocean. First severe weather warnings were issued on 16 January. The strengthened storm reached Ireland at 6 a.m. UTC on 18 January. Winds already gusted at 120 km/h (75 mph) along southwestern England's shores. The storm reached the Continent by noon, now reaching 960 mbar and by evening having a pressure gradient across Germany not seen in 20 years. Air pressure dropped by 51 mbar over a distance from the upper Rhine valley to the coast (approx. 600 km/373 mi). The storm continued through the following night. Wind speeds in Vienna, Austria clocked at 148 km/h (92 mph), breaking the 1946 record of 146 km/h (91 mph). The highest wind speeds of the storm were measured at the Aletsch Glacier, Switzerland (225 km/h at an elevation of 2850 m/9348 ft). Kyrill was the third "storm of the century" within only 17 years and it is not clear if the term has to be redefined.

Table 13.5 U.K. Met Office Weather Warning Levels(29)

Color	Warning Level
green	no severe weather
yellow	be aware
amber	be prepared
red	take action(o)

(o) warnings are issued for the following 5 categories: rain, snow, wind, fog, ice

Extratropical Cyclone Ulli (also named Emil by the Norwegian Weather Service), that formed in 31 December 2011 and dissipated on 6 January 2012, has an interesting history and dramatically illustrates how global weather is interconnected(29). On 30 December, a low-pressure system had formed over the U.S. Midwest that had already caused some damage. This low moved into the Atlantic Ocean off the coast of New Jersey on the following day when it was named Extratropical Cyclone Ulli by the Freie Universität Berlin, Germany. The storm then moved northward and strengthened. By the time its centered just east of Newfoundland. Starting at 6 p.m. UTC in 2 January, the barometric pressure dropped by 13 mbar to 970 mbar in only 6 hours by which time the storm was already located northwest of Scotland. It made landfall on Scotland in the early morning of 3 January, with a central pressure of 952 mbar, lower than that of Extratropical Cyclone

Kyrill. A day later, Ulli, now Emil, reached southern Norway before it dissipated over Finland in 6 January. On 1 January, the Irish meteorological service Met Éireann issued a national severe weather warning for Connacht and Ulster, predicting 87-mph winds (140 km/h) and heavy driving rain. A day later, UK's Met Office issued an amber weather warning for most of Scotland (Table 13.5), predicting heavy snow and strong winds.

Leading the passage of Ulli, many parts of the UK experienced heavy squally downpours in 2 January. High winds closed the Kingston, Erskine Tay and Forth bridges and bus, rail and ferry traffic was disrupted. Winds gusted at 169 km/h (105 mph) at Malin Head, Donegal, Ireland and 164 km/h (102 mph) in Edinburgh, Scotland. 10,000 people in Northern Ireland and 140,000 people in Scotland lost power. A tornado touched down in Hainault, London, uprooting trees and damaging roofs. Another tornado was reported 50 km (31 mi) to the east in Clacton-on-Sea 30 min later. Torrential downpours affected parts of southern England and France. High winds of up to 100 km/h (62 mph) were also experienced in Netherlands, Germany, Denmark and Sweden. Two people were killed and one went missing in the UK. The storm caused $306 million (2012 dollars) in damages. The Met Office received criticism for upgrading the warning from amber to red too late.

The Nor'Easter

Northeasters, commonly called nor'easters are large, powerful extratropical cyclones that develop near or off the east coasts of continents and then migrate north along the coast, steered by the jet stream. They get their name from the strong northeasterly winds that blow on its northern side (Fig. 13.29). Nor'Easters occur along some east coasts in the Northern Hemisphere (e.g. Japan) but the term is most widely known for some of the most violent storms along the U.S. eastern seaboard. Nor'Easters often feed on the warm Gulf Stream to intensify rapidly, and some actually feed off the remnants of dissipating late-season hurricanes. Studies suggest that even some nor'easters themselves that batter the coastline in winter possess characteristics of a tropical hurricane. A rapid drop in air pressure occurs in the center and some storms even develop something like a hurricane's eye. For example, the winds at the center of the storm in Fig. 13.29 were calm when it moved over Atlantic City, NJ.

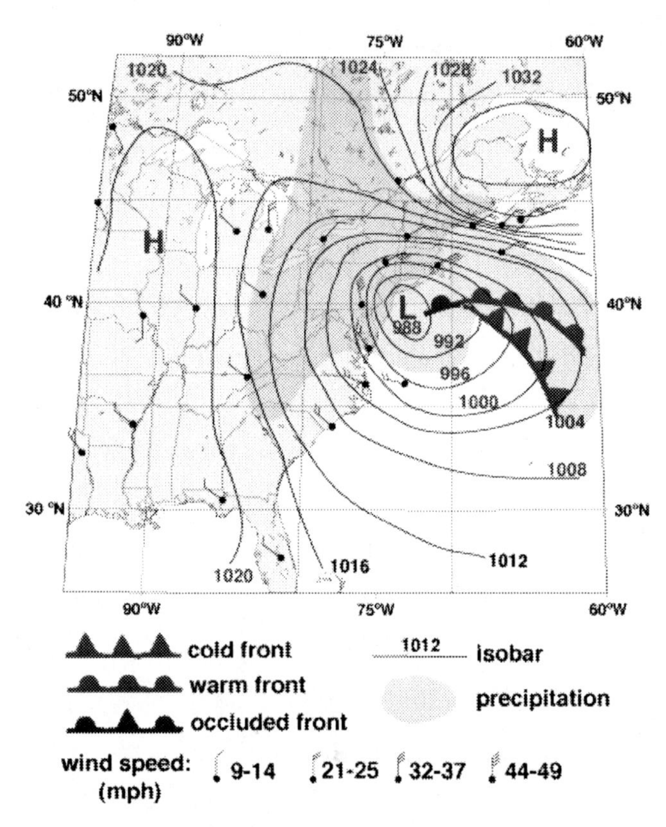

Figure 13.29 The surface weather map at 7:00 a.m. (EST) on 11 December 1992. An intense low-pressure area (central pressure 988 mbar/29.18 in) generates strong northeasterly winds and heavy precipitation (shaded area) from the mid-Atlantic states into New England. This nor'easter devastated a wide area of the eastern seaboard, causing damage in the hundreds of millions of dollars.

Nor'Easters usually occur in late winter and early spring, but they can develop as early as October and as late as April. They are often most intense off the coast of New England. The main characteristics of a nor'easter include gale force northeasterly winds on the northern side of the storm's center. Secondly, because of its large extent over the open ocean, a nor'easter typically generates high waves that cause significant coastal erosion. Thirdly, heavy precipitation falls. Coupled with high wind speeds, colder storms often cause blizzard conditions (e.g. the Great Blizzard of 1888).

What is typical of a nor'easter and makes it different from other storms is the fact the counterclockwise flowing cyclone is fueled along the northern perimeter by bringing in cool but moist mP air from the northeast. This allows the development of heavy precipitation. Despite some similarities to tropical hurricanes, nor'easters are not hurricanes, the main difference being that nor'easters are cold-core low-pressure systems, thriving on cold air, not warm tropical air as hurricanes do.

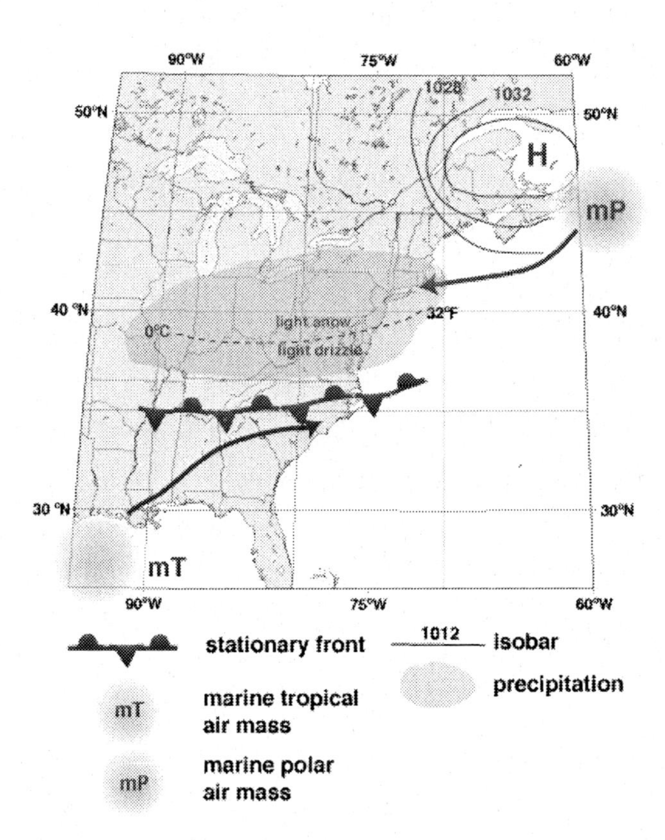

Figure 13.30 A weather situation that often precedes the development of a nor'easters. A nearly stationary front is located over the southern states while a high sits over Newfoundland. It delivers cold, moist marine air from the north to the south to collide with warm, moist marine air from the south along the stationary front. A wave cyclone then develops.

A basic ingredient for the development of a strong nor'easters is a slow-moving, cold continental anticyclone (high-pressure system) that may originally form somewhere else drifts to park over Newfoundland (Fig. 13.30). The High then captures the cool, moist mP air from the Atlantic Ocean and brings it ashore along its southern perimeter. A stationary front separates this air from warmer air further south (mT air brought in from the Gulf of Mexico). North of the front, northeasterly winds provide damp weather with drizzle, rain and snow. When upper atmosphere conditions are just right, storms develop along the front. They can then be pushed eastward by the jet stream and intensify to become nor'easters.

With a minimum central pressure of 992 mbar, the **December 1992 Nor'Easter** shown in Fig. 13.29 produced strong northeasterly winds from Maryland to Massachusetts. Huge waves called **storm surge** were pushed by hurricane-force winds. In Wildwood, NJ wind speeds clocked at 78 kn (144 km/h, 89 mph). The high storm surge caused extensive damage to beaches, beachfront homes, seawalls and boardwalks in New Jersey. Coupled with high tides, the surge caused extensive coastal flooding, engulfing coastal highways. In Atlantic City, NJ winds also gusted at 90 mph and the storm surge reached 8.8 ft (2.68 m) above mean low water, tying the mark left by Hurricane Gloria in 1985, and just inches below the all-time record of 9 ft (2.74 m) set by the 1944 hurricane. The Red Cross estimated that 3,200 homes in New Jersey were severely damaged or destroyed. The high surge also caused the Delaware River to back up, and with 5 in (127 mm) of rain, sent tributaries including the Rancocas Creek over they banks. In metro area of New York City, winds gusted to hurricane force and tide-enhanced storm surge rose 3-5 ft above normal, flooding low-lying areas. On 11 December, power outages caused a 5-hour closure of the subway system. Heavy snow and rain fell for several days. The storm killed between 4 and 19 people (uncertain because sources given conflicting numbers) and caused damages estimated at \$0.5 – \$2 billion dollars[29].

Figure 13.31 A street in New York during the Great Blizzard of 1888. Many overhead power lines broke and presented a hazard to city dwellers. (source: Wikipedia, The New York Historical Society)

In American folklore, no blizzard quite matches the fame of the **Great Blizzard of 1888**, often also called the **Great White Hurricane** (Fig. 13.31). This blizzard came less than two months after the Great Plains Howler than killed some 237 people in the Midwest. That stormed seemed mythically large, so "civilized" East Coast dwellers were much surprised to live through such a storm themselves only a short time later. Contributing to the surprise also was the fact that the weather had been unseasonably mild just before the blizzard hit. This incredible 11 – 14 March nor'easter dumped 50 in of snow (127 cm) in Connecticut and Massachusetts, while New Jersey and the state of New York received 40 in (102 cm). Drifts 40 – 50 ft high (12 – 15 m) buried homes and trains. The storm reached a central low pressure of 982 mbar and sustained wind speeds topped 72 km/h (45 mph). New York's Central Park measured a minimum temperature of -14.4°C (6°F), with a daytime average of -12.8°C (9°F) on 13 March, the coldest ever for March. With 400 fatalities, 200 of them in New York City alone, the Great Blizzard of 1888 has been the deadliest and one of the most severe blizzards in U.S. history. The storm paralyzed the North American east coast from Chesapeake Bay to Maine and into Canada. The telegraphy infrastructure was disabled, isolating Montreal and most of the large northeastern U.S. cities from Washington, D.C. to Boston for days. In New York, neither rail nor road transportation was possible anywhere for days. From Chesapeake Bay to Nantucket, 200 ships sunk[24], leaving 100 seamen dead. At the time, damages were estimated at $25 million, which is 1.2 billion in today's U.S. dollars (2008 dollars). Following the storm, New York began placing its telecommunication infrastructure underground to prevent damage from future such storms. The paralyzing effects from the storm were also partially responsible that Boston opened the first subway system in the U.S. nine years later.

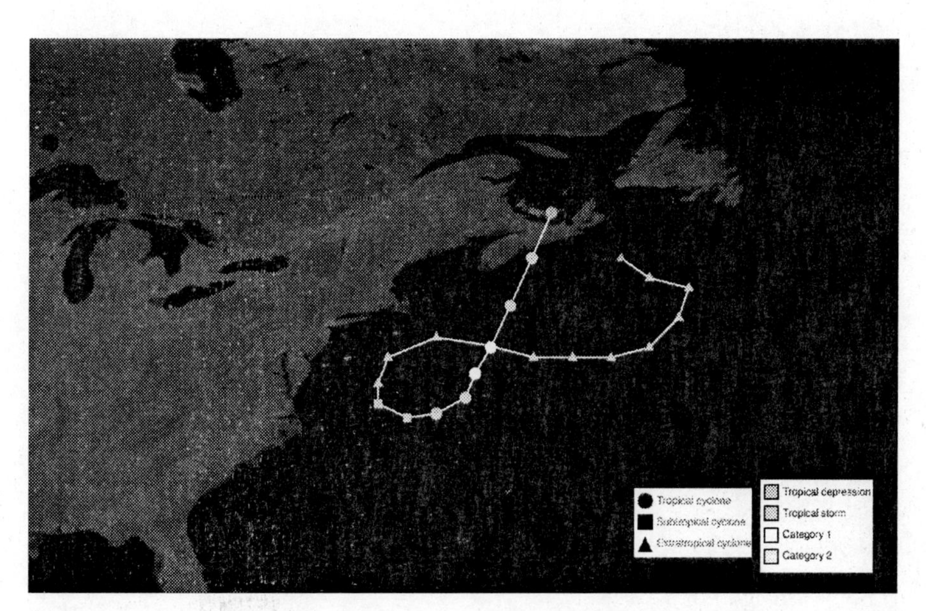

Figure 13.32 The unique track of the Halloween 1991 Perfect Storm. The nor'easter moved southwest from the Canadian coast in the open ocean and joined with remnants of Hurricane Grace. Fueled by the hurricane and the warm Gulf Stream, the storm strengthened into a hurricane, while turning around counterclockwise, moved north and eventually made landfall in Nova Scotia. (source: NOAA, Wikipedia).

Perhaps the most well-known nor'easter in recent history is the **Halloween 1991 Perfect Storm** which is described in the book "The Perfect Storm" by Sebastian Junger and in the movie of the same name. The storm initially developed as a typical nor'easter. But it then captured Hurricane Grace, the last one in the 1991 season. The storm then moved on a unique track eastward, and turned around while fueling on the warm Gulf Stream (Fig. 13.32). This happens only once every 50 to 100 years or so. It developed into a category 1 (tropical) hurricane itself before making landfall in Nova Scotia. The central low pressure reached 972 mbar and sustained wind speeds clocked at 120 km/h (75 mph). However, most of the damage was done while this storm still was an extratropical cyclone. A unique and disastrous aspect of this storm was that it caused incredible waves of up to 30 m in height in the open ocean. A Canadian buoy reported a wave height of 30.7 m (100.7 ft). NOAA buoys 100-150 km further south reported sustained wind speeds of 91 km/h (49 kn) and 117 km/h (63 kn), and significant wave heights (average high waves) of 12 m (39 ft) and 9 m (31 ft). The fishing vessel Andrea Gail was no match for this onslaught and the ship perished with its 6 sailors onboard. Near the coast, the storm surge reached a height of more than 10 m (33 ft), from Canada to Florida and southeast to Puerto Rico. The storm surge caused extensive coastal erosion, mostly in Massachusetts and New Jersey. Heavy rain caused extensive flooding inland. In all, this storm killed 13 people and caused damages estimated at more than $200 million (1991 dollars).

Box 1: Pressure and Temperature Extremes

air pressure[1,2]

lowest recorded air pressure at sea level: 870 mbar (25.7 in) during Typhoon Tip, W. Pacific Ocean (12 Oct 1979)

lowest in western hemisphere: 888 mbar (26.22 in) during Hurricane Gilbert, W. Pacific Ocean (Sep 1988)

highest air pressure at sea level: 1085.6 mbar (32.06 in) in Tosontsengel, Zavkhan Province, Mongolia; equivalent pressure at sea level (Tosontsengel is located at 1,723 m above sea level) (19 Dec 2001)

highest air pressure in North America: 1080 mbar (31.88 in) in Dawson, Yukon (2 Feb 1989)

U.S.: 1079 mbar (31.85 in) in Northway, AK (31 Dec 1989)

contiguous U.S.: 1064 mbar (31.42 in) in Miles City, MT (24 Dec 1983)

smallest difference in pressure extremes:

U.S.: Honolulu, HI: low at 994 mbar (29.34 in), high at 1027 mbar (30.32 in) continental U.S.: San Diego, CA: low at 995 mbar (29.37 in), high at 1034 mbar (30.53 in)

largest difference in pressure extremes:
U.S.: St. Paul Island, AK: low: 926 mbar (27.35 in), high: 1045 mbar (30.86 in)
contiguous U.S.: Charleston, SC: 936 mbar (27.64 in), high: 1045 mbar (30.85 in)

NB: the record high at Tosontsengel may have been surpassed on 29 Dec 2004 as the pressure was above 1086 mbar, as graphs available online indicate[6]. However, this has not been confirmed in the literature nor on Wikipedia.

air temperature[1,2]
lowest T (world): -89.2°C (-128.6°F) Vostok, Antarctica (21 Jul 1983)
lowest T (North America): -66.1°C (-87°F) North Ice, Greenland (9 Jan 1954)
lowest T (North American Continent): -63°C (-81°F) Snag, Yukon, Canada (3 Feb 1947)
lowest T (U.S.): -62°C (-80°F) Prospect Creek, AK (23 Jan 1971)
lowest T (Europe): -58.1°C (-72.6°F) Ust' Shchugor, Russia (31 Dec 1978)
lowest T (Germany): -45.9°C (-50.6°F) Funtensee, Bavaria (24 Dec 2001)

highest T (world): 57.8°C (136°F) Al Aziziyah, Libya (13 Sep 1922)
highest T (North America): 56.7°C (134°F) Death Valley, CA (10 Jul 1913)
highest T (Europe): 48.0°C (118.4°F) Athens, Greece (10 Jul 1977)
highest T (Germany): 40.2°C (104.4°F) Grämersdorf bei Amberg/Karlsruhe and Freiburg (27 Jul 1983/13 Aug 2003)

fastest temperature rise[3]: 27°C (48.6°F) in 2 minutes; Spearfish, SD (22 Jan 1943)
most consecutive days above 100°F (37.8°C)[3]: Marble Bar, W. Australia (31 Oct 1923 – 7 Apr 1924)
fastest temperature drop[3]: 27.2°C (49°F) in 15 minutes; Rapid City, SD (10 Jan 1911)

Box 2: Other Weather Extremes

wind speed
highest wind speed ever measured (with Doppler On Wheels radar unit): 484±32 km/h (301±20 mph); near Oklahoma City, OK; 3-s gust in tornado (3 May 1999)

highest wind speed measured with anemometer: 407 km/h (253 mph); Barrow Island, W. Australia; 3-s gust in Cyclone Olivia (10 Apr 1996)

highest wind speed outside of a tropical cyclone (measured with anemometer): 372 km/h (231 mph); Mt. Washington, NH; sustained 1-min average (12 Apr 1934)

highest 24-h average: 174 km/h (108 mph); Port Martin, Antarctica 21-22 Mar 1951

rain

least annual rate: Antofagasta, Atacama Desert, Chile (none in recorded history)

most in 1 min: 31.2 mm (1.23 in) Unionville, MD (4 Jul 1956)

most in 60 min: 305 mm (12 in) Holt, MO (22 Jun 1947)

most in 12 h: 1,144 mm (45 in) Foc-Foc, La Réunion (8 Jan 1966, tropical cyclone Denise)

most in 24 h: 1,825 mm (71.9 in) Foc-Foc, La Réunion (7-8 Jan 1966, tropical cyclone Denise)

most in 1 year: 26,470 mm (1,042 in) Cherrapunji, India (1860-61)

highest annual average(2): 13,299 mm (523.6 in) Lloro, Colombia (estimated)

highest annual average(3): 11,872 mm (467.4 in) Mawsynram, India (measured)

highest annual average U.S.(2): 11,684 mm (460 in) Mt. Waialeale, Kauai, HI

snow

most in 1 year: 31.1 m (102 ft) Mt. Rainier, WA (19 Feb 1971 – 18 Feb 1972)

most in 1 season (1 Jul – 30 Jun) 29.0 m (95 ft) Mt. Baker, WA (1998 – 1999)

Box 3: Extreme Cities in the U.S.

the 5 hottest U.S. cities (1970-2000), in °F(2)

average annual T: Key West, FL (78.0°); Honolulu, HI (77.5°); Miami, FL (76.6°); Fort Lauderdale, FL (75.7°); West Palm Beach, FL (75.3°)

average July maximum T: Palm Springs, CA (108.3°); Yuma, AZ (107.0°); Phoenix, AZ (106.0°); Las Vegas, NV (104.17deg;); Tucson, AZ (101.0°)

the 5 coldest U.S. cities (1970-2000), in °F(2)

average annual T: Fairbanks, AK (26.7°); Anchorage, AK (36.2°); International Falls, MN (37.4°); Duluth, MN (39.1°); Caribou, ME (39.2°)

average January minimum T: Fairbanks, AK (-19.0°); International Falls, MN (-8.4°), Grand Forks, ND (-4.3°); Alamosa, CO (-3.7°), Williston, ND (-3.3°)

the 5 driest U.S. cities (annual rainfall, in mm/inches)[(2)]

Yuma, AZ (76.2/3.0); Las Vegas, NV (114.3/4.5); Bishop, CA (127/5.0); Palm Springs, CA (132.1/5.2); Bakersfield, CA (165.1/6.5)

the 5 wettest U.S. cities (annual rainfall, in mm/inches)[(2)]

Aberdeen, WA (2,126/83.7); Astoria, OR (1,704/67.1); Mobile, AL (1,684/66.3); Miami, FL (1,676/66.0); Baton Rouge, LA (1,653/65.1)

the 5 snowiest U.S. cities (annual average snowfall)[(2)]

Truckee, CA (203.4"); Marquette, MI (179.8"); Steamboat Springs, CO (173.3"); Oswego, NY (153.3"); Sault Ste. Marie, MI (131.2")

References and Recommended Reading

1. Ahrens, C.D., 2003. Meteorology Today, 7th edition. Brooks/Cole-Thomson Learning, Pacific Grove, CA, pp. 544.
2. Burt, C.C. 2004. Extreme Weather. Norton&Company, New York, pp. 304.
3. Wikipedia webpage on temperatures extremes: http://en.wikipedia.org/wiki/Temperature_extremes
4. Wikipedia webpage on European winterstorms: http://en.wikipedia.org/wiki/List_of_European_windstorms
5. Alfred-Wegener-Institut für Polar- und Meeresforschung, 12 Jan 2012, press release: http://www.awi.de/en/news/press_releases/detail/item/folgt/?cHash=1b63419646171417ea4df6002b4d1b5d
6. Weather Online news feature on Tosontsengel air pressure record. The site also provides access to current data at the weather station (click on Tosontsengel): http://www.weatheronline.co.uk/feature/aa211201.htm
7. The Free Dictionary online encyclopedia: Karaburan at http://encyclopedia2.thefreedictionary.com/karaburan
8. Wikipedia webpage on list of local winds: http://en.wikipedia.org/wiki/List_of_local_winds
9. Maps of average temperatures and other weather parameters in the U.S. can be obtained at the NOAA-sponsored site of the High Plains Regional Climate Center at the University of Nebraska Lincoln. The user can choose from a list of weather parameters as well as time

COPYRIGHTED MATERIAL — DO NOT DUPLICATE, DISTRIBUTE, OR POST

periods. http://www.hprcc.unl.edu/maps/current/ global temperature anomaly maps can also be obtained at NOAA at http://www.ncdc.noaa.gov/ghcnm/maps.php

10. on a dust storm: Los Angeles Times, 18 Oct 2011. Dust storm shrouds Texas city. Online at http://articles.latimes.com/2011/oct/18/nation/la-na-dust-storm-20111019 on the drought: New York Times, 29 Nov 2011. As water levels drop, Texas Drought Reveals Secrets of the Deep. Online at http://www.nytimes.com/2011/11/30/us/texas-drought-is-revealing-secrets-of-the-deep.html?_r=2&scp=2&sq=texas&st=cse
11. individual climate normal data for U.S. stations can be obtained at NOAA directly at http://www.ncdc.noaa.gov/oa/climate/normals/usnormals.html though the data need to be processed. More readily available graphs and climate history can be obtained at http://www.wunderground.com
12. NOAA's web page on the heat index and heat safety at http://www.nws.noaa.gov/os/heat/heatindex.shtml
13. "Luke Howard: The Man Who Named The Clouds" on the website of the Weather Doctor: http://www.islandnet.com/~see/weather/history/howard.htm this website sites Howard, Luke, 1833. The Climate of London, Vol. I-III, Harvey and Darton, London.
14. National Weather Service website on the urban heat island in changing environments; part of the Professional Development Series online training, unit 6: http://www.nws.noaa.gov/om/csd/pds/PCU6/IC6_2/tutorial1/Factors_environment.htm
15. Billion Dollar U.S. Weather/Climate Disasters. Website of NOAA's National Climate Data Center at http://lwf.ncdc.noaa.gov/oa/reports/billionz.html
16. the data can be obtained from the National Climate Data Center (NCDC) at http://www.ncdc.noaa.gov/oa/climate/stationlocator.html heat index and fatality data were taken from: NOAA/U.S. Department of Commerce/NWS, 1995. Natural Disaster Survey Report: July 1995 Heat Wave. NWS, Silver Spring, MD, pp. 74.
17. Klinenberg, E., 2002. Heat Wave: A social autopsy of disaster in Chicago, Univ. Chicago Press, Chicago, 305 pp.; see also a NOAA summary of heat wave effects in Iowa at http://www.crh.noaa.gov/dvn/?n=07171995_summerof1995heatwave NOAA assessment of heat waves in Wisconsin at http://www.crh.noaa.gov/mkx/?n=heatflyer this site also gives tips on what to do during heat waves and how to ease heat-related illnesses
18. Wikipedia and NASA Earth Observatory web pages of the 2003 European heat wave at: http://en.wikipedia.org/wiki/2003_European_heat_wave and http://earthobservatory.nasa.gov/IOTD/view.php?id=3714

19. WMO, 11 Aug 2010. Unprecedented sequence of extreme weather events. At http://www.preventionweb.net/english/professional/news/v.php?id=14970
20. meteorological data at Clermont-Ferrand can be retrieved through freemeteo.com at http://freemeteo.com/default.asp?pid=15&gid=3024635&la=3
21. Münchener Rückversicherungs-Gesellschaft: TOPICS geo. Tabelle Hitzetote in Europa im Sommer 2003, pp. 25.
22. NASA Earth Observatory MODIS image and description at http://earthobservatory.nasa.gov/IOTD/view.php?id=77126 German and English Wikipedia webpages on the 2012 European cold wave at: http://de.wikipedia.org/wiki/K%C3%A4ltewelle_in_Europa_2012 and http://en.wikipedia.org/wiki/2012_European_cold_wave as well as Dr. Jeff's Master's blog at wunderground http://www.wunderground.com/blog/JeffMasters/comment.html?entrynum=2030
23. NOAA and NWS webpage on wind chill at http://www.wrh.noaa.gov/fgz/safety/windchil.php?wfo=fgz the NWS has a wind chill calculator at http://www.nws.noaa.gov/om/windchill/
24. NOAA webpage on blizzards at http://www.wrh.noaa.gov/fgz/science/blizzard.php?wfo=fgz and NWS Weather Forecast Office website on largest snowstorms in the U.S. at http://www.crh.noaa.gov/mkx/?n=biggestsnowstorms-us
25. Wikipedia page on the 2011 North American blizzard at http://en.wikipedia.org/wiki/Chicago_Blizzard_of_2011#Illinois and Dr. Jeff Masters' weather blog "Great Blizzard pounding Chicago" at http://www.wunderground.com/blog/JeffMasters/comment.html?entrynum=1739
26. online Britannica entry for front at http://www.britannica.com/EBchecked/topic/220786/front and NOAA/NWS website on weather map legend at http://www.hpc.ncep.noaa.gov/html/fntcodes2.shtml
27. current weather maps issued by the NOAA/NWS Hydrometeorological Prediction Center can be accessed at http://www.hpc.ncep.noaa.gov/national_forecast/natfcst.php
28. English, German and French Wikipedia webpages on Storms Daria, Lothar and Kyrill as well as own accounts
29. Wikipedia webpage on Cyclone Ulli at http://en.wikipedia.org/wiki/Cyclone_Ulli the official website of U.K.'s Met Office is at http://www.metoffice.gov.uk/

30. Wikipedia page on 1992 Nor'Easter as well as on hurricanes-blizzards-noreasters.com at http://en.wikipedia.org/wiki/User:Bobby122/December_1992_Nor'easter and http://www.hurricanes-blizzards-noreasters.com/1992noreaster.html

National Weather Service Severe Weather Forecast. Virtually all commercial weather websites and news agencies depend on the NWS for data:
http://www.weather.gov/

Glossary of the National Weather Service (NWS) at
http://weather.gov/glossary/

National Weather Service Regional Office, Central Region Headquarters:
http://www.crh.noaa.gov/crh/

NOAA's Annual State of the Climate Global Analysis (since 1997):
http://www.ncdc.noaa.gov/sotc/global/

The Wikipedia list of weather records provides easy and ready access to extremes in temperature, pressure, wind speeds, various types of precipitation. It also lists severe tornadoes, tropical cyclones and other severe weather. NB, there are two Tosontsengel in Mongolia and the one in Khüovsgöl Province is the wrong one:
http://en.wikipedia.org/wiki/List_of_weather_records

The Wikipedia list of extreme weather events is also very helpful:
http://en.wikipedia.org/wiki/List_of_extreme_weather_events

Chapter 14: Windstorms, Thunderstorms and Tornadoes

Sometimes, severe wind conditions are not related to any typical storm that produces precipitation. Being quite the opposite, prolonged droughts can cause disastrous conditions in which dust storms remove the top soil from large agricultural areas. Large dust storms can carry this soil across large oceans. Certain configurations of high- and low-pressure systems can cause damaging windstorms, also a type of storm without precipitation. Heat waves often end with violent thunderstorms that extend over a large area. Violent derecho winds can occur along long lines of thunderstorms. Very large, hail-producing thunderstorms can also spawn devastating tornadoes.

14.1 Droughts and Dust Storms

Table 14.1 Palmer Drought Severity Index

Palmer Index	**Soil Moisture**
Above +4	extremely moist
+3 to +4	very moist
+2 to +3	moist
-2 to +2	average, normal
-3 to -2	dry
-4 to -3	very dry
below -4	extremely dry

Droughts occur after long periods (a month or more) with less than 30% of normal precipitation. The actual amount of precipitation depends on the climate zone, however, and someone's drought may be "normal climate" for someone else. In Southern California, summer is the dry season. In San Diego, CA a total of only 0.12 in (3.1 mm) of rain from June through August is considered normal. In Seattle, WA, the summer also is the dry season but so little rain would be considered a drought condition, because the normal for Seattle is 3.2 in (81.3 mm). The severity of a drought is often described through the Palmer Drought Severity Index (Table 14.1). This index was developed in 1965 by meteorologist Wayne Palmer. It is based on a complex supply-and-demand model for the soil moisture and works best for longer droughts that last several months. Critics of the scale point out, however, that it does not take into account snow and frozen ground. But the scale is nevertheless widely used today.

Figure 14.1 May 1992 view from the space shuttle Endeavour of a huge dust storm in the Sahara desert, which covered hundreds of miles in Libya and Algeria. (source: The Guardian/NASA)

Short-term droughts may be closely related to heat waves, as the 2003 European heat wave showed. But not every heat wave leads to a drought, and not every drought is related to a heat wave. Long-term droughts last for years, maybe for decades. Some of the worst droughts, and related famines, have occurred in the sub-Sahara Sahel where rainfall has stayed below normal for several decades. Droughts almost always go hand in hand with dying vegetation. This exposes the soil beneath which is now vulnerable to wind erosion, or to erosion by floods when the drought ends. If droughts persist for long times, an area becomes **desertified**.

In large desertified regions, powerful dust storms can pick up the exposed soil and carry dust and other particles high into the atmosphere. High-altitude winds can then easily carry these particles across oceans. Such large dust storms can be seen from space (Fig. 14.1). The frequency of Saharan dust storms has increased approximately ten-fold during the half-century since the 1950s, causing loss of top soil in Niger, Chad, northern Nigeria and Burkina Faso. Dust storms decrease visibility, sometimes severely. The sand blasted by strong winds is highly abrasive and intrudes buildings through even then smallest cracks. Air quality is largely reduced which causes health problems. Prolonged and unprotected exposure of the respiratory system in a dust storm can cause Silicosis, a condition that can lead to asphyxiation if untreated. There is also increasing evidence that pathogens travel on the dust particles across oceans. This aspect of dust storms is examined in more detail in Chapter 19.

Figure 14.2 Effects from the U.S. Dust Bowl in the 1930ies. A dust storm approached Stratford, TX in 1935. (source: Wikipedia/USDA)

Figure 14.3 Effects from the U.S. Dust Bowl in the 1930ies. Buried machinery in a barn lot in Dallas, South SD in May 1936. (source: Wikipedia/USDA)

The Dust Bowl

In the U.S., droughts can occur anywhere when precipitation are less than normal. But their impact is most devastating in the Midwest because this region depends heavily on non-irrigated agriculture (dry farming). Also called the **Dust Bowl**, or the Dirty Thirties, the most memorable drought in U.S. history occurred in the 1930s (Figs. 14.2, 14.3). Starting in 1929, several years of below-normal rainfall dried out the soil to the point where farming collapsed. The dry, now exposed soil became airborne during severe dust storms. These "Black Blizzards" often reduced visibility to only a few feet. The soil was carried high enough to catch the jet stream and travel across the Atlantic Ocean into Europe. The severe soil erosion affected 100 million acres (400,000 km^2) of farmland in 756 counties in 19 states. The worst affected areas centered on the panhandles of Texas and Oklahoma, and adjacent parts of New Mexico, Colorado and Kansas. With no farmland left, hundreds of thousands of people had to leave their homes. Some migrated to California and other states, while some stayed on as migrant workers to earn pitiful wages. Only very few were probably able to improve their situation as the economy was slow during the Great Depression.

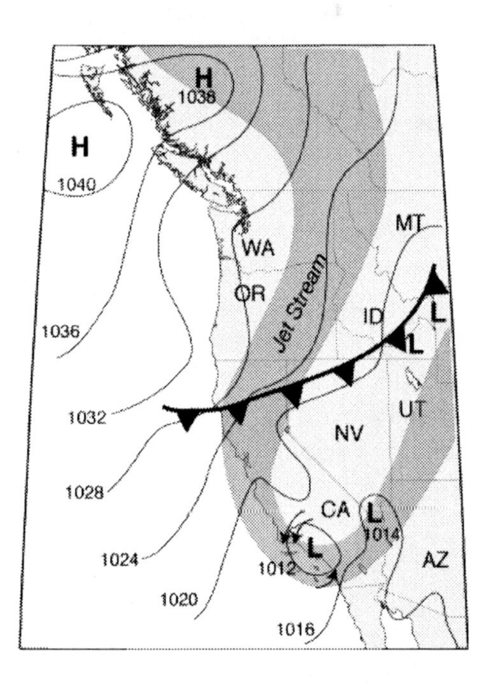

Figure 14.4 Basic features on the weather map of the 30 November – 1 December inside slider. High-pressure systems formed in the northwest where a cold front passed on the previous day. The cold front moved south and replaced high-pressure systems in Idaho. A strong local low-pressure system formed in the Southern California Bight on 30 November. The jet stream developed a far south-reaching loop that wrapped around this L. The insider slider brought no precipitation to California but

high winds to California, particularly the greater Los Angeles area, and several states to the east. (source: data taken from NOAA and the jet stream archive at SFSU [12])

14.2 Windstorms

The California Insider Slider

High atmospheric pressure is usually associated with fair weather though some high-pressure systems with strong pressure gradients can cause strong, damaging winds. The Nevada High associated with Southern California's Santa Ana winds is an example. A recent event with strong Santa Ana-like winds (that was not related with devastating wildfires) occurred during an **inside slider** on 30 November 2011 (Fig. 14.4). This event was also called a "cold Santa Ana" because the winds blew like during a Santa Ana but temperatures were relatively low. The glossary webpage of NWS Forecast Office of the San Francisco Bay Area defines an inside slider as a term used to describe a weather system that moves into California from the northwest, with the bulk of the system's energy moving inland toward the Sierra Nevada and the Great Basin[11]. During a Santa Ana, on the other hand, a strong high-pressure system forms over the Great Basin. Inside sliders usually bring cool, breezy weather to Northern and Central California. A north-south blowing jet stream allows the inside slider to move south to affect Southern California as well. Since the associated winds do not pass over the ocean, an inside slider normally produces only light precipitation, if any. On some occasions, surface high pressure builds strongly into the Great Basin, causing an offshore weather pattern similar to that of a Santa Ana.

The 30 November – 1 December 2011 brought fierce winds to the greater Los Angeles area[12]. The NWS called this a "once-in-a-decade" event. On the afternoon of 30 November, two runways had to be closed at Los Angeles International Airport (LAX) because of wind-blown debris. During the evening hours, twenty flights had to be diverted to Ontario International Airport, CA because of severe crosswinds. The strong winds knocked out power in areas including El Segundo, La Cañada Flintridge, Venice, Westchester, Monrovia and Pasadena. Some 370,000 customers lost power, including LAX and residents nearby. This widespread weather event also caused 18,000 customers to lose power in along California's Central Coast in Santa Cruz and Monterey counties, and thousands more as far as the state of Utah, and high-wind advisories to Nevada, Arizona and even to Wyoming and New Mexico. Hurricane-force winds gusted at 97 mph (156 km/h) at Whitaker Peak in the Los Angeles National Forest. Gale-force winds reached 47 mph (76 km/h) at LAX. In nearby Venice, wind speeds clocked at 36 mph (58 km/h). In the state of Utah, winds reached 100 mph (161 km/h) a day later. Griffith Park was closed out of concern that downed power lines might spark fires in the piles of hundreds of dry, fallen trees. The foothills of the San Gabriel Valley were hit hardest by the windstorm. A state of emergency was declared in Pasadena, where hurricane-force winds gusting at 123 mph (198 km/h) damaged 200 buildings, more than 3 dozen so badly that they were "red-tagged" (deemed unsafe to use). One tree fell on a gas station. In South Pasadena, residents were urged to conserve water because a pump to the city's reservoir failed after the power outage. In Arcadia, east of Pasadena, the streets were lined with fallen trees. The Los Angeles County Arboretum and Botanical Garden was closed for three weeks after 700 trees were damaged, including a loss of 235 trees. At least one person was killed when a tree fell on a wildlife biologist in Monterey county. Dry, more classical Santa Ana winds followed two days later with storm-force winds gusting at 73 mph (118 km/h) on a mountain near Acton in north central Los Angeles County. Intriguingly, only mountainous areas in San Diego's east county recorded wind gusts nearing 50 mph (81 km/h), while this inside slider was a "no-show" in the city.

Figure 14.5 A haboob approaching Al Asad, Iraq, just before nightfall on 27 April 2005. (source: Wikipedia/U.S. Marine Corps)

The Haboob and other Sandstorms

The terms dust storm and sandstorm are often used synonymously for storms in arid (desert) and semi-arid regions that pick up and transport sand. Dust storm is more inclusive and is also used for storms taking dust high into the atmosphere. Here, we will focus on effects closer to the surface. Winds for powerful sandstorms have to be strong enough to keep the sand suspended. Initially picked up by winds, sand particles start to leap around or **saltate**. The sand particles break up into smaller pieces that are then easier to go into suspension. Recent research suggests that the friction caused by saltating sand induces a static electric field where the sand particles are negatively charged relative to the ground. This in turn loosens more sand particles from the ground, allowing the sandstorm to grow. Strong winds can form in areas with strong pressure gradients, and quite often are associated with advancing dry cold fronts (i.e. not producing any precipitation), as occurred in the Dust Bowl in the 1930s.. Many powerful sandstorms are also related to large thunderstorms.

A spectacular example of a sandstorm is the haboob (from Arabic *hebbe* for "blown"). Photos of haboobs in Iraq circulate widely on the internet (Fig. 14.5). These sandstorms are most common in the African Sudan and in the desert of the U.S. Southwest, particularly in Arizona but they also occur elsewhere, e.g. in Australia. Haboob winds form as cold downdrafts along the leading edge of thunderstorms and usually occur in the warm season. Dust or sand is lifted into a huge, tumbling dark cloud that may extend horizontally for over 150 km (93 mi) and rise vertically to the base of the thunderstorm. Spinning whirlwinds of dust frequently form along the turbulent cold air boundary, giving rise to huge dust devils (called willy-willy in Australia) and even tornadoes. Haboob winds can travel at 35-100 km/h (22-62 mph) and may approach with little warning.

In Arizona, several haboobs formed from July through October in 2011. A haboob on 5 July caused a 45-min shut-down of Sky Harbor International Airport in Phoenix(2). Winds gusted at 53 mph (85 km/h) at the airport and visibility fell to less than 1/8 mi (200 m). Haboobs also occur in deserts and drought-stricken regions elsewhere in the U.S.. Severe drought conditions in Texas during the summer of 2011 paved the way for a large haboob that engulfed the city of Lubbock in mid-October(3). In this case, the driving force was not a thunderstorm but a cold front. Meteorologists are concerned that conditions there have been very reminiscent of the Dust Bowl in the 1930s. Lubbock, a city of 230,000, is used to minor dust storms but the one in 2011 was very large and dark, with wind speeds reaching 75 mph (121 km/h). The sandstorm knocked out power, tipped small planes, toppled trees and damaged and airport hangar.

The **shamal** is a northwesterly wind blowing over Iraq and the Persian Gulf states can cause sandstorms. Shamal comes from the Farsi word for "north" and "wind". A shamal typically lasts for 3 to 5 days. The winds are strong during the day and weaken at night. The strong winds are funneled through the mountains of Turkey and Iraq. The winds are usually strongest in spring and summer. During these times, the winds are associated with strong cold fronts passing of the mountains. Shamals in winter are associated with the strengthening of high pressure over the Arabian peninsula after the passage of a cold front. These sandstorms occur a few times a year. They are several thousand feet deep and so travel by air and ground comes to a standstill. A notable shamal occurred on 8 August 2005 and covered Baghdad with sand. Thousands of people with respiratory distress had to be treated at hospitals.

A **simoom** (or simoon) is a very rare sandstorm with local cyclonic winds. The name comes from the Arab word for "to poison". This very short-lived sandstorm (a few hours at most) is very dry and hot, with rel. humidity reaching into the single digits and temperatures exceeding 54°C (129°F). It has a suffocating effect on humans and animals and also causes heat strokes. The Simoom blows in the Sahara, Palestine, Israel, Jordan, Syria and the Arabian Peninsula. There is also a 19th-century account of a simoom in Egypt. The only storm remotely meeting this description that has ever occurred in North America is a sandstorm on 17 June 1859 in Goleta and Santa Barbara, CA. While morning temperatures were normal on that day (24–27°C/75-80°F) hot winds blew from the Santa Ynez Mountains by early afternoon. At 2 p.m., temperatures reached 56°C (133°F) and by 7 p.m. they were down back to normal. However, the great length of the heat and the origin of the hot air (from an inland valley) are not typical for a simoom.

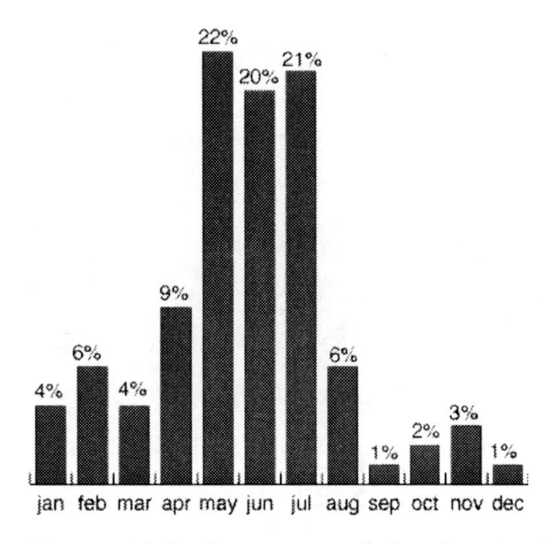

Figure 14.6a Occurrence of derechos in the U.S. throughout the year. (source: data taken from NOAA[1])

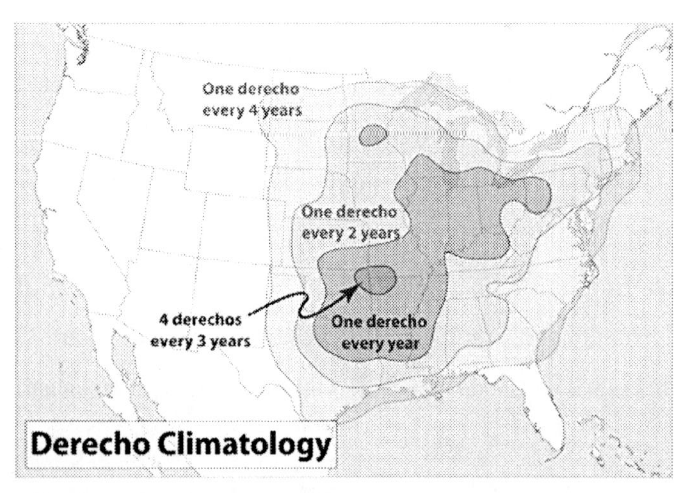

Figure 14.6b Geographic distribution and occurrence of derechos in the U.S.. (source: NOAA[1])

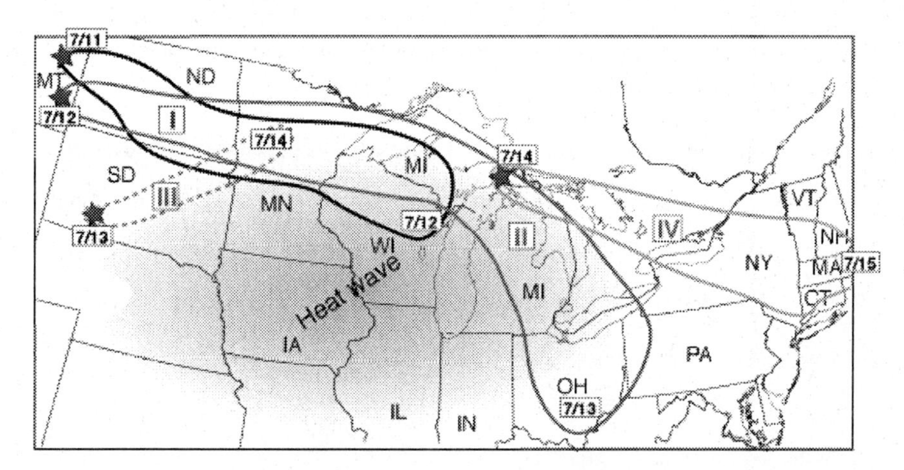

Figure 14.7 Outlines areas that were affected by the four derechos that occurred during the "Chicago heat wave" on 11 – 15 July 1995. The sequence of derechos are marked by Roman numerals. The derechos started at the western end on the evening of the days marked by a star and the date. Derechos I ended by mid-morning on the day after it started (date box in the eastern corner). Derecho II (also called the "right-turn" derecho) ended in southern Ohio by late evening on the day after it started. The relatively short-lived Derecho III lasted only until dawn. Derecho IV moved off the coast of southern New England by mid-morning of the day after it started. (modified from NOAA[(1)])

Derechos

Strong downburst winds associated with a cluster of severe thunderstorms can form so-called straight-line winds that can exceed wind speeds of 165 km/h (104 mph; hurricane-force according to Beaufort Scale, Table 13.2). These winds are not associated with any rotational motion, such as tornadoes are, and blow in the direction the thunderstorms are moving. If the damage from these winds extends for several hundreds of kilometers, these winds are called **derechos**, after the Spanish word for "straight ahead". Typically, derechos form in the early evening and last throughout the night.

In the U.S., derechos are most common during late spring and early summer, where 63% of the derechos develop from May through July (Fig. 14.6a). They usually form along two principal axes. One extends along the "Corn Belt" from the upper Mississippi Valley southeast into the Ohio Valley, and the other from the southern Plains northeast into the mid Mississippi Valley (Fig. 14.6b). The NOAA website on derechos[(1)] hosts an extensive list of past derecho events, as recent as the 8 May 2009 **Super Derecho** that devastated parts of Kansas, Missouri and Illinois.

During the **11 – 15 July 1995 "Chicago heat wave"**, a very warm and humid air mass over parts of the Midwest contributed to the development of several derechos that progressed within 18 hours or less from the west to the east along a stationary west-east front along the northern fringes of the heat wave region (Fig. 14.7). People in Minnesota experienced damaging derechos during three consecutive nights. Derecho II, also called the **Right Turn Derecho**, produced wind speeds of 113 km/h (70 mph) in Bismark, ND and 146 km/h (91 mph) at the Fargo, MN airport. Damage was extreme in Minnesota, with over 5 million trees blown down and many buildings damaged and some destroyed. Damage totaled well over $30 million (1995 dollar). Millions of dollars of damage and 3 fatalities were the result on the next day in Michigan and 400,000 customers lost power. On Lake Erie, many boaters were caught by surprise by the approaching winds and dozens of boats were capsized. At one point over Canada, the wind speeds clocked at 120 km/h (75 mph). In Ohio, where winds speeds of 142 km/h (88 mph) were recorded, 3 people were killed. Most fatalities occurred from fallen trees. Derecho IV (the **Ontario-Adirondacks Derecho**) reached similar wind speeds, with 136 km/h (85 mph) clocking at the Buttonville Airport near Toronto. The most serious damage was caused along and near the northeastern corner of Lake Ontario. Several tornadoes, including an EF2, were also observed. The damage was estimated at $35 million (1995 Canadian dollars) in insured property and Canada. In the state of New York, five people were killed and 900,000 acres (3,640 km^2) of forest damaged. The loss in timber was estimated at $200 million (1995 U.S. dollars). At the Syracuse Airport, winds pushed a parked Boeing 727 into another plane. The Ontario-Adirondacks Derecho was one of the most costly severe thunderstorm events to occur in eastern North America during the 20th century causing nearly $500 million in damage. The derecho raced along a nearly 1300-km (800 mi) long path, at an average speed of 105 mk/h (67 mph). Seven people lost their lives. In total, the four derechos killed 14 people and injured nearly 100.

Derechos develop also elsewhere in the world where the meteorological conditions allow it. On 10 July 2002, a derecho occurred over eastern Germany and adjacent countries in Europe. In the greater Berlin area, 8 people perished and 39 were injured, mainly from falling trees. In Bangladesh and neighboring India, some windstorms also exhibit the characteristics of derechos.

14.3 Thunderstorms

Thunder and lightning are probably the type of severe weather that most of us can relate to. A lightning bolt is frightening but also awesome to watch. And chances are pretty good that it will not strike you while you watch. Thunderstorm clouds often grow into beautiful towering, cauliflower-like cumulus clouds. Nevertheless, the risk from thunderstorms is significant. Lightning is the #1 natural cause for wildfires. Large thunderstorms are relatively short-lived, but they can cause devastating flash floods through heavy rain. Strong gusty winds are usually associated with thunderstorms and severe thunderstorms produce hail. A few very large thunderstorms, so-called supercell thunderstorms can spawn tornadoes. Thunderstorms able to produce tornadoes are called tornadic. About 2,000 thunderstorms occur worldwide at any given time, and over 100,000 occur in the U.S. each year.

About 10% of these are classified as severe. A typical thunderstorm is 15 mi (24 km) across and lasts 30 min(17).

Thunderstorms form in three basic ways:

1) A cold front moves into a region of a particularly warm air mass. This occurs quite often in mid-latitude North America when cold polar air collides with warm air from the Gulf of Mexico.
2) Solar radiation drives convective lifting in moisture rich regions. In tropical regions, this often leads to nearly daily thunderstorms. Examples for this are tropical rain forest but also Florida.
3) Orographic lifting causes warm moist air to rise. Condensation and the release of latent heat warm the surrounding air, causing further lifting. Orographic lifting occurs particularly in summer along mountains, such and the Rockies and the Sierra Nevada, when the heated surface air has no other way to go but up. Thunderstorms then form in the later afternoon.

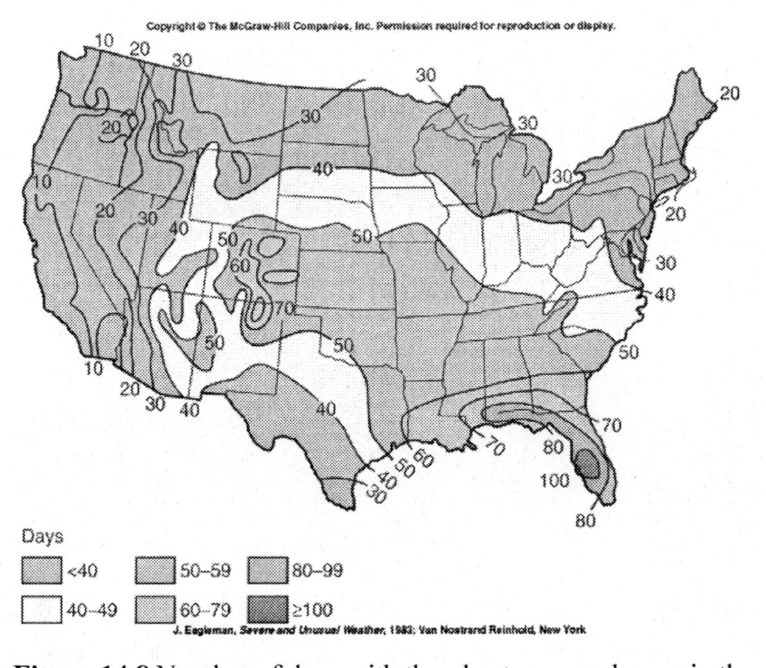

Figure 14.8 Number of days with thunderstorms each year in the contiguous U.S.. (source: AB)

In the U.S. thunderstorms occur most often in Florida were convective lifting provides the energy for thunderstorms (Fig. 14.8). In some areas in Florida, thunderstorms occur nearly every third day. The Rocky Mountains provide ample orographic lifting so that Colorado is the state with the second most days on which thunderstorms occur. Most of these days are during the summer.

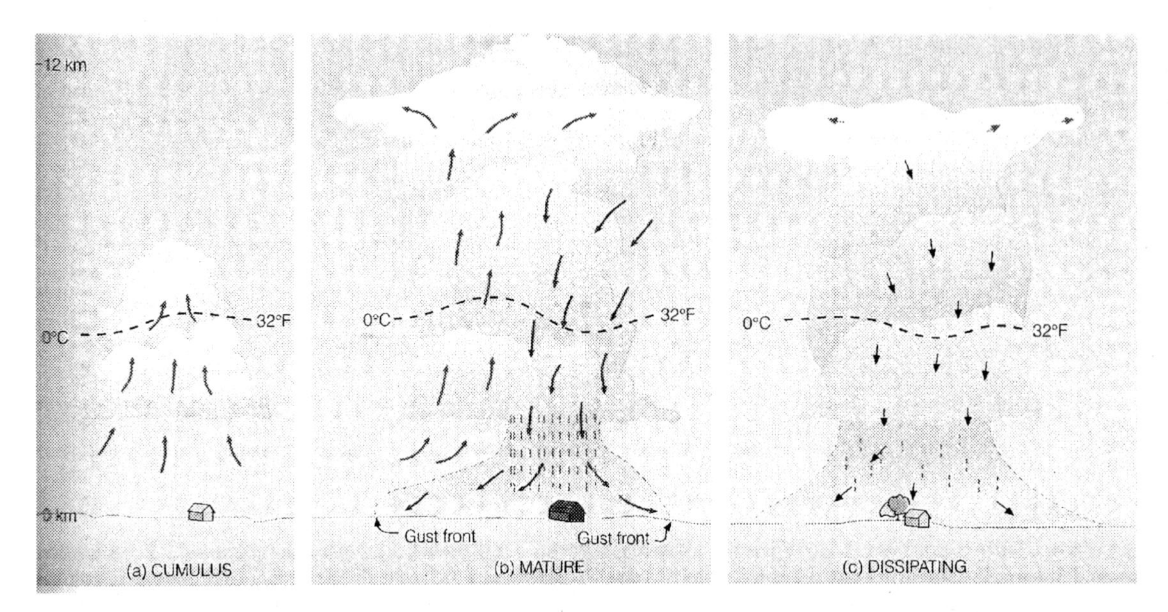

Figure 14.9 Simplified model of the three stages in the life cycle of a thunderstorm. The thunderstorm shown here is nearly stationary (stays in place) or moves slowly to the right. The thunderstorm is most intense in the mature stage. (source: MT)

Figure 14.10 The towering anvil cloud from a thunderstorm in its mature state over Swifts Creek, Victoria. (source: wikipedia)

The **life cycle of a thunderstorm** has three basic stages (Fig. 14.9): the towering cumulus stage, the mature stage and the dissipating stage. In the towering cumulus stage, the formation of a thunderstorm is often similar to that of an ordinary storm, except that the cumulus cloud reaches particularly high into the atmosphere. Strong updraft and the growth of the towering cloud are fed by latent heat released during condensation of large amounts of warm rising air. When the clouds reach the tropopause, they usually spread to form anvil clouds because conditions in the stratosphere prevent further ascend (Fig. 14.10). Updrafts within the clouds are particularly strong so that electric charges can separate (H^+ and OH^-). This sets the stage for lightning to occur.

As the cloud builds well above the freezing level, the cloud particles grow larger and heavier. Eventually, the rising air is no longer able to keep them suspended and they begin to fall. A thunderstorm is in its mature stage once rain starts to fall (Fig. 14.11). The rain triggers a strong downdraft that, together with the updraft constitutes a (convection) **cell.** In most storms, there are several cells, each of which may last for an hour or so. Thunderstorms are most severe during this stage. The cloud may now extend to an altitude of over 12 km (40,000 ft) and be several km in diameter at its base. At this point, the thunderstorm can disturb even high-flying air traffic, so pilots usually change routes to avoid the thunderstorm. Updraft and downdraft reach their greatest strength in the middle of the cloud, creating severe turbulence. Lightning and thunder are now present. Heavy rain and occasionally small hail falls from the leading end of the cloud. At the surface, there is often a down-rush of cold air with the onset of precipitation. A gust front forms on the ground where the colder air intrudes into the warmer surrounding air. Along the gust front, winds can change rapidly in speed and direction. To an observer on the ground, the passage of the gust front resembles that of a cold front as temperatures drop sharply.

Figure 14.11 The out flow from a thunderstorm's core shows up as sheets of wind-drive rain that spreads from right the left. (source: NOAA Digital Image Library[13])

After the storm enters the mature stage, it begins to dissipate within 15 – 30 min, when the downdraft overpowers the updraft. When this happens, the thunderstorm is cut off from its supply of fueling warm, moist surface air. The updraft weakens and the gust front moves away from the storm.

Light precipitation now falls, accompanied by only weak downdraft. As the storm dies, the lower-level cloud particles disappear rapidly, sometimes leaving only the cirrus anvil as the reminder of a once mighty thunderstorm.

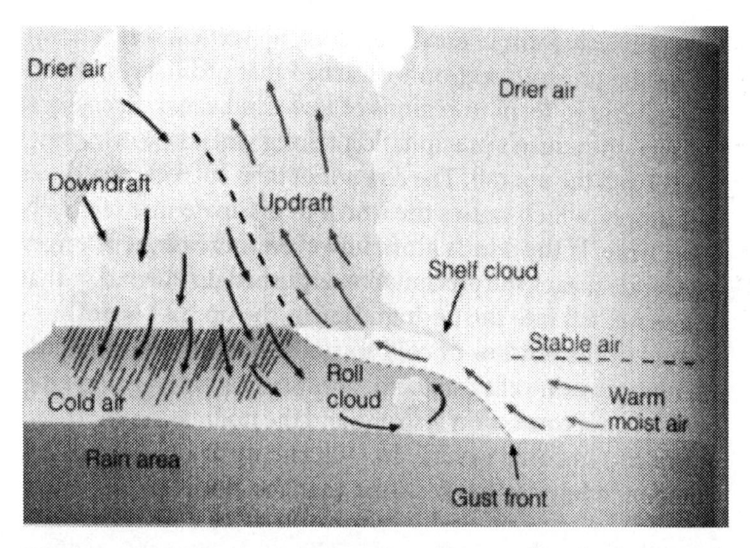

Figure 14.12 The lower part of an intense thunderstorm and some of the features associated with it. (source: MT)

Figure 14.13 Shelf cloud at the base of a supercell thunderstorm. Shelf clouds are sometimes mistaken for wall clouds from which tornadoes emerge. (source: NOAA[(13)])

Figure 14.14 A roll cloud associated with a severe thunderstorm over Racine, WI. (source: Wikipedia)

By definition, a **severe thunderstorm** is any thunderstorm that produces hail at least ¾ inches in diameter and/or surface wind gusts of 50 kn (93 km/h; 56 mph) or greater and/or spawns a tornado. Winds are particularly strong in the gust front (Fig. 14.12). Along the leading edge of the gust front, strong winds can pick up loose dust and soil and lift them into a huge tumbling haboob. As warm, moist air rises along the forward edge of the gust front, a **shelf cloud** forms (also called an arcus cloud) (Fig. 14.13). Such clouds occur where the atmosphere is very stable near the base of the thunderstorm. Beneath the shelf cloud, warm moist air is drawn into the thunderstorm, while cold air is blasted along the surface beneath. Occasionally, a roll cloud can form that appears to be slowly rotation about a horizontal axis (Fig. 14.14).

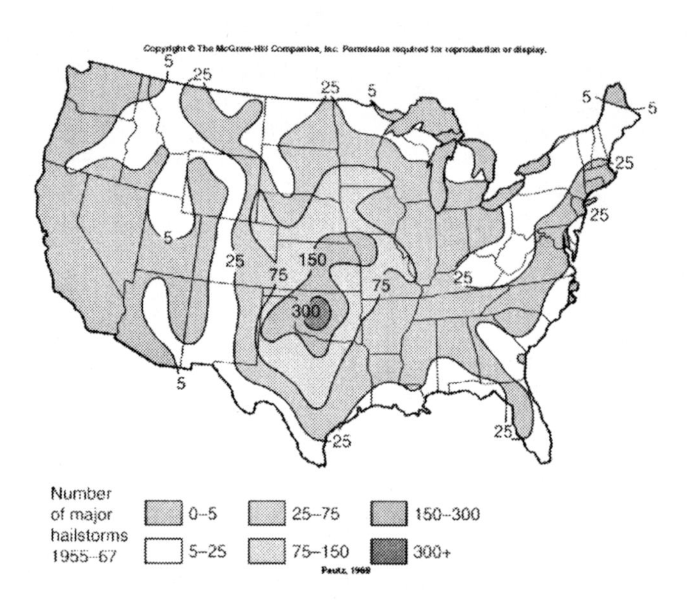

Figure 14.15 Major hailstorm frequency, 1955–67 in the contiguous U.S.. Contours indicate the total number of storms producing hail with a diameter of 3/4 in or larger.(source: AB)

In the contiguous U.S., hail-producing thunderstorms occur mostly east of the Rocky Mountains (Fig. 14.15). They are extremely rare in Hawaii and Alaska though two freak storms on 7 March 2012 and two days later produced the most intense hail on Oahu, HI since records began in 1950. Some hailstones were reportedly the size of tennis ball, and a rare tornado struck Oahu's east coast at Kailua[14] (NB: the EF0 tornado was relatively small. But then, Hawaii is not known for having tornadoes!). The area with a large number of hailstorms is a broad corridor from Texas north across Nebraska and into southern Minnesota, and east across Missouri. Oklahoma records the highest number hailstorms, indicative that thunderstorms there are typically severe. Recall from Figure 14.8 that Oklahoma typically records 50 thunderstorms a year. With an extrapolated 600 thunderstorms between 1955 and 1967, this means that every other thunderstorm is a severe thunderstorm! We will later see that this "hailstorm corridor" roughly coincides with Tornado Alley, the corridor most notorious for devastating tornadoes.

The likelihood of a thunderstorm to become severe increases with the length of time the thunderstorm survives. Ordinary thunderstorms form in regions of low wind shear that allows the rain (and the downdraft) to fall into the updraft. In severe thunderstorms, stronger winds aloft create a moderate wind shear the precipitation is pushed downwind so that it does not fall into the updraft. The updraft is now not suppressed and the storm can continue to grow. The cloud top can now intrude well into the stratosphere, up to 18 km (60,000 ft) in altitude. This violent updraft keeps hailstones suspended long enough to grow to considerable size. Once the hailstones are large enough, they can fall out of the bottom of the cloud with the downdraft, but a strong updraft may also toss them out the side of the cloud, or even from the base of the anvil. Airplanes have actually encountered such hail in clear air several kilometers away from a storm.

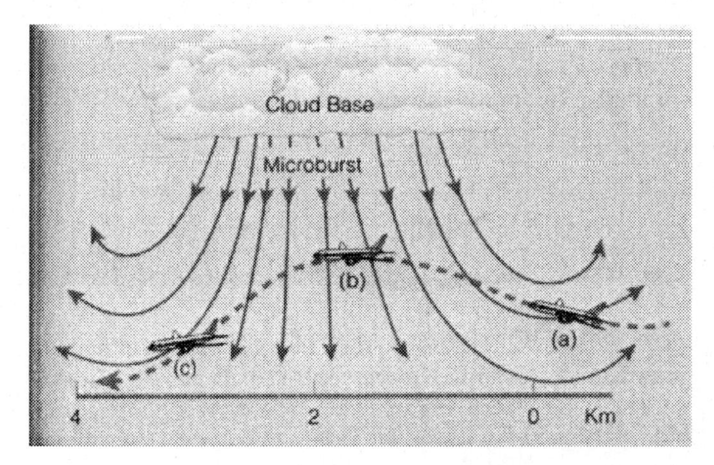

Figure 14.16 An airplane flying into a microburst. At position (a), the pilot encounters a headwind; at position (b), a strong downdraft; and at position (c), a tailwind that reduces lift and causes the aircraft to lose altitude. (source: MT)

Beneath an intense thunderstorm, the downdraft may locally hit the ground as a downburst and spread out horizontally in a burst of wind. Some of these downbursts may be quite localized and are called **microbursts,** if the affected area is less than 4 km in diameter (2.5 mi). Some microbursts are only a few 10s of meters large. Despite of their small size, microbusts can pack winds reaching 146 kn (270 km/h; 168 mph). They are usually associated with intense thunderstorm but they can also occur beneath ordinary thunderstorms. Microbursts are particularly dangerous to aviation and are thought responsible for several airline crashes (Fig.14.16). Low-flying planes on approach for landing may gain lift when flying into a microburst but then lose lift rapidly within a matter of seconds. If the pilot noses the plane in response to the initial headwinds, the plane accelerates to the ground upon entering the center of the microburst. An accident attributed to a microburst occurred near Dallas-Forth Worth International Airport on 2 August 1985. Just as the aircraft from Los Angeles (Delta Flight 191; a Lockheed L-1011) made its final approach, it encountered severe wind shear beneath a small but intense thunderstorm. The plane crashed to the ground, killing 8 of 11 crew members, 128 of 152 passengers and one person on the ground. Nowadays, pilots are held in a holding pattern to wait out the thunderstorm or are diverted to other airports. Several airports have Low-Level Wind Shear Alert Systems (LLWSAS) to warn air traffic control and pilots. These systems measure wind speed and directions at 6 to 32 remote stations around the airport and were developed by the Federal

Aviation Administration (FAA) in 1976 in response to the Eastern Air Lines Flight 66 wind shear accident at New York's John F Kennedy International Airport. On 24 June 1975, a Boeing 727 from New Orleans flew through a microburst, struck approach lights and skidded along the ground until it eventually burst into flames. Of 124 people on board, 113 people died. Nine passengers and two crew members near the rear of the aircraft survived. The subsequent investigation revealed that an earlier plane reported encountering severe wind shear along the runway involved in the accident and nearly crashed. Two more planes landed safely before the fatal accident. The recovered cockpit voice recorder showed that the pilot of Flight 66 was aware of the wind shear problem but decided to land anyway.

Figure 14.17 During the late afternoon and early evening of 3 April 2004, this supercell thunderstorm dropped hail 2 in diameter over Chaparral, NM, causing widespread damage. (source: wikipedia)

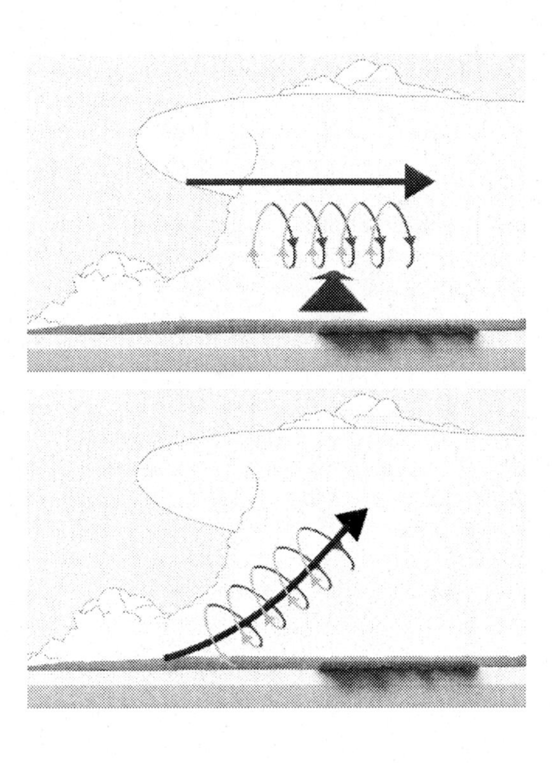

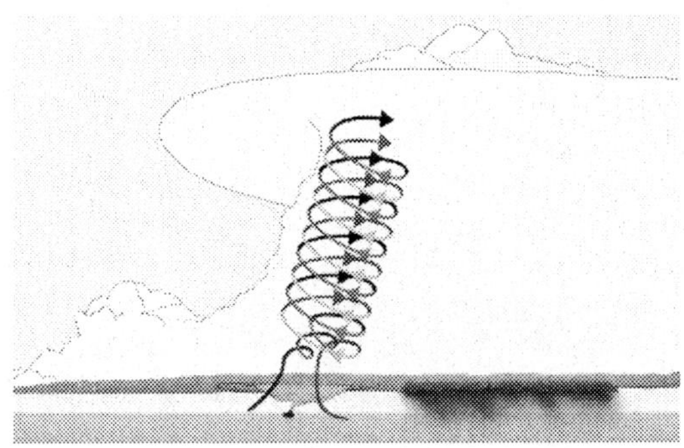

Figure 14.18 Evolution of a thunderstorm into a supercell thunderstorm featuring a mesocyclone. Top: wind from different directions at varying altitudes set up a winds shear that cause air to rotate about a horizontal axis. Middle: the horizontal roll of air is drawn into the updraft of the storm. The roll starts to tilt upward. bottom: The strong updraft starts to rotate with the roll, forming a **mesocyclone**. Tornadoes can emerge from the mesocyclone. (source: wikipedia)

If the winds aloft become even stronger (strong wind shear) and change direction with height (e.g. from more southerly at the surface to more westerly aloft), the outflow of cold air from the downdraft may never cut off the updraft. Often, the jet stream provides the strong winds aloft. The wind shear may be strong enough to create a horizontal roll, which when tilted into the updraft causes it to rotate. The horizontal roll is also called a **vortex tube**. The thunderstorm may now grow into a larger, long-lasting **supercell thunderstorm** that has a violent, rotating updraft in a single cell (Fig. 14.18). This rotating updraft is called a **mesocyclone**. It is this rotational aspect of supercells that can lead to the formation of tornadoes. A supercell thunderstorm is capable of maintaining itself for several hours. Strong updrafts can exceed 90 kn (167 km/h; 104 mph), hail can grow to the size of a grapefruit, damaging surface winds blow and large tornadoes can form. Precipitation normally does not form in the region of updraft and if it does, it is swept outward by the rapidly rotating air.

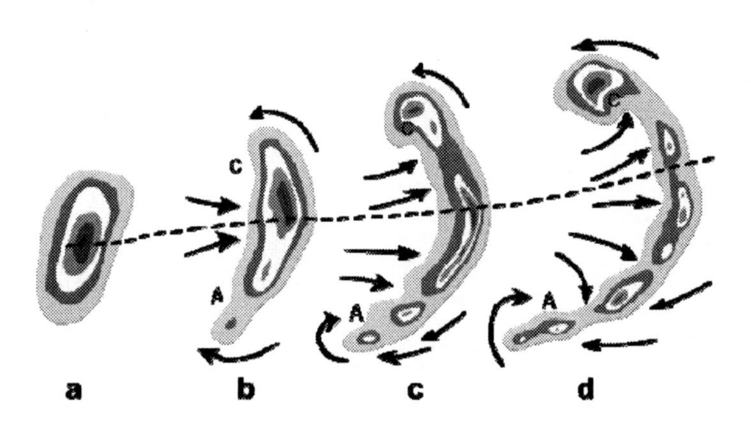

Figure 14.19 Typical evolution of a thunderstorm radar echo (a) into a **bow echo** (b,c) and into a comma echo (d). The dashed line traces the axis of greatest potential for downbursts. Arrows indicate wind flow relative to the storm. Wind flow can be cyclonic (C) as well as anticyclonic (A); both regions but particularly (C) are capable of supporting tornado development in some cases. (source: Wikipedia, NOAA)

Under favorable atmospheric conditions, thunderstorms can form as a line called a **squall line** or as a cluster of storms, called a **mesoscale convective system (MCS)**. The squall line forms along a cold front but it can also form in the warm air 100 – 300 km (62 – 186 mi) ahead of the front. In mid-latitudes, these pre-frontal squall-line thunderstorms are the largest and most severe type of squall line. It can extend over 1000 km (620 mi). The derechos during and ending the 1995 "Chicago heat wave" occurred along such a squall line (Fig. 14.7). On radar images, the weakly organized squall lines of MCSs are often shaped like an archer's bow. This feature is called a **bow echo**. Especially strong bow echoes are often indicative of derechos. The appearance of bow echos appears to indicate a lower likelihood of tornados to form within the thunderstorms itself, but they can occur at the perimeter of the bow echo (Fig. 14.19). Severe weather prediction forecasters use the formation of **bow echo** to issue warnings.

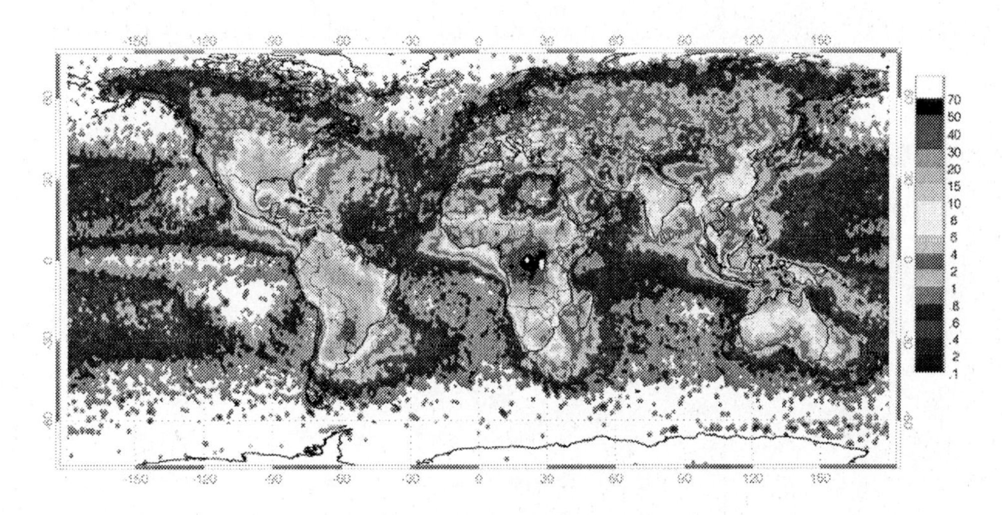

Figure 14.20 Global map of annual lightning flash rate, per km^2. (source: Wikipedia, NASA)

Figure 14.21 Lightning strikes over Albuquerque, NM. This thunderstorm on 4 October 2004 severely compromised air traffic. Airplanes were in a 1-hour holding pattern when the thunderstorm moved in. The famous Balloon Festival that was scheduled for the next day had to be canceled.

Figure 14.22 Lightning within a thunderstorm over the Great Plains on 12 May 2009. (source: NOAA[13])

Lightning

Thunderstorms and consequently lightning is not evenly distributed on Earth (Fig. 14.20). The area with the highest number of lightning strikes include parts of Central Africa, particularly the Democratic Republic of Congo where typically 70 lightning strikes per km^2 occur per year[5]. Second in lightning frequency the Bolivar area in northwestern Colombia, the Zuila area in western Venezuela and Kashmir and North-West Frontier foothills in Pakistan, with at least 50 strikes per year. With about 35 strikes per year, Florida stands out in North America, as already discussed in the thunderstorm section.

The exact causes and mechanisms of thunderstorm-related lightning are not yet fully understood and are an area of intense research. But in principle, lightning is simply a discharge of electricity, a giant spark, which occurs in mature thunderstorms. Lightning may take place within the cloud, from cloud to cloud, or from a cloud to the ground (Figs. 14.21; 14.22). The majority of lightning strikes occur within the cloud, while only about 20% or so occur between a cloud and the ground. The lightning flash heats the surrounding air instantaneously to temperatures of 8,000 – 33,000°C (14,000 – 59,400°F). The air then expands explosively, creating the sound of **thunder**. By observing the lag time between lightning and thunder, an observer can estimate the distance of the lightning strike. Since the speed of light is so high ($3 \cdot 10^8$ m/s), we can assume that the light arrives instantaneously. But sound waves travel much slower, at 330m/s. So if we count roughly 3 seconds between the lightning and the thunder, the lightning strike was 1km away. Or, for observers thinking in miles, a 5-second count places a lighting strike at 1 mi away.

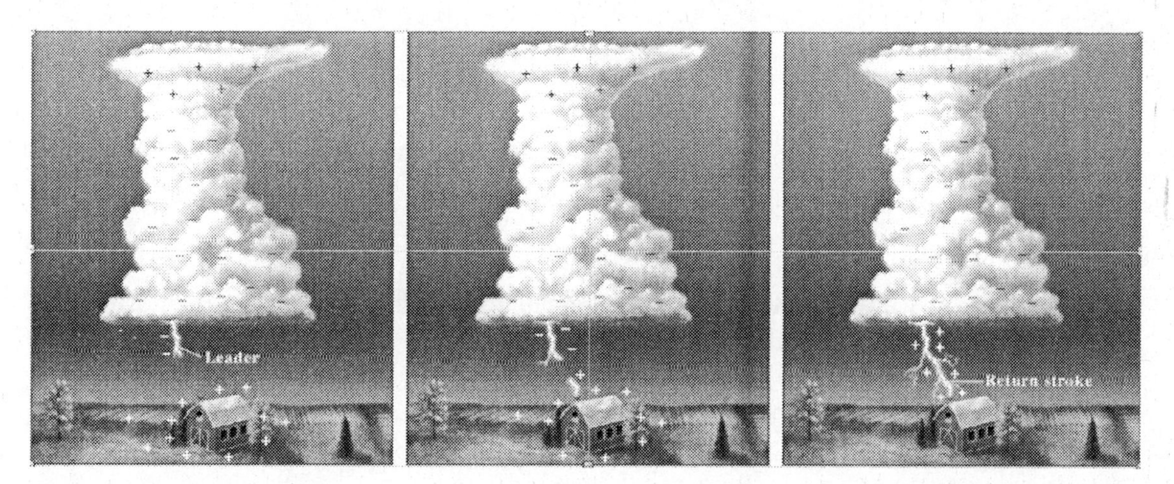

Figure 14.23 Left: A charge separation develops in a cloud, with a negative charge at the base and a positive charge at the top. The negative charge repels negative charges on the ground, so positive charges develop on the ground. A leader begins to descend from the cloud. Middle: As the leader grows downward, positive charges begin to flow upward from an object on the ground. Right: When the connection is complete, the return stroke carries positive charges rapidly from the ground to the cloud, creating the main part of the flash. (source: ME)

But what process prepares a thundercloud for lightning? One theory is that the strong updraft within a rapidly growing cumulus cloud separates charges when ice crystals rebound off graupel or hail, much like charges separate when a balloon is rubbed against one's hair. Graupel forms when the updraft carries water droplets upward that then supercool to temperatures between -10 and -40ºC (14 and -40ºF). These droplets collide with ice crystals to form the soft ice-water mixture called graupel. The collision between graupel and ice crystals transfers negative charges to the graupel, leaving the ice crystals positively charged. The lighter ice crystals are then pushed upward in the cloud while the negatively-charged graupel falls toward the middle and bottom of the cloud (Fig. 14.23). Another school of thought proposes that falling ice and rain become electrically polarized as they fall through Earth's natural electric field. Yet another idea is that during the formation of precipitation, regions of separate charge already exist within tiny cloud droplet and large precipitation particles, with the upper part becoming positively charged and the middle and lower part negatively charged.

As charges of like polarity repel each other, the negative charge beneath the cloud moves away, leaving the ground positively charged. The space between the cloud and the ground can be likened to that between the plates in a plate capacitor. The air beneath the cloud is a good electrical insulator, so no spark occurs while the electrical field between the cloud and the ground increases as more charges separate. At some point, the electrical field overcomes the insulating capacity of the air and a very faint **leader** develops, invisible to the human eye, and begins to descend from the cloud.

The leader covers about 50 – 100 m at a time, for about 50 millionths of a second, and so steps downward as a stepped leader. As the leader descends, an invisible flow of positive charges, the **streamer**, ascends from the ground, usually from and along elevated objects (Fig. 14.23 middle). After they meet, a large amount of negatively charged electrons flows downward and a bright **return stroke**, several cm in diameter, surges upward to the cloud along the path of the leader (Fig. 14.23 right). This all happens within 1/10,000 s. A single lightning stroke may involve a current of 100,000 amperes.

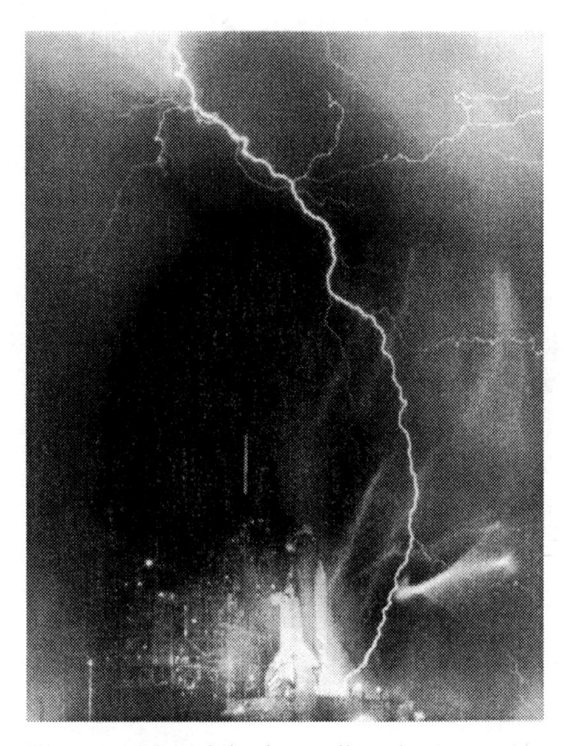

Figure 14.24 Lightning strikes the Space Shuttle Challenger at the Kennedy Space Center, FL on 30 August 1983, before the launch of STS-8. The photo was taken by a remote camera. (source: Wikipedia, NASA)

The reason why the streamer usually ascends from an elevated point is because the electrical field is strongest between the cloud and a point connected to the ground but that has the shortest distance to the cloud. It is therefore most likely that the streamer will flow through this point (e.g. Fig. 14.24). People therefore should never seek shelter under a tree in the open field because lightning is most likely to strike this tree. In fact, many lightning fatalities occur near relatively isolated trees, either through direct electrocution or by being struck by falling branches. In the U.S., nearly 14% of lightning deaths occur under trees (Table 14.2). There have been reports of people surviving lightning strikes but many have not been as fortunate.

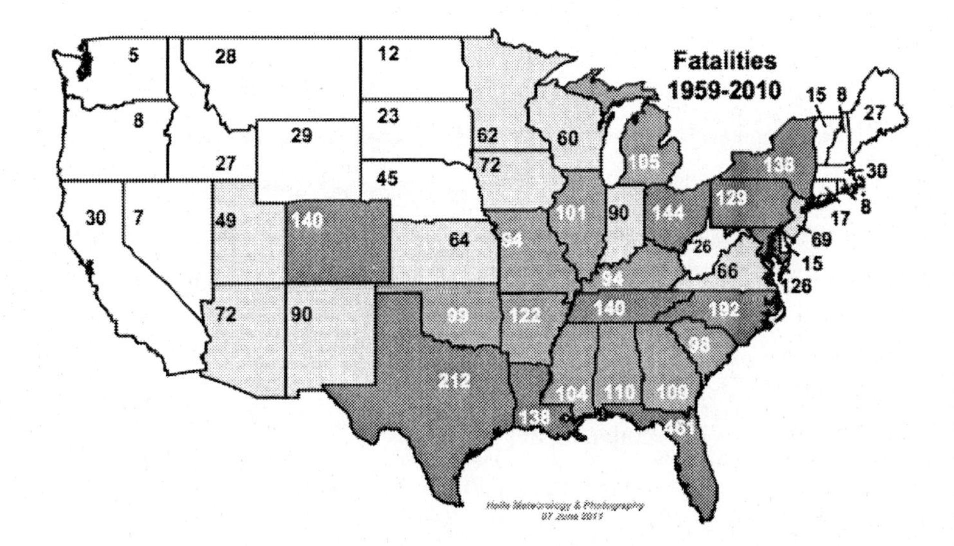

Figure 14.25 Lightning deaths in the contiguous U.S. from 1959 through 2010. In the 52 years of data, only two states have not reported any lightning deaths: Hawaii and Alaska. Darker shades indicate higher fatality rates. (source: NWS [17])

The NWS website lists recommendations on what to do during a lightning storm [16]. To quote the site, "when thunder roars, go indoors!". Nevertheless, about as many people in the U.S. get killed each year by lightning (44) as by wind (47) and winter storms (46). These numbers are based on the 10-year average between 1997-2006 (see Chapter 1). By far, most lightning deaths occur in Florida were most thunderstorms occur (Fig. 14.25). In the 52 years from 1959 through 2010, this state counted 461 fatalities. Texas and North Carolina follow with 212 and 192 fatalities. Relatively few deaths are reported in the western states and along the northeastern seaboard were thunderstorms are less common.

Table 14.2 Lightning deaths by locations in the U.S., 1959 – 1994[15]

Location	**Fatalities [%]**
open field, ballpark	26.8
under tree	13.7
fishing, boating, swimming	8.1
golfing (incl. under trees)	4.9
farming, heavy road equipment	3.0
telephone-related	2.4
radios, transmitters, antennas	0.7

unknown/not specified	40.4
total	100

People working and playing, standing upright in the open field, riding on a horse or playing golf are most vulnerable to lightning strikes and should seek proper shelters immediately (Table 14.2). In the U.S. more than 30% of lightning deaths occur during such activities. Since an electric current flows through the ground after the lightning strikes, it is also not a good idea to lie down flat, because then a large electrical potential (voltage difference) can form between the toes and the head and a deadly current can flow through the body. It is best to hunker down, with the feet close together, since this minimizes the electrical potential the human body is exposed to. It is for this reason that birds can sit on power lines unharmed but large flying birds become electrocuted when the wings touch power lines. There are some warning signs to alert you that a strike is imminent. If your hair begins to stand on end, your skin begins to tingle, or you hear clicking sounds, lightning may be about to strike. Lightning can also hit water which is a good conductor. So swimmers and surfers should come ashore. Small boats with no cabin are also not good places to be during a thunderstorm.

A safe place, on the other hand, is the inside of a car (but not a gold cart!) which acts like a Faraday cage. The same applies to large boats with cabins, especially those with lightning protection systems installed. When lightning hits a car, the charges will flow along the outside (metal casing) of the car to the bottom. The charges will then jump to the road through the air or through the tires. Nowadays, lightning rods and other lightning protection systems guide lightning strikes along the outside of structures and buildings to avoid damage. The main point here is that these systems are more conductive than the structures they protect and so the electric current is more likely to flow through them. The lightning rod was invented by Benjamin Franklin in 1749. His famous but extremely dangerous kite experiment proved that lightning is some form of electricity.

When a strong electric field forms near a pointed object such as the tip of a lightning rod, a mast, a spire, chimney or even aircraft wings, a bright blue or purple glow known as **St. Elmo's Fire** has been observed. The phenomenon also appears on ships at sea. The electric field around the object causes the ionization of nitrogen and oxygen molecules which produces the faint colored glow.

Figure 14.26 Lightning strikes observed during the 3 December 1982 eruption of Galungung volcano in Java, Indonesia. The eruption caused 68 fatalities. Lightning was first observed, and documented by Pliny the Younger, during the 79 A.D. eruption of Mt. Vesuvius, Italy. (source: Wikipedia, USGS available through NOAA)

In general, separation of charges and subsequent lightning can also occur without water molecules in the atmosphere. In such cases aerosols such as volcanic ash take on the role of water molecules. Such clouds are called pyrocumulus clouds, and lightning associated with volcanic eruptions has been observed (Fig. 14.26). Pyrocumulus clouds can also form in severe wildfires where strong convection can create fire weather. Sometimes rain falls and helps extinguish the fire but sometimes tornadoes that intensify the fire, and lightning can lightning can start new fires. The term **dry lightning** is used in North America for lightning that occurs when no significant precipitation falls. This is somewhat of a misnomer because the "dry" refers to precipitation, not the lightning. The type of lightning is particularly dangerous because a fire started by lightning is not extinguished by subsequent rain.

Heat Bursts

Heat bursts are rare but interesting atmospheric events during which the temperature rises extremely quickly, often lasting only a few minutes[4,5]. Most of them occur during the night, when relatively cool air contrasts strongly with a sudden burst of hot air. Heat bursts are usually accompanied by gusty winds and are associated with decaying thunderstorms where the downburst of air replaces a shallow pool of cool air hugging the ground. Heat bursts can also be caused by winds blowing down a mountain slope (katabatic winds) thereby funneling hot air through a mountain valley. During heat bursts that reach temperatures well over 32°C (90°F) (at night!), the air temperature typically rises by

10°C (20°F) though much larger jumps have been reported. Associated wind speeds often exceed 80 km/h (50 mph). The relative humidity can briefly decline into single digits, often desiccating plants. Heat burst can cause significant, though localized damage. For example, someone's garden furniture lands in a neighbor's backyard, but no damage occurs at homes further down the road.

The first well-documents heat burst goes back to 1909 in Oklahoma. On 11 July at 3 a.m., the temperature briefly rose to 136°F (57.8°C). Crops in a small area at the center of the heat burst were reportedly instantly desiccated. Though less specific, there are earlier reports of heat burst. The Minneapolis Tribune reported on 10 Jul 1879 that a "blast of hot air passed from south to north through portions of New Ulm and Renville County last Sunday evening. It lasted only a minute or two, but so intense was the heat that people rushed out of their houses believing them to be on fire."[7] And earlier yet, in 1860, Scientific American (vol. 3:106) gave the following report on two heat bursts though not specifying time and location: [7]"A hot wind extending about 100 yards in width, lately passed through middle Georgia, and scorched up the cotton crops on a number of plantations. A hot wind also passed through a section of Kansas; it burned vegetation in its track and several persons fell victim to its poisonous blast. It lasted for a very short period, during which the thermometer stood at 120°F" (48.9°C).

At least three heat bursts have also been reported in the U.S. in 2011[8]. In Wichita, KS, temperature rose from 85°F (29.4°C) to 102°F (38.9°C) on 9 June between 12:22 a.m. and 12:42 a.m.. Associated winds with speeds up to 50 mph (80 km/h) caused some damage. Similar wind speeds were observed during a heat burst on 3 July around 1:30 a.m. in Indianapolis, IN. A person reported that his neighbor's garden furniture ended up in his backyard. For more such events, the interested reader is referred to the Wikipedia webpage on heat bursts[7].

Heat burst occur worldwide, not only in the U.S.. Some extreme, and not scientifically documented, examples include temperature reports of 66.7°C (152°F) at Antalya, Turkey, on 10 July 1977[6]. A temperature rise from 37.8°C (100°F) to 70°C (158°F) in two minutes was reported near Lisbon, Portugal on 6 July 1949. A nearly unbelievable heat burst to 86.7°C (188°F) was reported in June 1967 in Abadan, Iran, where press reports said dozens of people died and asphalt streets melted. A heat burst reported by a meteorologist in Kimberley, South Africa, raised the temperature at 9:00 p.m. from 67°F (19.4°C) to 110° (43.3°C) within 5 minutes during a thunderstorm and fell back to 67°F by 9:45 p.m.[6] though it is not reported on which day this event occurred. A recent documented heat burst occurred on 29 October 2009 in Buenos Aires, Argentina after an unusually hot and day[8]. Around 10 p.m., temperatures rose from 31°C (87.8°F) to 34.6°C (94.3°F) within a few minutes, the relative humidity dropped to 8% (the lowest relative humidity in at least 20 years) and winds gusted at over 60 km/h (37.3 mph).

Figure 14.27 The 2 June 1995 tornado south of Dimmitt, TX. The F3 tornado was watched by the Probe 1 Vortex team, making it the most closely studied tornado at the time. (source: NOAA image library [13])

Figure 14.28 Before an after Google street views at East 24th Street and South Pennsylvania Ave, Joplin, MO. The area was hit 22 May 2011 by an EF5 tornado. (source: New York Times [18])

Figure 14.29 Aerial view of the path of destruction of the 27 April 2011 tornado that struck Tuscaloosa and Birmingham, AL during the **2011 Super Outbreak**. The path of destruction can be seen across the lower right. (source: NOAA[(19)])

14.4 Tornadoes

Tornadoes are fast-spinning near-vertical funnel-shaped clouds that form in violent thunderstorms (Fig. 14.27). Still to-date, maximum wind speeds in a tornado are estimated based on damage on the ground, because anemometers to obtain wind speeds *in situ* (in place) break before the fastest winds go through. Tornadoes do very odd things. They pick up things that we would never expect to travel through the air. Some fish and frogs "raining" from the skies, for example, were likely picked up elsewhere (e.g. in a nearby lake) by a whirlwind and then dropped once the winds became too weak to carry the unusual freight. In L. Frank Baum's famous "The Wizard of Oz", Dorothy's house in Kansas is picked up by a tornado and lands intact in the land of Oz. This is fiction. However, there have been reports of baby's in their cribs being picked up by a tornado and dropped some distance away, with the crib being intact and the baby unharmed. There is also the story of a man sitting in the bathroom. He survived while the house around him was destroyed. In 1991, a tornado in Kansas sucked a mother and her 7-year-old son out of their house in a bath tub[(4)]. The tub with its occupants hit the ground hard, rose into the air, and hit the ground again, tossing its passengers into the neighbor's backyard. Mother and son survived, with a few bruises and scratches, and a lump. The bathtub, which disappeared into the tornado, was never found.

Unfortunately, a similar story on 4 March 2012 in the San Diego Union Tribune did not have such a happy ending. Two days earlier, violent storms raged through at least a dozen states from Georgia to Illinois, killing at least 38 people. Kentucky, Indiana and Ohio were worst hit, with 4 twisters being the worst in the region in 24 years. In West Liberty, KY no house was untouched, and two police cruisers were picked up from the ground and tossed into City Hall. Henryville, IN, the birthplace of Kentucky Fried Chicken founder "Colonel" Harland Sanders, was demolished losing 3 schools and a church made of brick. About 20 mi (32 km) to the west, a tornado in Pekin, IN wiped out a family in their mobile home, including a 2-month-old girl and a 3-year-old boy. Their 15-month-old sister ended up in a field and was the sole survivor of the family, only to die two days later from severe brain injury. In fact, 2011 was one of the worst tornado years in history. With 550 fatalities, tornadoes in 2011 killed nearly as many people as in the previous 11 years taken together (605), obviously skewing the 11-year statistics of about 61 fatalities a year. Taken over a longer time period of 30 years, prior to 2011, 420 to 1100 tornadoes occurred each year causing on average 80 casualties, though a single F5 tornado may kills hundreds. The year 2011 counted nearly 1900 tornadoes. April set a new record with 771 confirmed tornadoes, upsetting the previous record of 552 in May 2003. As devastating as the 2011 tornado season was, with some 322 during the 25 – 28 April Super Outbreak alone the most fatal tornado did not occur in 2011. That infamous claim belongs to the 18 March 1925 tornado which killed 695 people.

So how do tornadoes work? Tornadoes rotate extremely rapidly around the axis of the funnel. Their wind gusts must count as the most destructive on earth. The highest wind speeds ever observed on Earth were those within tornadoes (Chapter 13; box 2), and it is likely that wind speeds exceed 500 km/h (311 mph) in the strongest tornadoes. The strong updraft within the funnel causes the air pressure to drop at the center, often by as much as 100 mbar. It is the **high winds** and the **low air pressure** at the center that make a tornado very destructive (Fig. 14.28). It is a common conception that the low pressure in the tornado sucks out the windows of a house. But this is probably an urban myth. Instead, it is more likely that the strong winds shatter the windows. However, the extremely low pressure at the center helps a tornado to achieve the high wind speeds.

Tornadoes vary in size and strength and are categorized using the (Enhanced) Fujita Scale (Tables 14.3 and 14.4). The diameter of the base is only 5 – 1500 m (16 – 4,900 ft) wide, making tornadoes extremely local phenomena. But the destruction is total and the path of destruction can be 1 – 500 km (0.6 – 310 mi) long (Fig. 14.29). In the U.S., tornadoes travel at 0 – 100 km/h (0 – 62 mph), generally from southwest to northeast due to prevailing winds. But tornadoes occasionally take a different direction. Tornadoes occur worldwide, including Eastern India and Bangladesh, South Africa, Australia, the Philippine and Europe. But no area is struck as often as "Tornado Alley" in the

U.S. that stretches from Texas north toward Nebraska and east into Indiana. Tornadoes are also common in Florida (Fig. 14.30).

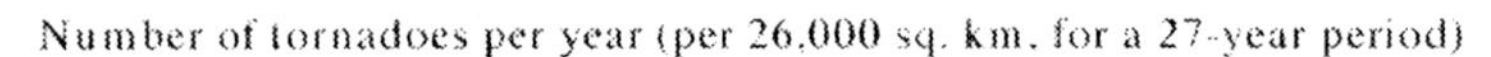

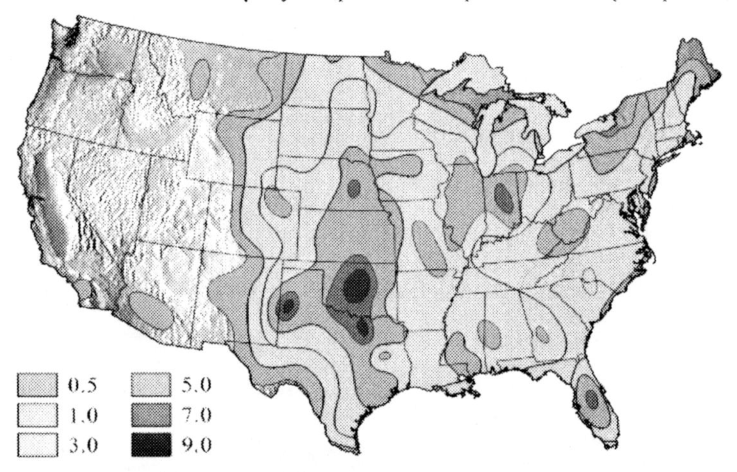

Figure 14.30 North American tornadoes are most common in "Tornado Alley", a band extending from Texas north to Nebraska and east into Indiana, where cold polar air masses collide with warm, moist maritime air from the Gulf of Mexico. The storms are pushed eastward by the jet stream and related high-altitude westerlies. (source: ME)

The Fujita Scale

The Fujita scale was introduced in 1971 by Tetsuya Fujita at the University of Chicago. The scale categorizes tornadoes by wind speed and destruction into 6 categories (F0-F5). The wind speeds are estimated as no one has yet succeeded in measuring winds directly in a tornado with a classical anemometer. The measurement of the highest wind speed ever measured involved a Doppler On Wheels radar unit. On 3 May 1999, this unit measured winds at 484±32 km/h (301±20 mph) during a 3-second gust in a tornado near Oklahoma City, OK. Wind speeds on the Fujita Scale represent that of 3-second gusts, not wind speeds that are gusting or sustained over longer times. Because of the difficulty to measure wind speeds in a tornado, the Fujita Scale is essentially a damage scale. The scale actually extends to F12 but no tornado larger than F5 has ever been observed. F2 tornadoes and larger are also classified as "significant" and F3 and large are termed "intense". In 2007, the Fujita Scale was superseded by the **Enhanced Fujita Scale** (EF). This scale was introduced after scientific research suggested that wind speeds required to inflict the described damage on the Fujita Scale was underestimated for smaller tornadoes but overestimated for intense tornadoes. The EF scale also has an improved damage description (with 28 damage indicators). But the scale reflects construction practices in the U.S. and may not accurately represent the strength of tornadoes elsewhere.

The new damage indicators make several distinctions:

- is a damaged mobile home a single- or double-wide unit?
- is “well-constructed” buildings made of masonry or other material (metal, glass)?
- number of stories
- category of building (school, strip mall, office buildings, warehouse)
- type of residential buildings (single or multi-family homes, apartment complex)
- free-standing structures (power line, utility tower, light and flag poles)
- type of tree (hardwood, softwood)

Table 14.3 The Old Fujita Scale

Scale	Category	Wind Speed [km/h]	Wind Speed [mph]	Typical Damage (old Fujita Scale)
F0	weak	72 – 126	45 – 78	light: branches and sign boards broken
F1	moderate	127 – 188	79 – 117	moderate: trees snapped; windows broken; shingles peeled off; mobile homes moved off foundations
F2	strong	189 – 259	118 – 161	considerable: large tree broken; mobiles homes destroyed; roofs torn off
F3	severe	260 – 336	162 – 209	severe: trees leveled; cars overturned; well-constructed roofs and walls removed
F4	violent	337 – 420	210 – 261	devastating: strong houses destroyed; buildings torn off foundations; cars thrown; trees carried away
F5	very violent	421 – 510	262 – 317	incredible: cars and trucks carried more than 90m; strong houses disintegrated; bark stripped off trees; asphalt peeled off roads

Table 14.4 Wind Speeds in the Enhanced Fujita Scale and Typical Tornado Sizes

Scale	Wind Speed [km/h]	Wind Speed [mph]	Path Length [km]	Path Width [m]
EF0	105 – 137	65 – 85	0 – 1.6	0 – 17

EF1	138 – 177	86 – 110	1.6 – 5.0	18 – 55
EF2	178 – 217	111 – 135	5 – 16	56 – 175
EF3	218 – 266	136 – 165	16 – 50	176 – 556
EF4	267 – 322	166 – 200	50 – 160	560 – 1500
EF5	> 322	> 200	160 – 500	1500 – 5000

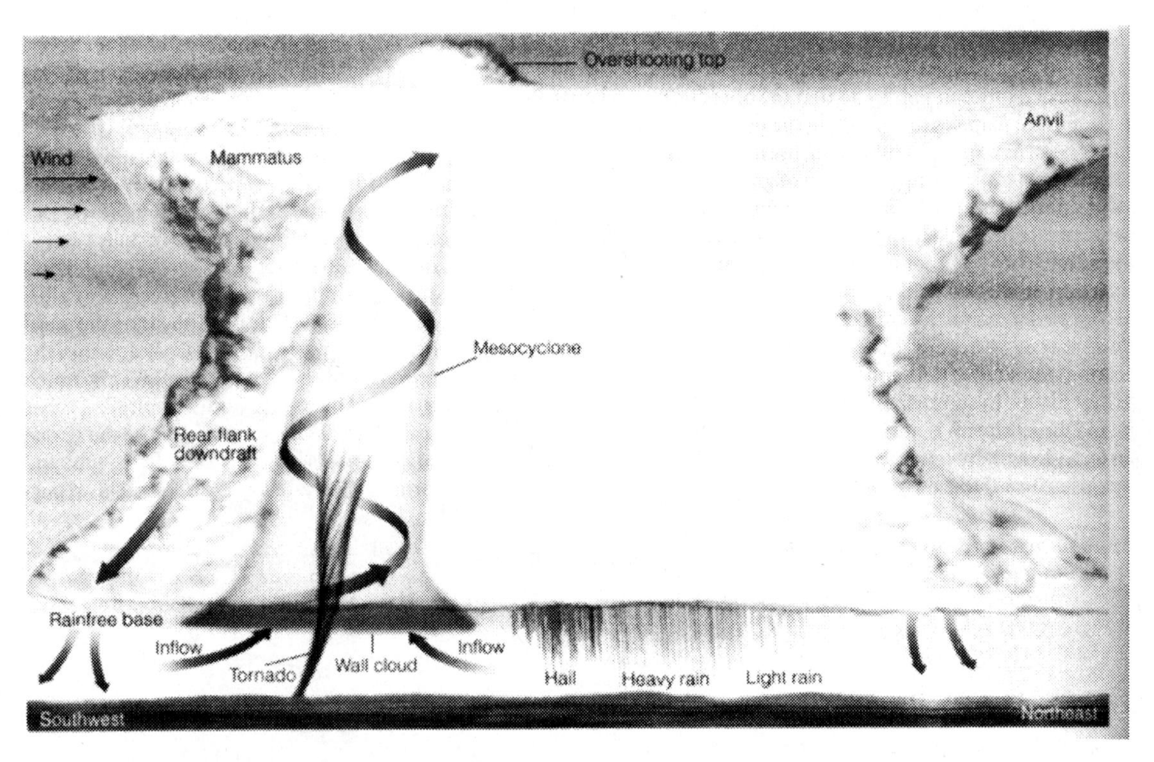

Figure 14.31 Some of the features associated with a tornado-spawning supercell thunderstorm. In the U.S., most such storms move from southwest to northeast. In this case, the viewpoint for this figure is in the southeast. (source: MT)

Tornadoes and Thunderstorms

In principle, tornadoes can form in any thunderstorm that exhibits a strong enough wind shear. So-called **gustnadoes** can form near the gust front (Fig. 14.12). Gustnado is short for gust front tornado. They are typically short-lived and involved low-level rotating clouds and usually do not extend from the surface all the way to the base of the thunderstorm. Their strength are typically in the EF0 – EF1 range. **Dust devils** are whirlwinds that form due to superheated, rotating surface air on sunny warm days in light winds. **Landspouts** have an intensely rotating column of air that is in contact with both the surface and the cloud base. The term may describe a tornado in its early stage but landspouts are not necessarily associated with mesocyclones. **Waterspouts** are essentially landspouts occurring over water. More so than their counterpart on land, water spouts are usually not associated with mesocyclones but they can be. When water or landspouts are associated with mesocyclones, they turn into tornadoes. The term **whirlwind** is a general description of phenomena with a vortex (spinning flow) of wind and includes all of the above. Most violent tornadoes occur in supercell thunderstorms, where a violent thunderstorm can spawn several tornadoes. However, tornadoes are relatively rare and only 15% of supercell thunderstorms actually spawn tornadoes[(4)].

Supercell tornadoes typically occur near the rear end of violent thunderstorms, where the winds, the updraft and the downdraft are strongest (Fig. 14.31). The heaviest precipitation, on the other hand, occurs near the center front of the storm. Supercell thunderstorms in the U.S. typically move in a northeasterly direction so tornadoes occur at the southwestern end of the storms. Counter to popular perception, the ultimate rotation direction (clockwise or counterclockwise) of a tornado is not influenced by the Coriolis Effect. Recall from Fig. 14.19 that the rotation direction depends on where the tornado forms relative to the storm center. It also matters where the cold front related to the storm is coming from (west or east) relative to the location of the updraft and downdraft in the thunderstorm.

Figure 14.32 A wall cloud forming at the base of a thunderstorm on 19 June 1980 in Miami, TX. (source: NOAA[(13)])

Figure 14.33 The 22 May 1981 Alfalfa, OK tornado emerging from a wall cloud. The rear flank downdraft manifests itself visually as a drying out of clouds. (source: NOAA[13])

The Spawning of A Tornado

The exact way how tornadoes form and why when the do is not fully understood. But certain atmospheric conditions are favorable for tornadoes to form. 1) A **conditionally unstable atmosphere** is essential for tornado development. In most cases, this involves cold air sitting above warm air that wants to rise. 2) **Strong wind shear** is also necessary to force air into a rotating motion. 3) Favorable to the formation of large tornadoes is the rotating updraft in a **mesocyclone.** In the U.S., the jet stream often blows blowing directly above the unstable atmosphere and provides the perfect recipe for strong, lasting tornadoes. This is why large tornadoes tend to form in intense, supercell thunderstorms.

But how does a tornado form from a mesocyclone? The updraft, the swirling precipitation and the surrounding air may all interact to produce a **rear flank downdraft (RFD)**[4] (Fig. 14.31). When this downdraft strikes the ground, it may interact with the region of surface inflow into the mesocyclone. Under favorable conditions, a supercell tornado can form. A first sign that a tornado is about to form is the sight of a **wall cloud**, a rotating cloud at the base of the storm near the downdraft/updraft interface (Fig. 14.32). As a result of conservation of angular momentum, air spiraling inward accelerates. It also spirals upward, toward the low-pressure core of the mesocyclone. Air cools and condenses and contributes to cloud formation. The wall cloud will lower toward the ground to form a **funnel cloud**. The rapidly rotating funnel cloud will then extend from the wall cloud down to the surface. The newly born supercell **tornado touches down** and manifests its dark self by drawing in dirt and debris (Fig. 14.33). After the tornado touched down, air outside of the funnel spirals upward, while air inside descends toward the extreme low pressure near the surface. The resulting evaporation during warming makes the core free of clouds.

The timing between the formation and the formation of a tornado (tornadogenesis) is uncertain and can involved anywhere between 1 and 60 minutes, but tornadogenesis usually occurs after about 15 min. Although it is rotating wall clouds that contain most strong tornadoes, many rotating wall clouds do not produce tornadoes. Tornadoes very rarely occur without the rear flank downdraft that manifests itself visually through a rain- and cloud-free section of the storm, so the RFD is an essential trigger for tornado-spawning (Fig. 14.33). The wall cloud withers and will often be gone by the time the tornado lifts off the ground. If conditions are favorable, another wall cloud (and tornado) may form downwind from the old wall cloud.

It is a common perception that **mammatus clouds** (Fig. 12.15g) that can be seen beneath thunderstorm anvils are a harbinger of tornadoes. These clouds form as cold air in the anvil sinks into warmer air beneath it. But mammatus clouds are not exclusive to supercells and can in fact be associated with the development of any thunderstorm or cumulonimbus cloud.

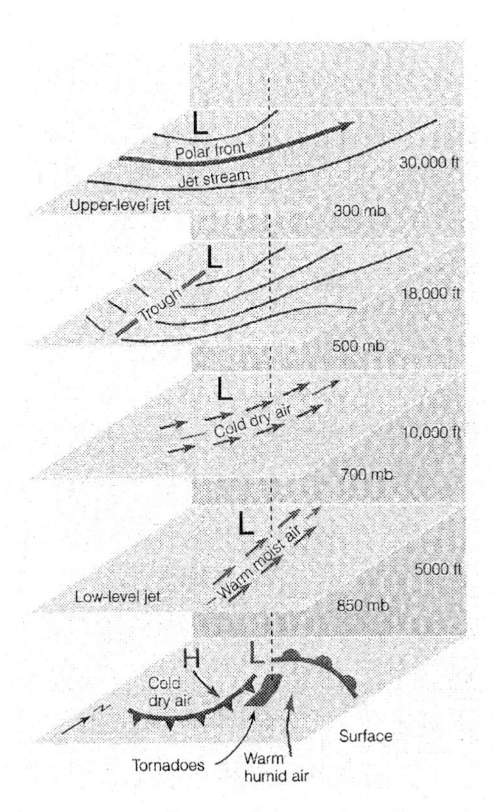

Figure 14.34 Conditions in "Tornado Alley" in the U.S. leading to the formation of severe thunderstorms that can spawn deadly tornadoes. (source: MT[(4)])

14.5 Tornado Alley

Tornado Alley (Fig. 14.30) holds the infamous world record on annual tornado counts. Unfortunately, conditions are often just right for tornadic supercell thunderstorms to form, particularly from late winter through late spring. During this time of year, the jet stream still makes extensive loop that can allow cold polar air from Canada to penetrate deeply to the south and form a cold front. On the other hand, warm, moist marine air tends to move in from the Gulf of Mexico to replace colder air to form a warm front. Such warm fronts are less likely to occur in summer because continental air by then is likely hotter than the mT air. This is a situation typical of a wave cyclone to form, with a cold front advancing on a warm front, and low pressure forming near the center. However, a unique situation now is that the jet stream returning to the north after a long excursion to the south provides the strong winds aloft necessary to increase the likelihood for mesocyclones to form in a thunderstorms. In Tornado Alley, then, a complex profile of conditionally unstable air forms where winds blow from different directions with increasing altitude (Fig. 14.34):

- at low altitudes (about 5000 ft/1525 m), a northerly flow of mT air from the Gulf of Mexico is accompanied by southerly winds
- at mid-altitudes (about 10,000ft/3050 m), cold and dry cP air moves south from Canada or east from the Rockies and is accompanied by strong westerly winds
- at mid-to-high altitude (about 18,000 ft/5490 m), a trough of low pressure exists to the west of the surface low as a result of wave cyclone formation
- at high-altitudes (about 30,000ft/9150 m), jet-stream winds racing northeastward, along the cold front and across the low-pressure trough center

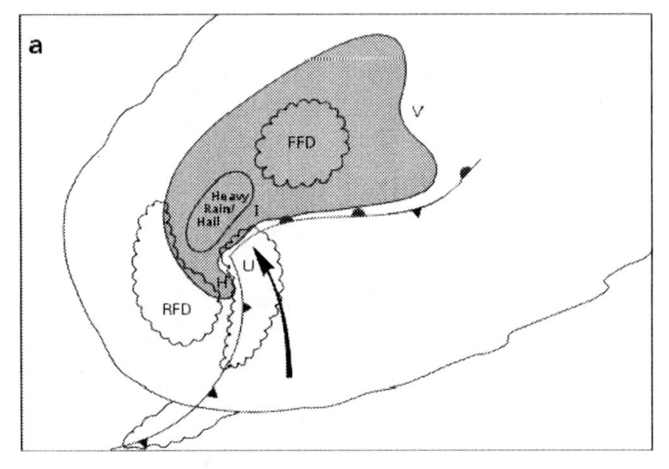

Figure 14.35 View from above on a **tornadic** supercell thunderstorm in Tornado Alley. A cold front chases a warm front around the low-pressure center. Heavy rain falls to the northwest, leading the

warm front. The RFD (rear flank downdraft) behind the cold front will trigger tornadoes in the main updraft (U); FFD: front flank downdraft; V: V-notch; I: updraft-downdraft interface; H: hook echo (in satellite radar image) (source: wikipedia)

The most likely location for violent tornadoes is near the rear of the developing low-pressure center and in front of the cold front, typically to the southeast of the low-pressure center (Figs. 14.34 and 14.35). A tornadic thunderstorm develops a characteristic hook echo in weather radar images just before tornadoes develop. The hook echos are used to issue tornado warnings (Box 1). The weather map in Fig. 14.36 shows the forecast situation for 27 April 2011 the day that became the day of the 2011 Super Outbreak (of tornadoes). In the days leading to the outbreak, a Low formed over Texas, with a leading and a trailing cold front. The jet stream aloft started to blow over the Low and along the leading cold front. By the time the Low advanced into Tennessee, the leading front became quasi-stationary, taking on the characteristics of a warm front locally. The jet stream still blew over the Low and aligned with the fronts. This system spawned numerous tornadoes, including 4 devastating EF5 tornadoes in Philadelphia and Smithville, Mississippi and in Hackleburg and Rainsville, Alabama (Case Study 1).

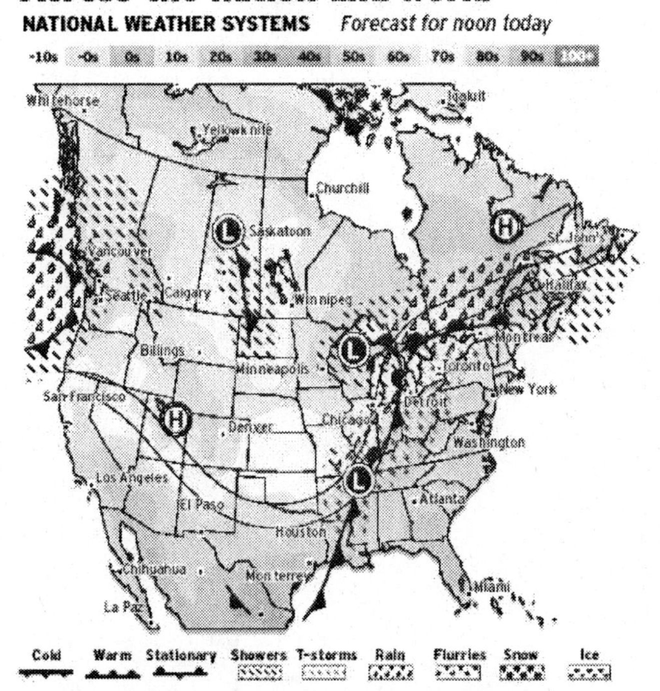

Figure 14.36 AccuWeather weather forecast map for 27 April 2011, the day of the 2011 tornado Super Outbreak. A little over a day earlier, an L moved north to park over the Great Lakes and another L formed in western Texas. A cold front was leading the Texas L and another cold front followed. For 2 days, the jet stream had been blowing over the advancing Texas L and along the leading cold front. Severe tornadic thunderstorms formed along the way but particularly so when the L and the trailing cold front advanced into the Tennessee area, spawning 4 EF5 tornadoes in Mississippi and Alabama on 27 April. Tornadoes also touched down in Arkansas and Tennessee. (source: San Diego Union Tribune)

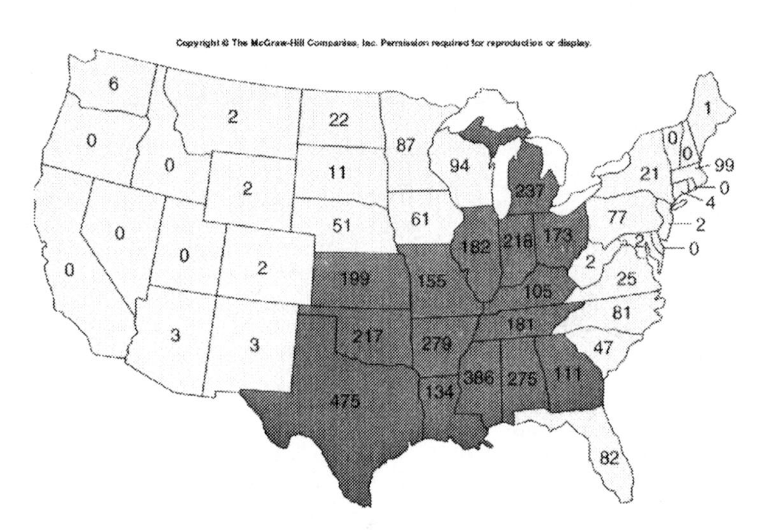

Figure 14.37 Number of tornado fatalities by state (cumulative 1950 – 1994) (source: AB)

In the U.S., most tornado fatalities occur in states in or around Tornado Alley (Fig. 14.37). With 475 cumulative deaths between 1950 and 1994, Texas records the highest fatality rate. Despite having the highest number of tornadoes, Oklahoma records less than half this rate. The state with the second-highest fatality rate is Mississippi with 386 deaths, and numbers in Alabama and Arkansas are also higher than in Oklahoma. Curiously, Kansas is not near the top even though L. Frank Baum's "The Wizard of Oz" may prompt one to think it is. Also note that the number of fatalities is also quite high, with Michigan and Indiana recording 237 and 218 fatalities.

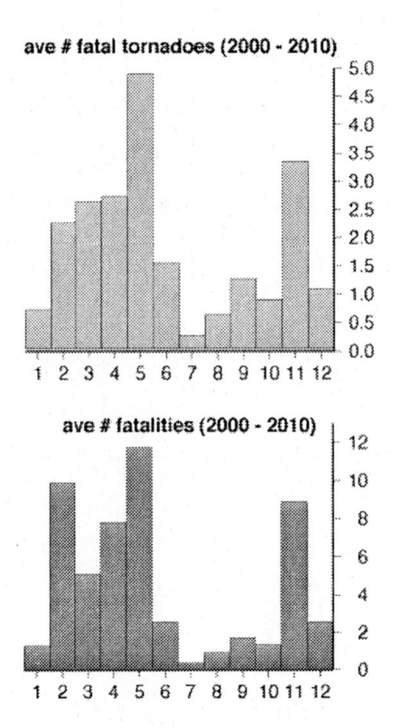

Figure 14.38 Top: Annual average number of fatal tornadoes in the U.S. by month. Bottom: Number of fatalities by month, 2000 – 2010. (data from NOAA)

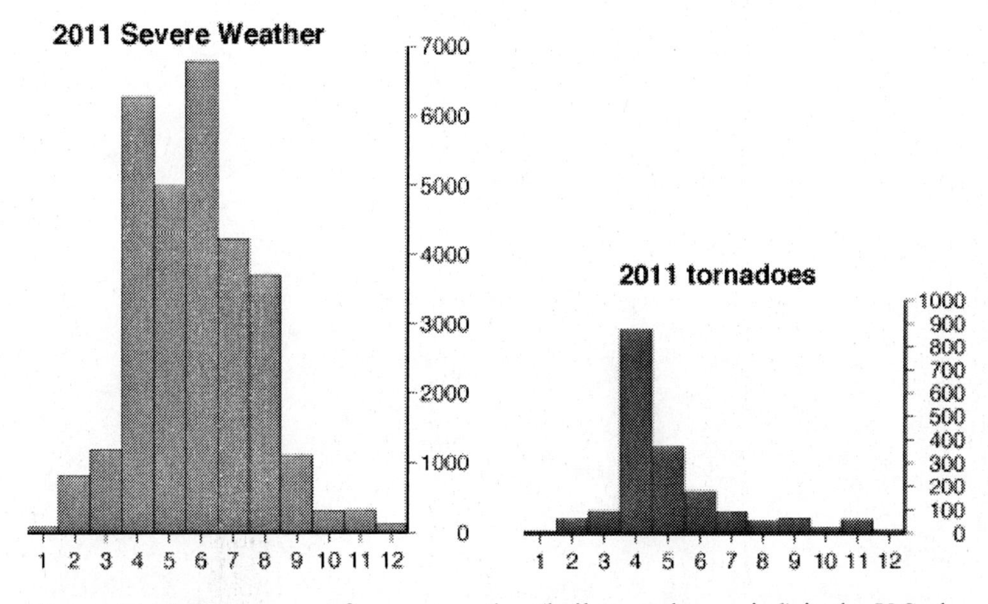

Figure 14.39 Occurrence of severe weather (hail, tornadoes, wind) in the U.S. throughout the year of 2011. Of the total of nearly 30,000 events, the vast majority of 62% were wind events (most of them thunderstorm-related), 31.5% were hail events, and 6.5% were tornadoes. Tornadoes occurred early, peaking in April. Hail events were more spread out and peaked in May, while wind events spread mostly from April through August, peaking in June. There were 26 lightning deaths in 2011, only a little more than half of a typical year, with a peak in July. (data from NOAA[20])

Tornadoes can occur throughout the year but they are not spread out evenly (Fig. 14.38). In the U.S., they typically occur from winter through late spring and early summer. Following the influx of mT air from the Gulf of Mexico, tornadoes in the south tend to occur earlier in the spring, while tornadoes in the north are more likely to occur later. This is very different from the typical statistics of thunderstorms which occur mostly throughout the summer months when fatalities from lightning also peaks (Fig. 14.39). Recall that only a small fraction of thunderstorms are tornadic. Disturbingly, fatalities peak earlier in the year (in February) than tornadoes do, suggesting that either tornadoes earlier in the year are more destructive or people are not prepared for tornadoes in the middle of winter. It turns out that in 2008, February had 12 fatal F3 and F4 tornadoes which is quite unusual.

Table 14.5 U.S. Tornado Deaths by Decade, 1920 – 2012[20]

Decade	Recorded Deaths
1920 – 1929	3,169
1930 – 1939	1,944
1940 – 1949	1,788
1950 – 1959	1,409
1960 – 1969	935
1970 – 1979	987
1980 – 1989	522
1990 – 1999	580
2000 – 2009	547
2010 – 2012	614 (2 years, 2.5 months)

90-year average 1920-2009: 132 per year

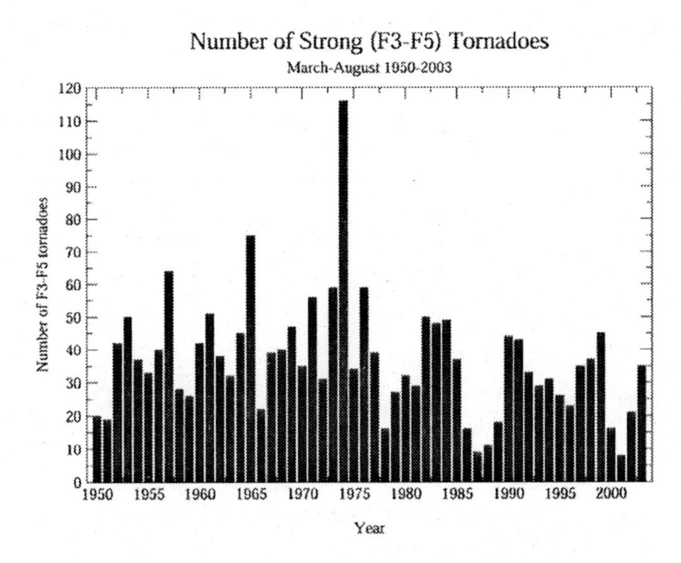

Figure 14.40 Yearly distribution of strong tornadoes (F3 and greater) 1950 – 2003. (source: NOAA)

Figure 14.41a A traditional storm cellar on the Texas plains. Such shelters are common in "Tornado Alley" and the Southeast, where tornadoes are particularly frequent and the low water table permits underground structures. Common in the 1930s and 1940s, such shelters are becoming relics of the past. (source: Wikipedia, USDA)

Figure 14.41b Storm shelter sign in Chicago's O'Hare International Airport. Restrooms near the inside of the terminal buildings serve as storm shelters. (source: Wikipedia, USDA)

At the beginning of the tornado section, we learned that in the last 30 years prior to 2011, 80 people were killed by tornadoes each year. The good news is that the long-term average in higher (132 in Table 14.5) and that fatality rates have gone down at least 5-fold. This is mainly due to the construction of tornado shelters but also due to improved tornado warning (see next section) because there has not been a decline in strong tornadoes that would otherwise explain the decline in fatalities (Fig. 14.40). Tornado shelters do not necessarily have to be of the traditional kind we are familiar with from the "Wizard of Oz" movie (Fig. 14.41a). Cellars beneath solid buildings also give protection. Storm shelters can also be found within public buildings. For example some major airports, such as O'Hare in Chicago and the Denver International Airport harbor such shelters (Fig. 14.41b).

Nevertheless, fatality rates seemed to have leveled off since the end of the 1970s and it appears hard to bring down death rates any further. In fact, it appears that fatality rates may actually be on the increase. With 550 fatalities in 2011 alone, the current decade definitely has an upward trend. At the same time, there is no indication that there has been a recent increase in strong tornadoes (Fig. 14.40), and agencies as well as the population may have to rethink disaster preparedness. It is worth considering where people died during a tornado. Traditionally, people have been killed either in the open or poorly constructed structures. In the 11 years prior and including 2010, the majority of 63% have been killed either outside, in vehicles or in relatively flimsy mobile homes (Table 14.6), while less than a third of fatalities occurred in permanent homes. It therefore appears that permanent homes give ample protection, especially when people heed to the advise to seek shelter toward the inside of sturdy buildings. However, in 2011 nearly as many people were killed in permanent homes (422) as in mobile homes (426). Shortly after the devastating 22 May 2011 Joplin, MO tornado, the

San Diego Union Tribune lamented that less and less homes have tornado shelters[9]. People then seek shelter in closets instead. Experts warn that even basements are not safe if they have windows and lack an integrated concrete roof. But, of course, the construction of such shelters, or even a basement, increases the construction costs of permanent homes. Given that cities are increasingly strapped for cash, residents should probably not trust that there will be ample space in public shelters. One therefore has to wonder whether it is a matter of not much time until the record number of fatalities of the 1925 tornado will be broken.

Table 14.6 U.S. Fatalities by location, 2000-2010[20]

Location	Fraction of Fatalities [%]
mobile home	53
permanent home	31
vehicle	7
business	6
outside	3

Some misconceptions about tornadoes

- opening windows of a building reduces damage risk. The reality: It may actually increase.
- highway underpasses provide proper shelter
- southwest corners in building provide proper shelter
- tornadoes cannot strike downtown areas. The reality: A tornado in 2000 caused damage in downtown Fort Worth, TX and there are many other examples, including Oklahoma City.
- escaping in a vehicle is the safest way to avoid a tornado. The reality: vehicles are easy targets to be picked up.
- tornadoes always move in a predictable way
- tornadoes skip houses. The reality: this occurs rarely.
- smaller tornadoes are always the weaker ones
- tornadoes always visibly reach from the ground to the thundercloud
- tornadoes only occur in North America
- tornadoes do not occur in winter
- tornadoes are attracted to mobile home parks. The reality: tornadoes strike anywhere but mobile homes are easy targets.

- some areas are protected through rivers, mountains or tall buildings or other geographical or man-made features

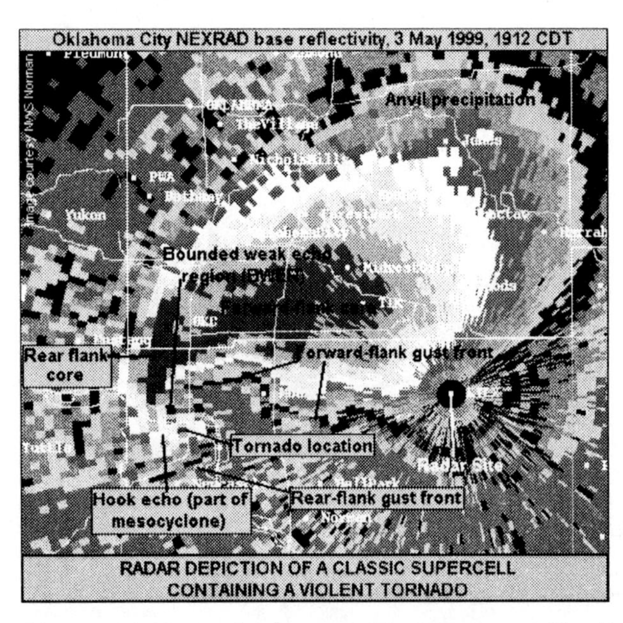

Figure 14.42 A radar image of a violent tornadic classic supercell near Oklahoma City, OK on 3 May 1999. (source: Wikipedia, NOAA)

BOX 1: Tornado Forecast

Up until the early 1980s, local tornado spotters looked out for suspicious cloud formations, such a wall cloud or rotating clouds, and a Tornado Alert was issued. However, tornado spotting declined in the aftermath of the 1974 Super Outbreak. It became obsolete after the advent of satellite imagery that has been more reliable and enabling forecasters the extend the warning times significantly.

Meteorologists watch out for tornado-suspicious weather conditions on a large scale, using radar imagery and other satellite methods. Classical radar imagery displays precipitation in particular. A **tornado watch** is issued by the Storm Prediction Center (SPC) to indicated that conditions are favorable for the formation of tornadoes. Local NWS offices issue a **tornado warning** if a hook echo can be identified (Fig. 14.42). Both, the NWS and the SPC are offices under the NOAA umbrella. A tornado warning means that there is imminent danger of a tornado striking. The updraft in the mesocyclone of a supercell thunderstorm is typically so strong that precipitation (rain and hail) cannot fall. Southwesterly winds aloft blow the precipitation northeastward, in front of the mesocyclone. If the mesocyclone lives long enough, it can swing the precipitation around it in a counterclockwise direction. While the mesocyclone itself does not show up in a Doppler Radar image, the hook-like swirl of precipitation (the hook echo) is a sign that a mesocyclone may be present.

The first documented tracking of a hook echo was on 9 April 1953 by the Illinois State Water Survey. But hook echoes are not always obvious. In the U.S. southern states, thunderstorms tend the produce heavier rainfall which obscures the hook shape. Doppler Radar such as NEXRAD (**Next**-Generation **Rad**ar) allows the detection of tornadoes even when the hook echo is not present, and for greater certainty when it is. Doppler Radar allows the detection of relative wind velocities in different parts of the storm, and so also the rotation in a mesocyclone.

Locally, tornado sirens may be activated after a tornado warning has been issued. Thanks to modern monitoring technology, the warning time for tornadoes could be increased substantially to 15 minutes (sometimes even up to 1 hour). But the exact path of tornadoes currently cannot be predicted and for many, a warning of 15 minutes may not be perceived as being enough. After the high number of fatalities in 2011 tornado preparedness may also have to be reconsidered and improved.

An excellent website to get information about imminent severe weather is the website of NOAA's storm prediction center at http://www.spc.noaa.gov.

References and Recommended Reading

1. NOAA website on derechos at http://www.spc.ncep.noaa.gov/misc/AbtDerechos/derechofacts.htm and NOAA, The Mid-July 1995 derechos: series overview. online at http://www.spc.ncep.noaa.gov/misc/AbtDerechos/casepages/jul1995derechopage.htm
2. World Weather Post, 6 July 2011. Massive haboob dust storm sweeps through Phoenix. Online at http://www.worldweatherpost.com/2011/07/06/massive-haboob-dust storm-sweeps-through-phoenix/#.T071Z8y27kw with link to weather data at wunderground.com and The Huffington Post, 6 July 2011. Phoenix Dust Storm: Arizona Hit With Monstrous 'Haboob'. Online (with video footage)at http://www.huffingtonpost.com/2011/07/06/phoenix-dust-storm-photos-video_n_891157.html
3. Los Angeles Times, 18 Oct 2011. Dust storm shrouds Texas city. Online at http://articles.latimes.com/2011/oct/18/nation/la-na-dust-storm-20111019
4. "Meteorology Today" by C. Donald Ahrens, 2003, Thomson Brooks/Cole, ISBN: 0-534-39771-9
5. (5)Wikipedia webpage on temperatures extremes: http://en.wikipedia.org/wiki/Temperature_extremes
6. Burt, C.C. 2004. Extreme Weather. Norton&Company, New York, pp. 304.
7. Wikipedia webpage on heat bursts: http://en.wikipedia.org/wiki/Heat_burst

8. online report on the 29 Oct 2009 heat burst in Argentina: http://foro.tiempo.com/meteorologia+general/bolsa+de+aire+caliente+en+buenos+aires+345ordm+c+y+8+de+hr+a+las+10+pm-t109635.0.html;msg2192335#new
9. San Diego Union Tribune, 28 May 2011. Shelters become relics of past.
10. Wikipedia page in Enhanced Fujita Scale at http://en.wikipedia.org/wiki/Enhanced_Fujita_Scale And NOAA's webpage on the Enhanced Fujita Scale published through the Storm Prediction Center at http://www.spc.noaa.gov/efscale/ef-scale.html
11. Glossary of the NWS Forecast Office of the San Francisco Bay Area at http://www.wrh.noaa.gov/mtr/afd-guide.php
12. Archive of jet stream maps for North America at San Francisco State University at http://virga.sfsu.edu/crws/jetstream_fcsts.html and NOAA weather map archive http://www.hpc.ncep.noaa.gov/dailywxmap and News stories at the Los Angeles Times at http://latimesblogs.latimes.com/lanow/2011/11/winds-lax-.html and related links
13. UCAR/NCAR online digital image library at http://www.fin.ucar.edu/ucardil/ and NOAA online digital photo library at http://www.photolib.noaa.gov
14. Hawaii khon2 news channel on 9 March 2012: Record-sized hail pounds windward Oahu at http://www.khon2.com/news/local/story/Record-sized-hail-pounds-windward-Oahu/EeuTUFC8iUiYn_JH-6LU3A.cspx and Rare tornado causes damage in Kailua at http://www.khon2.com/news/local/story/Rare-tornado-causes-damage-in-Kailua/jLXb71410kuMfxfDHaKKAQ.cspx and http://photoblog.msnbc.msn.com/_news/2012/03/09/10628559-a-rare-tornado-touches-down-on-oahu
15. Curran, E.B. and Holle, R.L., 1997. Lightning Fatalities, Injuries, And Damage Reports In The United States From 1959- 1994, NOAA *Technical Memorandum NWS SR-193* online version at http://www.nssl.noaa.gov/papers/techmemos/NWS-SR-193/techmemo-sr193.html
16. NOAA and NWS: Lightning Risk Reduction Outdoors at http://www.lightningsafety.noaa.gov/outdoors.htm
17. http://www.lightningsafety.noaa.gov/statistics.htm and http://www.noaawatch.gov/themes/severe.php
18. Google before and after images of the 2011 Joplin Tornado at the New York Times, retrieved March 2012 at http://www.nytimes.com/interactive/2011/05/27/us/joplin-panoramas.html and aerial before and after photos http://www.nytimes.com/interactive/2011/05/25/us/joplin-aerial.html

19. aerial views retrieved from the NOAA April 2011 Tornado Response Imagery Viewer at http://ngs.woc.noaa.gov/storms/apr11_tornado/
20. U.S. tornado statistics available on line at NOAA's Storm Prediction Center at http://www.spc.noaa.gov/ and http://www.spc.noaa.gov/climo/online/monthly/2011_annual_summary.html# and http://www.lightningsafety.noaa.gov/fatalities11.htm and http://www.spc.noaa.gov/climo/torn/fataltorn.html
21. 3 April 1974 F5 tornado striking Xenia, OH on YouTube at http://www.youtube.com/watch?v=r-9HBpHN_uY

COPYRIGHTED MATERIAL — DO NOT DUPLICATE, DISTRIBUTE, OR POST

Chapter 16: Long- and Short-term Climate Variations (draft)

Climate Changes - Why Do We Care?

Mounting scientific evidence that temperature changes can happen fairly rapidly (e.g. warming by 10°C in Greenland in only few decades) (from oxygen isotope ratios in ice cores)

- mounting evidence that most recent change is caused by human impact
- current level of CO_2 highest in last 800,000 years (evidence from ice cores)
- climate change may have negative impact on human life (financially and otherwise)

Examples for Possible Causes for Climate Change

- external: impacts, change in solar output, changes in Earth's position relative to Sun
- internal: changes in atmosphere and ocean chemistry, ocean circulation, volcanism, continental drift

16.1 Earth's Greenhouse Revisited

Greenhouse Gases and Their Relevance

There is almost 200 times more CO_2 in the atmosphere than CH_4, yet the warming by CH_4 is relatively important because its ability to trap heat is 20 times stronger. The contribution of each greenhouse gas to warming is [2]:

Gas	relative contribution	ability to trap heat
CO_2	60%	1
CH_4	16%	21
CFCs	11%	12,000
tropospheric O_3	8%	2000
N_2O	5%	310

COPYRIGHTED MATERIAL — DO NOT DUPLICATE, DISTRIBUTE, OR POST

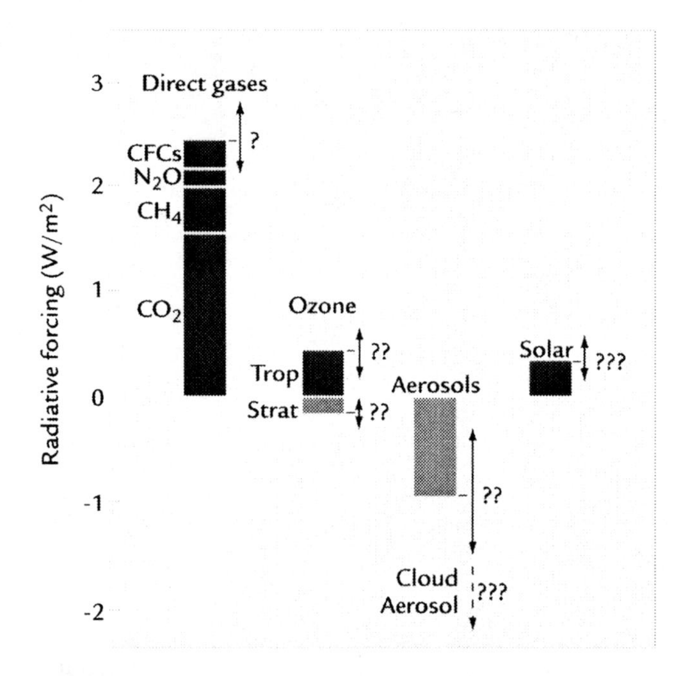

Figure 16.1 Radiative forcing of natural and anthropogenic greenhouse gases and aerosols. (source: RU)

Radiative Forcing[1]

Radiative forcing is interpreted as an increase (positive) or a decrease (negative) in net radiant energy observed over an area at the tropopause. All factors being equal, an ***increase in radiative forcing*** may induce surface ***warming***, whereas a ***decrease*** may induce surface ***cooling***. Examples:

- **greenhouse gases** have positive radiative forcing
- since the 1700s, changes in the sun's energy output may have produce a small positive forcing (~0.3 W/m^2)
- **ozone in the troposphere** has positive radiative forcing (greenhouse gas)
- **ozone in the stratosphere** has negative forcing (captures UV radiation that otherwise would reach into the troposphere and Earth's surface)
- **water vapor** is a greenhouse gas and has positive radiative forcing
- **clouds (condensed water vapor)** have negative radiative forcing because they reflect sunlight back to space that would otherwise reach Earth's surface
- **aerosols in the lower atmosphere** absorb heat, thus have positive radiative forcing
- **aerosols in the upper atmosphere** scatter and reflect sunlight, thus have negative radiative forcing
- large volcanic eruptions that inject sulfur-rich particles into the stratosphere that cause negative forcing for several years (estimates to more than offset the positive forcing by sun's increased output)

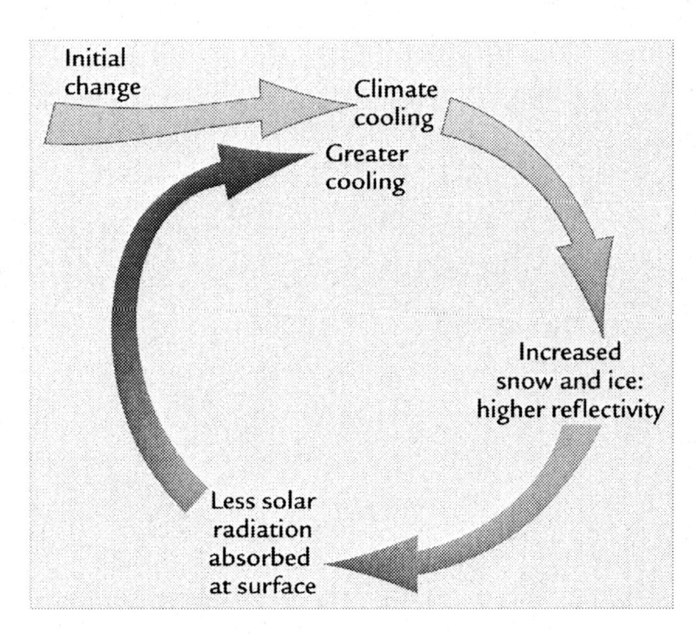

Figure 16.2 Example of a positive climate feedback. When climate cools, the increased extent of reflective snow and ice increases the albedo of Earth's surface in high-latitude regions, causing further cooling by positive feedback. (source: RU)

16.2 Climate Feedback Mechanisms

- a secondary process that responds to and influences a change
- positive feedback mechanism: process that enhances change
    - during global cooling, more snow and ice fall. This increases the albedo and enhances cooling.
    - during global warming, snow and ice melt. This decreases the albedo and enhances warming.
    - During warming, methane that resides below seafloor in ice lenses may melt. Methane gets released into the atmosphere and, being a potent greenhouse gas, accelerates warming.
    - during global warming, warming air can hold more water vapor which is a greenhouse gas, enhancing warming
- negative feedback mechanism: process that counteracts and mutes change
    - during warming, ice sheets melt. Cold fresh water floats on warmer ocean water. Warm ocean currents are diverted or slowed down which counteracts warming.
    - during warming, climate gets wetter. More clouds form which increases albedo and counteracts warming.
    - During cooling, climate gets drier. Less snow falls, so cooling may slow.

NB: Climate modeling shows that doubling the concentration of CO_2 and allowing the atmospheric water vapor content to rise (positive feedback), will result in an increase in average surface air temperature of more than 2.5°C (4.5°F) - a much larger rise in temperature than that produced by CO_2 alone[1].

16.3 Earth's Long-term Climate History

Table 16.1 Atmospheres of Mars, Earth and Venus

Constituent	Venus	Early Earth	Mars	Earth Today
CO_2	96.5%	98%	95.3%	0.038%
N_2	3.4%	1.9%	2.7%	78%
Ar	trace	trace	0.13%	21%
T [ºC]	477	290	-53	16
Pressure [bars]	92	60	0.006	1

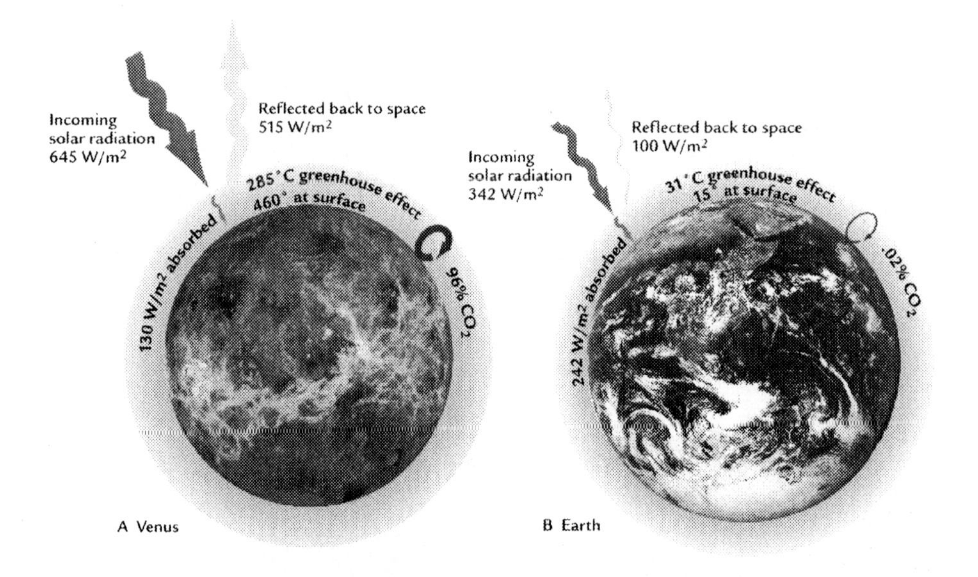

Figure 16.3 Venus (A) receives almost twice as much solar radiation as Earth (B), but its dense cloud cover permits less total radiation to penetrate to the surface. Yet Venus is much hotter than Earth because its CO2-rich atmosphere creates a stronger greenhouse effect and traps much more heat. (source: RU)

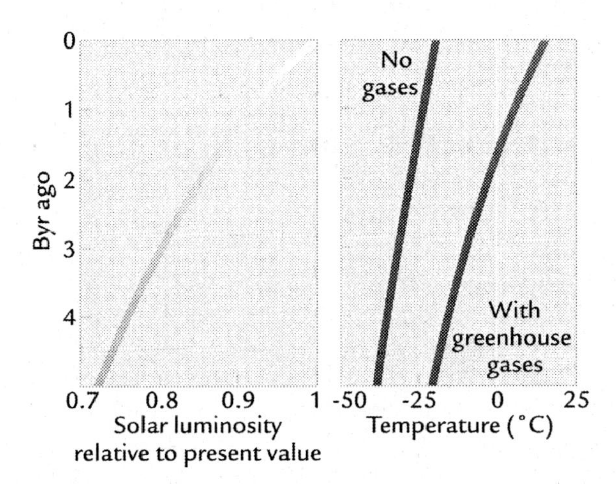

Figure 16.4 Astrophysical models of the Sun's evolution indicate it was 25% to 30% weaker early in Earth's history (left). Climate models show this situation would have produced a completely frozen Earth for well over half its early history if the atmosphere had the same composition as it has today (right). (source: RU)

Climate of the Early Earth

- Earth's early atmosphere is thought to have been similar to Venus' and Mars' current atmosphere that have changed very little over time; early CO_2 content caused T to be an unpleasant 290°C, nearly half as hot as Venus is today
- Earth's atmosphere underwent dramatic change from CO_2 rich to CO_2 poor
- much is thought to have occurred through chemical weathering
- most CO_2 is stored in limestone from chemical weathering, fossil shells, corals etc. as carbonate (e.g. Calcium Carbonate, $CaCO_3$).
- much CO_2 is now stored in the deep oceans (see Carbon Cycle; Appendix D)
- plants removed some CO_2 and respired O_2

A few textbooks emphasizes that the major change in the Earth's atmosphere has been the decrease in the greenhouse gas CO_2 to make this a habitable planet. Many other textbooks emphasize that the important change in the Earth's atmosphere was the increase in O_2, which was sole produced by early photosynthesis. Hence, early life (also) caused a major change in Earth's atmosphere. Both is probably correct. The removal of CO_2 lessened the greenhouse effect to allow Earth to be more habitable.

The Run-away Greenhouse

In a run-away greenhouse, warming continues until all water is gone.

- warming atmosphere causes water to evaporate
- atmospheric water vapor increases greenhouse effect, accelerating evaporation
- without moderating negative feedback, Earth's temperature wound increase until all oceans are evaporated
- this chain-reaction is called the run-away greenhouse
- there is no evidence that Earth ever had a run-away greenhouse

  NB: Venus' dense CO_2 atmosphere that keeps its surface at 480 deg C (900 deg F) is an example of a run-away greenhouse in which positive feedback mechanisms enhanced the warming. Mars' atmosphere is too thin to produce a significant greenhouse.

Climate History of the Earth: Time Scale in Millions of Years

The evidence for long-term past climates comes from **stratigraphic indicators** (geologic features and fossils in sedimentary rocks, see also Appendices B,C,D):

- **rock profiles**; e.g. different layers (strata) at one site inform us about local environmental fluctuations.
- **local environmental change over time**: can be result of climate fluctuation, continental drift, or tectonics; e.g. a place can be a marine environment at some point, a desert at another, and a glacial area at yet another point in time
- **tell-tale indicators in sedimentary rocks** to look for: fossilized bones; plant casts; corals; shells; coal; fossil dunes and soils
- Comparison of many profiles around the globe (corrected for continental drift!) give us an idea about the ancient global climate.

Stratigraphic Indicators

Stratigraphic Indicators for warm climate:

- tropical fossil reefs and micro-shells (most limestones)
- aluminum ore (bauxite) only found in tropical soil
- kaolinite (porcelain; a white clay that forms during weathering of feldspar - a mineral found in Granite - in a humid environment)
- evaporite minerals (halites and gypsum)

Indicators for cold climate:

- erosion of glaciers sculpt tell-tale landscapes (U-shaped valleys)
- glaciers leave polished and grooved surfaces (striation)
- massive piles of debris (moraines)

- o lack of fossils(?)

Figure 16.4 Glacial striation. Sediment rubble carried in the bottom layers of ice sheets and glaciers about 430 million years ago gouged grooves, so-called striations, in North African bedrock similar to those in modern-day ice in Alaska, shown here. (source: RU)

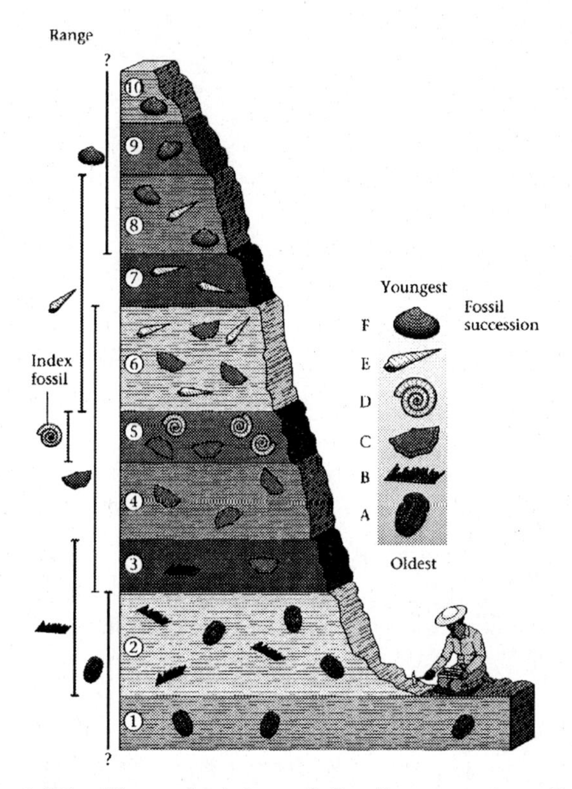

Figure 16.5 The principles of fossil succession. Each species has only a limited range in a succession. A mass extinction occurred when many species disappear at the same time. (source: ME)

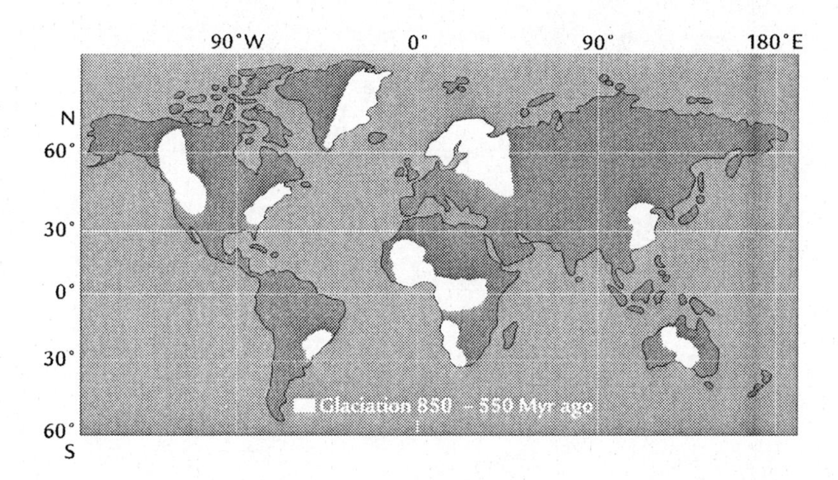

Figure 16.6 A snowball Earth? Evidence of several glaciations between 850 and 550 million years ago is found on Earth's modern-day continents. If these glaciated regions were located at or near the polar regions, climate may have been little different from what it is today. But if the were located in the tropics, a snowball Earth may have existed. (source: RU)

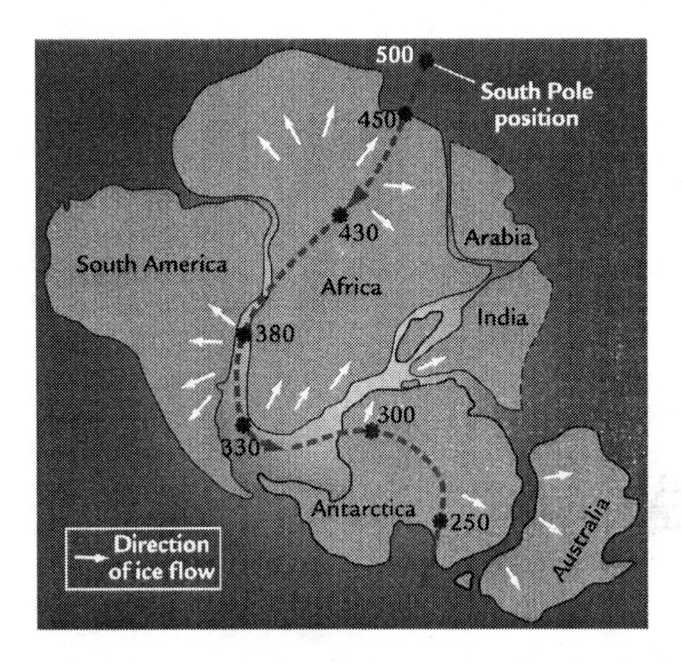

Figure 16.7 Gondwana and the South Pole. Changes in the position of the south magnetic pole in relation to the Gondwana continent were caused by the slow movement of Gondwana across a stable pole. Glaciation occurred in the northern Sahara about 430 million years ago ind southern Gondwana (South Africa, Antarctica, India, South America and Australia) 325-240 million years ago. The water shown between the modern-day continental outlines was land during Pangaean times. (source: RU)

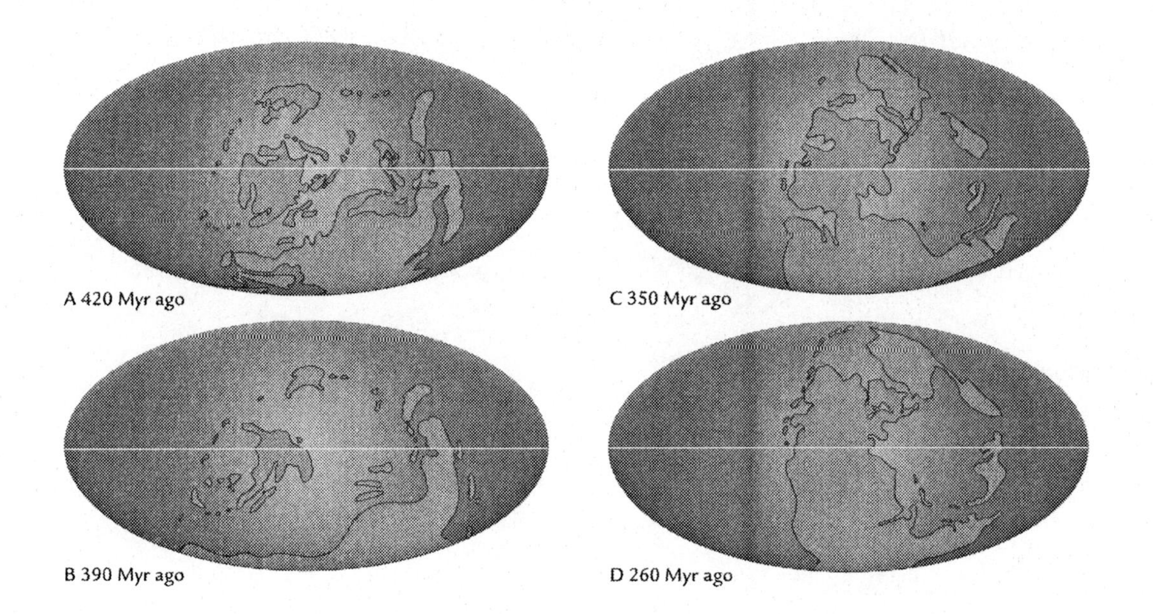

Figure 16.8 Moving continents. (A-C) After 450 million years ago, plate tectonic activity carried the southern continent of Gondwana across the South Pole on a path headed toward continents scattered across the northern hemisphere. Subsequent collisions formed the supercontinent Pangaea (D). (source: RU)

Glaciation

- Local past glaciation does not necessarily indicate that global climate has been colder: e.g. striation/glacial scratch marks on rocks found in Sahara desert can mean that either past climate was an ice house or that the Sahara was near the pole when the striations formed (the later was the case).
- Glaciation depends on:
    - position of continent relative to pole
    - precipitation
    - global climate
- Glaciation has been relatively rare in Earth's history.
- More often than not, Earth has had ice free climates.
- The interplay of different factors into Earth's climate are non-linear and complicated. There is therefore currently no completely satisfactory theory that can account for Earth's past and present glaciation.

talk about Rodinia before Pangaea.

Possible Causes of Climate Change on this Time Scale

Sun

- changes in solar energy output on scales of millions to billions of years
- sun used to be fainter; its output increases 10% every 1 billion years
- predictions are that in 1 billion years Sun will be too hot for liquid water to exist on Earth

Plate Tectonics

- position of land masses change over time so one spot can go through different climate stages
- extended rifting and volcanism influences regional to global climate
- variation in spreading rates influences global sea level
- large landmasses near pole leads to global cooling
- supercontinents covering large range of latitudes inhibit latitude-parallel currents; causes mixing of polar and tropical water -> more balanced climates

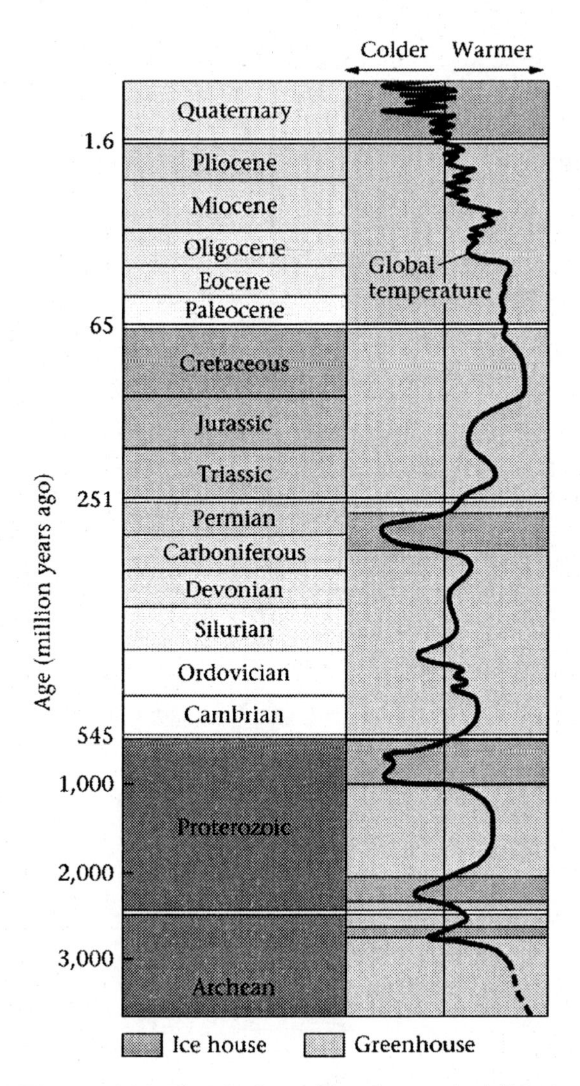

Figure 16.9 The timing of ice-house (cold/dry) and greenhouse (warm/humid) periods during Earth's history. Ice-house conditions have been relatively rare. (source: ME)

Past Long-term Cold Periods

- pre-Cambrian (700-550 Mio yrs ago)
- late Carboniferous and early Permian (320-270Mio yrs ago) this is the "late paleozoic ice age" referred to in Appendix E.
- Pleistocene (last 2.5 Mio yrs)this is the "late cenozoic ice age" referred to in Appendix E (cooling trend started 40 Mio yrs ago)this period has been warmer than the other two cold periods

Possible Causes for Long-term Cold Climate:

- one or more large (super)continents near poles lead to higher albedo (positive feedback)
- few large continents do not disturb west-east ocean currents, so less exchange between latitudes

- disappearance of warm shallow seas during sea level drop
- less ocean evaporation leads to less water vapor (greenhouse gas!)
- less water vapor would also lead to less precipitation (dry!)

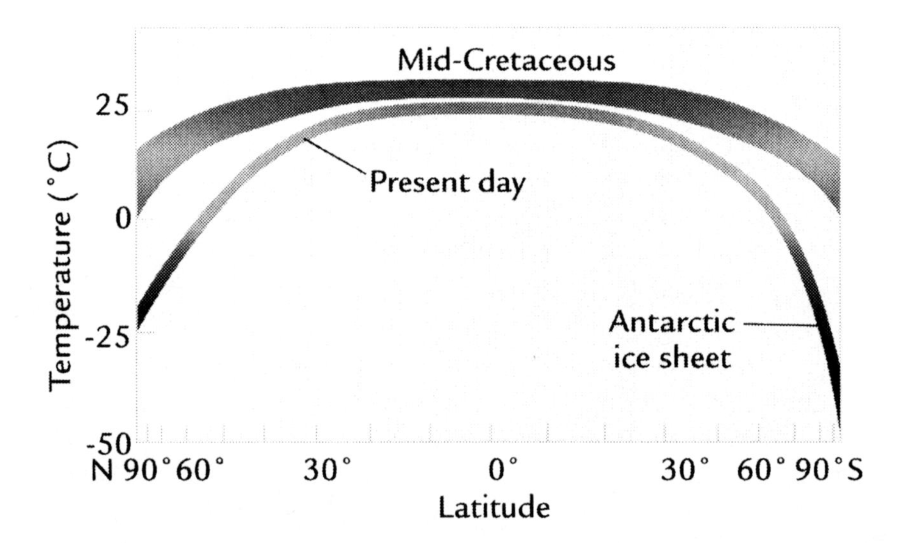

Figure 16.10 Comparison of today's temperatures with those in the Mid-Cretaceous, as function of latitude. While in low-latitude regions, the temperatures were not so different from today's temperatures, to polar regions were by more than 25°C (45°F) warmer. This is too warm for ice sheets to exist. (source: RU)

Past Long-term Warm Periods

- mid-Triassic (240-220 Mio yrs ago)
- late Jurassic through Cretaceous (170-65 Mio yrs ago)
- Eocene (50 - 40 Mio yrs ago); hottest climate in last 3 Ga this is the "late paleocene torrid age" referred to in Appendix F.

  NB:Ga stands for billion years.

Possible Causes for Long-term Warm Climate:

- smaller continents have less extreme climates
- smaller continents divert ocean currents north-south; drive global heat-conveyor (though some closure of passages (e.g. Isthmus of Panama), blocking off warm currents, can have opposite effect)
- melting ice sheets decrease albedo (positive feedback)

- extended large-scale volcanism injects greenhouse gases into atmosphere

Outstanding Episodes[2]:

- **Late Paleozoic Ice Age (360 - 260 Mio yrs ago)**: major ice age; one or more large supercontinent near poles lead to higher albedo, less water vapor, less precipitation, less greehouse gases
- **Late Paleocene Torrid Age (65-40 Mio yrs ago)**:equatorial zones slightly higher rainfalls than today but higher latitudes much warmer; southern oceans 10-15°C (18-27°F); less temperature difference in shallow and deep oceans and geographically led to less circulation; possibly catastrophic release of ice gas hydrates (methanes) from ocean floor
- **Late Cenozoic Ice Age (last 40 Mio yrs)**: Earth undergoes long-term cooling trend; continental ice sheets 5-3 Mio yrs ago

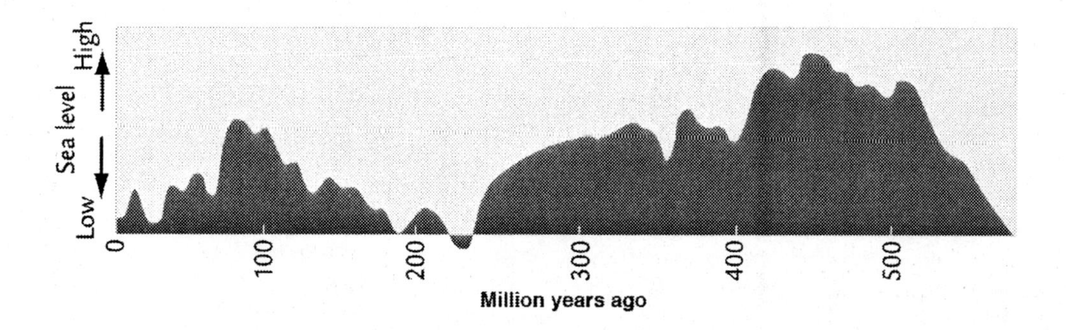

Figure 16.11 Sea level in the last 100 million years. Quantitative estimates of sea level change over the last 100 million years vary widely, but all show a progressive drop to the low sea level of today. The blue area defines the range of estimates based on different methods. (source: RU)

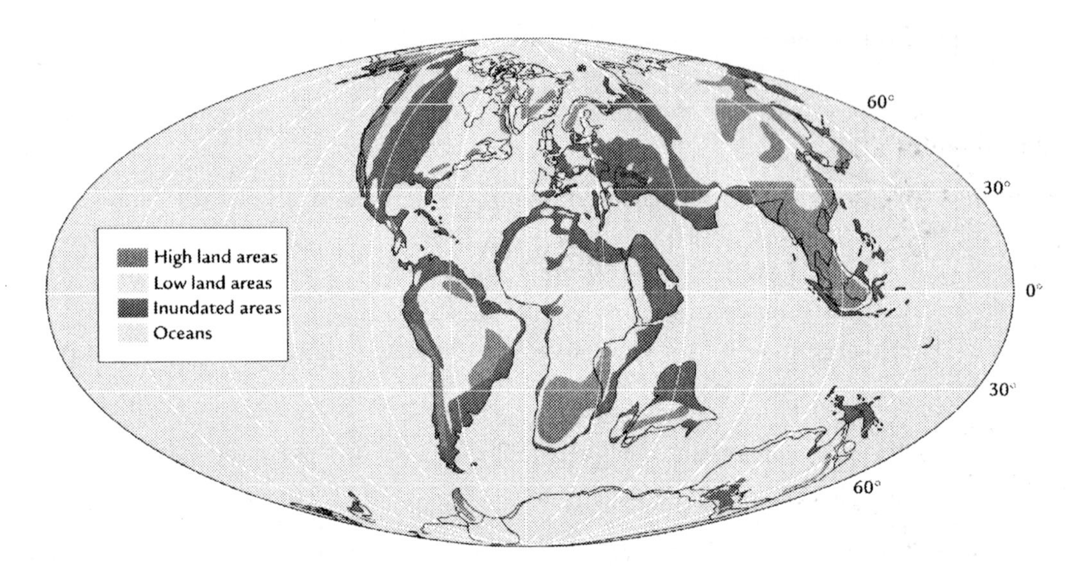

Figure 16.12 The world 100 million years ago. By this time, plate tectonic processes had broken the supercontinent Pangaea into smaller continents, and flooding of these continents by shallow seas further reduced the extent of dry land. (source: RU)

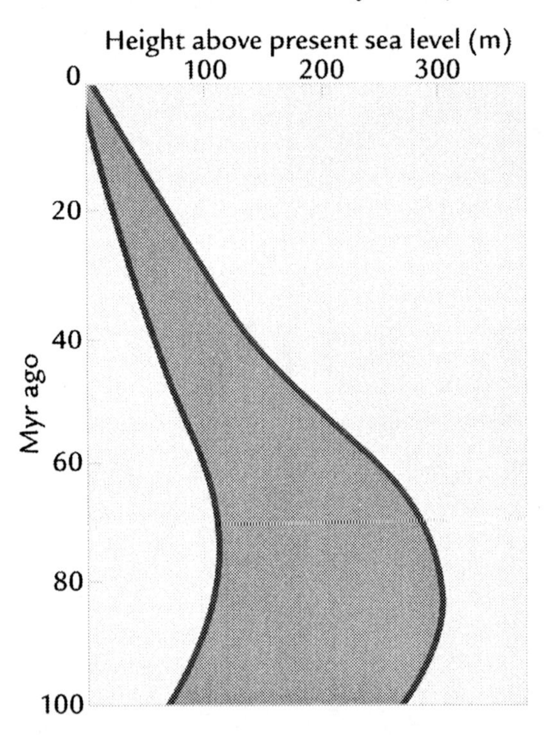

Figure 16.13 Sea level in the last 100 million years. Quantitative estimates of sea level change over the last 100 million years vary widely, but all show a progressive drop to the low sea level of today. The blue area defines the range of estimates based on different methods. (source: RU)

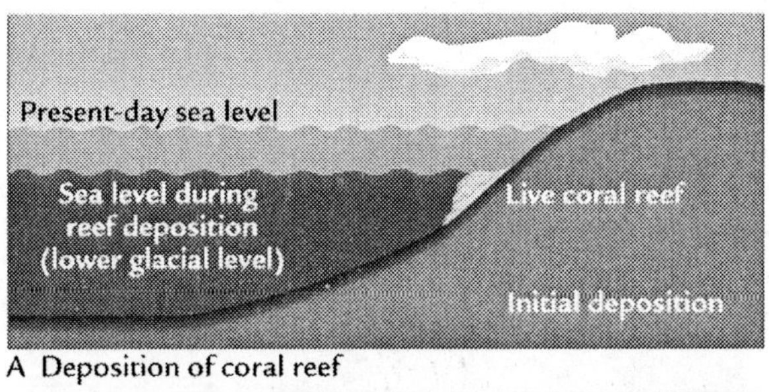

A Deposition of coral reef

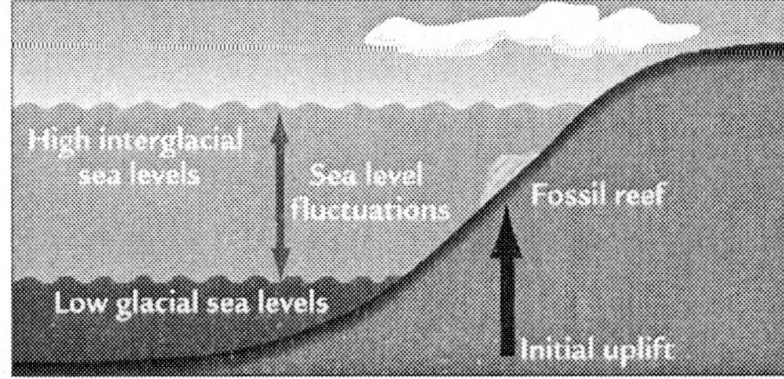

B Subsequent changes

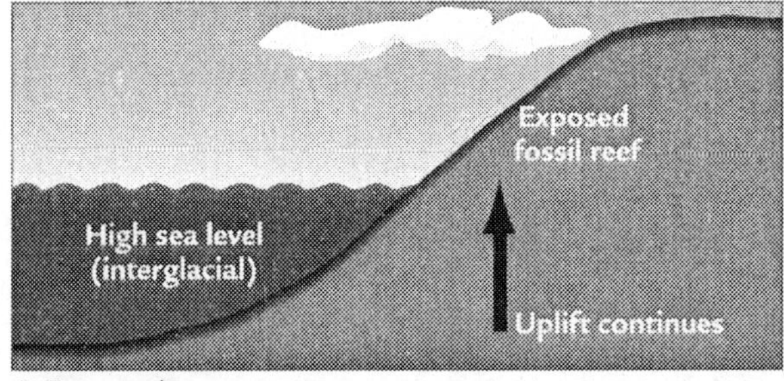

C Present day

Figure 16.14 Gradual uplift of coral reef. Coral reefs may initially form on the edge of an uplifting island at times when sea level lies below its modern position (A). As time passes, uplift steadily raises the island and the fossil reef toward higher elevations, while sea level moves up and down against the island in response to changes in ice volume (B). Today old fossil reefs may lie well above sea level as a result of uplift (C). (source: RU)

16.4 Sea Level Changes

- on scales of millions to 100s of years
- **sea level changes with global climate but also with plate tectonics**
- Local sea level depends on:
  - sediment input, compaction and subsidence (e.g. Mississippi Delta)
  - isostatic post-glacial rebound (e.g. Hudson Bay and coast along Canadian Shield)
  - plate tectonics (e.g. fast spreading causes relatively shallow oceans, raising sea level)
  - global climate

climate related examples:

- rises in warm climate as ice sheets melt
- falls in cold climates as water gets trapped in ice sheets
- on scale of 100s of years to 100s millions of years

tectonics related:

- rises and falls with changes in ocean spreading rate
- falls during coastal tectonic uplift (e.g. in subduction zone)
- on scale of millions of years

see also next chapter!

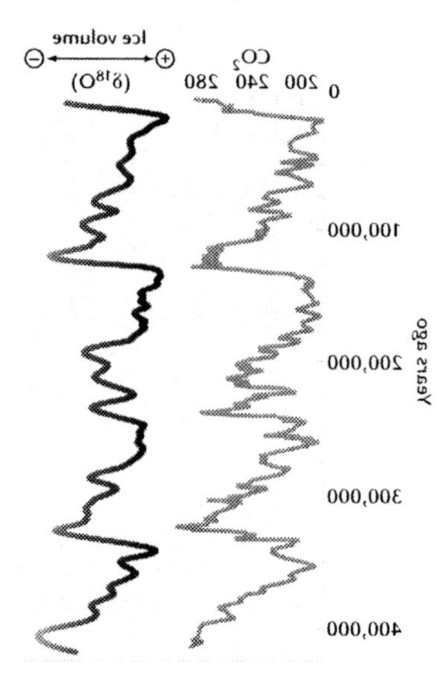

Figure 16.15 Long-term CO_2 and Ice volume changes. A 400,000-year record of CO_2 from Vostok ice in Antarctica shows four long cycles at a period of 100,000 years similar to those in the marine $d^{18}O$ (ice volume) record. At the beginning of a cycle, CO_2 slowly declines and ice volume/marine $d^{18}O$ increases leading into an ice age. A sharp increase in CO_2 and a sharp loss of ice volume marks the abrupt end of an ice age. (source: RU)

COPYRIGHTED MATERIAL — DO NOT DUPLICATE, DISTRIBUTE, OR POST

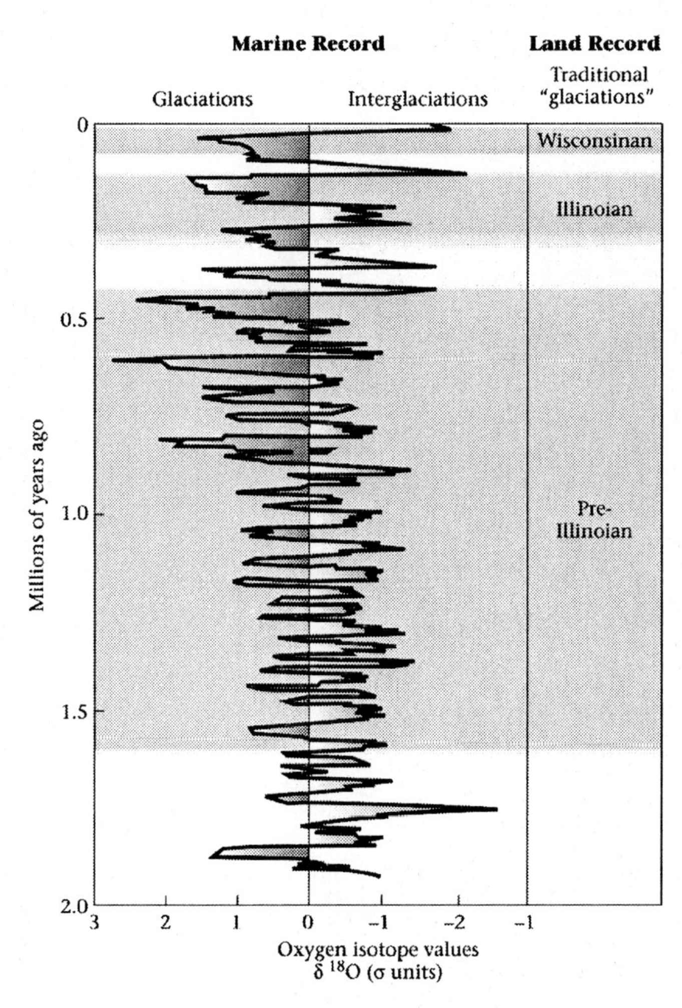

Figure 16.16 This time column shows the variations in oxygen-isotope ratios from marine sediment that define twenty to thirty glaciations and interglacials during the Pleistocene Epoch (most recent 2 million years). Darker (green) bands represent the approximate boundaries of the principal glacial stages recognized on land in the Midwestern USA. The traditional names "Kansan" and "Nebraskan" are no longer used, and have been replaced by "Pre-Illinoian". (source: ME)

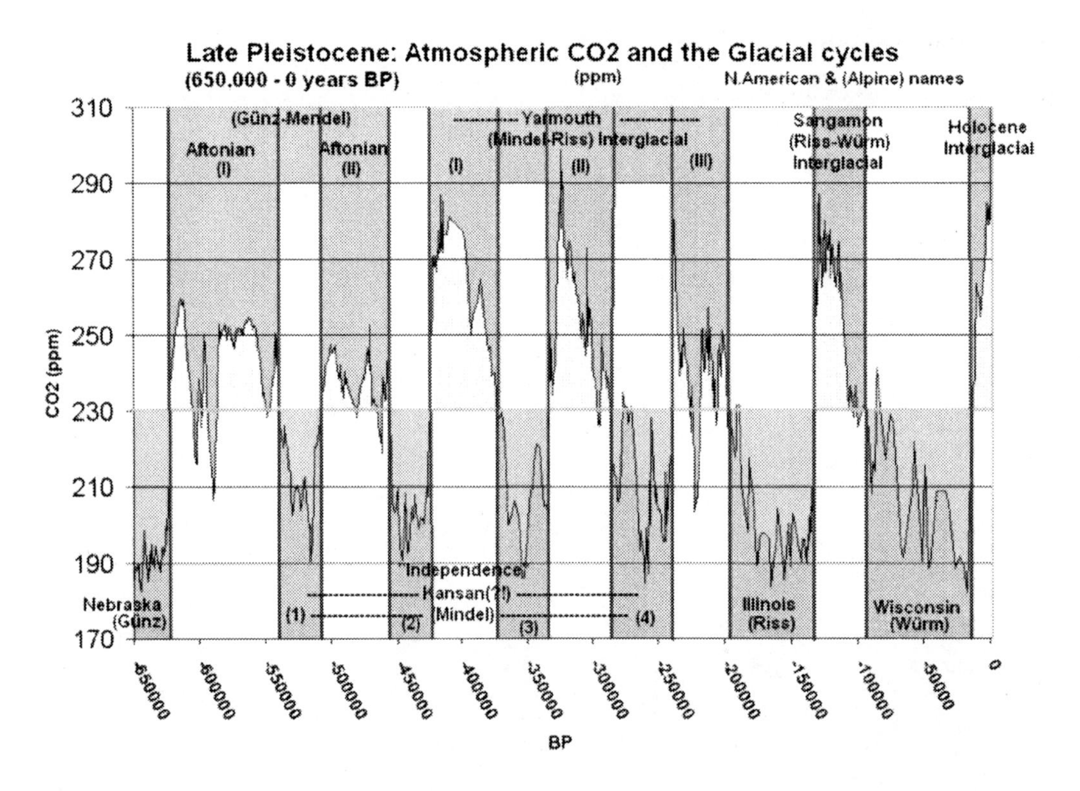

Figure 16.17 Glacial and interglacial cycles of the late Pleistocene epoch within the Quaternary glaciation, as represented by atmospheric CO_2, measured from air bubbles trapped in ice core samples going back 650,000 years. The stage names are part of the North American and the European Alpine subdivisions. The correlation between both subdivisions is tentative. Also note that the terms "Nebraskan" and "Kansan" are no longer used. The onset of an ice age is assumed during the slow descent of CO_2 when CO_2 falls below a certain threshold (here 230 ppm). (source: Wikipedia; Tom Ruen)

16.5 Climate Variation on Scales of 10,000s of Years

The evidence for past climates on the scale of 10,000s of years comes from **environmental climate indicators**

- pollen composition
- tree rings
- sea level changes
- ice cores (amount of CO2 in bubbles); data for about last 800,000 years
- coral rings, oxygen isotopes in corals, marine sediments and ice sheets/glaciers

Oxygen Isotope Ratios in Corals and Ice Cores:

- $^{16}O/^{18}O$ ratios; ^{18}O has 2 neutrons more than ^{16}O so is slightly heavier
- ^{18}O evaporates more easily in warm water than in cold water
- cold water evaporating from oceans removes relatively more ^{16}O, leaving behind ^{18}O enriched water to include in corals and marine sediments; so $^{18}O/^{16}$ **ratio in corals is higher in cold climates**
- glaciers trap more ^{16}O so $^{18}O/^{16}$ **ratio in glaciers and ice sheets is lower in cold climates**
- The situation is reversed in warm climate.

Possible Causes of Climate Change on this Time Scale

- changes on time scales of less than a few Mio years cannot be caused by long-term plate tectonics
- changes in solar energy output; on scales of years to 100,000s of years
- volcanism; on scales of years (SO_2 output causing cooling) to possibly 1000s of years (CO_2 output during very long-lasting, large-scale eruptions of flood basalts)
- **Milankovitch Cycles** (changes in Earth's orbital parameters, see below) most important factor; control amount of sunlight that Earth receives

Climates

- several ice ages and warm interglacial periods in Pleistocene epoch (last 2 Mio yrs)

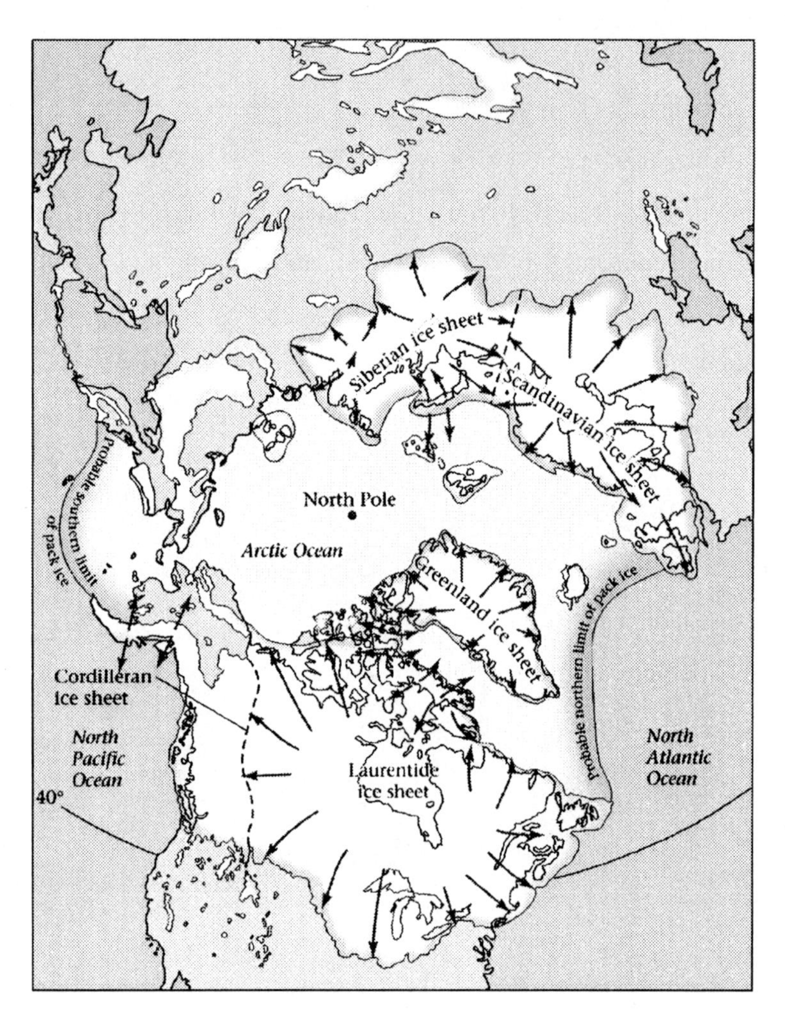

Figure 16.18 A simplified map of the distribution of major ice sheets during the Pleistocene Epoch. The arrows indicate the flow trajectories of the ice sheets. (source: ME)

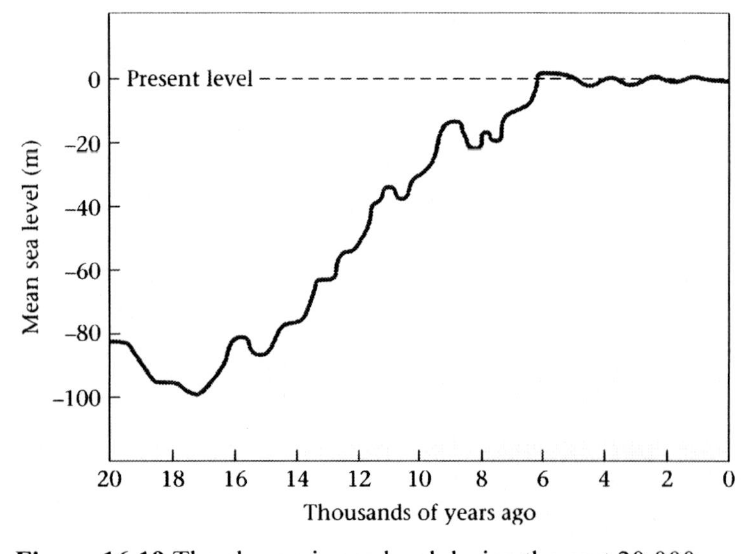

Figure 16.19 The change in sea level during the past 20,000 years. (source: ME)

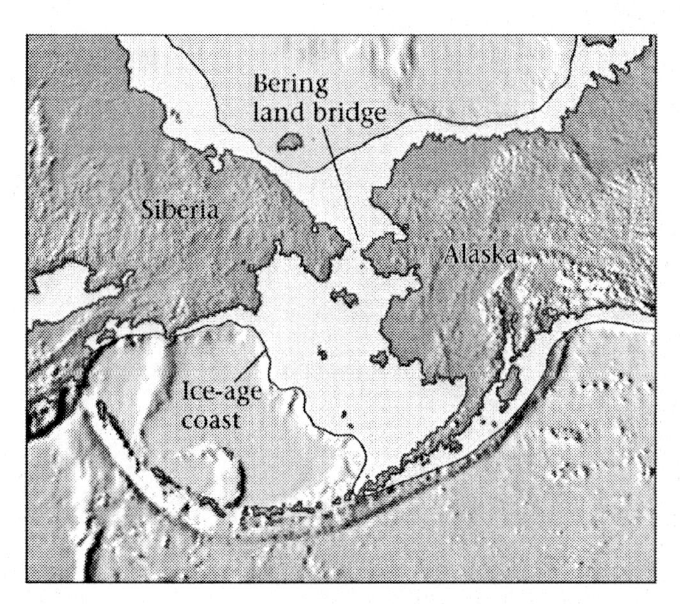

Figure 16.20 A land bridge existed across the Bering Strait between Asia and North America during the last ice age. (source: ME)

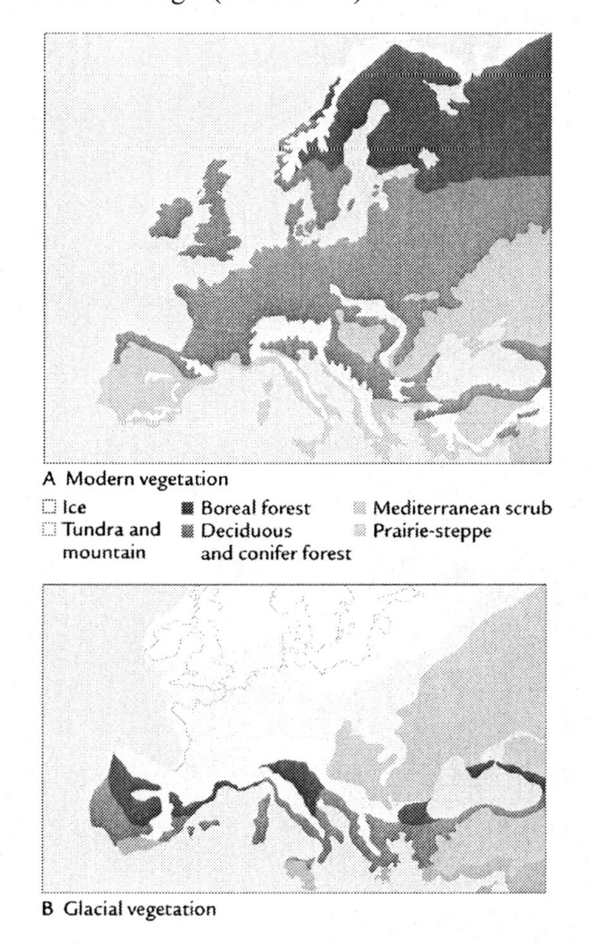

Figure 16.21 During the last ice age, glacial north-central Europe was treeless. Vegetation in present-day Europe (A) is dominated by forest, with conifers in the north and deciduous trees to the south. At

the glacial maximum (B), Arctic tundra covered a large area south of the ice sheet, with grassy steppe farther south and east, and forests reduced to patches near the southern coasts. (source: RU)

The last major ice age

- 18,000-10,000 years ago
- ice sheet expanded over North America and Fennoscandia
- more midlatitudinal rain
- sea level dropped considerably (130m/425ft)
- land bridge across Bering Strait facilitated migration from Eurasia into New World
- currently long-term cooling interglacial period but short-term warming
- if current ice sheets were to melt, sea level would rise by 65m/210ft

19.1 Dust Storms

- Earth can experience massive dust storms that transport sediments over long distances
- **Loess**: up to 30m high loess banks (very fine sand/silt formed during glaciation by glaciers grind on rocks) in Europe, Asia and America formed in and since the last ice age, when strong winds transported the loess across continents

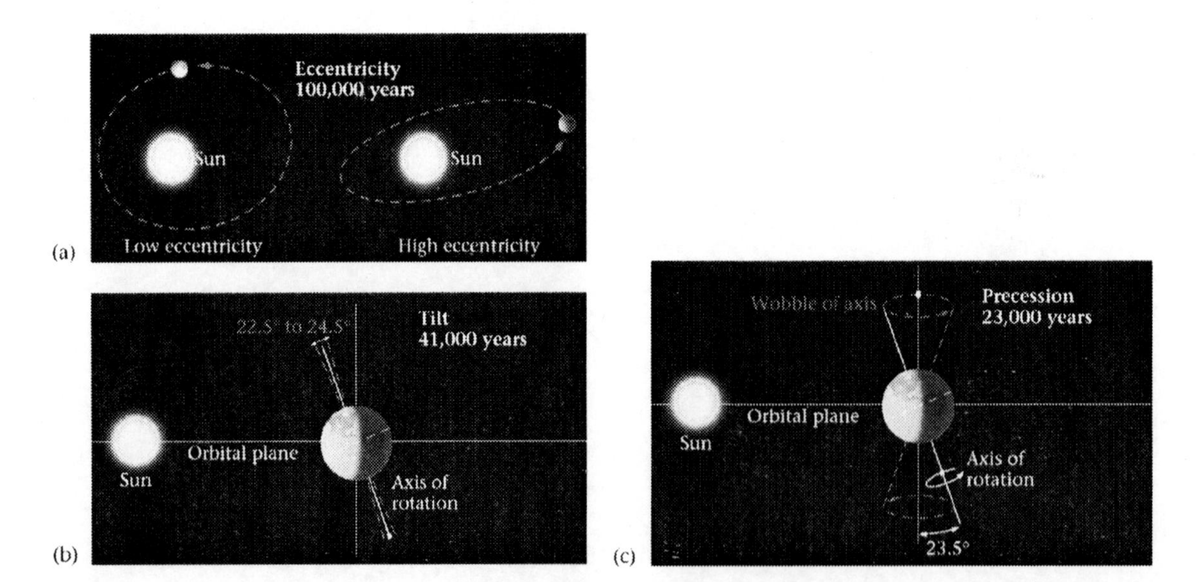

Figure 16.22 The Milankovitch cycles affect the amount of insolation at high latitudes. (a) Variations in insolation caused by changes in orbital shape; (b) variations caused by changes in the tilt angle of Earth's axis; (c) variations caused by the precession of Earth's axis. (source: ME)

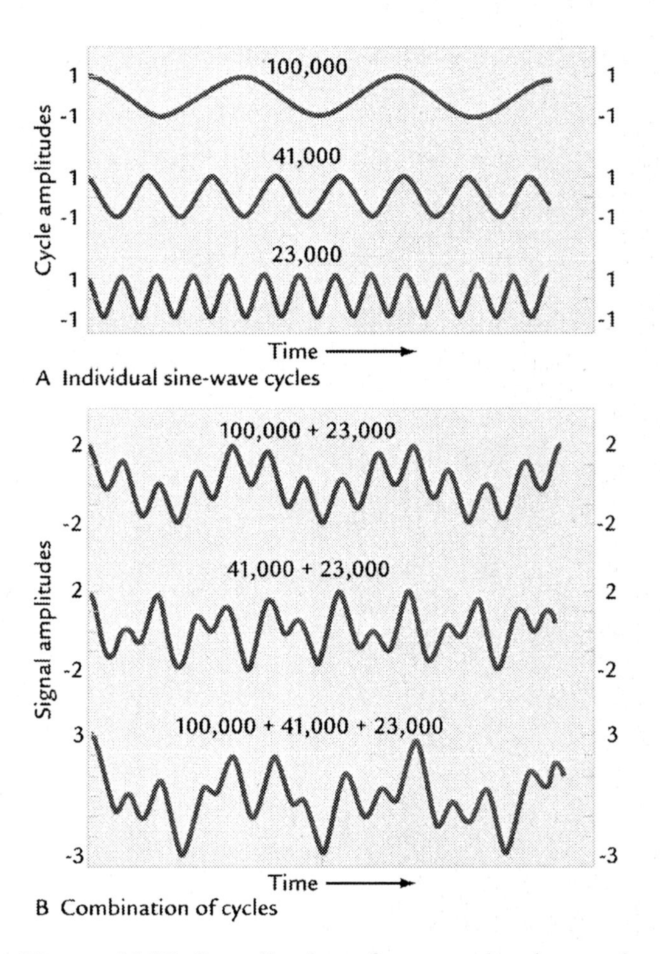

Figure 16.23 Complications from overlapping cycles. If perfect sine-wave cycles with periods of 100,000, 41,000 and 23,000 years are added together, the original cycles are very hard to detect by eye in the combined signal, if at all. (source: RU)

The Milankovitch Cycles

21. eccentricity of Earth's orbit around sun (100,000 years)
22. tilt of Earth's axis (41,000 years) (21.5° to 24.5°; 23.5° at present)
23. precession of Earth's spin axis (like a spinning top) (23,000 and 19,000 years); NB: most textbooks give one period at 20,000 years. This is good enough for this class!

- the combined effect of all 3 parameters is a complicated function where not every decline in insolation triggers an ice age
- The Milankovitch cycles alone cannot account for the temperature variations associated with ice ages. The change in orbital parameters can cause temperature variations of up to 4 °C. However, observed changes are larger (5-7° C in coastal areas and 10-13° C inland). Therefore other mechanisms (e.g. including positive feedback) must be at work.

- Difference in season more important than total amount of insolation (need cool summers to avoid snow melt and mild winters that cause extensive snow fall)
- glacial ice volume over time has **saw-tooth pattern** with warming happening much faster than cooling; extensive warming can occur over only few years!
- warming accelerated through **positive feedback mechanism** of melting high-albedo snow and ice

16.6 Climate Variation on Scales of Millennia to Decades

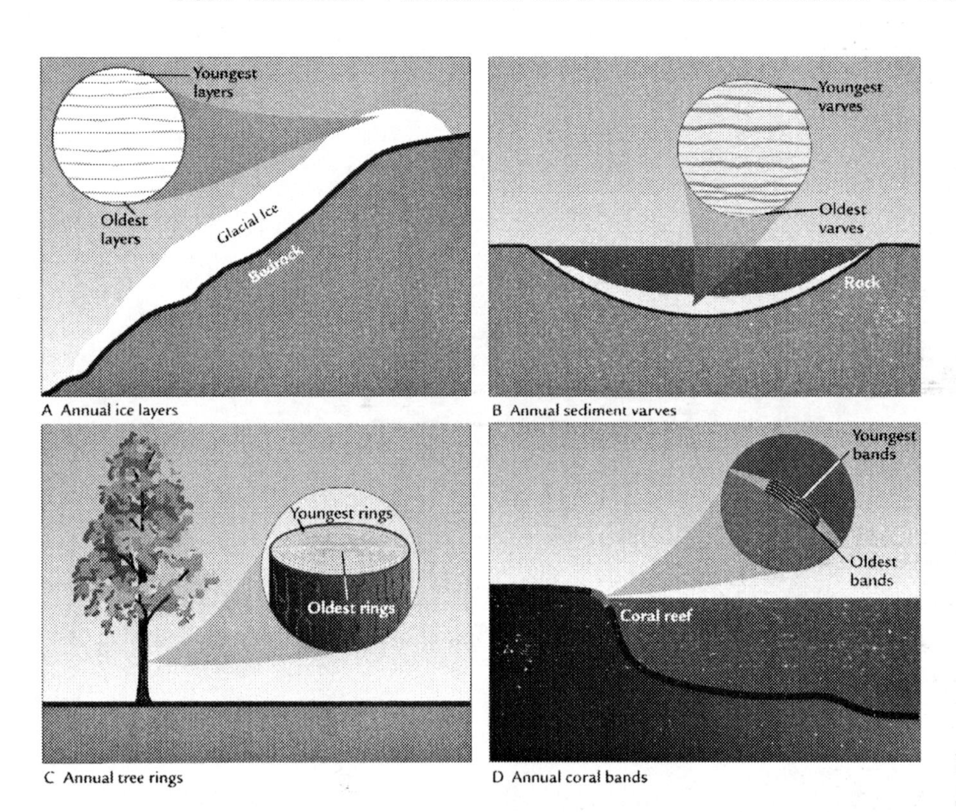

Figure 16.24 Annual layering. Four kinds of climate archives have annually deposited layers that can be used to date the climate records the contain: ice, varved lake sediments, trees, and corals. (source: RU)

The evidence for past climates on the scale of 10s to 1000s of years comes from **historical (recorded) climate indicators**

- tax records for crops
- paintings
- songs, epics and stories
- length of seasonal ice around Iceland
- names (e.g. Greenland due to warm period 800-1000)

- wine growth in England
- recorded advance and retreat of mountain glaciers
- modern measurements

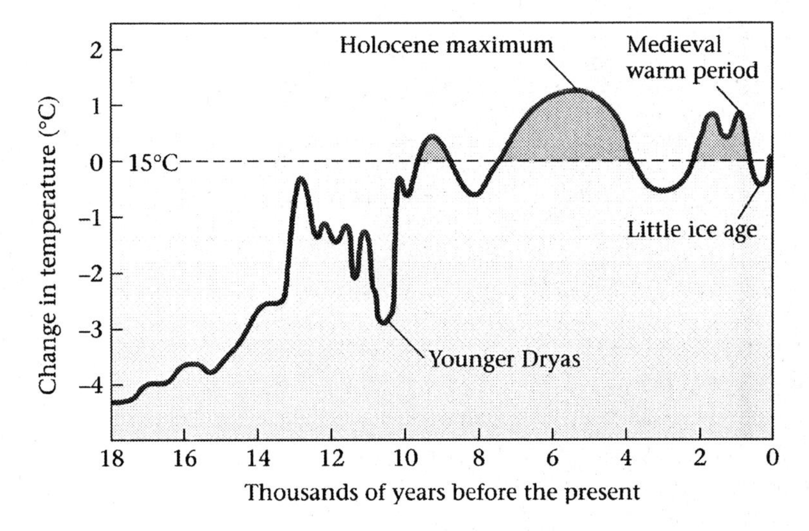

Figure 16.25 The past 15,000 years (the Holocene Epoch) experienced several periods of warming and cooling. (source: ME)

Figure 16.26 Coastal Greenland was settled by the Vikings during one of the warmer periods, when the region could support agriculture. (source: ME)

Figure 16.27 Life during the Little Ice Age. Skaters on the frozen canals of the Netherlands (painting by Hendrick Avercamp, ca. 1608). (source: ME/wikipedia)

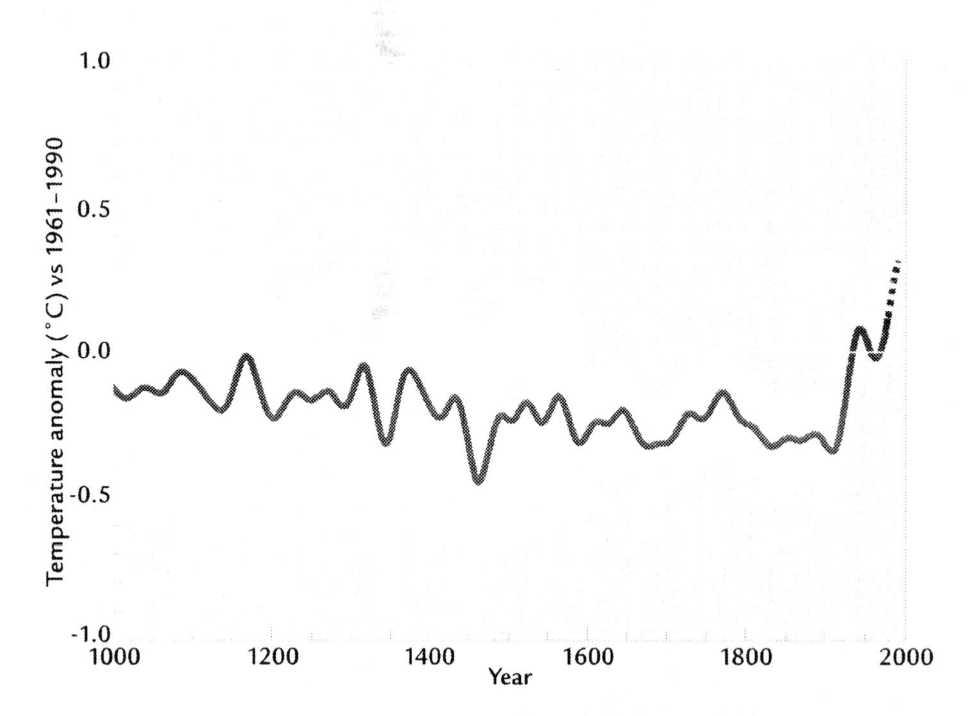

Figure 16.28 Temperatures in the northern hemisphere during the last millennium. A synthesis of high-resolution climate records spanning all or part of the last millennium shows a gradual cooling in the northern hemisphere for 900 years, followed by an abrupt warming in the 20th century to temperatures higher than any in the earlier record. Light shading indicates uncertainty in estimated temperature. (source: RU)

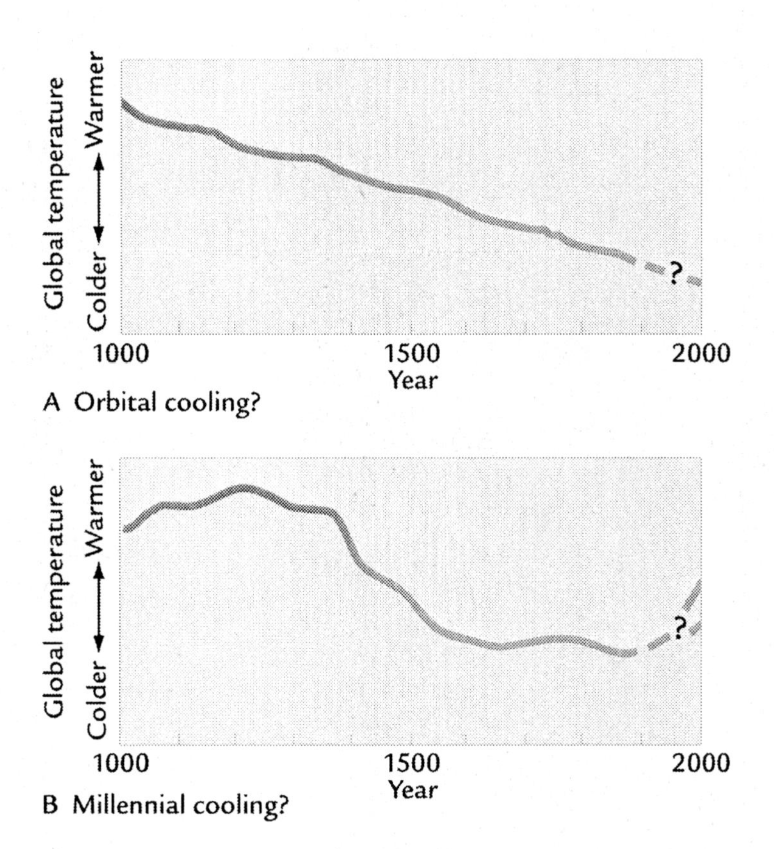

Figure 16.29 Two interpretations of the Little Ice Age. The cool Little Ice Age interval in Europe between 1400 and 1900 could have resulted from (A) slow orbital-scale cooling or (B) a millennial-scale oscillation. (source: RU)

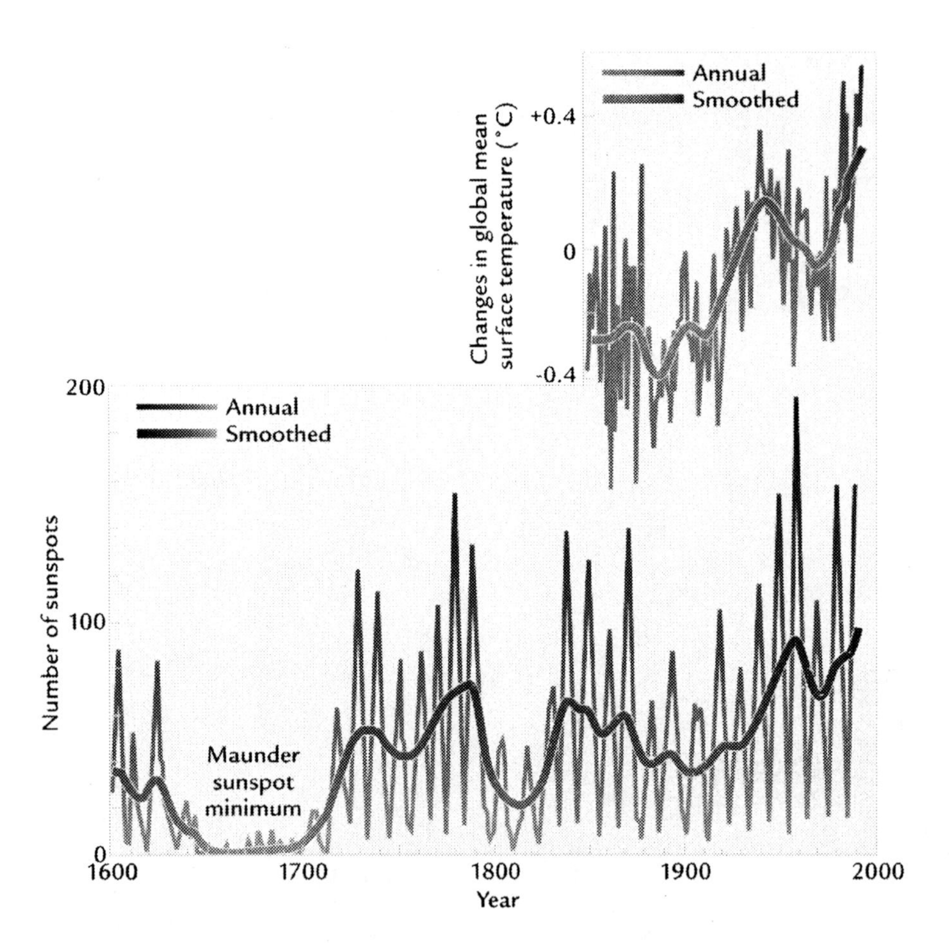

Figure 16.30 Sunspot history from telescopes. Measurements made with telescopes over the last several hundred years show an 11-year sunspot cycle, as well as earlier intervals such as the Maunder minimum, when sunspots were absent for several decades. The longer-term average number of sunspots resembles observed temperature changes during the twentieth century (top). (source: RU)

Possible Causes of Climate Change on this Time Scale

- changes in ocean currents, decadal oscillations in Pacific and Atlantic (decades to hundreds of years?)
- interactions between oceans, atmosphere, ice sheets (years to decades?)
- volcanism; on scales of years (SO_2 output causing cooling) to possibly 1000s of years (CO_2 output during very long-lasting, large-scale eruptions of flood basalts)
- fertile crescent in Near East (5000-6000years ago) could be due to more solar activity
- variations in solar output (10s - 100s of years): some decades have less solar activity
- sun spot cycles on a scale of 11 years
- El Niño Southern Oscillation (ENSO) and La Niña on a scale of few to 10 years (include global impact maps?)

Short-term Cold Climates on 100-year Time Scale

- Maunder Minimum (~1700) during "Little Ice Age" (1400-1850); advance of glaciers; sea surrounding Iceland froze; rivers and canals in Europe froze; could have been caused by 0.25% less solar output/50-year absence of sunspot
- cooler first half of 20th century (prior to 1950s); correlates with less solar activity/sunspots

Short-term Warm Climates on 100 to 1000-year Time Scale

- for last 10,000 yrs, general interglacial period (warming) after last ice age
- 5000-6000 years ago (fertile crescent in Near East; global T was 2°C higher than today; possible cause: variation in solar output
- Greenland was green in AD 900, hence warmer than today!!
- current trend is warming on short scale (10s-100s of years) but cooling on long scale (millions of years)
- most recent trend: rapid warming at rate not seen in last 650,000 years

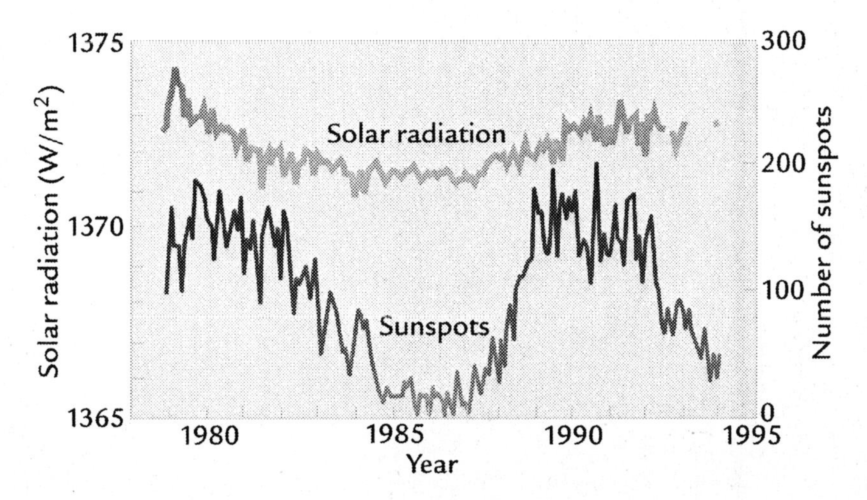

Figure 16.31 Solar radiation and sunspots. Since 1978, satellites have measured changes in the solar radiation arriving at the top of the atmosphere (top) and these changes correlate with the observed numbers of sunspots (bottom). (source: RU)

Sunspot Cycle: Climate change on the Scale of 10 years

- periodicity: about 11 years
- observed since 1749
- main source of periodic solar variation driving variations in space weather

- sunspots are darker but surrounding regions are hotter/brighter than average -> # of sunspots correlates with intensity of solar radiation
- variations are small (0.1%) compared to solar constant (solar output; 1366 W/m^2)
- last maximum in 2000; last minimum in 2008 culminating with nearly no sunspot activity in August (see Earthwatch), nearly breaking a 100-year record
- overall, the sun is currently at a heightened level of sunspot activity and was last similarly active over 8000 years ago

El Niño Southern Oscillation (ENSO) and La Niña: Climate change on the Scale of 10 years

Observations:

- the main modern, satellite-based observation is that during an El Niño, equatorial waters in the East Pacific ocean are up to 5°C warmer than normal, whereas during a La Niña, surface waters are colder than normal
- El Niño causes rainfall in California but also droughts (e.g. rain in north and dry in south; rain in strong events but droughts in weak events).
- El Niño can change the **global weather** creating both more severe droughts and rainfall than normal (e.g. droughts in East Africa in the 1997 El Niño).

Mechanisms:

- for details in mechanisms, see lecture on air circulation (lecture 15)
- ENSO events have a periodicity of 4-7 years but the strength of events vary. A possible cause is the interaction with other oscillations (e.g. the Pacific Decadal Oscillation with a periodicity of 20-30 years.)

Recommended Reading

- [1]"Meteorology Today" by C. Donald Ahrens, Brooks/Cole Thomson Learning, 2003, ISBN: 0-534-39771-9

- "Earth's Climate, Past and Future" by William F. Ruddiman, 2000. W.H. Freeman and Company, ISBN: 0-7167-3741-8

Also, food for thought to discuss climate change and human impact in this and future classes "State of the World" by The Worldwatch Institute, W.W. Norton and Company. Topics change from year to year but the 2004 book (ISBN: 0-393-32539-3) was on consumerism and globalization, waste/recycling of resources, catch-up of developing countries, water productivity and increasing shortage. Earlier books were on energy resources, greenhouse effect, Kyoto Protocol and the spread of and fight against diseases.

BOX 1: The Carbon Cycle

- path of Carbon of special interest because CO_2 is a greenhouse gas
- Carbon cycle describes the exchange of C between different reservoirs

  RESERVOIRS[1]:

  - atmosphere: 720 Gt (gigatons)
  - oceans: 38,400 Gt
    - total inorganic: 37,400 (surface layer: 670; deep layer: 36,730)
    - total organic: 1000
  - terrestrial biosphere: 2000 Gt
    - living biomass: 600-1000
    - dead biomass: 1200
  - sediments and rocks: 75,000,000 Gt
    - sedimentary carbonates: > 60,000,000
    - kerogens (geologically early stage of source rock for oil): 15,000,000
  - aquatic biosphere: 1-2
  - fossil fuels: 4130

  PATHWAYS INTO ATMOSPHERE[2] (according to coursebook):

  - atmosphere-ocean exchange: 100-115 Gt
  - decay of organic material: 50-60
  - plant respiration: 40-60
  - burning of fossil fuels: 6
  - land clearing: 1-2.5
  - volcanism: unknown

  PATHWAYS OUT OF ATMOSPHERE[2] (according to coursebook):

  - photo synthesis: 100-120
  - atmosphere-ocean exchange: 105-120
  - weathering of rock: unknown, perhaps 0.6; most important in early Earth

- amount of CO_2 in atmosphere is only 0.037%; but if CO_2 and water vapor were absent in the atmosphere, Earth would be 33°C (59°F) colder; current average surface T is 15 °C/59°F.
- human activity, especially in the U.S. (25% of world's CO_2 production!) is now increasing atmospheric CO_2 at an alarming rate

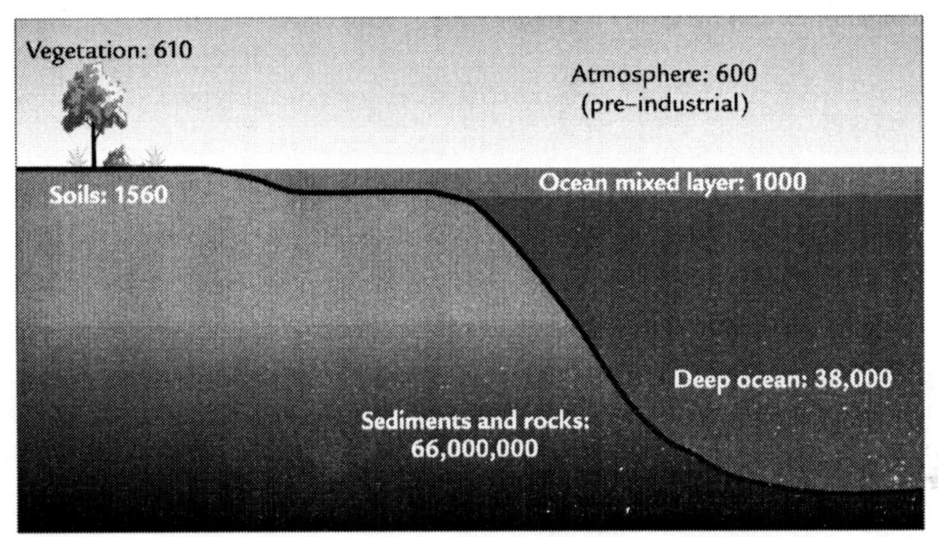

Figure 16.33 The largest reservoir of carbon on Earth lies in its rocks, while the smallest was the pre-industrial atmosphere but now is vegetation (atmospheric carbon increased by 40% by 2011). (source: RU)

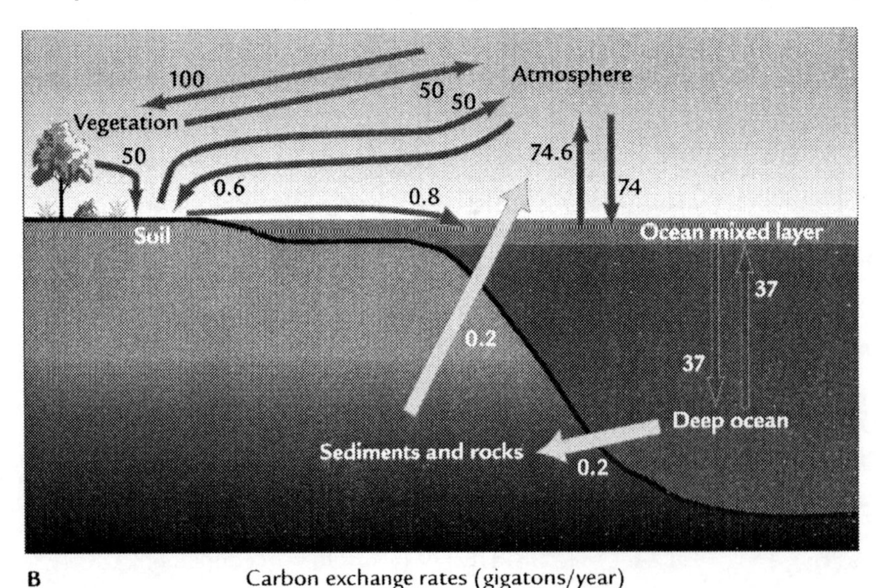

Figure 16.34 Over intervals of millions of years, slow exchanges among the rock and ocean/vegetation/soil/atmosphere reservoirs can cause large changes in atmospheric CO_2 levels. (source: RU)

BOX 2: Laws and Fossil Records

Geologists use stratigraphic profiles, the layering of sediments and the fossils they find within the sediments, to estimate past climates and environments. This can only be done under certain circumstances that are formulated in laws. These include:

- **uniformitarianism**: processes that happen today are the same as those that happened in ancient times
- **law of superposition**: in an undisturbed stratigraphic sequence younger strata lie over older strata
- **law of fauna assemblage**: strata of like age can be recognized by the like assemblage of fossils they contain

The validity of the last law was particularly important before physical dating methods (i.e. radiometric dating) was available.

REFERENCES

1. Falkowski et a., 2000. "The Global Carbon Cycle: A Test of Our Knowledge of Earth as a System". Science, vol 290 (13 October), 291-296
2. Patrick Abbott, 2005. "Natural Disasters". McGraw Hill

COPYRIGHTED MATERIAL — DO NOT DUPLICATE, DISTRIBUTE, OR POST

Chapter 17: Life and Mass Extinctions (draft)

Figure 17.1 (source: NASA/wikipedia)

see chapter 19 for human-induced recent extinctions!!

What are Mass Extinctions?

The natural disasters we have discussed so far have killed individuals of one or more species. They are not considered events that triggered an extinction.

- **Extinction**: so many individuals die that reproduction fails
- **Mass Extinction**: all members of one or more species die suddenly or within a relatively short time. Also, a mass extinction that involves many species can be triggered by removal of just one member in the food chain (e.g. large-bodied herbivores). Today, more than 99% of all species that have ever existed are extinct.

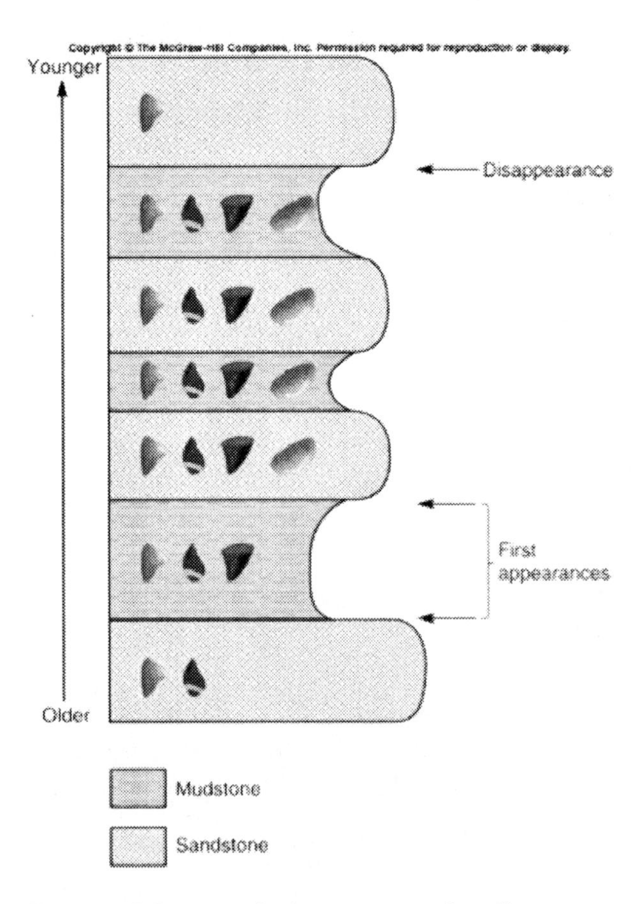

Figure 17.8 A vertical sequence of sedimentary rock layers illustrates how fossils emerge and disappear/become extinct. A time when many fossils disappear marks a mass extinction. (source: AB)

17.1 Why does Earth have Life?

Figure 17.2a Mars as seen by the Hubble Space Telescope. (source: NASA/wikipedia)

COPYRIGHTED MATERIAL — DO NOT DUPLICATE, DISTRIBUTE, OR POST

Figure 17.2b Europa's trailing hemisphere, as seen by the Galileo spacecraft. Darker regions are areas where Europa's primarily water ice surface has a higher mineral content. (source: NASA/wikipedia)

Figure 17.2c Mosaic of Titan from spacecraft Cassini's February 2005 flyby. (source: NASA/wikipedia)

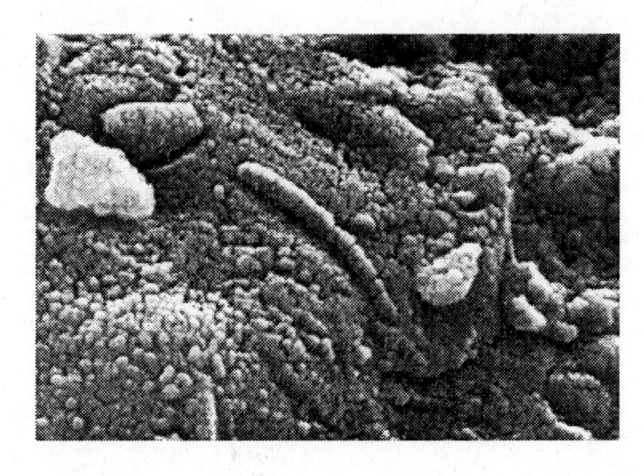

Figure 17.3 An electron microscope reveal chain structure in Martian meteorite fragment ALH84001. The meteorite was fund in Allan Hills, Antarctica on 27 December 1984. The structures are thought to be remains of bacteria but contamination from handling the sample on Earth may also explain the structures. (source: NASA/wikipedia)

Astronomy

Earth is uniquely located in the solar system to have the right conditions:

- Earth's distance to Sun allows it to maintain an atmosphere and bearable temperatures
- Earth's closeness to Sun allows Earth to receive enough sunlight for bearable temperature and water to exist in liquid form
- Moon is a relatively large satellite that contributes to relatively stable orbital parameters for Earth
- changes to Earth's orbital parameters are relatively small, causing relatively mild climate changes (see Milankovitch cycles in Lecture 15)
- comparison with Mars:
    - Mars' distance from Sun is 1.52 AU (astronomical units; average Earth-Sun distance)
    - Earth's eccentricity is 1-5%, Mars' is 9.3%
    - Earth's obliquity (tilt of spin axis) changes between 21.5 and 24.5°: Mars' is currently 25.2° but varies immensely over time
    - Mars' atmosphere: major constituent: CO_2; no greenhouse because atmosphere is thin
- comparison with Venus:
    - Venus' distance from Sun is 0.72 AU
    - Venus' current eccentricity is 0.7%, so not anomalous
    - Venus' obliquity is 177.4° so Venus' revolution around Sun is retrograde (spin and orbital revolution have opposite sense); practically no seasons
    - Venus' length of day is 243 Earth days; Venus' orbital period is 225 Earth days; a Venus "year" is therefore longer than a Venus "day"
    - Venus' atmosphere: major constituent: CO_2; extremely dense causing extreme atmospheric pressure at surface (92 times Earth's); extreme greenhouse causing surface temperatures around 450°C (surface rocks glow red)

COPYRIGHTED MATERIAL — DO NOT DUPLICATE, DISTRIBUTE, OR POST

Biochemistry

The origin of life involves the sequence of many complicated steps but Earth has all the chemical ingredients for life to form. Some are listed here:

- after the initial cooling, early Earth had traces of CH_4, NH_3, H_2 and - above all - H_2O
- lab experiments show that exposing such a mixture to lightning caused condensation to organic compounds (made of C, N, H)
- addition of water is the next step to make amino acids and other building blocks of life
- polymerization leads to macromolecules such as proteins, nucleic acids, polysaccharides
- three additional conditions: 1) supply of self-reproducing molecules; 2) copying of these molecules must be subject to errors (mutations); 3) perpetual supply of free energy and partial isolation from general environment (membranes)
- ..... eventually formation of DNA (protein, hereditary molecule) from "proto" RNA (nucleic acid)

Classification of Life

In a somewhat outdated but simple scheme, different life forms on Earth are classified like a **family tree**. Life on Earth has **5 (or 6) kingdoms** which then branch out into subcategories (**phyla, classes, orders, families, genera and species**).

Monera are now obsolete in the three-domain system.

Examples:

- **kindom**:animals, plants, fungi, protista (eukaryotes), monera (prokaryotes) Monera are sometimes sub-divided into two kingdoms: Eubacteria (true bacteria) and Archaeobacteria or Archea (ancient bacteria).
- **phylum**: e.g. the animal kingdom branches out into vertebrates (chordata) and invertebrates
- **class**: e.g. the vertebrate phyllum branches out into: mammals, fish, dinosaurs (that includes birds in a sub-class), amphibians and reptiles

  The word **protista** comes from the Greek word "protos" for first. Protista include all simple unicellular and multicellular forms. For example, tracing the lineage of today's humans through the tree of life: kingdom: *animals*; phyllum: *vertebrates*;

class: *mammals*; order: *primates*; family: *hominids*; genus: *homo*; species: **homo sapiens**.

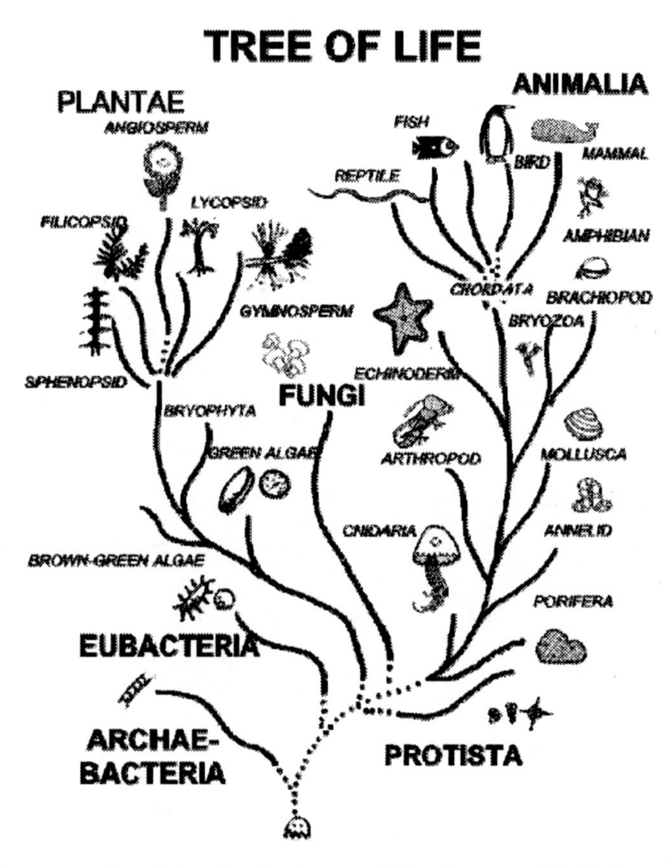

Figure 17.3 An example of the classical tree of life. (source: NASA)

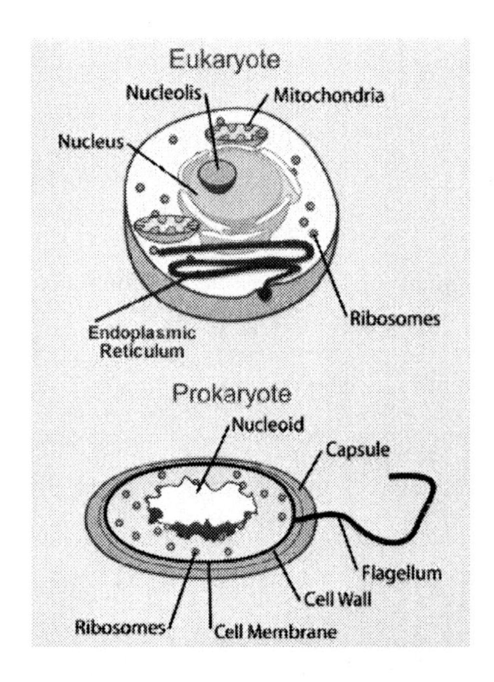

Figure 17.4 A simplified comparison between prokaryotes and eukaryotes. (source: NASA)

Some Basic Life Forms

The distinction between prokaryotes and eukaryotes comes from the fact whether a cell has a nucleus or not (from Greek word "karyon" for nut).

- **Prokaryotes**: from Greek "pro" for before; have no nucleus; basically all bacteria (kingdom: monera); DNA as strands in cytoplasm; two main sub-categories that are sometimes divided into two different kingdoms:
  - **Archaeobacteria**: can live in extreme environments, e.g. within the Earth several km down, hot, anaerobic environment such as active vents near mid-ocean ridges. In fact, O_2 kills archaea and their life is based on S instead. The processes involved to collect and use energy are relatively inefficient.
  - **Eubacteria**: some eubacteria are capable of doing photosynthesis, e.g. cyanobacteria, or "green-blue algae" (this name is actually misleading as cyanobacteria are bacteria and not algae)
- **Eukaryotes**: cell has a nucleus in which DNA is stored; kingdom protista and higher; various organelles distributed through the cell that perform different functions; e.g. chloroplast (photosynthesis); mitochondria (power source)
- **viruses**: not included in the traditional classification of life. Viruses are basically simple strands of DNA sheathed in proteins; they can't live independently and often need a host to reproduce. In contrast to bacteria, can't be cultured in artificial media. Many viruses are inactivated by heat. High mutation rate (eg. polio virus mutates 2% during its three-day passage through the human intestinal system; humans took 8 Mio yrs for the same level of mutation). Bacteria produce a new generation within 20min, viruses accomplish this is less time.
- **prions**: not included in the traditional classification of life. Examples include the agent causing BSE (bovine songiform encephalopathy or mad cow disease). Proteins folded in an abnormal way. No traditional reproduction. Turn other proteins into prions upon contact. Evoke no immune response in infected bodies. Resist heat, UV, radiation, sterilization.

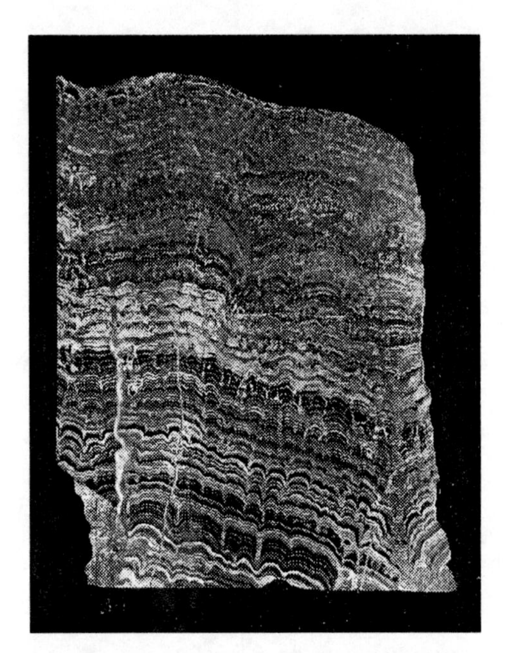

Figure 17.5a Fossilized stromatolites from Bolivia. (source: wikipedia)

Figure 17.5b Stromatolites growing in Yalgorup national park in western Australia. (source: wikimedia)

Traces from Early Life Forms

Traces of bacteria can be examined under the microscope (often bacteria products, not the bacteria themselves!). The geologist in the field relies on **stromatolites** which are macroscopic features that are left behind from thriving colonies of bacteria. Fossil stromatolites can be found e.g. in Glacier National Park. Living forms can be found everywhere, e.g. Mexico and New Zealand. Many of these grow in the intertidal zone. Cyanobacteria grow long fibers which can trap mud and sand and form mats. Bacteria then grow into the trapped sediments to form a thick crust. This happens repeatedly, layer after layer, until mounds up to 1m in height and diameter form, the stromatolites.

17.2 Early Life and its Timeline

About 3/4 of Earth's existence saw only "simple life". It is only the last 600Mio years or so, that we have complex multi-cellular life forms, i.e. life as we know it. Approximate timeline of early life forms:

- 3.5 billion yrs ago: appearance of Archaeobacteria
- shortly thereafter: appearance of Eubacteria (some doing photosynthesis)
- 1.1 billion yrs ago: simple Eukaryotes; single-celled organisms with a nucleus

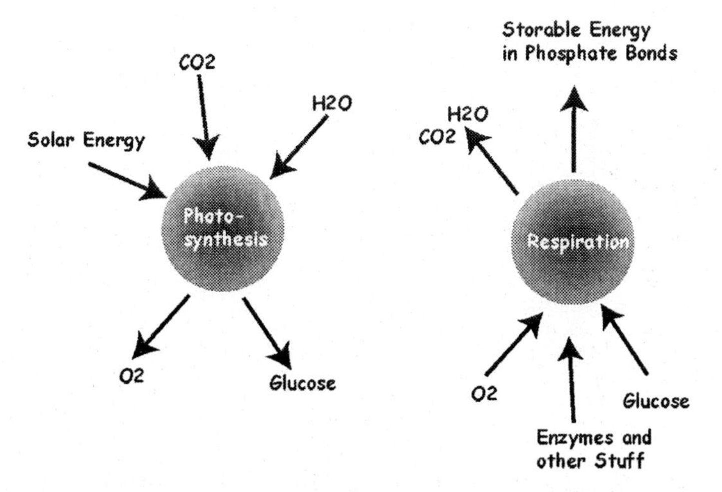

Figure 17.6 The concept of photosynthesis and respiration.

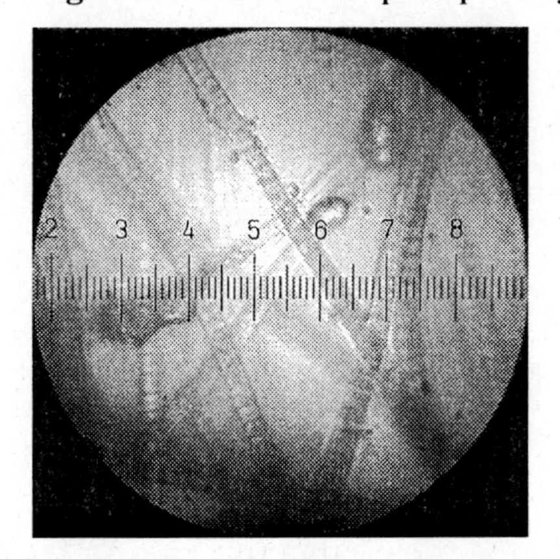

Figure 17.7 Cyanobacteria dramatically changed the composition of life forms on Earth through photosynthesis, leading to the near-extinction of oxygen-intolerant organisms. (source: Wikipedia)

Photosynthesis and Respiration

Photosynthesis is the most important process by which Earth's atmosphere gained oxygen. (see APPENDIX B). Living creatures absorb energy in some form and store it through some form of metabolisms for later use when needed. Some organisms collect energy through **photosynthesis**. This process was particularly important to the early Earth because it provided the oxygen that we now have in the atmosphere. Oxygen intolerant organisms use a less effective process called **chemisynthesis** (used by archaea). **Respiration** is a process in which the sugars from photosynthesis are transformed in storable energy in form of phosphate bonds. **Fermentation** a process similar to respiration but does not require oxygen and is less efficient (used by many bacteria).

- Photosynthesis: uses sunlight, CO_2 and H_2O to produce sugars/cellulose and O_2
- Chemisynthesis: uses, H_2S instead of H_2O to produce S instead of O_2; slower metabolism and less efficient than photosynthesis
- Respiration: uses enzymes and O_2 to break down sugars to storable energy in phosphate bonds; CO_2 and H_2O as by-products; in regard to oxygen, this process is a "reversed photosynthesis" though respiration typically does not require as much O_2 as is produced during photosynthesis
- Fermentation: low-efficiency equivalent to respiration; no O_2 required; produces alcohols in intermediate step

Oxygen sources and Sinks in the Atmosphere

Oxygen Producers and Users

Process	Production (+)/Usage (-); kg/yr
photochemistry	$+ 10^8$
weathering of rock	$- 10^{11}$
volcanism	$- 10^4$
photosynthesis	$+ 10^{14}$
respiration/decay	$- 10^{14}$
burial of carbon from organisms	$+ 10^{11}$
recycling of buried sediments	$- 10^{11}$
fossil fuel combustion	$- 10^{12}$

Comments (**see also section on photosynthesis**):

- Photochemistry is the process in the atmosphere when H_2O molecules are broken down after absorbing UV radiation. H_2 escapes into space.
- Carbon in decaying organisms uses oxygen from the atmosphere so oxygen is "gained" when burying these organisms.
- The recycling (bringing back to surface) of sediments is in balance with the burial.
- Respiration is now in balance with photosynthesis but was less in early geologic times.
- The atmosphere has currently $6x10^{17}$kg oxygen. If photosynthesis was shut off, weathering and volcanism would destroy the oxygen in only 6 Mio years.
- **photosynthesis is the most significant process to increase oxygen in the atmosphere**

Atmospheric Oxygen over Time

Approximate Timeline

Time before Present	Amount of Oxygen
3.5 - 2 billion years	prob. less than 1%
2 - 1.5 billion years	significant increase (prob. up to 5%)
	Eukaryotes (that require aerobic environment)
	formation of ozone layer
1.5 billion years	oxygen level near 5%
0.6-0.4 billion years	tremendous increase in oxygen level (> 30%)
0.3 billion years	oxygen level nears today's level (> 80%)

How do we know ancient Oxygen Levels?

In early times, the oxygen level could have been not higher than 2% because we find old rocks that could not have been formed or would have oxidized in an aerobic environment.

- The first case includes the iron band formations (BIF). These are typically 2-3 billion years old and consist of alternating red (30% Fe) and light (silica) layers. These deposits worldwide and make up 90% of the commercial supply. They are thought of have formed in marine environment. Iron was dissolved in ocean water are ferrous iron (FeO) and the absence of oxygen kept the iron from precipitating on the ocean floor. In today's atmosphere, FeO combines with oxygen to ferric iron, Fe_2O_3, which has relatively more oxygen. This precipitates to the ocean floor. The conclusion is that the early atmosphere must have been depleted in oxygen. The episodic events that precipitated the ferric iron in ancient times are not well understood. One theory includes upwelling from iron rich waters near active vents to bring it to the surface where photosynthetic organisms provided the necessary oxygen. There was no more BIF production after 2 billion years.

- Another piece of evidence for the early oxygen depletion comes from finds of 2.7 billion year old Pyrite (FeS) and Uraninite (UO_2) that show no oxidation, which would have happened in an oxygen-rich atmosphere.
- A lower limit of 10% oxygen after 2 billion years comes from multi-celled microbes
- oxygen levels found in100-200 Mio year old charcoal records suggest an oxygen level of at least 62% of today's level

17.3 Evolution of Life and the Geologic Time Scale

Rough Geologic Timescale

Time Span (Mio yrs before present)	**Eon**	**Era**	**Period**	**Major Appearance**
2.6 - 0			Quaternary	Humans (1.0)
65 - 2.6		Cenozoic	Tertiary	Human ancestors (3.5)
142 - 65		^	Cretaceous	
206 - 142		|	Jurassic	birds
251 - 206		Mesozoic	Triassic	dinosaurs and mammals
290 - 251			Permian	
354 - 290			Carboniferous	Reptiles (325)
417 - 354	^	^	Devonian	Amphibians
443 - 417	|	|	Ordovician	Land Plants
495 - 443	|	|	Silurian	
544 - 495	Phanerozoic	Paleozoic	Cambrian	vertebrates; Fishes

2500 - 544		Proterozoic	sexual reproduction (1000)
4000 - 2500	Precambrian	Archean	oldest fossil (3600); oldest rock (4000)
4570 - 4500	Hadean		

If the evolution of life on Earth were projected onto 1 hour, then life with no or very little oxygen would take up almost 30 min. Oxygen-based life under a protecting ozone layer was possible only within the last 20 min. Multicellular marine life started roughly 660 Mio years ago, just before the Cambrian, with invertebrate creatures that left little traces. This translates to the last 8.5 min. Life as we know it, with creatures that have hard parts, did not start until the Cambrian, 550 Mio years ago, the last 7.5 min. During this time many species evolved and died either as being part of the background extinction or a mass extinction. Dinosaurs that reigned from the Triassic through the Cretaceous would live in the last 3min20s, only to die after 2.5min. The last 52 seconds correspond to the Cenozoic, the Tertiary and Quaternary, that began 65 Mio years ago. And ancestors of humans appeared within the last 3 seconds, or so, 3.5 Mio years ago. Homo sapiens has been around for the last 1.5 Mio yrs, or 1.3s.

- 3 major faunas have dominated animal life: Cambrian, Paleozoic, Modern
- only Cambrian explosion brought forth new lineages, while the other two only enhanced the number of species
- size and complexity of organisms have increased
- predators have become more efficient
- human-induced extinctions used to affect only large animals but now reach into microscopic scales

- **Fossils**

  Nicolaus Steno discovered in the 1600s that in undisturbed sequences of sedimentary rock, the oldest rocks are near the bottom and the youngest near the top. Subsequently, geologists have used this to date fossils found in these layers. They found that: 1) similar fossils can be found at similar times but different locations; 2) certain fossils are always found in younger strata but missing in older ones; 3) younger fossils resemble more closely still existing species than older fossils; 4) some lineages are documented by time-consecutive fossil finds [e.g. whale

(Basilosaurus, 42 mya) evolved from land mammal (Mesonychid, 55 mya) through Ambulocetus (52 mya) and Rodhocetus (46 mya)].

Figure 17.9a Fossil dinosaur bones exposed on a tilted bed of sandstone in Dinosaur National Monument, UT. (source: ME)

Figure 17.9b Molds and casts from organisms, including those that have no hard parts. (source: ME)

Figure 17.9c The carbonized impression of fern fronds. (source: ME)

Figure 17.9d Petrified wood from the Petrified National Forest, AZ. Petrified wood is much harder than the surrounding volcanic tuff and thus remains after the tuff has eroded away. (source: ME)

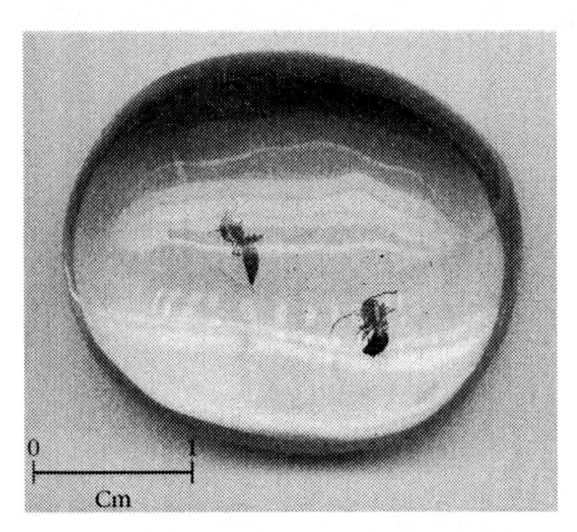

Figure 17.9e A piece of amber (fossilized resin) containing two fossil insects. (source: ME)

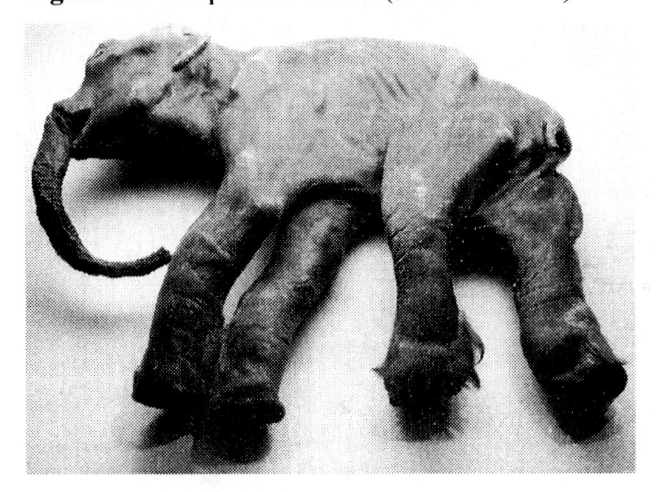

Figure 17.9f A mummified 100,000-year-old frozen mammoth, found in the permafrost of Siberia. It still had flesh and fur. (source: ME)

Figure 17.9g Foot prints of a dinosaur preserved in hardened (lithified) mud. (source: ME)

Figure 17.9h Worm burrows on siltstone. (source: ME)

Figure 17.10a A soft-bodied invertebrate excavated at Ediacara Hills in the Flinders Ranges of South Australia, formed 600 million years ago, in the pre-Cambrian. The period between 635 and 542 million years ago is also called the Ediacaran Period. (source: LIFE)

Figure 17.10b Cast of an assemblage of soft-bodied organisms (a "Dickinsonia") from the Ediacaran Period. (source: ME)

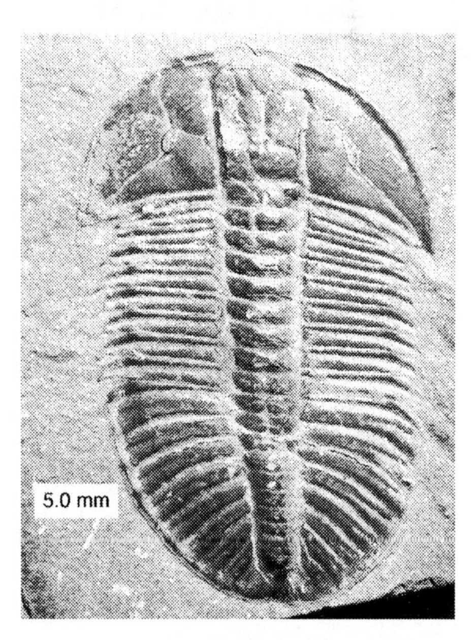

Figure 17.11 Trilobites from the Mt. Stephen Trilobite Beds near Field, B.C., Canada. First appeared in the Cambrian. Trilobites ranged in length from 1 mm (0.04 in) to 72 cm (28 in). The closest extant relatives of trilobites may be the horseshoe crab. (source: wikipedia)

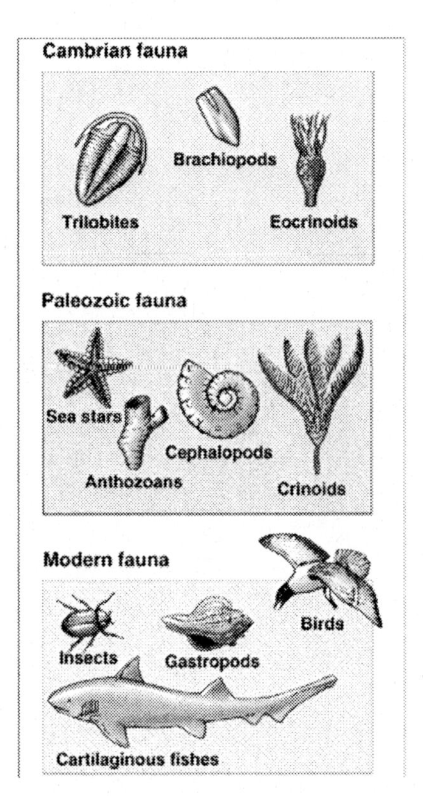

Figure 17.12 Representatives of the three major evolutionary faunas (source: LIFE)

The Cambrian Explosion of Life

Starting at about 550 Mio years ago, a 40 Mio year long burst of evolutionary changes. Nature experimented with species that could survive. Many died out after a short while but others survived. Animals start to have hard parts (vertebrates) and so are better preserved. This tremendously simplifies paleontologists' jobs to trace live in the past. Examples of species that appeared (as were shown in class) include the snale-like Lapworthellid, the miniature-limpet like Mobergella and the Trilobite came in all sizes, from 1mm to 72cm (1/25 - 28").

Figure 17.13 The Meganeura lived in the Carboniferous. It resembles the present-day dragonfly but had a wingspan of more than 75 cm (2.5 ft)! (source: wikipedia)

Life in the Carboniferous through the Cretaceous Periods

(times in Mio years before present)

- **Carboniferous (354 - 290)**: relatively cold; large glaciers on Gondwanaland; swamps in tropics (tree ferns, horsetail, lycopod); worldwide coal deposits; first tetrapods; reptiles; examples shown in class include a plant cast of horsetail and a dragon fly whose wing span was 60cm (24in)! Dragon flies are predators! It is still debated why and how gigantism works. Research indicates that higher oxygen levels promoted growth. This study reports that oxygen levels were higher than today (30-35% instead of the 21%) [(1).]

Figure 17.14 Etching depicting some of the most significant plants of the relatively cold Carboniferous. Characteristic of the flora were tree ferns and lycopod trees. (source: Wikipedia/Meyers Konversationslexikon)

- **Permian (290 - 251)**: Pangaea; further cooling; sea level drop;
- swamp-loving lycopod trees of the Carboniferous replaced by more advanced seed ferns and early conifers; more reptiles than amphibians; lineage for mammals from reptiles; bony fish

Figure 17.15 Dimetrodon and Eryops of the early Permian. (source: Wikipedia)

- **Triassic (251 - 206)**: frogs, turtles, dinosaurs; conifers, seed ferns; examples shown in class include the Pterodactyl (flying lizard) and the Kuehneosaurus whose wing span was 60cm (24in); his little cousin, the Draco Volans still lives in Southeast Asia;

Figure 17.16 The Kuehneossuchus and the Kuehneosaurus lived in the Triassic in the U.K.. The Kuehneosaurus was 72 cm (28 in) long and its wings extended 14.5 cm (5.7 in) from its body. (source: Wikipedia)

- **Jurassic (206-142)**: bony fish; salamanders; lizards; flying reptiles; bipedal predatory and quadrupedal herbivores dinosaurs; mammals; examples shown in class include the Ammonite, a sea creature that left us beautiful curly fossils; the stegosaur (a plant-eating dinosaur with spiky plates on the back) and a mosquito preserved in amber (fossil tree sap). This gave the author of Jurassic Park (Michael Crichton) the idea that dinosaurs could be "revived" by placing the DNA from dinosaur blood preserved in the mosquito in the amber in frog cells (or was it chicken??). IS this scientifically plausible? Check out recommended reading below! An interesting "relict" from the Jurassic is the Coelacanth, a fish living in the Indian ocean and thought extinct until it was discovered by a fisherman in 1939.

Figure 17.17 The Ammonite first appeared in the Devonian and disappeared in the Cretaceous. The Asteroceras shown here lived in the Jurassic in England. All close relatives are extinct. They are more closely related to octopi living today than to nautili even though the resemblance to the latter is stronger. (source: Wikipedia)

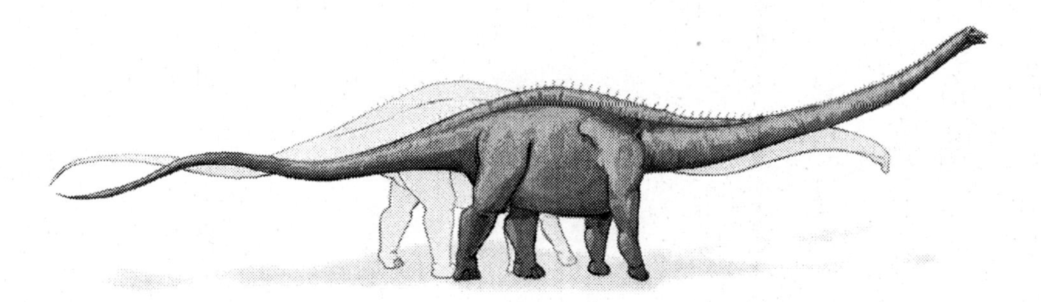

Figure 17.18 The supersaurus, found in Colorado, was one of the largest dinosaurs and lived in the Cretaceous. (source: Wikipedia)

Figure 17.19 The Tyrannosaurus rex lived in the late Cretaceous. (source: Wikipedia; Nobu Tamura)

- **Cretaceous (142 - 65)**: high sea level; marine invertebrates diversified; snakes; flowering plants; many small mammals; some of the largest dinosaurs including the supersaurus (30m/100 ft high) and the seismosaurus (36m/120ft high); the predatory Tyrannosaurus Rex lived in the late Cretaceous

Evolution

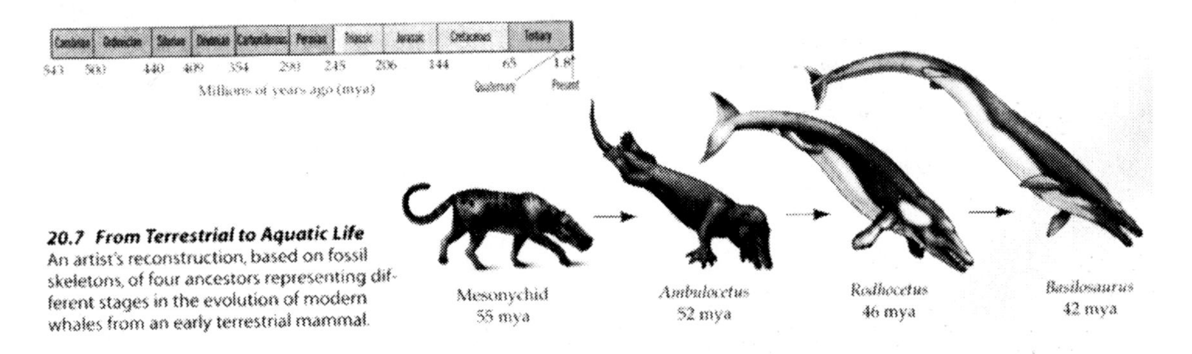

Figure 17.20 From Terrestrial to Aquatic Life. An artist's reconstruction, based on fossil skeletons, of four ancestors representing different stages in the evolution of modern whales from an early terrestrial mammal. (source: LIFE)

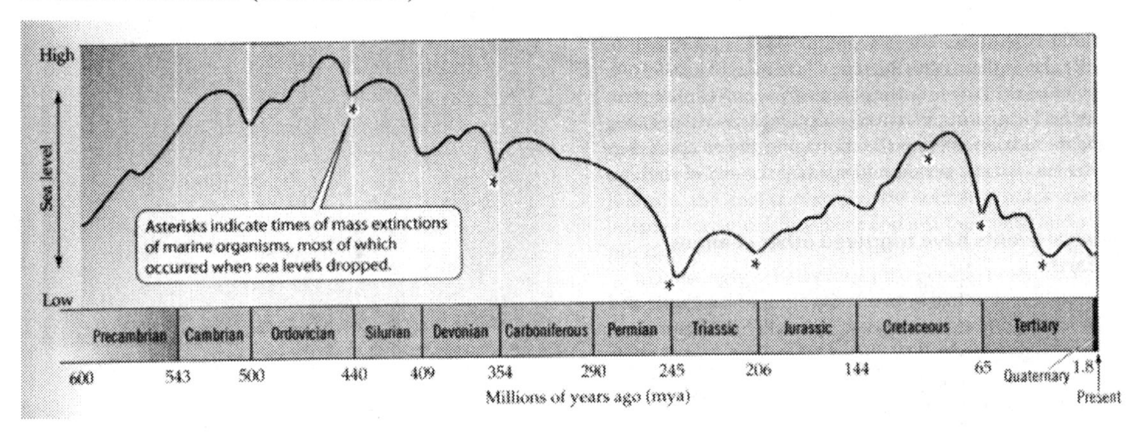

Figure 17.21 Sea levels have changed repeatedly and most mass extinctions of marine organisms have coincided with major changes in sea level. (source: LIFE)

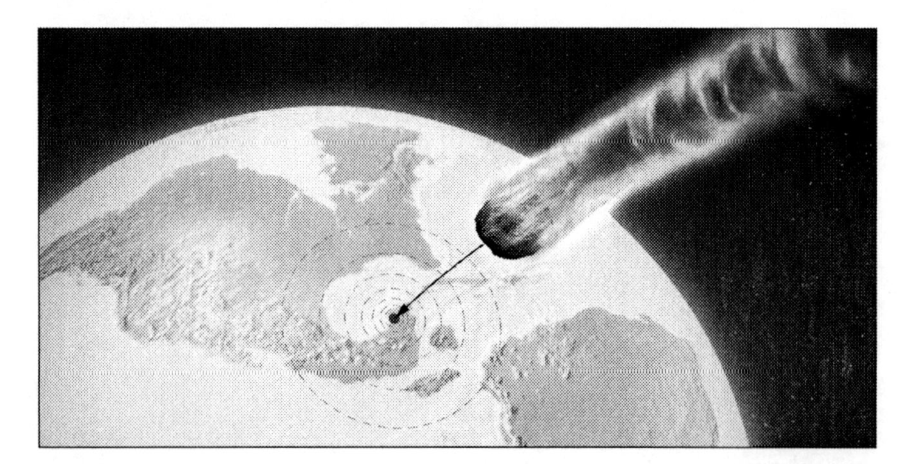

Figure 17.21 Artist's rendering of the Chicxulub impact, a possible cause for the mass extinction at the K-T boundary 65 million years ago. (source: ME)

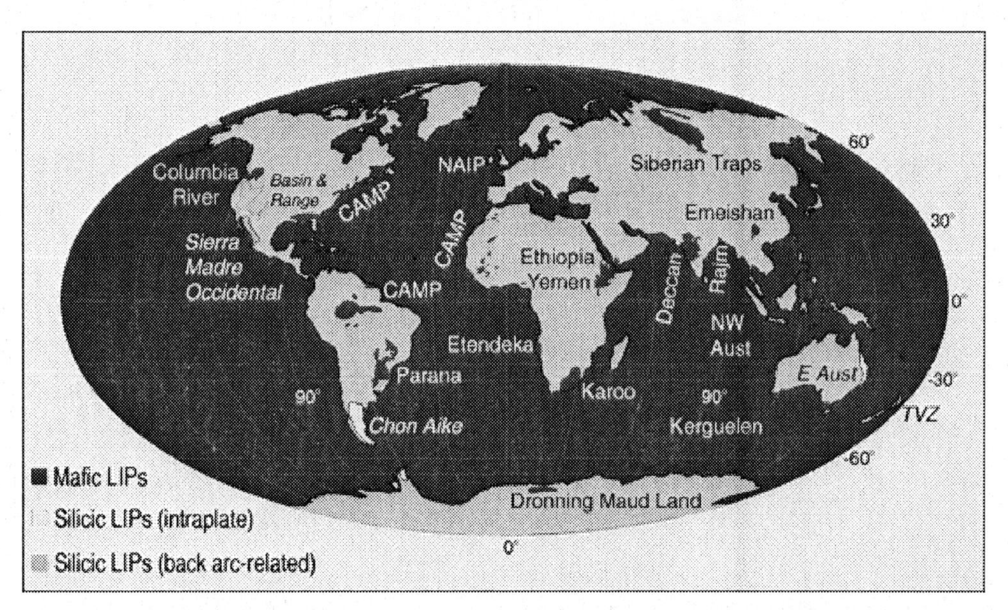

Figure 17.21 Large igneous provinces (LIP) are areas of extensive flood-basalt volcanism. Some are linked to mass extinctions. The Deccan traps in India are linked to the mass extinction at the K-T boundary, while the Siberian LIP is linked to the mass extinction at the end of the Permian period. (source: mantleplume.org)

17.4 Causes of Mass Extinctions

Geological Processes

- climate changes (warming as well as cooling though more extinctions appear to be related to warming)
- bolides/large impacts: global cooling (aerosols block sunlight), fire, tsunami, earthquakes
- flood basalt volcanism: changes atmospheric chemistry -> global cooling (SO_2) or warming (CO_2)
- sea level changes: both drop and increase has impact on shallow coral reefs and seas
- sea floor spreading: changes sea level as well as output of volcanic gases -> climate, ocean chemistry
- plate tectonics: geometry (size and location) of continents influences ocean currents/climate
- change in ocean processes/ocean chemistry; e.g. acidification leads to collapse of corals/shell forming species; lack of deep ocean mixing/loss in oxygen in deep ocean
- release of methane/gas hydrates -> accelerated warming

Species-Related Processes

- o interruption of food chain
- o new predator or competitor
- o new disease
- o habitat loss

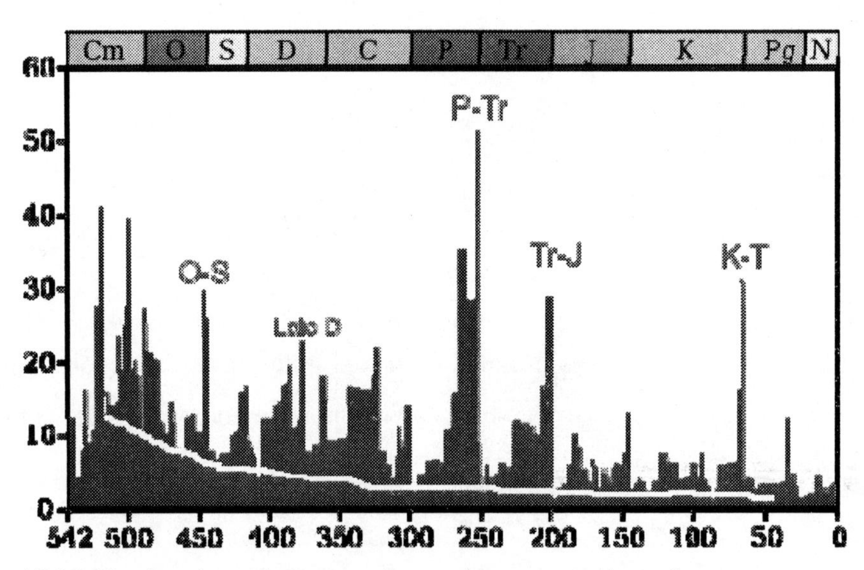

Figure 17.22 The fraction of genera going extinct at any given time, as reconstructed from the fossil record. The recent Holocene extinction is not included. A line marks the declining level of background extinctions. (source: modified from wikipedia)

The Five Major Mass Extinctions [(2),(3)]

Geologists designed the geologic time scale roughly where major changes in the stratigraphic profiles occurred. Some of these involved mass extinctions. Extinctions happen all the time as background extinction in which up to 10% of the genera disappear. In serious mass extinctions up to 65% of all genera die out. Over all, 99% of all species that ever lived are now extinct. The coursebook states that 99.9% of all animal and plant species (which does not include the other 3 kingdoms) are now extinct. The five most prominent extinctions in the phanerozoic eon were at the end of the Ordovician, Devonian, Permian, Triassic and Cretaceous periods, the most devastating of them being at the end of the Permian.The discussion of whether an impact or volcanic activity triggered mass extinction is still ongoing. Evidence for either of these comes from the sudden appearance of **Iridium**, an extremely rare element. Iridium on the Earth's surface comes either from meteorite impacts or from deep inside the Earth through the extensive volcanism of large igneous provinces/flood basalts.

A nice summary of the five mass extinctions was given in the "Quest" section of the San Diego Union Tribune on November 12, 2003. A similar summary can be found in the biology textbook "Life" (see reference below).

- **Ordovician**: marine life with more than 600 different families was not destined to last; possible causes: pronounced global cooling and decrease in precipitation according to Figure 10.25 in Abbott's book; significant drop in sea level (by as much as 1000ft/330m, according to Atlas referred to below); transition into an ice age; shallow seas on continental shelves turned into dry plains

- **Summary**: second greatest extinction of marine organisms
- **Time of period**: 510 - 438Mio years ago
- **Setting**:Shallow tidal seas in tropical regions brought about extensive diversification of marine organisms, including new types of cephalopods (the forerunners of squid), corals, the mosslike bryozoans and bivalves. Nautilus may have reached several feet. *(more than 600 families of marine life)*
- **Probable Cause of Extinction**: Rapid cooling of surface waters. As seas retreated into ice sheets, marine habitats drastically changed. *(decrease in precipitation according to Abbott's figure 10.25. According to atlas significant drop in sea level. It also cites a, contradictory, increase in precipitation.)*
- **Species affected**: More than 100 families of marine invertebrates wiped out.

- **Devonian**:"age of the fishes" accelerated evolution; featured abundance of diverse aquatic life forms; tree ferns; first forest soil; possible causes: reason for extinction not clear; long period when many species died; some impacts; lowering in sea level by 300ft; destruction of shallow seas by collision of continents

- **Summary**: as much as 70% of marine invertebrates lost. *"Age of the Fishes". Extinction was not a sudden one but seemed to have taken several tens of Mio years.*
- **Time of period**: 410 - 360Mio years ago
- **Setting**:Shallow seas flooded great areas of the continent. First appearance of *primitive* sharks, bony fish, *air-breathing lungfishes* and the coiled mollusks called ammonoids *(the atlas placed ammonites (?) into the Jurassic)*. World's oceans were dominated by corals. On land, first amphibians, insects and great swamp forests.

- **Probable Cause of Extinction**: Meteor impact or glaciation in the south polar area that cooled surface water. (*no evidence for this in Abbott's Figure 10.25, sea level did drop by 300ft after the extinction, according to atlas*).
- **Species affected**: Corals were so severely affected the they remained relatively uncommon for 150 Mio years. Brachiopods, trilobites, condodonts and jawless fish suffered. Little impact on freshwater and land animals and plants.

  - **Permian**: the period when Pangaea formed; cool climate; most serious mass extinction; 90% of all species died out (63% tetrapods, 33% amphibians, 60% marine life including trilobites marine diversity reduced from 250,000 species to just 10,000);

    possible causes: sea level dropped, laying bare most coral reefs - all-time low in last 545 Mio yrs and below today's sea level-; weather warming at the end of Permian though precipitation decreased (Figure 10.25); possible bolide impact; large flood basalts in Siberia (largest in the world); new evidence suggests massive belching of methane bubbles from deep ocean floor causing global warming

- **Summary**: greatest mass extinction known
- **Time of period**: 286 - 250 Mio years ago
- **Setting**:Formation of supercontinent Pangea and all-time low sea level (*only time sea level was below today's level!*) created for the first time more continents than oceans. Extensive development of terrestrial animals, including insects, amphibians, reptiles and mammal-like reptiles.
- **Probable Cause of Extinction**: Geologic events may have triggered methane hydrate release from sea floor, dramatically reducing atmospheric oxygen levels. *Other causes include extensive production of flood basalts in Siberia (largest on Earth). Meteor impacts are also discussed). Extensive global warming associated with the break-up of Pangea, rise in sea level though no particular change in precipitation have also been considered. The release of methane hydrates is an interesting idea but I wonder why this would happen in a cold global climate. In Abbott's chapter on global warming, the gas hydrate release is discussed as a consequence of global warming (the gas hydrates are frozen lenses in submarine sediments).*
- **Species affected**: 96% of marine species (*250,000 down to 10,000*), including trilobites, and 75% of terrestrial vertebrates decimated. *(63% tetrapods, 30% amphibians)*

  - **Triassic**:

    possible causes: not clear; no obvious impact crater found so far; Iridium find points

COPYRIGHTED MATERIAL — DO NOT DUPLICATE, DISTRIBUTE, OR POST

toward volcanic cause; drop in sea level; slight cooling but probably not enough for a mass extinction

- **Summary**: three extinction events spanned the period
- **Time of period**: 250 - 206 Mio years ago
- **Setting**:Warm and hot, with vast areas of desert. huge marine reptiles probably swam in small schools, feeding on mollusks. Dinosaurs became abundant and diversified. Appearance of first true mammals.
- **Probable Cause of Extinction**: Climate change (*cooling, drop in sea level*) or meteor impact. One large impact crater exists in Canada, although new dates place it out of time with extinction events. *(increased volcanism seems out of phase with extinctions though Iridium finds point toward increased volcanism near the extinction)*
- **Species affected**: Great devastation to sponges, marine snails, bivalves, cephalopods, brachiopods and marine reptiles. On land, 35 families of insects and 8 families of reptiles died out.

- **Cretaceous**:

  the mass extinction at the K-T boundary (K from German Kreide for chalk) marks the demise of the dinosaurs; mammals survived and took over; the cause for the mass extinction is the most hotly debated; Iridium finds and the recently imaged impact crater at Chicxulub point toward death by impact; but the massive Deccan Trap flood basalts in India also formed at this time, suggesting that death came by a combination of factors; sea level dropped sharply just before and temperature and precipitation increase after the proposed time of the impact

- **Summary**: known as the dinosaur extinction
- **Time of period**: 144 - 65 Mio years ago
- **Setting**: Continents drifted apart to approach their current position. Emergence of flowering plants. Major diversity in marine reptiles, bivalves, ammonoids and corals.
- **Probable Cause of Extinction**: meteorite impact created gigantic dust cloud that shrouded Earth, halting photosynthesis and leading to massive animal die-offs. *Alternatives include massive productions of flood basalt (Deccan Traps in India). End of cretaceous also became drier and cooler. Dramatic drop in sea level (could imply increase in oceanic spreading rates??).*
- **Species affected**: 75% to 85% of species, including 50% of vertebrates. Non-avian dinosaurs, pterosaurs, marime reptiles perished. The ammonites disappeared. Mammals, birds, reptiles and amphibians were largely unaffected.

Surviving Relics

Coelacanth

Figure 17.23 A present-day coelacanth. (source: University of Chicago News Office)

Figure 17.24 A horseshoe crab on Ana Maria Island, FL. (source: wikimedia)

17.5 Mass Extinctions in Modern Times

Figure 17.25 Mammals of the last ice age. Many mammals that are now extinct once roamed the glacial maximum world of 21,000 years ago, including woolly mammoths, saber-toothed tigers, and giant ground sloths. (source: RU/Smithsonian Institution; painted by Jay Matternes)

Figure 17.26 A cave painting of the glacial era. Despite that harsh ice age climate, our ancestors left striking paintings on cave walls in southern Europe. Some of these paintings document how humans interacted with wildlife at the time. (source: RU)

Figure 17.27 A yurt built by humans in Eurasia during the last ice age. These domed dwelling were made of hides draped over intricately linked mammoth bones. Other bones served as anchors. (source: RU)

The Ice Age and More Recent Extinctions

Recent extinctions are likely caused by human interaction. We will revisit this issue in Lectures 25 through 27. Many species did not survive the last ice age. These include

- woolly mammoths
- mastodons (related to elephant)
- wooly rhinos
- giant ground sloths (lived in South America)
- musk oxen
- saber-toothed tigers
- cave lions
- cave bears
- giant deer
- steppe bison
- wild horses
- scrats (rat-like creatures)

NB: the "wild" mustang in North America was "reintroduced" by the Spaniards.

- Though there may be natural components as well, humans had a major role in the reduction of mammoths. In Eurasia, mammoths were hunted down to build huts (about 80 mammoths provided bones for just one hut). Recent studies suggest that climate changes (and associated changes in the food source) also had an impact.

- **Flightless Birds**: Many victims of recent extinctions include flightless birds. Humans either hunted them for food, pleasure or fashion statements or introduced predators such as cats and mongooses (a big problem in Hawaii right now!). Large birds have disappeared from Madagaskar (elephant bird; 3.4m high; weighing 500kg) and New Zealand (moa; 4m high), while smaller cousins still survive in Africa (ostrich), Australia (emu) and South America (rhea).

Figure 17.28 Early 20th century reconstruction of a moa hunt. (source: Wikipedia)

Figure 17.29 Dodo reconstruction reflecting modern research at Oxford University Museum of Natural History. (source: Wikipedia)

The DODO on Mauritius: A particularly sad example was the DODO a flightless bird on Mauritius in the Indian Ocean. The DODO stood up to two feet high. In the animated movie "Ice Age" we can watch the last female jumping over the cliffs (thereby terminating reproduction) but the bird died out much later. It was discovered on Mauritius in 1598 by sailors and was extinct within less than 100 years (1681) by overhunting and the introduction of predators.

Possible Causes for Mass Extinction - Human Induced

- climate changes
- overhunting/overharvesting
- interruption of food chain
- introduction of predator or competitor (e.g. cats, rats, rabbits)
- introduction of disease/pest/poison
- habitat destruction

Ongoing Extinctions

Extinction of species is going on on virtually all continents, and in the oceans. We will address this issue in the remaining lectures. Here are some examples:

Figure 17.30 A Black Rhinoceros in Tanzania. (source: Wikipedia)

Black Rhino (critically endangered): its ancestors roamed the planet for the last 50 Mio yrs; in the early 1900s there were several 100,000 in Africa; by 1960, there were still more than 100,000; by 1995, there were just 2410; the rhinos were decimated through poaching in the 70s/80s for their horn; poaching is still going on though the remaining rhinos are heavily protected; the problem here is not (only) the poachers but also the market that provides the demand for the horn; the western black rhino has 10 individuals left and has been declared extinct (see Gambler's Ruin); the northern white rhino is also declared extinct, with 4 individuals left; the number of species is now down to 5 (from 30).

Figure 17.31 A California Condor juvenile at the San Diego Wild Animal Park. (photo: Gabi Laske)

California Condor (critically endangered): this mighty bird was decimated by the early 1900s through hunting, habitat loss and lead poisoning; by 1987, the condor was extinct in the wild, with 27 individuals captured; captive breeding programs (such as that of the San Diego Wild Animal Park) and reintroduction are problematic as there are only few places left that are adequate as habits; one of these is the Grand Canyon; recently, 40 individuals were returned to the wild where some died.

Threats to released condors are manifold. At least one was electrocuted when flying into power lines. One got shot. Some dead condors were found with lead bullets in their stomach that they ingested with dead animals. Most likely they died from lead poisoning. A new California law prohibits using lead ammunition in Condor land. Some of the 2008 fires in California burned through important condor habitat and at least two condors and a rearing station fell victim to the fires. The Wild Animal Park lost its condor station in the Guejito fire in 2007 but no condor was killed and the station has been replaced.

Buffalo (extinct in the wild; conservation dependent): A recent example in North America is the American Buffalo (*bison bison*). At one point an estimated 50 Mio buffaloes roamed through the prairies. By 1875, they were virtually extinct in the wild. There are now two places, the Yellowstone National Park and Wood Buffalo, CN where buffaloes can run free, in a protected area.

Figure 17.32 North Island Brown Kiwi in captivity. (source: wikipedia)

Kiwi (vulnerable): New Zealand's State Bird (endangered): a flightless bird that has a 40-year life-span; the only bird whit nostrils (sniffs for worms) and burrows; nocturnal; very vulnerable to predators (95% of the chicks are killed by cats); 1000 yrs ago there were 12 Mio kiwis; by 1923, this number was down to 5 Mio; 70,000 were left by the 1990s; according to a BBC report, the bird has already vanished on the North Island and is predicted to be extinct by 2010.

Talk about Grey Wolves!

The extinction of species has recently accelerated and it is estimated that if the extinction rate of the last 200 years continues for the next 200 years, we will face the greatest mass extinction the Earth has ever seen

References:

1. http://en.wikipedia.org/wiki/Meganeura; based on Chapelle, G. and Peck, L.S., 1999. Polar gigantism dictated by oxygen availability. Nature, 399, 114-115.
2. San Diego Union Tribune, 12 November 2003. Quest article on the five major mass extinctions.
3. Abbott, P.

Recommended Literature

- "Atlas of the Prehistoric World" by Douglas Palmer, 1999, Discovery Books, ISBN: 1-56331-829-6; some pictures shown in class come from this references
- "Life: The Science of Biology" by Purves, Sadava, Orians and Heller, 2001, Freeman & Co., ISBN: 0-7167-3873-2; some comments on evolution and life come from this comprehensive college textbook
- "The Science of Jurassic Park and the Lost World" by Rob DeSalle and David Lindley, 1998, Harper Perennial, ISBN: 0-465-07279-4; can dinosaurs be "revived" from Mosquitos preserved in amber?
- "Secret Agents: the menace of emerging infections" by Madeline Drexler, 2003, Penguin Books, ISBN: 0-1420.0261-5; interesting read on bacteria, viruses and prions

APPENDIX D: Evolution

Basics on Evolution

Changes to species came through evolution as the tried to adapt to changes in the environment. For details, see APPENDIX D. Some major points:

- changes to species through adaptation/survival of the fittest
- natural selection is not a random process
- changes to gene pool may be random but this is a rare process
- need large enough gene pool for survival (e.g. to preserve traits in the species)
- evolution may be a long process (e.g. the evolution of the land-living mesonychid to the basilosaurus, the ancestor of our whales, took from 55 Mio yrs ago to 42 Mio yrs ago)

- **Charles Darwin (1844)**

  Based on Charles Darwin's theory that organisms adapt to environmental changes by natural selection. Based on observations that species in South America are different from those in Europe, that species on the Galapagos Islands are similar to those in S. America but that there are differences even between islands. Natural selection: differential reproductive success and survival, i.e. slight variations among individuals significantly affect the chance that a given individual will survive and the number of offspring it will produce. Also, selective breeding can produce specific kinds of pigeons.
- **Gregor Mendel (contemporary to Darwin)**

  Biological evolution is due to changes in genetic composition of a population over time.
  - gene: unit of heredity
  - allele: alternate forms of a genetic character (e.g. A,B) (dominant and recessive allele)
  - trait: one form of a character (e.g. eye color)
  - genotype: genetic constitution that governs a trait (e.g. pairs AA, AB, BB); an organism typically has many of these
  - gene pool: sum of all alleles found in a population

  - Hardy-Weinberg Equilibrium: population is not changing genetically (i.e. same allele and genotype frequencies from generation to generation)

- conditions for H-W E.: 1) mating is random; 2) population very large; 3) no migration between populations; 4) ignore mutations; 5) natural selection does not affect alleles
- factors that change genetic structure of population: 1) mutations (rare!); 2) gene flow (between populations); 3) non-random mating; 4) random genetic drifts (can have serious consequences in small populations)

- most populations are genetically variable

  Example: wild mustard that does not change over generations is in H.-W. equilibrium.

  However, through genetic manipulations (selection of genotype frequency), farmers have derived cabbage, Brussel sprouts, kohlrabi, kale, broccoli and cauliflower.
- small populations affected greatly by environmental changes (reduced variability in gene pool; "survival genes" might have been lost)

Chapter 18: Anthropogenic Changes: The Atmosphere (draft)

http://sealevel.colorado.edu/content/2012rel2-global-mean-sea-level-time-series-seasonal-signals-removed

Climate in The Twentieth and Twenty-first Centuries

- global temperature rose 0.6°C in last century
- greatest increases in 1910-1945 and since 1976
- first may be due to hotter sun and lack of global volcanism (according to textbook)
- second is due to injection of greenhouse gases
- warming trend was temporarily offset by 1991 Mt. Pinatubo eruption
- most glaciers are retreating, including equatorial ice caps and glaciers
- Antarctic ice sheet calves huge ice masses (i.e. of the size of Delaware)
- if the Greenland and Antarctica ice sheets melted completely, global sea level would rise by 65m
- causes for some ice melting are complex, e.g. Kilimanjaro's disappearing ice cap could be due to global warming but also due to local reduction in precipitation as a consequence of local deforestation

- (though the ice cap has decreased by >80% since about 1935)

Figure 18.1a Rhone glacier in the Swiss Alps in 1870. The Rhone Glacier is easily accessible so its evolution has been observed since the 19th century. The glacier lost ca. 1300 m in length during the last 120 years. (source: Wikipedia)

Figure 18.1b Rhone glacier in 1900. (source: Wikipedia/Library of Congress)

Figure 18.1c Rhone glacier in 2005. (source: Wikipedia)

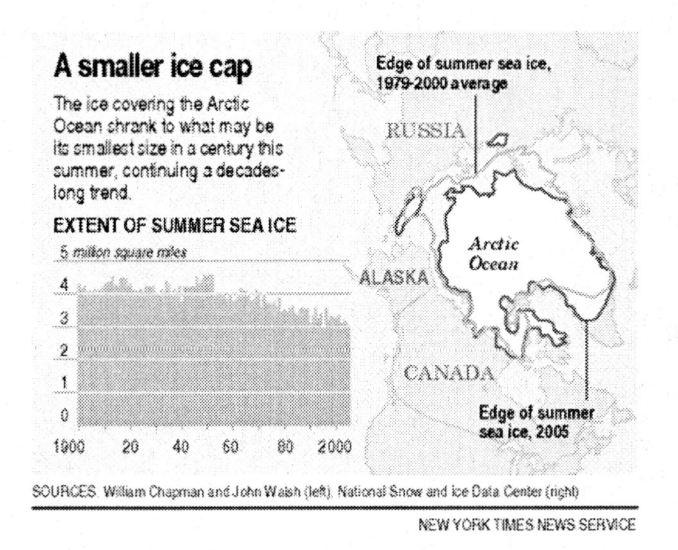

Figure 18.2 The loss of Arctic sea ice. (source: SD Union Tribune, 29 Sep 05)

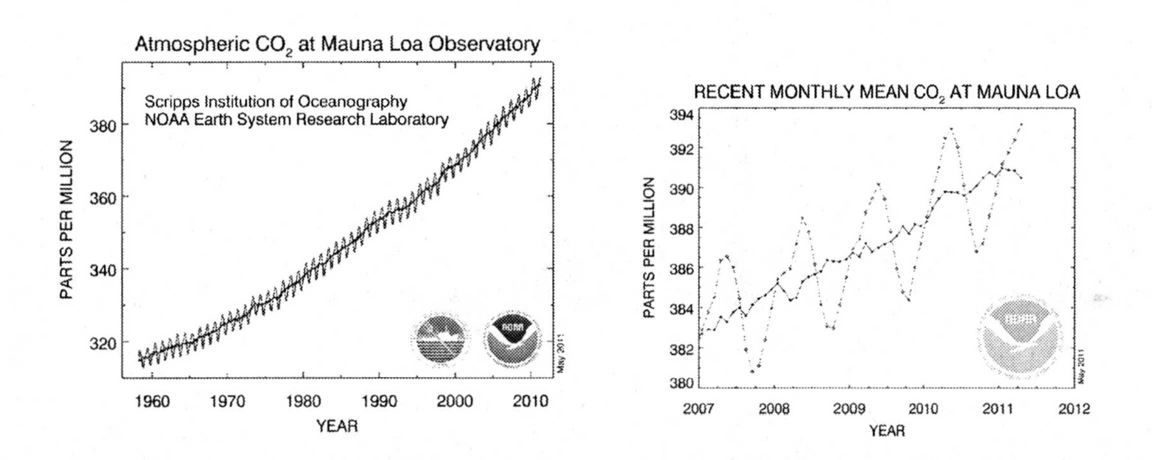

Figure 18.3 The Keeling Curve: In 1958, Charles Keeling from SIO started recording atmospheric CO2 at Mauna Loa, HI. The exponential increase over the last 5 decades by 75 ppm (24%) dwarfs the seasonal changes caused by changes in photosynthesis (right). (source: NOAA, SIO)

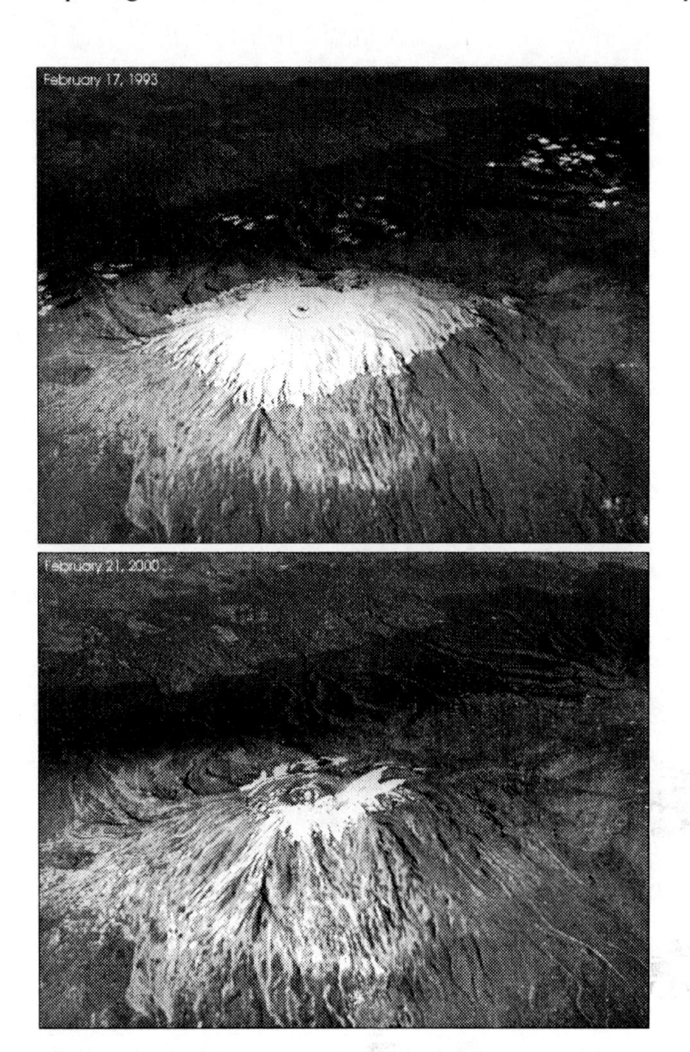

Figure 18.4 This image showing the disappearing snow and ice cap of Kilimanjaro in Tansania and Kenya was first published on NASA's Earth Observatory on 20 December 2002. The top snapshot was taken on 17 February 1993 while the bottom one is from 21 February 2000. With an elevation of 5895 m (19,341 ft), it is Africa's highest mountain and has had a remarkably persistent ice cap throughout the last 11,000 years. The fast disappearance – Kili lost 85% of its ice in the last century - has often been attributed to recent global warming. (source: Wikipedia/NASA's Earth Observatory)

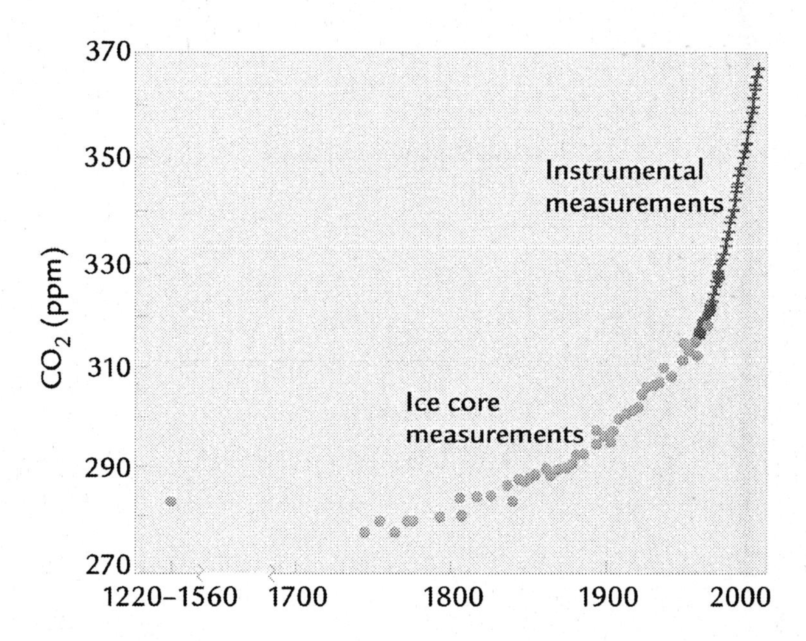

Figure 18.5 Preindustrial and recent exponential increase in atmospheric CO_2 from the combined records from ice core air bubbles and from instrument measurements since 1958. The increase in atmospheric CO_2 has accelerated in the last 200 years. In 2000, it was 32% above the natural baseline of 280 ppm. (source: RU)

In April 2011, atmospheric CO_2 at the Mauna Loa, HI station was measured at 393 ppm. Subtracting the seasonal contribution of about 2.5 ppm, this leaves 392.5ppm, 40% above the pre-industrial baseline of 280 ppm. The most recent measurements can be accessed at http://www.esrl.noaa.gov/gmd/ccgg/trends/.

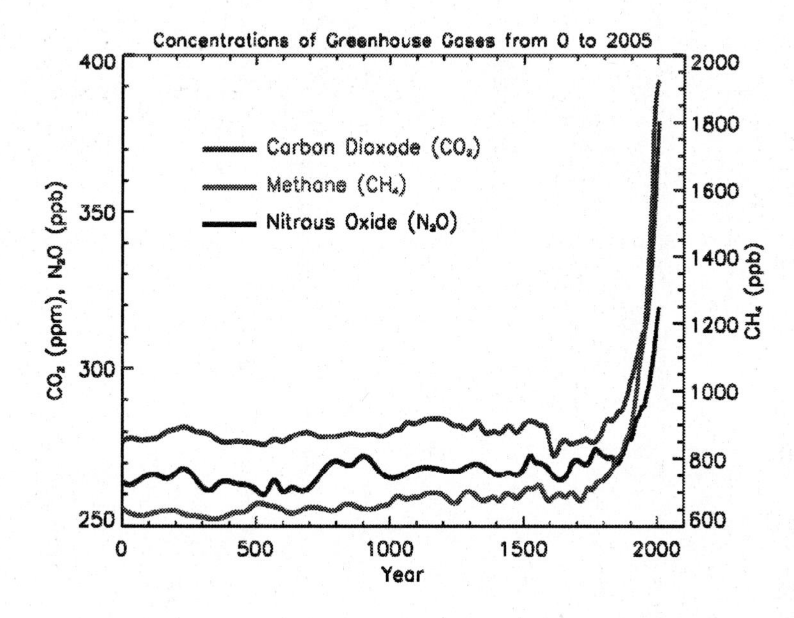

Figure 18.6 Atmospheric concentrations of important long-lived greenhouse gases over the last 2000 years. Increases since about 1740 are attributed to human activities in the industrial era. Concentrations units are parts per million (ppm) or part per billion (ppb), indicating the number of molecules in a sample. (source: IPCC report)

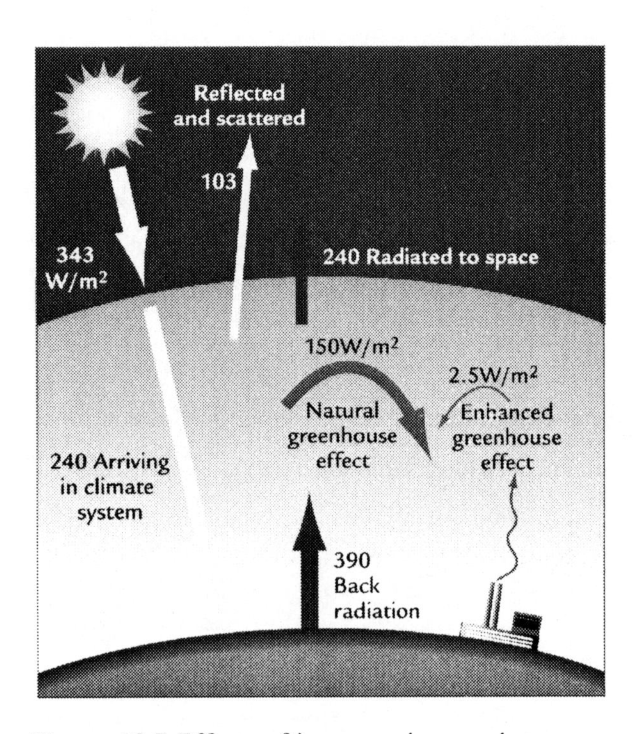

Figure 18.7 Effects of increases in greenhouse gases on radiation. Human activities since the start of the industrial era have increased greenhouse gas concentrations enough to enhance the natural greenhouse effect by more that 1%. (source: RU)

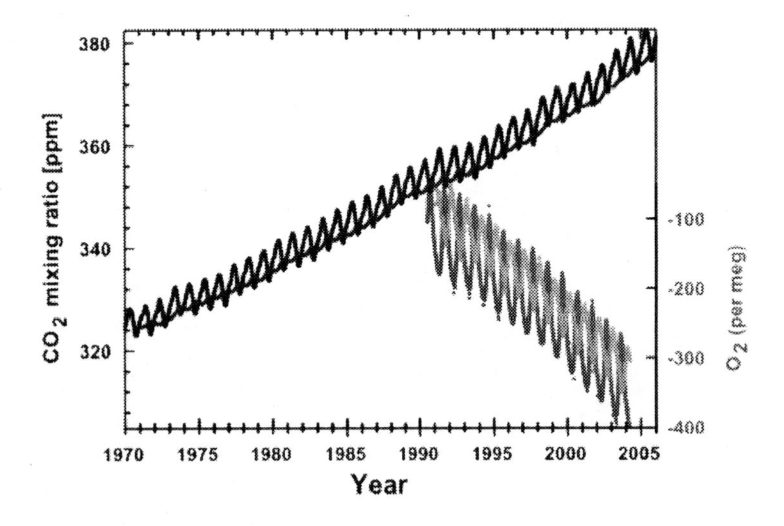

Figure 18.8 The recent rise in CO_2 at Mauna Loa, HI (black, large oscillations) and Baring Head, New Zealand (blue, small oscillations). The Mauna Loa record shows a stronger seasonal cycle because of the larger biosphere in the Northern Hemisphere. Also shown is the recent decline in atmospheric oxygen since the 1990s (pink: Alert, Canada; cyan: Cape Grim, Australia). Oxygen is measured in per meg (like per mile but 1000 times smaller) with respect to an arbitrary O_2/N_2 baseline. (source: IPCC report)

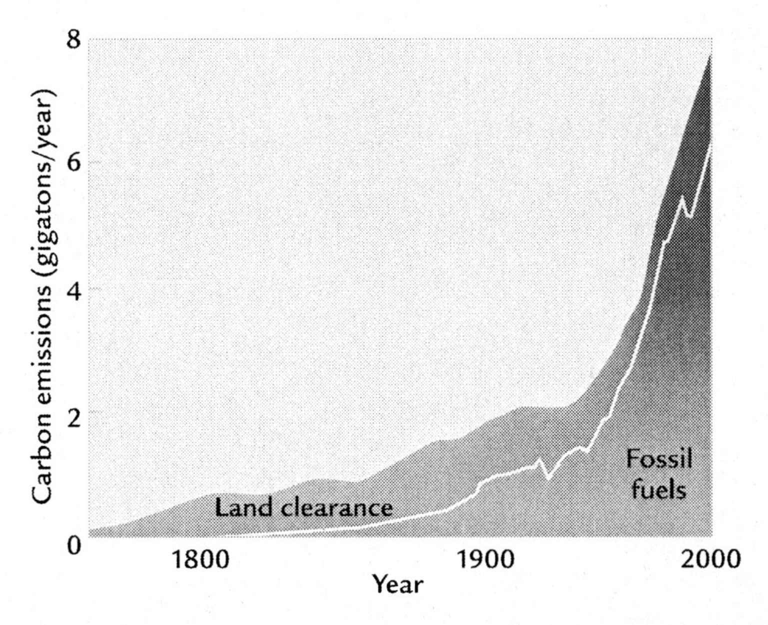

Figure 18.9 Human production of CO_2. Two factors account for the increase in atmospheric CO2 caused by human activities in the last 250 years: (1) clearing land (reduction in photosynthesis) and burning trees (release of CO_2) for agriculture and burning of carbon in fossil fuels (coal, oil, natural gas). (source: RU)

Greenhouse Gases

- **H_2O** (most important greenhouse gas; huge variations due primarily to natural causes, i.e. climate and weather)

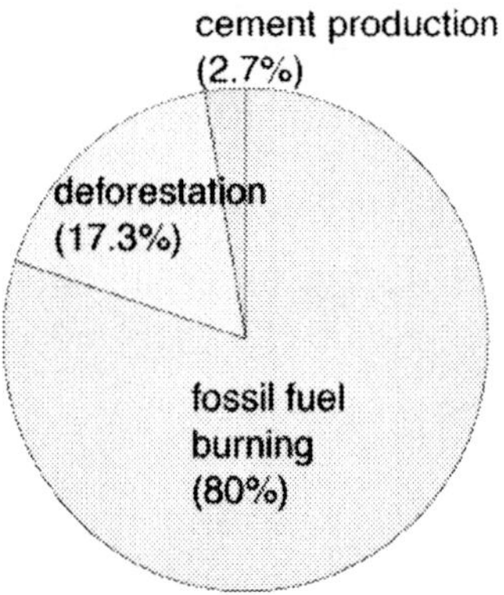

Figure 18.10 Anthropogenic sources of the greenhouse gas carbon dioxide.

- **CO_2**: amount of CO_2 in atmosphere has increased by 39% from 280ppm in 1800 to 389ppm in 2010.

Causes for change:

- Natural: temporarily also from volcanic activity, but the contribution in historical times has been insignificant
- Human: burning of biological mass and fossil fuels (80%)
- Human: removal of rainforest (though the contribution of this is relatively small) (17.3%)
- Human: cement production (smaller yet) (2.7%)

NB: almost all of the current increase in CO_2 is anthropogenic.

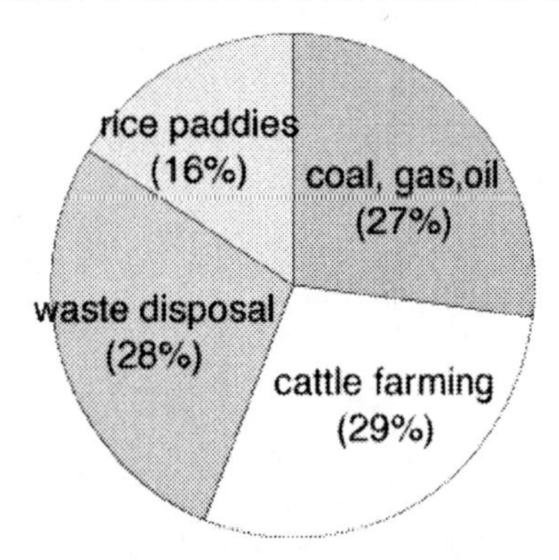

Figure 18.11 Anthropogenic sources of the greenhouse gas methane.

- **CH_4**: has increased 150% from 0.7ppm in 1750 to 2ppm in 2000; more effective greenhouse gas than CO_2 but less relevant due to slower current increase. This problem will become more prominent as the population increase requires more farming.

Causes for change:

- Natural: wetlands (72%)
- Natural: termites (12.5%)
- Natural: ocean (6%)
- Natural: other (9.5%)
- Natural: recent global warming also releases methane in the tundra that has been "locked" in frozen ground (positive climate feedback)
- Natural: potential release of large amounts of gas hydrates from submarine icy lenses could happen, if current warming persists (at which point this actually happens is currently unknown)
- Natural: a natural gas seep off-shore Santa Barbara releases significant amounts though most may be dissolved in water, unless it is transported very quickly to the sea surface
- Human: coal mining, natural gas, petroleum industry (27%)
- Human: rice paddies (16%)
- Human: cattle farming (enteric fermentation, animal wastes)(29%)
- Human: waste disposal sites (sewage, landfill, biomass burning) (28%)

NB: anthropogenic release is currently estimated to be 2.5 times larger than natural release.

- **N_2O**:

  Causes for change:

  - Human: agriculture (chemical fertilizer)
  - Human: car and truck engines

- **O_3**: protects from UV radiation in stratosphere but is a greenhouse gas and a pollutant in the troposphere

  Causes for change:

  - Natural: changes in solar activity (in stratosphere)
  - Human: cars and industry (troposphere)

- **CFCs**: exclusively of human origin, does not occur naturally; greenhouse gas and destroyer of protecting stratospheric ozone layer; long lived; a single CFC molecule is therefore capable of destroying large amounts of ozone molecules

Sources:

- air conditioning and refrigerators
- foam insulations
- until recently: propellant in aerosol spray cans
- most countries have banned CFCs and other ozone killers but some developing countries still produce CFCs

The increase in greenhouse gases since the Industrial Revolution (mid 1700s) caused an increase of radiative forcing by greenhouse gases by 2.43 W/m^2. Compare this to a possible increase of 0.3 W/m$_2$ due to changes in solar output.

18.1 The Industrial Age and CO2

- CO_2 measured from trapped bubbles in ice cores for times prior to 1958 and direct measurements in atmosphere since then
- industrial revolution at the end of 1700s started a steady increase in CO_2
- increase from 280ppm in 1800 to 370ppm in 2000 to 387ppm in 2008; the increase has accelerated within last 50 years
- current CO_2 level is the highest in last 650,000 years

- the U.S. have been the major contributor (25% of world's CO_2 production!)
- China and India catching up rapidly, with China surpassing U.S. in 2009

18.2 The Global Increase in Temperature

The link between atmospheric CO_2 increase and temperature increase is difficult to assess since global temperature depends on many factors, not just the atmospheric content of greenhouse gases.

- global temperature rose 1.0°C within last 140 yrs

  Possible Reasons:

  - millennial warming (<0.02°C)
  - solar warming (0.2°C)
  - industrial revolution (>0.8°C)
  - volcanism (insignificant)
- greatest increases in 1910-1945 and since 1976
- Abbott speculates that the first increase is due to increased sun activity (but sunspot maximum was actually in 1950s to 1960s) and lack of volcanism

- temperatures decreased between 1940 and 1970; the minimum in 60ies-70ies coincides with drop in solar activity
- second increase is most likely due to injection of greenhouse gases
- global average temperature rose by 0.4°C between 1970 and 1990; the current trend is up

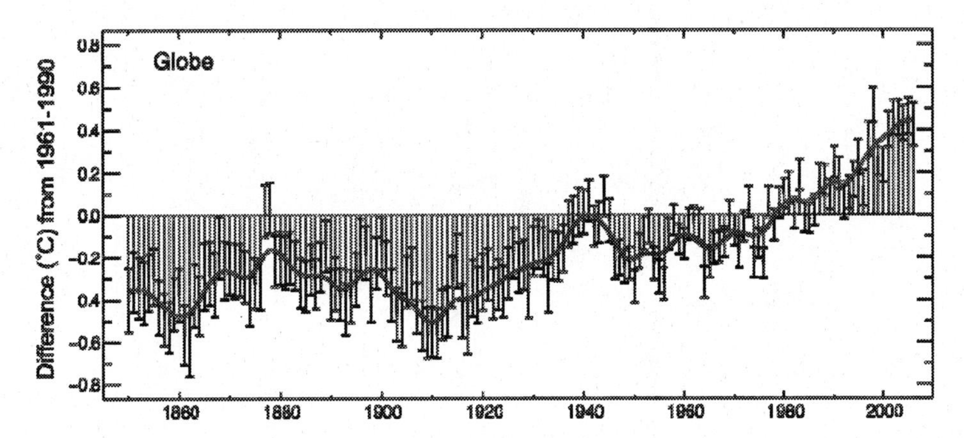

Figure 18.12 Global annual combined land-surface air temperature and SST anomalies (°C) for 1850 to 2006 relative to the 1961 to 1990 mean, along with 5 to 95% error bar ranges. The smooth curve shows decadal variations. (source: IPCC report)

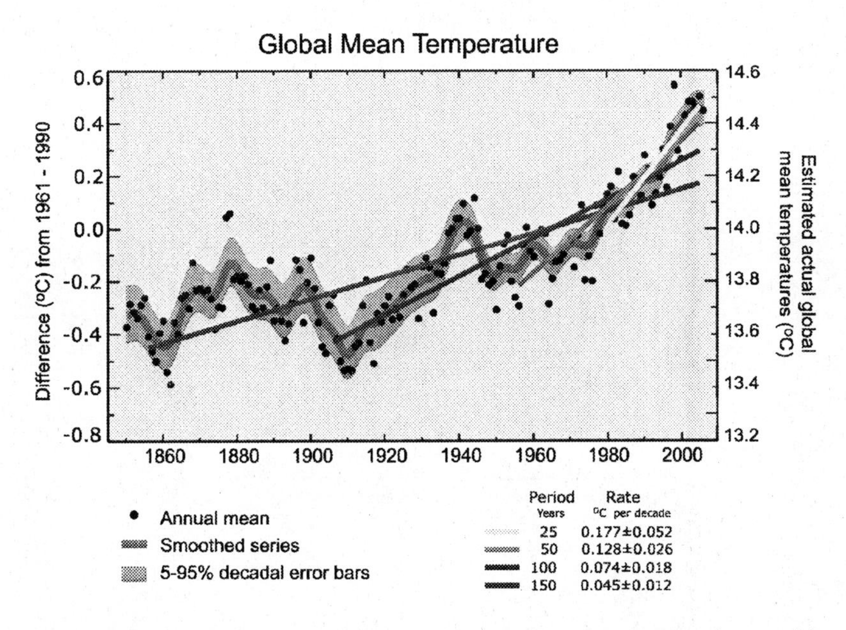

Figure 18.13 Global annual mean observed temperatures along with trends to fit the data. The warming trend in the last 50 years (0.128°C per decade) is three times larger than that over the last 150 years (0.045°C per decade). Hence, warming is accelerating. (source: IPCC report)

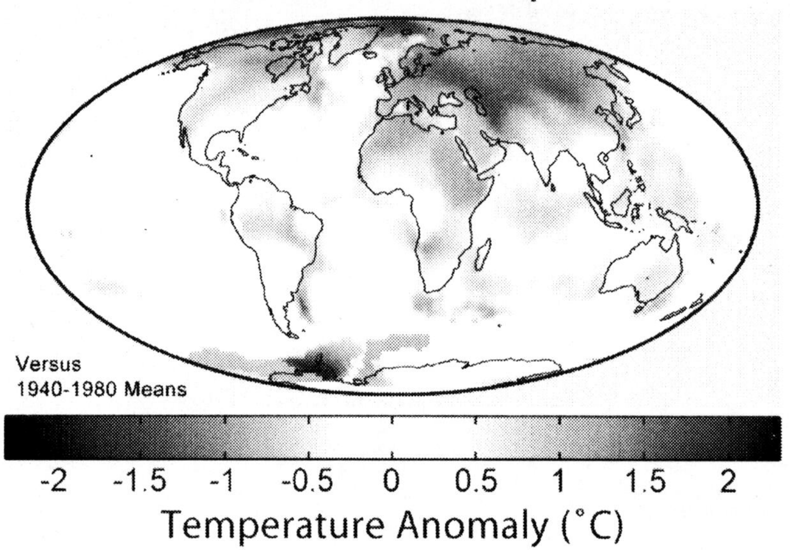

Figure 18.14 Difference in instrumentally determined surface temperatures in the period January 1999 through December 2008 and "normal" temperatures at the same locations, as defined to be the average over the interval January 1940 to December 1980. The average increase on this graph is 0.48°C. Widespread temperature increase is observed primarily at high latitudes in the Northern Hemisphere. (source: Wikipedia/NASA)

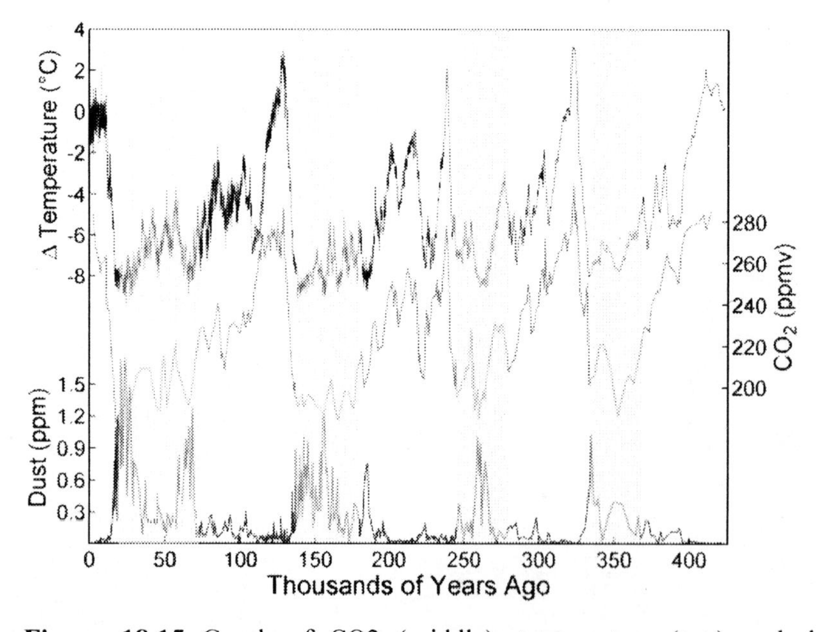

Figure 18.15 Graph of CO2 (middle), temperature (top) and dust concentration (bottom) measured from the Vostok, Antarctica ice core. Higher dust levels are believed to be caused by cold, dry periods. Unlike with current warming, the rise in temperature led the rise in CO_2. Under

current warming, CO_2 rises faster and has an earlier onset. Temperature and CO_2 follow a saw-tooth pattern, with sharp rises and slow decreases. (source: modified from Wikimedia/NOAA)

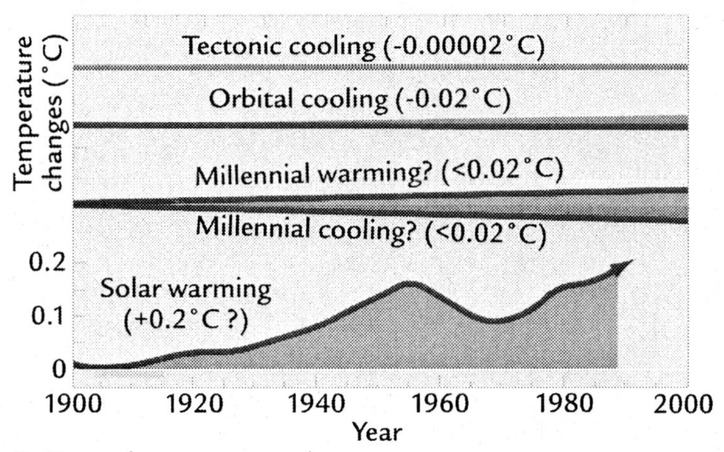

Figure 18.16 Of all natural processes, only solar warming may contribute to the current warming. But the amount predicted is only 0.2ºC. Warming from other causes is insignificant. (source: RU)

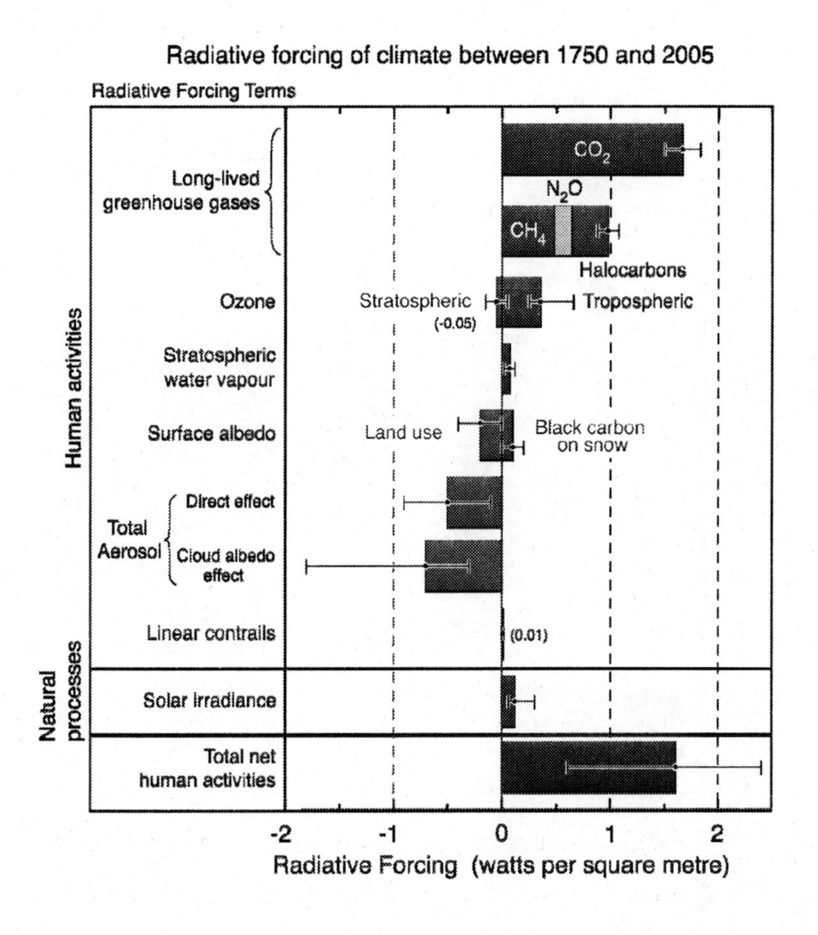

Figure 18.17 Summary of the principal components of the radiative forcing of climate change. All these radiative forcings result from one or more factors that affect climate and are associated with human activity or natural processes. The values represent the forcings in 2005 relative to the start of the industrial era (about 1750). Human activities cause significant changes in long-lived gases, ozone, water vapor, surface albedo, aerosols and contrails. The only increase in natural forcing of any significance between 1750 and 2005 occurred in solar irradiance. Positive forcings lead to warming and negative forcings lead to cooling. (source: IPCC report)

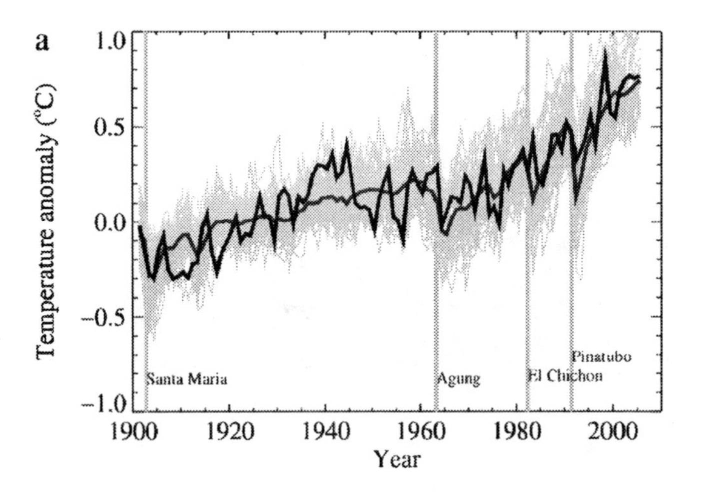

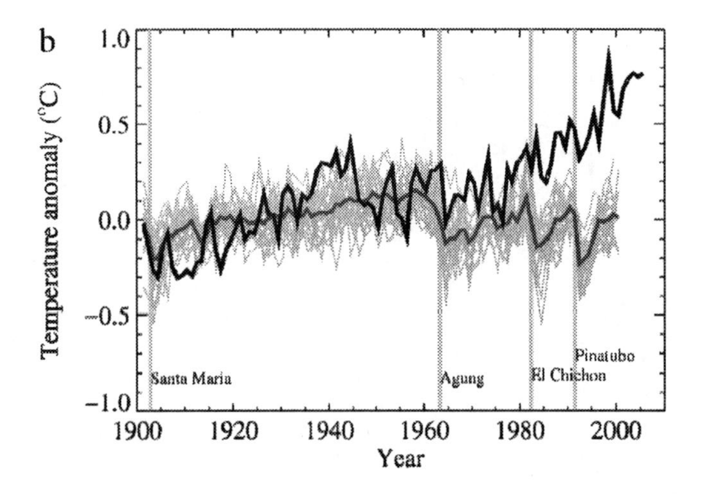

Figure 18.18 Comparison between global mean surface temperature anomalies (°C) from observations (black) and climate simulations (light curves) a) with anthropogenic and natural forcings and (b) natural forcings only. Data are shown as global mean temperature anomalies relative to the average from 1901 to 1950. (source: IPCC report)

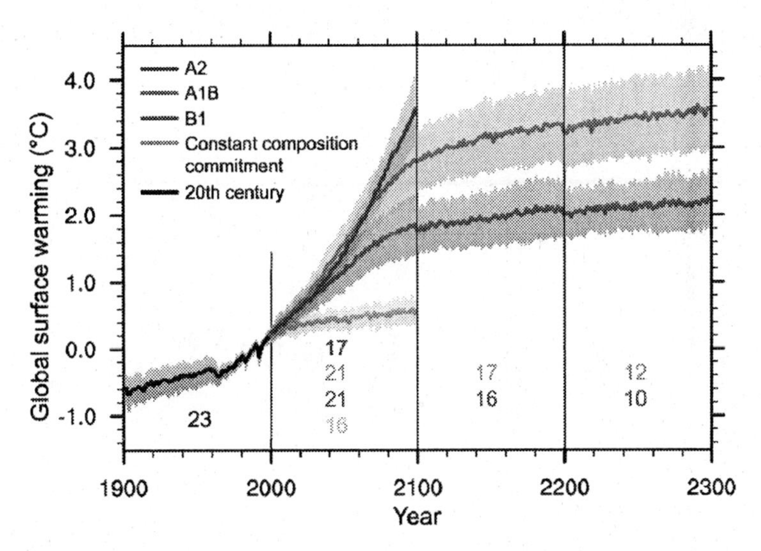

Figure 18.19 Multi-model means of surface warming (relative to 1980-1999) for the scenarios A2, A1B and B1, shown as continuations of the 20th-century simulations. (source: IPCC report)

Special Report on Emission Scenarios (SRES):

A1: very rapid economic growth, global population peaking in mid-century then declining, rapid introduction of new and more efficient technologies. A1FI: fossil-intensive; AIT: non-fossil energy sources; A1B: balance across all sources.

A2: very heterogeneous world, self-reliance and preservation of local identities, economic development regionally oriented and per capita economic growth and technological change fragments and slower than other scenarios.

B1: convergent world with same global population as in A1 but with rapid change in economic structures toward service and information economy, reduction of material intensity, clean resource-efficient technologies; solutions with economic, social and environmental sustainability;

B2: local solutions to economic, social and environmental sustainability; world population growing slower than A2; economic development with less rapid and more diverse technological change than in B1 and A1.

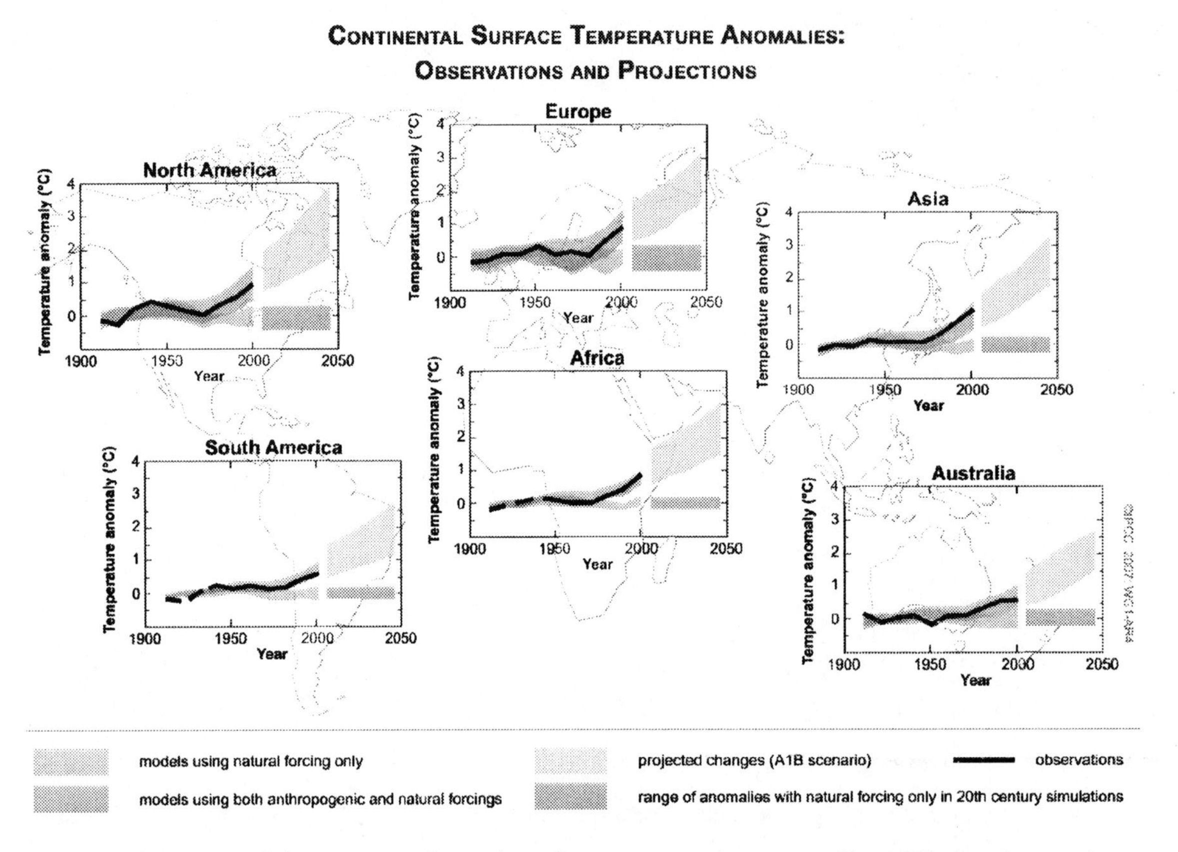

Figure 18.20 Decadal mean continental surface temperature anomalies (ºC) in observations and simulations for the period 1906 to 2005 and in projections for 2001 to 2050. Anomalies are calculated from the 1901 to 1950 average. The bold line represents observations while shaded bands represent the outcome of simulations, with and without anthropogenic forcings. (source: IPCC report)

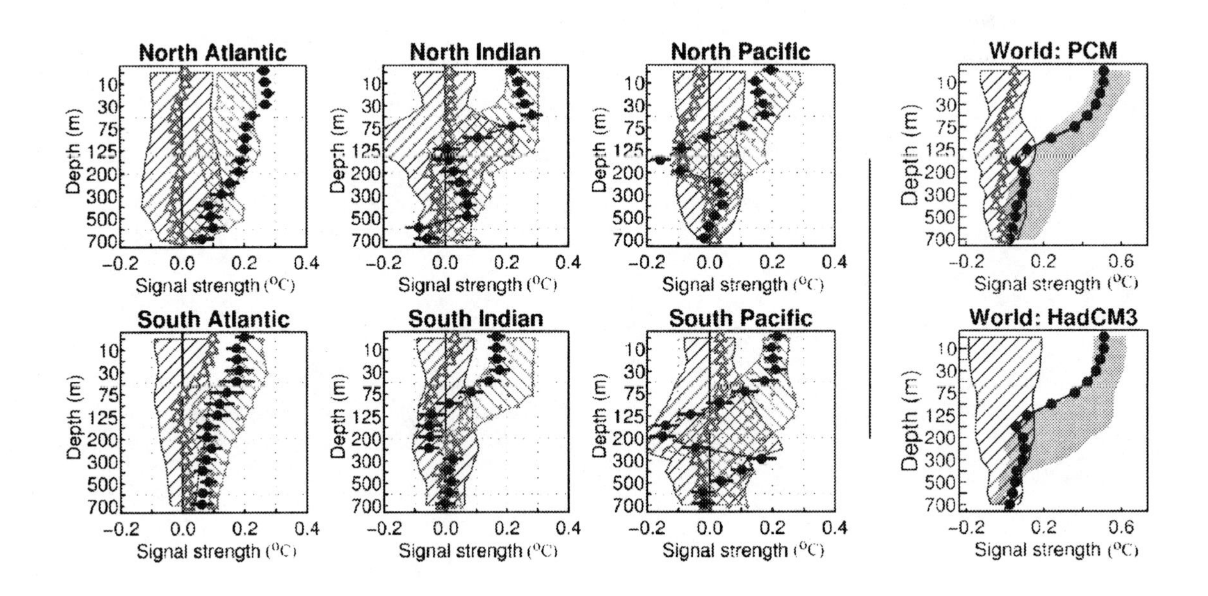

Figure 18.21 Strength of observed and model-simulated ocean warming signal as function of depth for the World ocean and for each ocean individually. Predictions are for two different models (PCM and HadCM3) which account for anthropogenic contributions differently. Triangles mark predictions using only natural forcings. (source: IPCC report)

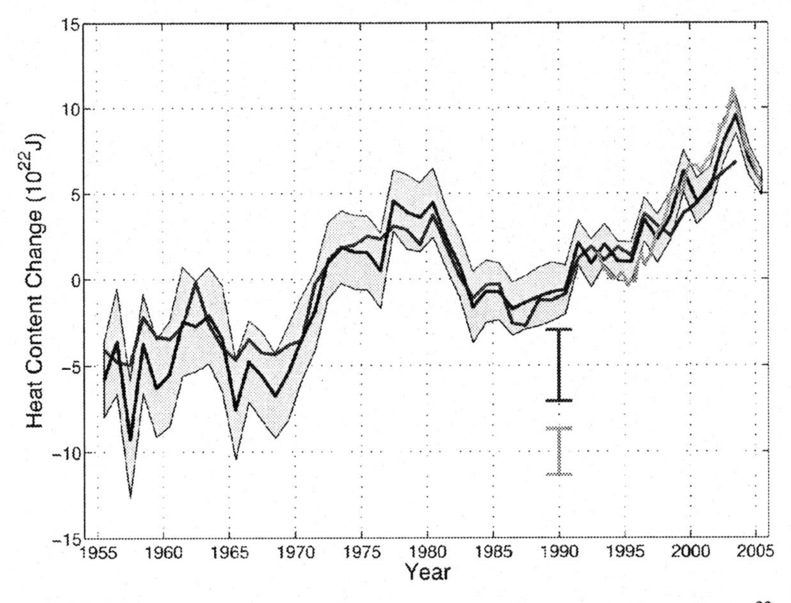

Figure 18.22 Time series of global annual ocean heat content (in 10^{22} J) for the top 700 m. Different curves indicate the results from different studies as deviation from the 1961 to 1990 average. (source: IPCC report)

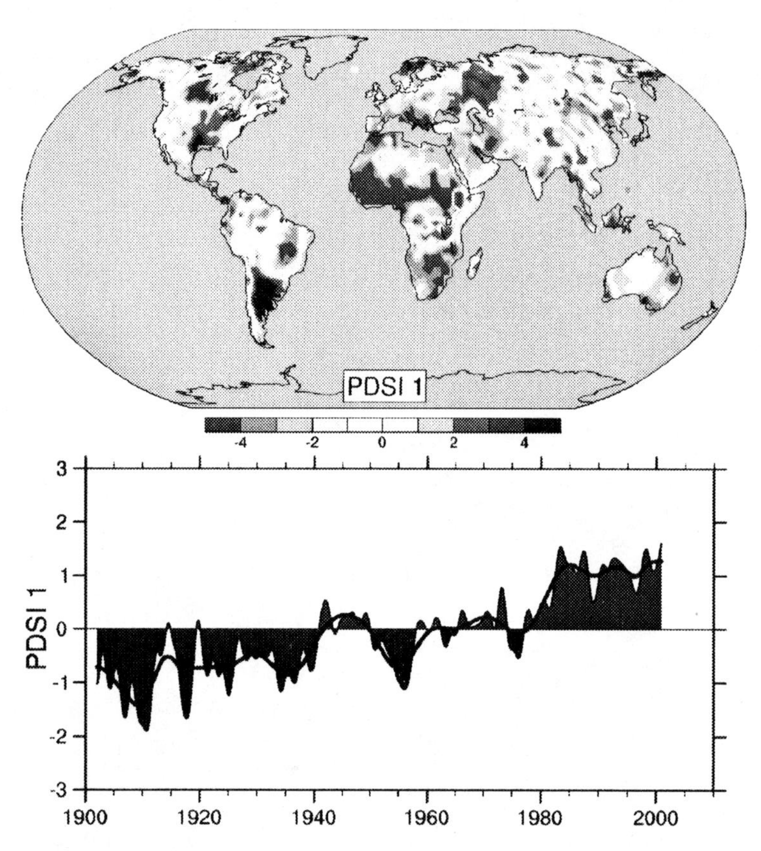

Figure 18.23 Global map of the Palmer Drought Index for 1900 to 2002. The Palmer index is a prominent index of cumulative deficit relative to local mean conditions. The bottom panel shows how the sign and strength of this pattern has changed since 1900. The smooth curve shows decadal variations. Some areas (e.g. Texas in North America and Paraguay in South America) have become wetter but much of Africa has become drier, especially the sub-Saharan Sahel. (source: IPCC report)

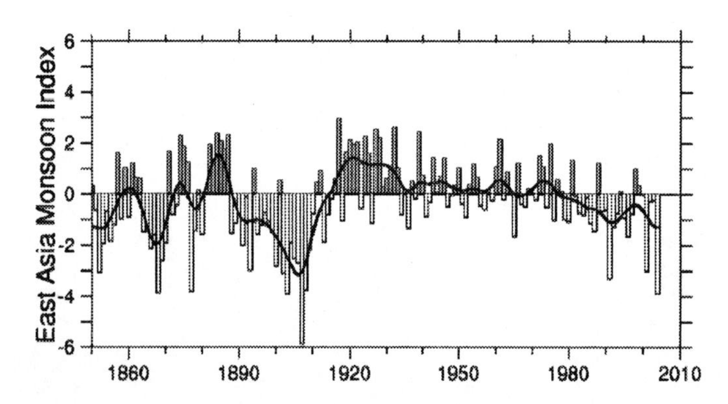

Figure 18.24 Annual values of the East Asia summer monsoon index derived from mean sea level pressure (MSLP) gradients between land and ocean in the Asia region. The smooth line shows decadal variations. (source: IPCC report) also mention sahel in text (refer to next chapter)

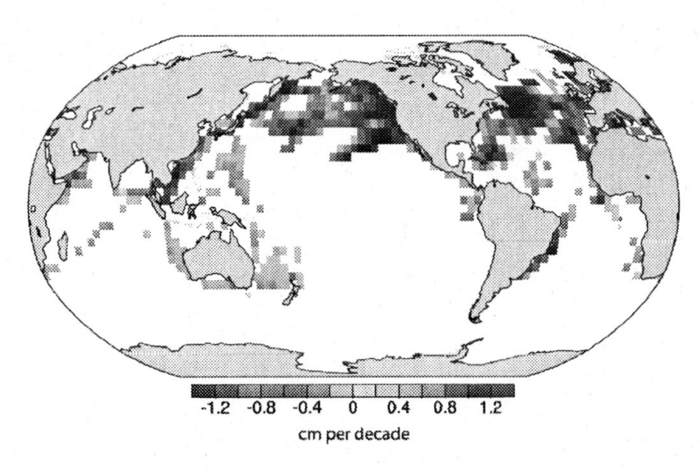

Figure 18.25 Estimates of linear trends in significant wave height (cm per decade) for regions along the major ship routes of the global ocean for 1950 to 2002. (source: IPCC report)

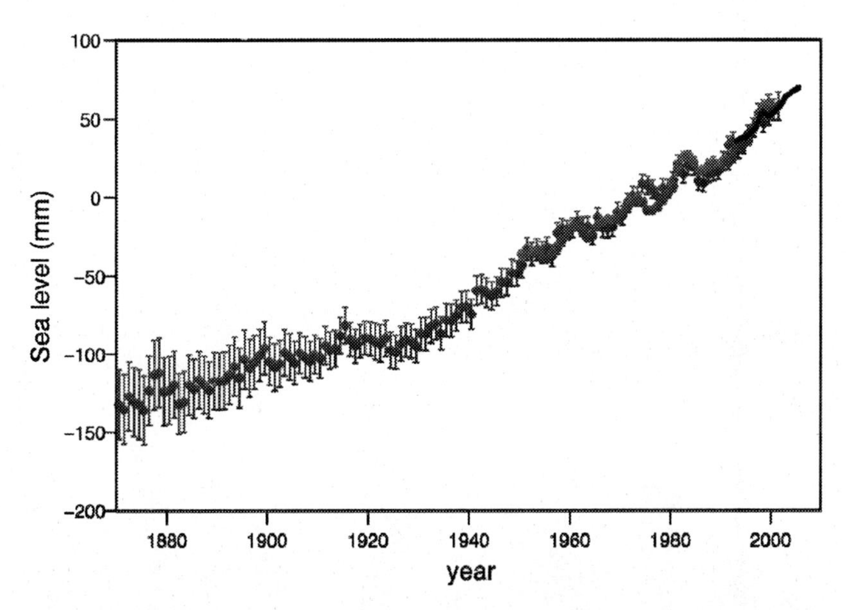

Figure 18.26 Annual averages of the global mean sea level (mm). Red/light-colored data show reconstructed sea level fields since 1870. Blue/dark data are from tide gauge records since 1950. The solid line since the 1990s is from satellite altimetry. The data before satellite are deviations from the average for 1961 to 1990. (source: IPCC report)

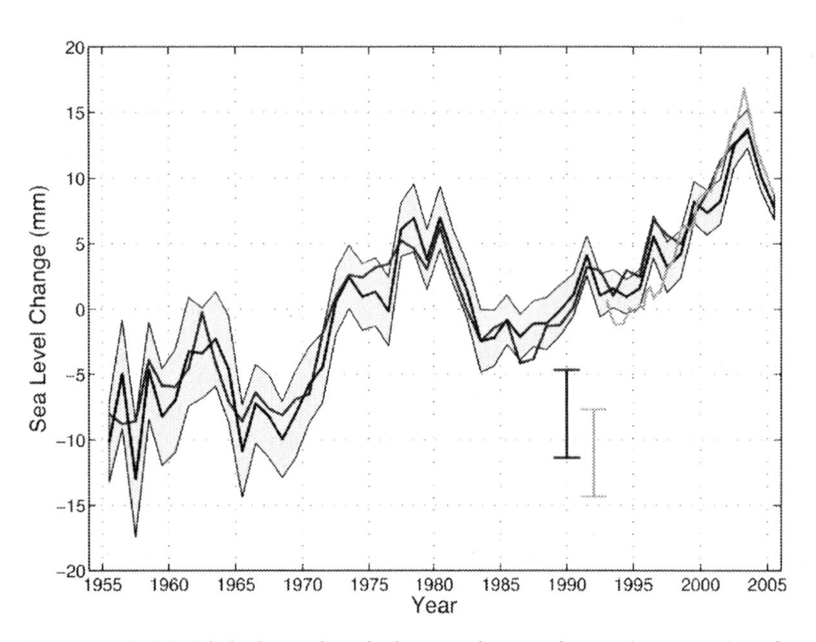

Figure 18.27 Global sea level change due to thermal expansion for 1955 to 2003, based on the heat content estimates for the upper 700m. The shaded area marks 90% confidence limits. Values are with respect to the average from 1961 to 1990. (source: IPCC report)

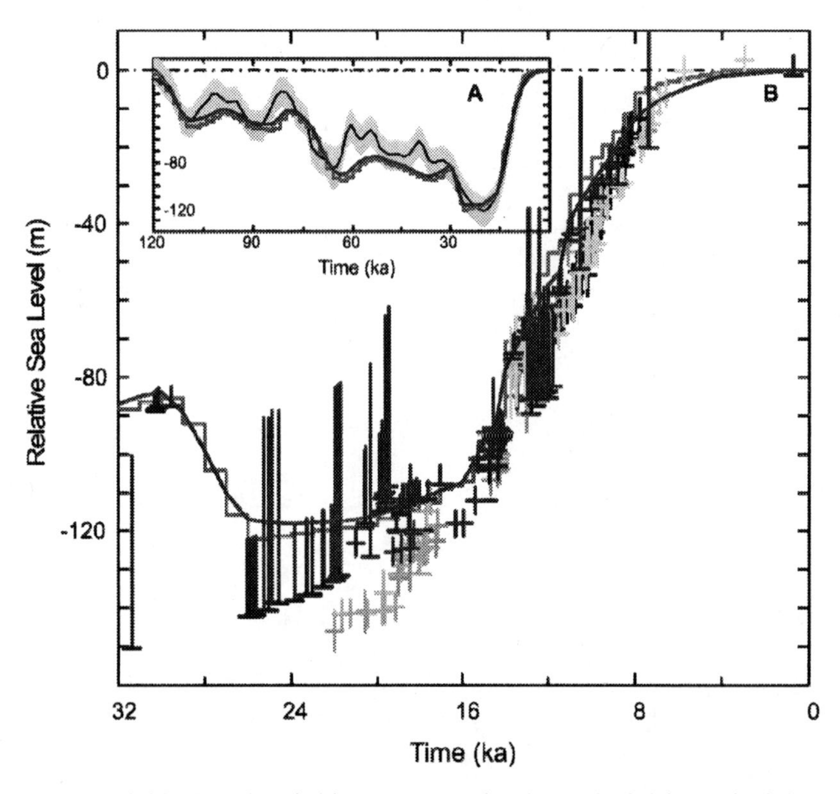

Figure 18.28 Sea level history over the last glacial-interglacial cycle. The black smooth line represents smoothed data while the red line is from model predictions. Crosses mark observations from coral samples in different coral reefs. (source: IPCC report)

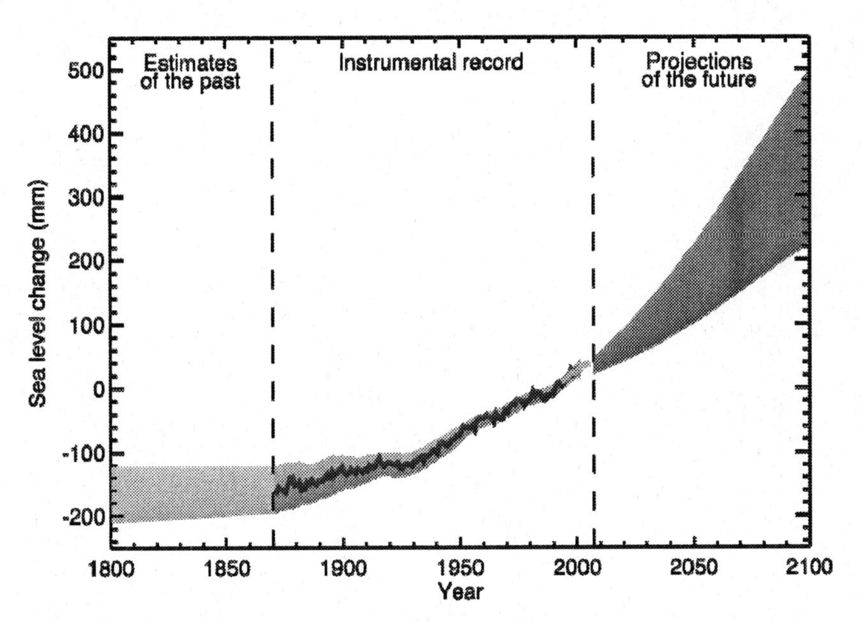

Figure 18.29 Time series of global mean sea level (deviation from the 1980-1999 mean) in the last and as projected for the future. For the period before 1870, global measurements of sea level are not available. The grey shading shows the uncertainty in the estimated long-term rate of sea level change. Instrumental records are from tide gauge records and, since the 1990s, from satellite altimetry. Uncertainties in future projections result from the assumption of different emission scenarios for greenhouse gases, mainly CO_2. (source: IPCC report)

18.3 Global Warming and Sea Level Rise

Rising temperatures lead to thermal expansion of the water in the oceans and melting of glaciers and ice sheets. Both results in a rise in sea level.

- In last 100 years, global sea level has risen by 15cm

Estimated contributions to sea level rise:

- thermal expansion: 30%
- melting of glaciers: 20%
- melting of Greenland ice sheet: 20%
- melting of Antarctic ice sheet: unknown but likely large

NB: the melting of sea ice (icebergs) does not contribute to sea level rise! The icebergs floating in seawater already displace the seawater of equivalent weight. So when the icebergs melt, they displace less seawater but the meltwater replaces the "displaced" seawater.

- various model predictions range from 2-5°C warming and a 40 - 200cm rise in sea level in next 100 years
- this is of the same order as the typical tidal reach and sea surface elevation due to low pressure in a hurricane
- consequences: flooding of coastal areas; landward migration of estuaries; displacement of millions of people

18.4 Global Warming and Some Effects

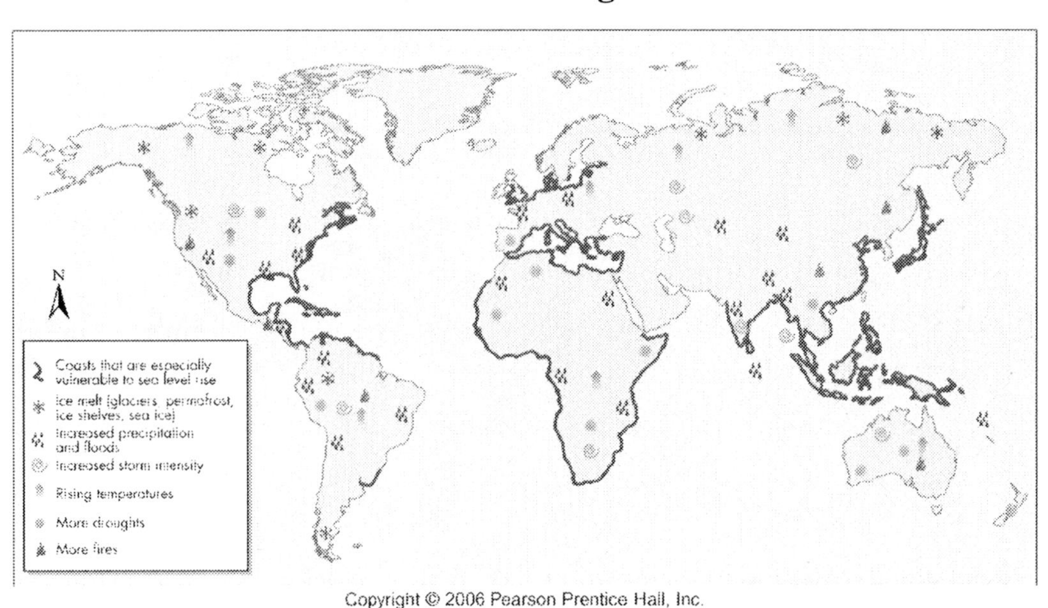

Figure 18.30 Predictions for increases in natural hazards as a result of global warming. Events will be spread around the globe and include sea-level rise, ice and snow melt, increases in precipitation and storm intensity, flooding, rising temperatures, and more droughts and fires. (source: KB)

Global Warming and the Oceans - Effects on The Carbon Cycle

- Carbon reservoirs on Earth are coupled (see Lecture 22 APPENDIX E)
- changes in atmospheric CO_2 therefore affect other reservoirs, such as oceans
- human CO_2 production has actually risen faster than content in atmosphere -> some CO_2 had to have gone somewhere else
- evidence points toward oceans
- some argue that increase in CO_2 in oceans could be beneficial to increase photosynthesis and plankton production

- current estimates are that atmosphere stores about 46% of human produced CO_2, oceans take up 29%, northern hemisphere forest regrowth 7% and other parts in the biosphere 18%
- and experiment off Hawaii therefore looks at the possibility of *CO_2 sequestration* into the deep ocean
- however, recent findings indicate that oceans instead become more acidic
- profound negative impact for coral growth and shell producing life forms

18.5 The Kyoto Protocol

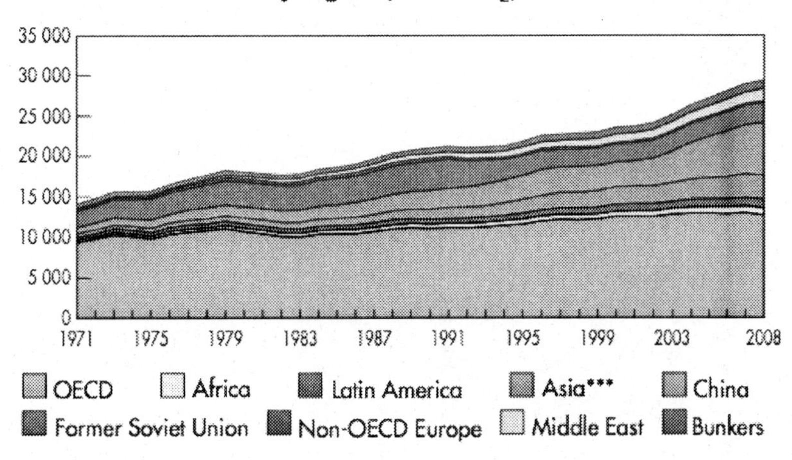

Figure 18.31 CO_2 emissions by region. * Includes international aviation and international marine bunkers, which are shown together as Bunkers. **Calculated using IEA's energy balances and the Revised 1996 IPCC Guidelines. CO_2 emissions are from fuel combustion only. ***Asia excludes China[13]. (source: IEA)

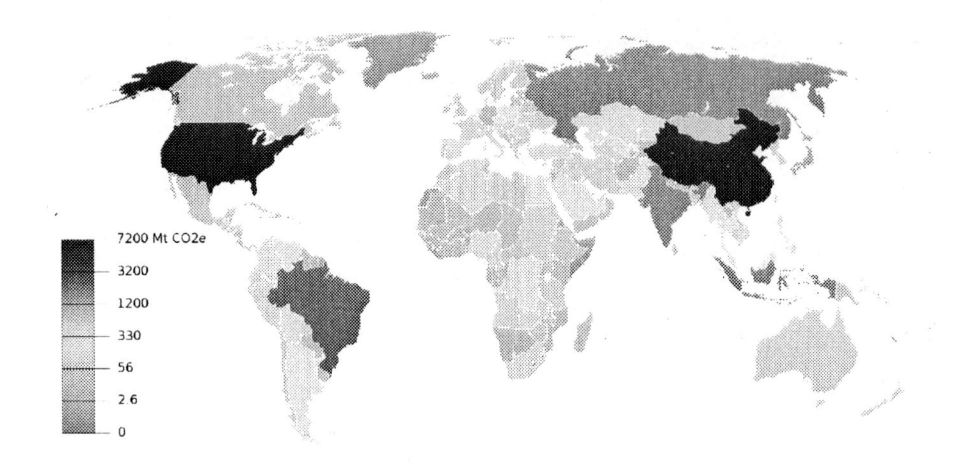

Figure 18.32 Greenhouse gas emission by country in millions of metric tons CO_2 Equivalent in 2005. CO_2 equivalent emissions from land use change and emissions of CO_2, CH_3, N_2O, PFC, HFC and SF6 are included. Bunker fuel (i.e. ships) is not. (source: wikipedia)

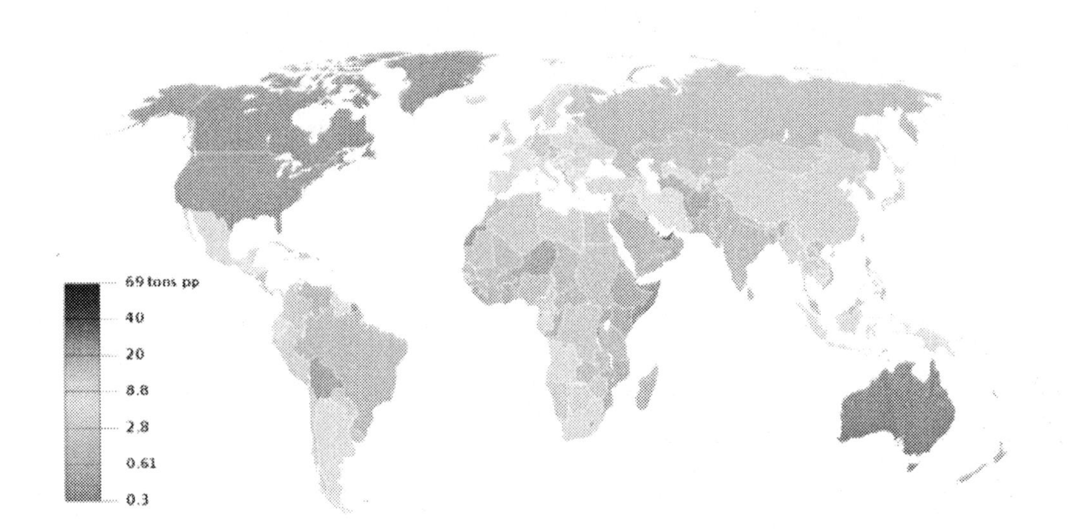

Figure 18.33 Greenhouse gas emission per capita in metric tons per person for each country in 2005. CO_2 Equivalent. CO_2 equivalent emissions from land use change and emissions of CO_2, CH_3, N_2O, PFC, HFC and SF6 are included. Bunker fuel (i.e. ships) is not. (source: wikipedia)

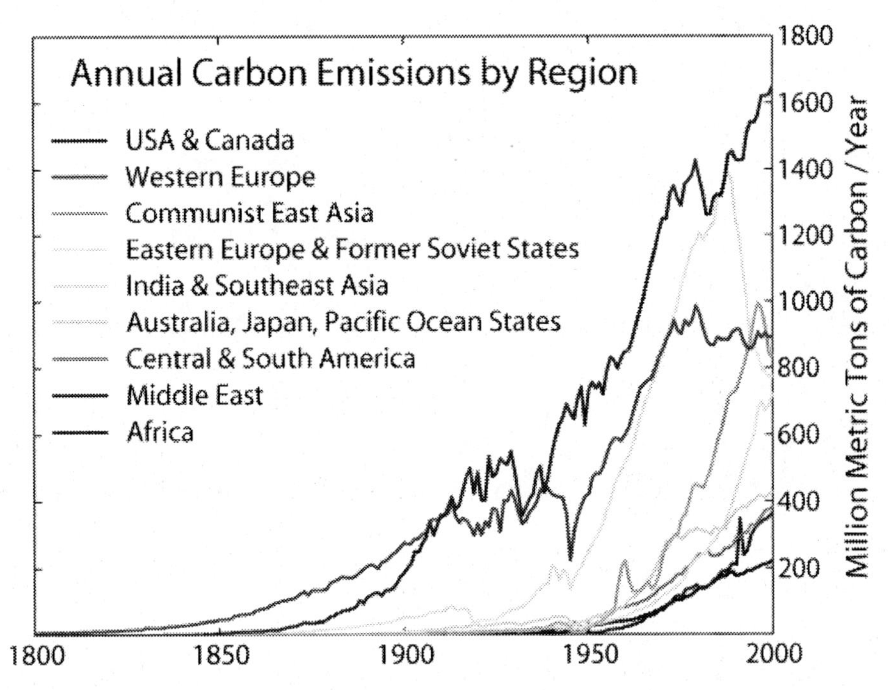

Figure 18.34 Annual fossil fuel carbon dioxide emission, in million metric tons of carbon, for a variety of non-overlapping regions in the World. Emissions in North America are highest in the World and account for 25% of total emission (with only 5% of thc World's population). Emissions in Eastern Europe nearly reached those of North America but dropped dramatically after the collapse of the USSR at the beginning of the 1990s. On the other hand, emission in Communist East Asia (incl. China) and India& Southeast Asia skyrocketed. (source: wikipedia)

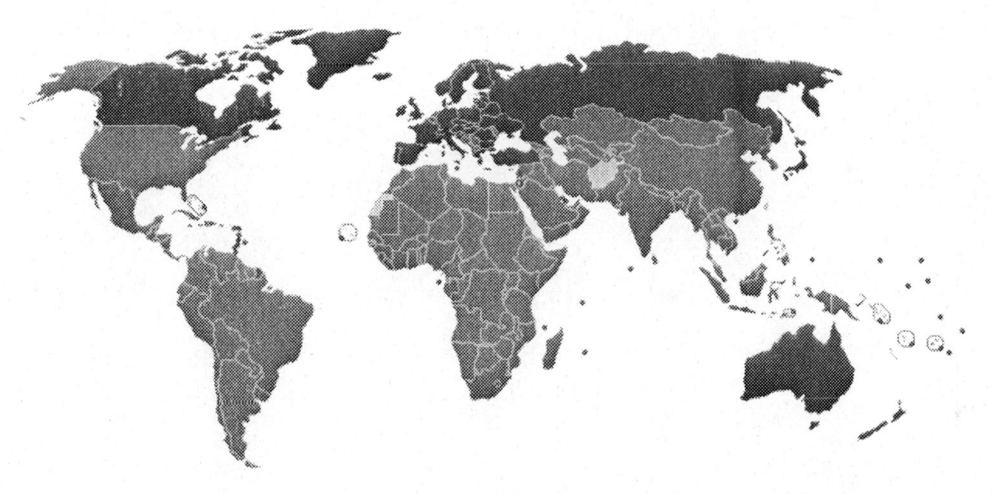

Figure 18.35 Kyoto Protocol participation map 2010. Green indicates countries that have ratified the treaty. Dark green are Annex I and II countries that have ratified the treaty. Light grey is not yet decided. Brown is not intention of ratifying (essentially the U.S.). (source: wikipedia)

- established in 1997; to be replaced by the Copenhagen protocol in 2009
- realization that, in order to stabilize CO_2 at 450 ppm, emission would have to fall below 1990 level
- industrial countries have to reduce to 5.2% below 1990 level by 2012
- 96 countries ratified, incl. E.U. and Japan which make up 37.4% of production in industrial countries
- need countries responsible of > 50% to ratify
- U.S. (25% of total CO_2 and 36.4% of production in industrial countries) bailed out; major argument against ratification: "this would hurt national economy"
- Russia (17.4%) and Poland (3%) initially pledged to ratify the protocol, which would account for 57.8% of production
- jeopardized when Russia bailed out in 2003, but ratified when it joined in October 2004
- European Union and Brazil proposed phasing out fossil fuel burning; Europe has increased the use of alternative energy (e.g. wind), while Brazil has a strong biodiesel program
- the Kyoto protocol is opposed by:
    - fossil fuel industry
    - oil-producing countries
    - major fossil fuel users (U.S., China)
- new threat: developing countries are catching up (e.g. China, India) though some was accounted for in Kyoto Protocol
- check out news clip

A Table from the "State of the World" (2006) issued by the World Watch Institute:

Country	Carbon Emission (Mio tons)	per person (tons)	Emission per GDP,PPP (tons per $Mio)	increase 1990-2004 (%)
China	1,021	0.8	158	+67%
India	301	0.3	99	+88%
Europe	955	2.5	94	+6%
Japan	338	2.7	95	+23%
United	1,616	5.5	147	+19%

States				

NB: China surpassed U.S. emissions in 2009 (see "Industrial Age" above).

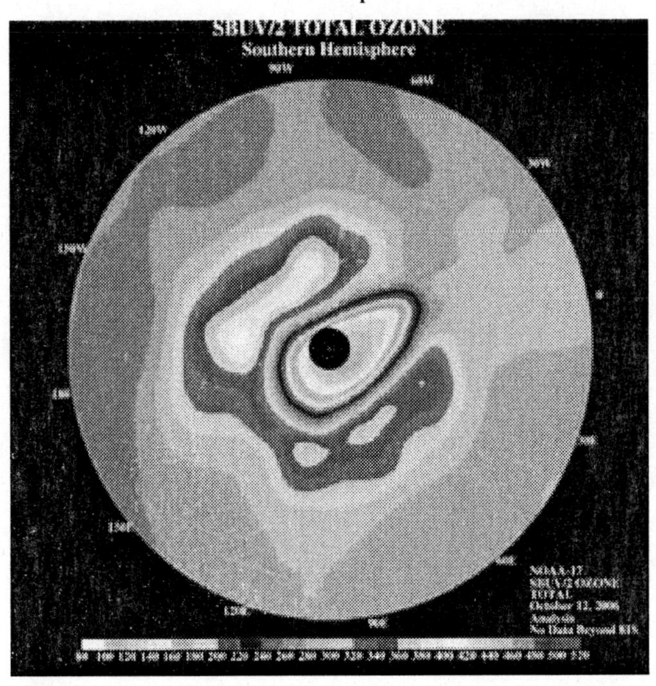

Figure 18.36 The ozone hole of 2006 broken records in both size and depth. (source: NOAA)

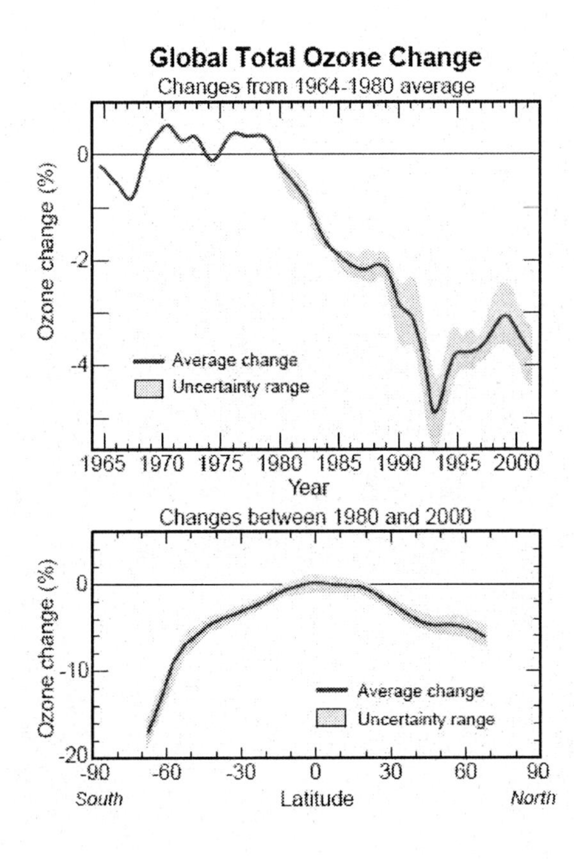

Figure 18.37 Decline in global ozone. Top: Global total ozone values decreased by an average of a few percent in the last two decades. Between 1980 and 2000, the largest decreases occurred following the 1991 eruption of Mt. Pinatubo. Bottom: changes in ozone for different latitudes. The greatest losses that occur each year are over Antarctica. (source: NOAA)

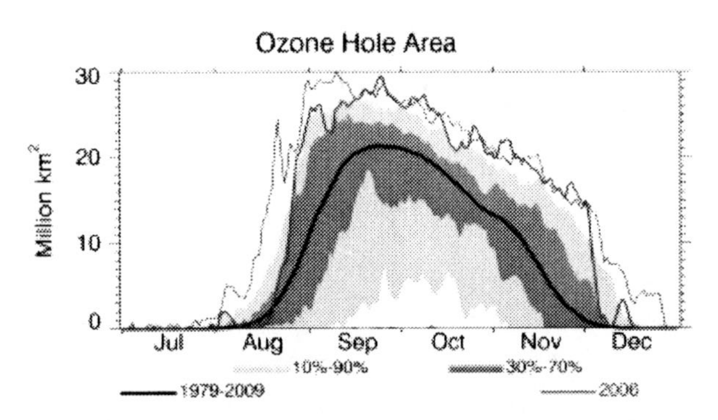

Figure 18.38 The area of the ozone hole in 2006, compared to the years 1979-2009. The ozone hole is the region of ozone levels below 220 Dobson Units (UD) located south of 40°S. Values below 220 DU represent anthropogenic ozone losses over Antarctica. (source: NASA)

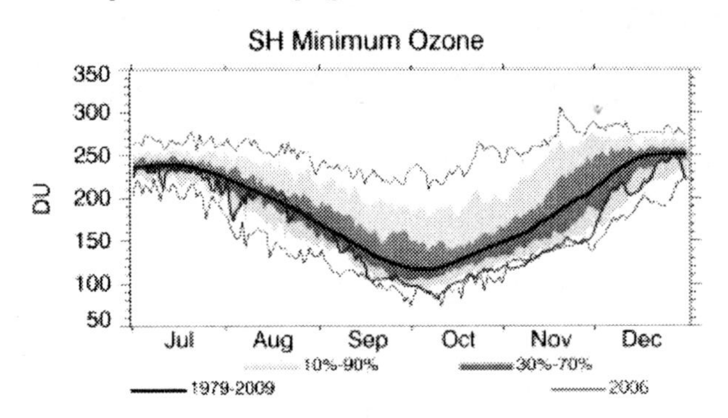

Figure 18.39 The minimum ozone is found from total ozone satellite measurements south of 40°S. (source: NASA)

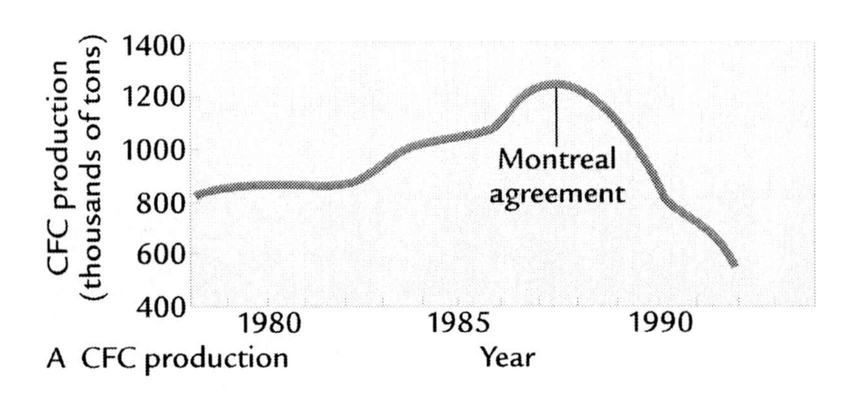

Figure 18.40 Anthropogenic CFC. Because CFCs do not occur naturally in the atmosphere, increasing CFC production by humans in recent decades has caused rising concentrations in the atmosphere. Production has decreased after the Montreal agreement but CFCs was projected to increase before declining because of the residence time in the atmosphere. (source: RU)

18.6 The Ozone Hole

- Ozone: solar radiation converts O_2 to O_3. Protective layer in 10-50km altitude
- lack in stratospheric ozone lets carcinogenic UV radiation reach the Earth's surface, causing skin cancer, cataracts, impaired immune system, reduced crop yields
- stratospheric ozone attacked by human-made CFC and other halogen compounds
- ozone "holes": sharp decrease in thickness of protective ozone layer in stratosphere
- process most effective at low temperatures and in presence of sunlight
- ozone holes therefore near poles; seasonal changes
- ozone destruction most severe in early spring
- decline first noticed in 1980s
- ozone hole caused by human action alone; no natural process known to destroy ozone layer
- chlorofluorocarbons (CFCs) attack and destroy ozone molecules in stratosphere
- sunlight breaks down CFCs; Cl recombines with one of the oxygen atoms of O_3
- CFCs have very long lifetime so can destroy ozone over a long time
- it has been conjectured that the ozone hole will likely increase before the ozone layer recovers, despite recent efforts to reduce CFC input
- **Montreal Protocol** established in 1987: CFC phased out by 1996 by 140 countries, incl. U.S.
- unlike Kyoto, Montreal was ratified by many because the (sole) human cause of CFC production was more obvious
- effects from CFC reduction not expected to be effective before 2025
- in 2006, the Antarctic ozone hole was the largest ever
- another problem: other halocarbons - like hydrochloro-flourocarbons (HCFC) replaced CFCs. These are less destructive to the ozone layer but are nevertheless greenhouse gases. These will be phased out by 2030. HFCs (hydroflourocarbons) are also greenhouse gases but their lifetime in the atmosphere is significantly shorter than CFCs and HCFCs.

NB: the ozone hole does not contribute to global warming/global warming and the ozone hole are two phenomena that are not related. Ozone absorbs in the infrared (IR) part of the spectrum, but much more so in the ultraviolet (UV) part. The ozone hole lets more UV light to Earth's surface which does not directly warm the planet.

Figure 18.41 A house in Los Angeles tented for termite fumigation. (source: Wikipedia)

Until 2000, an often used fumigant to eradicate termites from homes was methyl bromide (or bromomethane, CH_3Br). Like nearly every other fumigant used in this business, it is toxic and cumbersome preparations include the removal of all foods, toiletry and clothing as well as plants and animals from the homes. Methyl bromide depletes the ozone layer and so was phase out after the Montreal Protocol. It has since been replaced in termite control by sulfuryl fluoride (SO_2F_2). The problem with this agent, however, is that it is a greenhouse gas. Its residence time in the atmosphere was originally estimated at 5 years but more recent studies suggest a lifetime of 30-40 years. Sulfuryl fluoride is 4000-5000 times more efficient that CO_2 in trapping IR radiation. Currently however, the relatively small amount (about 2000 metric tons/year) injected into the atmosphere is considered irrelevant in the global warming discussion as CO_2 is injected at a rate of 30 billion tons/year.

Figure 18.42 (source: SD Union Tribune, 23 May 11; photo: Gosia Wozniacka)

methyl bromide as ozone hole killer! (SD UT 052311)

not using new fumigant could cost industry $1.7 billion

However, methyl bromide is still used today after the U.S. has lobbied successfully for critical-use exemptions from the Montreal Protocol. In 1999, an estimated 71.500 tonnes were used annually worldwide [(3)]. Over 75% is used by developed nations, with the U.S. leading the usage with 43% compared to Europe with 24% as the next biggest user. Say something about heath risks!

Methyl bromide has widely been used as soil sterilant for production of seed but also for some crops such as strawberries (including the local strawberries in San Diego county) and almonds[(5)].

18.7 Fossil Fuels and Pollution

- atmospheric pollutants:

  primarily through burning of fossil fuels

  - sulfates (SO_4^{-2}; also from volcanoes; causes acid rain)
  - nitrates (NOx; interaction with sunlight produces ozone; causes acid rain)
  - CO (an odorless toxin from combustion engines)

  - CFCs (destroys stratospheric ozone layer; from leaking cooling systems; not naturally occurring)

- **Consumption of Energy in the U.S., 1997**

Energy Resource	Fraction 1997	Fraction 2002
Petroleum	41%	41%
Natural Gas	24%	22%
Coal	22%	24%
Nuclear	7%	9%
Hydroelectric	5%	3%
others	1%	1%

Primary Pollutants from Burning of Fossil Fuels

Pollutant	Fraction
Carbon Monoxide	49.1%
Sulfur Oxides	16.4%
Nitrogen Oxides	14.8%
Volatile Organics	13.6%
Particulates	6.0%

- NB: the greenhouse gas CO_2 is not considered a pollutant and is therefore not included in this table!
- NB: Ozone near the surface is actually a pollutant. Surface ozone is the major constituent of smog and is most severe on sunny days when sunlight interacts with nitrous oxides from car engines to form ozone. Ozone is unhealthy and causes crop damage and corrodes material.

Sources of the Primary Pollutants

Process	Fraction
Transportation	46.2%
Stationary Source Fuel Combustion	27.3%
Industrial Processes	15.0%
Miscellaneous	9.0%

Solid Waste Disposal	2.5%

Given the fact that the burning of fossil fuels causes global warming and pollutes the atmosphere, and given the fact that some fossil fuels will be used up within a century, it seems clear that we need to cut back our dependency on fossil fuels. Nevertheless the World Summit on Sustainable Development in Johannesburg/South Africa in the summer of 2002 displayed the great dilemma. The European Union and Brazil proposed the adoption of specific numerical targets for the phasing out of fossil fuels and the increase in usage of renewable energy worldwide. This was strongly opposed by the fossil fuel industry, the governments of most oil producing countries and major fossil fuel consumers, including the U.S. and China.

- **Most Polluted Cities by Particulate Matter (PM)[1]**

PM μg/m^3(2004)	**City**
169	Cairo, Egypt
150	Delhi, India
128	Calcutta, India
125	Taiyuan, China
123	Chongqing, China
109	Kanpur, India
109	Lucknow, India
104	Jakarta, Indonesia
101	Shenyang, China

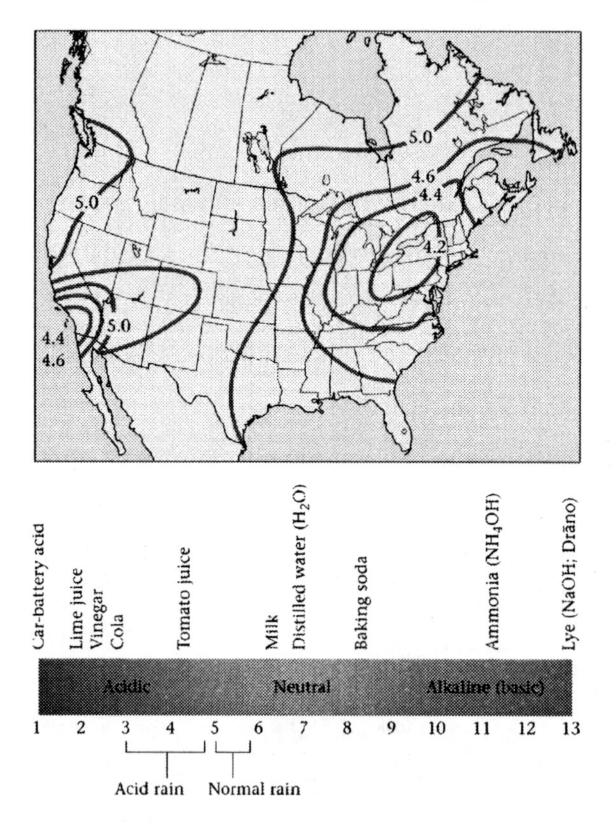

Figure 18.43 Acid rain has affected large portions of the U.S.. The map contours pH numbers, indicating the concentration of hydrogen ions in a solution. Very acidic rain falls in the Northeast and Southwest. The scale gives a sense of what the numbers mean. A solution with a pH of 7 is neutral; acidic solutions have a pH less than 7, while alkaline solutions have a pH greater than 7. Note that "normal" rain is slightly acidic. (source: ME)

Figure 18.44 Acid rain affects statues and building facades made of many materials such as limestone, sandstone and marble, such as this limestone gargoyle on the new town hall in Munich, Germany. (source: wikimedia)

Figure 18.45 Dead fir trees photographed in 1998 in the Erzgebirge, Germany. The vast "waldsterben" (dying of forest) was caused by aerosols from outdated Czech power plants burning lignite (low-grade, dirty coal). (source: wikipedia)

Acid Rain

- main cause in Eastern U.S.: burning of coal, especially low-grade coal that contains large amounts of Pyrite (FeS_2) produces SO_2
- SO_2 mixes with rain in air to form sulfuric acid, a major contributor to acid rain; much more aggressive than carbonic acid
- main cause in Southern California: combustion engines (cars and trucks) emit nitrogen oxides (NO_x)
- some NO_x mix with rain to form nitric acid, a major contributor to acid rain
- Acid rain is responsible for massive fish kills in freshwater lake (e.g. Scandinavia, Canada, northeastern U.S.)
- Survey of 1000 lakes and thousands of km of streams showed that 75% of lakes and 50% of streams were acidified
- in U.S. the pH value is lowest along the coal mining and burning areas in the Eastern U.S. (sulfuric acid), and in Southern California, where traffic related acid fog is found (nitric acid)
- acid rain is responsible for massive tree kills, especially conifers
- mountain forests damaged, particularly at higher elevations
- $1 billion of damage each year to buildings and monuments
- Relationship between acid rain and coal burning firmly established as damage is most severe downwind from coal burning power plants

- About 50% of Canada's fallout has source in U.S.
- **Clean Air Act**: passed by U.S. congress in 1991. Legislature requires power plants to reduce annual SO_2 production by 10Mio tons and annual NO_2 production by 2 Mio tons by 2000 (see also air quality index, APPENDIX B).
- Deregulation of U.S. railroad system lowered transport costs to make transport of higher grade western U.S. coal to eastern U.S. more attractive.

Air Quality Index (AQI)

In the US, air quality index (AQI) is assessed by calculating the AQI for 5 major pollutants regulated by the Clean Air Act: ground-level O_3, PM, CO, SO_2 and NO_2. The AQI runs from 0 to 500.

AQI	Level of Health Concern
0-50	good
51-100	moderate
101-150	unhealthy for sensitive people
151-200	unhealthy
201-300	very unhealthy
301-500	hazardous

In San Diego, the AQI is usually around 50 and is a little higher inland (e.g. El Cajon, Alpine). In summer, values can exceed 100 when ozone levels are high. After the fires in 2003 and 2007, values briefly exceeded 200, due to high particulate counts.

Summary

There are basically four problems with a fossil fuel-based society like ours:

- hydrocarbons, such as oil, are non-renewable and resources are probably exhausted within a few decades but 100 years at most; natural gas will last a little longer; so will coal (300 years)
- burning fossil fuels pollutes the environment (gas burns "cleaner" but still pollutes)
- burning fossil fuels accelerates the greenhouse effect
- burning coal causes acid rain

Action is Needed NOW - Some Suggestions

- temporary transition to "cleaner burning" natural gas instead of oil and coal is ok as short-term solution, but only as that
- nuclear power "clean" but poses waste disposal, accident and public security problems
- "traditional" alternative energy (incl. hydroelectric power, geothermal, solar, wind, biomass burning) each have disadvantages, the most serious being that they cannot replace the massive amount of energy produced by fossil fuels
- "modern" alternate energy (incl. fuel cells, hydrogen-based society, fusion) is in its infancy; some of which may never be realized
- more efficient use of energy: more fuel-efficient cars; technically feasible but repeatedly discouraged by oil-industry, especially in U.S.; replace incandescent light bulbs with fluorescent tubes; more energy-efficient household appliances; turn computer off instead of putting it in sleep mode
- societal changes/conservation

OECD: Organization for Economic Co-operation and Development

OPEC: Organization of the Petroleum Exporting Countries

IEA: International Energy Agency

EIA: Energy Information Administration, Department of Energy (DOE)

References

1. Wikipedia
2. air now
3. http://en.wikipedia.org/wiki/Methyl_bromide
4. http://en.wikipedia.org/wiki/Sulfuryl_fluoride
5. San Diego Union Tribune, 23 May 2011, Fumigant worry in strawberry fields.

Related Web Sites

- Intergovernmental Panel on Climate Change
- Real Climate
- Stephen Schneider at Stanford
- Worldwatch Institute

http://ozonewatch.gsfc.nasa.gov/meteorology/ozonetab_2006.html

Recommended Reading

- "Earth's Climate, Past and Future" by William F. Ruddiman, 2000. W.H. Freeman and Company, ISBN: 0-7167-3741-8
- "State of the World" by The Worldwatch Institute, W.W. Norton and Company. Topics change from year to year but here is a short list on recent topics: 2003 (ISBN: 0-393-32386-2): special 20th Anniversary Edition on spread of and fight against diseases, decline of birds, energy resources, mining and ecosystems, greenhouse effect, Kyoto Protocol, sustainable Earth 2004 (ISBN: 0-393-32539-3): consumerism and globalization, waste/recycling of resources, catch-up of developing countries, water productivity and increasing shortage. 2005 (ISBN: 0-393-32666-7): global security, infectious diseases, food security, water, oil economy, postwar societies 2006 (ISBN: 0-393-32771-X): China and India, global meat industry, freshwater, renewable energy, Mercury, sustainable development, corporations 2007 (ISBN: 0-393-32923-0): urban future, water and sanitation, urban farming, transportation, energy, natural disaster risks, public health, poverty 2008 (ISBN: 0-393-33031-1): toward a sustainable global economy, rethinking progress, sustainable lifestyle, low-carbon economy, water, biodiversity, human energy, trade governance
- check out Worldwatch Institute web site
- "Earthshock" by Andrew Robinson, 1993, Thames and Hudson, ISBN: 0-500-27738-9
- "Global Warming" by John Houghton, 1999, Cambridge Univ. Press, ISBN: 0-521-62932-2

Chapter 19: Anthropogenic Changes: The Ground (draft)

Today, we have better tools and knowhow to protect soil from erosion. But nevertheless, such sandstorms happen outside of desert areas. In the first week of April 2011, a freak sandstorm causes a massive pileup on a 4-lane autobahn near Rostock, Germany. Some 110 people in 80 cars were involved in the fiery accident, killing 8 people and injuring at least 41. The cars driving into the sandstorm encountered a nearly instantaneous drop in visibility. Strong winds with speeds of 100 km/h had picked up sand and soil from fields that were recently plowed nearby. An extended period of extremely dry weather also contributed. Experts agree that the sandstorm in northern Germany may have been a freak storm but they warn that smaller storms do occur. Measures against such a disaster are two-fold: 1) plant wind-breaking hedges and reduce the size of the fields to make them less vulnerable to soil erosion; 2) adjust driving to current conditions.

America's Dust Bowl: Excessive broad-scale farming in the Midwest without windbreakers or other measures to diminish wind erosion. Continued waves of immigrants from Europe brought settlers to the plains at the beginning of the 20th century. Unusually wet weather conditions encouraged people to start large-scale agriculture in areas with normally drier climates. Technological advances such as mechanized plowing also helped increase the management of larger fields. World War I led to an increase in agricultural prices which in turn encouraged a dramatic increase in cultivated land, thereby doubling or tripling the managed acreage in the space of only a decade. Finally, farmers used practices that allowed soil erosion. To start with, they removed the native grasses that once covered the prairie lands for centuries, holding the soil in place and maintaining its moisture. Farmers did not plant windbreakers. Cotton farmers left the fields bare over the winter months, when winds in the High Plains are strongest. They burned the stubble thereby depriving the soil of organic nutrients and increasing exposure to erosion.

DID YOU KNOW? It takes 30 years to grow 1 inch of U.S. soil!

Dust Bowl

To protect crops and soil, windbreaks (also called shelterbelts) are planted. These are usually rows of mixed conifer and deciduous trees planted perpendicular to the prevailing winds. They greatly reduce wind speeds on the lee side. It is thereby important to plant trees not too close together because this could generate unwanted turbulent eddies that could swirl the soil about, or strong downdrafts can occur in very windy conditions. The use of properly designed windbreaks has been greatly beneficial to agriculture. But many of the windbreaks planted after the Dust Bowl have been removed for various reasons, including that they occupy valuable crop land and they interfere with large center pivot sprinkler systems. There is no guarantee that weather conditions during the Dust Bowl will not repeat themselves and one is left to wonder if collective memory fades after a generation or so.

19.1 Dust Storms

- Earth can experience massive dust storms that transport sediments over long distances

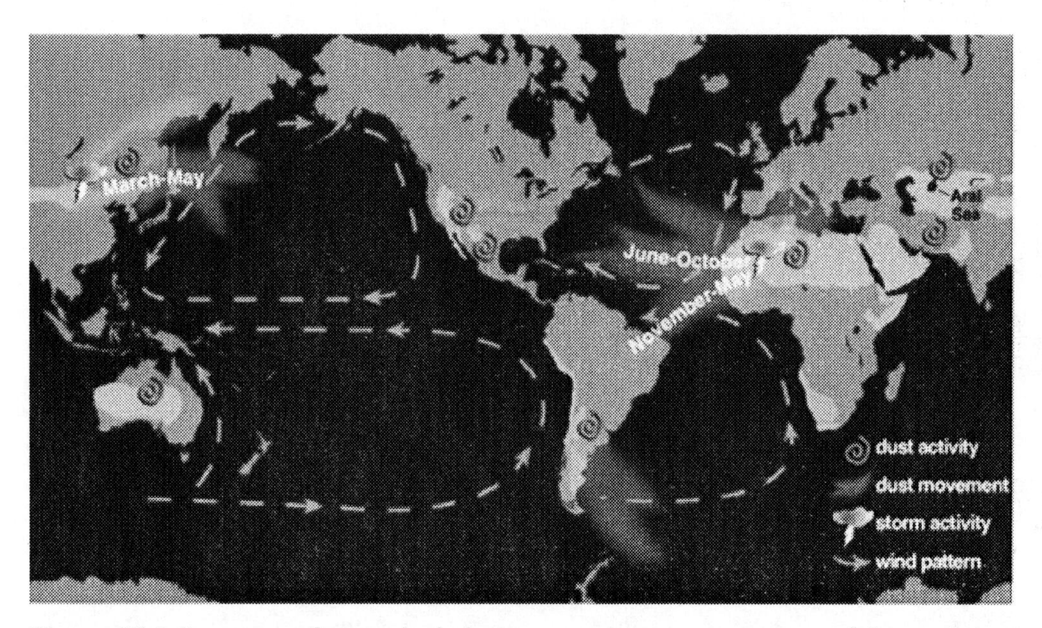

Figure 19.1 Summary of general wind patterns, primary source areas of dust and dust activity. In the time from June through October, strong winds can carry dusts, toxins and pathogens from the western Sahara to the Caribbean. Similarly, dust from the Gobi desert can reach North American carried by strong storms in March through May. (source: Griffin et al., 2002).

Figure 19.2 A plume of dust off the coast of northwestern Africa carried away by a storm in early April 2008. Dust storms from the Sahara provide minerals to the Mediterranean and Black Seas, sometimes causing blooms of plant-like marine organisms called phytoplankton. The latter acts as the base of the food web but overgrowth and decay can create dead zones where all the dissolved oxygen has been consumed. Farther away, Saharan dust sometimes damages Caribbean corals while providing soil for Caribbean islands. (source: NASA Earth Observatory).

COPYRIGHTED MATERIAL — DO NOT DUPLICATE, DISTRIBUTE, OR POST

Figure 19.3 Dust from East Asia reaching Alaska on 13 April 2002. The photo shows Alaska in the upper right (white patch) and the Bering Sea (center). The dust (grayish compared to the with clouds) travels southeast along Alaska's west coast. (source: NASA Earth Observatory).

Dust Storms transport bacteria: example: sub-Saharan dust storms can pick up bacteria from poorly developed sewer systems; the bacteria can caused diseases near the coastline across the Atlantic. **Observation**: It has been a puzzle for some time why the Caribbean Islands experience "dusty skies" despite their tropical climate. The health of corals, sea fans, sea urchins and other marine organisms has been deteriorating. It turned out that the microorganisms responsible for this are pathogenic (e.g. bacteria of human origin) and come from open sewage ditches in Western Africa (e.g. Mali).

- reference: the paper by Griffin given below

TOXINS see Case Study Aral Sea

- **Alaska dust storm**: A satellite image showing a November 2006 dust storm that transported loess from Alaska far into the Pacific Ocean can be found at NASA's Earth Observatory. This dust storm was pushed by Chinook winds that are caused by a similar mechanism as our Santa Ana winds.
-
- **Dust Clouds reaching the U.S.**: can be generated as far away as the Gobi Desert. A storm driven dust cloud in April 2001 reached the West Coast only two days later and crossed the continents in 7 days. Implications of this is that pollution anywhere on the planet can be carried elsewhere to cause major damage. An example of this is the

South East Asian Brown Cloud that is caused by industry and slash-and-burn deforestation.

- the following example of dying corals in the Caribbean illustrates just how far-reaching local action/mistakes can be to the global environment

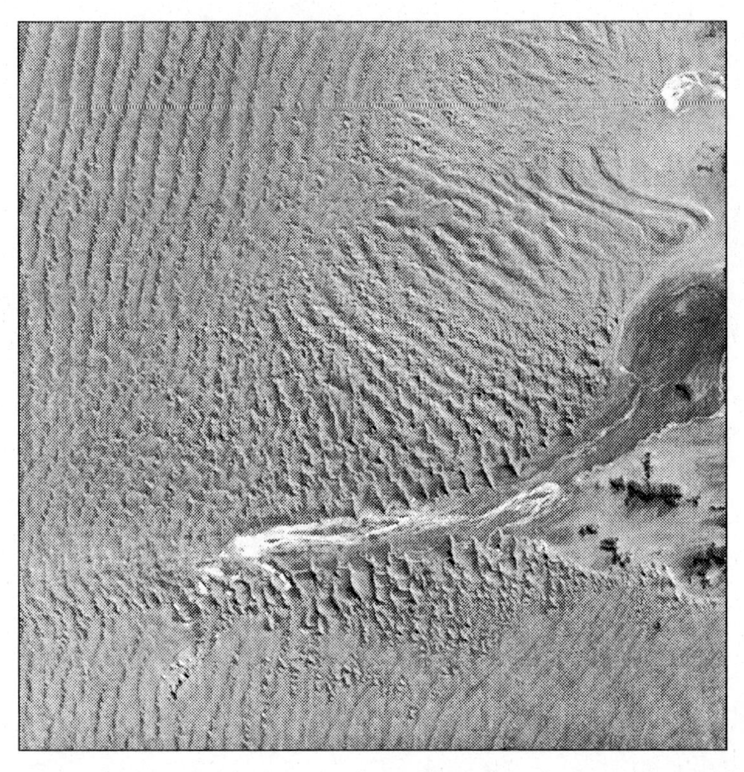

Figure 19.4 Landsat image of sand dunes in the Namib-Naukluft National Park. Coastal winds create the tallest sand dunes in the world here, with some dunes reaching 300 m (980 ft) in height.
. (source: Wikipedia/NASA).

Figure 19.5 Wreck of the ship Eduard Bohlen that ran aground in thick fog at Conception Bay off the coast of the Namib Desert, Namibia on 5 September 1909. Advancing sand dunes have buried the ship which is now 400 m (1/4 mi) inland. (source: Wikimedia).

Figure 19.6 Landsat image of san dunes advancing on Nouakchott, the capital of Mauretania. (source: Wikimedia/NASA).

Figure 19.7 Dune fixation in the Sahel. Most attempts to control the movement of African dunes have failed. (source: Robinson, 1993).

19.2 Desertification

Deserts

- a desert is defined as a region where vegetation is supported on no more than 15% of its surface area; it contains no permanent streams except for rivers that may bring water from a different climate belt (e.g. Nile); desert are typically associated with annual rainfalls of less than 25mm (10in);
- factors that also play a role: evaporation rate, characteristics of precipitation
- precipitation levels alone can be misleading; e.g. an area may have more than the 25mm of annual precipitation typical for a desert but precipitation could fall in one or few heavy rain storms once every or every few years, with extremely high run-off rates (i.e. useless for cultivation of land); there may be long dry spells/droughts in between
- the extent of deserts is nowadays assessed by estimating the abundance of vegetation using satellite imagery; in terms of inhabitable land, this is sometimes controversial as satellites do not distinguish between vegetation that is more or less "useful" to humans
- the gravity of recent desertification is difficult to assess as "dry" and "wet" phases in deserts may have oscillated in the past (e.g. Sahara)
- natural causes of advancing deserts include climate shifts and fast-advancing sand dunes
- drought problems especially severe in semi-arid regions where little shifts in climate can cause great droughts

- severe sand storms can reach high into the atmosphere and carry dust to other continents; some West African sandstorms reach the Caribbean and North America; diseases were carried by such sand storms at least as far as the Caribbean

DID YOU KNOW? Antarctica is strictly speaking a desert; it has as little as 2cm of annual precipitation making it one of the driest places on Earth

According to the Eden Foundation, land covers 14.9 billion hectares of Earth's surface. A UNEP (United Nations Environmental Programme) study shows that 41% of this is dryland of which 1 billion hectares are naturally hyperarid desert (7% of the total land). The rest of the dryland has either become desert or is being threatened by desertification. About 25% of world's population inhabit the drylands. While true deserts have a relatively stable environment, drylands can experience profound short-term climate changes. Extended droughts can caused devastating famines.

  - deserts are probably the climatic zone that is least understood so the definition of "desertification" appears somewhat controversial
  - some associate "deserts" with degradation of land

Figure 19.8 Rio Puerco, New Mexico a tributary to the Rio Grande. (source: USGS).

Figure 19.9 An example of overgrazing in western New South Whales, Australia. On the right of the fence, vegetation has been removed due to overgrazing while on the left, vegetation has been untouched. (modified from NASA Earth Observatory).

Figure 19.10 Goats are very undemanding. On the other hand, they can eat plants down to point beyond recovery. (photo: Robinson, 1993)

Grazing goats and cattle, collection firewood, the erosion of land around waterholes, greatly exacerbate the vanishing of vegetation, leading to sandstorms and the spread of deserts.

Human Impact

- overgrazing: particularly goats eat anything, even small poorly growing plants
- overpopulation: serious problem in semi-arid regions and deserts which could support nomadic people who adapted to desert life, but not permanent settlements
- slash-and-burn technique: in tropical to semi-desert climates; clear rainforest or bushland for annual farming;
- additional stress to environment by burning remaining crop stalks after harvest
- after slash-and-burn event, land is often quickly depleted of its nutrients, forcing farmers to move on to new locations
- land now particularly vulnerable to wind erosion; wind can blow top soil away, uproot seedlings and suffocate plants where soil later accumulates
- once top soil is removed, land can no longer be used for farming
- abandoned land also subject to massive water erosion, especially on slopes. The removal of soil results in severely furrowed badlands that cannot be used by humans
- drought problems especially severe in Sahel where little shifts in climate cause great famine

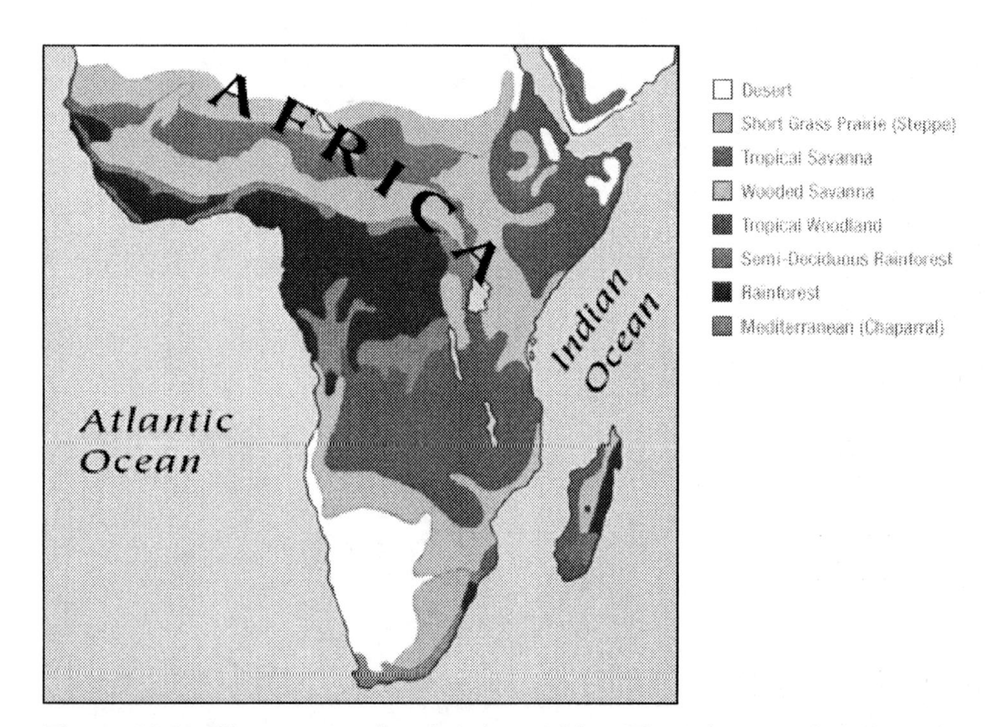

Figure 19.11 Climate zones in sub-Sahara Africa. The Sahel stretches along the Short Grass Prairie. (source: NASA Earth Observatory)

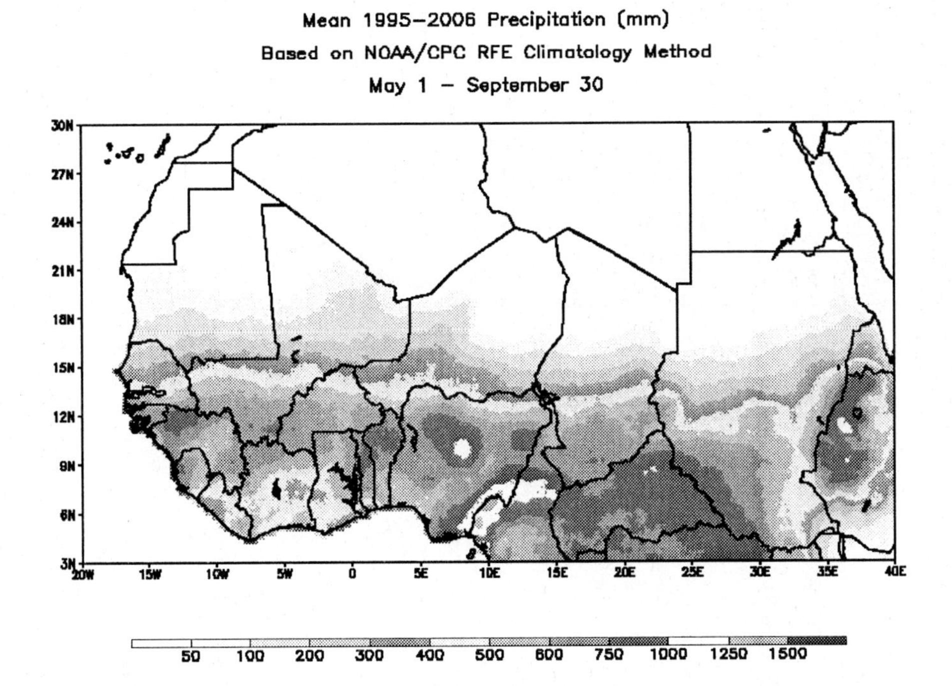

Figure 19.12 Ten-year average over dry seasonal rainfall in sub-Sahara (May through September). (source: Wikimedia/NOAA)

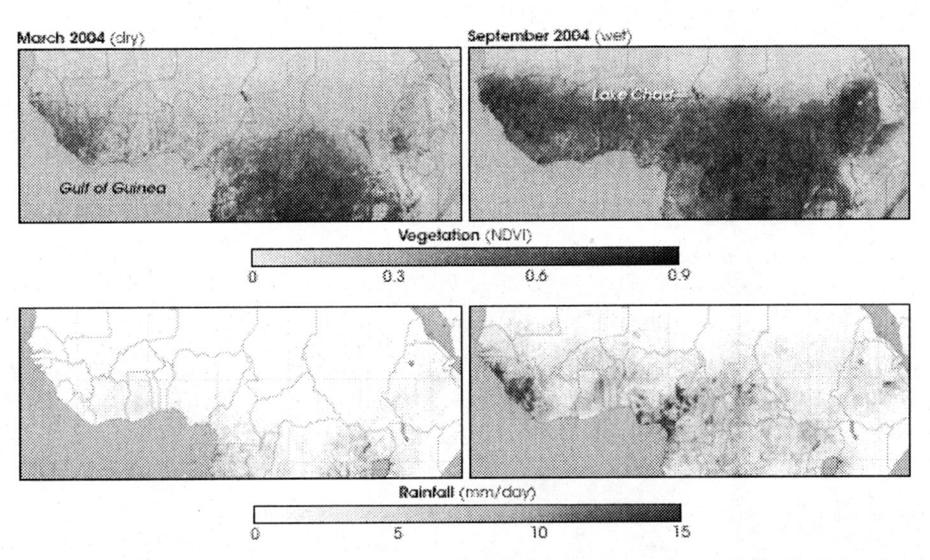

Figure 19.13 Vegetation in the Sahel follows seasonal rainfall. In March, during the dry season, rainfall and lush vegetation do not extend north of the Gulf of Guinea. September brings rain and vegetation into the Sahel as far north as the northern edge of Lake Chad. (source: NASA Earth Observatory)

Figure 19.14 A photo taken in Senegal showing the difference in vegetation between the dry (left) and wet (right) seasons. (source: NASA Earth Observatory)

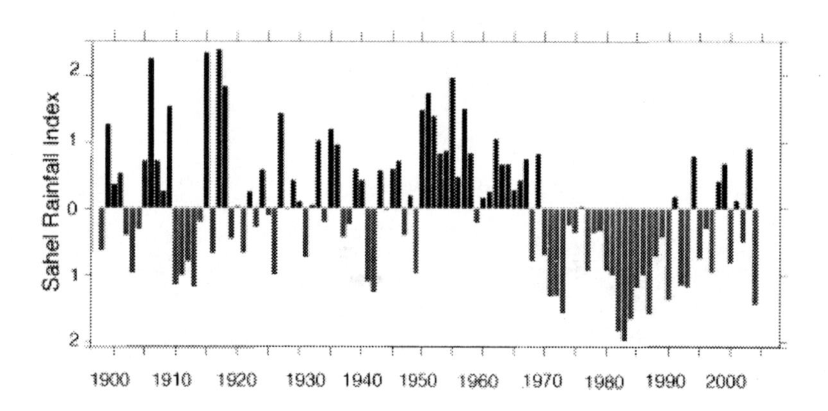

Figure 19.15 More than a century of rainfall data in the Sahel show an unusually wet period from 1950 until 1970 (positive index values), followed by extremely dry years from 1970 to 1990 (negative index values). From 1990 until 2004, rainfall returned to levels slightly below the 1898-1993 average, but year-to-year variability was high. (modified from NASA Earth Observatory).

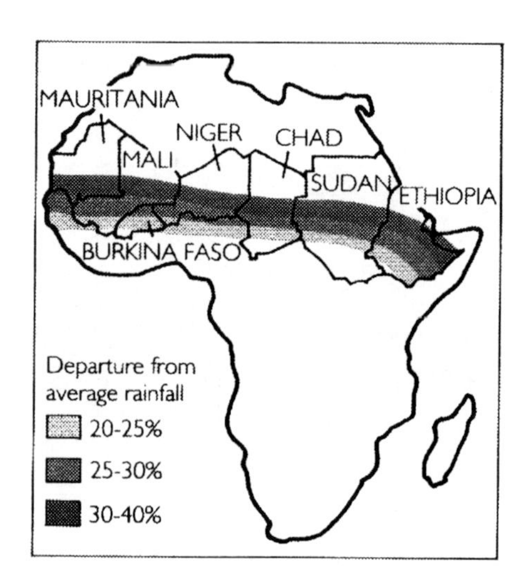

Figure 19.16 Deviation of annual rainfall from normal between 1968 and 1993. [(1)] Rainfall deficit were particularly severe in the Sahel, the region closest to the Sahara desert.

The Sahel:

- the Sahel is a belt of semi-arid dryland/savanna south of the Sahara desert. The Sahel receives 15-50mm (6-20in) of annual rainfall, primarily in the monsoon season between June and September. During summer, the northward shift of the intertropical convergence zone allows moisture rich air move in from the southwest to bring the rain (see Lecture 15). Soils in the Sahel are mostly acidic (which results in aluminum toxicity to plants), and are very low in nitrogen and phosphate (i.e. not very nutritious). Most people living in the Sahel are semi-nomadic, farming and raising cattle. This way of temporary land use (transhumance) is probably the most sustainable way of utilizing the Sahel. People move to the usually drier north during wetter years because the soil there is more nutritious, having higher quality plants for grazing. People then move south for several hundred km where plants are more abundant but of less nutritional quality. Recently, the ratio of permanent settlements and farming has increased, which stresses the environment and leads to conflicts. A long wet period in the 1950s and 1960s was followed by a serious drought in the 1970s and 1980s. Grazing and farming in the north became unsustainable which led to large-scale famine, killing a million people and afflicting more than 50 million people. Had it not been for massive humanitarian efforts, many more people would have starved to death.
- *The critical role of the green belt*: in some places, the northern part of the Sahel pastoral land is protected from the Sahara desert by a natural green belt (e.g. in Niger). Several species grow large despite low rainfall, inhibiting wind and water erosion (desertification). According to a UNEP study, the greenbelt has survived because it is closer to the desert, hence drier, than the northern Sahel pastoral land. The greenbelt is too dry for sustainable millet production (a type of grain). However, careless use could destroy this zone and expose the northern pastoral Sahel to intense desertification. In some places, the northern Sahel pastoral land appears to be degrading (see Eden website). Near the town of Tanout in Niger vegetation thinned in the last 40 years and millet harvest is only a small fraction of what it used to be 40 years ago (the Eden website quotes 1/7 of the harvest 40 years ago, on fields three times larger). Sand dunes are advancing and can reach roof level. These sand dunes are not advancing from the Sahara desert but form from regional soil erosion.
- *pollution and the last Sahel drought*: recent research suggests that the cause for the Sahel drought of the 1970s and 1980s had a human component. There is evidence that pollution in

the atmosphere caused "global dimming" and generated enough aerosols to inhibit the formation of clouds necessary for extensive rainfall. It was initially suggested that local slash-and-burn practices were responsible for the pollution but recent studies suggest that the pollution was caused by coal burning in North America.

Some Numbers

- an estimated 2,500 km^2 are being lost to desertification each year in Niger alone (the size of Luxembourg, twice the size of New York city)
- according to the Eden website, peanut production in the Tessaoua area in Central Niger rose from 4,500 tons in 1928 to 78,900 in 1970 before it declined due to lower prices and disease. Peanuts were replaced by millet (72,000 hectares in 1970 to 162,000 hectares in 1980), occupying up to 80% of the area by 1981. Intense agriculture resulted in rapid decline of stable perennial vegetation and desertification over wide areas. Species rich woodlands that harbored monkeys disappeared by 1981.

Fig. 19.17 Removal of a rainforest for agriculture in southern Mexico (Lacanja, Chiapas). (source: Wikipedia; Jami Dwyer)

Fig. 19.18 Large clear-cut areas of temperate rainforest in Valdivia, Chile. (source: Nature Conservancy)

Fig. 19.19 Betsiboka River, Madagascar estuary as seen from space. Deforestation leads to severe erosion. The soil is washed into rivers and eventually ends up in the oceans. In tropical latitudes, the huge input of sediments can bury and choke coral reefs. (source: NASA)

The 525-km-long Betsiboka River carries enormous amounts of reddish-orange silt to the sea. It is dramatic evidence of the catastrophic erosion of northwestern Madagascar. Removal of the native forest for cultivation and pastureland in the last 50 years has led to massive soil erosion of 250 metric tones per hectare (112 tons per acre) in some areas, the largest amount recorded anywhere in the world. [3]

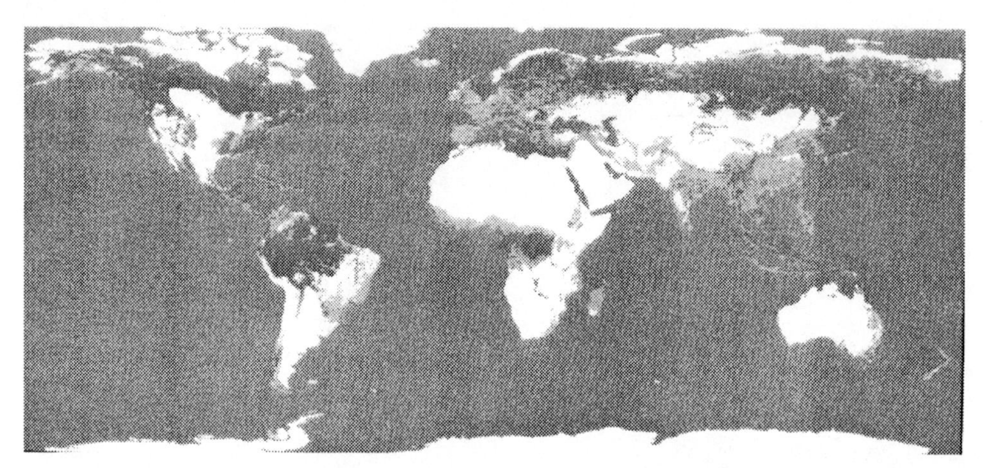

Fig. 19.20 The area of Earth covered by rain forest shrank steadily during the past century. One this map, the dark areas indicate existing forest (including high-latitude scrub forest), while the light colored areas indicate regions that were forested 8,000 years ago. Note how tropical rain forests are shrinking. (source: Marshak, 2005)

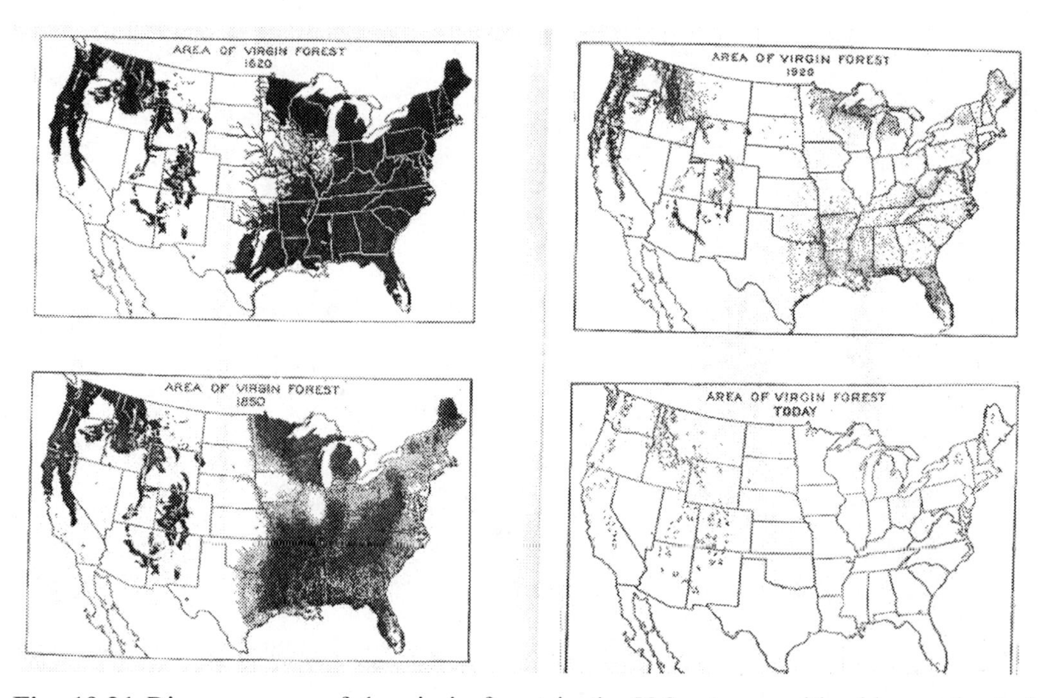

Fig. 19.21 Disappearance of the virgin forest in the U.S.. Areas with old-growth virgin forest are shown for 1620, 1850 (left), 1920 and today (right). Some of the areas have been reforested but not at the same rate as losses occur. (source: library.wixi.org)

19.3 Deforestation

There is no natural cause for complete deforestation, except for dramatic shifts in climate or the introduction of a new disease or pest.

- o Forests hold soil to the ground
- o Forests regulate the water supply to the surrounding region
- o Forests are part of the climate system
- o Forests are home to a rich variety of wildlife. Of all ecosystems, the tropical rainforest has the richest biodiversity.
- o forests are a major producer of oxygen in Earth's atmosphere
- o Logging is especially damaging to the environment. Without a strategy for replacing trees, permanent damage is inflicted.

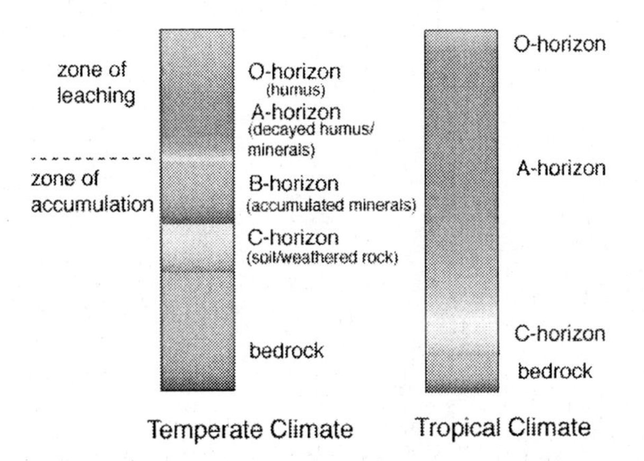

Fig. 19.22 simplified soil profiles for temperate and tropical climate zones. The soil in the tropical rain forest is very thick but has a very thin humus layer making tropical soil not very fertile.

Soil

- A typical soil profile has several distinct zones. The thickness of soil and its zones depend on the prevailing climate.

Zone of leaching and zone of accumulation: similar to the coffee grounds in a filter, nutrients are in the upper regions of the soil profile. When rain percolates through the soil, the nutrients are leached from the top and accumulate further down

- horizons: a soil profile is divided into various horizons. On the top, the O-horizon include the humus layer that contains partially decayed organic material. This is the horizon most important to plant growth. Below is the A-horizon in which the humus has decayed and mixed with minerals. The B-horizon comprises the subsoil into which the nutrients leach. The B-horizon also contains weathered rock from below.
- The richest soil with large amounts of nutrients can be found in the temperate climate zone
- In the tropical climate zone, the soil is very thick, comprised mainly of the A-horizon from which minerals are leached out. The B-horizon is missing. The humus layer is extremely thin. The reason is that in the warm humid environment, organic material decays so fast that humus cannot accumulate. The rainforest adapted by taking up nutrients through the roots faster than plants in temperate climates.

Fig. 19.23 The tropical rainforest in Borneo has recently also suffered from deforestation. (source: Nature Conservancy)

Rainforest

- cover less than 2% of Earth's total surface area, but are home to 50% of Earth's plants and animals
- many plants (e.g. orchids) are unique to rainforest and provide base for important medical remedies, incl. some that have anti-cancer properties

- rainforest can be found all over the world from as far north as Alaska and Canada (temperate rainforest) to Latin America, Asia and Africa (tropical rainforest)
- the largest temperate rainforests are found on North America's Pacific Coast
- rainforests regulate the world's temperatures and weather patterns
- rainforests are critical in maintaining the Earth's limited supply of drinking and fresh water
- 20% of the world's fresh water is found in the Amazon Basin
- rainforests provide important products, e.g. timber, coffee, cocoa
- a typical 4 square mile patch of rainforest contains as many as 1500 flowering plants, 750 species of trees, 400 species of birds and 150 species of butterflies

Human Impact

- deforestation has been going on during the last 8000 years
- between the 1960s and 1990s, 20% of the world's tropical forest was destroyed
- clear-cutting of temperate forests
- temperate forests are often replenished with monocultures that can't sustain great biodiversity
- today, only 50% of temperate rainforests remain (75 Mio acres)
- monocultures are more vulnerable to pest infestation than mixed cultures
- slash-and-burn technique in rainforest for fuel, timber or farming
- removal of forest especially problematic in tropical rainforest as nutritious humus layer is extremely thin and easy to erode and lost
- additional stress to environment by burning remaining crop stalks after harvest
- in tropical soil, only one year of agriculture can leave the soil depleted of its nutrients, forcing farmers to move on to new locations
- land now particularly vulnerable to erosion
- on slopes, abandoned land subject to massive erosion, resulting in mass movements and severely furrowed badlands that cannot be used by humans (e.g. Madagascar)
- every second, rainforest of an area the size of a football field is lost (area the size of Iowa per year)
- 57% of the originally 6 Mio square mi of tropical rainforest are now gone
- tropical deforestation results in the loss of 100 species per day

Impact on Humans

- 57% of the world's forests, incl. most tropical forests, are located in developing countries

- burning of tropical rainforest creates thick smoke that stresses the environment as well as people; particularly bad examples can be found in and around Indonesia where the smoke in far away large cities can be so bad that people have to wear masks
- destruction of forest also has an impact on indigenous people. E.g. before 1500A.D., approx. 6 Mio indigenous people lived in the Brazilian Amazon. In the early 1900s, there were less than 250,000.
- countries with the highest annual losses in natural forest: Mauritania, Niger, Nigeria, Sierra Leone, Côte d'Ivoire, Togo, Rwanda, Burundi, El Salvador, Nicaragua, Haiti
- The rate at which forest is being lost is slowing, partly because of international concern, but partly because in many places there is little forest left to be cut down [1]
- For many poor countries, hardwood is the main reliable source of income to repay international debts

DID YOU KNOW?

- The Amazon rainforest produces 20% of the world's oxygen [2]
- 70% of the plants identified by the U.S. National Cancer Institute as useful in cancer treatments are found only in rainforests [2]

Once the humus layer is removed, it takes 1000s of years for a tropical rainforest to grow back

19.4 Farming

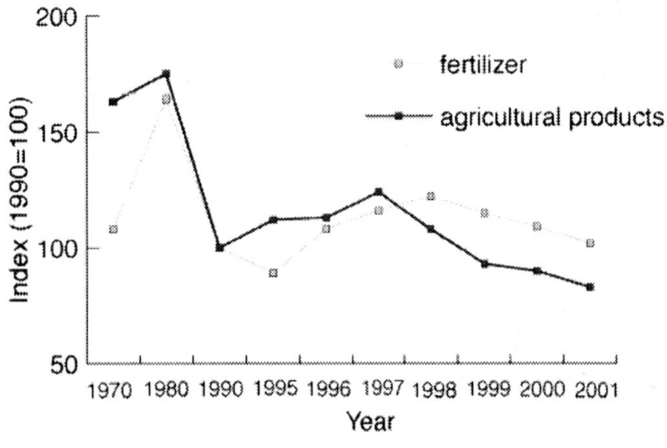

Figure 19.24 The base of an ordinary well penetrates below the water table. If groundwater is extracted faster that it can be replenished, a cone of depression forms around the well. Pumping by the big well may lower the water table sufficiently to cause the small well to become dry. (source: xx)

- o big-style farming as harmful to environment as monocultural temperate forests
- o fertilizer gets into groundwater and contaminates it: some contains so much nitrate that water is unsuitable for feeding babies
- o monocultures, herbicides and pesticides do not allow biodiversity
- o growing corn that needs massive amount of herbicides is one of worst agricultural culprits
- o farming and pest control have unexpected, far-reaching effects; e.g. some Sea Birds go to near-extinction (e.g. see California Brown Pelican, Lecture 26)

Figure 19.25 Bee on a honeycomd from a hive in Germany. (source: wikipedia)

Pumping

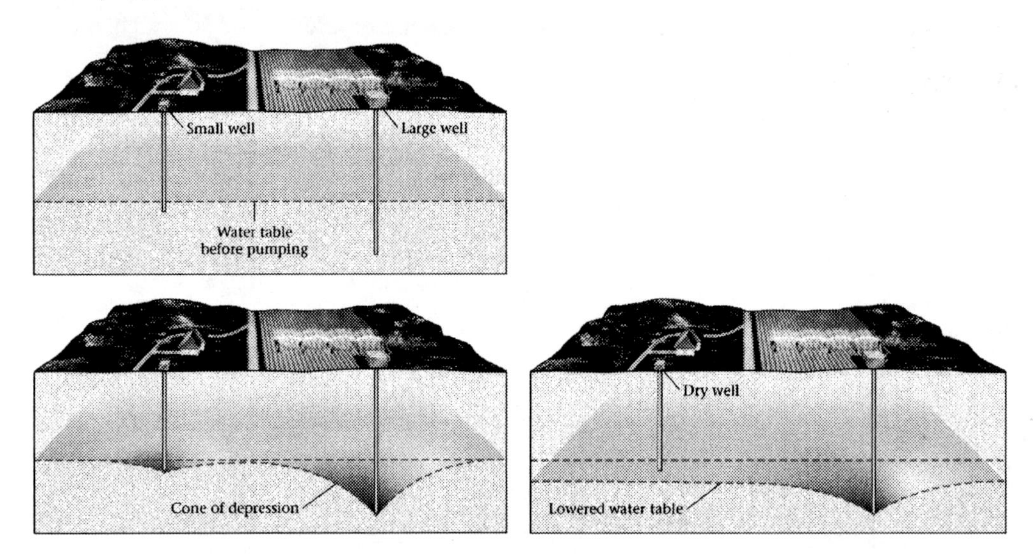

Figure 19.26 The base of an ordinary well penetrates below the water table. If groundwater is extracted faster that it can be replenished, a cone of depression forms around the well. Pumping by the big well may lower the water table sufficiently to cause the small well to become dry. (source: Marshak, 2005)

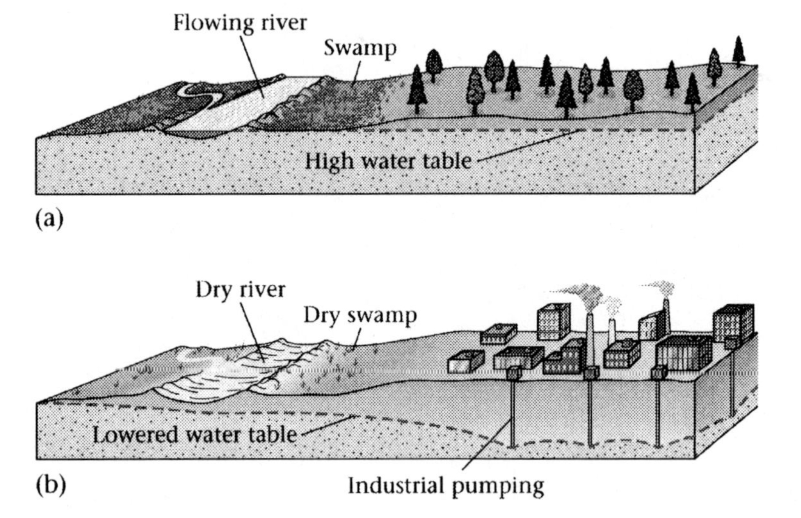

Figure 19.27 (a) Before a water table is lowered, a large swamp exists. (b) Pumping by a nearby city causes the water table to sink, so the swamp dries up. (source: Marshak, 2005)

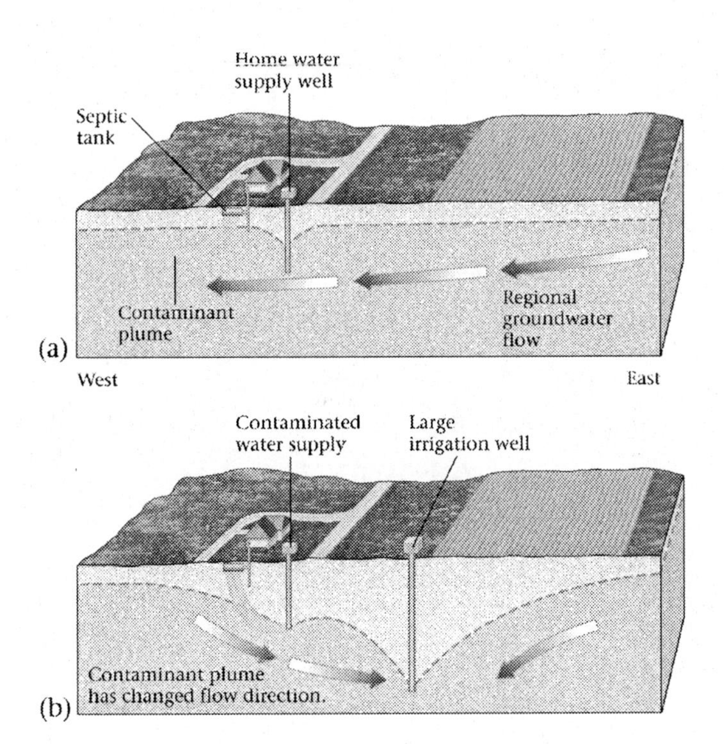

Figure 19.28 (a) Before pumping, effluent from a septic tank drifts west with the regional groundwater flow. (b) After pumping, it drifts east into the well, in response to the local slope of the water table. (source: Marshak, 2005)

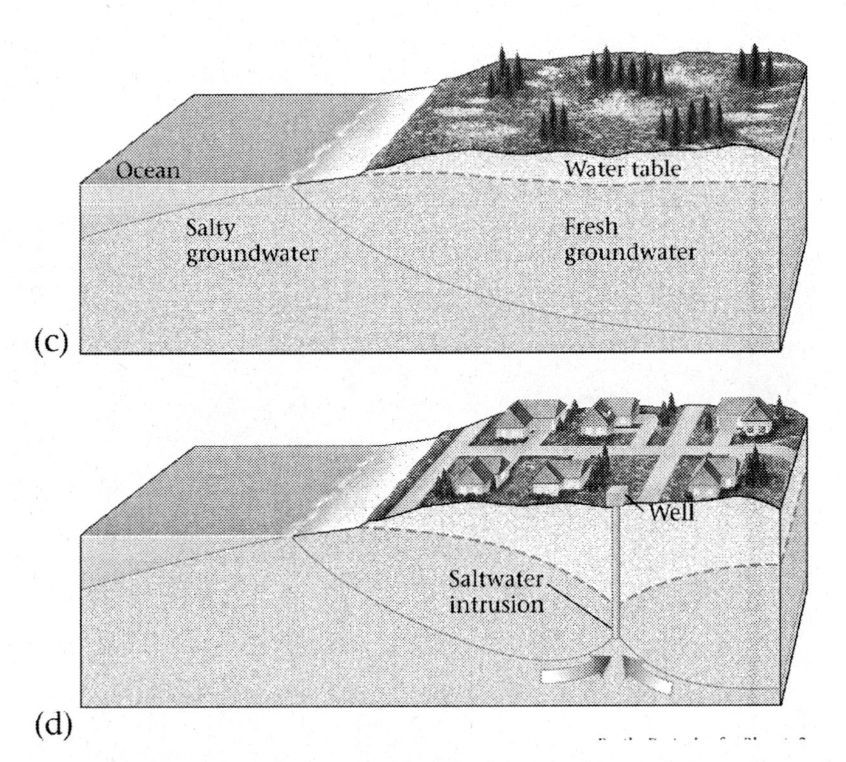

Figure 19.29 (c) Before pumping, fresh groundwater forms a large lens over salty groundwater. (d) Pumping too fast sucks saltwater from below into the well. (source: Marshak, 2005)

- much farming done in adverse climates, e.g. High Plains in U.S. that gets very little annual rainfall (see Ogallala)
- lowering of groundwater table due to overdrawing water (pumping exceeds rates of replenishment) causing shallow-reaching wells to go dry
- excess removal of groundwater leads to significant subsidence (see San Joaquin)
- agriculture in desert climates subject to salinization; extremely high evaporation rates leave salt behind; some land so salty that nothing grows anymore
- excess diversion of water to irrigation channels leave many streams devoid and shrink inland lakes; the most extreme example is the Aral Sea
- ignorant farming techniques lead to soil erosion (see Dust Bowl)
- poorer countries typically depend on farming for export
- compared to 1990, fertilizer has become more expensive while prices of agricultural products have declined

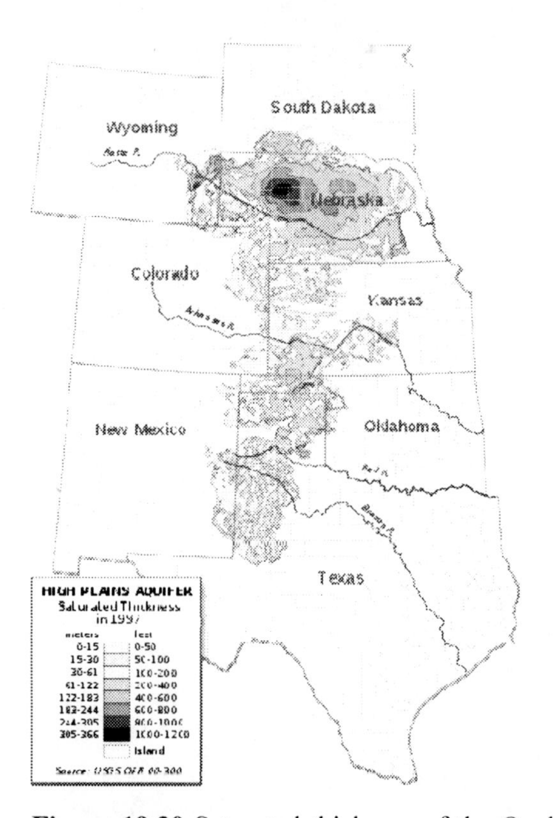

Figure 19.30 Saturated thickness of the Ogallala Aquifer in 1997 after several decades of intensive withdrawal: The breadth and depth of the aquifer generally decrease from north to south. (source: wikipedia)

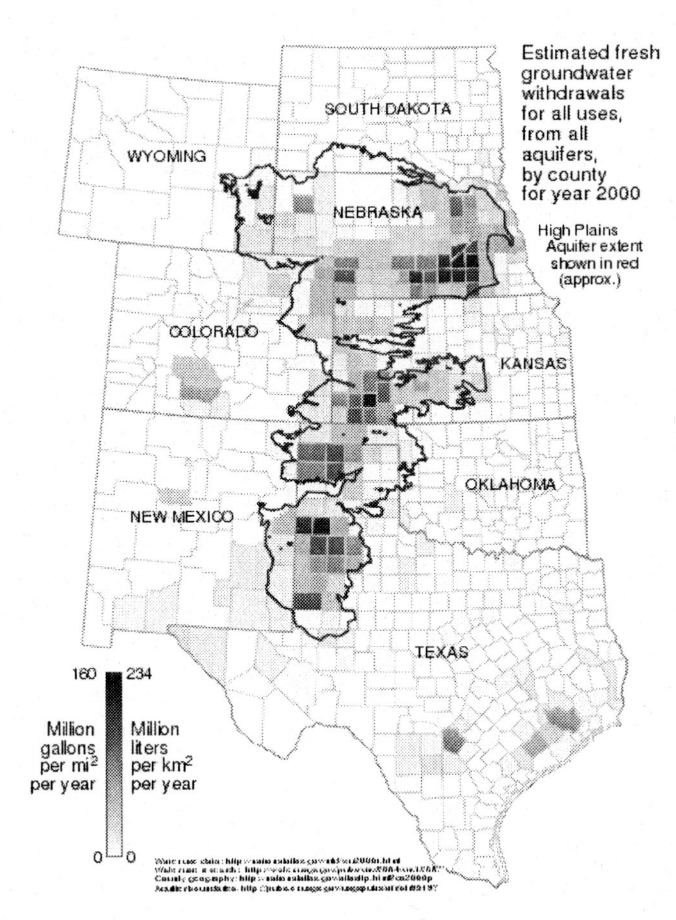

Figure 19.31 Groundwater withdrawal rates (fresh water, all sources) by county in 2000. (source: Wikipedia; National Atlas)

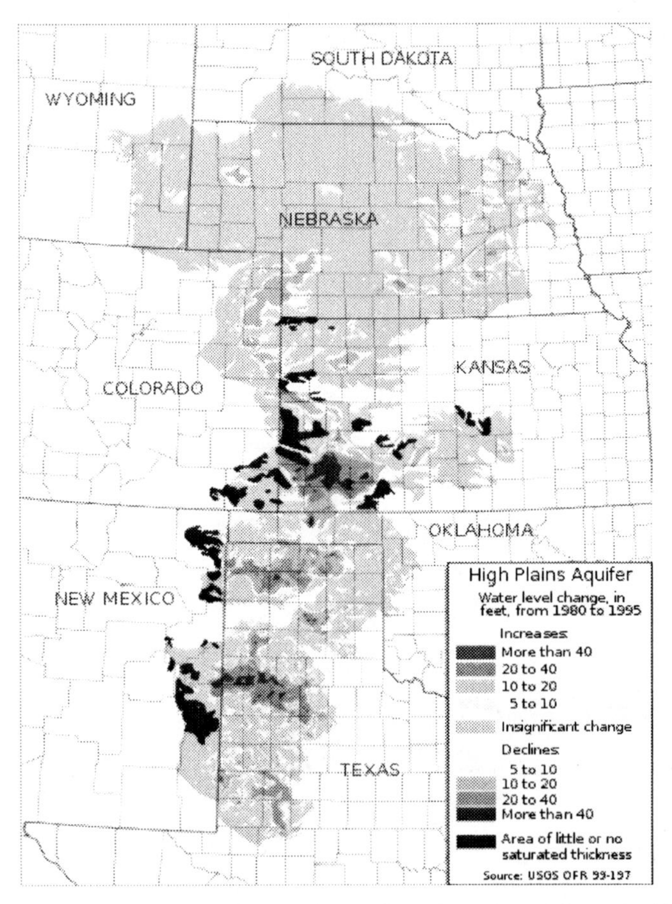

Figure 19.32 Regions where the water level has declined in the period 1980-1995 are shown in yellow and red; regions where it has increased are shown in shades of blue. (source: Wikipedia; USGS)

Case Study Ogallala Aquifer, U.S.:

Ogallala Formation (largest aquifer in U.S.) underlies about 450,000km^2 (175,000mi^2) of the High Plains (one of the most agricultural regions in U.S.). The connection between Rockies and aquifer (naturally) severed so all replenishment must come from meager rainfalls. Aquifer first used for agriculture in 1800s. Nowadays, 170,000 wells are being used to irrigate 65,000 km^2 (16Mio acres) of land which far exceeds rates of replenishment. Beginning in 1980s irrigated acreage has declined due to higher costs for pumping water out of greater depth (groundwater level dropped).

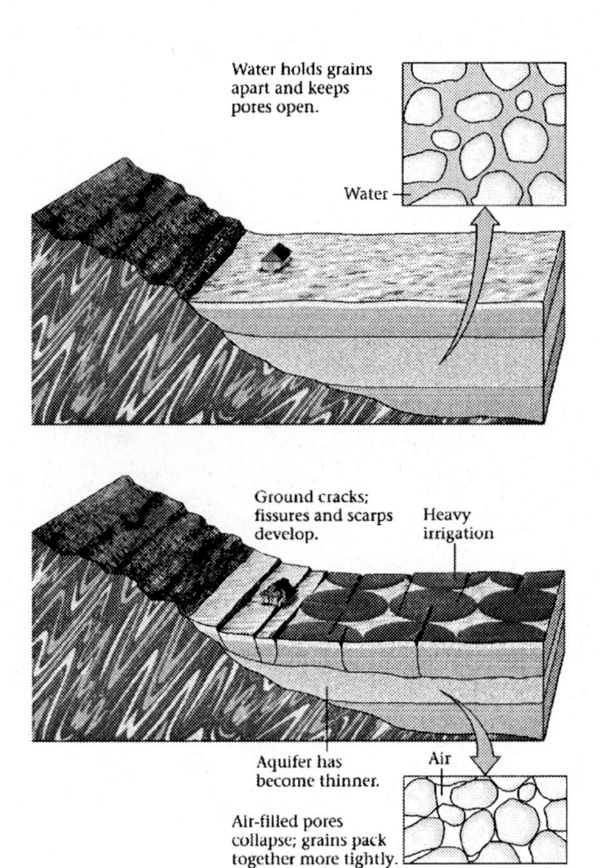

Figure 19.33 (top) Before pumping, the pore pressure in the aquifer maintains an equilibrium with the gravity of the layer above. (bottom) Pumping faster than the ground water is replenished increases the risk that pore space collapses. This causes the land to subside, causing fissures, scarps and cracked houses on the surface. (source: Marshak, 2005)

Figure 19.34 The photograph illustrates subsidence in the San Joaquin Valley, California. In the photo, USGS scientist, Joe Poland shows subsidence between 1925 and 1977 due to fluid withdrawal and soil consolidation. (source: USGS)

Case Study San Joaquin, CA:

Southern 2/3 of California's Central Valley; thick fill of sediments (870m/0.5mi on average); climate arid to semiarid (12-35cm/5-14in per year); strong agricultural economy requires extensive irrigation; 50% of this was met by groundwater; by 1970s groundwater levels had declined by up to 120m (400ft) and resulting ground subsidence exceeded 8.5m (29ft). One half of valley was affected (water level raised for short time only to subside faster after a drought in 1976/77 after sediment compacted).

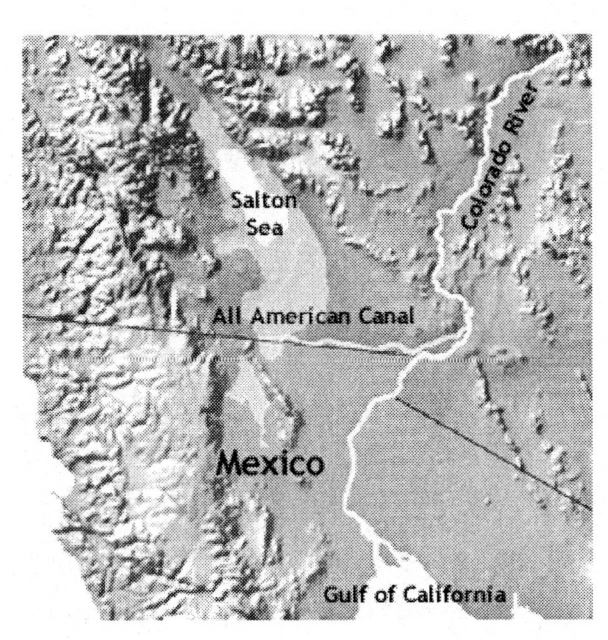

Figure 19.35 The All-American Canal transports much-needed water from the Colorado River to irrigate farmlands in Imperial Valley. (source: Wikimedia/USGS/National Atlas)

Figure 19.36 The All-American canal near Calexico. After the magnitude 7.2 2010 Easter earthquake in the area, FEMA inspected the canal for damage and potential threats to the community. (source: Wikimedia/FEMA)

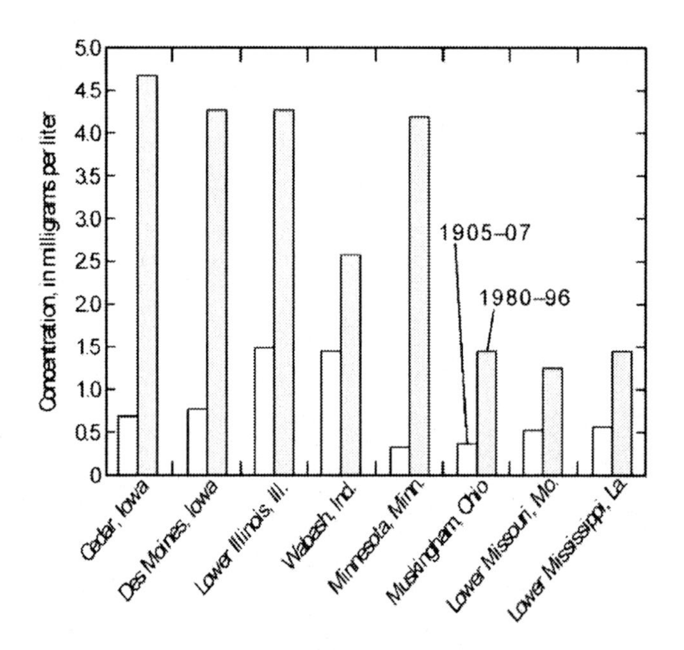

Figure 19.37. Average annual nitrate concentrations in selected rivers during 1905-07 and 1980-96. (source: USGS)

Case Study All American Canal, CA:

Hunting, Farming and Extinctions:

By introducing pesticides, fungicides and herbicides, farming and other practices (such as vector control) can lead to deliberate or unintended extinctions. The California Condor was extinct in the wild due to hunting but also unintentional lead poisoning. The California Brown Pelican was driven to near extinction by the use of DDT. See appendix for details.

Human Impact on Biodiversity

Farming, mining, industry and urbanization (see also Lectures 23, 25 and 26) have an impact on the natural flora and fauna. Logging is especially damaging. According to the IUCN (International Union for the Conservation of Nature and Natural Resources), large fractions of species are now threatened.

Threatened Species (as of 2000)[1]

Kingdom/Class/Phylum/Kingdom	Fraction of Threatened Species
mammals	24%
birds	12%
reptiles	25%

amphibians	21%
fishes	30%
invertebrates	29%
plants	49%

DID YOU KNOW?

- 1.7 Mio species have been identified - less than 10% of the likely total

Recent Extinctions (see also Lecture 21)

If current rates of extinctions will continue for any other 200 years, experts have estimated that never in Earth's history have so many species disappeared in such a short time as now. Some of this may be natural (especially the extinctions of many animals during/after the last ice age) but most of it now is due to human action.

- it is now accepted that overfishing and hunting depletes the oceans; in some areas, fishing is done with dynamite which does not distinguish between the desirable individuals (big edible fish) and the undesirable ones (inedible but perhaps in the food chain of the desirable fish and young edible fish); this type of fishing though still practiced is basically like cutting off the branch of a tree the fisherman is sitting on (why is this done? beats me!); the number of fish has dramatically decreased, especially the deep sea fish (have you looked at fish prices in the grocery store lately?); the size of animals caught along California's coast has dramatically decreased and some have disappeared within the last 10 years (e.g. abalone) (see Lecture 26)
- though the case is not (yet) proven, amphibians are declining at an alarming rate; some mutations are thought to be due to the thinning of the ozone layer
- some hatcheries (e.g. salmon) jeopardize and push out the native population
- perhaps surprisingly, the most severely hit class in the animal kingdom is probably that of birds
- birds are hunted and eaten (e.g. ducks everywhere and; sadly enough still, songbirds in Europe)
- birds are hunted for fashion statements (e.g. Dodo; Peacock?)
- birds, especially flightless ones are endangered by introduced predators (e.g. Dodo, Kauai O'o)
- bird are subject to introduced diseases (e.g. Kauai O'o)
- most birds probably go extinct due to loss of habitat (e.g. California Gnatcatcher, Kauai O'o)

- some birds go extinct due to human competition for food (anywhere in the food chain, not just at the end) (e.g. Canary Oystercatcher)
- some birds go extinct only short while after we discovered them (e.g. Aldabra Warbler)
- some birds go extinct due to poisoning (e.g. California Condor, California Brown Pelican)
- **Example Dodo**: Portuguese name of a turkey-size (22kg) flightless birds on Mauritius. It was discovered in 1598, only to become extinct in 1681, less than 100 years later. Believed to be related to pigeons. Though some were sent to museums no complete specimen now exists. Inhabited forests, laid one egg in large pile of grass. Vulnerable to imported hogs which ate the eggs and young.
- **Example Kauai O'o**: extinct in 1987 due to habitat loss, predation by introduced black rat, disease by exotic mosquito
- **Example Canary Islands Oystercatcher**: extinct in 1981 due to loss of mollusk prey by human overharvest, predation by introduced cats and rats, disturbance by people in coastal habitat
- **Example California Gnatcatcher**: not yet extinct but disappearing at an alarming rate right in our backyard due to habitat loss
- **Example Aldabra Warbler**: discovered on island in Indian Ocean in 1967; extinct by 1983 due to rat predation and habitat degradation by introduced goats
- **California Condor**: condors are vultures and feed on carcasses, so they don't kill animals. It is often reported that the bird got extinct in the wild in the 1980s after the use of DDT made egg shells so thin that chicken could no longer hatch alive (see also Brown Pelican, Lecture 26). The Bald Eagle, the national bird of the U.S., also was seriously affected by DDT. However, DDT is not the main reason for the extinction of the California Condor. Its decline started earlier in the century due to shooting, pesticide poisoning and habitat disappearance (150 individuals left in 1939; 50-60 by 1967). Less than 30 individuals survived in the wild by 1980. Desperate breeding programs started in zoos (e.g. San Diego Zoo) with 27 surviving individuals (small gene pool!). The condor seems on its road to success as the number of living condors has increased to 219 by 2003. A re-introduction is extremely difficult but attempts started in 1993 in the Los Padres National Forest and are being done in Baja (2002) and the Grand Canyon (1996). The efforts often controversial as local people object. The year 2002 has seen the first chicken hatch in the wild since 1984 but quite a few birds did not survive their second year in freedom. Some were shot by ignorant individuals (either out of pure hunting instinct or they asserted that they kill sheep --- Condors are vultures and do not kill animals!) and some flew into powerlines. Some carcasses contained significant amounts

of (the poisoning) lead, some birds couldn't find food. As of July 2006, there are 289 California condors and 138 of these live in the wild (want to know more? go to Condor Ridge at the San Diego Zoo.

References and Recommended Reading

1. Robinson, A., 1993. Earthshock, 1993, Thames and Hudson, London, 300 pp.
2. Marshak, S.,2005. Earth: Portrait of a Planet. 2nd edition. Norton & Company, New York, pp. 748.
    - "State of the World 2003" by Chris Bright et al., 2003, The Worldwatch Institute, Norton & Company, ISBN: 0-393-32386-2
    - Griffin, D.W., Kellogg, C.A., Garrison, V.H. and Shinn, E.A., May 2002. "The global transport of dust: an intercontinental river of dust, microorganisms and toxic chemicals flows through the Earth's atmosphere. American Scientist, volume 90, pages 228 -235.
    - (1)"The Penguin State of the World Atlas" by Dan Smith, 2003. Penguin Books, ISBN: 0-14.200318-2
    - (2)Nature Conservancy web site on rainforests
    - IUCN web site (International Union for the Conservation of Nature and Natural Resources)
    - (3) wikipedia.org on Betsiboka River

John Steinbeck's “The Grapes of Wrath” and “Of Mice and Men” describes the lives of the 1930s' Dust Bowl people.

Chapter 20: Anthropogenic Changes: Resources (draft)

20.1 Energy Resources – Overview

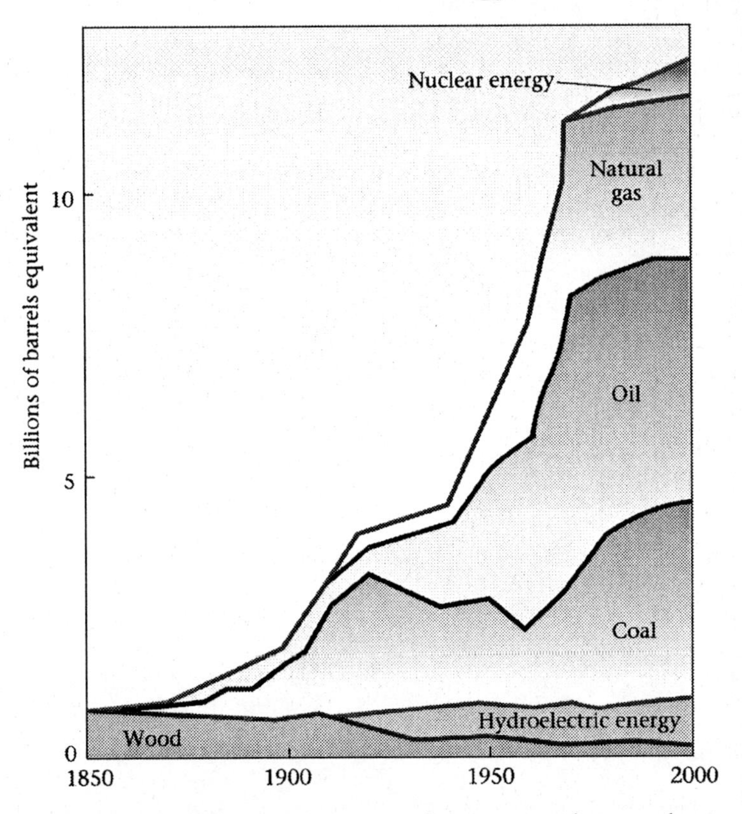

Figure 20.1a The steep increase of energy needs over the past 150 years, and how different energy resources have been used to fill those needs. Oil and natural gas together now account for more than half the world's energy usage. (source: ME)

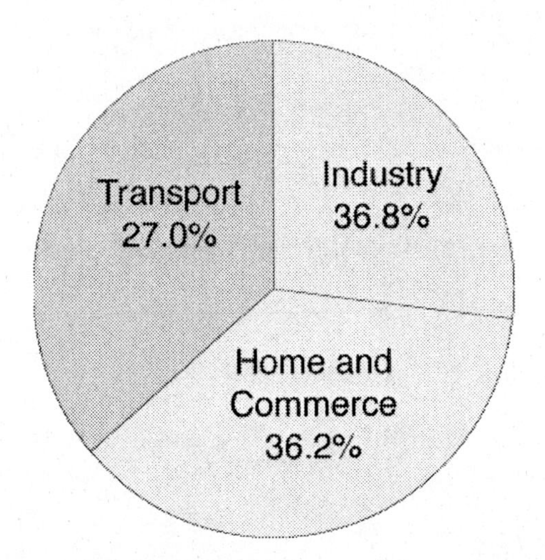

Figure 20.1b Breakdown of the consumed energy into the three main sectors. (source: ME)

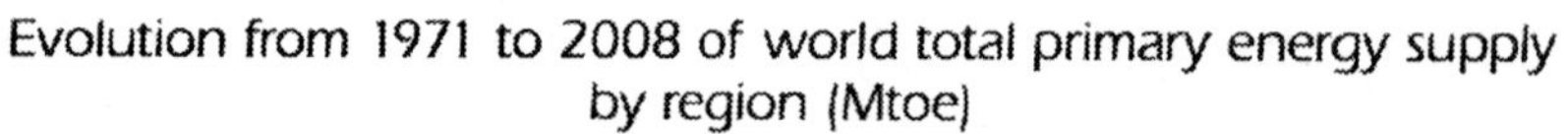

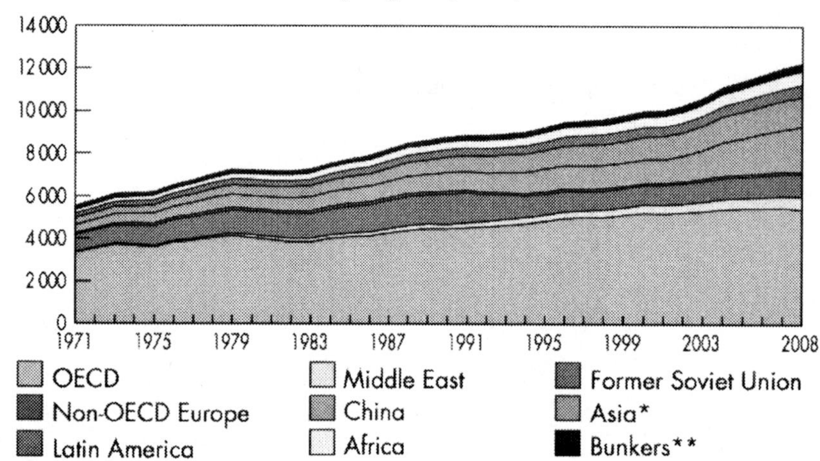

Figure 20.2 World suppliers of energy. * Asia excludes China. ** Includes international aviation and international marine bunkers. [13] (source: IEA)

The most widely used energy resources are fossil fuels (oil, gas, coal), nuclear power (nuclear fission) and moving water. The first two are nonrenewable energy resources, while moving water, wind and Earth's internal heat are renewable energy resource.

- **Fossil fuels:** oil and coal are made of organisms that lived a long time ago. Burning fossil fuels has the same effects as burning plants.
- **more than 85% of energy comes from fossil fuel, 63% from oil and gas**
- **non-renewable resources**: resources that take long to be replaced relative to the human life span
- download key world energy statistics published by the International Energy Agency
- petroleum is raw material for many chemical products, including solvents, fertilizers, pesticides and plastics
- 84% (37 out of 42 gallons in a typical barrel) of petroleum is processed as fuel, incl. gasoline, diesel and kerosene
- at the current rate of consumption, oil reserves will be depleted within a few decades

Consumption of Energy in the U.S., 1997

Energy Resource	**Fraction 1997**	**Fraction 2002**
Petroleum	41%	41%
Natural Gas	24%	22%

Coal	22%	24%
Nuclear	7%	9%
Hydroelectric	5%	3%
others	1%	1%

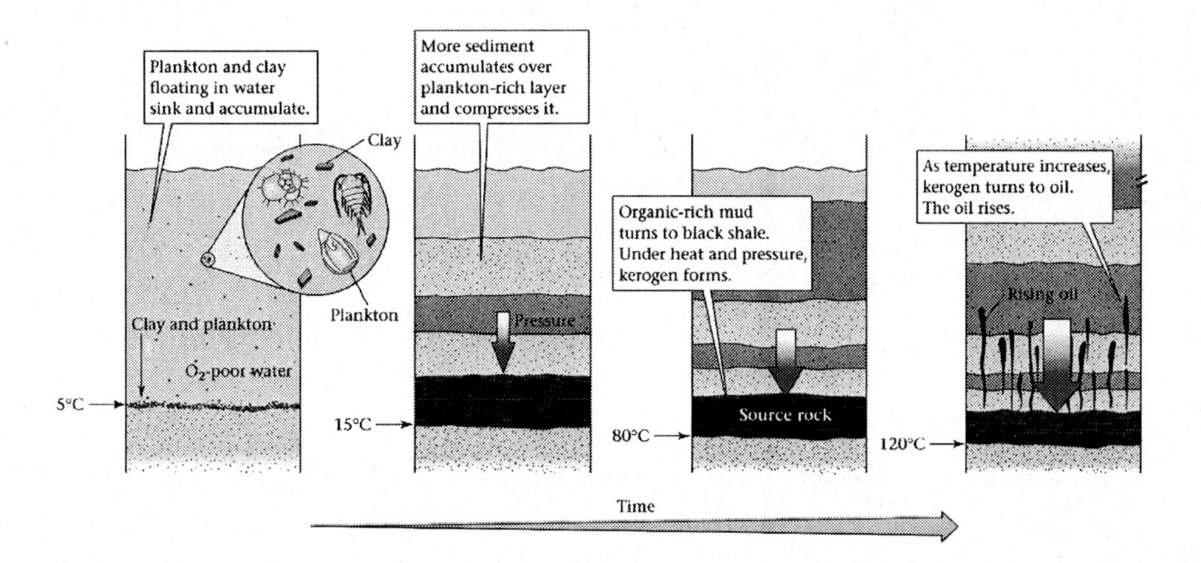

Figure 20.3 Plankton, algae, and clay settle out of water and become progressively buried and compacted, gradually being transformed into black organic shale. When heated for a long time, the organic matter in black shale is transformed into oil shale, which contains kerogen. Eventually, the kerogen transforms into oil and gas. The oil may then start to seep upward out of the shale. The large arrow indicates pressure, which increases as more sediment accumulates above. (source: ME)

20.2 Oil and Natural Gas

- industrialized societies today rely primarily on oil and natural gas for their energy needs
- Oil and gas are hydrocarbons, ring or chainlike molecules of C and H, and are organic chemicals
- Hydrocarbons have varying viscosity and the volatility depends on the molecule size
- form primarily from dead algae and plankton
- require a long (geologic times) process to form, involving burial and increase in T and P

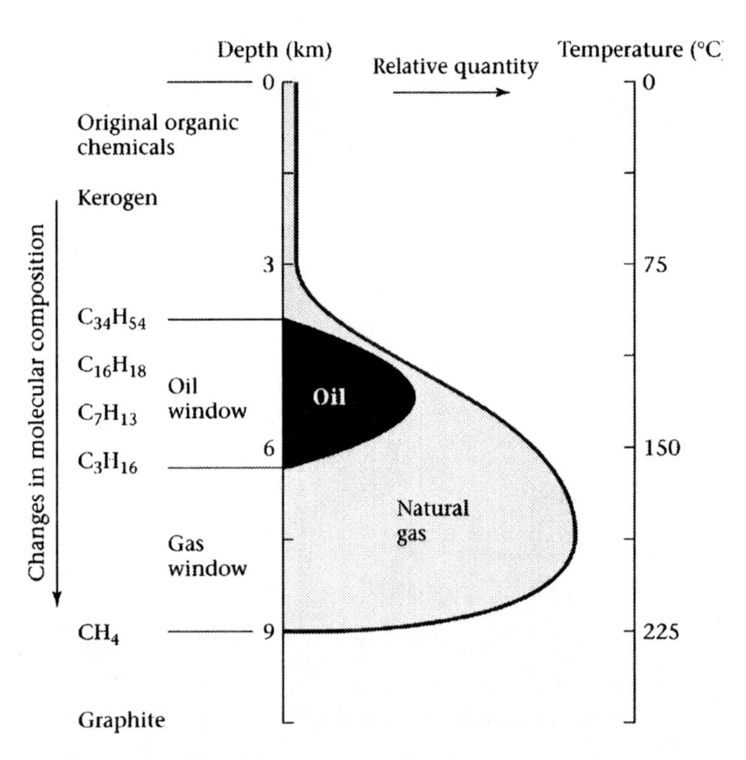

Figure 20.4 The oil window is the range of temperature conditions (i.e., depth) at which hydrocarbons form. In regions with a geothermal gradient of 25°C/km, oil occurs only at depths less that about 6.5 km. In regions with a shallow gradient (<25°C/km), oil may be found to greater depths. Natural gas can generally be found at greater depths than oil. The length of hydrocarbon chains decreases with increasing depth, because longer chains break to form smaller ones as temperature increases. (source: ME)

- **oil window**: oil and natural gas exist only under certain T and P conditions. At T>160°C, any remaining oil breaks down to form gas and at T>250° the remaining organic matter transforms into graphite. Under normal conditions, oil exists at depths only down to 6.5km.

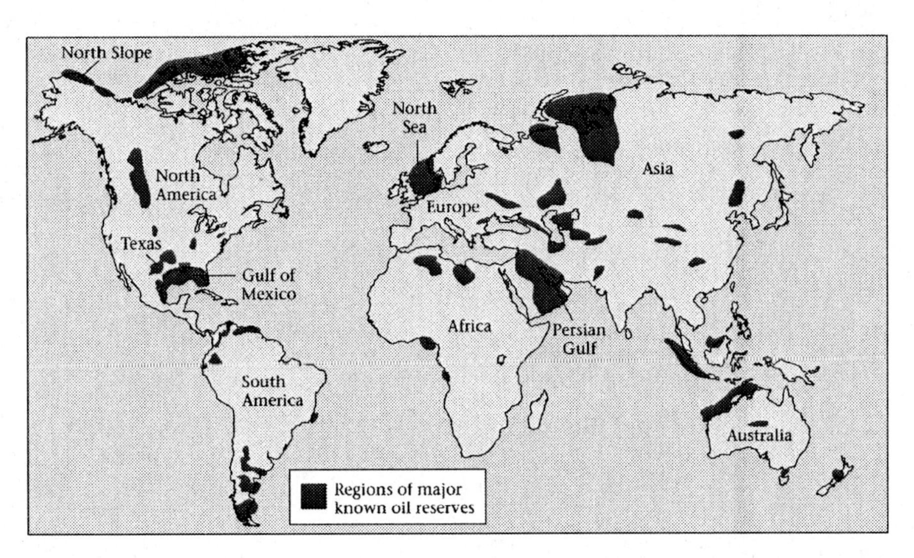

Figure 20.5a The distribution of oil reserves around the world. The largest fields are located in the region surrounding the Persian Gulf. (source: ME)

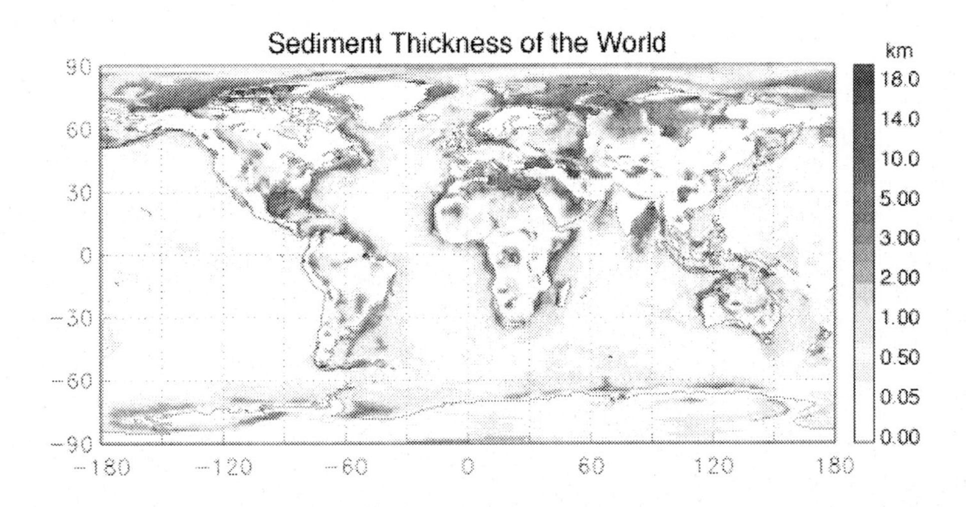

Figure 20.5b Since oil originates in marine sediments, rich oil reserves today are tied to large and deep sedimentary basins, such as the Gulf of Mexico, Canada's North Slope, the Caspian Sea and the Persian Gulf. (source: Gabi Laske/ http://igppweb.ucsd.edu/~gabi/sediment.html)

- **oil reserves**: oil can be found in sedimentary deposits. Large oil fields can be found in the Canadian Arctic Ocean, Gulf of Mexico, North Sea, Kara Sea/Northern Russia and the Persian Gulf (largest oil fields)
- Huge natural gas reserves can be found in Siberia (1/3 of the world's reserves)

- Two oil reserves that are very important U.S. domestic resources are the Gulf of Mexico and off-shore Alaska. The latter currently provides up to 20% of U.S. domestic oil and there are plans to expand oil production into the Arctic National Wildlife Refuge (see more).
- With 7 Mio barrels per day (1 barrel = 42 Gallons), the U.S. is the largest consumer worldwide, using about 25% of the produced oil (and contributing 25% of the world's CO_2 production)
- In the 1970ies, when the U.S. oil production passed its peak

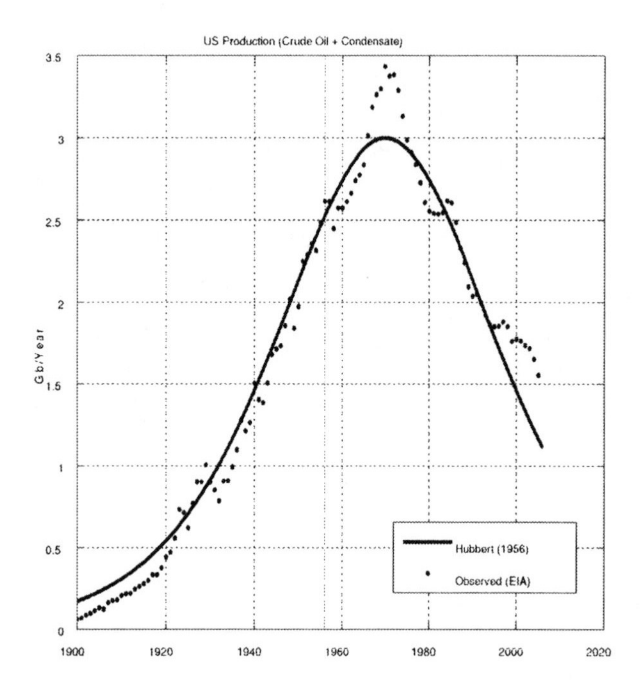

Figure 20.6 Hubbert's peak for U.S. oil production. (source: wikimedia)

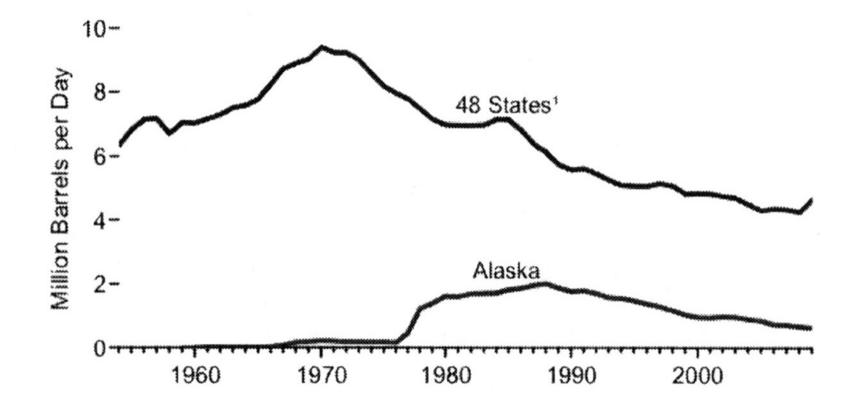

Figure 20.7 Crude oil production peaked in the 48 States at 9.4 million barrels/day in 1970. As production fell in the 48 States, Alaska's production came online and helped supply U.S. needs. Alaskan production peaked at 2.0 million barrels/day in 1988; in 2009, Alaska's production stood at 32% of its peak level, or 0.6 barrels/day. [1]U.S. excluding Alaska and Hawaii. (source: EIA/DOE [14])

Petroleum production on Alaska's north slope has caught new media attention in recent years as claims are circulated that Alaska's oil is exported to Japan rather than for domestic use. At least up to 2000, some of Alaska's oil was indeed sold to Korea, Japan and China, but it was never more than 7% of Alaska's crude oil[15].

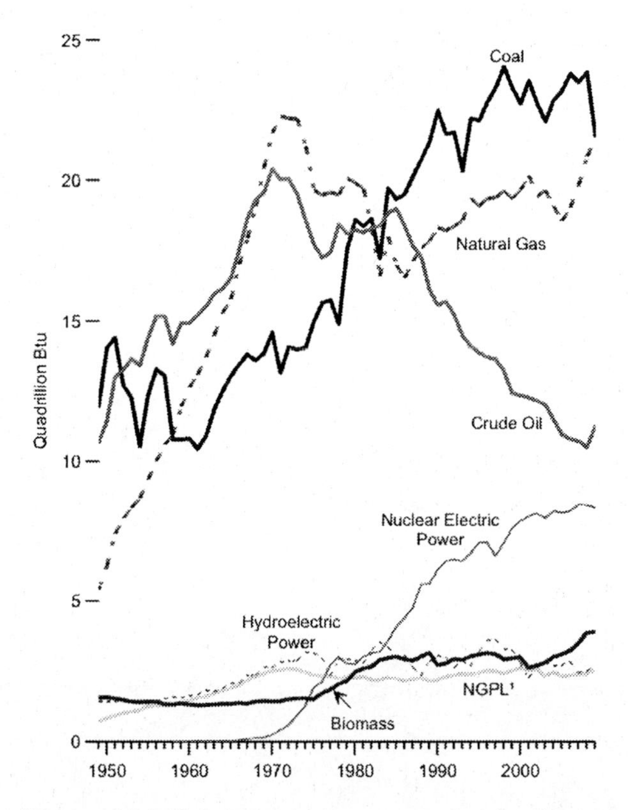

Figure 20.8 U.S. primary energy production by major source. Most energy produced in the U.S. came from fossil fuels. Coal, the leading source in the mid-20th century was surpassed in the 1950s by oil and natural gas but by the mid-1980s, coal again became the leading energy source and oil production declined sharply, while energy from nuclear fuel increased. [1]Natural gas plant liquids (source: EIA/DOE [14])

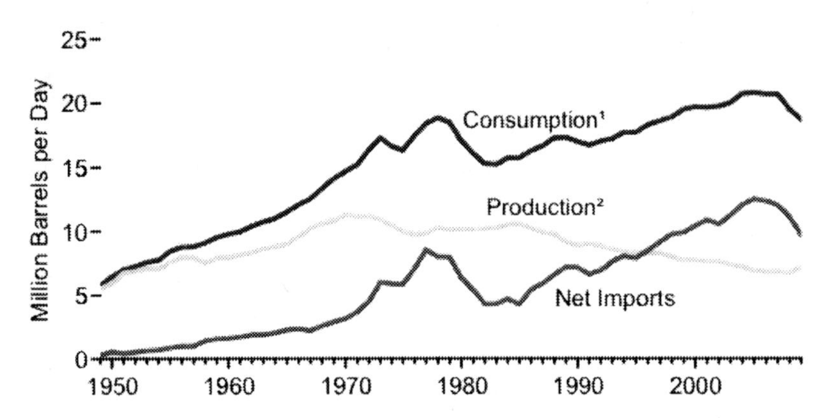

Figure 20.9 U.S. oil and natural gas production and consumption. Oil and gas production peaked at 11.3 million barrels/day in 1970. Net imports stood at 3.2 million barrels/day. In 2009, production was 7.2 million barrels/day and net imports were 9.7 million barrels/day. The steep decline in oil production is somewhat muted by a steady production of natural gas plant liquids (NGPL). Before the recent decline in consumption, the U.S. had to import more than 2/3 of the oil it consumed. [1]Petroleum products supplied is used as an approximation for consumption. [2]Crude oil and natural gas plant liquids production. (source: EIA/DOE [(14)])

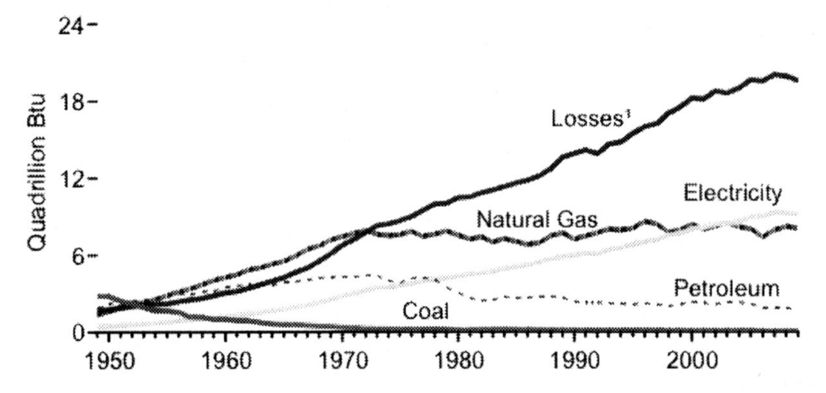

Figure 20.10 Residential and Commercial total energy consumption by major sources. The two big other sector, industrial and transportation, consume 60% of the total energy produced and are not considered here. [1]Energy lost during generation, transmission and distribution of electricity. (source: EIA/DOE [(14)])

The generation of electricity costs energy from resources, i.e. the efficiency of electric energy is always far less than 100%. The generation of electricity has increased dramatically since the 1950s and so have the losses. In 2009, 40% of the energy produced for residential and commercial energy was lost during electricity generation, transmission or distribution.

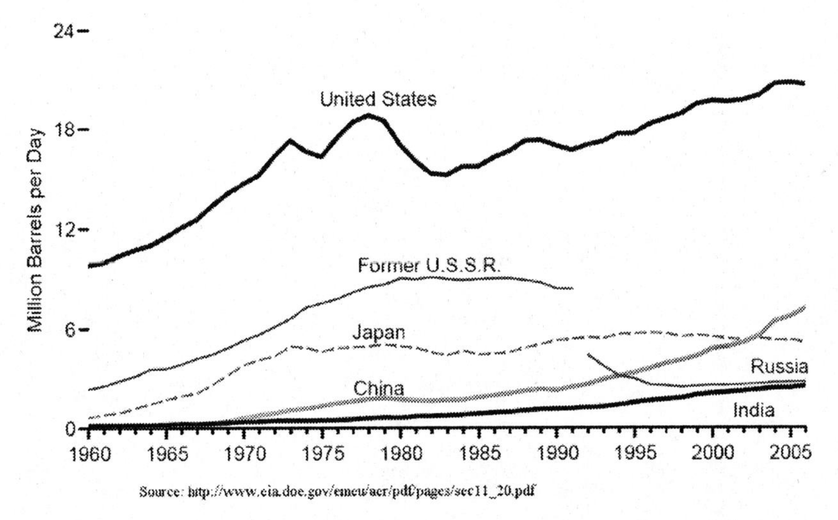

Figure 20.11 Oil consumption in selected countries from 1960 to 2006. (source: Wikipedia/EIA)

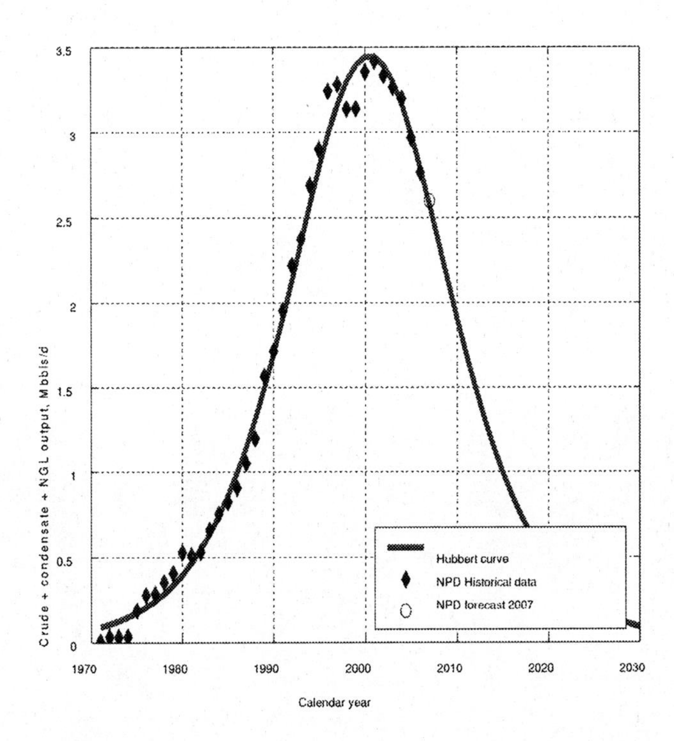

Figure 20.12 Hubbert's peak for Norwegian oil production. (source: wikimedia)

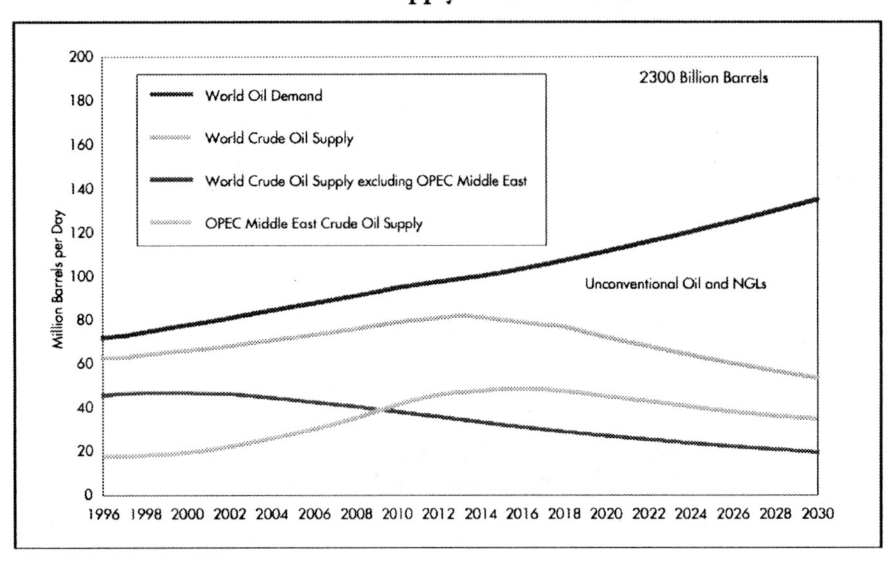

Figure 20.13 The World's growing demand for oil outpaces World oil production which is predicted to peak in the early 21st century. (source: IEA[12])

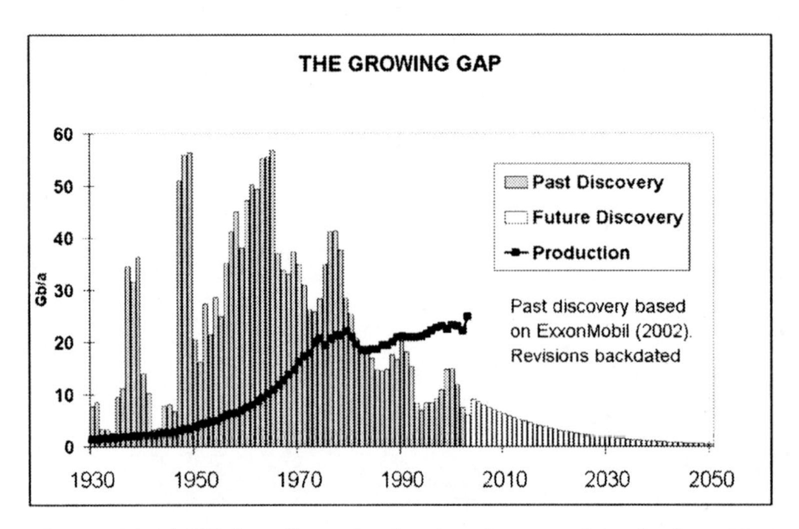

Figure 20.14 While oil production has increased in the last 60 years, the discovery of new reserves peaked in the 1960s and has been declining steadily[11]. (source: wikipedia)

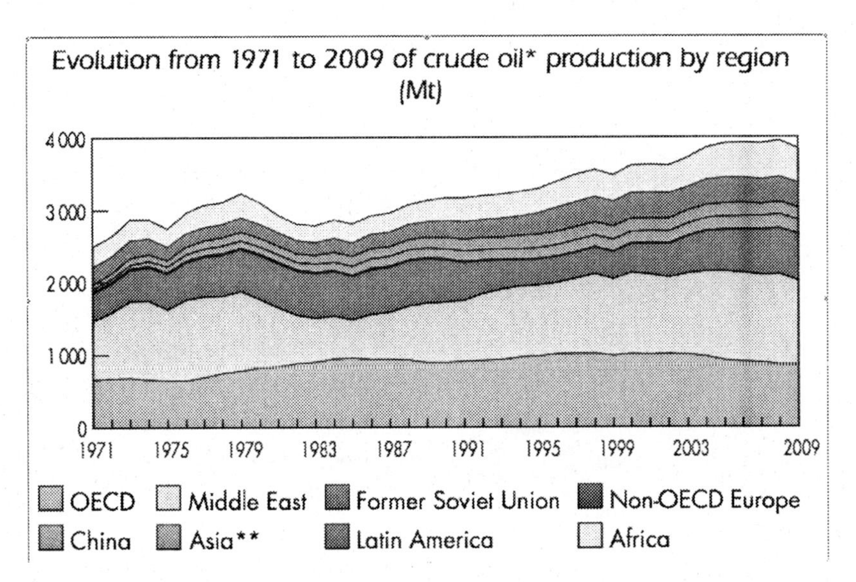

Figure 20.15 Oil production by region. *Includes crude oil, NGL, refinery feedstocks, additives and other hydrocarbons. **Asia excludes China[13]. (source: IEA)

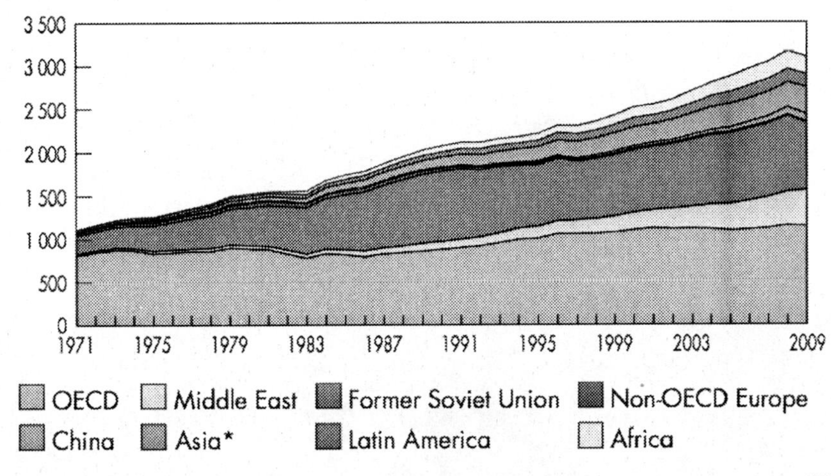

Figure 20.16 Natural gas production by region. *Asia excludes China[13]. (source: IEA)

… in WW 2005, oil production is in decline in 33 of the 48 largest oil-producing countries. Peak oil production has not been reached in the following nations but is expected to occur in the following years[10]: (11)

Country	Year
Algeria	2012
Azerbaijan	2013

India	2015
Iraq	2036
Kazakhstan	2020
Kuwait	2033
Saudi Arabia	2027
Qatar	2019

Hubbert's Peak, the U.S. lost its position as largest producer

- U.S. oil reserves now account for only 4% of the world's reserves
- the U.S. must now import more than 50% of its used oil.
- The world consumption now exceeds the rate of discovery of oil (by factor 3)
- at the current rate of consumption, know oil resources will be depleted within a few decades
- predictions place Hubbert's Peak of world oil production around 2005
- natural gas is more abundant than oil
- Gas burns more cleanly than oil. Burning gas produces only CO_2 (the least amount of C of all fossil fuels) and water, while burning oil produces more complex organic pollutants.
- Gas transportation requires expensive high-pressure pipelines and containers making gas less attractive as energy source than oil

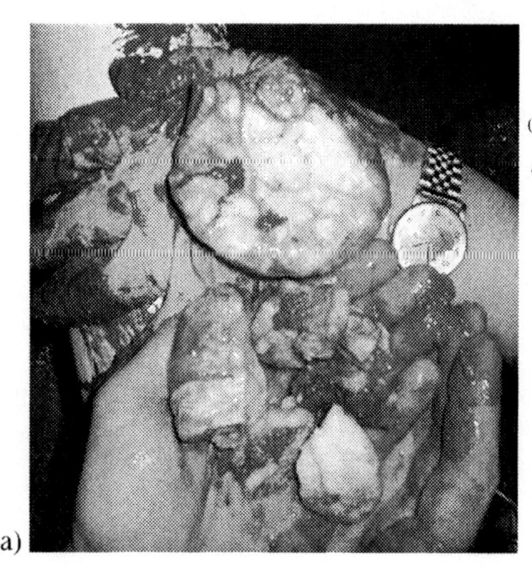
(a)

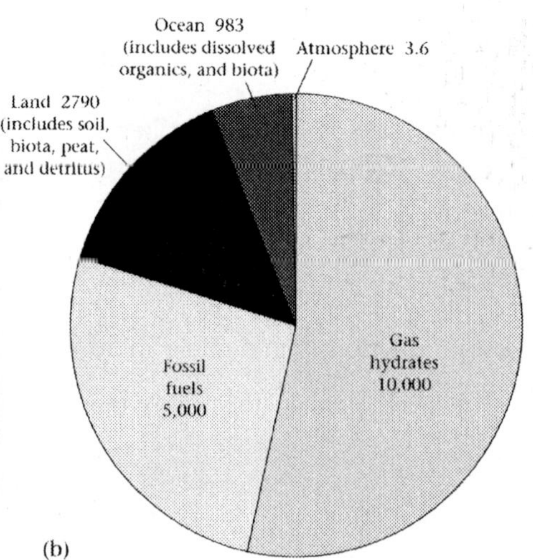

(b)

Figure 20.17 (a) Photo of gas hydrates samples (the white, icy material) brought up from the sea floor. (b) The proportion of organic matter in different materials. Gas hydrates may contain the most organic carbon (numbers are x10^5 tons of carbon). (source: ME)

Gas Hydrates (check 14.7 ME)

Figure 20.18 Chunks of coal. Since 1000 C.E., coal has been a major source of energy. (source: ME)

Figure 20.19 This museum diorama depicts a Carboniferous coal swamp. (source: ME)

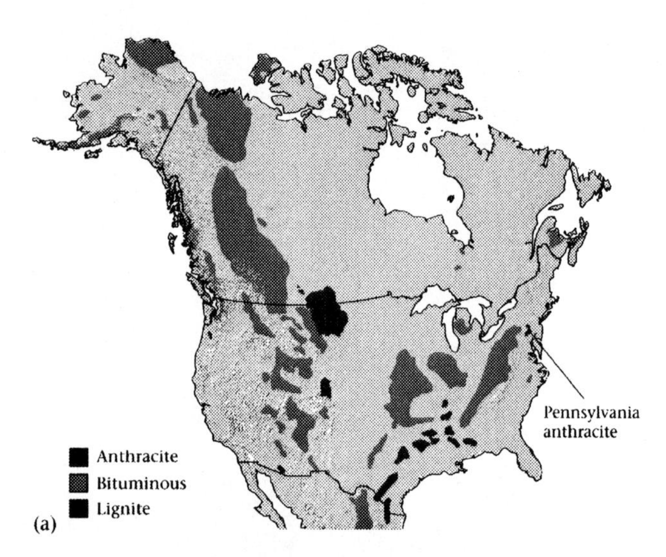

Figure 20.20 The distribution of coal reserves in the U.S.. The high-grade, clean-burning Anthracite coal is unfortunately relatively rare. (source: ME)

20.3 Coal

- Coal forms from plant material (wood, stems, leaves) in swamps and adjacent forests
- Most coal was formed in the late Carboniferous and Permian Periods (320-245 Mio years ago) when the continents were locked together to form Pangaea
- Coal is the most polluting fossil fuel but reserves will last 3 times as long as the oil reserves (300 years).
- The U.S. burn 1 Billion tons of coal per year, mostly at electrical power plants that generate ~23% of U.S. electricity.
- classification of coal:
  - **Peat**: 50% Carbon (C); not yet coal
  - **Lignite**: from peat (soft dark-brown coal) (70% carbon)
  - **Bituminous Coal**: from lignite at higher T (100-200°C) (85% C)
  - **Anthracite Coal**: from bituminous coal at even higher T (200-300^{0}C); formed at 8-10 km depth (95% C); burns most efficiently and clean
  - **Graphite**: at even higher T and P
- anthracite coal is the rarest type of coal. In the US, it is found at two places in Pennsylvania and Arizona
- More than half of the world's coal can be found in the United States and Former Soviet Union.

The Global Distribution of Coal Reserves

Country	Fraction
United States	28%
Former Soviet Union	26%
China	11%
Western Europe	10%
Australia	10%
Eastern Europe	6%
other	9%

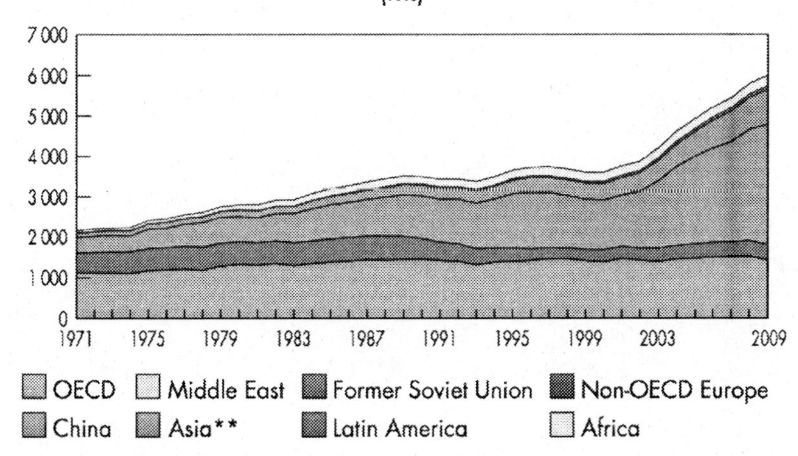

Figure 20.21 Coal production by region. * Includes recovered coal. **Asia excludes China[13]. (source: IEA)

20.4 The Burning of Fossil Fuel and Air Pollution (see also Lecture 23)

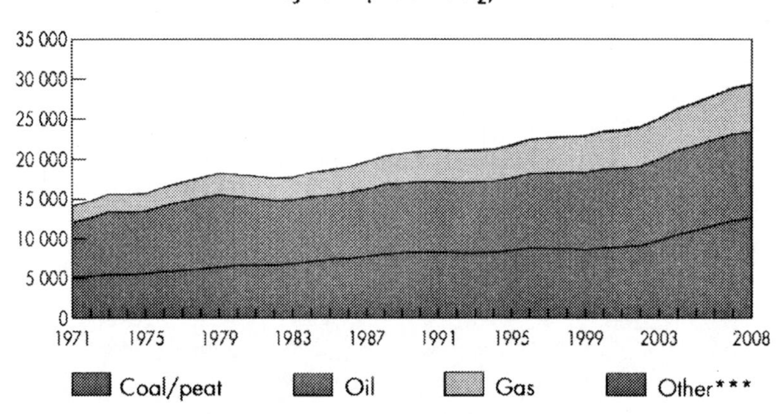

Figure 20.22 CO_2 emissions by fuel. * Includes international aviation and international marine bunkers. **Calculated using IEA's energy balances and the Revised 1996 IPCC Guidelines. CO_2 emissions are from fuel combustion only. ***Includes industrial waste and non-renewable municipal waste[13]. (source: IEA)

Primary Pollutants from Burning Fossil Fuels

Pollutant	**Fraction**
Carbon Monoxide	49.1%
Sulfur Oxides	16.4%
Nitrogen Oxides	14.8%
Volatile Organics	13.6%
Particulates	6.0%

NB: the greenhouse gas CO_2 is not considered a pollutant and is therefore not included in this table! CO_2 in the atmosphere has increased from 290 ppm in 1850 to 370 ppm in 2000.

Sources of the Primary Pollutants

Process	**Fraction**
Transportation	46.2%
Stationary Source Fuel Combustion	27.3%

Industrial Processes	15.0%
Miscellaneous	9.0%
Solid Waste Disposal	2.5%

The pollution per car was even greater before catalytic converters were introduced in the 1970ies (see more).

The Carbon Footprint

- the net amount of greenhouse gases (amount emitted minus amount consumed) a process, product or entity releases to the environment
- serves as a measure to estimate the net contribution of a process or production to global warming
- a carbon-neutral process produced as much greenhouse gases as it consumes

Some companies try to reduce their overall carbon footprint by adding processes that emit less greenhouse gases. E.g.

- electricity-generating companies adding renewable energy to their portfolio
- some airlines offer a surcharge to customers that goes toward regrowing rainforest to reduce the carbon footprint of an airline passenger
- the carbon footprint of individually bottled water is much higher than that of tap drinking water (transportation!)

Cap and Trade

A market-based approach of emission trading of pollutants and greenhouse gases to control net pollution of a company or entity. Typically, some economic incentives (such as reduced taxes) are provided to make this approach attractive.

- a central authority (state or country) sets emission limits (cap) on emission
- a company can buy permits (carbon credits) for higher emissions
- the total number of carbon credits cannot exceed the cap
- companies with higher emissions must buy credits from company with lower emissions, so that total cap is not exceeded (trade)
- in theory, this should reduce the pollution at the lowest cost to society, as companies who can reduce their emission cheaply will do so

Summary on Fossil Fuels and Consumption

COPYRIGHTED MATERIAL — DO NOT DUPLICATE, DISTRIBUTE, OR POST

- global energy demand doubled in last 30 years
- hydrocarbons now account for > 50% world's energy production
- fossil fuels account for nearly 80%
- **fossil fuels are non-renewable**
- U.S. oil production peaked in 1970ies
- U.S. has to import 50% of its oil
- peak in global oil production by 2005
- oil consumption now exceeds rate of discovery by factor 3
- oil reserves exhausted within 50 years
- gas reserves probably within 100 years
- coal reserves probably within 300 years
- **burning fossil fuels pollutes the environment**
- **burning fossil fuels accelerates the greenhouse effects**
- **burning coal causes acid rain**

20.5 Nuclear Power

- Nuclear power is considered an immense energy source because a tiny amount of matter transforms into a great amount of thermal energy
- nuclear power is considered as an alternative energy source to fossil fuels because it does not produce green house gases
- at current estimates nuclear fuel reserves will last about three times as long as oil reserves (about 300 years)
- Uranium is the only naturally occurring element that undergoes fission and is the last of the 92 naturally occurring elements. Uranium occurs naturally in granitic plutons
- Most of the naturally occurring Uranium is the relatively stable isotope 238U (half life: 4.5 Billion years), while the unstable 235U used in nuclear reactors accounts for only 0.7%
- Uranium enrichment for fuel rods requires energy
- by-products of uranium production, use and reprocessing produces Cesium (Cs) and Strontium (Sr) that can be absorbed by human tissue and do damage. Another by-product is Plutonium (Pu), which is weapons material and the most toxic element on earth. Pu also has a long-half life (24,000 years), i.e. stays around for a long time.
- a nuclear bomb requires only a few km of weapons-grade material. The world currently stockpiles more than 1000 tons.
- **the three main disadvantages in nuclear power**

- long half life of nuclear waste requires politically and geologically safe storage
- accidents contaminate the environment for a long time (e.g. after the 1986 explosion in Chernobyl, the area around the reactor is still uninhabitable)
- civil security is at risk by terrorist attacks

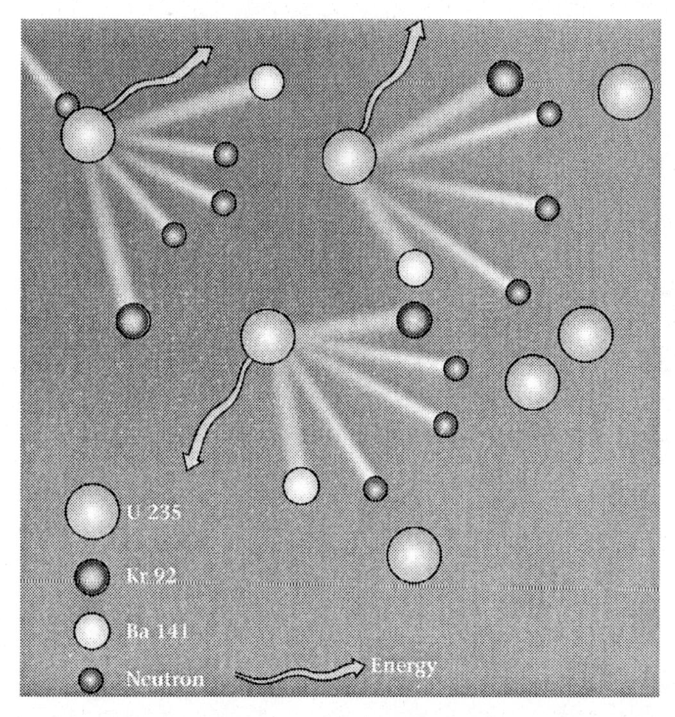

Figure 20.23 A chain reaction involving the radioactive decay of uranium. (source: ME)

Figure 20.24 The San Onofre, CA nuclear power plant. The plant with its signature domes was commissioned in 1968 and has pressurized water reactors. It uses seawater for cooling and so does not have cooling towers. In 2010, 92,687 people lived within 16 km (10 mi) of the plant. San Diego lies within a 50-mi radius within which nearly 8.5 million people lived in 2010. The plant was "built to withstand a magnitude 7.0 earthquake directly under the plant". (source: KPBS, Nuclear Regulatory Commission and wikipedia)

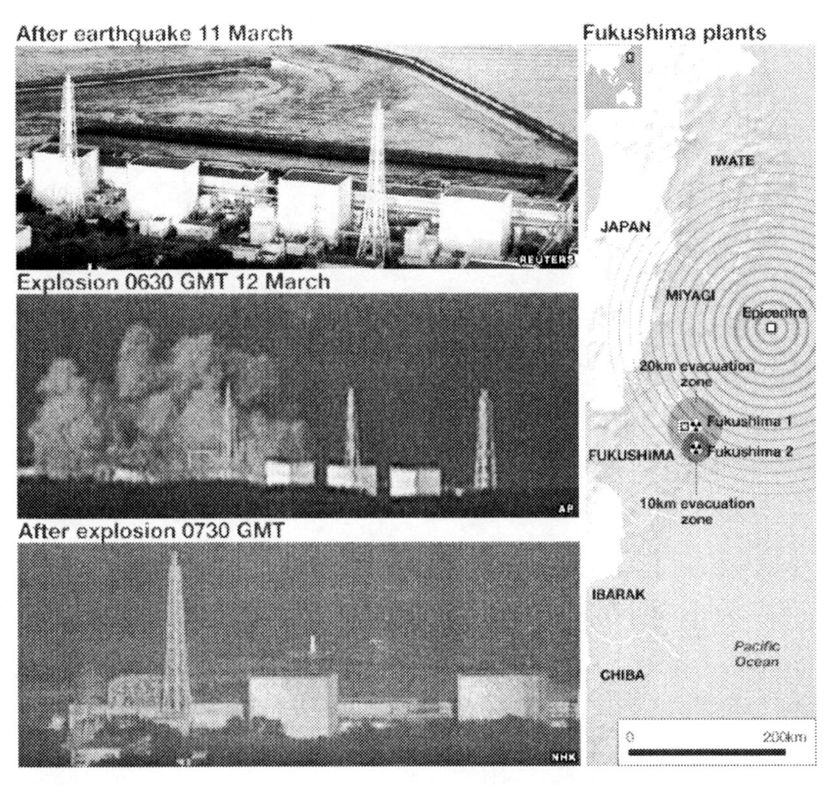

Figure 20.25 Explosions at the Fukushima nuclear power plants two days after the 11 March 2011 Tohoku earthquake and tsunami struck. (source: BBC online news)

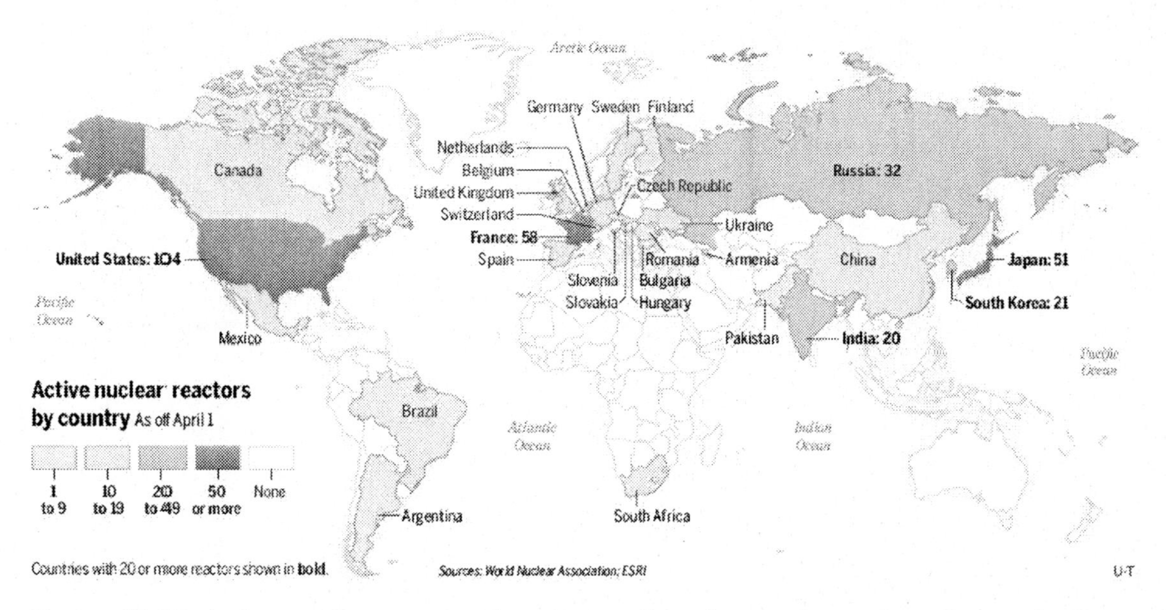

Figure 20.25 Active nuclear reactors by country. Reacting to the nuclear disaster in Japan, the German government announced in May 2011 that the country would phase out nuclear energy by 2022. (source: SD Union Tribune, 31 May and 2 June 2011)

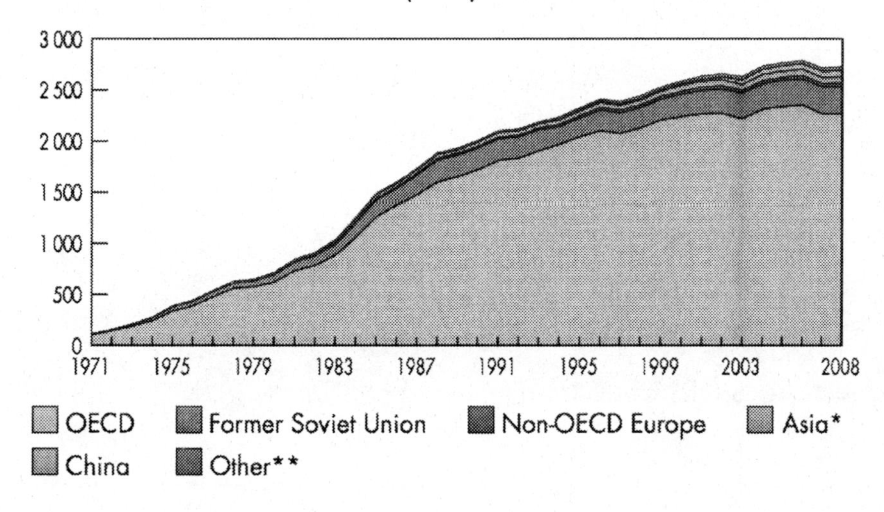

Figure 20.25 Nuclear power generation by region. *Asia excludes China. **Other includes, Africa, Latin America and the Middle East[13]. (source: IEA)

Installed Capacity (GW)	Current Production (TWh)	% of World total	% of domestic el. generation

COPYRIGHTED MATERIAL — DO NOT DUPLICATE, DISTRIBUTE, OR POST

U.S.	101	838	30.7	19.3
France	63	439	16.1	77.1
Japan	48	258	9.4	24.0
Russian Fed.	23	163	6.0	15.7
Germany	20	148	5.4	23.5
Korea	18	151	5.5	34.0
Canada	13	94	3.4	14.4
Ukraine	13	90	3.3	46.7
U.K.	11	-	-	-
Sweden	9	64	2.3	42.6
Rest of world	53			11.9*
China		68	2.5	2.0
World	**372**	**2731**	**100**	

* excludes countries with no nuclear power production.[13]

20.6 Renewable Energy Resources

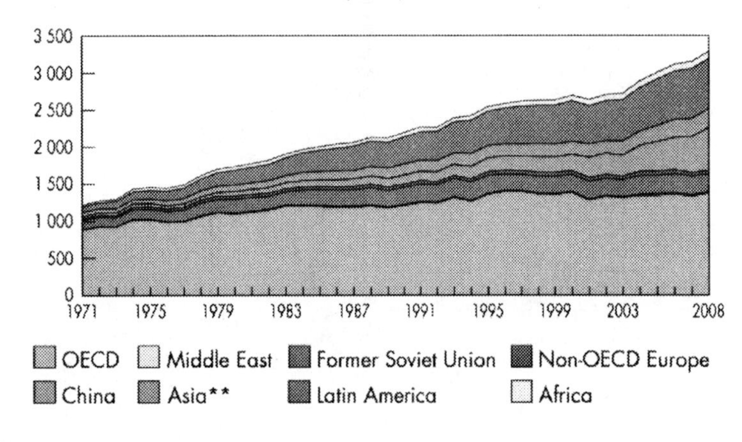

Figure 20.26 Hydro-electric power generation by region. *Includes pumped storage. **Asia excludes China[13]. (source: IEA)

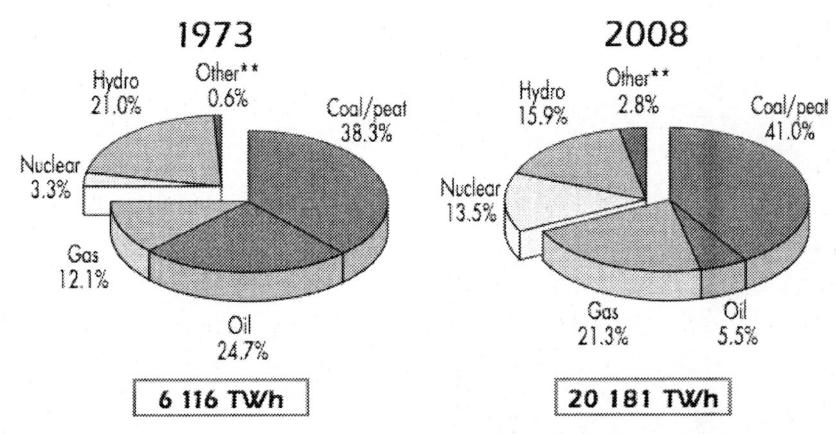

Figure 20.27 Electricity generation by type of fuel. *Excludes pumped storage. **Other includes geothermal, solar, wind, combustible renewables and waste, and heat[13]. (source: IEA)

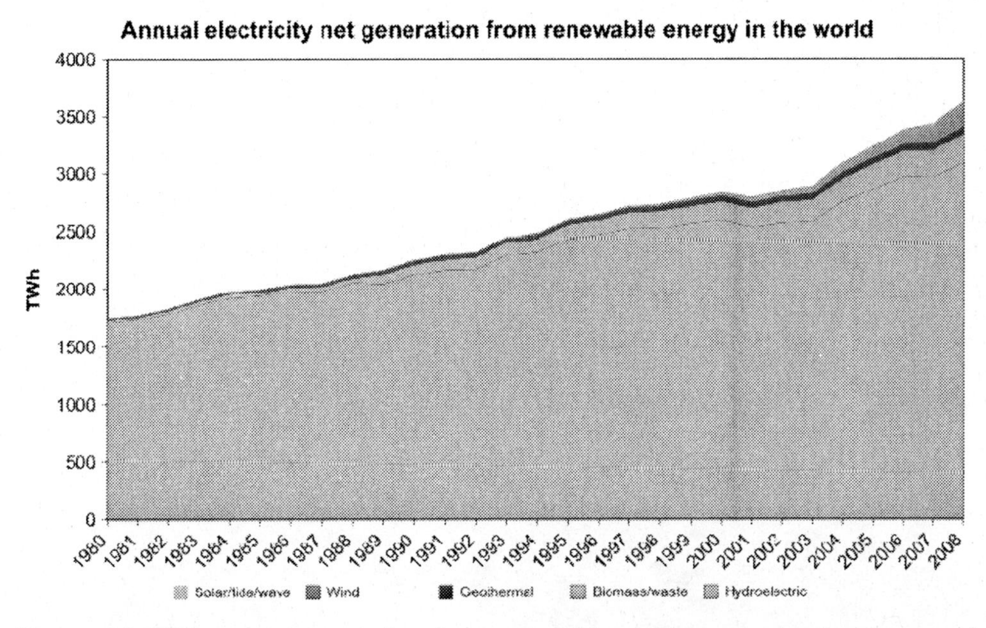

Figure 20.28 Electricity generation from renewable resources: hydroelectric comprises the lion share while minor contributors are biomass, wind, geothermal and solar. (source: Wikipedia/IEA)

Hydroelectric

Figure 20.29 The Hoover Dam when completed in 1936 was both the World's largest electric-power generating stations and the World's largest concrete structure. (source: wikipedia)

Figure 20.30 The Three Gorges Dam in China can provide large amounts of hydroelectric power; it will have a 22.5 GW capacity. But its construction also was controversial. (source: wikipedia)

Figure 20.31 The Rance Tidal Power Station, France is the World's first and largest tidal power station. The barrage across the Rance River is 750 m (2461 ft) long. The power plant portion is about half that long and has 25 turbines and the tidal basin measures 22.5 km^2 (9 sq mi). The power plat usually runs at 40% capacity. The tidal range is between 8 m (26 ft) and 13.5 m (44.3 ft). (source: wikipedia)

Figure 20.32 Top: The Rheinfelden, Germany hydroelectric plant, located in the High Rhine River, about 20 km (12 mi) upstream (east) of Basel, Switzerland. Top: the old plant was built in 1898. It had 20 turbines to produce 10 MW of electricity and was the oldest large hydroelectric plant of still in operation, after the 37 MW Adams Powerhouse at Niagara Falls closed in 1961 to become a National Historic Landmark. The Rheinfelden plant ran alongside the river on one side and collected the water in a side channel. A new plant that now runs across the river was built in 2010, 130 m upstream from the old plant. It has a capacity of 100 MW (4 times the most recent capacity of the old plant). Due to its environmental impact the project was not without controversy. Bottom: Despite its historical significance and attempts to make it a historical landmark, the old plant was demolished to give way to a fish ladder for salmon. (source: wikipedia)

Tidal Power Stations in Operation Aug 2010

Power Plant and Country	**Year**	**Capacity (MW)**
Rance Tidal Power Station, France	1966	240
Annapolis Royal Generating Station, Canada	1984	20
Jiangxia Tidal Power Station, China	1980	3.2
Kislaya Guba Tidal Power Station, Russia	1968	1.7
Strangford Lough SeaGen, U.K.	2008	1.2
Uldolmok Tidal Power Station, S. Korea	2009	1.0

Wind

Figure 20.33 Wind-powered turbines in Aalborg, Denmark used in electricity generation. (source: wikipedia)

Figure 20.34 Older windmills from the Tehachapi Pass Wind Farm in Southern California. Other major wind farms in California include that Altamont Pass Wind Farm near Livermore and the San Gorgonio Pass Wind Farm near Palm Springs. (source: wikipedia)

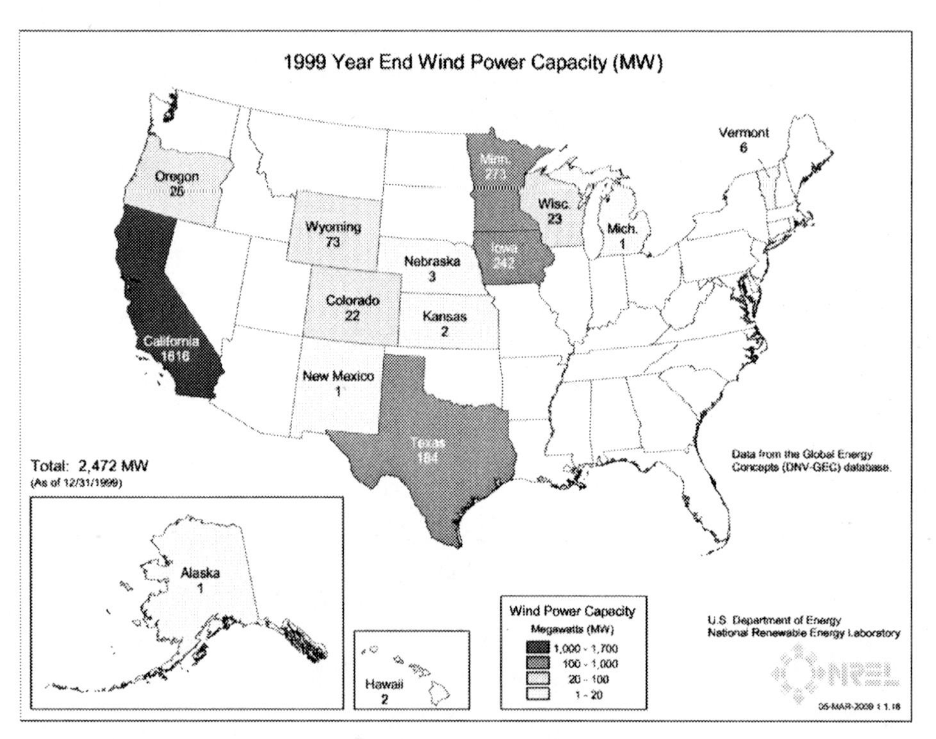

Figure 20.35a Installed wind farm capacity in the U.S. in 1999. With 65% of the total capacity, California dominated the market. (source: Wikipedia/windoweringamerica.gov)

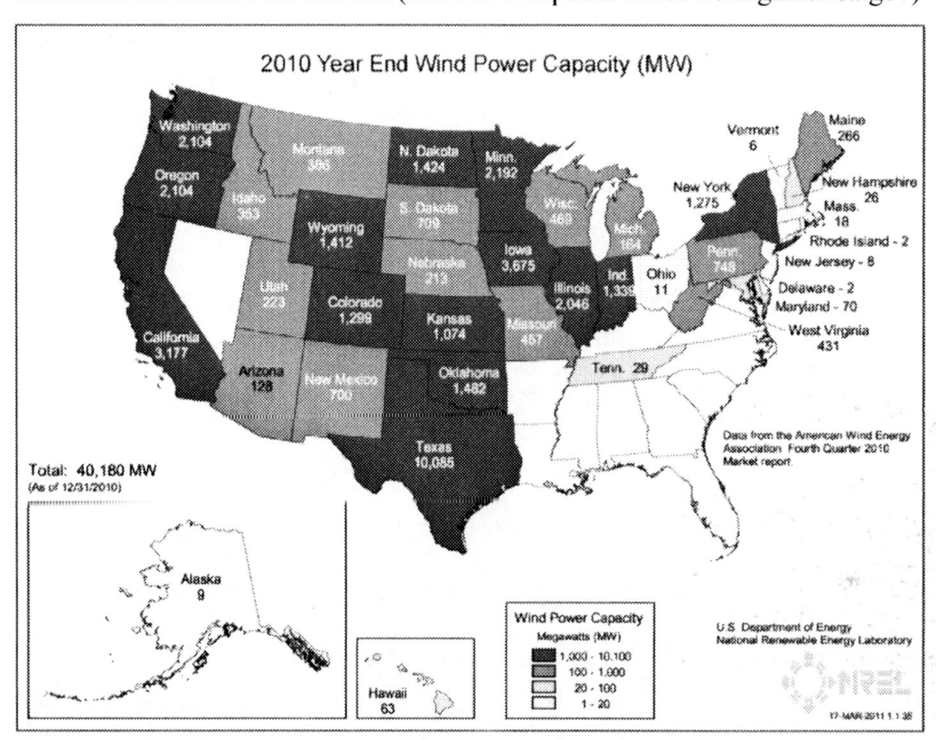

Figure 20.35b Installed wind farm capacity in the U.S. in 2010. In the last 11 years, wind capacity in the U.S. has seen a more than 15-fold increase, with Texas now providing 25% of the U.S. wind energy. (source: Wikipedia/windoweringamerica.gov)

Geothermal

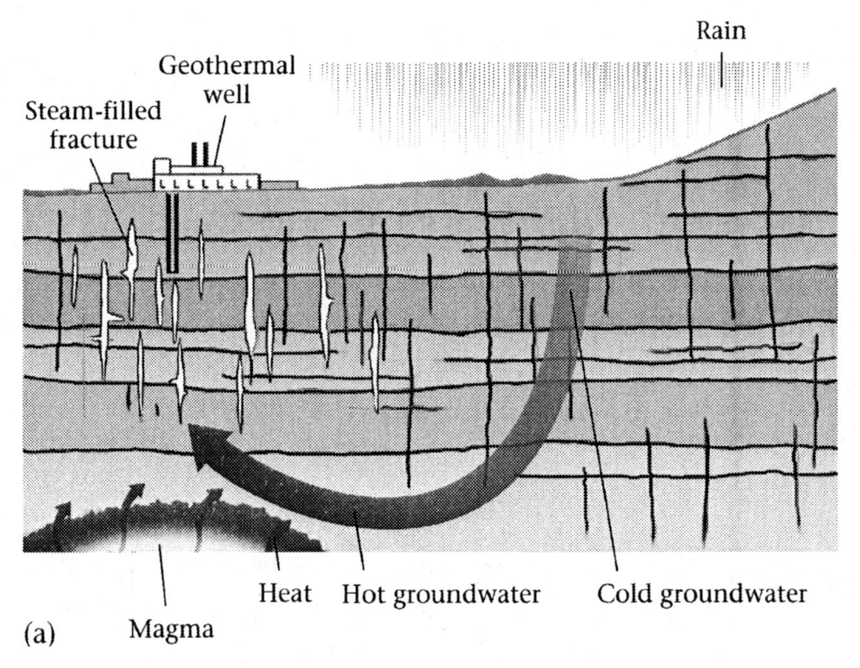

Figure 20.36 The principle of harnessing geothermal energy. Surface water sinks down into the ground as groundwater and gets heated by the hot ground (possibly by magma below). The hot water rises and, when it reaches shallow depths, turns to steam (as a result of decompression). Yhe water may be pumped out. It turns to steam under ground surface pressure, and runs turbines. (source: ME)

Figure 20.37 The Nesjavellir Geothermal Power Station in Iceland. (source: wikipedia)

Figure 20.38 The Leathers geothermal power plant in Calipatria in California's Imperial Valley. (source: Energy Efficiency&Renewable Energy, DOE)

Imperial Valley has become a "hot bed" of renewable energy project, both solar and geothermal. This is driven in part by California's mandate to generate 20% of its power from renewable sources by the end of 2010. CalEnergy currently runs one of 10 geothermal plants in the Salton Sea area. They generate 340 MW (megawatt), enough power for 300,000 homes, and could tap into more for up to 2.5 million homes. .... check out SDREADER article for more.

Solar

Solar Thermal Power.

Homes in Germany and Chula Vista.

Figure 20.39 Solar Two Power Tower Project in the Mojave Desert near Barstow, CA. The Solar project generated power from 1982 through 1999. (source: wikipedia)

Solar One was a pilot solar-thermal project built in the Mojave Desert just east of Barstow, CA. It was the first test of a large-scale solar thermal power tower plant. Solar One's method of collecting energy was based on concentrating the sun's energy onto a common focal point to produce heat to run a steam turbine generator. It had hundreds of large mirror assemblies (heliostats) that reflected the solar energy onto the central tower and operated between 1982 and 1989. In 1995 the system was renamed Solar Two after a second ring of 108 mirror systems was added. Using 1818 40-m^2 (430 ft^2) mirrors, the project was installed on a 126-acre (0.5 km^2) site. The project produced 10 MW of electricity until 1999 when it was decommissioned. Solar power generation then had a long hiatus in the U.S.. A similar plant, Nevada Solar One, started producing electricity in 1907 and generates 64 MW of power. This system uses a solar parabolic trough system.

Figure 20.40 Point focus parabolic mirror with Stirling engine at its center and its solar tracker at Plataforma Solar de Almeria in Spain. (source: wikipedia)

In Imperial Valley, Tessera Solar is currently working on the 750 MW Imperial Valley Solar Project (formerly Stirling Energy Systems Solar Two) to build one of the world's largest solar thermal plants, 26 km^2 (10 sq mi) with 30,000 "sun catchers", it will power up to 600,000 homes once it is fully operational by around 2015. Not to be confused with photo-voltaic systems, each sun catcher consists of a solar receiver heat exchanger and a close-cycle, high-efficiency Solar Stirling Engine specifically designed to convert solar power to rotary power to drive electrical generators.

Figure 20.41 The Nellis Solar Power Plant at Nellis Air Force Base in Nevada produces 15 MW of electricity. The solar panels track the sun around one axis. (source: wikipedia)

Solar Photovoltaic Power.

In photovoltaic (PV) solar panels, solar energy is converted directly to electricity.

As of 2010, solar photovoltaics generates electricity in more than 100 countries. Though yet comprising a tiny fraction of the 4.8 TW total global power-generating capacity from all sources, solar photovoltaic is the fastest growing power generation technology in the world. Between 2004 and 2009, grid-connected PV capacity increased at an annual average rate of 60%, to some 21 GW. Such installations may be ground-mounted, sometimes integrated with agriculture, or built into the roof or walls of a building, known as Building Integrated Photovoltaics (BIPV). Off-grid PV accounts for an additional 3-4 GW. Many solar PV power stations have been built, mainly in Europe. Perhaps oddly so, Germany us the current leader in solar production in Europe even though the solar electricity potential is relatively unfavorable, due to its high-latitude location between 47 and 55°N.

Largest PV power plants in the world (Dec 2010)

Power Plant and Location	**Capacity (MW)**
Sarnia PV Power Plant, Canada	97
Montalto di Castro PV Power Station, Italy	84.2
Finsterwalde Solar Park, Germany	80.7
Rovigo PV Power Plant, Italy	70
Olmedilla PV Park, Spain	60

Strasskirchen Solar Park, Germany	54
Lieberose PV Park, Germany	53

Currently the largest solar photovoltaic power plant in the U.S., the Copper Mountain Solar Facility is a 48 MW power plant in Boulder City, NV. The plant took about one year to be built and started generating electricity on 1 December 2010. It has 775,000 First Solar panels on an 380-acre site (1.5 km^2). The 25 MW DeSoto Next Generation Solar Energy Center in Florida consists of over 90,000 solar panels. The 15 MW Nellis Solar Power Plant and Nellis Airforce Force Base in Nevada was completed in 2007 and has about 70,000 solar panels.

Many schools, businesses and private homes now have photovoltaic solar panels on their roof. Most of these are grid connected and use net metering laws to allow use of electricity in the evening that was generated during the daytime. New Jersey leads the nation with the least restrictive net metering law, while California leads in total number of homes which have solar panels installed. Many were installed because of the million solar roof initiative. California has decided that it is not moving forward fast enough on photovoltaic generation and is considering to introduce feed-in tariffs. Washington state has a feed-in tariff of $0.15/kWh which increases to $0.54/kWh if components are manufactured in the state. Hawaii and Michigan also consider introducing feed-in tariffs.

20.7 Mineral Resources

Non-metallic Mineral Resources

E.g. building stone, gravel, sand, gypsum, phosphate (for fertilizer), salt.

- **dimension stone**: intact slabs and blocks of rock (e.g. granite or marble) cut out in a quarry (a quarry provides stone, a mine provides ore).
- **crushed stone**: for concrete, cement and asphalt.

  cement forms by precipitation of minerals out of slurry of water, lime (CaO) (66%), silica (SiO2) (25%), aluminum oxide (Al2O3) and iron oxide (Fe2O3). CaO comes from the calcite in limestone that is roasted in a furnace at high temperature. The other elements come from shale and sandstone. ***Concrete*** is a mixture of cement with sand and gravel. ***Bricks*** are made from dried and bakes clay (this process is a metamorphic process that recrystallize the clay and make it relatively impermeable). ***Glass*** comes from the melting of quartz and quartz sand. ***Drywall*** is made from a slurry of water and gypsum (CaSO4.H2O); ***asbestos*** is made from serpentine; ***plastics*** from oil; ***fertilizers*** such as phosphate (PO_4^{3-}) and potash (K2CO3) come from evaporites.

Figure 20.42 A native copper sample from Ray mine in the Scott Mt. area in Arizona. (source: wikipedia)

Figure 20.43 A bauxite sample, with a penny for scale. The three largest bauxite reserves are in Guinea, Australia and Vietnam. (source: Wikipedia/USGS)

20.8 Metallic Mineral Resources

Metallic resources come in *native* form and as *ore*. *Native metals* are easy to mine because they occur in pure form (e.g. gold nuggets, copper lumps). Examples for native metals: gold, copper, silverIn *ores* the metals occur in metallic compounds such as oxides. The extraction of the metals require elaborate techniques that use high temperature smelting (e.g. iron ore) or toxic chemicals (e.g. need mercury for some gold compounds) to extract the metal from the compound.Many metals are not found in native form, e.g. iron.

- some metals are not found in native form, e.g. iron
- a rock qualifies as an ore, if:

  - the metal content is high
  - **and** the process to extract this metal is cost-effective dependent on the market value we distinguish between base and precious metals:
- **base metals**: e.g. copper, iron, lead, zinc, tin
- **precious metals**: e.g. gold, silver, platinum
- of the 63 metals in use today, only 9 were known before 1700 (Gold, Silver, Copper, Mercury, Lead, Tin, Antimony, Iron, Arsenic), typically the metals found in native form

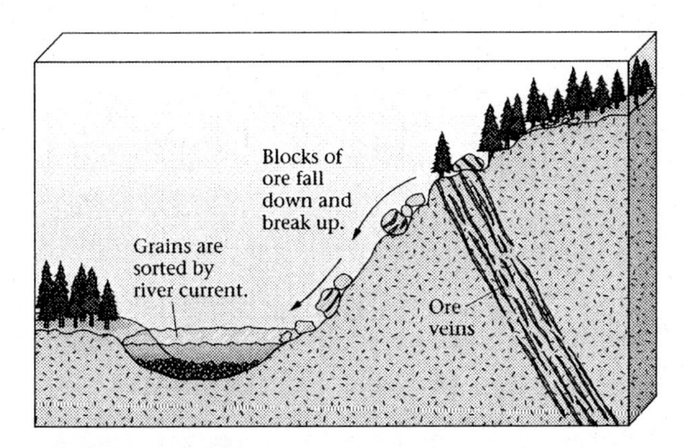

Figure 20.44 The process of forming a placer deposit. Ore-bearing rock is eroded, and blocks containing native metals fall into a stream. Sorting by the stream concentrates the metals. (source: ME)

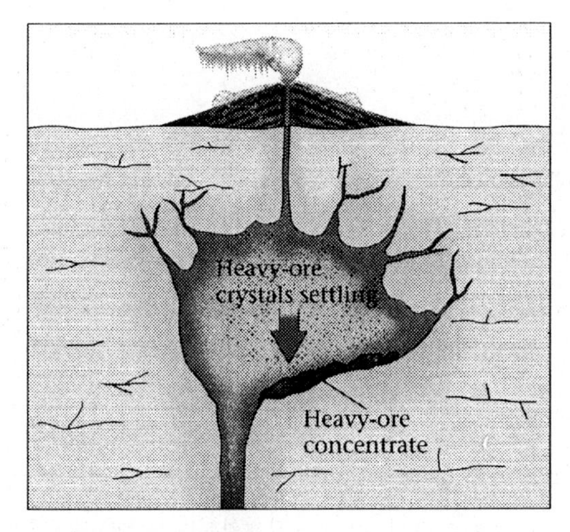

Figure 20.45 Heavy, metal-containing crystals can sink to the bottom of a magma chamber to form a massive-sulfide deposit. The sulfide concentrate may become an ore body in the future. (source: ME)

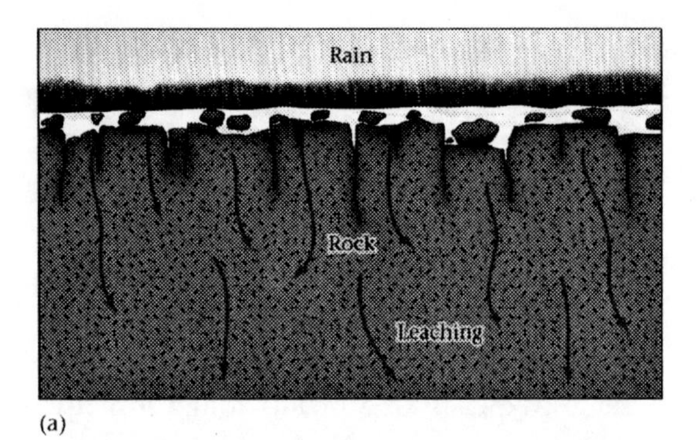

(a)

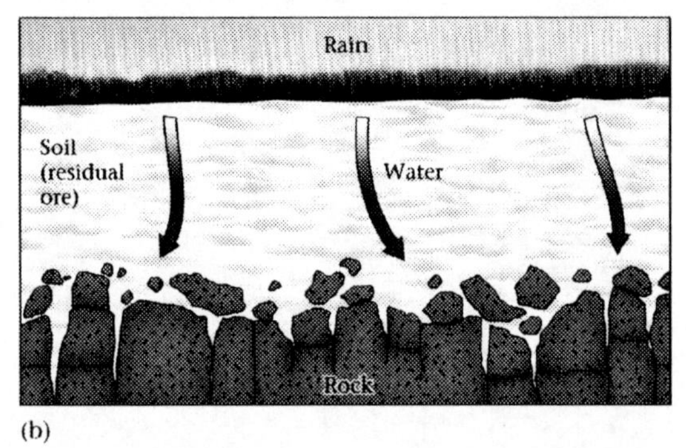

(b)

Figure 20.46 (a) When rainwater percolates through the soil, it dissolves and removes many elements. (b) A thick soil forms, containing a residuum of iron or aluminum. Similarly, rain can leach ore-minerals from a low-grade ore body that is buried in the sediment. The leached mineral are transported downward and can accumulate in a secondary enrichment zone to form a high-grade ore. (source: ME)

Where do Metallic Resources Come From?

- **magmatic deposits**: bottom of a magma chamber
- **hydrothermal deposits**: circulation of hot-water solutions
  - disseminated deposit: deposit throughout an intrusion
  - vein deposit: deposit in a preexisting crack
  - copper porphyry deposit: copper in a two-stage melted igneous rock
  - black smoker: precipitated nearly pure metal sulfides around black smokers
- **secondary-enrichment deposits**: transport and re-deposition of ore minerals, typically at higher concentrations
- **sedimentary deposits**: for example
  - banded-iron formation (BIF) in deep ocean, when environment suddenly became oxygen-rich

- manganese nodules (manganese-oxide, typically about 10 cm diameter) on the ocean floor. Rich in manganese and copper (20% Mn + Fe, Cu, Ni). Technology to mine vast deposits underway (e.g. copper supply would last 720 years).

- **residual mineral deposits**: tropical environments, iron and aluminum-rich soil residuum in the soil as consequence of severe leaching (E.g. bauxite is a aluminum-bearing residual from extreme leaching of granite).
- **placer deposits**: native metals eroded from original deposit and settled as flakes in rivers (recovery similar to what happens during panning).

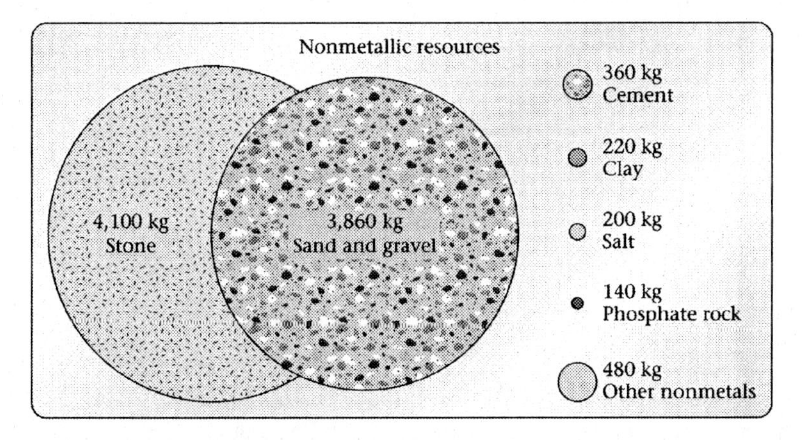

Figure 20.47 The weight of the annual per capita usage of non-metallic mineral resources, in a typical industrialized country. (source: ME)

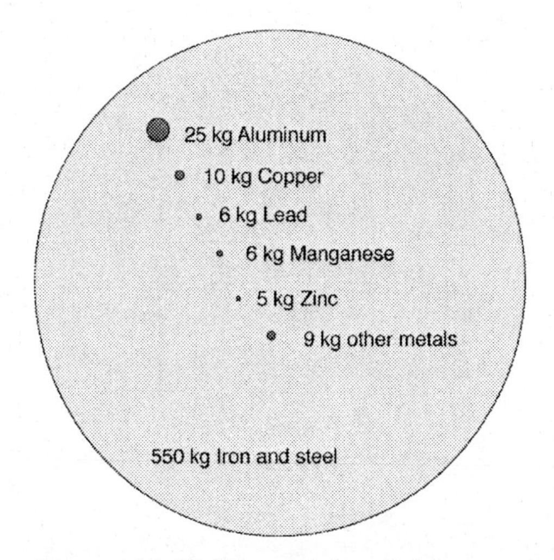

Figure 20.48 The weight of the annual per capita usage of metallic mineral resources, in a typical industrialized country. (data from ME)

20.8 Global Mineral Needs and Reserves

- Mineral resources are non-renewable resources (i.e. the lifetime is limited). Combined with energy resources, the U.S. mine, quarry and pump 18 billion tons/year. For comparison, the Mississippi transports 190 million tons of sediments into the Gulf of Mexico.
- How Much Do We Need?
  In "Physical Geology", Carlson et al., refer to the Mineral Information Institute to estimate per capita/year needs in the U.S.:
  - 18,000 kg total are mined, not including energy resources
  - 4,400 kg stone
  - 3,500 kg sand and gravel
  - 325 kg limestone for cement
  - 160 kg clays
  - 165 kg salt
  - 760 kg other nonmetals
  - 545 kg iron
  - 19 kg aluminum
  - 9 kg copper
  - 5 kg lead
  - 5 kg zinc
  - 3 kg manganese
  - 11 kg other metals

  for comparison, some numbers for energy resources[6]:
  - 3050 kg petroleum
  - 2650 kg coal
  - 1900 kg natural gas

The U.S.'s Dependence on Importing Minerals

- 99% of gemstones are imported (e.g. from Israel, Belgium and India)
- **Strategic metals**: metals that are alloyed with iron to make special-purpose steels needed in the aerospace industry; many need to be imported[6]: e.g. manganese (100%), cobalt (95%), chromium (73%) and platinum (92%).

- many other raw materials are 100% imported

Metals used in the U.S. and Where They Come From

metal	Metal's Origin
arsenic trioxide	China, Chile, Mexico
bauxite and alumina	Australia, Jamaica, Brazil
bismuth	Belgium, Mexico, China
strontium	Mexico, Germany
thallium	Mexico, Belgium, Germany
thorium	UK, France

China's Emergence on the Rare Earth Market

Element	Atomic No.	Commercial Use
Scandium	21	stadium lights
Yttrium	39	lasers
Lanthanum	57	electric car batteries
Cerium	58	lens polishes
Praseodymium	59	searchlights, aircraft parts
Neodymium	60	high-strength magnets
Promethium	61	portable X-ray units
Samarium	62	glass
Europium	63	compact fluorescent bulbs
Gadolinium	64	neutron radiology
Terbium	65	high-strength magnets
Dysprosium	66	high-strength magnets
Holmium	67	glass tint
Erbium	68	metal alloys
Thulium	69	lasers
Ytterbium	70	stainless steel
Lutetium	71	none

(source: ref #16)

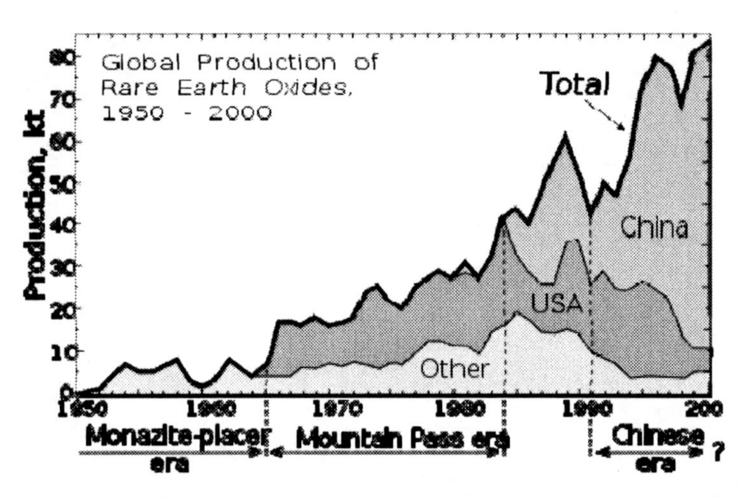

Figure 20.48 Global rare earth production 1950-2000. (data from ME)

Lifetime (in years) of Currently Known Ore Resources

metal	World Resources	U.S. Resources
Iron	120	40
Aluminum	330	2
Copper	65	40
Lead	20	40
Zinc	30	25
Gold	30	20
Platinum	45	1
Nickel	75	<1
Cobalt	50	<1
Manganese	70	0
Chromium	75	0

THE BOTTOM LINE: mineral resources are non-renewable and many global resources will last a few more decades. The U.S. has to import large fractions of the metals used.

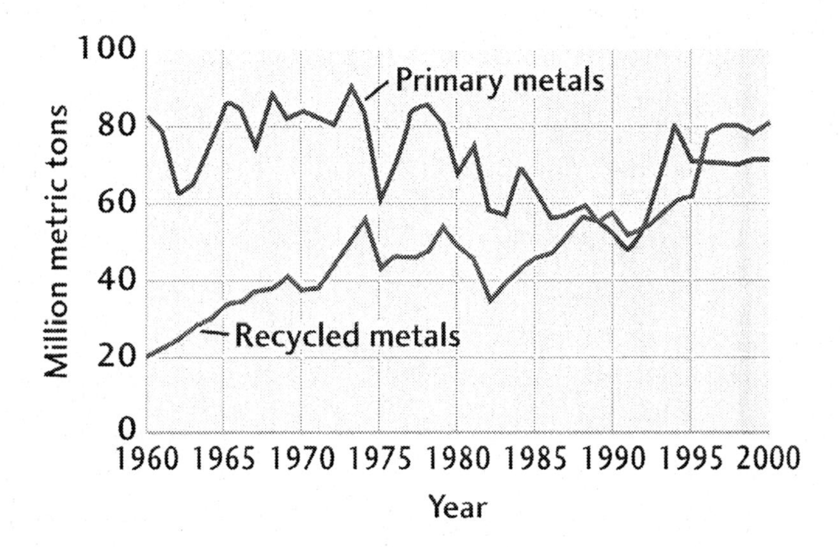

Figure 20.49 Estimate (by weight) of the amount of primary metals and recycled metals consumed in the U.S.. (source: UE/USGS)

20.9 Recycling

Estimates in the 1960ies and 1970ies predicted a shortfall of available minerals. But mineral consumption has slowed in the meantime and current estimates predict that resources will last about 100 years. The rate of mineral consumption has slowed especially in industrialized nations due to a shift of economics from construction and manufacturing to service and technology. The demand for raw minerals has decreased because of recycling and substitution by low-cost ceramics, composites and plastics.

Fraction of Recycled Metals

Metal	Fraction
Au (Gold)	45%
Pt (Platinum)	45%
Al (Aluminum)	45%
Pb (Lead)	73%
Cu (Copper)	60%
Fe (Iron), Steel	56%

Issues that determine whether a mineral is to be recycled or not: costs, environmental impact, availability.

Environmental effects of recycling over primary resource[17,18]

Material	Energy Savings	Air pollution savings
Aluminum	95%	95%
Cardboard	24%	-
Glass	5-30%	20%
Paper	40%	73%
Plastics	70%	-
Steel	60%	-

There is some debate over whether recycling is economically efficient. Some recycling programs may simply be driven by landfill costs. One Danish study reported that recycling is the most efficient method to dispose of household waste in 83% of cases but another suggests that incineration is most cost-effective to dispose of drinking containers, including aluminum cans. [nore]

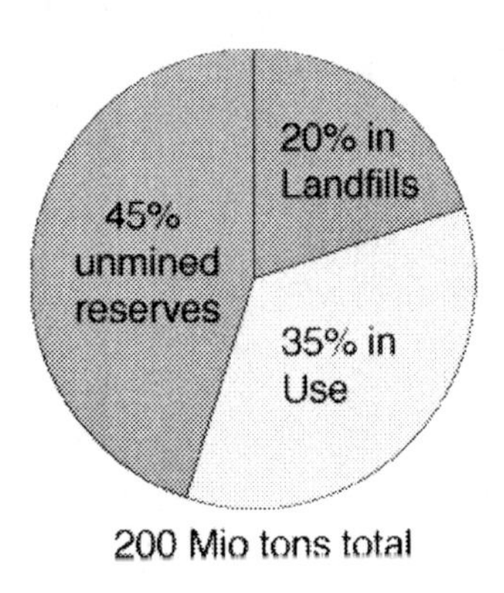

Figure 20.50 Reserves and usage of U.S. copper stocks in the 1990s.

Fluorescent Light Bulbs and Compact Fluorescent Light Bulbs (CFL)

Disposal is problematic because such bulbs contain mercury (Hg) vapor. The amount of Hg varies between 3 and 46 mg. A typical 2006-era 4-ft tube contains about 12 mg, CFLs that use 25-40 W of electricity have about 6 mg or less. Users are supposed to bring old CFL and fluorescent tubes to a recycling center (or hazardous waste disposal place. The San Diego/Miramar landfill is said to collect bulbs but only quantities that are consistent with "single household usage". Home Depot is said to have just started a take-back program.

20.10 Plastic [2]

- made from oil; many different kinds (e.g. from clear plastic bottles to styrofoam)
- plastics are durable, lightweight, and can be made into virtually anything
- plastics are not biodegradable
- only 3.5% of plastics are recycled in any way
- in the U.S., 63 pounds of plastic packaging goes into landfills per person per year
- broken, degraded plastic pieces outweigh surface zooplankton in the Central North Pacific by 6 to 1

SOME PLASTIC MILESTONES

- 1869: John Wesley Hyatt invents celluloid, the first plastic product given a trade name
- 1909: Bakelite introduced as the "first thermoset plastic" (once set it is set for life)
- 1939:Nylon stockings debut at the World's Fair
- 1946: Earl S. Tupper produced a 7-ounce polyethylene tumbler, the first Tupperware product
- 1955: the Corvette is the first car to use plastic for body panels
- 1957: the Hula Hoop creates a surge in demand for polyethylene
- 1983: microwave ovens open up a new market for plastic packaging
- 2000: in the U.S., pre-production plastics reaches 100 billion pounds of virgin resin pellets per year

20.11 Mining and the Environment

Resource Exploration and Production

In the past, prospectors looked for "shows" of ore (exposures of ore minerals at the surface). They looked out for quartz veins or stained rock (green or red by oxidizing metal-containing minerals). Today's geologists look at tectonic settings and measure the magnetic and gravity fields to find ore bodies (they are typically denser and more magnetic than ordinary rock).

- **open-pit mines**: explosions separate blocks from bedrock; waste rock is separated from ore and brought to a "tailings pile". The ore is crushed, ore minerals extracted and smelted or treated with acidic solutions to separate the metal. The metal is then melted and poured into molds to make ingots (brick-shaped blocks).
- **underground mines**: drive vertical shafts and horizontal adits into the ore-bearing rock. The deepest mine currently reaches 3.5 km where the temperature exceeds 55°C.

Mining and the Environment

- open-pit mining leaves landscape scars
- mining leaves waste rock in tailings piles (artificial hills)
- acid mine runoff (sulfides released in ore-mining)
- smelting in ore-processing creates acidic air and rain

Figure 20.51 The El Chino open-pit copper mine near Silver City, NM. It was started as the Chino Copper Company in 1909 and is perhaps the oldest mining site still being used in the American Southwest. Until it was surpassed by the Chuquicamata mine in Chile, it was also the largest open-pit copper mine in the World. Operations in the mine have declined drastically in recent years.(source: wikipedia)

Figure 20.52 Coal strip mining in Wyoming.(source: wikipedia)

Figure 20.53 A coalbed fire beneath Centralia, Pennsylvania, produces noxious gas. (source: ME)

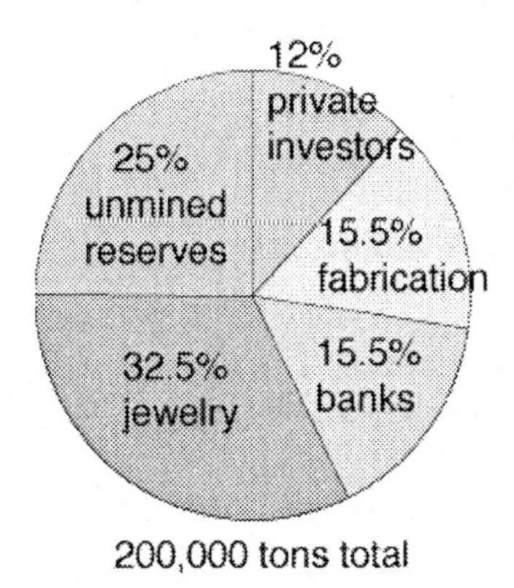

Figure 20.54 Reserves and usage of World gold stocks in 2000.

Recommended Reading and Websites

"Coal" by Barbara Freese, 2004, Penguin Books, ISBN: 01-42000981:a good book on coal, its history and how it changed mankind

U.S. Energy Information Administration: http://www.eia.doe.gov

References

1. "The Penguin State of the World Atlas" by Dan Smith, 2003. Penguin Books, ISBN: 0-14.200318-2
2. Flyer from the Algalita Marine Research Foundation

3. United Nations Environmental Programme (UNEP) web site
4. State of the World 2003" by Chris Bright et al., 2003, The Worldwatch Institute, Norton & Company, ISBN: 0-393-32386-2
5. "State of the World 2004" by Erik Assadourian et al., 2004, The Worldwatch Institute, Norton & Company, ISBN: 0-393-32539-3
6. Marshak, S., 2005. Earth, Portrait of a Planet, 2nd edition. Norton & Company Ltd., New York, 748 pp.
7. Earth, An Introduction to Physical Geology, 2005. E.J. Tarbuck and F.K. Lutgens, 8th ed., Prentice Hall, Upper Saddle River, N.J., ISBN: 0-13-114865-6
8. "Earthshock" by Andrew Robinson, 1993, Thames and Hudson, ISBN: 0-500-27738-9
9. Wikipedia web page on petroleum
10. Wikipedia page "Predicting the timing of peak oil"
11. Wikipedia page "Peak oil"
12. International Energy Agency (IEA), 1998. World Energy Outlook. Downloadable on the IEA website www.iea.org
13. International Energy Agency (IEA), 2010. Key World Energy Statistics. Downloadable on the IEA website www.iea.org
14. Energy Information Agency of the DOE, 2009. Annual Energy Review 2009. www.eia.gov/aer
15. Kumins, L., 2000. Alaska Oil Exports, CRS (Congressional Research Service) Report for Congress. National Library the Environment. http://ncseonline.org/nle/crsreports/natural/nrgen-25.cfm
16. SD Union Tribune, 1 November 2010, China pays a price as leader in rare earths production
17. http://en.wikipedia.org/wiki/Recycling
18. The Economist, 7 June 2007. The truth about recycling.
19. The Economist, 7 June 2007. The price of virtue.
20. The Economist, 7 June 2007. Plastics of evil.

APPENDIX A: The Special Role of Traffic

- transportation takes up 1/3 of the world's energy consumption
- transportation is responsible for almost half of the pollution produced
- finding alternatives current transportation is therefore essential!
- public transportation effective alternative in densely populated areas
- cars need to be made more fuel efficient (technology exists since the 1980s!)

- finding alternatives is crucial. Hybrid-automobiles and biodiesel are only temporary alternatives as hybrid cars still use fossil fuels and biodiesels contribute to greenhouse effect

•

General conversion factors for energy

To:	TJ	Gcal	Mtoe	MBtu	GWh
From:	*multiply by:*				
TJ	1	238.8	2.388×10^{-5}	947.8	0.2778
Gcal	4.1868×10^{-3}	1	10^{-7}	3.968	1.163×10^{-3}
Mtoe	4.1868×10^{4}	10^{7}	1	3.968×10^{7}	11630
MBtu	1.0551×10^{-3}	0.252	2.52×10^{-8}	1	2.931×10^{-4}
GWh	3.6	860	8.6×10^{-5}	3412	1

Conversion factors for mass

To:	kg	t	lt	st	lb
From:	*multiply by:*				
kilogramme (kg)	1	0.001	9.84×10^{-4}	1.102×10^{-3}	2.2046
tonne (t)	1 000	1	0.984	1.1023	2 204.6
long ton (lt)	1 016	1.016	1	1.120	2 240.0
short ton (st)	907.2	0.9072	0.893	1	2 000.0
pound (lb)	0.454	4.54×10^{-4}	4.46×10^{-4}	5.0×10^{-4}	1

Conversion factors for volume

To:	gal U.S.	gal U.K.	bbl	ft^3	l	m^3
From:	*multiply by:*					
U.S. gallon (gal)	1	0.8327	0.02381	0.1337	3.785	0.0038
U.K. gallon (gal)	1.201	1	0.02859	0.1605	4.546	0.0045
barrel (bbl)	42.0	34.97	1	5.615	159.0	0.159
cubic foot (ft^3)	7.48	6.229	0.1781	1	28.3	0.0283
litre (l)	0.2642	0.220	0.0063	0.0353	1	0.001
cubic metre (m^3)	264.2	220.0	6.289	35.3147	1000.0	1

Unit abbreviations

bcm	billion cubic metres	**kWh**	kilowatt hour
Gcal	gigacalorie	**MBtu**	million British thermal units
GCV	gross calorific value	**Mt**	million tonnes
GW	gigawatt	**Mtoe**	million tonnes of oil equivalent
GWh	gigawatt hour	**PPP**	purchasing power parity
kb/cd	thousand barrels per calendar day	**t**	metric ton = tonne = 1000 kg
kcal	kilocalorie	**TJ**	terajoule
kg	kilogramme	**toe**	tonne of oil equivalent = 10^7 kcal
kJ	kilojoule	**TWh**	terawatt hour

OECD: Organization for Economic Co-operation and Development

OPEC: Organization of the Petroleum Exporting Countries

IEA: International Energy Agency

EIA: Energy Information Administration, Department of Energy (DOE)

Chapter 21: Anthropogenic Changes: Water and the Oceans

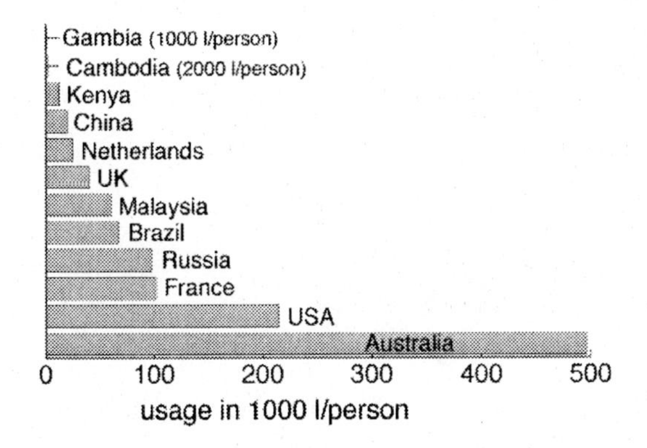

Figure 21.1 Annual domestic water for selected countries. Use estimated in 2000. (data: reference 1)

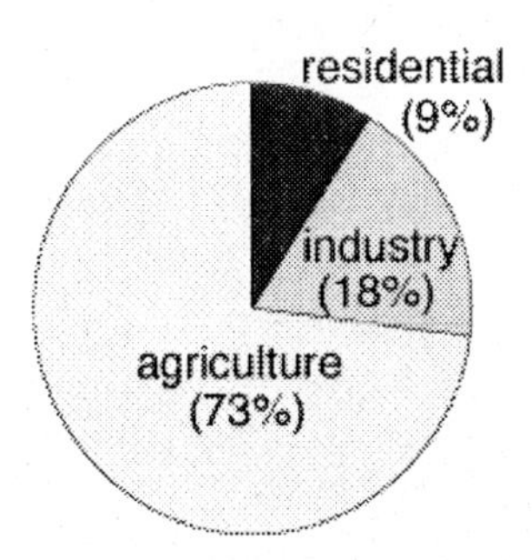

Figure 21.2 Worldwide average water use in the three main areas. (data: reference 1)

Table 21.1 Daily Wage spent on water[1]

Country	Usage [%]
USA	0.006
UK	0.013
Pakistan	1.1
Uganda	3.2
Tanzania	5.7

Table 21.2 The cost of hygiene[1]

Purpose	Usage [l]
1 flush of toilet	10
minimum healthy daily requirement	11.5

COPYRIGHTED MATERIAL — DO NOT DUPLICATE, DISTRIBUTE, OR POST

1 shower or 1 dishwasher cycle	35
1 bath	80

21.1 Global Water Supply and Use1

Some Tidbits

- the Amazon is the second longest river in the world [19] with the largest basin and volume of water[18]; with an average discharge greater than the next seven largest rivers combined[19]
- the largest river system in North America is the Mississippi[18]
- the Great Lakes are the largest group of freshwater lakes in the world[18]
- by 2025 two-thirds of the world's population will be short of water[1]
- water is becoming more and more scarce as a result of population increase and climate change
- global water use increased six-fold between 1900 and 1995; it has more than tripled since 1950; the rate of increase is twice as large as the rate of population growth (UNEP[3] estimate)
- at the same time, freshwater supplies and some freshwater lakes have shrunk dramatically, e.g. Aral Sea, Lake Chad and the marshlands of Mesopotamia
- global water use now stands at 4340 km^3 per year, 8 times the annual flow of the Mississippi River
- annual irretrievable water losses have increased about 7-fold between 1900 and 1995 and now stands at 2900 km^3; about 86% of this loss occurs in agriculture while the rest is lost in industry, municipal supply and reservoirs[6]
- several shortages affecting at least 400 Mio people today could affect 4 billion people (more than half the world's population) within 50 years; main culprits: waste and inadequate management
- 2.4 billion people (40% of world population) lack adequate sanitation facilities
- one person in three already lives in 'water stressed' countries where consumption exceeds 10% of total supply; if current trends continue, 2 out of 3 will live in such conditions by 2025
- one person in six has no regular access to safe drinking water
- water is used for domestic and industrial purposes but most (75%) is used in agriculture

- in 2000, agriculture and domestic each wasted 800 km^3 of water and industry 400 km^3; by 2025, a UNEP[3] estimate put these numbers to 1000, 1100 and 500
- drinking water supplies for poor people would be doubled with just a 10% improvement in the efficiency of irrigation
- countries with the highest rates of water usage (liters per person per day): Kazakhstan, Uzbekistan, Kyrgyzstan, Tajikistan, Azarbaijan, Armenia, Iraq and Guyana
- countries in which more than 50% have no access to improved water sources, such as a protected well, spring or tap: Haiti, Oman, Cambodia, Fiji, Papua New Guinea, Madagascar, Eritrea, Ethiopia, Kenya, Uganda, Rwanda, Democratic Republic of Congo, Angola, Equatorial Guinea, Chad, Sierra Leone, Guinea, Guinea-Bissau, Mauritania
- by 2025 two-thirds of the world's population will be short of water
- commercial companies are buying exclusive rights to rivers and aquifers, often in areas where the local people are chronically short of water
- power struggles over water are increasingly likely to be the cause of conflict
- the destruction or contamination of water supplies is itself a weapon of war

Examples:

- in the 1991 Gulf War, Iraq destroyed Kuwait's desalination capacity
- in the 1991 Gulf War, the U.S.-led coalition bombed Baghdad's water system
- from Angola to Timor, enemies have been killed and their bodies thrown in wells to poison local water supplies

DID YOU KNOW? by 2025 two-thirds of the world's population will be short of water[1]

Some Good News

- the number of people with some form of improved water supply rose from 4.1 billion (79% of world population) to 4.9 billion (82%) in 2000

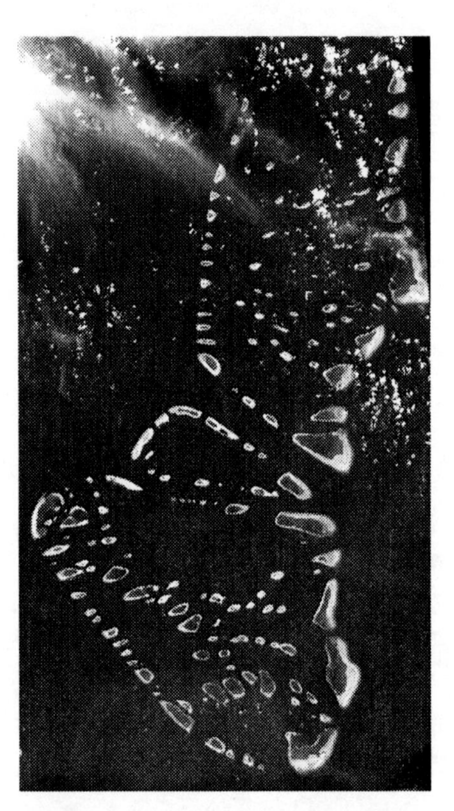

Figure 12.1a Malhosmadulhu Atoll of the Maldives in the India Ocean seen from space. The image was acquired on 22 December 2002. (source: Wikipedia/NASA)

Figure 12.1b Malé, the capital of the Maldives in the Indian Ocean. (source: Wikipedia)

21.2 Ocean Warming and Life along Ocean Coasts

- oceans cover 70.8% of Earth's surface
- oceans hold 95% of Earth's water

- 60% of population lives near coasts
- 200 Mio people depend on fishing

The Maldives, Indian Ocean. The Maldive Islands consist of 26 atolls with 1192 islands spread over an area of about 90,000 km^2, making it one of the most dispersed countries in the world. Two-hundred of the islands are inhabited by about 400,000 people. Malé, the capital, has a population of 104,000 (2006). Arguably he lowest-lying country in the world, the average elevation of the Maldives above sea level is only 1.5 m (4 ft 11 in) and the highest point is at 2.3 m (7 ft 7 in). The waves from the 26 December 2004 Sumatra-Andaman tsunami spilled over sea walls to flood Malé with sand-clouded water and then swept out just as suddenly. Residents fear that this was a foreboding of disasters to come if global sea levels are continuing to rise. Malé is 590 km (365 mi) from the nearest point in southwest India and 750 km (465 mi) from Sri Lanka. If sea level should rise by 1 m within the next 100 m, the Maldives will seize to exist, and with it probably its language and culture as its inhabitants will be relocated and dispersed as refugees.

Level of Flooding of Venice, Italy during Acqua Alta (Flood Stage)[10]

Sea Level above Normal [cm]	Area of Venice Submerged [%]
90	2
100	5
110	14
120	29
130	43
140	54
150	63
160	69
170	74
180	78
190	82
200	86
>200	100

Exceptionally High Floods in Venice, Italy Since 1923 [10]

Sea Level above Normal [cm]	Date
194	4 Nov 1966
166	22 Dec 1979
158	1 Feb 1986
156	1 Dec 2008
151	12 Nov 1951
147	16 Apr 1936
147	16 Nov 2002
145	25 Dec 2009
145	15 Oct 1960
144	23 Dec 2009
144	3 Nov 1968
144	6 Nov 2000
142	8 Dec 1992
140	17 Feb 1979

Figure 21.2 Piazza San Marco, Venice, Italy during an acqua alta (high water) in 2005. Floods like these are quite common. Note that the 2005 acqua alta did not make the top-14-list (Table 21.2). (source: Wikimedia)

Figure 21.3 Tourists taking photos on Piazza San Marco, Venice Italy during the 1 December 2008 acqua alta. (source: boston.com)

Figure 21.4 Aerial view of the MOSE Project to protect Venice, Italy from acqua alta. (source: wikipedia)

Venice Floods. The threat of rising sea levels becomes particularly clear in Venice. The city is essentially built on a sinking lagoon and the flood risk is increasing with time. A flood in Venice, the acqua alta, is not caused by extensive rainfall but by a combination of unusually high tides combined with wind-whipped surf pushing into the lagoon. The highest acqua alta in 1966 was causes by a maximum high tide. It takes only 1.5m of water above normal to flood more than 65% of the city. Such a flood occurred on 1 December 2008 when an exceptionally high tide, strong winds and rainfall combined. It was the 4th highest flood since 1872 and submerged 66% of the city. Of the 14 highest floods since 1923, 5 occurred in the last 11 years, with the 2008 flood being the most severe in 22 years. To save the city from future, more severe flooding, the Modulo Sperimentale Elettromeccanico project [11] (MOSE, engl. experimental electromechenical module) is a defense system that consists of rows of mobile gates. When the tide reaches higher than 110 cm - a level when significant flooding starts to occur (Table 21.1) - the gates close and isolate the lagoon from the Adriatic Sea. The gates are constructed to operate to a highest level on 3 m. Work on the MOSE system began in 2003 but temporarily stopped in 2006 to review the budget. The work is expected to be completed in 2012 and the structure will cost € 4.3 billion ($ 6.1 billion US). However, this structure alone will not save Venice. In addition, complementary measures such as coastal reinforcement, the raising of quaysides and improvement of the lagoon environment will also be necessary. The project is not without controversy as environmental impact and high costs are downsides of the project. For these reasons, alternative proposals were presented to the government in 2006. But experts expressed that alternative measures would be ineffective and inappropriate so the government gave the go-head for MOSE to resume construction.

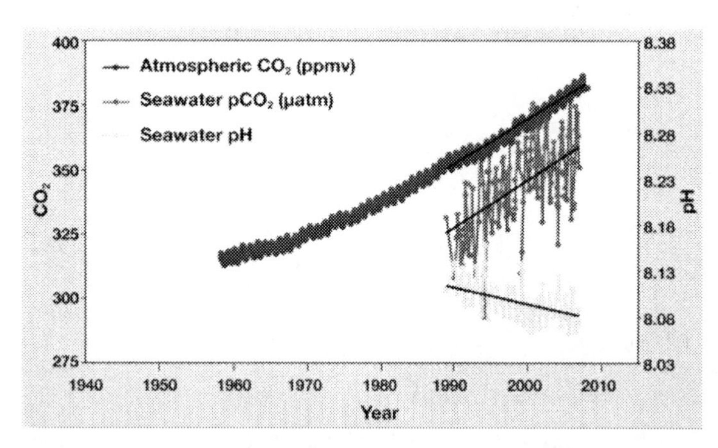

Figure 21.5 The correlation between rising levels of CO_2 in the atmosphere at Mauna Loa with rising CO_2 levels in the nearby ocean at Station Aloha, north of Oahu. As more CO_2 accumulates in the ocean, the pH of the ocean decreases. (source: NOAA)

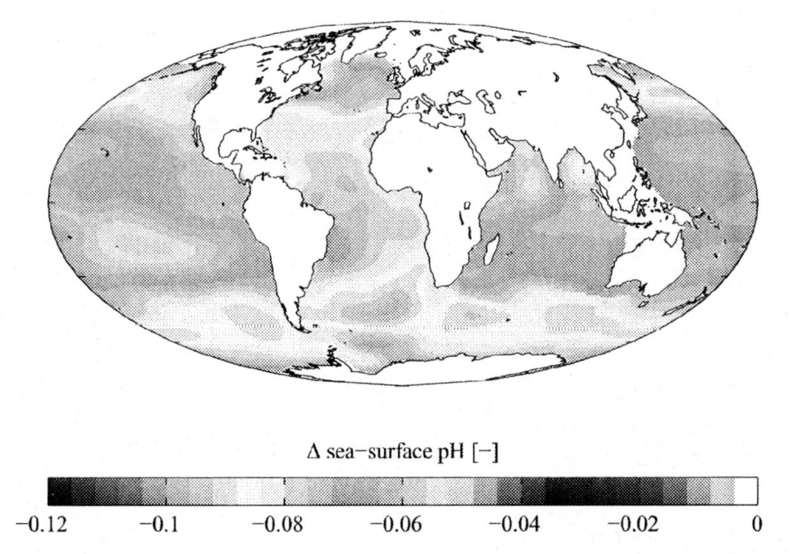

Figure 21.6 Change in sea surface pH caused by anthropogenic CO_2 between the 1700s and the 1990s. (source: wikipedia)

Figure 21.7 A rosette deployed from a research ship to collect ocean water samples off Puget Sound. (source: NOAA)

Material	Acidity [pH]
beer, vinegar	3
wine, tomatoes	4
black coffee	5
milk	6
rain	6.5
pure water	7

COPYRIGHTED MATERIAL — DO NOT DUPLICATE, DISTRIBUTE, OR POST

seawater today	8.069
baking soda	9
milk of magnesia	10
household ammonia	11
household bleach	12

Seawater Acidity and Change Relative to pre-Industrial Level[13]

Time	**Acidity [pH]**	**Change [%]**
seawater by 2100	7.82	+126.5
seawater by 2050	7.95	+69.8
seawater today	8.069	+28.8
seawater 1990s	8.104	+18.9
seawater 1700s	8.179	0

sea life unlikely to be harmed by changing ocean acidity: algae, cyanobacteria, sea grasses

sea life that may suffer from a more acidic ocean: soft-shelled clams, corals, calcareous plankton, pteropods

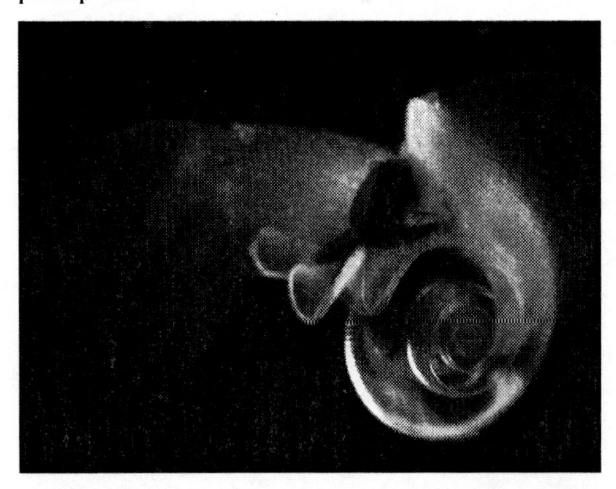

Figure 21.8 A pelagic (living in the open ocean) pteropod. (source: Wikimedia/NOAA)

12.3 Ocean Acidification

- see Lecture 14 for coupling between atmosphere and oceans
- current estimates are that the atmosphere stores about 46% of human produced CO_2, the oceans take up 29%, the northern hemisphere forest re-growth 7% and other parts in the biosphere 18%

COPYRIGHTED MATERIAL — DO NOT DUPLICATE, DISTRIBUTE, OR POST

- according to NOAA (National Ocean and Atmosphere Administration) studies, the oceans' chemistry has changed 100 faster since the industrial revolution began than in the previous 650,000yrs
- the amount of CO_2 dissolved in ocean water has increased from 398.1ppb in 1850 to 529.9 ppb in 2005 (30% increase in acidity); predictions put this number to 926.7 ppb by 2100
- the pH has changed from 8.16 in 1850 to 8.05 in 2005 and could be as low as 7.85 by 2100
- sea life that is likely impacted by a more acidic ocean: soft-shelled clams, corals, calcareous plankton and pteropods

Pteropods. Pteropods are small snails that build calcium carbonate shells; they are a critical part of the food chain in polar and near-polar seas; they are the preferred food for herring, pollock, cod, salmon and baleen whales

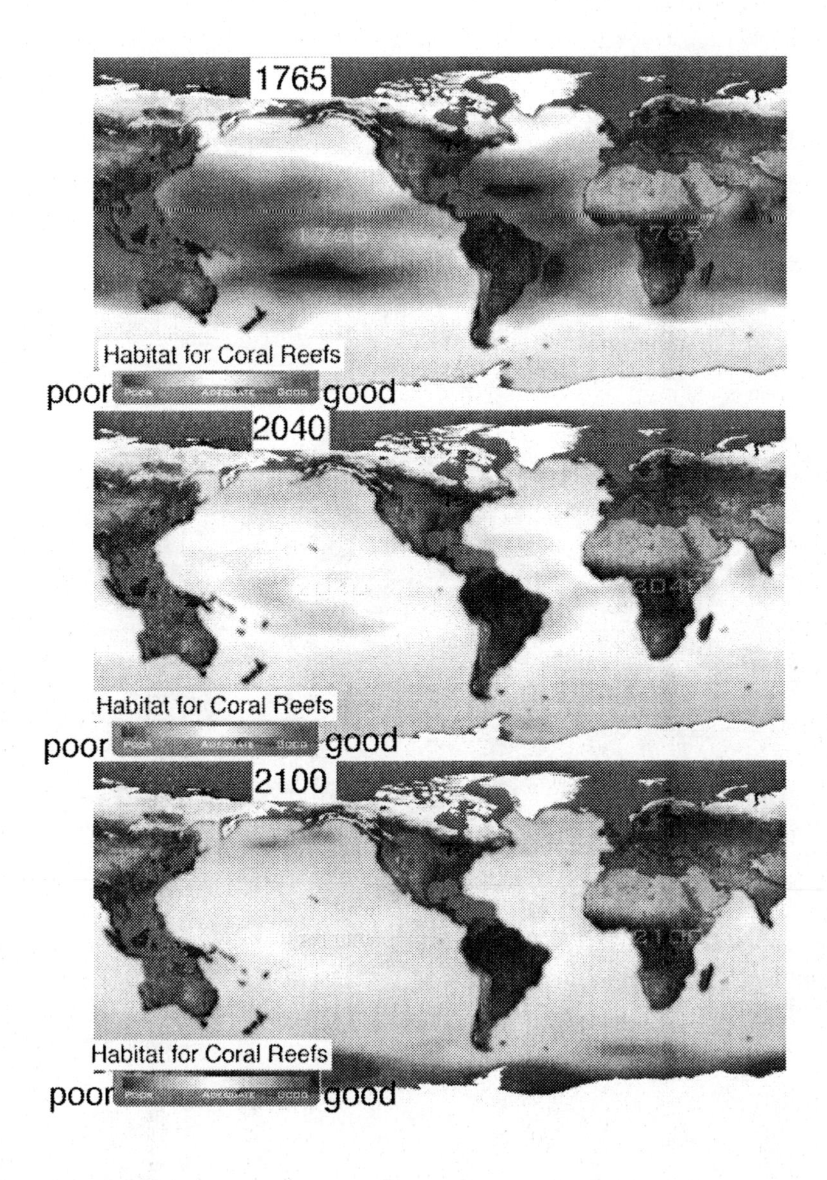

Figure 21.8 Change in conditions for coral reefs to thrive, 1765 and projected for 2040 and 2100, as a result of ocean acidification. (source: NOAA)

Figure 21.9 Example of the biodiversity of a coral reef in the Great Barrier Reef off the coast of Australia. A blue starfish in the foreground is resting on hard Acropora coral. (source: Wikipedia)

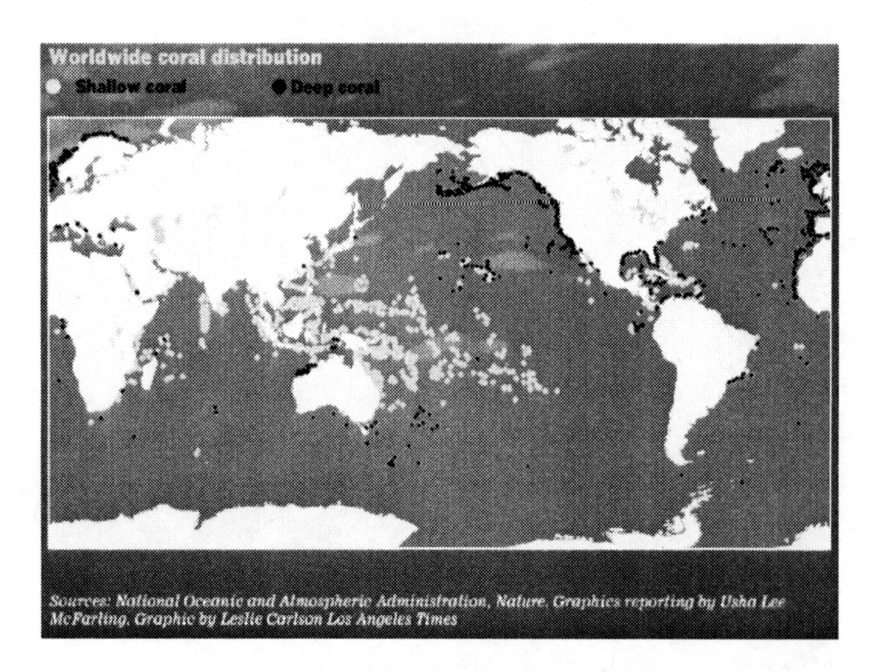

COPYRIGHTED MATERIAL — DO NOT DUPLICATE, DISTRIBUTE, OR POST

Figure 21.10 Location of shallow (mostly warm-water) and deep (mostly cold-water) coral reefs. For most of the warm-water coral reefs Landsat images are available at NASA's Millennium Coral Reefs Landsat Archive (http://seawifs.gsfc.nasa.gov/cgi/landsat.pl). (source: LA Times, November 2006)

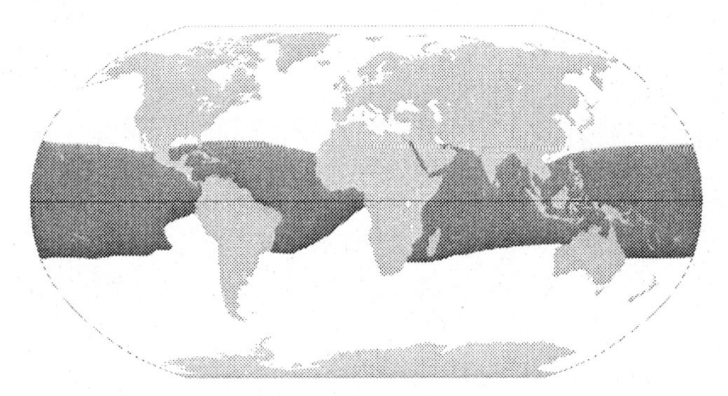

Figure 21.11 Boundary for 20°C isotherms. Most warm-water corals live within this boundary, at latitudes less than 30°. (source: Wikipedia/CIA World Factbook)

Figure 21.12 Bleaching corals expel the algae they depend on and lose their color. The photo was in the Moofushi coral reef, Maldives eight years after it was affected by the strong 1997/98 El Niño. (source: Wikipedia; Bruno de Giusti)

COPYRIGHTED MATERIAL — DO NOT DUPLICATE, DISTRIBUTE, OR POST

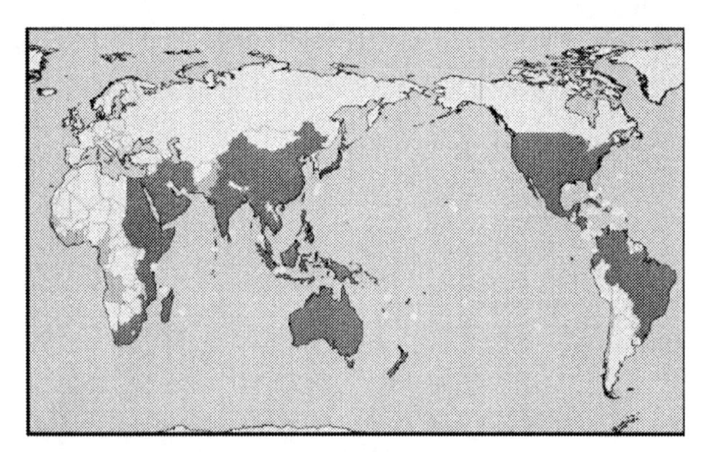

Figure 21.13 Locations of major bleaching events during the 15 years prior to the major 1997/98 El Niño bleaching event. (source: marinebiology.org; Jason Buchheim)

Get more info from http://www.marinebiology.org/coralbleaching.htm

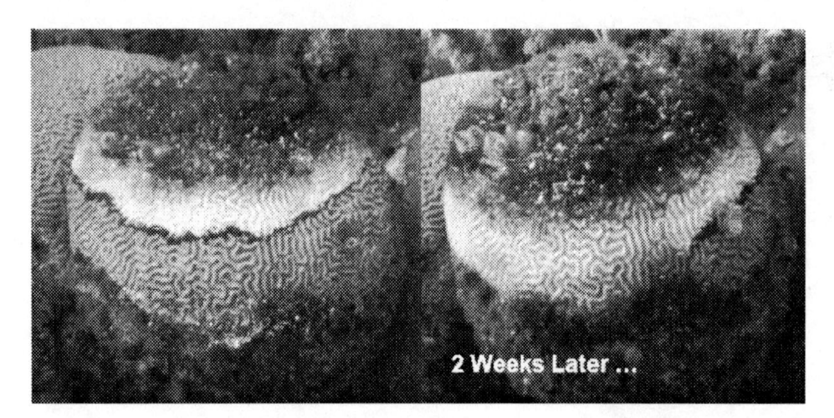

Figure 21.14 Example of Black band disease progression on a colony of brain corals. The disease is caused by a mix of pathogenic bacteria. (source: Wikipedia; Jon Shrives)

A bleached brain coral in the Caribbean is gradually infected with black band disease. Although the remaining bleached portion recovered, the diseased portion did not. (source: Science Daily, 3 Oct 2009). Some coral diseases come from fungi introduced by agriculture and water treatment plants

21.4 Coral Reefs and Coral Bleaching

- the health state of coral reefs is often used to estimate the oceans' health in general
- coral reefs are important source of revenue for tourist industry
- **impact on corals** has many sources [6]:

- physical damage: explosive fishing, anchor damage, construction, boat grounding, reef walking, removal of protective vegetation from islands
- freshwater runoff: from cities and cleared land
- introduction of invasive predators
- climate change: warming, sea level rise, hurricane activity
- exploitation: fishing, collecting corals, coral mining
- sedimentation: mining, dredging, filling, construction, soil erosion
- pollution: industry, oil spills, sewage, agricultural chemicals, fish farming, waste dumping, anti-fouling paint, etc

Coral Bleaching

- during coral bleaching the corals discard the algae living on them in symbiosis, leaving white skeletons behind
- since 1980, a quarter of the world's coral reefs have died
- possible causes for bleaching:
  - increased water temperatures
  - starvation due to decline in zooplankton
  - change in solar radiation
  - changes in water chemistry (see acidification below)
  - pathogenic infections (see Lecture 24)
  - salinity
- even though many corals thrive in warm water, they may bleach when the water gets warmer by just 1-2 °C
- major bleaching events occurred in the Caribbean in 2005 and during the 1997/98 El Niño
- different corals react differently to environmental factors; e.g. all of the hydrozoan *millepora boschmai* colonies near Panama were affected during the 1997/98 El Niño and died within 6 years. The species is now thought to be extinct
- the Great Barrier Reef east off Australia experienced massive bleaching events in 1998 (42% of corals affected), 2002 (54% affected) and the southern part also in 2006

Figure 21.15 Locations of kelp forests along the California coast. (source: imaging.geocomm.com)

Figure 21.16 A California Sheephead (marketed in Spanish as "vieja de California") in a local Giant Kelp forest. (source: LA Times November 2006; Rick Loomis)

The kelp is prime habitat for many sea creatures, providing hiding places for juvenile fish and food for abalone. Scientists believe that excessive harvesting of lobster and sheephead, which eat sea urchins, have resulted in an explosion of kelp-munching purple urchins.

21.5 The Kelp Forest

kelp: one of the oceans' most important resource

kelp forest: one of the ocean's most important habits

- large seaweeds (belonging to the brown algae); does not belong to the kingdom of plants but in the kingdom of protista (see Lecture 21)
- requires nutrient rich water below about 20°C (68°F); grows up to 30cm/day to a total length of over 60m (180ft)
- grows to form underwater forests and offers habitat for wide variety of sea creatures
- rich in iodine and alkali
- kelp ash used in soap and glass production
- alginate, a kelp-derived carbohydrate, used to thicken products such as ice cream, jelly, salad dressing, toothpaste, manufactured goods

Figure 21.17 Left: Nitrogen fertilizer being applied to growing corn on a field in Iowa; right: A commercial meet chicken production house in Florida. (source: wikimedia)

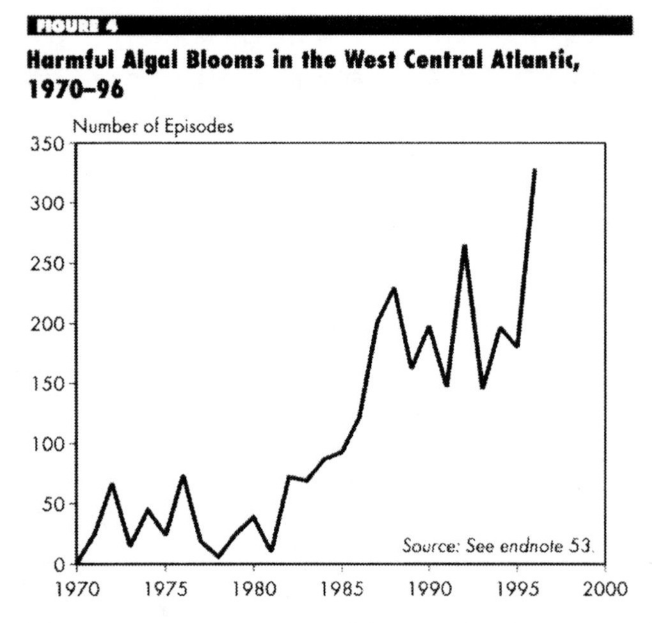

Figure 21.18 An example of the alarming increase in the occurrence of harmful algal blooms(14).

21.6 Impact of Farming and Industry on Ocean Wildlife

- costal habitats changed by pollution, dredging, filling, draining
- deforestation increases runoff into oceans (and polluting agricultural chemicals with it, see example of California Brown Pelican below)
- deforestation increases soil erosion and more sediments reach coastal areas
- mining runoff; e.g. mercury is used to extract gold from dirt
- waste disposal, untreated or improperly treated sewage, nuclear waste

- even pollution of atmosphere is taking toll on ocean life (see section on "Impact of Farming" below)
- **coastal dead zones**
    - : provoked by nitrogen, phosphorus, and other nutrients from fertilizer, large livestock farms, and septic systems
    - dramatic increase in recent decades
    - there are now 146 dead zones worldwide
    - most severe dead zones: (> 20,000 km^2) Bay of Bengals, Gulf of Mexico, Arabic Sea, Baltic Sea, East China Sea
- **beach cleanups**[7]: 6.2 Mio volunteers have removed nearly 50,000 tons of debris from world's beaches and waterways
- a small number of objects account for a large share of this debris:

- 60% is from recreational activity (fishing lines, nets, beach toys, food wrappers)
- 29% is from cigarette butts and filters

Many species in the oceans are affected by farming and other actions on land where the connection between the two are sometimes not immediately obvious.

Figure 21.19 California Brown Pelicans in spring plumage roosting at La Jolla Cove. (photo: Gabi Laske)

The California Brown Pelican was brought to near extinction in the 1970s/80s, due to the use of DDT on land. Among other things, DDT was widely used against mosquitoes which are vectors for a variety of diseases (e.g. malaria). Land birds greatly suffered by direct poisoning and sterility. They laid eggs with shells so fragile that they would break when the birds sat on them for incubation. Intriguingly, the California Brown Pelican, which only fishes in the oceans, was also affected. This emphasizes that actions on land can have far-reaching consequences for the oceans as well. The California Brown Pelican was affected by run-off in storm drains that started in farming areas and lagoons that were treated for pest control.

A little success story: The Brown Pelican has come back after the use of DDT was outlawed. The pelican now enjoys soaring just inches above the ocean waves in much greater numbers and drops like a rock to fish close to the beach. Its increase in numbers has prompted recent discussions to take the Brown Pelican off the endangered species list.

DDT: dichlorodiphenyltrichloroethanol; very effective poison that is used as insecticide; initially regards as wonder treatment against disease-carrying mosquitos because it is completely harmless to humans; organisms store this in fat cells where it accumulates; the insects became resistant to DDT but birds greatly suffered by direct poisoning, sterility, fragile egg shells; DDT was banned by many countries in the late 1970s/early 1980s after a lot of damage has been done

Figure 21.19 Predatory fish that are high in the food chain, such as the Yellowfin tuna shown here, are most likely of bioaccumulation of Mercury. (source: Wikipedia)

Tuna and Mercury

The release of Mercury (by human action) usually occurs far inland (e.g. gold mining and burning of low-grade coal). It is therefore intriguing why Tuna should have increased levels of Mercury in their tissue. It turns out that Mercury can travel long distances when airborne; e.g. power plants in the eastern and central U.S. pollute lakes in Nova Scotia, Canada (see special page).

In animals (and humans), Mercury is not reduced during metabolism. It is accumulated and often stored in the bones instead of Calcium. The Mercury accumulates as animals are eaten along the food chain, with tuna being near the top. Level of Mercury in tuna are now so high that certain groups in the population are advised not to consumer tuna at all or in only small, infrequent portions.

Figure 21.20 Children in Manila Bay, Philippines, collecting litter (to sell) in the harbor area. (source: UNEP)

Figure 21.21 Cruise ships anchoring at the San Diego, CA B Street Pier of the Embarcadero. (source: Wikimedia)

21.7 Marine Litter

Cruise ships:

- lax regulations allow world's growing fleet of more than 200 large cruise ships to dump untreated sewage into oceans (each day, a standard ship generates 95,000l of sewage from toilets, 540,000l sewage from sinks, galleys and showers, > 6 tons of garbage and solid waste, 56l of toxic chemicals, 26,500l oily bilge water)

Though not planned as such, this problem may be mitigated in San Diego to some extent in the near future. Until recently, San Diego's cruise ship industry was the second largest in California, generating an estimated $2 million annually from the purchase of food, fuel, supplies and maintenance services. At its peak in 2008 the number of ship calls was 255. In 2010, the Port of San Diego hosted an estimated 152 ship calls and more than 515,000 passengers. But due to drug violence, cruise destinations "south of the border" have become less attractive and numbers for 2011 are as low as 260,000 passengers. One of the three main cruise operators recently announced that they will relocate to Australia. By 2013, ship calls could therefore decline to only 76.[12]

Get more info on http://marine-litter.gpa.unep.org/facts/facts.htm
And
http://www.unep.org/regionalseas/marinelitter/about/effects/default.asp

Figure 21.22 Trash collected from a beach in the Los Angeles area. (source: Algalita.org)

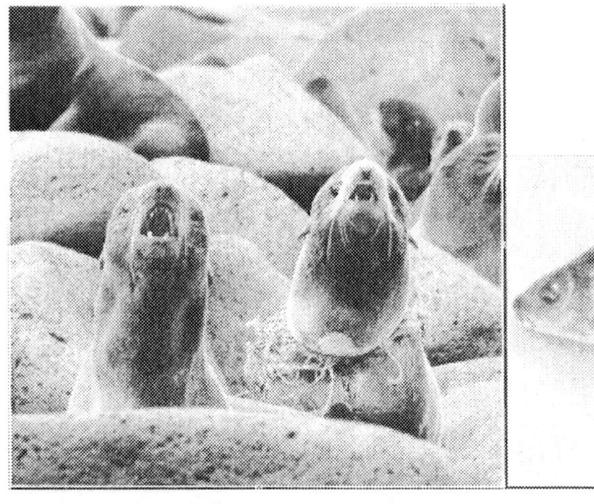

Figure 21.23a Marine life entangled in human trash. Most of them will not survive. (source: UNEP)

Figure 21.23b This common snapping turtle is at least 15 years old. She crawled into a plastic bottle ring as a youngster and her shell was then constricted by the ring. Her spinal cord is now partially unprotected. (source: Algalita.org and audubonistitute.org/zoo; Photo: Dino Ferri)

Figure 21.24 Albatrosses breeding and roosting on Midway Island and Kure Atoll amid trash. (source: Algalita.org)

COPYRIGHTED MATERIAL — DO NOT DUPLICATE, DISTRIBUTE, OR POST

Figure 21.25 Albatross carcasses exposing their most recent content in their stomach. It is estimated that more than 200,000 Laysan Albatrosses die per year loaded with plastic trash. More than 98% of albatross chicks have plastic in the stomachs[14]. (source: Algalita.org)

Plastic in the Oceans[2]

Recently, plastic has become one of the worst pollutants in the oceans and have become a very serious threat to numerous marine species.

- plastic is not bio-degradable; it degrades by chemical processes over a long time. For example, under sunlight it breaks up into ever smaller pieces (photodegradation). But it is still plastic for a very long time. It takes 450 years for a plastic bottle to degrade. It takes longer for plastic to degrade in the oceans than on land, because the water prevents heat buildup.
- UNEP estimates that 18,000 pieces of plastic litter are floating on every km^2 of the ocean. The San Diego Tribune wrote on November 30 that 65% of litter in the north Pacific Ocean is plastic.
- along shipping lane between Iceland and Scotland, 3 times more plastic found in 1990s than in 1960s
- broken, degraded plastic pieces outweigh surface zooplankton in the Central North Pacific (CNP gyre) by 6 to 1
- toxic chemicals in plastics can make marine animals sick. Over 80 species of seabirds mistakenly ingest plastic because they think it is food.
- more than 100,000 mammals and sea turtles die each year from entanglement in, or ingestion of, plastics
- 90% of Laysan Albatross chick carcasses contain plastic. 200,000 Laysan Albatross chicks die loaded with plastic each year. The San Diego Tribune wrote on November 30 that 98% of albatross chicks have plastic in their digestive system. Plastic stays there and makes the birds feel full. They eventually starve to death.
- in turtles, plastic can block the digestive system and make the animals float so that they cannot dive for food

The Great Pacific Garbage Patch

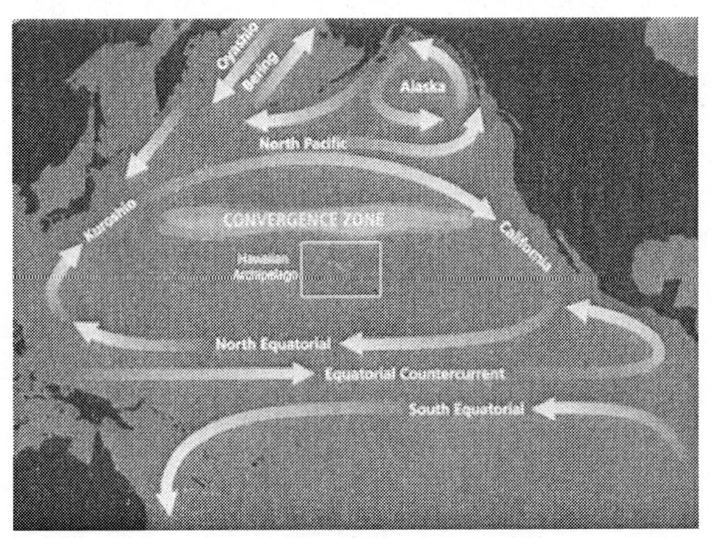

Figure 21.26 The Great Pacific Garbage Patch or the Pacific Trash Vortex may be twice the size of Texas and lies within the North Pacific Gyre.

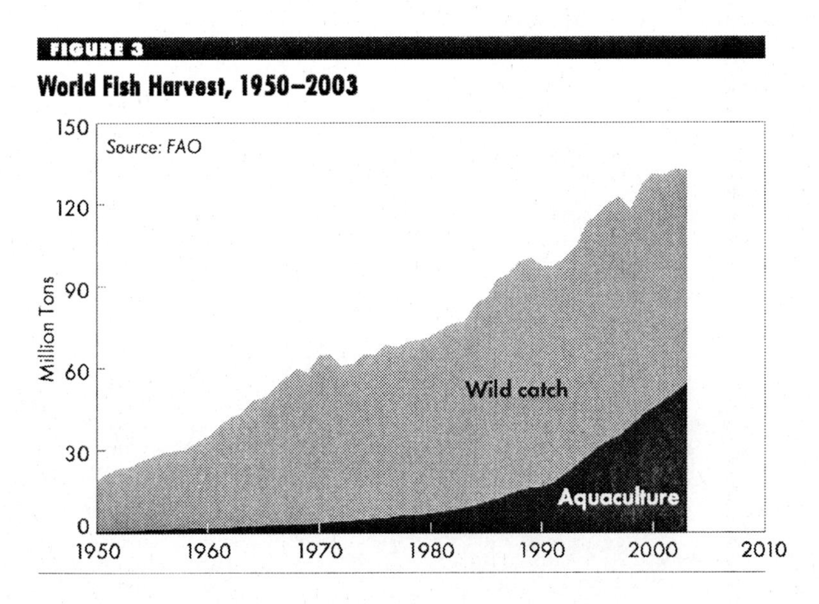

Figure 21.27 The increase in fish harvest since 1950. The serious state of wild fisheries is masked by increasing contribution from aquaculture.[7]

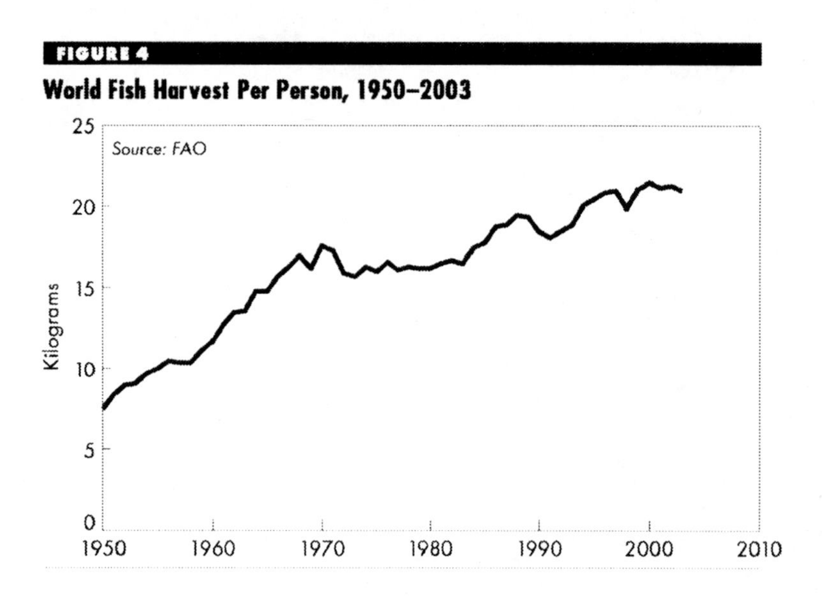

Figure 21.28 The increase in the per capita harvest of fish may not necessarily reflect the increase in per capita consumption as 23% of the harvest is for purposes other than food.(7)

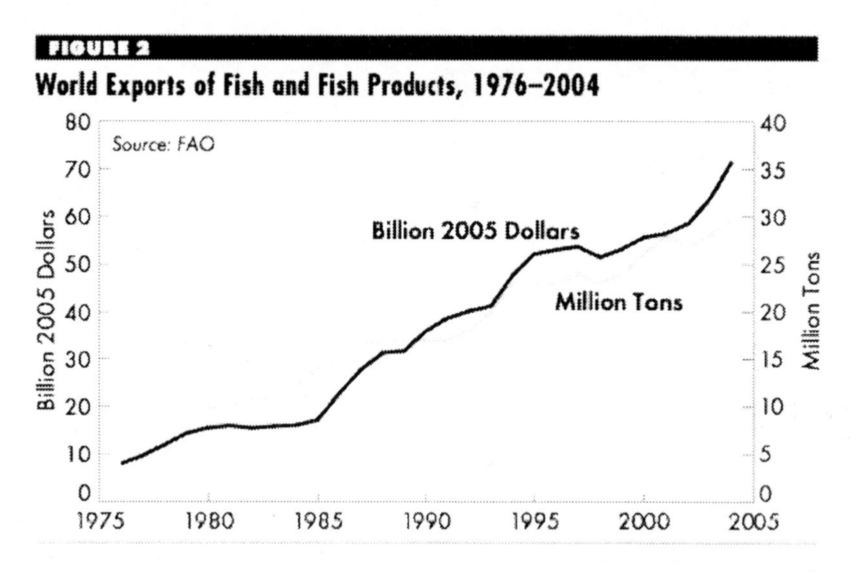

Figure 21.29 The increasing economic value of the fishing industry.(7)

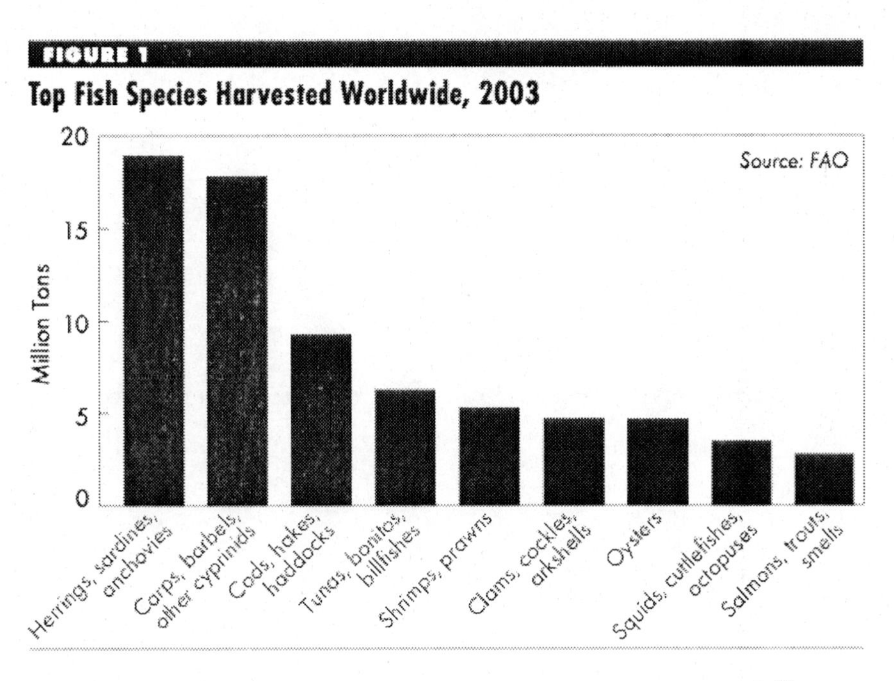

Figure 21.30 The top species harvested and marketed worldwide.[7]

Figure 21.31 A catch of orange roughy in 1988. (photo: Greenpeace; Westerskov)

The orange roughy or deep sea perch appeared in the mid 1980s on the market as cod and other fisheries disappeared. The fish was discovered as Slimehead along the Chatham Rise in the New Zealand exclusive economical zone (EEZ) and was renamed for marketing purposes. By 1982, just four years after its discovery, fishermen caught 35,000 tons. By the late 1980s, scientists learnt that it takes an orange roughy 30 years to mature, 10 times longer than a cod. Scientists calculated that the maximum sustainable yield was 7,500 tons. Just four years after its peak harvest in the 1989-90 season, orange roughy catches plummeted by 70%. Recently discovered substitute stocks are rapidly dwindling and scientists fear that this fish has been harvested beyond recovery. Orange roughy is a deep-sea fish caught with a deep-water trawl. (14)

21.8 The Global Fisheries – An Unlimited Renewable Resource?

- fish considered healthy food, with high levels in fatty acids and trace minerals; fish is good for neurological functions, reduced cancer risk, heart disease
- for more than 1 billion people (mostly in Asia), fish supplies 30% of proteins (6% on average globally)
- seafood has become scarcer as demand has grown; e.g. Chinese consumers now eat 5 times as much seafood per capita as in 1961; total fish consumption in China has increase 10-fold and in the U.S. 2.5-fold[7]
- people tend to remember fish populations only during own lifetime and are not aware what it was like only a generation ago, so many are not aware of creeping seafood decline
- experts may set false reference points for evaluating economic losses from overfishing and for recommending restoration; very large historical fish catches may be dismissed as myths/inaccurate anecdotes (this process is called "shifting baseline"); e.g. Atlantic cod was once plentiful but current generation may grow up thinking that cod never was in the Atlantic
- today, most of world's seafood, from tuna to salmon to bay scallops, is threatened with extinction
- for less-threatened species, like shrimp or farmed salmon, main issues are how they are raised or caught, which can have adverse impacts on the environment
- today's fishing technique's include sonar to locate fish, ever-longer longlines, ever-growing trawl and driftnets, larger and stronger winches, larger refrigerators
- today's fishing fleets now cover wide swaths, leaving little room for fish to hide and escape
- at same time, fuel consumption has increased dramatically; e.g. in 2000, burned 43 Mio tons of fuel to catch 80 Mio tons of fish; i.e. fishermen used 12.5 times as much energy to catch fish as fish provide to those who eat them; this is still more energy efficient than raising beef and raising salmon!
- there is no sign that commercial fishing companies will voluntarily change their practices, especially as the soaring demand for fish continues to push up its value

The Leading Nations in the 2001 Marine Capture Fisheries (million tones) [9]

China	14
Peru	8
US, Japan	5
Indonesia, Chile	4
Russia, India, Norway, Thailand	3
South Korea, Iceland	1.5-2
Philippines. Denmark	1.5-2
Vietnam	1-1.5
Mexico, Malaysia	1-1.5
Morocco, Spain, Canada	1-1.5

How we used the catch in 2001 [9]

canned food	10%
frozen food	19%
fresh food	40%
cured food	8%
non-food (oil, food, fertilizer)	23%

Declining Fisheries - Some Numbers

- of some 30,000 known fish species, only approx. 1000 are eaten by humans; only a small share of these make up most of the catch; e.g. Alaska pollock, Peruvian anchovy, Atlantic bluefin tuna, and Chilean jack mackerel make up 13% of global wild catch and carp, catfish, tilapia and salmon dominate aquaculture (fish farms)
- 7 nations take in 2/3 of the global catch: China (47.3 Mio tons), Peru (6.1), India (5.9), Indonesia (5.7), U.S. (5.5), Japan (5.5); Japan now world's largest importer
- in 2003, world's fish farmers and fishing fleets harvested 132.5 Mio tons of seafood, 7 times what it was in 1950
- 74% comes from marine areas, 26% come from inland freshwater bodies
- of the marine harvest, 88% come from shallow shelf areas (less than 120km from coast, no deeper than 200m)

- approx. 67% of world's major stocks have been fished at or beyond their capacity
- another 10% have been harvested so heavily that fish populations will take many years to recover
- since 1997, wild harvests have fallen 13% from the peak of approx. 87 Mio tons
- fish farming harvest increased 50% from 35.8 Mio tons the 54.8 Mio tons; now accounts for 40% of global fish harvest
- by 2004, industrial fleets had emptied oceans of at least 90% of all large predators (tuna, marlin, swordfish, sharks, cod, halibut, skates, flounder), within the last 50 years
- size and quality of fish declining
- some inland lakes may be affected even more; harvest from lakes and rivers has quadrupled since 1950s (now 8.7 Mio tons/year, excl. fish farms); e.g. catch of legendary Mekong catfish (3m long, 300kg) has fallen from 60 in 1995 to 4 in 2005
- endangered species (swordfish, Atlantic cod, Chilean sea bass) make comeback on restaurants' menus
- Patagonian toothfish renamed to Chilean sea bass; now again Patagonian toothfish

http://en.wikipedia.org/wiki/Forage_fish

The Peruvian anchoveta fishery is now the biggest in the world (10.7 million tonnes in 2004), while the Alaskan pollock fishery in the Bering Sea is the largest single species fishery in the world (3 million tonnes). The Alaskan pollock is said to be the largest remaining single species source of palatable fish in the world.[20] However, the biomass of pollock has declined in recent years, perhaps spelling trouble for both the Bering Sea ecosystem and the commercial fishery it supports. Acoustic surveys by NOAA indicate that the 2008 pollock population is almost 50 percent lower than last year's survey levels.[21] Some scientists think this decline in Alaska pollock could repeat the collapse experienced by Atlantic cod, which could have negative consequences for the entire Bering Sea ecosystem. Salmon, halibut, endangered Steller sea lions, fur seals, and humpback whales eat pollock and depend on healthy populations to sustain themselves.[22]

Use as animal feed

Eighty percent of the forage fish caught are fed to animals. Ninety percent is processed into fishmeal and fish oil. Of this, 46 percent was fed to farmed fish, 24 percent to pigs, and 22 percent to poultry

(2002).[4][23][24] Six times the weight of forage fish is fed to pigs and poultry alone than the entire seafood consumption of the U.S. market.

According to Turchini and De Silva (2008), another 2.5 million tonnes of the annual forage fish catch is consumed by the global cat food industry. In Australia, pet cats eat 13.7 kilograms of fish a year compared to the 11 kilograms eaten by the average Australian. The pet food industry is increasingly marketing premium and super-premium products, when different raw materials, such as the by-products of the fish filleting industry, could be used instead.

Environmental issues

A recent study (2008) by fisheries scientists Jacqueline Alder, Daniel Pauly and colleagues is the product of a nine-year Sea Around Us Project. The study concludes that...[4]

1. The composition of landings of forage fish fisheries have changed over the past 50 years with the trophic level of fish used in fishmeal increasing over the past 20 years.
2. Our understanding of the role of forage fish in marine ecosystem and the impact of fishing is still limited.
3. Landing of forage fish peaked by the 1970s, and these high levels are highly unlikely in the future, even if fisheries are managed sustainably.
4. The consumption of forage fish by seabirds and marine mammals is not likely to be onerous to fisheries, except in a few localized areas. By contrast, fisheries, by reducing the biomass of small pelagics, might pose a threat to these predators, particularly to those species for which stocks have been heavily depleted by human exploitation in the past.
5. Some forage fish species are consumed by many people with consumption patterns changing over the last 20 years.
6. Aquaculture continues to increase its consumption of fishmeal and fish oil.

Declining Fisheries Worldwide (9)

Fishing quotas. Quotas have exacerbated the problem of "illegal, unreported and unregulated (IUU) fishing" which, according to the FAO (Food and Agriculture Organization), is "growing both in scope and intensity" – in some localities the catch of commercially valuable species may exceed permitted levels by over 300%.

Figure 21.32a Atlantic cod. (photo: Wikipedia; Hans-Peter Fjeld)

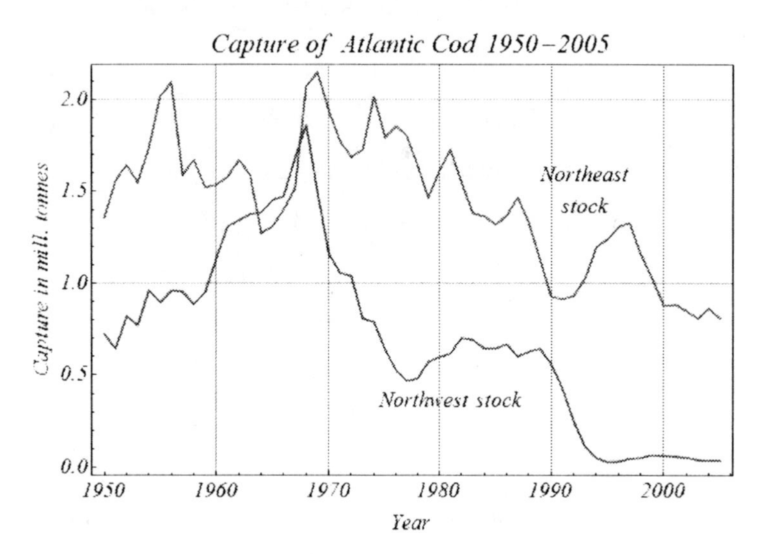

Figure 21.32b Atlantic cod capture 1950-2005. (source: Wikipedia from FAO data)

Disappearing cod. According to the World Wide Fund for Nature, the world's cod fisheries are disappearing fast, declining from 3.4 million tones in 1970 to a million tones in 2000. Particularly threatened is the world's largest remaining stock in the Barents Sea. North America's catch has declined by 90% since the early 1980s, while in European waters, the North Sea cod catch is now just a quarter of what it was two decades ago. Over- and illegal fishing together with oil exploration could "wipe out cod by 2020".

NW Atlantic overfishing. The total cod, hake and haddock catch in the Northwest Atlantic fell from 840,000 tonnes in 1980 to 148,000 tonnes in 2001.

Fall in abundance. The Georges Bank, off the coast of new England, US, was once one of the world's most productive fisheries. Stocks of flounder, haddock, and cod fell so low that fishing of these species was banned in 1994.

Club mackerel. Fished in wide areas of the Atlantic and Pacific, the catch has declined as a result of overfishing, falling from 2.4 million tonnes to 1.8 million tonnes between 1997 and 2001.

Southern African pilchard. Both local (South African and Angolan) and long-distance fleets (Polish) overexploited this fishery, causing its decline and collapse since 1970. Between 1995 and 2001, catch levels rose from 158,000 tonnes to 200,000 tonnes.

Indian mackerel. In the Gulf of Thailand, the change from traditional fishing to modern types of trwl in the 1960s initially produced an increase in catches of Indian mackerel and other species. Then it fell and the global catch continues to fall, from 304,000 tonnes in 1997 to 184,000 tonnes in 2001.

Indian Ocean shrimp. Factory fleets operating in shallow waters are scooping up the valuable shrimp, depriving traditional fishing peoples of their livelihood. The situation is aggravated by loss of nursery areas and coastal pollution.

The Anchoveta crisis. Until 1972, the enormously productive anchoveta fishery off Peru was the world's largest. The is spectacularly crashed due to years of overfishing combined with that year's strong El Niño, when a stable layer of warm surface water prevented the upwelling of nutrient-rich deep water. The catch continues to fluctuate: in 1997, it reached 7.7 million tonnes, then in 1998 only 1.7 million tonnes (though this was another strong El Niño cycle), before rising to 8.7 million tonnes in 1999 and 11.3 million tones in 2000. By 2001, it was down again to 7.2 million tonnes.

Pacific Ocean perch. From a 1965 harvest of almost 0.5 million tonnes, overfishing led to the continued fall in annual catches of Pacific Ocean perch to a low of 13,000 tonnes in 1979. Ten years later, the catch was only 33,000 tonnes.

California sardine. In the 1930s, annual catches of sardine exceeded 500,000 tonnes, providing prosperity to the fishing peoples of Monterey, immortalized in Steinbeck's Cannery Row. By the 1950s, overfishing resulted in stock collapse, with no significant recovery until the 1980s. Between 1998 and 2001, the catch rose from 380,000 tonnes to 685,000 tonnes.

Figure 21.33 The Giant Sea Bass reaches a size of 2.5 m and is critically endangered. Little is known about its biology and behavior. (source: Wikipedia)

Giant Sea Bass – a local example.

- critically endangered
- up to 2.5m long, and 255 kg
- very slow reproduction; fish stock doubles only once in 14 years
- lives in kelp forest off California and Japan
- once common, it supported commercial fishing industry in the 1800s in Southern California, with hundreds of thousands of kg annually
- by late 1970s if was found that local populations were in serious trouble
- since 1982 protected from commercial and sport fishing

Figure 21.34a Oceanic whitetip shark at Elphinstone Reef in the Red Sea. (source: Wikipedia; Oldak Quill)

Figure 21.34b Small purple-colored thresher shark caught at Pacifica Pier, CA. (source: Wikipedia; Paul Ester)

Figure 21.35 Shark fins drying on deck, after the de-finned shark was thrown back into the ocean left to die. (source: Greenpeace)

COPYRIGHTED MATERIAL — DO NOT DUPLICATE, DISTRIBUTE, OR POST

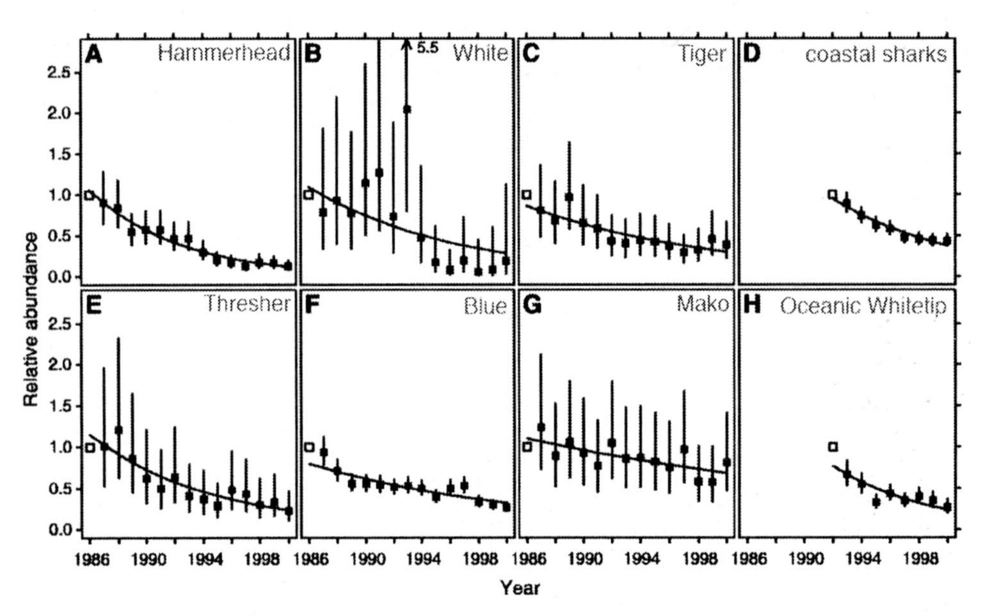

Figure 21.36 The decline in the Northwest Atlantic shark populations in the 15 years before 2003. Some species declined by 80% or more! (source: Baum et al., 2003 [(17)])

Declining Fish Stock - Example Sharks

- sharks have lived on this planet for 450 Mio years
- true fish (as opposed to whales which are mammals); largest fish: whale shark (up to 12.2 m/40 ft; lives up to 70 years)
- 350 species of variable size (a few inches to 15m), all of which are threatened
- dramatic decline in Northwest Atlantic in just 15 years[8]; e.g. Hammerhead has declined by 89% since 1986, White Shark has declined 79%, Tiger Shark by 65%, Thresher Shark by 80%, Blue Sharks by 60%
- some sharks in Gulf of Mexico declined by 90%
- nearly 50 shark species close to disappearing worldwide
- a bowl of shark fin soup can cost $200 -> shark fin industry booming
- approx. 100 Mio sharks are caught each year, yielding 8000 tons of fins
- particularly inhumane as only shark fins are harvested, while sharks are thrown back into ocean to die. Sharks do not have swim bladders so they need to constantly swim and move water over their gills in order to stay afloat and breathe. Once the fins are removed, they suffocate and "drown".
- several nations have banned such practice
- trade in several species in prohibited under international law but laws may not always be effective

- being on top of the food chain, shark meat and fins can contain concentrations of mercury higher than what is considered safe to eat
- finning has been banned from the Atlantic and Mediterranean but is still going on in the Indian and Pacific

Figure 21.37a By-catch from shrimp fishing: marine life not intended for marketing and consumption. (source: Wikipedia/NOAA)

Figure 21.37b A Dall's porpoise caught as by-catch in a fishing net. (source: Wikipedia/NOAA)

By-catch

Another 25% is discarded at sea as unwanted by-catch, e.g. birds, unwanted fish, marine mammals, turtles, juveniles of the targeted species, even fish sought after in other fisheries (e.g. after the 1992 Canadian moratorium to limit the catch of Atlantic cod, it can is now by-catch to other target fish); the by-catch in shrimp fishing is particularly high, due to the fine nets that are used. For each 1 kg of wild shrimp, there is 5 kg of by-catch.

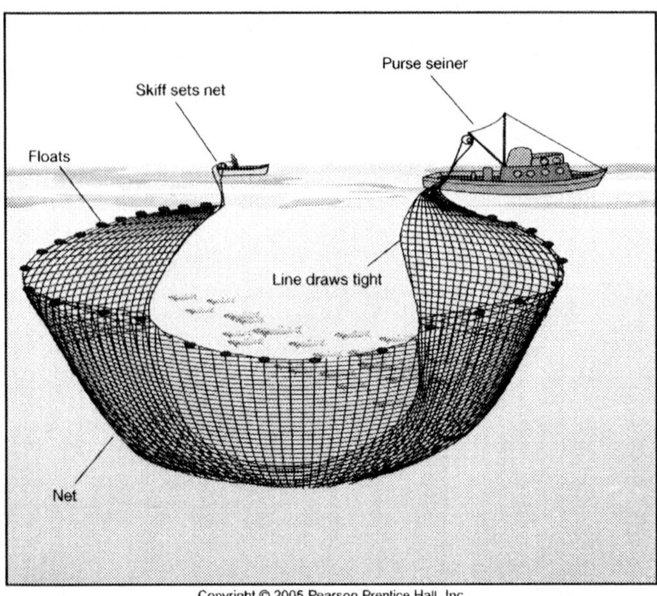

Figure 21.38 Fishing with a purse seine net. (source: Trujillo and Thruman, 2005 [16])

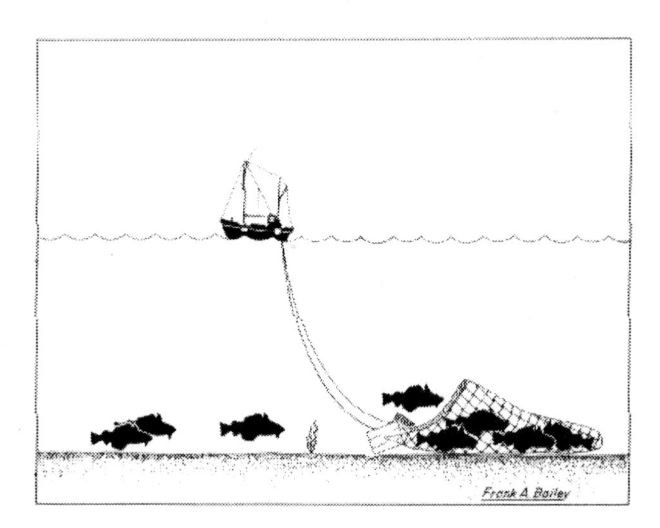

Figure 21.39 Concept drawing of fishing with a trawler. Because of its destructive impact on the ocean floor flora and fauna, trawl fishing is often likened to clear-cutting a forest. (source: wikipedia)

Some Bad Fishing Practices and an Alternative

- poisoning with sodium cyanide (e.g. S.E. Asia)
- dynamite (e.g. Mediterranean)
- gillnets/drift nets (trap/drown large mammals and birds)
- trawling; trawling clears virtually all continental shelved (within 120km of coast, up to 200m deep water) twice a year (equals 150 times area of global forest); likened to clear-cutting techniques in forests as everything along the floor is swept up/destroyed

- o alternative: aquaculture; raise fish in suitable fish farms (see also Appendix B)

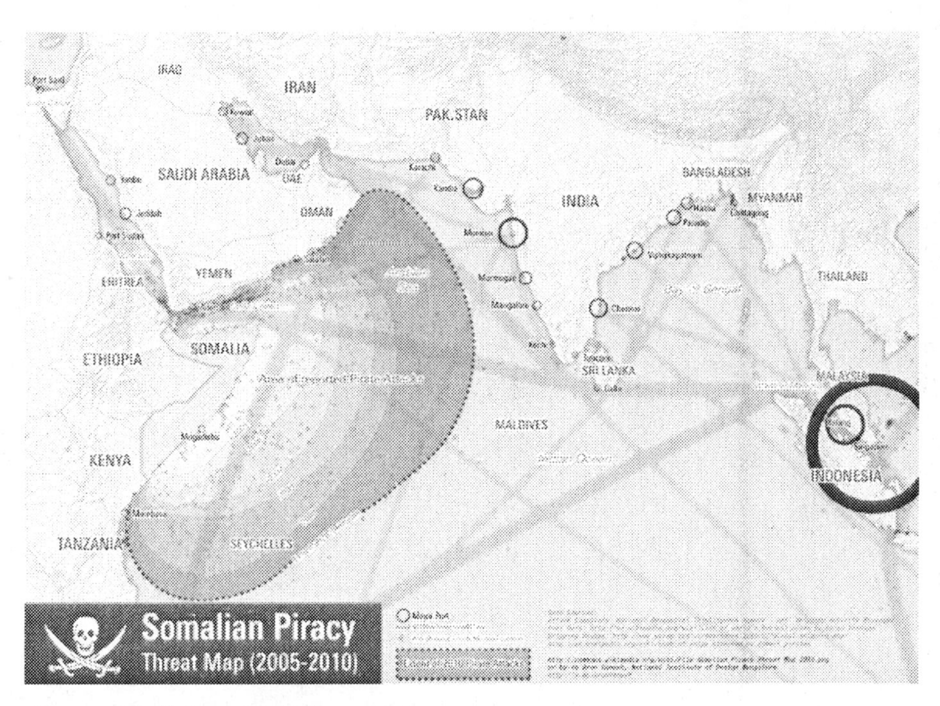

Figure 21.40 (source: Wikipedia)

Power struggle and competition: the ever-increasing appetite for marine fish and declining fisheries presses fishermen to venture out not only into the open ocean far away from home, sometimes even to invade each other's coastal economic zones. Tension has increased in 2011 between Australia and Japan as the latter kill whales "for scientific studies" off Australia's coasts. A very problematic and perhaps unethical situation has arisen in the waters around Africa. West Africa's coasts feature some prolific fisheries. To a large degree, local populations depend on subsistence fishing that leave fisheries at sustainable levels. Industrial fishing fleets from outside have exploited this zone with sophisticated catching gear and cold-storage facilities – long-distance equipment characteristic of advanced nations. In the 1960s, West Africans watched half their catch being taken away by developed nations. By the mid-1970s, the 22 African nations concerned had increased the catch substantially, but witnessed their share decline to one-third of all fish taken, the bulk being accounted for by the former USSR, Spain, France, Poland, Japan, and 14 other developed nations. One tonne of fishmeal fed to livestock in Europe produced less than half a tonne of pork or poultry, far less again if used as fertilizer to grow animal-feed grains. Had the West African fish resources been used for direct human consumption by the people of West Africa, they would have represented an additional 12 kg of animal protein per person per year, a 50% increase for many people.

Figure 21.41 A collage of armed Somali pirates who are typically 20-35 year-old men from the region of Puntland in northeastern Somalia. The East African Seafarer's Association estimates that there are at least five gangs, with a total of 1000 armed pirates. (source: Wikipedia)

Figure 21.42 Ships of the multinational fleet Combined Task Force 150 (CTF-150), March 2004. CTF-150 was established to monitor, inspect, board and stop suspect shipping to pursue the war on terrorism. Countries currently contributing to this fleet include Australia, Canada, France, Germany, Italy, Pakistan, New Zealand, Spain, United Kingdom and the United States. (source: Wikipedia)

A much more serious and despicable situation is escalating along Africa's east coast but it may show what could unfold elsewhere as resources become scarce. As along Africa's west coast, it started with subsistence fishing. But, according to a United Nations report, illegal fishing by outside nations and dumping of toxic waste from foreign vessels in Somalia waters have severely constrained the ability of local fishermen to earn a living and forced many into piracy instead. There is also a suggestion that many communities along the coast support privacy as a means of defending their coast. In the absence of an effective coast guard following the outbreak of the Somali Civil war, pirates may believe they are protecting their fishing grounds and compensate for the stolen marine resources. Now in 2011, after small sail boats are hijacked and killed it is obvious that the situation has escalated to the point that the original cause no longer justifies the means. In fact, there are reports that the pirates are no longer the original Somali fishermen but often mercenaries hired by terrorist groups. The increasing involvement of foreign military operations to free hostages and captured ships have made this situation a global crisis and the question is how to solve it. Clearly, a solution has to include reliably security measures to prevent illegal fishing toxic waste dumping in Somali waters but also employment opportunities and food security for the Somali people.

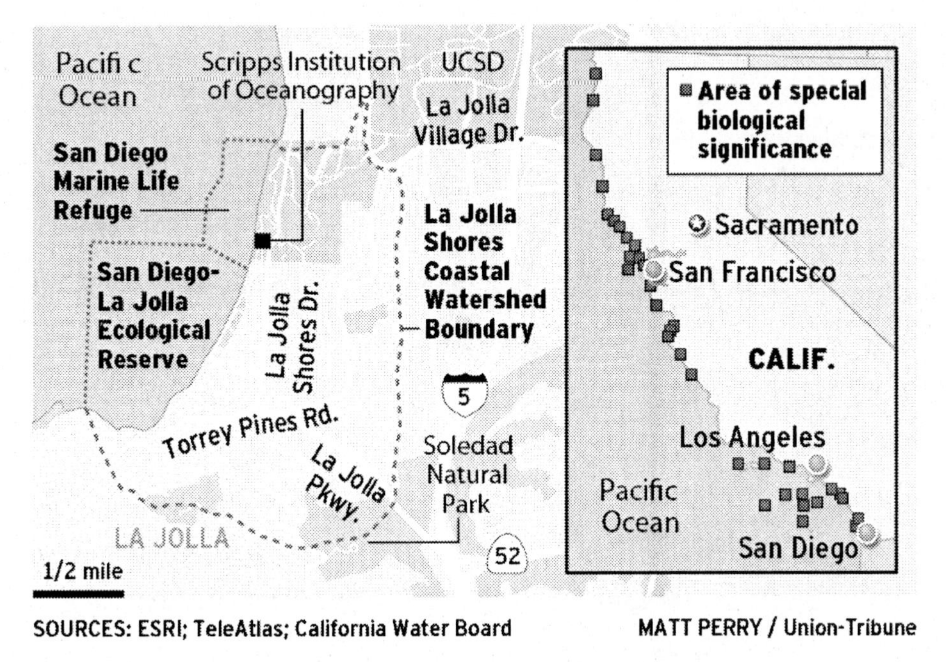

Figure 21.43 Boundaries of the marine reserve and refuge off the coast of La Jolla, CA. (source: SD Union Tribune, 12 Oct 2006)

21.9 Fishing Industry - Some Good News

- **marine reserves** have shown to be effective to rebuild depleted fish populations; have to be large enough; may not work in open ocean; have to include coastal area on land to effectively avoid/decrease pollution

North Sea herring. Industrialized fishing in the mid-1960s and 1970s caused a huge decline in annual catches. The fishery was closed between 1978 and 1982 to allow stock to recover, but in the late 1980s they declined again. A further recovery since the mid-1990s led to the highest level of stocks recorded for 40 years in 2003.

aquaculture: fish farms relieve strain on wild fisheries. The bulk of world marine aquaculture production, such as salmon, is in China, with 10 million tones of the total 18 million tones in 2001. Other major marine producers include Japan (0.7 million), Chile and Norway (0.5 million), Spain and

South Korea (almost 0.3 million). Inland aquaculture (e.g. catfish and trout) totaled 22 million tones, again dominated by China with 16 million tones.

- citizens are trained/help with seeding area with young fish/shellfish, e.g. Peconic Bay/ NY Long Island is seeded with scallops, clams and oysters
- London-based Marine Stewardship Council certifies certain seafood as "sustainable" and has granted its label to 18 fisheries worldwide, incl. North Sea herring, Australian mackerel, and Baja red rock lobster
- more than 370 products in nearly 30 nations now carry the "Fish Forever" logo
- certain food companies base their business on "the story behind the fish" (how it's raised, caught and processed)
- large chains like Unilever, Wal-Mart and Red Lobster have committed to source their seafood only from intact fish populations
- EcoFish identifies and markets seafood originating from environmentally sustainable fisheries (now in 1000 stores and 150 restaurants)
- CleanFish markets seafood caught by smaller-scale fishers that are less likely to harm the marine environment
- some high-profile celebrities and companies rally against shark-finning and Hong Kong university and Disneyland took shark fin soup off their menus
- groups like the Monterey Bay Aquarium in California and the Blue Ocean Institute in New York are issuing guise to inform people which seafood is the most "ocean friendly" (see below); guides encourage shoppers which species to avoid
- these fish cards are not yet widely used but could be effective to put pressure on fishing industry; e.g. campaigning for dolphin-safe tuna fishing in late 1980s was quite successful
- since becoming a U.S. state in 1959, Alaska's constitution requires to protect and conserve salmon habitat; local fishing industry worked with government to rebuild and not overfish stock; regulations strictly limit number of fishing permits; state biologists monitor spawning; since 1959, harvest increased from 20 Mio tons the 206 Mio tons, the third largest catch on record
- Cape Cod Hookers: local fishermen distinguish themselves by promoting "old-fashioned" hook line fishing that damage fish less that commercial techniques; by-catch more likely to survive; fish more expensive but story sells well

- Loch Duart, Scotland (farmed) Salmon: developed better farming techniques to avoid pollution; more room for fish avoids the application of antibiotics; uses only native fish species; now train fish farmers in Maine and Nova Scotia

on the downside: this does not include farmed salmon and Asian-farmed shrimp, which constitute the bulk of chains' seafood sales. Whilst aquaculture helps meet our appetites for fish and seafood, it might be confusing why farmed salmon is not in the green category but it turns out that salmon farming and other aquaculture can actually contribute to further decline of wild fisheries. More than two pounds of wild fish is needed to produce one pound of farmed salmon. Salmon food comprises ground sardines, anchovies, mackerel, herring, and other fish. In 2001, aquaculture demanded 1/3 of the produced fishmeal worldwide. Predictions have placed this number to ½ by 2010. As Peru's marine capture fisheries are largely dominated by sardine and anchovy catches, their second place in Table XX should be evaluated in this context. Peru's fish are caught to produce half their weight in salmon elsewhere. Apart from the 2-for-1 loss, Peru also loses out by exporting their catch instead of supporting domestic productivity. One of the problems is that sardines as food is not as popular as salmon. Ecologically, and socio-politically for Peru, it would make more sense to motivate people to (re-)develop a taste for sardines.

Aquaculture is causing problems for coastal ecosystems, particularly mangroves.

- suggested methods to improve situation[7]:
  - eliminate fisheries and energy subsidies (global fishing fleets are an estimated 250% larger than are needed to catch what oceans can sustainably produce)
  - reallocate payments to encourage the use of less destructive gear, direct marketing to consumers, and ecological fish farming
  - establish a global network of marine reserves
  - eliminate bottom trawling (dragging such nets has been likened to clear-cutting forests)
  - reduce wasted and illegal catches
  - encourage ecological fish farming

Good and Bad Fish to Buy

A Table Issued by the Audubon Society:

Enjoy	**be careful**	**avoid**
anchovies	cod (Pacific)	cod (Atlantic)
dungeness crab	lobster (American)	caviar (imported//wild-caught)
crawfish	shrimp (U.S. farmed or trawl-caught)	shrimp (imported)
catfish (farmed)	rainbow trout (farmed)	flounder and soles (Atlantic)
halibut (Pacific)	Mahi-Mahi	halibut (Atlantic)
mussels and clams (farmed)	scallops (bay and sea)	Chilean sea bass (toothfish)
oysters (Pacific farmed)	oysters (wild-caught)	monkfish
sablefish (Alaska, Brit. Columbia)	squid (calamari)	grouper
salmon (wild Alaskan)	swordfish (Atlantic)	salmon (farmed, incl. Atlantic)
sardines	tuna (canned)	orange roughy
tuna: ahi, yellowfin, bigeye, albacore (pole/troll-caught)	tuna: ahi, yellowfin, bigeye, albacore (longline caught)	tuna bluefin
striped bass (farmed)		red snapper
tilapia(U.S. farmed)		sharks

Related Web Sites

- Intergovernmental Panel on Climate Change
- California Ocean Protection Council
- check out Worldwatch Institute web site

References

1. "The Penguin State of the World Atlas" by Dan Smith, 2003. Penguin Books, ISBN: 0-14.200318-2
2. Flyer from the Algalita Marine Research Foundation
3. United Nations Environmental Programme (UNEP) web site
4. "State of the World 2003" by Chris Bright et al., 2003, The Worldwatch Institute, Norton & Company, ISBN: 0-393-32386-2
5. "State of the World 2004" by Erik Assadourian et al., 2004, The Worldwatch Institute, Norton & Company, ISBN: 0-393-32539-3
6. "The Human Impact on the Natural Environment" by Andrew Goudie, 2000, The MIT Press, ISBN: 0-262-57138-2
7. Halweil, B., 2006. Catch of the Day. from the World Watch Institute, publication #172, Worldwatch Institute. http://worldwatch.org
8. Baum et al, 2003. "Collapse and Conservation of Shark Populations in the Northwest Atlantic", Science, Vol 299, 389-392.
9. Myers, N. and Kent, J., 2005. The New Atlas of Planet Management. University of California Press. Berkeley and Los Angeles. 304 pp.
10. Wikipedia at http://en.wikipedia.org/wiki/Acqua_alta
11. Wikipedia at http://en.wikipedia.org/wiki/MOSE_Project
12. San Diego Union Tribune, 14 January 2011. Carnival pulling last ship out of San Diego.
13. Wikipedia at http://en.wikipedia.org/wiki/Ocean_acidification
14. Platt McGinn, A., 1999. Safeguarding the Health of Oceans. Worldwatch paper #145. Worldwatch Institute. http://worldwatch.org
15. San Diego Union Tribune, 29 November 2008. State panel floats ‘litter tax’ to curb debris along coast
16. Trujillo, A.P. and Thruman, H.V., 2005. Essentials of Oceanography. 8th edition. Pearson Prentice Hall, Upper Saddle River, NJ, 518 pp.
17. Baum, J.K., Myers, R.A., Kehler, D.G., Worm, B., Harley, S.J. and Doherty, P.A., 2003. Collapse and Conservation of Shark Populations in the Northwest Atlantic. Science, 299, 389-392.
18. David de Rothschild, 2008. "Earth Matters", DK, New York; ISBN: 978-0-7566-3435-3
19. wikipedia page on Amazon River

Recommended Reading

- "State of the World" by The Worldwatch Institute, W.W. Norton and Company. Topics change from year to year but the 2004 book (ISBN: 0-393-32539-3) was on consumerism and globalization, waste/recycling of resources, catch-up of developing countries, water productivity and increasing shortage. Earlier books were on energy resources, greenhouse effect, Kyoto Protocol and the spread of and fight against diseases.

- UNEP
- Algalita
- Greenpeace
- Oceansconservancy

Chapter 22: Epilogue

In this course, we encountered the awesome forces behind Earth processes. We also learn and the power and devastating global warming may increase frequency and severity of severe weather, coastal erosion, but the real problem Earth is facing is the exponential growth of human population; at the same time, we are looking at limited non-renewable resources; limited land suitable for farming, and limited food sources in the oceans;

Easter Island

In his book "Collapse", Jared Diamond mentions the rise and fall of the population of Easter Island. This Easter Island case is often used as a fable to reflect what could happen to our world if we continue business as usual and do not consider the limitation of the resources that we currently use. The story is also described in Chapter 1 in the course book. In a sense, Earth - the isolated habitable planet in the solar system - is like Easter Island, the most remote place on Earth. The story is summarized here:

- Easter Island is the most remote island on the planet, with the closest other island, Pitcairn, being > to the west 2000km and Chile > 3700km to the east. With only 165km^2 (64 mi^2), it is relatively small. At a latitude of 27°; S, the island is close to the subtropical divergence zone along which many deserts lie. But being an ocean island and Maunga Terevaka having a maximum elevation of 507m (it's a shield volcano), Easter Island receives about 1000mm (44") of rain annually (though this can vary). The island has high temperatures and humidity. In principle, one could expect that Easter Island has some kind of forest and water but it is a desolate place with poorly drained and marginal soils, no permanent stream, no terrestrial mammals. Some permanent water is available in little lakes within the volcano caldera. It turns out that the island has not always looked like this.
- Polynesian seafarers (about 25-50 or so) found a paradise with lush palm forests when they first arrived at the island around A.D. 400. They began to settle the island and use its resources. They brought some food staples, such as bananas and yams which appeared to be the only adequate staples for the relatively harsh conditions on the island. The Polynesians also brought chickens. The population thrived, using the wood of the palm trees to build boats for fishing, to cook, and to transport their Moais, large basalt statues that can still be found along the perimeter of the island.

- Pollen and other soil analysis revealed that a first sign of decline emerged in about 800AD and the last tree was gone by 1400AD, when about 10,000 people inhabited the island. By that time, first signs of a change in diet emerged in the waste disposal piles (e.g. discarded bones). The bones of porpoises were replaced by remains of shellfish, snails and bird eggs. Evidently, the people could not go out to sea far enough to catch the larger fish.
- Soon after, crop failures and soil erosion started to happen. The rats brought to the island by the settlers ate the seeds, and the people started to eat rats and insects, after all the chickens were gone and apparently unable to go out to sea anymore because there was no longer wood to build boats. The population declined dramatically after that.
- When Dutch explorers first visited the island in 1722, they found about 2000 people living in caves (no more wood to build houses) in a primitive society engaged in warfare. The island was completely desolated and people practiced cannibalism.
- It appears that human activities so overwhelmed the environment that it was no longer able to support the greatly enlarged human population.
- This example shows us what could happen if we are not carefully using our resources, some of which we take for granted but may last only a few more decades, after our ancestors have been using them for thousands of years.
- There is still scientific discussion going on about some details how things happened. Some scientists suggest that not the people are responsible for the demise of the palm forest but that the rats ate the seeds so fast that the forest couldn't grow back. Some scientists argue that the settling people didn't come from Polynesia but from Chile. Some scientists argue that climate changes may have caused droughts and triggered the crop failures. And some scientists argue that the timing of events is not exactly right. But the bottom line is ultimately the same, in that the island witnessed the collapse of a once thriving society, mostly due to unwise, or ignorant, management of the environment.

There is some indication, that with most of the most important fisheries depleted nearly beyond recovery we are now going down the food chain and collect smaller species that were once considered unworthy of harvesting....... squid anyone?

References

1. "Collapse" by Jared Diamond, 2005. Penguin Books, ISBN: 0-14303655-6; paperback/ 2004 ISBN:0-67003337-5; hardcover